Symbols

$=$	Is equal to
$\neq$	Is not equal to
$\approx$	Is approximately equal to
$>$	Is greater than
$\geq$	Is greater than or equal to
$<$	Is less than
$\leq$	Is less than or equal to
$a < x < b$	a is less than x and x is less than b
$\{a, b\}$	The set whose elements are a and b
$\varnothing$	Null set
$a \in B$	a is an element of set B
$a \notin B$	a is not an element of set B
$A \subseteq B$	Set A is a subset of set B
$A \nsubseteq B$	Set A is not a subset of set B
$A \cap B$	Set intersection
$A \cup B$	Set union
$\lvert x \rvert$	The absolute value of x
b^n	nth power of b
$\sqrt[n]{a}$	nth root of a
$\sqrt{a}$	Square root of a
i	Imaginary unit
$a + bi$	Complex number
$\pm$	Plus or minus
(a, b)	Ordered pair: first component is a and second component is b
f, g, h, etc.	Names of functions
$f(x)$	Functional value at x
$\log_b x$	Logarithm, to the base b, of x
$\text{antilog}_b\, x$	Antilogarithm, to the base b, of x
$\ln x$	Natural logarithm (base e)
$\log x$	Common logarithm (base 10)
S_n	Sum of n terms of a series
S_∞	Infinite sum

Intermediate Algebra

for College Students

The Prindle, Weber & Schmidt Series in Mathematics

Althoen and Bumcrot, *Introduction to Discrete Mathematics*
Boye, Kavanaugh, and Williams, *Elementary Algebra*
Boye, Kavanaugh, and Williams, *Intermediate Algebra*
Burden and Faires, *Numerical Analysis*, Fourth Edition
Cass and O'Connor, *Fundamentals with Elements of Algebra*
Cullen, *Linear Algebra and Differential Equations*, Second Edition
Dick and Patton, *Calculus, Volume I*
Dick and Patton, *Calculus, Volume II*
Dick and Patton, *Technology in Calculus: A Sourcebook of Activities*
Eves, *In Mathematical Circles*
Eves, *Mathematical Circles Adieu*
Eves, *Mathematical Circles Squared*
Eves, *Return to Mathematical Circles*
Fletcher, Hoyle, and Patty, *Foundations of Discrete Mathematics*
Fletcher and Patty, *Foundations of Higher Mathematics*, Second Edition
Gantner and Gantner, *Trigonometry*
Geltner and Peterson, *Geometry for College Students*, Second Edition
Gilbert and Gilbert, *Elements of Modern Algebra*, Third Edition
Gobran, *Beginning Algebra*, Fifth Edition
Gobran, *Intermediate Algebra*, Fourth Edition
Gordon, *Calculus and the Computer*
Hall, *Algebra for College Students*
Hall, *Beginning Algebra*
Hall, *College Algebra with Applications*, Third Edition
Hall, *Intermediate Algebra*
Hartfiel and Hobbs, *Elementary Linear Algebra*
Humi and Miller, *Boundary-Value Problems and Partial Differential Equations*
Kaufmann, *Algebra for College Students*, Fourth Edition
Kaufmann, *Algebra with Trigonometry for College Students*, Third Edition
Kaufmann, *College Algebra*, Second Edition
Kaufmann, *College Algebra and Trigonometry*, Second Edition
Kaufmann, *Elementary Algebra for College Students*, Fourth Edition
Kaufmann, *Intermediate Algebra for College Students*, Fourth Edition
Kaufmann, *Precalculus*, Second Edition
Kaufmann, *Trigonometry*
Kennedy and Green, *Prealgebra for College Students*
Laufer, *Discrete Mathematics and Applied Modern Algebra*
Nicholson, *Elementary Linear Algebra with Applications*, Second Edition
Pence, *Calculus Activities for Graphic Calculators*
Pence, *Calculus Activities for the TI-81 Graphic Calculator*
Plybon, *An Introduction to Applied Numerical Analysis*
Powers, *Elementary Differential Equations*

Powers, *Elementary Differential Equations with Boundary-Value Problems*
Proga, *Arithmetic and Algebra*, Third Edition
Proga, *Basic Mathematics*, Third Edition
Rice and Strange, *Plane Trigonometry*, Sixth Edition
Schelin and Bange, *Mathematical Analysis for Business and Economics*, Second Edition
Strnad, *Introductory Algebra*
Swokowski, *Algebra and Trigonometry with Analytic Geometry*, Seventh Edition
Swokowski, *Calculus*, Fifth Edition
Swokowski, *Calculus*, Fifth Edition (Late Trigonometry Version)
Swokowski, *Calculus of a Single Variable*
Swokowski, *Fundamentals of College Algebra*, Seventh Edition
Swokowski, *Fundamentals of Algebra and Trigonometry*, Seventh Edition
Swokowski, *Fundamentals of Trigonometry*, Seventh Edition
Swokowski, *Precalculus: Functions and Graphs, Sixth Edition*
Tan, *Applied Calculus*, Second Edition
Tan, *Applied Finite Mathematics*, Third Edition
Tan, *Calculus for the Managerial, Life, and Social Sciences*, Second Edition
Tan, *College Mathematics*, Second Edition
Trim, *Applied Partial Differential Equations*
Venit and Bishop, *Elementary Linear Algebra*, Third Edition
Venit and Bishop, *Elementary Linear Algebra*, Alternate Second Edition
Wiggins, *Problem Solver for Finite Mathematics and Calculus*
Willard, *Calculus and Its Applications*, Second Edition
Wood and Capell, *Arithmetic*
Wood and Capell, *Intermediate Algebra*
Wood, Capell, and Hall, *Developmental Mathematics*, Fourth Edition
Zill, *A First Course in Differential Equations with Applications*, Fourth Edition
Zill and Cullen, *Advanced Engineering Mathematics*
Zill, *Calculus*, Third Edition
Zill, *Differential Equations with Boundary-Value Problems*, Second Edition

The Prindle, Weber & Schmidt Series in Advanced Mathematics

Brabenec, *Introduction to Real Analysis*
Ehrlich, *Fundamental Concepts of Abstract Algebra*
Eves, *Foundations and Fundamental Concepts of Mathematics*, Third Edition
Keisler, *Elementary Calculus: An Infinitesimal Approach*, Second Edition
Kirkwood, *An Introduction to Real Analysis*
Ruckle, *Modern Analysis: Measure Theory and Functional Analysis with Applications*
Sieradski, *An Introduction to Topology and Homotopy*

Annotated Instructor's Edition

Intermediate Algebra

for College Students

Fourth Edition

Jerome E. Kaufmann

PWS-KENT Publishing Company
Boston

20 Park Plaza
Boston, Massachusetts 02116

PWS-KENT Publishing Company is a division of Wadsworth, Inc.

Library of Congress Cataloging-in-Publication Data

Kaufmann, Jerome E.
Intermediate algebra for college students/Jerome E. Kaufmann.—
4th ed.
p. cm.
Includes index.
ISBN 0-534-92853-6.—ISBN 0-534-92883-8
1. Algebra. I. Title.
QA154.2.K37 1991
512.9--dc20 91-35140
CIP

Sponsoring Editor: Timothy Anderson
Assistant Editor: Kelle Karshick
Production & Text Design: Susan Graham
Production Coordinator: Helen Walden
Cover Design: Helen Walden
Manufacturing Coordinator: Peter D. Leatherwood
Typesetter: Polyglot Pte Ltd
Printer/Binder: R. R. Donnelley
Cover Printer: Henry N. Sawyer Company
Cover Photo: Benjamin Mendlowitz

Printed in the United States of America
92 93 94 95 96—10 9 8 7 6 5 4 3 2

This book is printed on recycled, acid-free paper.

The cover photo is of the three-masted schooner *Natalie Todd* sailing on Frenchman's Bay, Maine.

Contents

Chapter 3

Chapter 4

Chapter 5

Quadratic Equations and Inequalities 262

Equations in Two Variables and Their Graphs 310

Functions 379

Chapter 9 Systems of Equations 426

Chapter 10 Exponential and Logarithmic Functions 493

Chapter 11 Sequences and Series 537

Preface

When I prepared this fourth edition, I attempted to preserve the features that made the previous editions successful; at the same time I wanted to incorporate a number of improvements that the reviewers suggested.

This text was written for college students who need an algebra course that bridges the gap between elementary algebra and the more advanced courses in precalculus mathematics. It covers topics that are usually classified as intermediate algebra topics.

The basic concepts of intermediate algebra are presented in a simple, straightforward manner. Algebraic ideas are developed in a logical sequence, but in an easy-to-read manner without excessive formalism. Concepts are frequently motivated by examples and they are continuously reinforced by additional examples. *In this fourth edition I added several examples to better illustrate the concept under discussion. Users of the previous editions were very helpful in suggesting such examples.

The examples demonstrate a large variety of situations, but I still leave many things for the students to think about in the problem sets. In the examples, I guide students to organize their work and to decide when a meaningful shortcut might be used.

A focal point of every revision is the problem sets. *Again in this edition I carefully analyzed the problem sets to make sure that variations of skill development exercises are contained in both the even- and odd-numbered problems. Users of the previous editions were very helpful in suggesting problems to be added, deleted, or changed.

---

* Specific references to this fourth edition

Some of the problem sets contain a category of problems called **Miscellaneous Problems**. These are either problems designed to give the students an opportunity to see what lies ahead in college algebra or problems designed to broaden the students' background in a particular topic. All of them could be omitted without breaking the continuity pattern of the text; however, I feel that they do add another flexibility feature. You may wish to turn to Problem Sets 5.6 and 7.5 to see examples of these problems.

Chapter Review Problem Sets provide a comprehensive vehicle for students to use to pull together the concepts of a chapter. **Cumulative Review Problem Sets** appear at the ends of Chapters 3, 5, 7, and 9.

In this edition, we have also included a few special problems at the end of most Chapter Review Problem Sets. These problems, titled **Thoughts into Words**, provide the students with an opportunity to express, in written form, their thoughts about various mathematical ideas.

There is a common thread throughout the book, which is *learn a skill*, then *use the skill to help solve equations and inequalities*, and then *use equations and inequalities to solve word problems*. This thread influenced some other decisions.

1. Numerous word problems are scattered throughout the text. Every effort was made to start with easy ones, then gradually increase the level of difficulty.
2. Many problem solving suggestions are offered throughout, with special discussions in several sections. The key is to work with various problem solving techniques; not to become overly concerned that all of the traditional types of problems are studied.
3. Newly acquired skills are used as soon as possible to solve equations and inequalities, which are, in turn, used to solve word problems. Therefore, the concept of solving equations and inequalities is introduced early and developed throughout the text. The concepts of factoring, solving equations, and solving word problems are tied together in Chapter 3.

As recommended by the American Mathematical Association of Two-Year Colleges, some basic geometric concepts used in subsequent courses were integrated in a problem solving setting as follows.

Section 2.2. Complementary and supplementary angles; sum of the angles of a triangle equals 180°

Section 2.4. Area and volume formulas

Section 3.4. More on area and volume formulas, perimeter and circumference formulas

Section 3.7. Pythagorean Theorem

Section 6.2. More on Pythagorean Theorem, including work with isosceles right triangles and 30°–60° right triangles

I tried to assign the calculator its rightful place in the study of mathematics, that is, as a tool, useful at times, unnecessary at other times. No special problems were created just so that we could use the calculator. Instead some of the usual intermediate algebra problems, which lend themselves to the use of a calculator, were labeled as calculator problems.

Specific Comments about Some of the Chapters

1. Chapter 1 was written so that it can be covered quickly, and on an individual basis if desired, by those needing only a brief review of some basic algebraic concepts. *In this fourth edition, Sections 1.2 (Addition and Subtraction of Integers) and 1.3 (Multiplication and Division of Integers) were combined to further facilitate a review situation.
2. Chapter 2 presents an early introduction to the heart of an intermediate algebra course. By introducing problem solving and the solving of equations and inequalities early, they can be used as unifying themes throughout the text.
3. In Chapter 5 exponents and radicals are developed separately and then merged at the end of the chapter to unify rational exponents and roots. The general concept of nth root is discussed, but in the simplification problems we concentrate on square root and cube root.
4. Chapter 6 was organized to give students an opportunity to learn on a day-to-day basis different techniques for solving quadratic equations. *At the request of users of the previous editions, I moved the work on complex numbers into the first section of the chapter. The other sections were rewritten to include complex numbers throughout the development. The process of completing the square is treated as a viable equation solving process for certain types of quadratic equations. The emphasis on completing the square here pays dividends in Chapter 7 when graphing parabolas and circles. Section 6.5 offers some guidance as to when to use a particular technique for solving quadratic equations. Furthermore, the often overlooked relationships involving the sum and products of roots are discussed and used as an effective checking procedure.
5. Chapter 7 was written on the premise that intermediate algebra students should become *very familiar* with straight lines, parabolas, and circles with only limited exposure to ellipses and hyperbolas. I intentionally kept the function concept out of Chapter 7. My personal experience indicates that these students need more work with coordinate geometry concepts—specifically graphing techniques—before being introduced to the idea of a function. *Section 7.1 was rewritten to place emphasis on finding intercepts and using the concept of symmetry for graphing purposes.
6. Chapter 8 is devoted entirely to functions and the issue is not clouded by jumping back and forth between functions and relations that are not functions. It includes some work on the composition of functions and the use of quadratic functions in problem solving situations.

* Specific references to this fourth edition

7. Chapter 9 contains the various techniques for solving systems of linear equations. It was organized so that instructors can use as much of the chapter as needed for their particular course. *I rewrote the first three sections so that we now begin with the substitution method in Section 9.1. Then in Section 9.2 the elimination-by-addition method is presented using an approach that emphasizes equivalent systems and sets the stage for future work with matrices.
8. Chapter 10 presents a modern day version of the concepts of exponents and logarithms. The emphasis is on making the concepts and their applications understood. The calculator is used as a tool to help with the messy computational aspects. *In this fourth edition, I have rewritten several parts of this chapter in order to place more emphasis on the applications of exponents and logarithms.

Supplements

For Instructors:

The following supplements are available to adopters:

1. An *Annotated Instructors's Edition* that contains answers for all of the problems in the text.
2. An *Answer Book* that contains answers for all of the problems except the Chapter Review Problem Sets, which are in the back of the book.
3. A printed *Test Bank* that contains two multiple choice and one short-answer test for each chapter.
4. An *EXPTest* computerized test bank for IBM PCs and compatibles that features over one thousand questions written specifically to accompany this text by Joan and Stuart Thomas of the University of Oregon. This program allows users to view and edit all tests, to add to, delete from, and modify existing questions, and to print any number of student tests.
5. A computerized testing program for the MacIntosh that features test questions written by Joan and Stuart Thomas. Questions can be stored by objectives, and the user can scramble the order of questions.

For Students:

1. A *Partial Solutions Manual* (which students may purchase) that contains solutions for most of the problems numbered 1, 5, 9, 13, etc.
2. A *TrueBASIC Algebra* software package available for both IBM PCs and compatibles and the MacIntosh contains a disk and user manual for

---

* Specific references to this fourth edition

self-study, free exploration of topics, and solution of problems. This package also includes a record and playback feature that makes the program ideal for classroom demonstrations.

3. A set of *Expert Algebra Tutor* disks for IBM PCs and compatibles by Sergei Ovchinnikov of San Francisco State University. These disks are tutorial software page-referenced to specific sections of the text. They define the level of tutoring needed by evaluating the user's need for further remediation or advancement in the tutoring session.
4. A set of *Videotapes* that can be used by students in an independent developmental math lab setting to review the topics in the textbook.

Acknowledgments

I would like to take this opportunity to thank all those who helped define and clarify the market for this text, including: Martin Clark, Montreat-Anderson College; Edward Doran, Front Range Community College; Janet Fiksdal, St. Mary's College; Linda Heithoff, Allegany Community College; Robert L. Horvath, El Camino College; James T. Hunter, Hopkinsville Community College; Glenn Kindle, Front Range Community College; Susann Mathews, Miami University—Middletown; Nancy L. Okalsky, St. Francis College; Thomas A. Reifenrath, Clark College; Wesley W. Tom, Chaffey College; John C. Van druff, Pierce College; and Betsey Whitman, Florida A & M University. Thanks are also in order for the following reviewers, who offered in-depth, specific comments on the working manuscript, including: Daniel S. Chesley, Anne Arundel Community College; Carleton H. Cowan, University of Maine; Irma T. Holm, Long Beach City College; Louise Hoover, Clark College; Richard Simmler, North Virginia Community College; and Larry D. Smith, Cypress College.

I am very grateful to the staff of PWS-KENT, especially Timothy Anderson and Kelle Karshick for their continuous cooperation and assistance throughout this project. I would also like to express my sincere gratitude to Helen Walden and Susan Graham. They continue to make my life as an author so much easier by carrying out the details of production in a dedicated and caring way. My thanks go out to Joan and Stuart Thomas for all of their hard work on the creation and programming of fresh questions for the computerized test banks.

Again, very special thanks are due my wife, Arlene, who spends numerous hours typing and proofreading manuscripts, answer keys, solutions manuals, and test banks.

Jerome E. Kaufmann

Chapter 1

Basic Concepts and Properties

Algebra is often described as a **generalized arithmetic**. That description may not tell the whole story, but it does indicate an important idea: A good understanding of arithmetic provides a sound basis for the study of algebra. Be sure that you thoroughly understand the basic concepts that we review in this first chapter.

1.1

Sets, Real Numbers, and Numerical Expressions

In arithmetic, symbols such as 6, $\frac{2}{3}$, 0.27, and π are used to represent numbers. The basic operations of addition, subtraction, multiplication, and division are commonly indicated by the symbols $+$, $-$, $\cdot$, and $\div$, respectively. Thus, we can form specific **numerical expressions**. For example, we can write the indicated sum of six and eight as $6 + 8$.

In algebra, the concept of a variable provides the basis for generalizing arithmetic ideas. For example, by using x and y to represent *any* numbers, we can use the expression $x + y$ to represent the indicated sum of *any* two numbers. The x and y in such an expression are called **variables** and the phrase $x + y$ is called an **algebraic expression**.

We can extend to algebra many of the notational agreements made in arithmetic, with a few modifications. The following chart summarizes these notational agreements pertaining to the four basic operations.

Operation	*Arithmetic*	*Algebra*	*Vocabulary*
addition	$4 + 6$	$x + y$	The **sum** of x and y
subtraction	$14 - 10$	$a - b$	The **difference** of a and b
multiplication	$7 \cdot 5$ or 7×5	$a \cdot b$, $a(b)$, $(a)b$, $(a)(b)$, or ab	The **product** of a and b
division	$8 \div 4$, $\frac{8}{4}$, or $4\overline{)8}$	$x \div y$, $\frac{x}{y}$, or $y\overline{)x}$	The **quotient** of x and y

Note the different ways to indicate a product, including the use of parentheses. The *ab form* is the simplest and probably the most widely used form. Expressions such as abc, $6xy$, and $14xyz$ all indicate multiplication. We also call your attention to the various forms to indicate division. In algebra, we usually use the fractional form $\frac{x}{y}$, although the other forms do serve a purpose at times.

Use of Sets

We can use some of the basic vocabulary and symbolism associated with the concept of sets in the study of algebra. A **set** is a collection of objects and the objects are called **elements** or **members** of the set. In arithmetic and algebra the elements of a set are usually numbers.

The use of set braces, { }, to enclose the elements (or a description of the elements) and the use of capital letters to name sets provide a convenient way to communicate about sets. For example, we can represent a set A, which consists of the vowels of the alphabet, as

$A = \{\text{vowels of the alphabet}\}$	word description,
$A = \{a, e, i, o, u\}$	list or roster description, or
$A = \{x \mid x \text{ is a vowel}\}$	set builder notation.

We can modify the listing approach if the number of elements is quite large. For example, all of the letters of the alphabet can be listed as

$$\{a, b, c, \ldots, z\}.$$

We simply begin by writing enough elements to establish a pattern, then the three dots indicate that the set continues in that pattern. The final entry indicates the last element of the pattern. If we write

$$\{1, 2, 3, \ldots\}$$

the set begins with the counting numbers 1, 2, and 3. The three dots indicate that it continues in a like manner forever; there is no last element. A set that consists of no elements is called the **null set** (written $\varnothing$).

The **set builder notation** combines the use of braces and the concept of a variable. For example, $\{x \mid x \text{ is a vowel}\}$ is read "the set of all x such that x is a vowel." Note that the vertical line is read "such that." We can use set builder notation to describe the set $\{1, 2, 3, \ldots\}$ as $\{x \mid x > 0 \text{ and } x \text{ is a whole number}\}$.

We use the symbol $\in$ to denote set membership. Thus, if $A = \{a, e, i, o, u\}$, we can write $e \in A$, which we read as "e is an element of A." The slash symbol, /, is commonly used in mathematics as a negation symbol. Therefore, $m \notin A$ is read as "m is *not* an element of A."

Two sets are said to be *equal* if they contain exactly the same elements. For example,

$$\{1, 2, 3\} = \{2, 1, 3\},$$

because both sets contain the same elements; the order in which the elements are written doesn't matter. The slash mark through the equality symbol denotes "not equal to." Thus, if $A = \{1, 2, 3\}$ and $B = \{1, 2, 3, 4\}$, we can write $A \neq B$, which we read as "set A is not equal to set B."

Real Numbers

Most of the algebra that we will study in this text is referred to as the **algebra of real numbers**. This simply means that the variables represent real numbers. Therefore, it is necessary for us to be familiar with the various terminology that classifies different types of real numbers.

$\{1, 2, 3, 4, \ldots\}$	natural numbers, counting numbers, positive integers,

$\{0, 1, 2, 3, \ldots\}$	whole numbers, nonnegative integers,
$\{\ldots -3, -2, -1\}$	negative integers,
$\{\ldots -3, -2, -1, 0\}$	nonpositive integers,
$\{\ldots -3, -2, -1, 0, 1, 2, 3, \ldots\}$	integers

We define a **rational number** as any number that can be expressed in the form $\frac{a}{b}$, where a and b are integers and b is not zero. The following are examples of rational numbers.

$-\frac{3}{4}$, $\frac{2}{3}$,

6 because $6 = \frac{6}{1}$,

-4 because $-4 = \frac{-4}{1} = \frac{4}{-1}$,

0 because $0 = \frac{0}{1} = \frac{0}{2} = \frac{0}{3} \ldots$,

0.3 because $0.3 = \frac{3}{10}$,

$6\frac{1}{2}$ because $6\frac{1}{2} = \frac{13}{2}$

We can also define a rational number in terms of a decimal representation. Before doing so, let's briefly review the different possibilities for decimal representations. We can classify decimals as **terminating, repeating**, or **nonrepeating**. Some examples of each are as follows.

$$\begin{bmatrix} 0.3 \\ 0.46 \\ 0.789 \\ 0.6234 \end{bmatrix} \text{ terminating decimals} \qquad \begin{bmatrix} 0.6666\ldots \\ 0.141414\ldots \\ 0.694694694\ldots \\ 0.2317171717\ldots \\ 0.5417283283283\ldots \end{bmatrix} \text{ repeating decimals}$$

$$\begin{bmatrix} 0.276314583\ldots \\ 0.21411811161111\ldots \\ 0.673183329333\ldots \end{bmatrix} \text{ nonrepeating decimals}$$

A repeating decimal has a block of digits that repeats indefinitely. This repeating block of digits may be of any number of digits and may or may not begin immediately after the decimal point. A small horizontal bar is commonly used to indicate the repeat block. Thus, 0.6666... is written as $0.\overline{6}$ and 0.2317171717... as $0.23\overline{17}$.

In terms of decimals, we define a **rational number** as a number that has either a terminating or repeating decimal representation. The following examples illustrate some rational numbers written in $\frac{a}{b}$ form and in decimal form.

$$\frac{3}{4} = 0.75, \quad \frac{3}{11} = 0.\overline{27}, \quad \frac{1}{8} = 0.125, \quad \frac{1}{7} = 0.\overline{142857}, \quad \frac{1}{3} = 0.\overline{3}$$

We define an **irrational number** as a number that *cannot* be expressed in $\frac{a}{b}$ form, where a and b are integers and b is not zero. Furthermore, an irrational number has a nonrepeating and nonterminating decimal representation. Some examples of irrational numbers and a partial decimal representation for each are as follows.

$\sqrt{2} = 1.414213562373095\ldots,\qquad \sqrt{3} = 1.73205080756887\ldots,$
$\pi = 3.14159265358979\ldots$

The entire set of **real numbers** is composed of the rational numbers along with the irrationals. The following tree diagram summarizes the various classifications of the real number system.

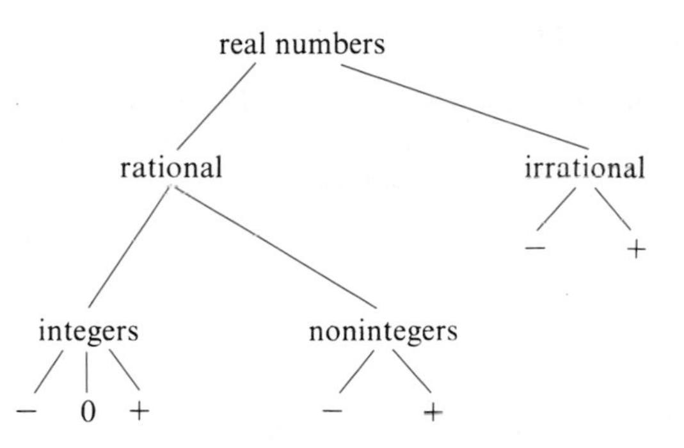

We can trace any real number down through the diagram as follows.

7 is real, rational, an integer, and positive,

$-\frac{2}{3}$ is real, rational, noninteger, and negative;

$\sqrt{7}$ is real, irrational, and positive;

0.38 is real, rational, noninteger, and positive.

REMARK We usually refer to the set of nonnegative integers, $\{0, 1, 2, 3, \ldots\}$, as the set of **whole numbers**, and the set of positive integers, $\{1, 2, 3, \ldots\}$, as the set of **natural numbers**.

The concept of subset is convenient to use at this time. A set A is a **subset** of a set B if and only if every element of A is also an element of B. This is written as $A \subseteq B$ and read as "A is a subset of B." For example, if $A = \{1, 2, 3\}$ and $B = \{1, 2, 3, 5, 9\}$, then $A \subseteq B$ because every element of A is also an element of B. The slash mark again denotes negation, so if $A = \{1, 2, 5\}$ and $B = \{2, 4, 7\}$, we can say that "A is not a subset of B" by writing $A \nsubseteq B$.

We can make the following kinds of statements about the real number system using the subset vocabulary.

1. The set of rational numbers is a subset of the set of real numbers.
2. The set of irrational numbers is a subset of the set of real numbers.
3. The set of integers is a subset of the set of rational numbers. The set of integers is also a subset of the set of real numbers.

Equality

The relation **equality** plays an important role in mathematics—especially when manipulating real numbers and algebraic expressions that represent real numbers. An equality is a statement in which two symbols, or groups of symbols, are names for the same number. The symbol $=$ is used to express an equality. Thus, we can write

$$6 + 1 = 7, \qquad 18 - 2 = 16, \qquad 36 \div 4 = 9.$$

(The symbol $\neq$ means *is not equal to*.) The following four basic properties of equality are self-evident, but we do need to keep them in mind. (This list will be expanded in Chapter 2 when we work with solutions of equations.)

Properties of Equality

Reflexive Property

For any real number a,

$$a = a.$$

Examples $14 = 14$; $x = x$; $a + b = a + b$.

Symmetric Property

For any real numbers a and b,

$$\text{if } a = b, \quad \text{then } b = a.$$

Examples If $13 + 1 = 14$, then $14 = 13 + 1$.
If $3 = x + 2$, then $x + 2 = 3$.

Transitive Property

For any real numbers a, b, and c,

$$\text{if } a = b \quad \text{and} \quad b = c, \quad \text{then } a = c.$$

Examples If $3 + 4 = 7$ and $7 = 5 + 2$, then $3 + 4 = 5 + 2$.
If $x + 1 = y$ and $y = 5$, then $x + 1 = 5$.

Substitution Property

For any real numbers a and b: if $a = b$ then a may be replaced by b, or b may be replaced by a, in any statement without changing the meaning of the statement.

Examples If $x + y = 4$ and $x = 2$, then $2 + y = 4$.
If $a - b = 9$ and $b = 4$, then $a - 4 = 9$.

Numerical Expressions

Let's conclude this section by **simplifying some numerical expressions** that involve whole numbers. The following examples illustrate several important ideas that pertain to this process. Study them carefully and be sure that you agree with each result.

EXAMPLE 1 Simplify $18 + 16 - 9 + 14 - 12 - 10$.

Solution The additions and subtractions are to be done from left to right in the order in which they appear. Thus, $18 + 16 - 9 + 14 - 12 - 10$ simplifies to 17. ■

EXAMPLE 2 Simplify $7 \cdot 4 \div 2 \cdot 3 \cdot 2 \div 4$.

Solution The multiplications and divisions are to be done from left to right in the order in which they appear. Thus, $7 \cdot 4 \div 2 \cdot 3 \cdot 2 \div 4$ simplifies to 21. ■

EXAMPLE 3 Simplify $5 \cdot 3 + 4 \div 2 - 2 \cdot 6 - 28 \div 7$.

Solution First, we do the multiplications and divisions in the order in which they appear. Then we do the additions and subtractions in the order in which they appear. Our work may take on the following format.

$$\begin{aligned} 5 \cdot 3 + 4 \div 2 - 2 \cdot 6 - 28 \div 7 &= 15 + 2 - 12 - 4 \\ &= 1 \end{aligned}$$

■

EXAMPLE 4 Simplify $(4 + 6)(7 + 8)$.

Solution We use the parentheses to indicate the *product* of the quantities $4 + 6$ and $7 + 8$. Perform the additions inside the parentheses first and then multiply.

$$(4 + 6)(7 + 8) = (10)(15) = 150$$

■

EXAMPLE 5 Simplify $(3 \cdot 2 + 4 \cdot 5)(6 \cdot 8 - 5 \cdot 7)$.

Solution First, we do the multiplications inside the parentheses.

$$(3 \cdot 2 + 4 \cdot 5)(6 \cdot 8 - 5 \cdot 7) = (6 + 20)(48 - 35)$$

Then, we do the addition and subtraction inside the parentheses.

$$(6 + 20)(48 - 35) = (26)(13)$$

Then, we find the final product.

$$(26)(13) = 338$$

■

EXAMPLE 6 Simplify $6 + 7[3(4 + 6)]$.

Solution We use brackets for the same purposes as parentheses. In such a problem we need to simplify *from the inside out*; that is, perform the operations in the innermost parentheses first. We would thus obtain

$$\begin{aligned} 6 + 7[3(4 + 6)] &= 6 + 7[3(10)] \\ &= 6 + 7[30] \\ &= 6 + 210 \\ &= 216. \end{aligned}$$

■

EXAMPLE 7 Simplify $\dfrac{6 \cdot 8 \div 4 - 2}{5 \cdot 4 - 9 \cdot 2}$.

Solution First, we perform the operations above and below the fraction bar. Then, we find the final quotient.

$$\begin{aligned} \frac{6 \cdot 8 \div 4 - 2}{5 \cdot 4 - 9 \cdot 2} &= \frac{12 - 2}{20 - 18} \\ &= \frac{10}{2} \\ &= 5 \end{aligned}$$

■

REMARK With parentheses we could also write the problem in Example 7 as $(6 \cdot 8 \div 4 - 2) \div (5 \cdot 4 - 9 \cdot 2)$.

Here is a summary of the ideas pertaining to simplifying numerical expressions. When evaluating a numerical expression perform the operations in the following order.

1. Perform the operations inside the symbols of inclusion (parentheses, brackets, and braces) and above and below each fraction bar. Start with the innermost inclusion symbol.
2. Perform all multiplications and divisions in the order in which they appear from left to right.
3. Perform all additions and subtractions in the order in which they appear from left to right.

Problem Set 1.1

For Problems 1–10, identify each statement as true or false.

1. Every irrational number is a real number. True
2. Every rational number is a real number. True
3. If a number is real, then it is irrational. False
4. Every real number is a rational number. False
5. All integers are rational numbers. True
6. Some irrational numbers are also rational numbers. False
7. Zero is a positive integer. False
8. Zero is a rational number. True
9. All whole numbers are integers. True
10. Zero is a negative integer. False

From the list 0, 14, $\frac{2}{3}$, π, $\sqrt{7}$, $-\frac{11}{14}$, 2.34, $3.2\bar{1}$, $6\frac{7}{8}$, $-\sqrt{17}$, -19, and -2.6 identify each of the following. (Problems 11–18)

11. The whole numbers 0 and 14
12. The natural numbers 14
13. The rational numbers 0, 14, $\frac{2}{3}$, $-\frac{11}{14}$, 2.34, $3.2\bar{1}$, $6\frac{7}{8}$, -19, and -2.6
14. The integers 0, 14, and -19
15. The nonnegative integers 0 and 14
16. The irrational numbers π, $\sqrt{7}$, and $-\sqrt{17}$
17. The real numbers All of them
18. The nonpositive integers 0 and -19

For Problems 19–32, use the following set designations.

$N = \{x \mid x \text{ is a natural number}\}$. $Q = \{x \mid x \text{ is a rational number}\}$.
$W = \{x \mid x \text{ is a whole number}\}$. $H = \{x \mid x \text{ is an irrational number}\}$.
$I = \{x \mid x \text{ is an integer}\}$. $R = \{x \mid x \text{ is a real number}\}$.

Place $\subseteq$ or $\nsubseteq$ in each blank to make a true statement.

19. R ______ N $\nsubseteq$

20. N ______ R $\subseteq$

21. I ______ Q $\subseteq$

22. N ______ I $\subseteq$

23. Q ______ H $\nsubseteq$

24. H ______ Q $\nsubseteq$

25. N ______ W $\subseteq$

26. W ______ I $\subseteq$

27. I ______ N $\nsubseteq$

28. I ______ W $\nsubseteq$

29. $\{1, 3, 5, 7, \ldots\}$ ______ I $\subseteq$

30. $\{0, 2, 4, \ldots\}$ ______ W $\subseteq$

31. $\{0, 3, 6, 9, \ldots\}$ ______ N $\nsubseteq$

32. $\{-2, -1, 0, 1, 2\}$ ______ W $\nsubseteq$

For Problems 33–42, list the elements of each set. For example, the elements of $\{x \mid x$ is a natural number less than $4\}$ can be listed as $\{1, 2, 3\}$.

33. $\{x \mid x$ is a natural number less than $3\}$ $\{1, 2\}$

34. $\{x \mid x$ is a natural number greater than $3\}$ $\{4, 5, 6, \ldots\}$

35. $\{n \mid n$ is a whole number less than $6\}$ $\{0, 1, 2, 3, 4, 5\}$

36. $\{y \mid y$ is an integer greater than $-4\}$ $\{-3, -2, -1, 0, \ldots\}$

37. $\{y \mid y$ is an integer less than $3\}$ $\{\ldots, -1, 0, 1, 2\}$

38. $\{n \mid n$ is a positive integer greater than $-7\}$ $\{1, 2, 3, 4, \ldots\}$

39. $\{x \mid x$ is a whole number less than $0\}$ $\varnothing$

40. $\{x \mid x$ is a negative integer greater than $-3\}$ $\{-2, -1\}$

41. $\{n \mid n$ is a nonnegative integer less than $5\}$ $\{0, 1, 2, 3, 4\}$

42. $\{n \mid n$ is a nonpositive integer greater than $3\}$ $\varnothing$

For Problems 43–50, replace each question mark to make the given statement an application of the indicated property of equality. For example, $16 = ?$ becomes $16 = 16$ because of the reflexive property of equality.

43. If $y = x$ and $x = -6$, then $y = ?$ (transitive property of equality) -6

44. $5x + 7 = ?$ (reflexive property of equality) $5x + 7$

45. If $n = 2$ and $3n + 4 = 10$, then $3(?) + 4 = 10$ (substitution property of equality) 2

46. If $y = x$ and $x = z + 2$, then $y = ?$ (transitive property of equality) $z + 2$

47. If $4 = 3x + 1$, then $? = 4$ (symmetric property of equality) $3x + 1$

48. If $t = 4$ and $s + t = 9$, then $s + ? = 9$ (substitution property of equality) 4

49. $5x = ?$ (reflexive property of equality) $5x$

50. If $5 = n + 3$, then $n + 3 = ?$ (symmetric property of equality) 5

For Problems 51–74, simplify each of the numerical expressions.

51. $16 + 9 - 4 - 2 + 8 - 1$ 26

52. $18 + 17 - 9 - 2 + 14 - 11$ 27

53. $9 \div 3 \cdot 4 \div 2 \cdot 14$ 84

54. $21 \div 7 \cdot 5 \cdot 2 \div 6$ 5

55. $7 + 8 \cdot 2$ 23

56. $21 - 4 \cdot 3 + 2$ 11

57. $9 \cdot 7 - 4 \cdot 5 - 3 \cdot 2 + 4 \cdot 7$ 65

58. $6 \cdot 3 + 5 \cdot 4 - 2 \cdot 8 + 3 \cdot 2$ 28

59. $(17 - 12)(13 - 9)(7 - 4)$ 60

60. $(14 - 12)(13 - 8)(9 - 6)$ 30

61. $13 + (7 - 2)(5 - 1)$ 33

62. $48 - (14 - 11)(10 - 6)$ 36

63. $(5 \cdot 9 - 3 \cdot 4)(6 \cdot 9 - 2 \cdot 7)$ 1320

64. $(3 \cdot 4 + 2 \cdot 1)(5 \cdot 2 + 6 \cdot 7)$ 728

65. $7[3(6-2)]-64$ 20

66. $12+5[3(7-4)]$ 57

67. $[3+2(4\cdot 1-2)][18-(2\cdot 4-7\cdot 1)]$ 119

68. $3[4(6+7)]+2[3(4-2)]$ 168

69. $14+4\left(\dfrac{8-2}{12-9}\right)-2\left(\dfrac{9-1}{19-15}\right)$ 18

70. $12+2\left(\dfrac{12-2}{7-2}\right)-3\left(\dfrac{12-9}{17-14}\right)$ 13

71. $[7+2\cdot 3\cdot 5-5]\div 8$ 4

72. $[27-(4\cdot 2+5\cdot 2)][(5\cdot 6-4)-20]$ 54

73. $\dfrac{3\cdot 8-4\cdot 3}{5\cdot 7-34}+19$ 31

74. $\dfrac{4\cdot 9-3\cdot 5-3}{18-12}$ 3

75. You should be able to do calculations like those in Problems 51–74 *with* and *without* a calculator. Different types of calculators handle the *priority-of-operations* issue in different ways. Be sure you can do Problems 51–74 with *your* calculator.

1.2

Operations with Real Numbers

Before we review the four basic operations with integers let's briefly discuss some concepts and terminology we commonly use with this material.

The symbol -1 can be read as *negative one*, *opposite of one*, or *additive inverse of one*. The *opposite of* and *additive inverse of* terminology is especially meaningful when we work with variables. For example, the symbol $-x$, read as "opposite of x" or "additive inverse of x," emphasizes an important issue. Since x can be any real number, $-x$ (opposite of x) can be zero, positive, or negative. If x is a positive number, then $-x$ is negative. If x is a negative number, then $-x$ is positive. If x is zero, then $-x$ is zero. These facts are illustrated as follows.

$-(6)=-6$	the opposite of six (additive inverse of six) is negative six;
$-(-4)=4$	the opposite of negative four is four;
$-(0)=0$	the opposite of zero is zero.

In general, it can be stated that **the opposite of the opposite of any real number is the real number itself**. This is symbolically expressed as $-(-a)=a$; we sometimes refer to it as the **double negative** property.

Absolute Value

We can use the concept of **absolute value** to precisely describe how to operate with positive and negative numbers. Geometrically, the absolute value of any number is the distance between the number and zero on the number line. For example, the absolute value of 2 is 2. The absolute value of -3 is 3. The absolute value of 0 is 0 (see Figure 1.1).

Figure 1.1

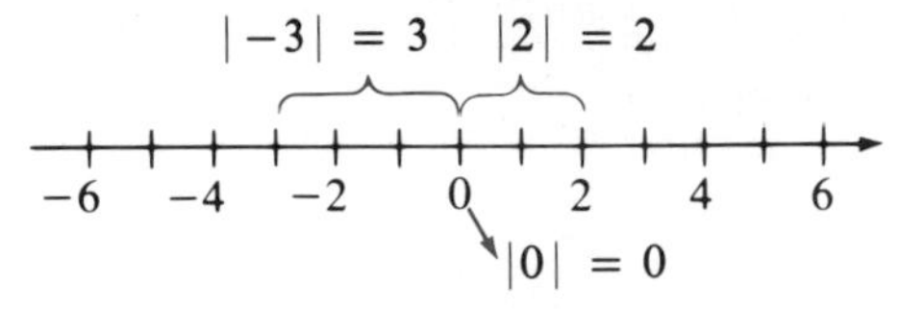

Symbolically, absolute value is denoted with vertical bars. Thus, we write

$$|2| = 2, \qquad |-3| = 3, \qquad |0| = 0.$$

More formally, we define the concept of absolute value as follows.

DEFINITION 1.1

For all real numbers a,

1. If $a \geq 0$, then $|a| = a$.

2. If $a < 0$, then $|a| = -a$.

According to Definition 1.1 we obtain

$\|6\| = 6$	by applying part 1 of Definition 1.1,
$\|0\| = 0$	by applying part 1 of Definition 1.1,
$\|-7\| = -(-7) = 7$	by applying part 2 of Definition 1.1.

Notice that the absolute value of a positive number is the number itself, but the absolute value of a negative number is its opposite. Thus, the absolute value of any number, except 0, is positive and the absolute value of 0 is 0.

Adding Integers

We can describe addition of integers using various physical models. For example, profits and losses pertaining to investments provide a meaningful model. A loss of \$25 (written as -25) on one investment along with a profit of \$60 (written as 60) on a second investment produces an overall profit of \$35. We could write this as $(-25) + 60 = 35$. Think in terms of profits and losses for each of the following examples.

$$50 + 75 = 125, \qquad 20 + (-30) = -10,$$
$$(-10) + (-40) = -50, \qquad (-50) + 75 = 25,$$
$$100 + (-50) = 50, \qquad (-50) + (-50) = -100$$

Although all problems that involve addition of integers could be solved using the profit-and-loss interpretation, it is sometimes convenient to give a more precise description of the addition process. For this purpose we can use the concept of absolute value. Suppose that we want to describe the process of adding -50 and 75. We could say "subtract the absolute value of -50 from the absolute value of 75." Thus,

$$\begin{aligned}(-50) + 75 &= |75| - |-50| \\ &= 75 - 50 \\ &= 25.\end{aligned}$$

We could describe the addition problem $20 + (-30)$ as: Subtract the absolute value of 20 from the absolute value of -30 and then take the opposite. Thus,

$$\begin{aligned}20 + (-30) &= -(|-30| - |20|) \\ &= -(30 - 20) \\ &= -(10) \\ &= -10.\end{aligned}$$

We could describe the addition problem $(-10) + (-40)$ as: Add the absolute value of -10 and the absolute value of -40 and then take the opposite. We would write

$$\begin{aligned}(-10) + (-40) &= -(|-10| + |-40|) \\ &= -(10 + 40) \\ &= -(50) \\ &= -50.\end{aligned}$$

In general, we can describe **addition of integers** as follows.

Two Positive Integers The sum of two positive integers is the sum of their absolute values.

Two Negative Integers The sum of two negative integers is the opposite of the sum of their absolute values.

One Positive and One Negative Integer We can find the sum of a positive integer and a negative integer by subtracting the smaller absolute value from the larger absolute value and giving the result the sign of the original number that has the larger absolute value. If the integers have the same absolute value, then their sum is 0.

Zero and Another Integer The sum of 0 and any integer is the integer itself.

Now consider the following examples in terms of the previous list concerning the addition of integers.

$$(-6) + (-8) = -(|-6| + |-8|) = -(6 + 8) = -14,$$

$$(-19) + (-11) = -(|-19| + |-11|) = -(19 + 11) = -30,$$

$$18 + (-14) = (|18| - |-14|) = 18 - 14 = 4,$$

$$(-17) + 25 = (|25| - |-17|) = 25 - 17 = 8,$$

$$14 + (-21) = -(|-21| - |14|) = -(21 - 14) = -7,$$

$$(-32) + 17 = -(|-32| - |17|) = -(32 - 17) = -15,$$

$$19 + (-19) = 0, \qquad -49 + 0 = -49,$$

$$-21 + 21 = 0, \qquad 0 + (-72) = -72$$

It is true that this *absolute value approach* precisely describes the process of adding integers, but don't forget about the profit-and-loss interpretation. The next problem set also includes other physical models that describe the addition of integers. Some people find such models very helpful.

Subtracting Integers

We describe subtraction of integers in terms of addition.

> **Subtraction of Integers** If a and b are integers, then $a - b = a + (-b)$.

It may be helpful for you to read $a - b = a + (-b)$ as "a minus b is equal to a plus the opposite of b." In other words, we can change every subtraction problem to an equivalent addition problem. Consider the following examples.

$$7 - 9 = 7 + (-9) = -2, \qquad -5 - (-13) = -5 + 13 = 8,$$

$$8 - (-12) = 8 + 12 = 20, \qquad -16 - (-11) = -16 + 11 = -5,$$

$$-9 - 6 = -9 + (-6) = -15$$

It should be apparent that addition of integers is a key operation. The ability to add integers effectively is a necessary skill for future algebraic work. To simplify numerical expressions that involve addition and subtraction of integers, first change all subtractions to additions and then perform the additions.

EXAMPLE 1 Simplify $7 - 9 - 14 + 12 - 6 + 4$.

Solution

$$\begin{aligned} 7 - 9 - 14 + 12 - 6 + 4 &= 7 + (-9) + (-14) + 12 + (-6) + 4 \\ &= -6 \end{aligned}$$

■

EXAMPLE 2 Simplify $-12 + 17 - (-10) + 9 - 3$.

Solution

$$\begin{aligned} -12 + 17 - (-10) + 9 - 3 &= -12 + 17 + 10 + 9 + (-3) \\ &= 21 \end{aligned}$$

■

It is helpful if you can *mentally* convert subtractions to additions. In the next two examples the work shown in the dashed boxes could be done mentally.

EXAMPLE 3 Simplify $4 - 9 - 18 + 13 - 10$.

Solution

$$\begin{aligned} 4 - 9 - 18 + 13 - 10 &= \boxed{4 + (-9) + (-18) + 13 + (-10)} \\ &= -20 \end{aligned}$$

■

EXAMPLE 4 Simplify $(3 - 7) - (4 - 9)$.

Solution

$$\begin{aligned} (3 - 7) - (4 - 9) &= \boxed{[3 + (-7)] - [4 + (-9)]} \\ &= (-4) - (-5) \\ &= \boxed{-4 + 5} \\ &= 1 \end{aligned}$$

■

Multiplying Integers

We can interpret multiplication of whole numbers as repeated addition. For example, $3 \cdot 2$ means three 2's; thus, $3 \cdot 2 = 2 + 2 + 2 = 6$. We can use this same "repeated addition" interpretation of multiplication to find the product of a positive integer and a negative integer as illustrated by the following examples.

$$2(-3) = -3 + (-3) = -6, \qquad 3(-2) = -2 + (-2) + (-2) = -6,$$

$$4(-5) = -5 + (-5) + (-5) + (-5) = -20$$

Note the use of parentheses to indicate multiplication. Sometimes both numbers are enclosed in parentheses, as with $(3)(-4)$.

When multiplying whole numbers we realize that the order in which we multiply two factors does not change the product. For example, $2(3) = 6$ and $3(2) = 6$. Using this idea, we can handle a negative integer times a positive integer as follows.

$$(-2)(3) = (3)(-2) = (-2) + (-2) + (-2) = -6,$$

$$(-3)(4) = (4)(-3) = (-3) + (-3) + (-3) + (-3) = -12,$$

$$(-4)(3) = (3)(-4) = (-4) + (-4) + (-4) = -12$$

Finally, let's consider the product of two negative integers. The following pattern helps with the reasoning.

$$4(-2) = -8,$$
$$3(-2) = -6,$$
$$2(-2) = -4,$$
$$1(-2) = -2,$$
$$0(-2) = 0, \quad \text{The product of zero and any number is zero.}$$
$$(-1)(-2) = ?$$

Certainly, to continue this pattern the product of -1 and -2 has to be 2. In general, this type of reasoning would help us realize that the product of any two negative integers is a positive integer.

Using the concept of absolute value we can describe the **multiplication of integers** as follows.

1. The product of two positive integers or two negative integers is the product of their absolute values.
2. The product of a positive and a negative integer (either order) is the opposite of the product of their absolute values.
3. The product of zero and any integer is zero.

The following examples illustrate this description of multiplication.

$$(-6)(-7) = |-6| \cdot |-7| = 6 \cdot 7 = 42,$$
$$(8)(-9) = -(|8| \cdot |-9|) = -(8 \cdot 9) = -72,$$
$$(-5)(12) = -(|-5| \cdot |12|) = -(5 \cdot 12) = -60,$$
$$(-19)(0) = 0, \qquad (0)(-31) = 0$$

The previous examples illustrated a step-by-step process for multiplying integers. In practice, however, the key issue is to remember whether the product is positive or negative. In other words, we need to remember that the product of two positive or two negative integers is positive and the product of a positive and a negative integer (either order) is negative.

Dividing Integers

The relationship between multiplication and division provides the basis for dividing integers. For example, we know that $8 \div 2 = 4$ because $2 \cdot 4 = 8$. In other words, the quotient of two numbers can be found by looking at a related multiplication problem. In the following examples we used this same type of reasoning to determine some quotients that involve integers.

$\frac{6}{-2} = -3$ because $(-2)(-3) = 6$;

$\frac{-12}{3} = -4$ because $(3)(-4) = -12$;

$\frac{-18}{-2} = 9$ because $(-2)(9) = -18$;

$\frac{0}{-5} = 0$ because $(-5)(0) = 0$;

$\frac{-8}{0}$ is undefined.   Remember that division by zero is undefined!

A precise description for **division of integers** follows.

1. The quotient of two positive or two negative integers is the quotient of their absolute values.
2. The quotient of a positive integer and a negative integer or a negative integer and a positive integer is the opposite of the quotient of their absolute values.
3. The quotient of zero and any nonzero integer (zero divided by any nonzero integer) is zero.

The following examples illustrate this description of division. Again, for practical purposes the key idea is to remember whether the quotient is positive or negative.

$$\frac{-16}{-4} = \frac{|-16|}{|-4|} = \frac{16}{4} = 4, \qquad \frac{28}{-7} = -\left(\frac{|28|}{|-7|}\right) = -\left(\frac{28}{7}\right) = -4,$$

$$\frac{-21}{3} = -\left(\frac{|-21|}{|3|}\right) = -\left(\frac{21}{3}\right) = -7, \qquad \frac{0}{-9} = 0$$

Now let's simplify some numerical expressions that involve the four basic operations with integers.

EXAMPLE 5 Simplify $-6 + 8(-3) - (-4)(-3)$.

Solution

$$-6 + 8(-3) - (-4)(-3)$$
$$= -6 + (-24) - 12 \quad \text{Do the multiplications first.}$$
$$= -42$$

■

EXAMPLE 6 Simplify $-24 \div 4 + 8(-5) - (-5)(3)$.

Solution

$$\begin{aligned} &-24 \div 4 + 8(-5) - (-5)(3) \\ &= -6 + (-40) - (-15) \quad \text{Do the multiplications and divisions first.} \\ &= -31 \end{aligned}$$

■

EXAMPLE 7 Simplify $-7 - 5[-3(6 - 8)]$.

Solution

$$\begin{aligned} &-7 - 5[-3(6 - 8)] \\ &= -7 - 5[-3(-2)] \quad \text{Start with the innermost parentheses.} \\ &= -7 - 5[6] \\ &= -7 - 30 \\ &= -37 \end{aligned}$$

■

EXAMPLE 8 Simplify $[3(-7) - 2(9)][5(-7) + 3(9)]$.

Solution

$$\begin{aligned} [3(-7) - 2(9)][5(-7) + 3(9)] &= [-21 - 18][-35 + 27] \\ &= [-39][-8] \\ &= 312 \end{aligned}$$

■

We have just reviewed the basic operations with *integers*; however, the descriptions we gave for manipulating integers apply to *real numbers* in general. For example, the sum of two negative real numbers is the opposite of the sum of their absolute values.

$$\begin{aligned} (-3.21) + (-4.47) &= -(|-3.21| + |-4.47|) \\ &= -(3.21 + 4.47) \\ &= -7.68 \end{aligned}$$

Likewise, the product of two negative real numbers is the product of their absolute values.

$$\begin{aligned} (-1.2)(-2.3) &= |-1.2| \cdot |-2.3| \\ &= (1.2)(2.3) \\ &= 2.76 \end{aligned}$$

Problem Set 1.2

Perform the following operations with integers.

1. $8 + (-15)$ -7 **2.** $9 + (-18)$ -9 **3.** $(-12) + (-7)$ -19

4. $(-7) + (-14)$ -21 **5.** $-8 - 14$ -22 **6.** $-17 - 9$ -26

7. $9 - 16$ -7 **8.** $8 - 22$ -14 **9.** $(-9)(-12)$ 108

10. $(-6)(-13)$ 78

11. $(5)(-14)$ -70

12. $(-17)(4)$ -68

13. $(-56) \div (-4)$ 14

14. $(-81) \div (-3)$ 27

15. $\dfrac{-112}{16}$ -7

16. $\dfrac{-75}{5}$ -15

17. $-24 + 52$ 28

18. $-19 + 38$ 19

19. $19 - (-14)$ 36

20. $11 - (-25)$ 36

21. $(-17)(12)$ -204

22. $(-15)(9)$ -135

23. $78 \div (-13)$ -6

24. $72 \div (-6)$ -12

25. $0 \div (-14)$ 0

26. $(-19) \div 0$ Undefined

27. $(-21) \div 0$ Undefined

28. $0 \div (-11)$ 0

29. $-21 - 39$ -60

30. $-23 - 38$ -61

31. $-42 - (-25)$ -17

32. $-53 - (-26)$ -27

33. $(-4)(-3)(6)$ 72

34. $(-5)(7)(-2)$ 70

35. $9(-8)(5)$ -360

36. $6(-8)(5)$ -240

37. $(-3)(-7)(-6)$ -126

38. $(-4)(-7)(-9)$ -252

39. $-\left(\dfrac{-56}{-7}\right)$ -8

40. $-\left(\dfrac{-48}{-6}\right)$ -8

Simplify each of the following numerical expressions.

41. $9 - 12 - 8 + 5 - 6$ -12

42. $6 - 9 + 11 - 8 - 7 + 14$ 7

43. $-21 + (-17) - 11 + 15 - (-10)$ -24

44. $-16 - (-14) + 16 + 17 - 19$ 12

45. $7 - (9 - 17)$ 15

46. $-7 - (8 - 15)$ 0

47. $16 - 18 + 19 - [14 - 22 - (31 - 41)]$ 15

48. $-19 - [15 - 13 - (-12 + 8)]$ -25

49. $[14 - (16 - 18)] - [32 - (8 - 9)]$ -17

50. $[-17 - (14 - 18)] - [21 - (-6 - 5)]$ -45

51. $6 - 5(-3)$ 21

52. $-12 - 2(-6)$ 0

53. $-5 + (-2)(7) - (-3)(8)$ 5

54. $-9 - 4(-2) + (-7)(6)$ -43

55. $5(-6) - (-4)(11)$ 14

56. $-7(9) + (-8)(12)$ -159

57. $(-6)(-9) + (-7)(4)$ 26

58. $(-7)(-7) - (-6)(4)$ 73

59. $3(5 - 9) - 3(-6)$ 6

60. $7(8 - 9) + (-6)(4)$ -31

61. $(6 - 11)(4 - 9)$ 25

62. $(7 - 12)(-3 - 2)$ 25

63. $-6(-3 - 9 - 1)$ 78

64. $-8(-3 - 4 - 6)$ 104

65. $56 \div (-8) - (-6) \div (-2)$ -10

66. $-65 \div 5 - (-13)(-2) + (-36) \div 12$ -42

67. $-3[5 - (-2)] - 2(-4 - 9)$ 5

68. $-2(-7 + 13) + 6(-3 - 2)$ -42

69. $\dfrac{-6 + 24}{-3} + \dfrac{-7}{-6 - 1}$ -5

70. $\dfrac{-12 + 20}{-4} + \dfrac{-7 - 11}{-9}$ 0

71. $[(-3)(4) - (-2)(1)][(-2)(-7) - (-8)(6)]$ -620

72. $[4(-6) - 3(-9)][(-1)(-6) - 7(-11)]$ 249

73. $(7 - 11)[-3 - (6 - 7)] - 12$ -4

74. $(-4 - 3)[-2 - (-1 + 6)] - 16$ 33

Perform the following operations with decimals.

75. $(-5.4) + (-7.2)$ -12.6

76. $(-1.8) + (-2.7)$ -4.5

77. $(12.2) + (-5.6)$ 6.6

78. $(13.7) + (-7.9)$ 5.8

79. $-17.3 + 12.5$ -4.8

80. $-16.3 + 9.6$ -6.7

81. $(-2.1)(-3.5)$ 7.35

82. $(-8.5)(-3.3)$ 28.05

83. $(5.4)(-7.2)$ -38.88

84. $(2.9)(-3.6)$ -10.44

85. $\dfrac{-1.2}{-6}$ 0.2

86. $\dfrac{-6.3}{0.7}$ -9

87. $(16.8) \div (-1.2)$ -14

88. $(16.5) \div (-1.5)$ -11

89. $-21.4 - (-14.9)$ -6.5

90. $-32.6 - (-9.8)$ -22.8

91. $1.42 - 7.29$ -5.87

92. $2.73 - 8.14$ -5.41

93. $-2.11 - 4.67$ -6.78

94. $-3.27 - 4.15$ -7.42

95. Do Problems 41–94 with your calculator.

Miscellaneous Problems

96. The number line is a convenient pictorial aid for interpreting addition of integers. Consider the following examples.

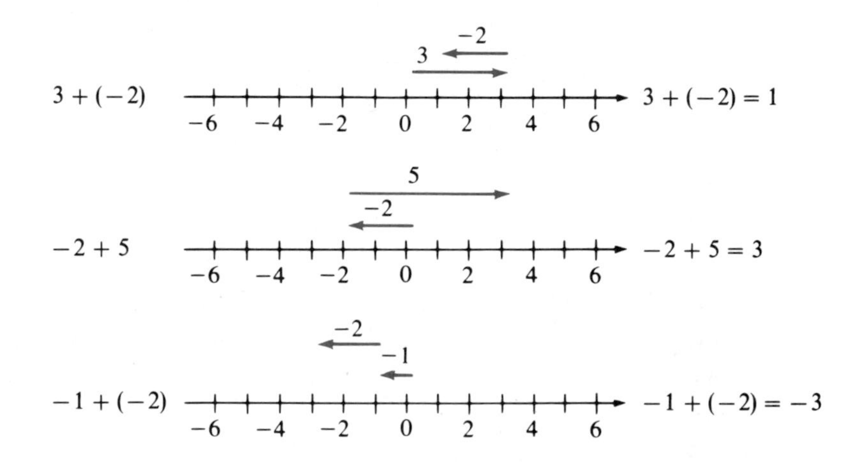

Once you develop a feeling for movement on the number line, a mental image of the movement is sufficient. Use the number line interpretation to find the following sums.

(a) $5 + (-9)$ -4

(b) $6 + (-7)$ -1

(c) $-8 + 5$ -3

(d) $-7 + 1$ -6

(e) $(-5) + (-6)$ -11

(f) $(-12) + (-4)$ -16

(g) $14 + (-23)$ -9

(h) $-18 + 27$ 9

(i) $(-14) + (-19)$ -33

97. A game like football can also be used to interpret addition of integers. A *gain* of 7 yards on one play followed by a *loss* of 3 yards on the next play places the ball 4 yards *ahead* of the original line of scrimmage. We can express this as $7 + (-3) = 4$. Use this "football interpretation" to find the following sums.

(a) $6 + (-4)$ 2
(b) $4 + (-3)$ 1
(c) $5 + (-8)$ -3
(d) $3 + (-9)$ -6
(e) $(-1) + (-5)$ -6
(f) $(-6) + (-4)$ -10
(g) $4 + (-12)$ -8
(h) $5 + (-17)$ -12
(i) $-14 + 23$ 9

1.3 Properties of Real Numbers and the Use of Exponents

At the beginning of this section we will list and briefly discuss some of the basic properties of real numbers. Be sure that you understand these properties, for not only do they facilitate manipulations with real numbers, they also serve as the basis for many algebraic computations.

Closure Property for Addition

If a and b are real numbers, then $a + b$ is a unique real number.

Closure Property for Multiplication

If a and b are real numbers, then ab is a unique real number.

We say that the set of real numbers is *closed* with respect to addition and also with respect to multiplication. That is, the sum of two real numbers is a unique real number and the product of two real numbers is a unique real number. We use the word *unique* to indicate *exactly one.*

Commutative Property of Addition

If a and b are real numbers, then

$$a + b = b + a.$$

Commutative Property of Multiplication

If a and b are real numbers, then

$$ab = ba.$$

We say that addition and multiplication are commutative operations. This means that the order in which you add or multiply two numbers does not affect the result. For example, $6 + (-8) = (-8) + 6$ and $(-4)(-3) = (-3)(-4)$. It is also important to realize that subtraction and division *are not* commutative operations; order does make a difference. For example, $3 - 4 = -1$ but $4 - 3 = 1$. Likewise, $2 \div 1 = 2$ but $1 \div 2 = \frac{1}{2}$.

Associative Property of Addition

If a, b, and c are real numbers, then

$$(a + b) + c = a + (b + c).$$

Associative Property of Multiplication

If a, b, and c are real numbers, then

$$(ab)c = a(bc).$$

Addition and multiplication are **binary operations**. That is, we add (or multiply) *two* numbers at a time. The associative properties apply if more than two numbers are to be added or multiplied; they are grouping properties. For example, $(-8 + 9) + 6 = -8 + (9 + 6)$; changing the grouping of the numbers does not affect the final sum. This is also true for multiplication, which is illustrated by $[(-4)(-3)](2) = (-4)[(-3)(2)]$. Subtraction and division *are not* associative operations. For example, $(8 - 6) - 10 = -8$, but $8 - (6 - 10) = 12$. An example showing that division is not associative is $(8 \div 4) \div 2 = 1$, but $8 \div (4 \div 2) = 4$.

Identity Property of Addition

If a is any real number, then

$$a + 0 = 0 + a = a.$$

Zero is called the identity element for addition. This merely means that the sum of any real number and zero is identically the same real number. For example, $-87 + 0 = 0 + (-87) = -87$.

Identity Property of Multiplication

If a is any real number, then

$$a(1) = 1(a) = a.$$

We call 1 the identity element for multiplication. The product of any real number and 1 is identically the same real number. For example, $(-119)(1) = (1)(-119) = -119$.

Additive Inverse Property

For every real number a, there exists a unique real number $-a$ such that $a + (-a) = -a + a = 0$.

The real number, $-a$, is called the **additive inverse of *a*** or the **opposite of *a***. For example, 16 and -16 are additive inverses and their sum is 0. The additive inverse of 0 is 0.

Multiplication Property of Zero

If a is any real number, then

$$(a)(0) = (0)(a) = 0.$$

The product of any real number and zero is zero. For example, $(-17)(0) = 0(-17) = 0$.

Multiplication Property of Negative One

If a is any real number, then

$$(a)(-1) = (-1)(a) = -a.$$

The product of any real number and -1 is the opposite of the real number. For example, $(-1)(52) = (52)(-1) = -52$.

Multiplicative Inverse Property

For every nonzero real number a, there exists a unique real number $\frac{1}{a}$, such that

$$a\left(\frac{1}{a}\right) = \frac{1}{a}(a) = 1.$$

The number $\frac{1}{a}$ is called the **multiplicative inverse** or the **reciprocal of** a. For example, the reciprocal of 2 is $\frac{1}{2}$ and $2\left(\frac{1}{2}\right) = \frac{1}{2}(2) = 1$. Likewise, the reciprocal of $\frac{1}{2}$ is $\frac{1}{\frac{1}{2}} = 2$. Therefore, 2 and $\frac{1}{2}$ are said to be reciprocals (or multiplicative inverses) of each other. Since division by zero is undefined, zero does not have a reciprocal.

Distributive Property

If a, b, and c are real numbers, then

$$a(b + c) = ab + ac.$$

The distributive property ties together the operations of addition and multiplication. We say that **multiplication distributes over addition**. For example, $7(3 + 8) = 7(3) + 7(8)$. Since $b - c = b + (-c)$, it follows that **multiplication also distributes over subtraction**. This can be symbolically expressed as $a(b - c) = ab - ac$. For example, $6(8 - 10) = 6(8) - 6(10)$.

Now let's consider some examples that illustrate the use of the properties of real numbers to facilitate certain types of manipulations.

EXAMPLE 1 Simplify $(74 + (-36)) + 36$.

Solution In such a problem it is much more advantageous to group -36 and 36.

$$(74 + (-36)) + 36 = 74 + ((-36) + 36)$$ by using the associative property for addition

$$= 74 + 0 = 74$$ ■

EXAMPLE 2 Simplify $[(-19)(25)](-4)$.

Solution It is much easier to group 25 and -4. Thus,

$$[(-19)(25)](-4) = (-19)[(25)(-4)]$$ by using the associative property for multiplication

$$= (-19)(-100)$$
$$= 1900$$ ■

EXAMPLE 3 Simplify $17 + (-14) + (-18) + 13 + (-21) + 15 + (-33)$.

Solution One could add in the order in which the numbers appear. However, since addition is commutative and associative we could change the order and group in any convenient way. For example, we could add all of the positive integers and add all of the negative integers and then find the sum of these two results. It might be convenient to use the vertical format as follows.

$$\begin{array}{rrr} & -14 & \\ 17 & -18 & \\ 13 & -21 & -86 \\ \underline{15} & \underline{-33} & \underline{45} \\ 45 & -86 & -41 \end{array}$$

■

EXAMPLE 4 Simplify $-25(-2 + 100)$.

Solution For this problem it might be easiest to first apply the distributive property and then to simplify as follows.

$$\begin{aligned} -25(-2 + 100) &= (-25)(-2) + (-25)(100) \\ &= 50 + (-2500) \\ &= -2450 \end{aligned}$$

■

EXAMPLE 5 Simplify $(-87)(-26 + 25)$.

Solution For this problem it would be better not to apply the distributive property, but instead to add the numbers inside the parentheses first and then find the indicated product.

$$\begin{aligned} (-87)(-26 + 25) &= (-87)(-1) \\ &= 87 \end{aligned}$$

■

EXAMPLE 6 Simplify $37(104) + 37(-4)$.

Solution Remember that the distributive property allows us to change from the form $a(b + c)$ to $ab + ac$ or from $ab + ac$ to $a(b + c)$. In this problem we want to use the latter change. Thus,

$$\begin{aligned} 37(104) + 37(-4) &= 37(104 + (-4)) \\ &= 37(100) \\ &= 3700 \end{aligned}$$

■

Examples 4, 5, and 6 illustrate an important issue. Sometimes the form $a(b + c)$ is more convenient, but at other times the form $ab + ac$ is better. In these cases, as well as in the cases of other properties, you should *think first* and decide whether or not the properties can be used to make the manipulations easier.

Exponents

Exponents are used to indicate repeated multiplication. For example, we can write $4 \cdot 4 \cdot 4$ as 4^3 where the "raised 3" indicates that 4 is to be used as a factor 3 times. The following general definition is helpful.

DEFINITION 1.2

> If n is a positive integer and b is any real number, then
>
> $$b^n = \underbrace{bbb \cdots b}_{n \text{ factors of } b}.$$

We refer to the b as the **base** and n as the **exponent**. The expression b^n can be read as "b to the nth power." We commonly associate the terms *squared* and *cubed* with exponents of 2 and 3, respectively. For example, b^2 is read "b squared" and b^3 as "b cubed." An exponent of 1 is usually not written, so b^1 is written as b.

The following examples illustrate Definition 1.2.

$$2^3 = 2 \cdot 2 \cdot 2 = 8, \qquad \left(\frac{1}{2}\right)^5 = \frac{1}{2} \cdot \frac{1}{2} \cdot \frac{1}{2} \cdot \frac{1}{2} \cdot \frac{1}{2} = \frac{1}{32},$$

$$3^4 = 3 \cdot 3 \cdot 3 \cdot 3 = 81, \qquad (0.7)^2 = (0.7)(0.7) = 0.49,$$

$$-5^2 = -(5 \cdot 5) = -25, \qquad (-5)^2 = (-5)(-5) = 25$$

Please take special note of the last two examples. Note that $(-5)^2$ means -5 is the base and is to be used as a factor twice. However, -5^2 means that 5 is the base and after it is squared then we take the opposite of that result.

Simplifying numerical expressions containing exponents creates no trouble if we keep in mind that exponents are used to indicate repeated multiplication. Let's consider some examples.

EXAMPLE 7 Simplify $3(-4)^2 + 5(-3)^2$.

Solution

$$\begin{aligned} 3(-4)^2 + 5(-3)^2 &= 3(16) + 5(9) && \text{Find the powers.} \\ &= 48 + 45 \\ &= 93 \end{aligned}$$

■

EXAMPLE 8 Simplify $[3(-1) - 2(1)]^3$.

Solution

$$\begin{aligned} [3(-1) - 2(1)]^3 &= [-3 - 2]^3 \\ &= [-5]^3 \\ &= -125 \end{aligned}$$

■

EXAMPLE 9 Simplify $2(2)^2(-3) - 3(2)(-3)^2 - (-3)^3$.

Solution

$$\begin{aligned} 2(2)^2(-3) - 3(2)(-3)^2 - (-3)^3 &= 2(4)(-3) - 3(2)(9) - (-27) \\ &= -24 - 54 + 27 \\ &= -51 \end{aligned}$$

■

Problem Set 1.3

For Problems 1–14, state the property that justifies each of the statements. For example, $3 + (-4) = (-4) + 3$ because of the commutative property of addition.

1. $[6 + (-2)] + 4 = 6 + [(-2) + 4]$ Associative property of addition
2. $x(3) = 3(x)$ Commutative property of multiplication
3. $42 + (-17) = -17 + 42$ Commutative property of addition
4. $1(x) = x$ Identity property of multiplication
5. $-114 + 114 = 0$ Additive inverse property
6. $(-1)(48) = -48$ Multiplication property of negative one
7. $-1(x + y) = -(x + y)$ Multiplication property of negative one
8. $-3(2 + 4) = -3(2) + (-3)(4)$ Distributive property
9. $12yx = 12xy$ Commutative property of multiplication
10. $[(-7)(4)](-25) = (-7)[4(-25)]$ Associative property of multiplication
11. $7(4) + 9(4) = (7 + 9)4$ Distributive property
12. $(x + 3) + (-3) = x + (3 + (-3))$ Associative property of addition
13. $[(-14)(8)](25) = (-14)[8(25)]$ Associative property of multiplication
14. $\left(\frac{3}{4}\right)\left(\frac{4}{3}\right) = 1$ Multiplicative inverse property

For Problems 15–26, simplify each numerical expression. Be sure to take advantage of the properties whenever they can be used to make the computations easier.

15. $36 + (-14) + (-12) + 21 + (-9) - 4$ 18
16. $-37 + 42 + 18 + 37 + (-42) - 6$ 12
17. $[83 + (-99)] + 18$ 2
18. $[63 + (-87)] + (-64)$ -88
19. $(25)(-13)(4)$ -1300
20. $(14)(25)(-13)(4)$ -18200
21. $17(97) + 17(3)$ 1700
22. $-86(49 + (-48))$ -86
23. $14 - 12 - 21 - 14 + 17 - 18 + 19 - 32$ -47
24. $16 - 14 - 13 - 18 + 19 + 14 - 17 + 21$ 8
25. $(-50)(15)(-2) - (-4)(17)(25)$ 3200
26. $(2)(17)(-5) - (4)(13)(-25)$ 1130

For Problems 27–54, simplify each of the numerical expressions.

27. $2^3 - 3^3$ -19
28. $3^2 - 2^4$ -7
29. $-5^2 - 4^2$ -41
30. $-7^2 + 5^2$ -24

31. $(-2)^3 - 3^2$ -17

32. $(-3)^3 + 3^2$ -18

33. $3(-1)^3 - 4(3)^2$ -39

34. $4(-2)^3 - 3(-1)^4$ -35

35. $7(2)^3 + 4(-2)^3$ 24

36. $-4(-1)^2 - 3(2)^3$ -28

37. $-3(-2)^3 + 4(-1)^5$ 20

38. $5(-1)^3 - (-3)^3$ 22

39. $(-3)^2 - 3(-2)(5) + 4^2$ 55

40. $(-2)^2 - 3(-2)(6) - (-5)^2$ 15

41. $2^3 + 3(-1)^3(-2)^2 - 5(-1)(2)^2$ 16

42. $-2(3)^2 - 2(-2)^3 - 6(-1)^5$ 4

43. $[4(-3) + 5(-2)^2]^2$ 64

44. $[-3(2) - 2(-1)^3]^3$ -64

45. $[3(-2)^2 - 2(-3)^2]^3$ -216

46. $[-3(-1)^3 - 4(-2)^2]^2$ 169

47. $2(-1)^3 - 3(-1)^2 + 4(-1) - 5$ -14

48. $(-2)^3 + 2(-2)^2 - 3(-2) - 1$ 5

49. $2^4 - 2(2)^3 - 3(2)^2 + 7(2) - 10$ -8

50. $3(-3)^3 + 4(-3)^2 - 5(-3) + 7$ -23

51. $5(-1)^4 - 4(-1)^3 - 3(-1)^2 + 8(-1) - 14$ -16

52. $-3(-1)^5 - 2(-1)^4 + 4(-1)^3 - 7(-1)^2 + 9(-1) + 12$ -7

53. $-3^2 - 2^3 + 4(-1)^4 - (-2)^3 + 11$ 6

54. $-2^4 + 2^3 - 2^2 + 2 - 15$ -25

55. Do Problems 41–54 with your calculator and be sure that your calculator can handle the use of exponents.

For Problems 56–64, use your calculator to evaluate each numerical expression.

56. 2^{10} 1024

57. 3^7 2187

58. $(-2)^8$ 256

59. $(-2)^{11}$ -2048

60. -4^9 $-262{,}144$

61. -5^6 $-15{,}625$

62. $(3.14)^3$ 30.959144

63. $(1.41)^4$ 3.9525416

64. $(1.73)^5$ 15.496389

1.4 Algebraic Expressions

Algebraic expressions such as

$$2x, \quad 8xy, \quad 3xy^2, \quad -4a^2b^3c, \quad \text{and} \quad z$$

are called **terms**. A term is an indicated product that may have any number of factors. The variables involved in a term are called **literal factors** and the numerical factor is called the **numerical coefficient**. Thus, in $8xy$, the x and y are literal factors and 8 is the numerical coefficient. The numerical coefficient of the term $-4a^2bc$ is -4. Since $1(z) = z$, the numerical coefficient of the term z is understood to be 1. Terms that have the same literal factors are called **similar terms** or **like terms**. Some examples of similar terms are

$$3x \quad \text{and} \quad 14x, \qquad 5x^2 \quad \text{and} \quad 18x^2,$$

$$7xy \quad \text{and} \quad -9xy, \qquad 9x^2y \quad \text{and} \quad -14x^2y,$$

$$2x^3y^2, \quad 3x^3y^2, \quad \text{and} \quad -7x^3y^2.$$

By the symmetric property of equality we can write the distributive property as

$$ab + ac = a(b + c).$$

Then the commutative property of multiplication can be applied to change the form to

$$ba + ca = (b + c)a.$$

This latter form provides the basis for simplifying algebraic expressions by **combining similar terms**. Consider the following examples.

$$\begin{aligned} 3x + 5x &= (3 + 5)x \\ &= 8x, \end{aligned} \qquad \begin{aligned} -6xy + 4xy &= (-6 + 4)xy \\ &= -2xy, \end{aligned}$$

$$\begin{aligned} 5x^2 + 7x^2 + 9x^2 &= (5 + 7 + 9)x^2 \\ &= 21x^2, \end{aligned} \qquad \begin{aligned} 4x - x &= 4x - 1x \\ &= (4 - 1)x = 3x \end{aligned}$$

More complicated expressions might first require some rearranging of the terms by applying the commutative property for addition.

$$\begin{aligned} 7x + 2y + 9x + 6y &= 7x + 9x + 2y + 6y \\ &= (7 + 9)x + (2 + 6)y \\ &= 16x + 8y, \end{aligned}$$

$$\begin{aligned} 6a - 5 - 11a + 9 &= 6a + (-5) + (-11a) + 9 \\ &= 6a + (-11a) + (-5) + 9 \\ &= (6 + (-11))a + 4 \\ &= -5a + 4 \end{aligned}$$

As soon as you feel that you understand the various simplifying steps you may want to do the steps mentally. Then you could go directly from the given expression to the simplified form as follows.

$$14x + 13y - 9x + 2y = 5x + 15y,$$

$$3x^2y - 2y + 5x^2y + 8y = 8x^2y + 6y,$$

$$-4x^2 + 5y^2 - x^2 - 7y^2 = -5x^2 - 2y^2$$

Sometimes an algebraic expression can be simplified by applying the distributive property to remove parentheses and then to combine similar terms as the next examples illustrate.

$$\begin{aligned} 4(x + 2) + 3(x + 6) &= 4(x) + 4(2) + 3(x) + 3(6) \\ &= 4x + 8 + 3x + 18 \\ &= 4x + 3x + 8 + 18 \\ &= (4 + 3)x + 26 \\ &= 7x + 26, \end{aligned}$$

$$\begin{aligned} -5(y+3)-2(y-8) &= -5(y)-5(3)-2(y)-2(-8) \\ &= -5y-15-2y+16 \\ &= -5y-2y-15+16 \\ &= -7y+1, \end{aligned}$$

$$\begin{aligned} 5(x-y)-(x+y) &= 5(x-y)-1(x+y) \\ &= 5(x)-5(y)-1(x)-1(y) \\ &= 5x-5y-1x-1y \\ &= 4x-6y \end{aligned}$$

Remember $-a = -1(a)$.

When multiplying two terms such as 3 and $2x$ the associative property for multiplication provides the basis for simplifying the product.

$$3(2x) = (3 \cdot 2)x = 6x$$

This idea can be put to use in the following example.

$$\begin{aligned} 3(2x+5y)+4(3x+2y) &= 3(2x)+3(5y)+4(3x)+4(2y) \\ &= 6x+15y+12x+8y \\ &= 6x+12x+15y+8y \\ &= 18x+23y \end{aligned}$$

After you are sure of each step, a more simplified format may be used, as the following examples illustrate.

Be careful with this sign.

$$\begin{aligned} 5(a+4)-7(a+3) &= 5a+20-7a-21 \\ &= -2a-1, \end{aligned}$$

$$\begin{aligned} 3(x^2+2)+4(x^2-6) &= 3x^2+6+4x^2-24 \\ &= 7x^2-18, \end{aligned}$$

$$\begin{aligned} 2(3x-4y)-5(2x-6y) &= 6x-8y-10x+30y \\ &= -4x+22y \end{aligned}$$

Evaluating Algebraic Expressions

An algebraic expression takes on a numerical value whenever each variable in the expression is replaced by a real number. For example, if x is replaced by 5 and y by 9, the algebraic expression $x + y$ becomes the numerical expression $5 + 9$, which simplifies to 14. We say that $x + y$ has a value of 14 when x equals 5 and y equals 9. If $x = -3$ and $y = 7$, then $x + y$ has a value of $-3 + 7 = 4$. The following examples illustrate the process of finding a value of an algebraic expression. We commonly refer to the process as **evaluating algebraic expressions**.

EXAMPLE 1 Find the value of $3x - 4y$ when $x = 2$ and $y = -3$.

Solution

$$\begin{aligned} 3x - 4y &= 3(2) - 4(-3), \quad \text{when } x = 2 \text{ and } y = -3 \\ &= 6 + 12 \\ &= 18 \end{aligned}$$

■

EXAMPLE 2 If $a = -1$ and $b = 4$, find the value of $2a^2 - 5b^2$.

Solution

$$\begin{aligned} 2a^2 - 5b^2 &= 2(-1)^2 - 5(4)^2, \quad \text{when } a = -1 \text{ and } b = 4 \\ &= 2(1) - 5(16) \\ &= 2 - 80 \\ &= -78 \end{aligned}$$

■

EXAMPLE 3 Evaluate $x^2 - 2xy + y^2$ for $x = -2$ and $y = -5$.

Solution

$$\begin{aligned} x^2 - 2xy + y^2 &= (-2)^2 - 2(-2)(-5) + (-5)^2, \quad \text{when } x = -2 \text{ and } y = -5 \\ &= 4 - 20 + 25 \\ &= 9 \end{aligned}$$

■

EXAMPLE 4 Evaluate $(3x + 2y)(2x - y)$ for $x = 4$ and $y = -1$.

Solution

$$\begin{aligned} (3x + 2y)(2x - y) &= (3(4) + 2(-1))(2(4) - (-1)), \quad \text{when } x = 4 \text{ and } y = -1 \\ &= (12 - 2)(8 + 1) \\ &= (10)(9) \\ &= 90 \end{aligned}$$

■

Simplifying by combining similar terms will help you evaluate some algebraic expressions. The following examples illustrate this idea.

EXAMPLE 5 Evaluate $7x - 2y + 4x - 3y$ for $x = -3$ and $y = 6$.

Solution Let's first simplify the given expression.

$$7x - 2y + 4x - 3y = 11x - 5y$$

Now we can evaluate for $x = -3$ and $y = 6$.

$$\begin{aligned} 11x - 5y &= 11(-3) - 5(6), \quad \text{when } x = -3 \text{ and } y = 6 \\ &= -33 - 30 \\ &= -63 \end{aligned}$$

■

EXAMPLE 6 Evaluate $2(3x + 1) - 3(4x - 3)$ for $x = -6$.

Solution

$$\begin{aligned} 2(3x + 1) - 3(4x - 3) &= 6x + 2 - 12x + 9 \\ &= -6x + 11 \end{aligned}$$

Substituting $x = -6$, we obtain

$$\begin{aligned} -6x + 11 &= -6(-6) + 11 \\ &= 36 + 11 \\ &= 47. \end{aligned}$$

■

EXAMPLE 7 Evaluate $2(a^2 + 1) - 3(a^2 + 5) + 4(a^2 - 1)$ for $a = 10$.

Solution

$$\begin{aligned} 2(a^2 + 1) - 3(a^2 + 5) + 4(a^2 - 1) &= 2a^2 + 2 - 3a^2 - 15 + 4a^2 - 4 \\ &= 3a^2 - 17 \end{aligned}$$

Substituting $a = 10$, we obtain

$$\begin{aligned} 3a^2 - 17 &= 3(10)^2 - 17 \\ &= 3(100) - 17 \\ &= 300 - 17 \\ &= 283. \end{aligned}$$

■

Translating from English to Algebra

To use the tools of algebra to solve problems, we must be able to translate from English to algebra. This translation process requires that we recognize key phrases in the English language that translate into algebraic expressions, which involve the operations of addition, subtraction, multiplication, and division. Some of these key phrases and their algebraic counterparts are listed in the following table. The variable n represents the number being referred to in each phrase.

English phrase	*Algebraic expression*
Addition	
the sum of a number and 4	$n + 4$
7 more than a number	$n + 7$
a number plus 10	$n + 10$
a number increased by 6	$n + 6$
8 added to a number	$n + 8$
Subtraction	
14 minus a number	$14 - n$
12 less than a number	$n - 12$
a number decreased by 10	$n - 10$
the difference between a number and 2	$n - 2$
a number subtracted from 13	$13 - n$

English phrase	*Algebraic expression*
Multiplication	
14 times a number	$14n$
the product of 4 and a number	$4n$
$\frac{3}{4}$ of a number	$\frac{3}{4}n$
twice a number	$2n$
multiply a number by 12	$12n$
Division	
the quotient of 6 and a number	$\frac{6}{n}$
the quotient of a number and 6	$\frac{n}{6}$
a number divided by 9	$\frac{n}{9}$
the ratio of a number and 4	$\frac{n}{4}$
Mixture of operations	
4 more than three times a number	$3n + 4$
5 less than twice a number	$2n - 5$
3 times the sum of a number and 2	$3(n + 2)$
2 more than the quotient of a number and 12	$\frac{n}{12} + 2$
7 times the difference of 6 and a number	$7(6 - n)$

An English statement may not always contain the key words *sum*, *difference*, *product*, or *quotient*. Instead, the statement may describe a physical situation and from this description we must deduce the operations involved. We make some suggestions for handling such situations in the following examples.

EXAMPLE 8 Sonya can type 65 words per minute. How many words will she type in m minutes?

Solution The total number of words typed equals the product of the rate per minute and the number of minutes. Therefore, Sonya should be able to type $65m$ words in m minutes. ■

EXAMPLE 9 Russ has n nickels and d dimes. Express this amount of money in cents.

Solution Each nickel is worth 5 cents and each dime is worth 10 cents. We represent the amount in cents by $5n + 10d$. ■

EXAMPLE 10 The cost of a 50-pound sack of fertilizer is d dollars. How much is the cost per pound for the fertilizer?

Solution We calculate the price per pound by dividing the total cost by the number of pounds. We represent the price per pound by $\frac{d}{50}$. ■

The English statement we want to translate to algebra may contain some geometric ideas. Tables 1.1 and 1.2 contain some of the basic relationships that pertain to linear measurement in the English and metric systems, respectively.

Table 1.1

English System
12 inches = 1 foot
3 feet = 1 yard
1760 yards = 1 mile
5280 feet = 1 mile

Table 1.2

Metric System
1 kilometer = 1000 meters
1 hectometer = 100 meters
1 dekameter = 10 meters
1 decimeter = 0.1 meter
1 centimeter = 0.01 meter
1 millimeter = 0.001 meter

EXAMPLE 11 The distance between two cities is k kilometers. Express this distance in meters.

Solution Since 1 kilometer equals 1000 meters, the distance in meters is represented by $1000k$. ■

EXAMPLE 12 The length of a rope is y yards and f feet. Express this length in inches.

Solution Since 1 foot equals 12 inches and 1 yard equals 36 inches, the length of the rope in inches can be represented by $36y + 12f$. ■

EXAMPLE 13 The length of a rectangle is l centimeters and the width is w centimeters. Express the perimeter of the rectangle in meters.

Solution A sketch of the rectangle may be helpful.

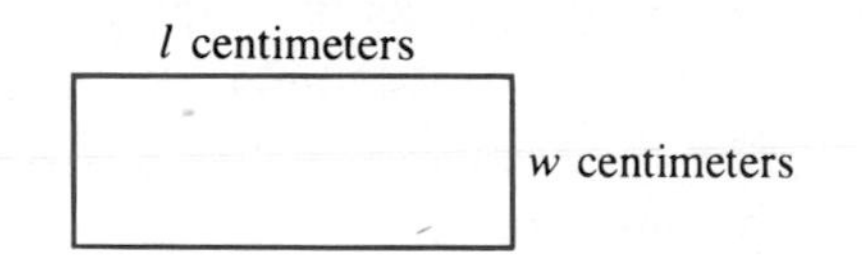

The perimeter of a rectangle is the sum of the lengths of the four sides. Thus, the perimeter in centimeters is $l + w + l + w$, which simplifies to $2l + 2w$. Now, since 1 centimeter equals 0.01 of a meter, the perimeter, in meters, is $0.01\,(2l + 2w)$. This could also be written as $\frac{2l + 2w}{100} = \frac{2(l + w)}{100} = \frac{l + w}{50}$. ■

Problem Set 1.4

Simplify each of the following algebraic expressions by combining similar terms.

1. $-7x + 11x$ $4x$

2. $5x - 8x + x$ $-2x$

3. $5a^2 - 6a^2$ $-a^2$

4. $12b^3 - 17b^3$ $-5b^3$

5. $4n - 9n - n$ $-6n$

6. $6n + 13n - 15n$ $4n$

7. $4x - 9x + 2y$ $-5x + 2y$

8. $7x - 9y - 10x - 13y$ $-3x - 22y$

9. $-3a^2 + 7b^2 + 9a^2 - 2b^2$ $6a^2 + 5b^2$

10. $-xy + z - 8xy - 7z$ $-9xy - 6z$

11. $15x - 4 + 6x - 9$ $21x - 13$

12. $5x - 2 - 7x + 4 - x - 1$ $-3x + 1$

13. $5a^2b - ab^2 - 7a^2b$ $-2a^2b - ab^2$

14. $8xy^2 - 5x^2y + 2xy^2 + 7x^2y$ $10xy^2 + 2x^2y$

Simplify each of the following algebraic expressions by removing parentheses and combining similar terms.

15. $3(x + 2) + 5(x + 3)$ $8x + 21$

16. $5(x - 1) + 7(x + 4)$ $12x + 23$

17. $-2(a - 4) - 3(a + 2)$ $-5a + 2$

18. $-7(a + 1) - 9(a + 4)$ $-16a - 43$

19. $3(n^2 + 1) - 8(n^2 - 1)$ $-5n^2 + 11$

20. $4(n^2 + 3) + (n^2 - 7)$ $5n^2 + 5$

21. $-6(x^2 - 5) - (x^2 - 2)$ $-7x^2 + 32$

22. $3(x + y) - 2(x - y)$ $x + 5y$

23. $5(2x + 1) + 4(3x - 2)$ $22x - 3$

24. $5(3x - 1) + 6(2x + 3)$ $27x + 13$

25. $3(2x - 5) - 4(5x - 2)$ $-14x - 7$

26. $3(2x - 3) - 7(3x - 1)$ $-15x - 2$

27. $-2(n^2 - 4) - 4(2n^2 + 1)$ $-10n^2 + 4$

28. $-4(n^2 + 3) - (2n^2 - 7)$ $-6n^2 - 5$

29. $3(2x - 4y) - 2(x + 9y)$ $4x - 30y$

30. $-7(2x - 3y) + 9(3x + y)$ $13x + 30y$

31. $3(2x - 1) - 4(x + 2) - 5(3x + 4)$ $-13x - 31$

32. $-2(x - 1) - 5(2x + 1) + 4(2x - 7)$ $-4x - 31$

33. $-(3x - 1) - 2(5x - 1) + 4(-2x - 3)$ $-21x - 9$

34. $4(-x - 1) + 3(-2x - 5) - 2(x + 1)$ $-12x - 21$

Evaluate each of the following algebraic expressions for the given values of the variables.

35. $3x + 7y$, $x = -1$ and $y = -2$ -17

36. $5x - 9y$, $x = -2$ and $y = 5$ -55

37. $4x^2 - y^2$, $x = 2$ and $y = -2$ 12

38. $3a^2 + 2b^2$, $a = 2$ and $b = 5$ 62

39. $2a^2 - ab + b^2$, $a = -1$ and $b = -2$ 4

40. $-x^2 + 2xy + 3y^2$, $x = -3$ and $y = 3$ 0

41. $2x^2 - 4xy - 3y^2$, $x = 1$ and $y = -1$ 3

42. $4x^2 + xy - y^2$, $x = 3$ and $y = -2$ 26

43. $3xy - x^2y^2 + 2y^2$, $x = 5$ and $y = -1$ -38

44. $x^2y^3 - 2xy + x^2y^2$, $x = -1$ and $y = -3$ -24

45. $7a - 2b - 9a + 3b$, $a = 4$ and $b = -6$ -14

46. $-4x + 9y - 3x - y$, $x = -4$ and $y = 7$ 84

47. $-9a^2 + 6 + 7a^2 - 14$, $a = -5$ -58

48. $9x - y - 3y + 8y$, $x = 6$ and $y = -7$ 26

49. $-2a - 3a + 7b - b$, $a = -10$ and $b = 9$ 104

50. $3(x - 2) - 4(x + 3)$, $x = -2$ -16

51. $-2(x + 4) - (2x - 1)$, $x = -3$ 5

52. $-4(2x - 1) + 7(3x + 4)$, $x = 4$ 84

53. $2(x - 1) - (x + 2) - 3(2x - 1)$, $x = -1$ 4

54. $-3(x + 1) + 4(-x - 2) - 3(-x + 4)$, $x = 2$ -31

55. $3(x^2 - 1) - 4(x^2 + 1) - (2x^2 - 1)$, $x = -4$ -54

56. $2(n^2 + 1) - 3(2n^2 - 3) + 3(5n^2 - 2)$, $n = -3$ 104

57. $5(x - 2y) - 3(2x + y) - 2(x - y)$, $x = -1$ and $y = -2$ 25

For Problems 58–63, use your calculator and evaluate each of the algebraic expressions for the indicated values. Express the final answers to the nearest tenth.

58. πr^2, $\pi = 3.14$ and $r = 2.1$ 13.8

59. πr^2, $\pi = 3.14$ and $r = 8.4$ 221.6

60. $\pi r^2 h$, $\pi = 3.14$, $r = 1.6$, and $h = 11.2$ 90.0

61. $\pi r^2 h$, $\pi = 3.14$, $r = 4.8$, and $h = 15.1$ 1092.4

62. $2\pi r^2 + 2\pi rh$, $\pi = 3.14$, $r = 3.9$, and $h = 17.6$ 526.6

63. $2\pi r^2 + 2\pi rh$, $\pi = 3.14$, $r = 7.8$, and $h = 21.2$ 1420.5

For Problems 64–78, translate each English phrase into an algebraic expression using n to represent the unknown number.

64. The sum of a number and 4 $n + 4$

65. A number increased by 12 $n + 12$

66. A number decreased by 7 $n - 7$

67. Five less than a number $n - 5$

68. A number subtracted from 75 $75 - n$

69. The product of a number and 50 $50n$

70. One-third of a number $\frac{1}{3}n$

71. Four less than one-half of a number $\frac{1}{2}n - 4$

72. Seven more than three times a number $3n + 7$

73. The quotient of a number and 8 $\frac{n}{8}$

74. The quotient of 50 and a number $\frac{50}{n}$

75. Nine less than twice a number $2n - 9$

76. Six more than one-third of a number $\frac{1}{3}n + 6$

77. Ten times the difference of a number and 6 $10(n - 6)$

78. Twelve times the sum of a number and 7 $12(n + 7)$

For Problems 79–99, answer the question with an algebraic expression.

79. Brian is n years old. How old will he be in 20 years? $n + 20$

80. Becky is n years old. How old was she 5 years ago? $n - 5$

81. Pam is t years old and her mother is 3 less than twice as old as Pam. What is the age of Pam's mother? $2t - 3$

82. The sum of two numbers is 65 and one of the numbers is x. What is the other number? $65 - x$

83. The difference of two numbers is 47 and the smaller number is n. What is the other number? $n + 47$

84. The product of two numbers is 98 and one of the numbers is n. What is the other number? $\frac{98}{n}$

85. The quotient of two numbers is 8 and the smaller number is y. What is the other number? $8y$

86. The perimeter of a square is c centimeters. How long is each side of the square? $\frac{c}{4}$

87. The perimeter of a square is m meters. How long, in centimeters, is each side of the square? $25m$

88. Eric has n nickels, d dimes, and q quarters in his bank. How much money, in cents, does he have in his bank? $5n + 10d + 25q$

89. Tina has c cents, which is all in quarters. How many quarters does she have? $\frac{c}{25}$

90. If n represents a whole number, what represents the next larger whole number? $n + 1$

91. If n represents an odd integer, what represents the next larger odd integer? $n + 2$

92. If n represents an even integer, what represents the next larger even integer? $n + 2$

93. The cost of a 5-pound box of candy is c cents. What is the price per pound? $\frac{c}{5}$

94. Larry's annual salary is d dollars. What is his monthly salary? $\frac{d}{12}$

95. Kim's monthly salary is d dollars. What is her annual salary? $12d$

96. The perimeter of a square is i inches. What is the perimeter expressed in feet? $\frac{i}{12}$

97. The perimeter of a rectangle is y yards and f feet. What is the perimeter expressed in feet? $3y + f$

98. The length of a line segment is d decimeters. How long is the line segment expressed in meters? $0.1d$ or $\frac{d}{10}$

99. The distance between two cities is m miles. How far is this, expressed in feet? $5280m$

Chapter 1 Summary

(1.1) A **set** is a collection of objects; the objects are called **elements** or **members** of the set. Set A is a subset of set B if and only if every member of A is also a member of B. The sets of **natural numbers, whole numbers, integers, rational numbers**, and **irrational numbers** are all subsets of the set of **real numbers.**

We can evaluate **numerical expressions** by performing the operations in the following order.

1. Perform the operations inside the parentheses and above and below fraction bars.
2. Find all powers or convert them to indicated multiplication.
3. Perform all multiplications and divisions in the order in which they appear from left to right.
4. Perform all additions and subtractions in the order that they appear from left to right.

(1.2) The **absolute value** of a real number a is a defined by (1) if $a \geq 0$, then $|a| = a$, and (2) if $a < 0$, then $|a| = -a$.

The sum of two positive integers is the sum of their absolute values. The sum of two negative integers is the opposite of the sum of their absolute values. The sum of a positive and a negative integer is given by: (a) If the positive integer has the larger absolute value, then their sum is the difference of their absolute values when you subtract the smaller absolute value from the larger; (b) If the negative integer has the larger absolute value, then their sum is the opposite of the difference of their absolute values when you subtract the smaller absolute value from the larger.

The statement $a - b = a + (-b)$ changes every subtraction problem to an equivalent addition problem.

The **product** of two positive or two negative integers is the product of their absolute values. The **product** of one positive and one negative integer is the opposite of the product of their absolute values.

The **quotient** of two positive or two negative integers is the quotient of their absolute values. The quotient of one positive and one negative integer is the opposite of the quotient of their absolute values.

(1.3) The following basic properties of real numbers help with numerical manipulations and serve as a basis for algebraic computations.

Closure Properties $a + b$ is a real number

ab is a real number

Commutative Properties $a + b = b + a$

$ab = ba$

Associative Properties $(a + b) + c = a + (b + c)$

$(ab)c = a(bc)$

Identity Properties $a + 0 = 0 + a = a$

$$a(1) = 1(a) = a$$

Additive Inverse Property $a + (-a) = (-a) + a = 0$

Multiplication Property of Zero $a(0) = 0(a) = 0$

Multiplication Property of Negative One $-1(a) = a(-1) = -a$

Multiplicative Inverse Property $a\left(\frac{1}{a}\right) = \left(\frac{1}{a}\right)a = 1$

Distributive Properties $a(b + c) = ab + ac$

$$a(b - c) = ab - ac$$

(1.4) Algebraic expressions such as $2x$, $8xy$, $3xy^2$, $-4a^2b^3c$, and z are called **terms**. A term is an indicated product and may have any number of factors. We call the variables in a term the **literal factors** and we call the numerical factor the **numerical coefficient**. Terms having the same literal factors are called **similar** or **like terms**.

The distributive property in the form $ba + ca = (b + c)a$ serves as the basis for **combining similar terms**. For example,

$$3x^2y + 7x^2y = (3 + 7)x^2y = 10x^2y.$$

To translate English phrases into algebraic expressions, we must be familiar with the standard vocabulary of *sum*, *difference*, *product*, and *quotient*, as well as other terms that express the same ideas.

Chapter 1 Review Problem Set

1. From the list $0, \sqrt{2}, \frac{3}{4}, -\frac{5}{6}, 8\frac{1}{3}, -\sqrt{3}, -8, 0.34, 0.2\overline{3}, 67$, and $\frac{9}{7}$

identify each of the following.

(a) The natural numbers 67

(b) The integers $0, -8$, and 67

(c) The nonnegative integers 0 and 67

(d) The rational numbers $0, \frac{3}{4}, -\frac{5}{6}, 8\frac{1}{3}, -8, 0.34, 0.2\overline{3}, 67$, and $\frac{9}{7}$

(e) The irrational numbers $\sqrt{2}$ and $-\sqrt{3}$

For Problems 2–10, state the property of equality or the property of real numbers that justifies each of the statements. For example, $6(-7) = -7(6)$ because of the *commutative property for multiplication*; and if $2 = x + 3$, then $x + 3 = 2$ is true because of the *symmetric property of equality*.

2. $7 + (3 + (-8)) = (7 + 3) + (-8)$ Associative property for addition

3. If $x = 2$ and $x + y = 9$, then $2 + y = 9$ Substitution property of equality
4. $-1(x + 2) = -(x + 2)$ Multiplication property of negative one
5. $3(x + 4) = 3(x) + 3(4)$ Distributive property
6. $[(17)(4)](25) = (17)[(4)(25)]$ Associative property for multiplication
7. $x + 3 = 3 + x$ Commutative property for addition
8. $3(98) + 3(2) = 3(98 + 2)$ Distributive property
9. $\left(\frac{3}{4}\right)\left(\frac{4}{3}\right) = 1$ Multiplicative inverse property
10. If $4 = 3x - 1$, then $3x - 1 = 4$. Symmetric property of equality

For Problems 11–22, simplify each of the numerical expressions.

11. $-8 + (-4) - (-6)$ -6
12. $9 - 12 + (-4) - (-1)$ -6
13. $-8 - 7 + 4 + 6 - 10 + 8 - 3$ -10
14. $4(-3) - 12 \div (-4) + (-2)(-1) - 8$ -15
15. $-3(2 - 4) - 4(7 - 9) + 6$ 20
16. $(48 + (-73)) + 74$ 49
17. $[5(-2) - 3(-1)][-2(-1) + 3(2)]$ -56
18. $-4^2 - 2^3$ -24
19. $(-2)^4 + (-1)^3 - 3^2$ 6
20. $2(-1)^2 - 3(-1)(2) - 2^2$ 4
21. $[4(-1) - 2(3)]^3$ -1000
22. $3 - [-2(3 - 4)] + 7$ 8

For Problems 23–32, simplify each of the algebraic expressions by combining similar terms.

23. $3a^2 - 2b^2 - 7a^2 - 3b^2$ $-4a^2 - 5b^2$
24. $4x - 6 - 2x - 8 + x + 12$ $3x - 2$
25. $ab^2 - 3a^2b + 4ab^2 + 6a^2b$ $5ab^2 + 3a^2b$
26. $3(x + 1) - 2(x - 4) + 5(x + 4)$ $6x + 31$
27. $3(2n^2 + 1) + 4(n^2 - 5)$ $10n^2 - 17$
28. $-2(3a - 1) + 4(2a + 3) - 5(3a + 2)$ $-13a + 4$
29. $-(n - 1) - (n + 2) + 3$ $-2n + 2$
30. $3(2x - 3y) - 4(3x + 5y) - x$ $-7x - 29y$
31. $4(a - 6) - (3a - 1) - 2(4a - 7)$ $-7a - 9$
32. $-5(x^2 - 4) - 2(3x^2 + 6) + (2x^2 - 1)$ $-9x^2 + 7$

For Problems 33–42, evaluate each of the algebraic expressions for the given values of the variables.

33. $-5x + 4y$ for $x = 3$ and $y = -4$ -31
34. $3x^2 - 2y^2$ for $x = -1$ and $y = -2$ -5
35. $-5(2x - 3y)$ for $x = 1$ and $y = -3$ -55
36. $(3a - 2b)^2$ for $a = -2$ and $b = 3$ 144
37. $a^2 + 3ab - 2b^2$ for $a = 2$ and $b = -2$ -16
38. $3n^2 - 4 - 4n^2 + 9$ for $n = 7$ -44
39. $3(2x - 1) + 2(3x + 4)$ for $x = 4$ 53
40. $-4(3x - 1) - 5(2x - 1)$ for $x = -2$ 53
41. $2(n^2 + 3) - 3(n^2 + 1) + 4(n^2 - 6)$ for $n = 3$ 6
42. $5(3n - 1) - 7(-2n + 1) + 4(3n - 1)$ for $n = -5$ -221

For Problems 43–50, translate each English phrase into an algebraic expression and use n to represent the unknown number.

43. Four increased by twice a number $4 + 2n$

44. Fifty subtracted from three times a number $3n - 50$

45. Six less than two-thirds of a number $\frac{2}{3}n - 6$

46. Ten times the difference of a number and 14 $10(n - 14)$

47. Eight more than five times a number $5n + 8$

48. The quotient of a number and three less than the number $\frac{n}{n-3}$

49. Three less than five times the sum of a number and 2 $5(n + 2) - 3$

50. Three-fourths of the sum of a number and 12 $\frac{3}{4}(n + 12)$

For Problems 51–60, answer the question with an algebraic expression.

51. The sum of two numbers is 37 and one of the numbers is n. What is the other number? $37 - n$

52. Tina can type w words in an hour. What is her typing rate per minute? $\frac{w}{60}$

53. Harry is y years old. His brother is 7 years less than twice as old as Harry. How old is Harry's brother? $2y - 7$

54. If n represents a multiple of three, what represents the next largest multiple of three? $n + 3$

55. Celia has p pennies, n nickels, and q quarters. How much, in cents, does Celia have? $p + 5n + 25q$

56. The perimeter of a square is i inches. How long, in feet, is each side of the square? $\frac{i}{48}$

57. The length of a rectangle is y yards and the width is f feet. What is the perimeter of the rectangle expressed in inches? $24f + 72y$

58. The length of a piece of wire is d decimeters. What is the length expressed in centimeters? $10d$

59. Joan is f feet and i inches tall. How tall is she, in inches? $12f + i$

60. The perimeter of a rectangle is 50 centimeters. If the rectangle is c centimeters long, how wide is it? $25 - c$

Chapter 2

Equations and Inequalities

Throughout this text we are developing algebraic skills, using the skills to help solve equations and inequalities, and then using equations and inequalities to solve applied problems. In this chapter we will review and expand concepts that are important for our problem solving development.

2.1 Solving First-Degree Equations

In Section 1.1 we stated that *an equality (equation) is a statement that two symbols, or groups of symbols, are names for the same number.* It should be further stated that an equation may be true or false. For example, the equation $3 + (-8) = -5$ is true, but the equation $-7 + 4 = 2$ is false.

Algebraic equations contain one or more variables. The following are examples of algebraic equations.

$$3x + 5 = 8, \qquad 4y - 6 = -7y + 9, \qquad x^2 - 5x - 8 = 0,$$
$$3x + 5y = 4, \qquad x^3 + 6x^2 - 7x - 2 = 0$$

An algebraic equation such as $3x + 5 = 8$ is neither true nor false as it stands and we often refer to it as an *open sentence.* Each time that a number is substituted for x, the algebraic equation $3x + 5 = 8$ becomes a numerical statement that is true or false. For example, if $x = 0$, then $3x + 5 = 8$ becomes $3(0) + 5 = 8$, which is a false statement. If $x = 1$, then $3x + 5 = 8$ becomes $3(1) + 5 = 8$, which is a true statement. **Solving an equation** refers to the process of finding the number (or numbers) that makes an algebraic equation a true numerical statement. We call such numbers the **solutions** or **roots** of the equation that **satisfy** the equation. We call the set of all solutions of an equation its **solution set**. Thus, $\{1\}$ is the solution set of $3x + 5 = 8$.

In this chapter we shall consider techniques for solving **first-degree equations in one variable**. This means that the equations contain only one variable and that this variable has an exponent of one. The following are examples of first-degree equations in one variable.

$$3x + 5 = 8, \qquad \frac{2}{3}y + 7 = 9,$$

$$7a - 6 = 3a + 4, \qquad \frac{x-2}{4} = \frac{x-3}{5}$$

Equivalent equations are equations that have the same solution. For example,

1. $3x + 5 = 8$

2. $3x = 3$

3. $x = 1$

are all equivalent equations since $\{1\}$ is the solution set of each.

The general procedure for solving an equation is to continue replacing the given equation with equivalent *but simpler* equations until we obtain an equation of the form *variable = constant* or *constant = variable.* Thus, in the example above, $3x + 5 = 8$ was simplified to $3x = 3$, which was further simplified to $x = 1$, from which the solution set $\{1\}$ is obvious.

To solve equations we need to use the various properties of equality. In addition to the *reflexive*, *symmetric*, *transitive*, and *substitution properties* we listed in Section 1.1, the following properties of equality play an important role.

Addition Property of Equality

For all real numbers a, b, and c,

$$a = b \quad \text{if and only if } a + c = b + c.$$

Multiplication Property of Equality

For all real numbers a, b, and c, where $c \neq 0$,

$$a = b \quad \text{if and only if } ac = bc.$$

The addition property of equality states that "any number can be added to both sides of an equation and an equivalent equation is produced." The multiplication property of equality states that "we obtain an equivalent equation whenever we multiply both sides of an equation by the same **nonzero** real number." The following examples demonstrate the use of these properties to solve equations.

EXAMPLE 1 Solve $2x - 1 = 13$.

Solution

$$2x - 1 = 13$$

$$2x - 1 + 1 = 13 + 1 \qquad \text{Add 1 to both sides.}$$

$$2x = 14$$

$$\frac{1}{2}(2x) = \frac{1}{2}(14) \qquad \text{Multiply both sides by } \frac{1}{2}.$$

$$x = 7$$

The solution set is $\{7\}$. ■

To check an apparent solution we can substitute it into the original equation and see if we obtain a true numerical statement.

CHECK $2x - 1 = 13$

$$2(7) - 1 \stackrel{?}{=} 13$$

$$14 - 1 \stackrel{?}{=} 13$$

$$13 = 13$$

Now we know that $\{7\}$ is the solution set of $2x - 1 = 13$. We will not show our checks for every example in this text, but do remember that checking is a way to detect arithmetic errors.

EXAMPLE 2 Solve $-7 = -5a + 9$.

Solution

$$-7 = -5a + 9$$

$$-7 + (-9) = -5a + 9 + (-9) \qquad \text{Add } -9 \text{ to both sides.}$$

$$-16 = -5a$$

$$-\frac{1}{5}(-16) = -\frac{1}{5}(-5a) \qquad \text{Multiply both sides by } -\frac{1}{5}.$$

$$\frac{16}{5} = a$$

The solution set is $\left\{\frac{16}{5}\right\}$. ■

Notice that in Example 2 the final equation is $\frac{16}{5} = a$ instead of $a = \frac{16}{5}$. Technically, the symmetric property of equality (if $a = b$, then $b = a$) would permit us to change from $\frac{16}{5} = a$ to $a = \frac{16}{5}$, but such a change is not necessary to determine that the solution is $\frac{16}{5}$. Notice that we could use the symmetric property at the very beginning to change $-7 = -5a + 9$ to $-5a + 9 = -7$; some people prefer working with the variable on the left side of the equation.

Let's clarify another point. We stated the properties of equality in terms of only two operations, addition and multiplication. We could also include the operations of subtraction and division in the statements of the properties. That is to say, we could think in terms of "subtracting the same number from both sides of an equation" and also in terms of "dividing both sides of an equation by the same nonzero number." For example, in the solution of Example 2, we could subtract 9 from both sides rather than add -9 to both sides. Likewise, we could divide both sides by -5 instead of multiplying both sides by $-\frac{1}{5}$.

EXAMPLE 3 Solve $7x - 3 = 5x + 9$.

Solution

$$7x - 3 = 5x + 9$$

$$7x - 3 + (-5x) = 5x + 9 + (-5x) \qquad \text{Add } -5x \text{ to both sides.}$$

$$2x - 3 = 9$$

$$2x - 3 + 3 = 9 + 3 \qquad \text{Add 3 to both sides.}$$

$$2x = 12$$

$$\frac{1}{2}(2x) = \frac{1}{2}(12) \qquad \text{Multiply both sides by } \frac{1}{2}.$$

$$x = 6$$

The solution set is $\{6\}$. ■

EXAMPLE 4 Solve $4(y-1)+5(y+2)=3(y-8)$.

Solution

$$4(y-1)+5(y+2)=3(y-8)$$

$$4y-4+5y+10=3y-24$$ Remove parentheses by applying distributive property.

$$9y+6=3y-24$$ Simplify left side by combining similar terms.

$$9y+6+(-3y)=3y-24+(-3y)$$ Add $-3y$ to both sides.

$$6y+6=-24$$

$$6y+6+(-6)=-24+(-6)$$ Add -6 to both sides.

$$6y=-30$$

$$\frac{1}{6}(6y)=\frac{1}{6}(-30)$$ Multiply both sides by $\frac{1}{6}$.

$$y=-5$$

The solution set is $\{-5\}$. ■

We can summarize the process of solving first-degree equations in one variable as follows.

Step 1. Simplify both sides of the equation as much as possible.

Step 2. Use the addition property of equality to isolate a term that contains the variable on one side and a constant on the other side of the equation.

Step 3. Use the multiplication property of equality to make the coefficient of the variable 1. That is, multiply both sides of the equation by the reciprocal of the numerical coefficient of the variable. The solution set should now be obvious.

Step 4. Check each solution by substituting it in the original equation and verify that the resulting numerical statement is true.

Use of Equations to Solve Problems

In order to use the tools of algebra to solve problems we must be able to translate back and forth between the English language and the language of algebra. More specifically, at this time we need to translate *English sentences* into *algebraic equations.* Such translations allow us to use our knowledge of *equation solving* to solve word problems. Let's consider an example.

PROBLEM 1 If we subtract 27 from three times a certain number, the result is 18. Find the number.

Solution

Let n represent the number to be found. The sentence "if we subtract 27 from three times a certain number, the result is 18" translates into the equation $3n - 27 = 18$. Solving this equation, we obtain

$$3n - 27 = 18$$

$$3n = 45 \qquad \text{Add 27 to both sides.}$$

$$n = 15. \qquad \text{Multiply both sides by } \frac{1}{3}.$$

The number to be found is 15. ■

We often refer to the statement "let n represent the number to be found" as **declaring the variable**. We need to choose a letter to use as a variable and indicate what it represents for a specific problem. This may seem like an insignificant idea, but as the problems become more complex the process of declaring the variable becomes even more important. Furthermore, it is true that you could probably solve a problem such as Problem 1 without setting up an algebraic equation. However, as problems increase in difficulty the translation from English to algebra becomes a key issue. Therefore, even with these relatively easy problems we suggest that you concentrate on the translation process.

The next example involves the use of integers. Remember that the set of integers consists of $\{\ldots -2, -1, 0, 1, 2, \ldots\}$. Furthermore, the integers can be classified as *even*, $\{\ldots -4, -2, 0, 2, 4, \ldots\}$ or *odd*, $\{\ldots -3, -1, 1, 3, \ldots\}$.

PROBLEM 2

The sum of three consecutive integers is 13 greater than twice the smallest of the three integers. Find the integers.

Solution

Since consecutive integers differ by 1, we will represent them as follows: let n represent the smallest of the three consecutive integers; then $n + 1$ represents the second largest, and $n + 2$ represents the largest.

$$\overbrace{n + (n + 1) + (n + 2)}^{\text{the sum of the three consecutive integers}} = \overbrace{2n + 13}^{\text{13 greater than twice the smallest}}$$

$$3n + 3 = 2n + 13$$

$$n = 10$$

The three consecutive integers are 10, 11, and 12. ■

To check our answers for Problem 2 we must determine whether or not they satisfy the conditions stated in the original problem. Since 10, 11, and 12 are consecutive integers whose sum is 33 and since twice the smallest plus 13 is also 33 ($2(10) + 13 = 33$), we know that our answers are correct. (Remember, when checking a result for a word problem it is *not* sufficient to check the result in the equation set up to solve the problem; the equation itself may be in error!)

In the two previous problems, the equation formed was almost a direct translation of a sentence in the statement of the problem. Now let's consider a situation where we need to think in terms of a *guideline* not explicitly stated in the problem.

PROBLEM 3 Danny received a car repair bill for \$106. This included \$23 for parts, \$22 per hour for each hour of labor, and \$6 for taxes. Find the number of hours of labor.

Solution Let h represent the number of hours of labor. Then $22h$ represents the total charge for labor. We can use a guideline of "charge for parts plus charge for labor plus tax equals the total bill" to set up the following equation.

$$\underset{\text{parts}}{23} + \underset{\text{labor}}{22h} + \underset{\text{tax}}{6} = \underset{\text{total bill}}{106}$$

Solving this equation we obtain

$$22h + 29 = 106$$

$$22h = 77$$

$$h = 3\frac{1}{2}.$$

Danny was charged for $3\frac{1}{2}$ hours of labor. ■

Problem Set 2.1

Solve each of the following equations.

1. $3x + 4 = 16$ $\{4\}$
2. $4x + 2 = 22$ $\{5\}$
3. $5x + 1 = -14$ $\{-3\}$
4. $7x + 4 = -31$ $\{-5\}$
5. $-x - 6 = 8$ $\{-14\}$
6. $8 - x = -2$ $\{10\}$
7. $4y - 3 = 21$ $\{6\}$
8. $6y - 7 = 41$ $\{8\}$
9. $3x - 4 = 15$ $\{\frac{19}{3}\}$
10. $5x + 1 = 12$ $\{\frac{11}{5}\}$
11. $-4 = 2x - 6$ $\{1\}$
12. $-14 = 3a - 2$ $\{-4\}$
13. $-6y - 4 = 16$ $\{-\frac{10}{3}\}$
14. $-8y - 2 = 18$ $\{-\frac{5}{2}\}$
15. $4x - 1 = 2x + 7$ $\{4\}$
16. $9x - 3 = 6x + 18$ $\{7\}$
17. $5y + 2 = 2y - 11$ $\{-\frac{13}{3}\}$
18. $9y + 3 = 4y - 10$ $\{-\frac{13}{5}\}$
19. $3x + 4 = 5x - 2$ $\{3\}$
20. $2x - 1 = 6x + 15$ $\{-4\}$
21. $-7a + 6 = -8a + 14$ $\{8\}$
22. $-6a - 4 = -7a + 11$ $\{15\}$
23. $5x + 3 - 2x = x - 15$ $\{-9\}$
24. $4x - 2 - x = 5x + 10$ $\{-6\}$

25. $6y + 18 + y = 2y + 3$ $\{-3\}$

26. $5y + 14 + y = 3y - 7$ $\{-7\}$

27. $4x - 3 + 2x = 8x - 3 - x$ $\{0\}$

28. $x - 4 - 4x = 6x + 9 - 8x$ $\{-13\}$

29. $6n - 4 - 3n = 3n + 10 + 4n$ $\{-\frac{7}{2}\}$

30. $2n - 1 - 3n = 5n - 7 - 3n$ $\{2\}$

31. $4(x - 3) = -20$ $\{-2\}$

32. $3(x + 2) = -15$ $\{-7\}$

33. $-3(x - 2) = 11$ $\{-\frac{5}{3}\}$

34. $-5(x - 1) = 12$ $\{-\frac{7}{5}\}$

35. $5(2x + 1) = 4(3x - 7)$ $\{\frac{33}{2}\}$

36. $3(2x - 1) = 2(4x + 7)$ $\{-\frac{17}{2}\}$

37. $5x - 4(x - 6) = -11$ $\{-35\}$

38. $3x - 5(2x + 1) = 13$ $\{-\frac{18}{7}\}$

39. $-2(3x - 1) - 3 = -4$ $\{\frac{1}{2}\}$

40. $-6(x - 4) - 10 = -12$ $\{\frac{13}{3}\}$

41. $-2(3x + 5) = -3(4x + 3)$ $\{\frac{1}{6}\}$

42. $-(2x - 1) = -5(2x + 9)$ $\{-\frac{23}{4}\}$

43. $3(x - 4) - 7(x + 2) = -2(x + 18)$ $\{5\}$

44. $4(x - 2) - 3(x - 1) = 2(x + 6)$ $\{-17\}$

45. $-2(3n - 1) + 3(n + 5) = -4(n - 4)$ $\{-1\}$

46. $-3(4n + 2) + 2(n - 6) = -2(n + 1)$ $\{-2\}$

47. $3(2a - 1) - 2(5a + 1) = 4(3a + 4)$ $\{-\frac{21}{16}\}$

48. $4(2a + 3) - 3(4a - 2) = 5(4a - 7)$ $\{\frac{53}{24}\}$

49. $-2(n - 4) - (3n - 1) = -2 + (2n - 1)$ $\{\frac{12}{7}\}$

50. $-(2n - 1) + 6(n + 3) = -4 - (7n - 11)$ $\{-\frac{12}{11}\}$

Solve each of the following problems by setting up and solving an algebraic equation.

51. If 15 is subtracted from three times a certain number the result is 27. Find the number. 14

52. If 1 is subtracted from seven times a certain number the result is the same as if 31 is added to three times the number. Find the number. 8

53. Find three consecutive integers whose sum is 42. 13, 14, and 15

54. Find four consecutive integers whose sum is -118. -31, -30, -29, and -28

55. Find three consecutive odd integers such that three times the second minus the third is 11 more than the first. 9, 11, and 13

56. Find three consecutive even integers such that four times the first minus the third is 6 more than twice the second. 14, 16, and 18

57. The difference of two numbers is 67. The larger number is 3 less than six times the smaller number. Find the numbers. 14 and 81

58. The sum of two numbers is 103. The larger number is 1 more than five times the smaller number. Find the numbers. 17 and 86

59. Dan is paid double time for each hour he works over 40 hours in a week. Last week he worked 46 hours and earned \$572. What is his normal hourly rate? \$11 per hour

60. Suppose that a plumbing repair bill, not including tax, was \$130. This included \$25 for parts and an amount for 5 hours of labor. Find the hourly rate that was charged for labor. \$21 per hour.

61. Suppose that Maria has 150 coins consisting of pennies, nickels, and dimes. The number of nickels she has is 10 less than twice the number of pennies; the number of dimes she has is 20 less than three times the number of pennies. How many coins of each kind does she have? 30 pennies, 50 nickels, and 70 dimes

62. Hector has a collection of nickels, dimes, and quarters totaling 122 coins. The number of dimes he has is 3 more than four times the number of nickels, and the number of quarters he has is 19 less than the number of dimes. How many coins of each kind does he have? 15 nickels, 63 dimes, and 44 quarters

63. The selling price of a ring is \$750. This represents \$150 less than three times the cost of the ring. Find the cost of the ring. \$300

64. In a class of 62 students, the number of females is one less than twice the number of males. How many females and how many males are there in the class? 41 females and 21 males

65. An apartment complex contains 230 apartments each having one, two, or three bedrooms. The number of two-bedroom apartments is 10 more than three times the number of three-bedroom apartments. The number of one-bedroom apartments is twice the number of two-bedroom apartments. How many apartments of each kind are in the complex? 20 three-bedroom, 70 two-bedroom, and 140 one-bedroom

66. Barry sells bicycles on a salary-plus-commission basis. He receives a monthly salary of \$300 and a commission of \$15 for each bicycle that he sells. How many bicycles must he sell in a month to have a total monthly income of \$750? 30 bicycles

Miscellaneous Problems

67. Solve each of the following equations.

(a) $5x + 7 = 5x - 4$ $\varnothing$

(b) $4(x - 1) = 4x - 4$ All reals

(c) $3(x - 4) = 2(x - 6)$ $\{0\}$

(d) $7x - 2 = -7x + 4$ $\{\frac{3}{7}\}$

(e) $2(x - 1) + 3(x + 2) = 5(x - 7)$ $\varnothing$

(f) $-4(x - 7) = -2(2x + 1)$ $\varnothing$

68. Verify that for any three consecutive integers, the sum of the smallest and largest is equal to twice the middle integer. [*Hint*: Use n, $n + 1$, and $n + 2$ to represent the three consecutive integers.]

69. Verify that no four consecutive integers can be found such that the product of the smallest and largest is equal to the product of the other two integers.

2.2 Equations Involving Fractional Forms

To solve equations that involve fractions it is usually easiest to begin by **clearing the equation of all fractions**. This can be accomplished by multiplying both sides of the equation by the least common multiple of all the denominators in the equation. Remember that the least common multiple of a set of whole numbers is the smallest nonzero whole number that is divisible by each of the numbers. For example, the least common multiple of 2, 3, and 6 is 12. When working with fractions, we refer to the least common multiple of a set of denominators as the **least common denominator** (LCD). Let's consider some equations involving fractions.

EXAMPLE 1 Solve $\frac{1}{2}x + \frac{2}{3} = \frac{3}{4}$.

Solution

$$\frac{1}{2}x + \frac{2}{3} = \frac{3}{4}$$

$$12\left(\frac{1}{2}x + \frac{2}{3}\right) = 12\left(\frac{3}{4}\right)$$ Multiply both sides by 12, which is the LCD of 2, 3, and 4.

$$12\left(\frac{1}{2}x\right) + 12\left(\frac{2}{3}\right) = 12\left(\frac{3}{4}\right)$$ Apply distributive property on left side.

$$6x + 8 = 9$$

$$6x = 1$$

$$x = \frac{1}{6}$$

The solution set is $\left\{\frac{1}{6}\right\}$.

CHECK $\frac{1}{2}x + \frac{2}{3} = \frac{3}{4}$

$$\frac{1}{2}\left(\frac{1}{6}\right) + \frac{2}{3} \stackrel{?}{=} \frac{3}{4}$$

$$\frac{1}{12} + \frac{2}{3} \stackrel{?}{=} \frac{3}{4}$$

$$\frac{1}{12} + \frac{8}{12} \stackrel{?}{=} \frac{3}{4}$$

$$\frac{9}{12} \stackrel{?}{=} \frac{3}{4}$$

$$\frac{3}{4} = \frac{3}{4}$$

EXAMPLE 2 Solve $\frac{x}{2} + \frac{x}{3} = 10$.

Solution

$$\frac{x}{2} + \frac{x}{3} = 10$$ Recall that $\frac{x}{2} = \frac{1}{2}x$.

$$6\left(\frac{x}{2} + \frac{x}{3}\right) = 6(10)$$ Multiply both sides by the LCD.

$$6\left(\frac{x}{2}\right) + 6\left(\frac{x}{3}\right) = 6(10) \qquad \text{Apply distributive property on left side.}$$

$$3x + 2x = 60$$

$$5x = 60$$

$$x = 12$$

The solution set is $\{12\}$. ■

As you study the examples of this section, pay special attention to the steps shown in the solutions. Certainly, there are no rules as to which steps should be performed mentally; this is an individual decision. When you solve problems show enough steps to allow the flow of the process to be understood and to minimize the chances of making careless computational errors.

EXAMPLE 3 Solve $\dfrac{x-2}{3} + \dfrac{x+1}{8} = \dfrac{5}{6}$.

Solution

$$\frac{x-2}{3} + \frac{x+1}{8} = \frac{5}{6}$$

$$24\left(\frac{x-2}{3} + \frac{x+1}{8}\right) = 24\left(\frac{5}{6}\right) \qquad \text{Multiply both sides by the LCD.}$$

$$24\left(\frac{x-2}{3}\right) + 24\left(\frac{x+1}{8}\right) = 24\left(\frac{5}{6}\right) \qquad \text{Apply distributive property to left side.}$$

$$8(x-2) + 3(x+1) = 20$$

$$8x - 16 + 3x + 3 = 20$$

$$11x - 13 = 20$$

$$11x = 33$$

$$x = 3$$

The solution set is $\{3\}$. ■

EXAMPLE 4 Solve $\dfrac{3t-1}{5} - \dfrac{t-4}{3} = 1$.

Solution

$$\frac{3t-1}{5} - \frac{t-4}{3} = 1$$

$$15\left(\frac{3t-1}{5} - \frac{t-4}{3}\right) = 15(1) \qquad \text{Multiply both sides by LCD.}$$

$$15\left(\frac{3t-1}{5}\right) - 15\left(\frac{t-4}{3}\right) = 15(1) \qquad \text{Apply distributive property to left side.}$$

$$
\begin{aligned}
3(3t-1)-5(t-4) &= 15 \\
9t-3-5t+20 &= 15 \qquad \text{Be careful with this sign!} \\
4t+17 &= 15 \\
4t &= -2 \\
t &= -\frac{2}{4} = -\frac{1}{2} \qquad \text{Reduce!}
\end{aligned}
$$

The solution set is $\left\{-\frac{1}{2}\right\}$. ■

Problem Solving

As we expand our skills for solving equations, we also expand our capabilities for solving word problems. There is no one definite procedure that will ensure success at solving word problems, but the following suggestions can be helpful.

Suggestions for Solving Word Problems

1. Read the problem carefully and make certain that you understand the meanings of all of the words. Be especially alert for any technical terms used in the statement of the problem.
2. Read the problem a second time (perhaps even a third time) to get an overview of the situation being described. Determine the known facts as well as what is to be found.
3. Sketch any figure, diagram, or chart that might be helpful in analyzing the problem.
4. Choose a meaningful variable to represent an unknown quantity in the problem (perhaps t, if time is an unknown quantity) and represent any other unknowns in terms of that variable.
5. Look for a *guideline* that you can use to set up an equation. A guideline might be a formula such as "distance equals rate times time," or a statement of a relationship such as "the sum of the two numbers is 28."
6. Form an equation containing the variable that translates the conditions of the guideline from English to algebra.
7. Solve the equation and use the solution to determine all facts requested in the problem.
8. Check all answers back into the **original statement of the problem**.

Keep these suggestions in mind as we continue to solve problems. We will elaborate on some of these suggestions at different times throughout the text. Now let's consider some problems.

PROBLEM 1 Find a number such that three-eighths of the number minus one-half of it is 14 less than three-fourths of the number.

Solution Let n represent the number to be found.

$$\frac{3}{8}n - \frac{1}{2}n = \frac{3}{4}n - 14$$

$$8\left(\frac{3}{8}n - \frac{1}{2}n\right) = 8\left(\frac{3}{4}n - 14\right)$$

$$8\left(\frac{3}{8}n\right) - 8\left(\frac{1}{2}n\right) = 8\left(\frac{3}{4}n\right) - 8(14)$$

$$3n - 4n = 6n - 112$$

$$-n = 6n - 112$$

$$-7n = -112$$

$$n = 16$$

The number is 16. Check it! ■

PROBLEM 2 The width of a rectangular lot is 8 feet less than three-fifths of the length. The perimeter of the lot is 400 feet. Find the length and width of the lot.

Solution Let l represent the length of the lot. Then $\frac{3}{5}l - 8$ represents the width.

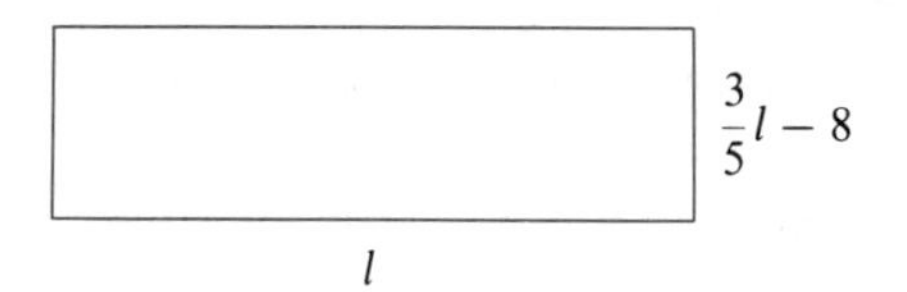

A *guideline* for this problem is the formula "perimeter of a rectangle equals twice the length plus twice the width ($P = 2l + 2w$)." Use this formula to form the following equation.

$$\begin{array}{cccc} P & = 2l & + & 2w \\ \downarrow & \downarrow & & \downarrow \\ 400 & = 2l & + & 2\left(\frac{3}{5}l - 8\right) \end{array}$$

Solving this equation we obtain

$$400 = 2l + \frac{6l}{5} - 16$$

$$5(400) = 5\left(2l + \frac{6l}{5} - 16\right)$$

$$2000 = 10l + 6l - 80$$

$$2000 = 16l - 80$$

$$2080 = 16l$$

$$130 = l.$$

The length of the lot is 130 feet and the width is $\frac{3}{5}(130) - 8 = 70$ feet. ■

In Problems 1 and 2 notice the use of different letters as variables. It is helpful to choose a variable that has significance for the problem you are working on. For example, in Problem 2 the choice of l to represent the length seems natural and meaningful. Certainly this is another matter of personal preference, but you might consider it.

In Problem 2 a geometric relationship ($P = 2l + 2w$) serves as a guideline for setting up the equation. The following geometric relationships pertaining to angle measure may also serve as guidelines.

1. **Two** angles whose sum of measures is 90° are called **complementary angles.**
2. **Two** angles whose sum of measures is 180° are called **supplementary angles.**
3. The sum of the measures of the three angles of a triangle is 180°.

PROBLEM 3 One of two complementary angles is 6° larger than one-half of the other angle. Find the measure of each of the angles.

Solution Let a represent the measure of one of the angles. Then $\frac{1}{2}a + 6$ represents the measure of the other angle. Since they are complementary angles, the sum of their measures is 90°.

$$a + \left(\frac{1}{2}a + 6\right) = 90$$

$$2a + a + 12 = 180$$

$$3a + 12 = 180$$

$$3a = 168$$

$$a = 56$$

If $a = 56$, then $\frac{1}{2}a + 6$ becomes $\frac{1}{2}(56) + 6 = 34$. The angles have measures of 34° and 56°. ■

PROBLEM 4 Find the measures of the three angles of a triangle if the smallest angle is one-sixth of the largest and the other angle is twice the smallest angle.

Solution Let a represent the measure of the largest angle. Then $\frac{1}{6}a$ represents the measure of the smallest angle and $2\left(\frac{1}{6}a\right)$, which simplifies to $\frac{1}{3}a$, represents the other angle. Since the sum of the measures of the angles of a triangle is 180°, we can set up and solve the following equation.

$$a + \frac{1}{6}a + \frac{1}{3}a = 180$$

$$6a + a + 2a = 1080$$

$$9a = 1080$$

$$a = 120$$

If $a = 120$, then $\frac{1}{6}a$ becomes $\frac{1}{6}(120) = 20$ and $\frac{1}{3}a$ becomes $\frac{1}{3}(120) = 40$. The angles have measures of 120°, 20°, and 40°. ■

Problem Set 2.2

Solve each of the following equations.

1. $\frac{3}{4}x = 9$ $\{12\}$
2. $\frac{2}{3}x = -14$ $\{-21\}$
3. $\frac{-2x}{3} = \frac{2}{5}$ $\left\{-\frac{3}{5}\right\}$
4. $\frac{-5x}{4} = \frac{7}{2}$ $\left\{-\frac{14}{5}\right\}$
5. $\frac{n}{2} - \frac{2}{3} = \frac{5}{6}$ $\{3\}$
6. $\frac{n}{4} - \frac{5}{6} = \frac{5}{12}$ $\{5\}$
7. $\frac{5n}{6} - \frac{n}{8} = \frac{-17}{12}$ $\{-2\}$
8. $\frac{2n}{5} - \frac{n}{6} = \frac{-7}{10}$ $\{-3\}$
9. $\frac{a}{4} - 1 = \frac{a}{3} + 2$ $\{-36\}$
10. $\frac{3a}{7} - 1 = \frac{a}{3}$ $\left\{\frac{21}{2}\right\}$
11. $\frac{h}{4} + \frac{h}{5} = 1$ $\left\{\frac{20}{9}\right\}$
12. $\frac{h}{6} + \frac{3h}{8} = 1$ $\left\{\frac{24}{13}\right\}$
13. $\frac{h}{2} - \frac{h}{3} + \frac{h}{6} = 1$ $\{3\}$
14. $\frac{3h}{4} + \frac{2h}{5} = 1$ $\left\{\frac{20}{23}\right\}$
15. $\frac{x-2}{3} + \frac{x+3}{4} = \frac{11}{6}$ $\{3\}$
16. $\frac{x+4}{5} + \frac{x-1}{4} = \frac{37}{10}$ $\{7\}$

17. $\frac{x+2}{2} - \frac{x-1}{5} = \frac{3}{5}$ $\{-2\}$

18. $\frac{2x+1}{3} - \frac{x+1}{7} = -\frac{1}{3}$ $\{-1\}$

19. $\frac{n+2}{4} - \frac{2n-1}{3} = \frac{1}{6}$ $\left\{\frac{8}{5}\right\}$

20. $\frac{n-1}{9} - \frac{n+2}{6} = \frac{3}{4}$ $\left\{-\frac{43}{2}\right\}$

21. $\frac{y}{3} + \frac{y-5}{10} = \frac{4y+3}{5}$ $\{-3\}$

22. $\frac{y}{3} + \frac{y-2}{8} = \frac{6y-1}{12}$ $\{-4\}$

23. $\frac{4x-1}{10} - \frac{5x+2}{4} = -3$ $\left\{\frac{48}{17}\right\}$

24. $\frac{2x-1}{2} - \frac{3x+1}{4} = \frac{3}{10}$ $\left\{\frac{21}{5}\right\}$

25. $\frac{2x-1}{8} - 1 = \frac{x+5}{7}$ $\left\{\frac{103}{6}\right\}$

26. $\frac{3x+1}{9} + 2 = \frac{x-1}{4}$ $\left\{-\frac{85}{3}\right\}$

27. $\frac{2a-3}{6} + \frac{3a-2}{4} + \frac{5a+6}{12} = 4$ $\{3\}$

28. $\frac{3a-1}{4} + \frac{a-2}{3} - \frac{a-1}{5} = \frac{21}{20}$ $\{2\}$

29. $x + \frac{3x-1}{9} - 4 = \frac{3x+1}{3}$ $\left\{\frac{40}{3}\right\}$

30. $\frac{2x+7}{8} + x - 2 = \frac{x-1}{2}$ $\left\{\frac{5}{6}\right\}$

31. $\frac{x+3}{2} + \frac{x+4}{5} = \frac{3}{10}$ $\left\{-\frac{20}{7}\right\}$

32. $\frac{x-2}{5} - \frac{x-3}{4} = -\frac{1}{20}$ $\{8\}$

33. $n + \frac{2n-3}{9} - 2 = \frac{2n+1}{3}$ $\left\{\frac{24}{5}\right\}$

34. $n - \frac{3n+1}{6} - 1 = \frac{2n+4}{12}$ $\left\{\frac{9}{2}\right\}$

35. $\frac{3}{4}(t-2) - \frac{2}{5}(2t-3) = \frac{1}{5}$ $\{-10\}$

36. $\frac{2}{3}(2t+1) - \frac{1}{2}(3t-2) = 2$ $\{-2\}$

37. $\frac{1}{2}(2x-1) - \frac{1}{3}(5x+2) = 3$ $\left\{-\frac{25}{4}\right\}$

38. $\frac{2}{5}(4x-1) + \frac{1}{4}(5x+2) = -1$ $\left\{-\frac{22}{57}\right\}$

39. $3x - 1 + \frac{2}{7}(7x-2) = -\frac{11}{7}$ $\{0\}$

40. $2x + 5 + \frac{1}{2}(6x-1) = -\frac{1}{2}$ $\{-1\}$

Solve each of the following problems by setting up and solving an algebraic equation.

41. Find a number such that one-half of the number is 3 less than two-thirds of the number. 18

42. One-half of a number plus three-fourths of the number is 2 more than four-thirds of the number. Find the number. -24

43. Suppose that the width of a certain rectangle is 1 inch more than one-fourth of its length. The perimeter of the rectangle is 42 inches. Find the length and width of the rectangle. 16 inches long and 5 inches wide

44. Suppose that the width of a rectangle is 3 centimeters less than two-thirds of its length. The perimeter of the rectangle is 114 centimeters. Find the length and width of the rectangle. 36 centimeters long and 21 centimeters wide

45. Find three consecutive integers such that the sum of the first plus one-third of the second plus three-eighths of the third is 25. 14, 15, and 16

46. Lou is paid $1\frac{1}{2}$ times his normal hourly rate for each hour he works over 40 hours in a week. Last week he worked 44 hours and earned $276. What is his normal hourly rate? $6 per hour

47. A board 20 feet long is cut into two pieces such that the length of one piece is two-thirds of the length of the other piece. Find the length of the shorter piece of board. 8 feet

48. Jody has a collection of 116 coins consisting of dimes, quarters, and silver dollars. The number of quarters is 5 less than three-fourths of the number of dimes. The number of silver dollars is seven more than five-eighths of the number of dimes. How many coins of each kind are in her collection? 48 dimes, 31 quarters, and 37 dollars

49. The sum of the present ages of Angie and her mother is 64 years. In eight years Angie will be three-fifths as old as her mother at that time. Find the present ages of Angie and her mother. Angie is 22 and her mother is 42

50. Donna's present age is two-thirds of Jessie's present age. In 12 years the sum of their ages will be 54 years. Find their present ages. Jessie is 18 and Donna is 12

51. Aura took three biology exams and has an average score of 88. Her second exam was ten points better than her first and her third exam was four points better than her second exam. What were her three exam scores? 80, 90, and 94

52. The average of the salaries of Tim, Ann, and Eric is \$24,000 per year. If Ann earns \$10,000 more than Tim, and Eric's salary is \$2000 more than twice Tim's salary, find the salary of each person. \$15,000 for Tim, \$25,000 for Ann, and \$32,000 for Eric

53. One of two supplementary angles is 4° more than one-third of the other angle. Find the measure of each of the angles. 48° and 132°

54. If one-half of the complement of an angle plus three-fourths of the supplement of the angle equals 110°, find the measure of the angle. 56°

55. If the complement of an angle is 5° less than one-sixth of its supplement, find the measure of the angle. 78°

56. In $\triangle ABC$, angle B is 8° less than one-half of angle A and angle C is 28° larger than angle A. Find the measures of the three angles of the triangle. A is 64°, B is 24°, and C is 92°

2.3 Equations Involving Decimals

Certainly we can solve the equation $x - 0.17 = 0.18$ by adding 0.17 to both sides. However, as equations that contain decimals become more complex, it is often easier to first **clear the equation of all decimals** by multiplying both sides by an appropriate power of 10. Let's consider some examples.

EXAMPLE 1 Solve $0.6x = 1.2$.

Solution

$$0.6x = 1.2$$

$$10(0.6x) = 10(1.2) \quad \text{Multiply both sides by 10.}$$

$$6x = 12$$

$$x = 2$$

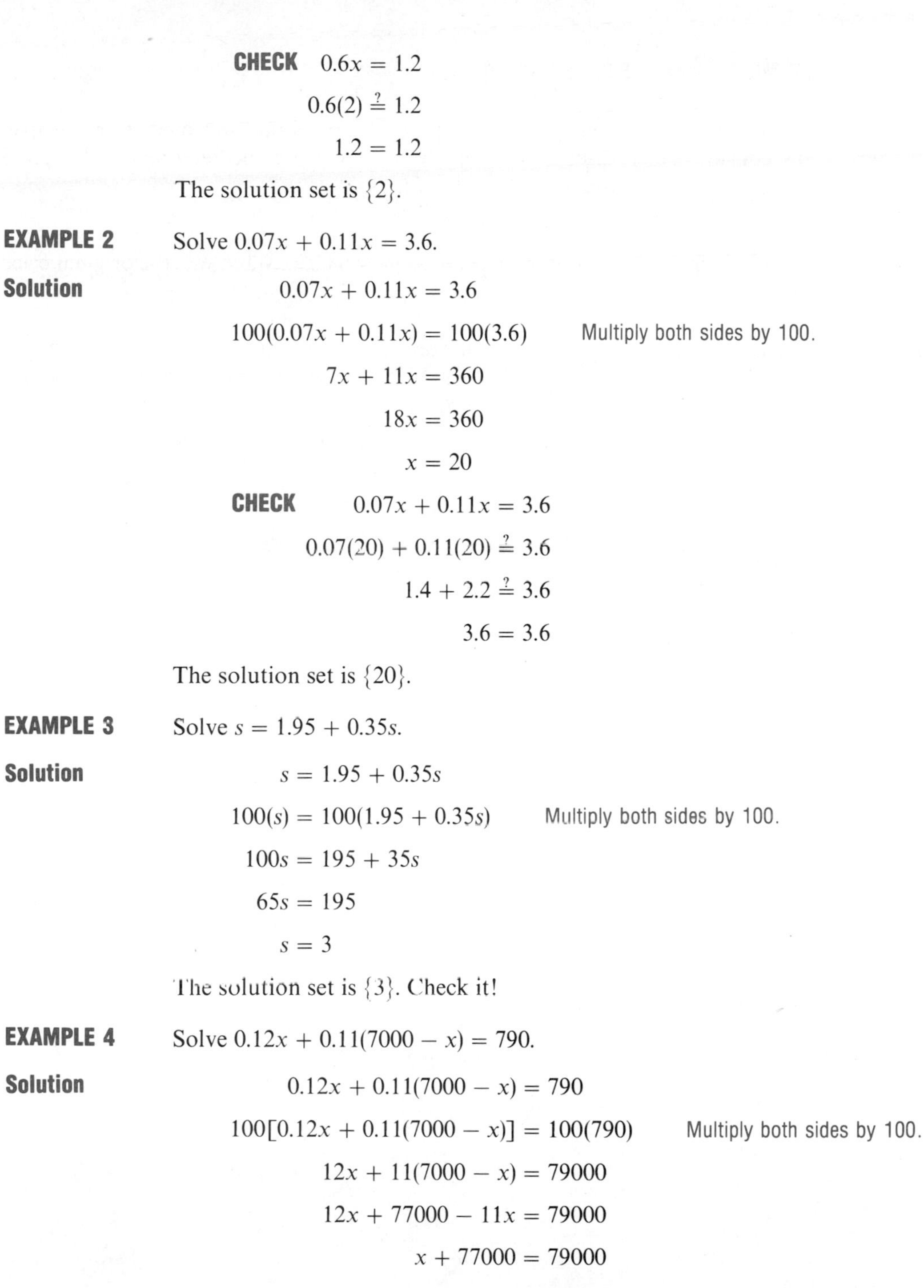

CHECK

$$0.6x = 1.2$$
$$0.6(2) \stackrel{?}{=} 1.2$$
$$1.2 = 1.2$$

The solution set is $\{2\}$. ■

EXAMPLE 2 Solve $0.07x + 0.11x = 3.6$.

Solution

$$0.07x + 0.11x = 3.6$$
$$100(0.07x + 0.11x) = 100(3.6) \quad \text{Multiply both sides by 100.}$$
$$7x + 11x = 360$$
$$18x = 360$$
$$x = 20$$

CHECK

$$0.07x + 0.11x = 3.6$$
$$0.07(20) + 0.11(20) \stackrel{?}{=} 3.6$$
$$1.4 + 2.2 \stackrel{?}{=} 3.6$$
$$3.6 = 3.6$$

The solution set is $\{20\}$. ■

EXAMPLE 3 Solve $s = 1.95 + 0.35s$.

Solution

$$s = 1.95 + 0.35s$$
$$100(s) = 100(1.95 + 0.35s) \quad \text{Multiply both sides by 100.}$$
$$100s = 195 + 35s$$
$$65s = 195$$
$$s = 3$$

The solution set is $\{3\}$. Check it! ■

EXAMPLE 4 Solve $0.12x + 0.11(7000 - x) = 790$.

Solution

$$0.12x + 0.11(7000 - x) = 790$$
$$100[0.12x + 0.11(7000 - x)] = 100(790) \quad \text{Multiply both sides by 100.}$$
$$12x + 11(7000 - x) = 79000$$
$$12x + 77000 - 11x = 79000$$
$$x + 77000 = 79000$$
$$x = 2000$$

The solution set is $\{2000\}$. ■

Back to Problem Solving

We can solve many consumer problems with an algebraic approach. For example, let's consider some discount sale problems involving the relationship "original selling price minus discount equals discount sale price."

PROBLEM 1 Karyl bought a dress at a 35% discount sale for \$32.50. What was the original price of the dress?

Solution Let p represent the original price of the dress. Using the discount sale relationship as a guideline, the problem translates into an equation as follows.

original selling price	minus	discount	equals	discount sale price
↓		↓		↓
$(100\%)(p)$	$-$	$(35\%)(p)$	$=$	\$32.50

Switching this equation to decimal form and solving, we obtain

$$(100\%)(p) - (35\%)(p) = 32.50$$
$$(65\%)(p) = 32.50$$
$$0.65p = 32.50$$
$$65p = 3250$$
$$p = 50.$$

The original price of the dress was \$50. ■

PROBLEM 2 A pair of jogging shoes that was originally priced at \$25 is on sale for 20% off. Find the discount sale price of the shoes.

Solution Let s represent the discount sale price.

original price	minus	discount	equals	sale price
↓		↓		↓
\$25	$-$	$(20\%)(\$25)$	$=$	s

Solving this equation we obtain

$$25 - (20\%)(25) = s$$
$$25 - (0.2)(25) = s$$
$$25 - 5 = s$$
$$20 = s.$$

The shoes are on sale for \$20. ■

REMARK Keep in mind that if an item is on sale for 35% off, then you are going to pay $100\% - 35\% = 65\%$ of the original price. Thus, in Problem 1 you could begin with the equation $.65p = 32.50$. Likewise, in Problem 2, you could start with the equation $s = .8(25)$.

Another basic relationship that pertains to consumer problems is "selling price equals cost plus profit." We can state profit, also called markup, markon, and margin of profit, in different ways. It may be stated as a percent of the selling price, a percent of the cost, or simply in terms of dollars and cents. We shall consider some problems for which the profit is calculated either as a percent of the cost or as a percent of the selling price.

PROBLEM 3 A retailer has some shirts that cost \$20 each. She wants to sell them at a profit of 60% of the cost. What selling price should be marked on the shirts?

Solution Let s represent the selling price. Use the "selling price equals cost plus profit" relationship as a guideline.

selling price	equals	cost	plus	profit
↓		↓		↓
s	$=$	\$20	$+$	(60%)(\$20)

Solving this equation yields

$$s = 20 + (60\%)(20)$$

$$s = 20 + (.6)(20)$$

$$s = 20 + 12$$

$$s = 32.$$

The selling price should be \$32. ■

REMARK A profit of 60% of the cost means that the selling price is 100% of the cost plus 60% of the cost, or 160% of the cost. Thus, in Problem 3 we could solve the equation $s = 1.6(20)$.

PROBLEM 4 A retailer of sporting goods bought a putter for \$18. He wants to price the putter to make a profit of 40% of the selling price. What price should he mark on the putter?

Solution Let s represent the selling price.

selling price	equals	cost	plus	profit
↓		↓		↓
s	$=$	\$18	$+$	(40%)(s)

Solving this equation yields

$$s = 18 + (40\%)(s)$$

$$s = 18 + .4s$$

$$10s = 180 + 4s \qquad \text{Multiply both sides by 10.}$$

$$6s = 180$$

$$s = 30.$$

The selling price should be \$30. ■

We can solve certain types of investment problems by using an algebraic approach. Consider the following examples.

PROBLEM 5 A man invests \$8000: part of it at 11% and the remainder at 12%. His total yearly interest from the two investments is \$930. How much did he invest at each rate?

Solution Let x represent the amount he invested at 11%. Then $8000 - x$ represents the amount he invested at 12%. Use the following guideline.

interest earned from 11% investment	+	interest earned from 12% investment	=	total amount of interest earned
↓		↓		↓
$(11\%)(x)$	+	$(12\%)(8000 - x)$	=	\$930

Solving this equation yields

$$(11\%)(x) + (12\%)(8000 - x) = 930$$

$$0.11x + 0.12(8000 - x) = 930$$

$$11x + 12(8000 - x) = 93000 \qquad \text{Multiply both sides by 100.}$$

$$11x + 96000 - 12x = 93000$$

$$-x + 96000 = 93000$$

$$-x = -3000$$

$$x = 3000.$$

Therefore, $8000 - x = 5000$. \$3000 was invested at 11% and \$5000 at 12%. ■

PROBLEM 6 A certain amount of money is invested at 8% and \$1500 more than that amount is invested at 9%. The annual interest from the 9% investment exceeds the annual interest from the 8% investment by \$160. How much is invested at each rate?

Solution Let x represent the amount invested at 8%. Then $x + 1500$ represents the amount invested at 9%. Use the following guideline.

$$\begin{array}{ccccc} \text{interest earned from 9\% investment} & = & \text{interest earned from 8\% investment} & + & \$160 \\ \downarrow & & \downarrow & & \downarrow \\ (9\%)(x + 1500) & = & (8\%)(x) & & + \$160 \end{array}$$

Solving this equation, we obtain

$$(9\%)(x + 1500) = (8\%)(x) + 160$$

$$0.09(x + 1500) = 0.08(x) + 160$$

$$9(x + 1500) = 8x + 16000 \qquad \text{Multiply both sides by 100.}$$

$$9x + 13500 = 8x + 16000$$

$$x = 2500.$$

Therefore, $x + 1500 = 4000$. \$2500 was invested at 8% and \$4000 at 9%. ■

Don't forget to check word problems; determine whether the answers satisfy the conditions stated in the original problem. A check for Problem 6 is as follows.

CHECK We claim that \$2500 was invested at 8% and \$4000 at 9% and this satisfies the condition "\$1500 more was invested at 9% than at 8%." The \$2500 at 8% produces \$200 of interest and the \$4000 at 9% produces \$360. Therefore, the interest from the 9% investment exceeds the interest from the 8% investment by \$160. The conditions of the problem are satisfied and our answers are correct.

Problem Set 2.3

Solve each of the following equations.

1. $.14x = 2.8$ {20}

2. $1.6x = 8$ {5}

3. $.09y = 4.5$ {50}

4. $.07y = .42$ {6}

5. $n + .4n = 56$ {40}

6. $n - .5n = 12$ {24}

7. $s = 9 + .25s$ {12}

8. $s = 15 + .4s$ {25}

9. $s = 3.3 + .45s$ {6}

10. $s = 2.1 + .6s$ $\{\frac{21}{4}\}$

11. $.11x + .12(900 - x) = 104$ {400}

12. $.09x + .11(500 - x) = 51$ {200}

13. $.08(x + 200) = .07x + 20$ {400}

14. $.07x = 152 - .08(2000 - x)$ {800}

15. $.12t - 2.1 = .07t - .2$ {38}

16. $.13t - 3.4 = .08t - .4$ {60}

17. $.92 + .9(x - .3) = 2x - 5.95$ {6}

18. $.3(2n - 5) = 11 - .65n$ {10}

19. $.1d + .11(d + 1500) = 795$ {3000}

20. $.8x + .9(850 - x) = 715$ {500}

21. $.12x + .1(5000 - x) = 560$ {3000}

22. $.10t + .12(t + 1000) = 560$ {2000}

23. $.08(x + 200) = .07x + 20$ {400}

24. $.09x = 1650 - .12(x + 5000)$ {5000}

25. $.3(2t + .1) = 8.43$ $\{14\}$

26. $.5(3t + .7) = 20.6$ $\{13.5\}$

27. $.1(x - .1) - .4(x + 2) = -5.31$ $\{15\}$

28. $.2(x + .2) + .5(x - .4) = 5.44$ $\{8\}$

Solve each of the following problems by setting up and solving an algebraic equation.

29. Judy bought a coat at a 20% discount sale for $72. What was the original price of the coat? $90

30. Jim bought a pair of slacks at a 25% discount sale for $24. What was the original price of the slacks? $32

31. Find the discount sale price of a $64 item that is on sale for 15% off. $54.40

32. Find the discount sale price of a $72 item that is on sale for 35% off. $46.80

33. A retailer has some skirts that cost $30 each. She wants to sell them at a profit of 60% of the cost. What price should she charge for the skirts? $48

34. The owner of a pizza parlor wants to make a profit of 70% of the cost for each pizza sold. If it costs $2.50 to make a pizza, at what price should each pizza be sold? $4.25

35. If a ring costs a jeweler $200, at what price should it be sold to make a profit of 50% on the selling price? $400

36. If a head of lettuce costs a retailer $.32, at what price should it be sold to make a profit of 60% on the selling price? $.80

37. If a pair of shoes costs a retailer $24 and he sells them for $39.60, what is his rate of profit based on the cost? 65%

38. A retailer has some skirts that cost her $45 each. If she sells them for $83.25 per skirt, find her rate of profit based on the cost. 85%

39. Robin's salary for next year is $34,775. This represents a 7% increase over this year's salary. Find Robin's present salary. $32,500

40. Don bought a car with 6% tax included for $15,794. What was the price of the car without the tax? $14,900

41. Cindy invested a certain amount of money at 10% interest and $1500 more than that amount at 11%. Her total yearly interest was $795. How much did she invest at each rate? $3000 at 10% and $4500 at 11%

42. A total of $4000 was invested, part of it at 8% interest and the remainder at 9%. If the total yearly interest amounted to $350, how much was invested at each rate? $1000 at 8% and $3000 at 9%

43. If $500 is invested at 6% interest, how much additional money must be invested at 9% so that the total return for both investments averages 8%? $1000

44. A sum of $2000 is split between two investments, one paying 7% interest and the other 8%. If the return on the 8% investment exceeds that on the 7% investment by $40 per year, how much is invested at each rate? $1200 at 8% and $800 at 7%

45. Suppose that Javier has a handful of coins consisting of pennies, nickels, and dimes worth $2.63. The number of nickels is one less than twice the number of pennies and the number of dimes is 3 more than the number of nickels. How many coins of each kind does he have? 8 pennies, 15 nickels, and 18 dimes

46. Sarah has a collection of nickels, dimes, and quarters worth $15.75. She has 10 more dimes than nickels and twice as many quarters as dimes. How many coins of each kind does she have? 15 nickels, 25 dimes, and 50 quarters

47. A collection of 70 coins consisting of dimes, quarters, and half-dollars has a value of \$17.75. There are three times as many quarters as dimes. Find the number of each kind of coin. 15 dimes, 45 quarters, and 10 half-dollars

48. Abby has 37 coins, consisting only of dimes and quarters, worth \$7.45. How many dimes and how many quarters does she have? 12 dimes and 25 quarters

Solve each of the following equations and express the solutions in decimal form. Check all of your solutions. Use your calculator whenever it seems helpful.

49. $1.2x + 3.4 = 5.2$ $\{1.5\}$

50. $0.12x - 0.24 = 0.66$ $\{7.5\}$

51. $0.12x + 0.14(550 - x) = 72.5$ $\{225\}$

52. $0.14t + 0.13(890 - t) = 67.95$ $\{-4775\}$

53. $0.7n + 1.4 = 3.92$ $\{3.6\}$

54. $0.14n - 0.26 = 0.958$ $\{8.7\}$

55. $0.3(d + 1.8) = 4.86$ $\{14.4\}$

56. $0.6(d - 4.8) = 7.38$ $\{17.1\}$

57. $0.8(2x - 1.4) = 19.52$ $\{12.9\}$

58. $0.5(3x + 0.7) = 20.6$ $\{13.5\}$

Miscellaneous Problems

59. The following formula can be used to determine the selling price of an item when the profit is based on a percent of the selling price.

$$\text{selling price} = \frac{\text{cost}}{100\% - \text{percent of profit}}$$

Show how this formula is developed.

60. A retailer buys an item for \$90, resells it for \$100, and claims that he is only making a 10% profit. Is his claim correct?

61. Is a 10% discount followed by a 20% discount equal to a 30% discount? Defend your answer.

2.4 Formulas

To find the distance traveled in 4 hours at a rate of 55 miles per hour we multiply the rate times the time; thus, the distance is $55(4) = 220$ miles. We can state the rule "distance equals rate times time" as a formula: $d = rt$. **Formulas** are rules we state in symbolic form, usually as equations.

The techniques we have considered for solving equations can be used to solve a formula for a specified variable if we are given numerical values for the other variables in the formula. Let's consider some examples.

EXAMPLE 1 If we invest P dollars at r percent for t years, the amount of simple interest i is given by the formula $i = Prt$. Find the amount of interest earned by \$500 at 7% for 2 years.

Solution By substituting \$500 for P, 7% for r, and 2 for t, we obtain

$$i = Prt$$

$$i = (500)(7\%)(2)$$

$$i = (500)(0.07)(2)$$

$$i = 70.$$

Thus, we earn \$70 in interest. ■

EXAMPLE 2 If we invest P dollars at a simple rate of r percent, then the amount A accumulated after t years is given by the formula $A = P + Prt$. If we invest \$500 at 8%, how many years would it take to accumulate \$600?

Solution Substituting \$500 for P, 8% for r, and \$600 for A, we obtain

$$A = P + Prt$$

$$600 = 500 + 500(8\%)(t).$$

Solving this equation for t yields

$$600 = 500 + 500(0.08)(t)$$

$$600 = 500 + 40t$$

$$100 = 40t$$

$$2\frac{1}{2} = t.$$

It will take $2\frac{1}{2}$ years to accumulate \$600. ■

When using a formula it is sometimes convenient to first change its form. For example, suppose we are to use the *perimeter* formula for a rectangle ($P = 2l + 2w$) to complete the following chart.

Perimeter **(*P*)**	32	24	36	18	56	80
Length **(*l*)**	10	7	14	5	15	22
Width **(*w*)**	?	?	?	?	?	?

all in centimeters

Since w is the unknown quantity, it would simplify the computational work if we first solved the formula for w in terms of the other variables as follows.

$$P = 2l + 2w$$

$$P - 2l = 2w \qquad \text{Add } -2l \text{ to both sides.}$$

$$\frac{P - 2l}{2} = w \qquad \text{Multiply both sides by } \frac{1}{2}.$$

$$w = \frac{P - 2l}{2} \qquad \text{Apply the symmetric property of equality.}$$

Now for each value for P and l, we can easily determine the corresponding value for w. Be sure you agree with the following values for w: 6, 5, 4, 4, 13, and 18.

We can also solve the formula $P = 2l + 2w$ for l in terms of P and w as follows.

$$P = 2l + 2w$$

$$P - 2w = 2l \qquad \text{Add } -2w \text{ to both sides.}$$

$$\frac{P - 2w}{2} = l \qquad \text{Multiply both sides by } \frac{1}{2}.$$

$$l = \frac{P - 2w}{2} \qquad \text{Apply the symmetric property of equality.}$$

Let's consider some other often used formulas and see how we can use the properties of equality to alter their forms. Throughout this section, we will identify formulas when we first use them.

EXAMPLE 3 Solve $A = \frac{1}{2}bh$ for h. (area of a triangle)

Solution

$$A = \frac{1}{2}bh$$

$$2A = bh \qquad \text{Multiply both sides by 2.}$$

$$\frac{2A}{b} = h \qquad \text{Multiply both sides by } \frac{1}{b}.$$

$$h = \frac{2A}{b} \qquad \text{Apply the symmetric property of equality.}$$

■

EXAMPLE 4 Solve $A = P + Prt$ for t.

Solution

$$A = P + Prt$$

$$A - P = Prt \qquad \text{Add } -P \text{ to both sides.}$$

$$\frac{A - P}{Pr} = t \qquad \text{Multiply both sides by } \frac{1}{Pr}.$$

$$t = \frac{A - P}{Pr} \qquad \text{Apply the symmetric property of equality.}$$

■

EXAMPLE 5 Solve $A = P + Prt$ for P.

Solution

$$A = P + Prt$$

$$A = P(1 + rt) \quad \text{Apply the distributive property to the right side.}$$

$$\frac{A}{1 + rt} = P \quad \text{Multiply both sides by } \frac{1}{1 + rt}.$$

$$P = \frac{A}{1 + rt} \quad \text{Apply the symmetric property of equality.}$$

■

EXAMPLE 6 Solve $A = \frac{1}{2}h(b_1 + b_2)$ for b_1. (area of a trapezoid)

Solution

$$A = \frac{1}{2}h(b_1 + b_2)$$

$$2A = h(b_1 + b_2) \quad \text{Multiply both sides by 2.}$$

$$2A = hb_1 + hb_2 \quad \text{Apply the distributive property to right side.}$$

$$2A - hb_2 = hb_1 \quad \text{Add } -hb_2 \text{ to both sides.}$$

$$\frac{2A - hb_2}{h} = b_1 \quad \text{Multiply both sides by } \frac{1}{h}.$$

$$b_1 = \frac{2A - hb_2}{h} \quad \text{Apply the symmetric property of equality.}$$

■

In Example 5, notice that we used the distributive property to change from a form of $P + Prt$ to $P(1 + rt)$. However, in Example 6 we used the distributive property to change $h(b_1 + b_2)$ to $hb_1 + hb_2$. In both problems the key issue is to *isolate the term* that contains the variable being solved for so that an appropriate application of the multiplication property of equality will produce the desired result. Also note the use of *subscripts* to identify the two bases of a trapezoid. Subscripts allow us to use the same letter b to identify the bases, but b_1 represents one base and b_2 the other.

Sometimes we are faced with equations such as $ax + b = c$, where x is the variable and a, b, and c are referred to as *arbitrary constants*. Again we can use the properties of equality to solve the equation for x as follows.

$$ax + b = c$$

$$ax = c - b \quad \text{Add } -b \text{ to both sides.}$$

$$x = \frac{c - b}{a} \quad \text{Multiply both sides by } \frac{1}{a}.$$

In Chapter 7 we will be working with equations such as $2x - 5y = 7$, which are called equations of *two* variables in x and y. Often we need to change the form of such equations by "solving for one variable in terms of the other variable." The properties of equality provide the basis for doing this.

EXAMPLE 7 Solve $2x - 5y = 7$ for y in terms of x.

Solution

$$2x - 5y = 7$$

$$-5y = 7 - 2x \qquad \text{Add } -2x \text{ to both sides.}$$

$$y = \frac{7 - 2x}{-5} \qquad \text{Multiply both sides by } -\frac{1}{5}.$$

$$y = \frac{2x - 7}{5}$$

Multiply the numerator and denominator of the fraction on the right by -1. (The final step would not be absolutely necessary, but usually we prefer to have a positive number as a denominator.) ■

Equations of two variables may also contain arbitrary constants. For example, the equation $\frac{x}{a} + \frac{y}{b} = 1$ contains the variables x and y, and the arbitrary constants a and b.

EXAMPLE 8 Solve the equation $\frac{x}{a} + \frac{y}{b} = 1$ for x.

Solution

$$\frac{x}{a} + \frac{y}{b} = 1$$

$$ab\left(\frac{x}{a} + \frac{y}{b}\right) = ab(1) \qquad \text{Multiply both sides by } ab.$$

$$bx + ay = ab$$

$$bx = ab - ay \qquad \text{Add } -ay \text{ to both sides.}$$

$$x = \frac{ab - ay}{b} \qquad \text{Multiply both sides by } \frac{1}{b}.$$

■

Formulas and Problem Solving

We often use formulas as *guidelines* for setting up an appropriate algebraic equation when solving a word problem. Let's consider an example to illustrate this point.

PROBLEM 1 How long will it take \$500 to double itself if we invest it at 8% simple interest?

Solution Let t represent the number of years it will take \$500 to earn \$500 in interest (double itself). We can use the formula $i = Prt$ as a guideline.

$$\overset{i}{500} = \overset{P}{500}\overset{r}{(8\%)}\overset{t}{(t)}$$

Solving this equation we obtain

$$500 = 500(0.08)(t)$$

$$1 = 0.08t$$

$$100 = 8t$$

$$12\frac{1}{2} = t.$$

It will take $12\frac{1}{2}$ years. ■

Sometimes we use formulas in the analysis of a problem but not as the main guideline for setting up the equation. For example, uniform motion problems involve the formula $d = rt$, but the main guideline for setting up an equation for such problems is usually a statement about either *times*, *rates*, or *distances*. Let's consider an example to demonstrate.

PROBLEM 2 Lori starts jogging at 5 miles per hour. One-half hour later, Karen starts jogging on the same route at 7 miles per hour. How long will it take Karen to catch Lori?

Solution First, let's sketch a diagram and record some information.

Lori at 5 mph.

Karen at 7 mph, but starts $\frac{1}{2}$ hour later.

If we let t represent Karen's time, then $t + \frac{1}{2}$ represents Lori's time. We can use the statement "Karen's distance equals Lori's distance" as a guideline.

Karen's distance Lori's distance

$$7t = 5\left(t + \frac{1}{2}\right)$$

Solving this equation we obtain

$$7t = 5t + \frac{5}{2}$$

$$2t = \frac{5}{2}$$

$$t = \frac{5}{4}.$$

Karen should catch Lori in $1\frac{1}{4}$ hours. ■

Note that in Problem 2 we used a simple arrow diagram to record and organize the pertinent information of the problem. Some people find it helpful to use a chart with uniform motion problems. We shall use a chart in Problem 3 but keep in mind that we are not trying to dictate a particular approach; you decide what works best for you.

PROBLEM 3 Two trains leave a city at the same time, one traveling east and the other traveling west. At the end of $9\frac{1}{2}$ hours, they are 1292 miles apart. If the rate of the train traveling east is 8 miles per hour faster than the other train, find their rates.

Solution If we let r represent the rate of the westbound train, then $r + 8$ represents the rate of the eastbound train. Now we can record the *times* and *rates* in a chart and then use the distance formula ($d = rt$) to represent the *distances*.

	Rate	*Time*	*Distance* ($d = rt$)
Westbound train	r	$9\frac{1}{2}$	$\frac{19}{2}r$
Eastbound train	$r + 8$	$9\frac{1}{2}$	$\frac{19}{2}(r + 8)$

Since the distance that the westbound train travels plus the distance that the eastbound train travels equals 1292 miles, we can set up and solve the following equation.

$$\frac{19r}{2} + \frac{19(r + 8)}{2} = 1292$$

$$19r + 19(r + 8) = 2584$$

$$19r + 19r + 152 = 2584$$

$$38r = 2432$$

$$r = 64$$

The westbound train travels at a rate of 64 miles per hour and the eastbound train at a rate of $64 + 8 = 72$ miles per hour. ■

Now let's consider a problem that is often referred to as a mixture-type problem. There is no basic formula that applies to all of these problems but we suggest that you "think in terms of a pure substance," which is often helpful in setting up a guideline. Also keep in mind that a statement such as "a 40% solution of some substance" means that the solution contains 40% of that particular substance and 60% of something else mixed with it. For example, a 40% salt solution contains 40% salt and the other 60% is something else, probably water. Now let's illustrate what we mean by the suggestion "think in terms of a pure substance."

PROBLEM 4 How many liters of pure alcohol must we add to 20 liters of a 40% solution to obtain a 60% solution?

Solution The key idea to solving such a problem is to recognize the following guideline.

$$\begin{pmatrix}\text{amount of pure}\\ \text{alcohol in the}\\ \text{original solution}\end{pmatrix} + \begin{pmatrix}\text{amount of pure}\\ \text{alcohol to be}\\ \text{added}\end{pmatrix} = \begin{pmatrix}\text{amount of pure}\\ \text{alcohol in the}\\ \text{final solution}\end{pmatrix}$$

Let l represent the number of liters of pure alcohol to be added and the guideline translates into the following equation.

$$(40\%)(20) + l = 60\%(20 + l).$$

Solving this equation yields

$$\begin{aligned} 0.4(20) + l &= 0.6(20 + l) \\ 8 + l &= 12 + 0.6l \\ 0.4l &= 4 \\ l &= 10. \end{aligned}$$

We need to add 10 liters of pure alcohol. (Perhaps you should check this answer back into the original statement of the problem!) ■

Problem Set 2.4

1. Solve $i = Prt$ for i, given that $P = \$300$, $r = 8\%$, and $t = 5$ years. \$120
2. Solve $i = Prt$ for i, given that $P = \$500$, $r = 9\%$, and $t = 3\frac{1}{2}$ years. \$157.50
3. Solve $i = Prt$ for t, given that $P = \$400$, $r = 11\%$, and $i = \$132$. 3 years
4. Solve $i = Prt$ for t, given that $P = \$250$, $r = 12\%$, and $i = \$120$. 4 years

5. Solve $i = Prt$ for r, given that $P = \$600$, $t = 2\frac{1}{2}$ years, and $i = \$90$. Express r as a percent.
6%
6. Solve $i = Prt$ for r, given that $P = \$700$, $t = 2$ years, and $i = \$126$. Express r as a percent.
9%
7. Solve $i = Prt$ for P, given that $r = 9\%$, $t = 3$ years, and $i = \$216$. \$800
8. Solve $i = Prt$ for P, given that $r = 8\frac{1}{2}\%$, $t = 2$ years, and $i = \$204$. \$1200
9. Solve $A = P + Prt$ for A, given that $P = \$1000$, $r = 12\%$, and $t = 5$ years. \$1600
10. Solve $A = P + Prt$ for A, given that $P = \$850$, $r = 9\frac{1}{2}\%$, and $t = 10$ years. \$1657.50
11. Solve $A = P + Prt$ for r, given that $A = \$1372$, $P = \$700$, and $t = 12$ years. Express r as a percent. 8%
12. Solve $A = P + Prt$ for r, given that $A = \$516$, $P = \$300$, and $t = 8$ years. Express r as a percent. 9%
13. Solve $A = P + Prt$ for P, given that $A = \$326$, $r = 7\%$, and $t = 9$ years. \$200
14. Solve $A = P + Prt$ for P, given that $A = \$720$, $r = 8\%$, and $t = 10$ years. \$400
15. Use the formula $A = \frac{1}{2}h(b_1 + b_2)$ and complete the following chart. 6 feet; 14 feet; 10 feet; 20 feet; 7 feet; 2 feet

Area (A)	98	104	49	162	$16\frac{1}{2}$	$38\frac{1}{2}$	square feet
Height (h)	14	8	7	9	3	11	feet
One base (b_1)	8	12	4	16	4	5	feet
Other base (b_2)	?	?	?	?	?	?	feet

16. Use the formula $P = 2l + 2w$ and complete the following chart. (You may want to change the form of the formula.) 8 centimeters; 6 centimeters; 4 centimeters; 10 centimeters; 20 centimeters

Perimeter (P)	28	18	12	34	68	centimeters
Width (w)	6	3	2	7	14	centimeters
Length (l)	?	?	?	?	?	centimeters

Solve each of the following for the indicated variable.

17. $V = Bh$ for h (volume of a prism) $h = \frac{V}{B}$
18. $A = lw$ for l (area of a rectangle) $l = \frac{A}{w}$
19. $V = \pi r^2 h$ for h (volume of a circular cylinder) $h = \frac{V}{\pi r^2}$

20. $V = \frac{1}{3}Bh$ for B (volume of a pyramid) $B = \frac{3V}{h}$

21. $C = 2\pi r$ for r (circumference of a circle) $r = \frac{C}{2\pi}$

22. $A = 2\pi r^2 + 2\pi rh$ for h (surface area of a circular cylinder) $h = \frac{A - 2\pi r^2}{2\pi r}$

23. $I = \frac{100M}{C}$ for C (intelligence quotient) $C = \frac{100M}{I}$

24. $A = \frac{1}{2}h(b_1 + b_2)$ for h (area of a trapezoid) $h = \frac{2A}{b_1 + b_2}$

25. $F = \frac{9}{5}C + 32$ for C (Celsius to Fahrenheit) $C = \frac{5}{9}(F - 32)$ or $C = \frac{5F - 160}{9}$

26. $C = \frac{5}{9}(F - 32)$ for F (Fahrenheit to Celsius) $F = \frac{9C + 160}{5}$ or $F = \frac{9}{5}C + 32$

For Problems 27–36, solve each equation for x.

27. $y = mx + b$ $x = \frac{y - b}{m}$

28. $\frac{x}{a} + \frac{y}{b} = 1$ $x = \frac{ab - ay}{b}$

29. $y - y_1 = m(x - x_1)$ $x = \frac{y - y_1 + mx_1}{m}$

30. $a(x + b) = c$ $x = \frac{c - ab}{a}$

31. $a(x + b) = b(x - c)$ $x = \frac{ab + bc}{b - a}$

32. $x(a - b) = m(x - c)$ $x = \frac{mc}{m + b - a}$

33. $\frac{x - a}{b} = c$ $x = a + bc$

34. $\frac{x}{a} - 1 = b$ $x = ab + a$

35. $\frac{1}{3}x + a = \frac{1}{2}b$ $x = \frac{3b - 6a}{2}$

36. $\frac{2}{3}x - \frac{1}{4}a = b$ $x = \frac{12b + 3a}{8}$

For Problems 37–46, solve each equation for the indicated variable.

37. $2x - 5y = 7$ for x $x = \frac{5y + 7}{2}$

38. $5x - 6y = 12$ for x $x = \frac{6y + 12}{5}$

39. $-7x - y = 4$ for y $y = -7x - 4$

40. $3x - 2y = -1$ for y $y = \frac{3x + 1}{2}$

41. $3(x - 2y) = 4$ for x $x = \frac{6y + 4}{3}$

42. $7(2x + 5y) = 6$ for y $y = \frac{6 - 14x}{35}$

43. $\frac{y - a}{b} = \frac{x + b}{c}$ for x $x = \frac{cy - ac - b^2}{b}$

44. $\frac{x - a}{b} = \frac{y - a}{c}$ for y $y = \frac{cx - ac + ab}{b}$

45. $(y + 1)(a - 3) = x - 2$ for y $y = \frac{x - a + 1}{a - 3}$

46. $(y - 2)(a + 1) = x$ for y $y = \frac{x + 2a + 2}{a + 1}$

Solve each of the following problems by setting up and solving an appropriate algebraic equation.

47. Suppose that the length of a certain rectangle is 2 meters less than four times its width. The perimeter of the rectangle is 56 meters. Find the length and width of the rectangle. 22 meters long and 6 meters wide

48. The perimeter of a triangle is 42 inches. The second side is 1 inch more than twice the first side and the third side is 1 inch less than three times the first side. Find the lengths of the three sides of the triangle. 7 inches, 15 inches, and 20 inches

49. How long will it take \$500 to double itself at 9% simple interest? $11\frac{1}{9}$ years

50. How long will it take \$700 to triple itself at 10% simple interest? 20 years

51. How long will it take P dollars to double itself at 9% simple interest? $11\frac{1}{9}$ years

52. How long will it take P dollars to triple itself at 10% simple interest? 20 years

53. Two airplanes leave Chicago at the same time and fly in opposite directions. If one travels at 450 miles per hour and the other at 550 miles per hour, how long will it take for them to be 4000 miles apart? 4 hours

54. Dave leaves city A on a moped traveling toward city B at 18 miles per hour. At the same time, Tina leaves city B on a bicycle traveling toward city A at 14 miles per hour. The distance between the two cities is 112 miles. How long will it take before Dave and Tina meet? $3\frac{1}{2}$ hours

55. Dennis starts walking at 4 miles per hour. An hour and a half later Cathy starts jogging along the same route at 6 miles per hour. How long will it take Cathy to catch up with Dennis? 3 hours

56. A car leaves a town at 60 kilometers per hour. How long will it take a second car traveling at 75 kilometers per hour to catch the first car if it leaves 1 hour later? 4 hours

57. Bret starts on a 70-mile bicycle ride at 20 miles per hour. After a time he becomes a little tired and slows down to 12 miles per hour for the rest of the trip. The entire trip of 70 miles took $4\frac{1}{2}$ hours. How far had Bret ridden when he reduced his speed to 12 miles per hour? 40 miles

58. How many cups of grapefruit juice must be added to 40 cups of punch that contains 5% grapefruit juice to obtain a punch that is 10% grapefruit juice? $2\frac{2}{9}$ cups

59. How many milliliters of pure acid must be added to 150 milliliters of a 30% solution of acid to obtain a 40% solution? 25 milliliters

60. How many gallons of a 12%-salt solution must be mixed with 6 gallons of a 20%-salt solution to obtain a 15%-salt solution? 10 gallons

61. Suppose that you have a supply of a 30% solution of alcohol and a 70% solution of alcohol. How many quarts of each should be mixed to produce 20 quarts that is 40% alcohol? 15 quarts of 30% solution and 5 quarts of 70% solution

62. A 16-quart radiator contains a 50% solution of antifreeze. How much needs to be drained out and replaced with pure antifreeze to obtain a 60%-antifreeze solution? $3\frac{1}{5}$ quarts

For Problems 63–70, use your calculator to help solve each formula for the indicated variable.

63. Solve $i = Prt$ for i, given that $P = \$875$, $r = 12\frac{1}{2}\%$, and $t = 4$ years. \$437.50

64. Solve $i = Prt$ for i, given that $P = \$1125$, $r = 13\frac{1}{4}\%$, and $t = 4$ years. \$596.25

65. Solve $i = Prt$ for t, given that $i = \$453.25$, $P = \$925$, and $r = 14\%$. 3.5 years

66. Solve $i = Prt$ for t, given that $i = \$243.75$, $P = \$1250$, and $r = 13\%$. 1.5 years

67. Solve $i = Prt$ for r, given that $i = \$356.50$, $P = \$1550$, and $t = 2$ years. Express r as a percent. 11.5%

68. Solve $i = Prt$ for r, given that $i = \$159.50$, $P = \$2200$, and $t = 0.5$ of a year. Express r as a percent. 14.5%

69. Solve $A = P + Prt$ for P, given that $A = \$1423.50$, $r = 9\frac{1}{2}\%$, and $t = 1$ year. \$1300

70. Solve $A = P + Prt$ for P, given that $A = \$2173.75$, $r = 8\frac{3}{4}\%$, and $t = 2$ years. \$1850

2.5 Inequalities

Just as we use the symbol $=$ to represent *is equal to*, we also use the symbols $<$ and $>$ to represent "is less than" and "is greater than," respectively. Thus, various **statements of inequality** can be made as follows.

$a < b$ means a is less than b;

$a \leq b$ means a is less than or equal to b;

$a > b$ means a is greater than b;

$a \geq b$ means a is greater than or equal to b.

The following are examples of **numerical statements of inequality**.

$$7 + 8 > 10, \qquad -4 + (-6) \geq -10,$$
$$-4 > -6, \qquad 7 - 9 \leq -2,$$
$$7 - 1 < 20, \qquad 3 + 4 > 12,$$
$$8(-3) < 5(-3), \qquad 7 - 1 < 0$$

Notice that only $3 + 4 > 12$ and $7 - 1 < 0$ are *false*; the other six are *true* numerical statements.

Algebraic inequalities contain one or more variables. The following are examples of algebraic inequalities.

$$x + 4 > 8, \qquad 3x + 2y \leq 4;$$
$$3x - 1 < 15, \qquad x^2 + y^2 + z^2 \geq 7,$$
$$y^2 + 2y - 4 \geq 0$$

An algebraic inequality such as $x + 4 > 8$ is neither true nor false as it stands and we call it an **open sentence**. For each numerical value we substitute for x, the algebraic inequality $x + 4 > 8$ becomes a numerical statement of inequality that is true or false. For example, if $x = -3$ then $x + 4 > 8$ becomes $-3 + 4 > 8$, which is false. If $x = 5$, then $x + 4 > 8$ becomes $5 + 4 > 8$, which is true. **Solving an inequality** refers to the process of finding the numbers that make an algebraic inequality a true numerical statement. We call such numbers the *solutions* of the inequality; the solutions *satisfy* the inequality.

The general process for solving inequalities closely parallels that for solving equations. We continue to replace the given inequality with *equivalent, but simpler*, inequalities. For example,

$$3x + 4 > 10 \qquad (1)$$

$$3x > 6 \qquad (2)$$

$$x > 2 \qquad (3)$$

are all equivalent inequalities; that is, they all have the same solutions. By inspection we see that the solutions for (3) are "all numbers greater than 2." Thus, (1) has the same solutions.

The exact procedure for simplifying inequalities so that we can determine the solutions is primarily based on two properties. The first of these is the **addition property of inequality**.

Addition Property of Inequality

For all real numbers a, b, and c,

$$a > b \quad \text{if and only if } a + c > b + c.$$

The addition property of inequality states that *we can add any number to both sides of an inequality to produce an equivalent inequality.* We have stated the property in terms of $>$, but analogous properties exist for $<$, $\geq$, and $\leq$.

Before we state the multiplication property of inequality let's look at some numerical examples.

$$2 < 5 \xrightarrow{\text{multiply by 4}} \underline{4}(2) < \underline{4}(5);$$

$$-3 > -7 \xrightarrow{\text{multiply by 2}} \underline{2}(-3) > \underline{2}(-7);$$

$$-4 < 6 \xrightarrow{\text{multiply by 10}} \underline{10}(-4) < \underline{10}(6);$$

$$4 < 8 \xrightarrow{\text{multiply by } -3} \underline{-3}(4) > \underline{-3}(8);$$

$$3 > -2 \xrightarrow{\text{multiply by } -4} \underline{-4}(3) < \underline{-4}(-2);$$

$$-4 < -1 \xrightarrow{\text{multiply by } -2} \underline{-2}(-4) > \underline{-2}(-1)$$

Notice in the first three examples that when we multiply both sides of an inequality by a *positive number* we get an inequality of the *same sense.* That means that if the original inequality is *less than,* then the new inequality is *less than*; and if the original inequality is *greater than,* then the new inequality is *greater than.* The last three examples illustrate that when we multiply both sides of an inequality by a *negative number* we get an inequality of the *opposite sense.*

We can state the multiplication property of inequality as follows.

Multiplication Property of Inequality

(a) For all real numbers a, b, and c, *with* $c > 0$,

$$a > b \quad \text{if and only if } ac > bc.$$

(b) For all real numbers a, b, and c, *with* $c < 0$,

$$a > b \quad \text{if and only if } ac < bc.$$

Similar properties hold if we reverse each inequality or if we replace $>$ with $\geq$ and $<$ with $\leq$. For example, if $a \leq b$ and $c < 0$, then $ac \geq bc$.

Now let's use the addition and multiplication properties of inequality to help solve some inequalities.

EXAMPLE 1 Solve $3x - 4 > 8$.

Solution

$$3x - 4 > 8$$

$$3x - 4 + 4 > 8 + 4 \qquad \text{Add 4 to both sides.}$$

$$3x > 12$$

$$\frac{1}{3}(3x) > \frac{1}{3}(12) \qquad \text{Multiply both sides by } \frac{1}{3}.$$

$$x > 4$$

The solution set is $\{x \mid x > 4\}$. (Remember that we read the set $\{x \mid x > 4\}$ as "the set of all x such that x is greater than 4.") ■

In Example 1, once we obtained the simple inequality $x > 4$, the solution set $\{x \mid x > 4\}$ became obvious. We can also express solution sets for inequalities on a number line graph. Figure 2.1 shows the graph of the solution set for Example 1. The left-hand parenthesis at 4 indicates that 4 is *not* a solution and the thickened portion to the right of 4 indicates that all numbers greater than 4 are solutions.

Figure 2.1

−4 −2 0 2 4

It is also convenient to express solution sets of inequalities using **interval notation**. For example, the notation $(4, \infty)$ also refers to the set of real numbers greater than 4. As in Figure 2.1, the left-hand parenthesis indicates that 4 is not to be included. The infinity symbol, ∞, along with the right-hand parenthesis, indicates that there is no right-hand endpoint. Following is a partial list of interval notations along with the sets and graphs they represent. We will add to this list in the next section.

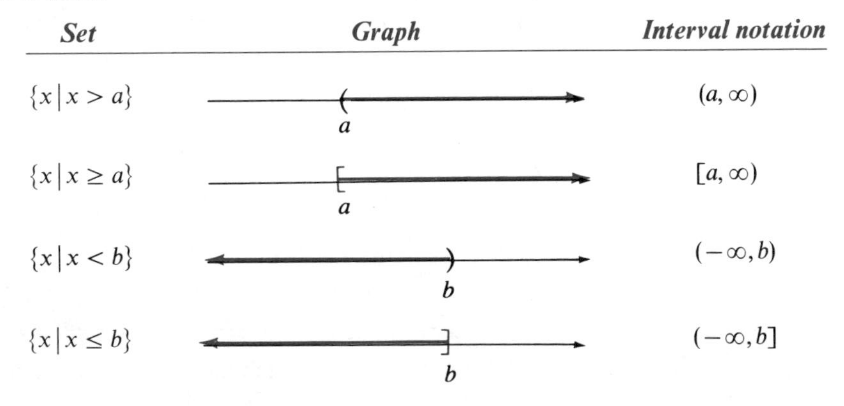

Set	*Graph*	*Interval notation*
$\{x \mid x > a\}$	a	(a, ∞)
$\{x \mid x \geq a\}$	a	$[a, \infty)$
$\{x \mid x < b\}$	b	$(-\infty, b)$
$\{x \mid x \leq b\}$	b	$(-\infty, b]$

Notice the use of square brackets to include endpoints. From now on, we will express the solution sets of inequalities using interval notation.

EXAMPLE 2 Solve $-2x + 1 > 5$ and graph the solutions.

Solution

$$-2x + 1 > 5$$

$$-2x + 1 + (-1) > 5 + (-1) \qquad \text{Add } -1 \text{ to both sides.}$$

$$-2x > 4$$

$$-\frac{1}{2}(-2x) < -\frac{1}{2}(4) \qquad \text{Multiply both sides by } -\frac{1}{2}.$$

$$x < -2$$

Notice that the sense of the inequality has been reversed.

The solution set is $(-\infty, -2)$, which can be illustrated on a number line as in Figure 2.2.

Figure 2.2

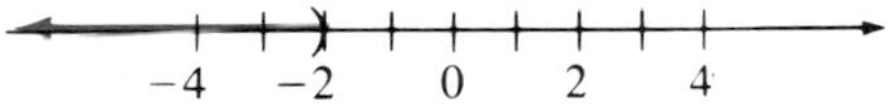

■

Many of the same techniques used to solve equations, such as removing parentheses and combining similar terms, may be used to solve inequalities. However, we must be extremely careful when using the multiplication property of inequality. Study each of the following examples very carefully. The format we used highlights the major steps of a solution.

EXAMPLE 3 Solve $-3x + 5x - 2 \geq 8x - 7 - 9x$.

Solution

$$-3x + 5x - 2 \geq 8x - 7 - 9x$$

$$2x - 2 \geq -x - 7 \qquad \text{Combine similar terms on both sides.}$$

$$3x - 2 \geq -7 \qquad \text{Add } x \text{ to both sides.}$$

$$3x \geq -5 \qquad \text{Add 2 to both sides.}$$

$$\frac{1}{3}(3x) \geq \frac{1}{3}(-5) \qquad \text{Multiply both sides by } \frac{1}{3}.$$

$$x \geq -\frac{5}{3}$$

The solution set is $\left[-\frac{5}{3}, \infty\right)$. ■

EXAMPLE 4 Solve $-5(x-1) \leq 10$ and graph the solutions.

Solution

$$-5(x-1) \leq 10$$

$$-5x + 5 \leq 10 \qquad \text{Apply the distributive property on left side.}$$

$$-5x \leq 5 \qquad \text{Add } -5 \text{ to both sides.}$$

$$-\frac{1}{5}(-5x) \geq -\frac{1}{5}(5) \qquad \text{Multiply both sides by } -\frac{1}{5}, \text{ which reverses the inequality.}$$

$$x \geq -1$$

The solution set is $[-1, \infty)$ and it can be graphed as in Figure 2.3.

Figure 2.3

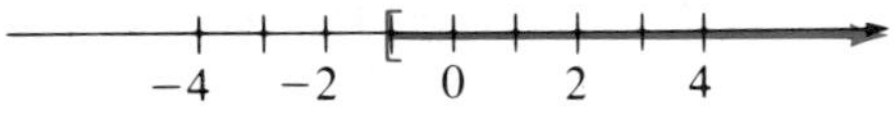

−4 −2 0 2 4

■

EXAMPLE 5 Solve $4(x-3) > 9(x+1)$.

Solution

$$4(x-3) > 9(x+1)$$

$$4x - 12 > 9x + 9 \qquad \text{Apply the distributive property.}$$

$$-5x - 12 > 9 \qquad \text{Add } -9x \text{ to both sides.}$$

$$-5x > 21 \qquad \text{Add 12 to both sides.}$$

$$-\frac{1}{5}(-5x) < -\frac{1}{5}(21) \qquad \text{Multiply both sides by } -\frac{1}{5}, \text{ which reverses the inequality.}$$

$$x < -\frac{21}{5}$$

The solution set is $\left(-\infty, -\dfrac{21}{5}\right)$. ■

The next example will solve the inequality without indicating the justification for each step. Be sure that you can supply the reasons for the steps.

EXAMPLE 6 Solve $3(2x+1) - 2(2x+5) < 5(3x-2)$.

Solution

$$3(2x+1) - 2(2x+5) < 5(3x-2)$$

$$6x + 3 - 4x - 10 < 15x - 10$$

$$2x - 7 < 15x - 10$$

$$-13x - 7 < -10$$

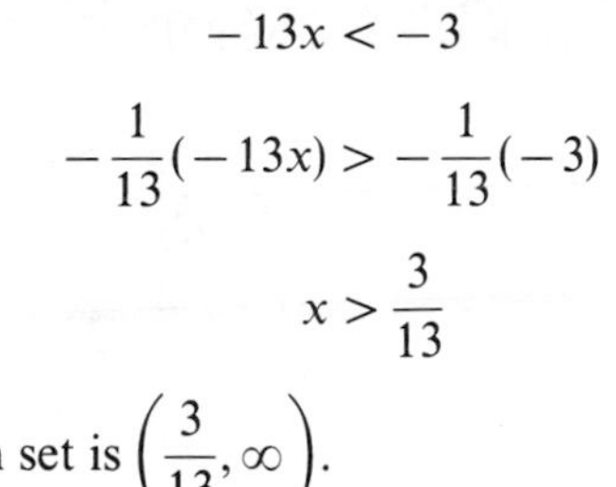

$$-13x < -3$$

$$-\frac{1}{13}(-13x) > -\frac{1}{13}(-3)$$

$$x > \frac{3}{13}$$

The solution set is $\left(\frac{3}{13}, \infty\right)$. ■

Checking solutions for an inequality presents a problem. Obviously, we cannot check all of the infinitely many solutions for a particular inequality. However, by checking at least one solution, especially when the multiplication property has been used, we might catch a mistake of forgetting to change the sense of an inequality. In Example 6 we are claiming that *all numbers greater than* $\frac{3}{13}$ will satisfy the original inequality. Let's check one such number, say 1.

$$3(2x + 1) - 2(2x + 5) < 5(3x - 2)$$

$$3(2(1) + 1) - 2(2(1) + 5) \stackrel{?}{<} 5(3(1) - 2)$$

$$3(3) - 2(7) \stackrel{?}{<} 5(1)$$

$$0 - 14 \stackrel{?}{<} 5$$

$$-5 < 5$$

Thus, 1 satisfies the original inequality. Had we forgotten to switch the sense of the inequality when both sides were multiplied by $-\frac{1}{13}$, then our answer would have been $x < \frac{3}{13}$, and we would have detected such an error by the check.

Problem Set 2.5

For Problems 1–8, express the given inequality in interval notation and sketch a graph of the interval.

1. $x > 1$ $(1, \infty)$
2. $x > -2$ $(-2, \infty)$
3. $x \geq -1$ $[-1, \infty)$
4. $x \geq 3$ $[3, \infty)$
5. $x < -2$ $(-\infty, -2)$
6. $x < 1$ $(-\infty, 1)$
7. $x \leq 2$ $(-\infty, 2]$
8. $x \leq 0$ $(-\infty, 0]$

For Problems 9–16, express each interval as an inequality using the variable of x. For example, we can express the interval $[5, \infty)$ as $x \geq 5$.

9. $(-\infty, 4)$ $x < 4$
10. $(-\infty, -2)$ $x < -2$
11. $(-\infty, -7]$ $x \leq -7$
12. $(-\infty, 9]$ $x \leq 9$
13. $(8, \infty)$ $x > 8$
14. $(-5, \infty)$ $x > -5$
15. $[-7, \infty)$ $x \geq -7$
16. $[10, \infty)$ $x \geq 10$

For Problems 17–40, solve each of the inequalities and graph the solution set on a number line.

17. $x - 3 > -2$ $(1, \infty)$

18. $x + 2 < 1$ $(-\infty, -1)$

19. $-2x \geq 8$ $(-\infty, -4]$

20. $-3x \leq -9$ $[3, \infty)$

21. $5x \leq -10$ $(-\infty, -2]$

22. $4x \geq -4$ $[-1, \infty)$

23. $2x + 1 < 5$ $(-\infty, 2)$

24. $2x + 2 > 4$ $(1, \infty)$

25. $3x - 2 > -5$ $(-1, \infty)$

26. $5x - 3 < -3$ $(-\infty, 0)$

27. $-7x - 3 \leq 4$ $[-1, \infty)$

28. $-3x - 1 \geq 8$ $(-\infty, -3]$

29. $2 + 6x > -10$ $(-2, \infty)$

30. $1 + 6x > -17$ $(-3, \infty)$

31. $5 - 3x < 11$ $(-2, \infty)$

32. $4 - 2x < 12$ $(-4, \infty)$

33. $15 < 1 - 7x$ $(-\infty, -2)$

34. $12 < 2 - 5x$ $(-\infty, -2)$

35. $-10 \leq 2 + 4x$ $[-3, \infty)$

36. $-9 \leq 1 + 2x$ $[-5, \infty)$

37. $3(x + 2) > 6$ $(0, \infty)$

38. $2(x - 1) < -4$ $(-\infty, -1)$

39. $5x + 2 \geq 4x + 6$ $[4, \infty)$

40. $6x - 4 \leq 5x - 4$ $(-\infty; 0]$

For Problems 41–70, solve each inequality and express the solution sets using interval notation.

41. $2x - 1 > 6$ $(\frac{7}{2}, \infty)$

42. $3x - 2 < 12$ $(-\infty, \frac{14}{3})$

43. $-5x - 2 < -14$ $(\frac{12}{5}, \infty)$

44. $5 - 4x > -2$ $(-\infty, \frac{7}{4})$

45. $-3(2x + 1) \geq 12$ $(-\infty, -\frac{5}{2}]$

46. $-2(3x + 2) \leq 18$ $[-\frac{11}{3}, \infty)$

47. $4(3x - 2) \geq -3$ $[\frac{5}{12}, \infty)$

48. $3(4x - 3) \leq -11$ $(-\infty, -\frac{1}{6}]$

49. $6x - 2 > 4x - 14$ $(-6, \infty)$

50. $9x + 5 < 6x - 10$ $(-\infty, -5)$

51. $2x - 7 < 6x + 13$ $(-5, \infty)$

52. $2x - 3 > 7x + 22$ $(-\infty, -5)$

53. $4(x - 3) \leq -2(x + 1)$ $(-\infty, \frac{5}{3}]$

54. $3(x - 1) \geq -(x + 4)$ $[-\frac{1}{4}, \infty)$

55. $5(x - 4) - 6(x + 2) < 4$ $(-36, \infty)$

56. $3(x + 2) - 4(x - 1) < 6$ $(4, \infty)$

57. $-3(3x + 2) - 2(4x + 1) \geq 0$ $(-\infty, -\frac{8}{17}]$

58. $-4(2x - 1) - 3(x + 2) \geq 0$ $(-\infty, -\frac{2}{11}]$

59. $-(x - 3) + 2(x - 1) < 3(x + 4)$ $(-\frac{11}{2}, \infty)$

60. $3(x - 1) - (x - 2) > -2(x + 4)$ $(-\frac{7}{4}, \infty)$

61. $7(x + 1) - 8(x - 2) < 0$ $(23, \infty)$

62. $5(x - 6) - 6(x + 2) < 0$ $(-42, \infty)$

63. $-5(x - 1) + 3 > 3x - 4 - 4x$ $(-\infty, 3)$

64. $3(x + 2) + 4 < -2x + 14 + x$ $(-\infty, 1)$

65. $3(x - 2) - 5(2x - 1) \geq 0$ $(-\infty, -\frac{1}{7}]$

66. $4(2x - 1) - 3(3x + 4) \geq 0$ $(-\infty, -16]$

67. $-5(3x + 4) < -2(7x - 1)$ $(-22, \infty)$

68. $-3(2x + 1) > -2(x + 4)$ $(-\infty, \frac{5}{4})$

69. $-3(x + 2) > 2(x - 6)$ $(-\infty, \frac{6}{5})$

70. $-2(x - 4) < 5(x - 1)$ $(\frac{13}{7}, \infty)$

Miscellaneous Problems

71. Solve each of the following inequalities.

(a) $5x - 2 > 5x + 3$ $\varnothing$

(b) $3x - 4 < 3x + 7$ $(-\infty, \infty)$

(c) $4(x + 1) < 2(2x + 5)$ $(-\infty, \infty)$

(d) $-2(x - 1) > 2(x + 7)$ $(-\infty, -3)$

(e) $3(x - 2) < -3(x + 1)$ $(-\infty, \frac{1}{2})$

(f) $2(x + 1) + 3(x + 2) < 5(x - 3)$ $\varnothing$

2.6 More on Inequalities

To solve equations involving fractions we found that **clearing the equation of all fractions** by multiplying both sides by the least common denominator of all of the denominators in the equation was frequently an effective technique. This same basic approach also works very well with inequalities that involve fractions, as the next examples demonstrate.

EXAMPLE 1 Solve $\frac{2}{3}x - \frac{1}{2}x > \frac{3}{4}$.

Solution

$$\frac{2}{3}x - \frac{1}{2}x > \frac{3}{4}$$

$$12\left(\frac{2}{3}x - \frac{1}{2}x\right) > 12\left(\frac{3}{4}\right)$$

Multiply both sides by 12, which is the LCD of 3, 2, and 4.

$$12\left(\frac{2}{3}x\right) - 12\left(\frac{1}{2}x\right) > 12\left(\frac{3}{4}\right)$$

Apply the distributive property.

$$8x - 6x > 9$$

$$2x > 9$$

$$x > \frac{9}{2}$$

The solution set is $\left(\frac{9}{2}, \infty\right)$. ■

EXAMPLE 2 Solve $\frac{x+2}{4} + \frac{x-3}{8} < 1$.

Solution

$$\frac{x+2}{4} + \frac{x-3}{8} < 1$$

$$8\left(\frac{x+2}{4} + \frac{x-3}{8}\right) < 8(1)$$

Multiply both sides by 8, which is the LCD of 4 and 8.

$$8\left(\frac{x+2}{4}\right) + 8\left(\frac{x-3}{8}\right) < 8(1)$$

$$2(x + 2) + (x - 3) < 8$$

$$2x + 4 + x - 3 < 8$$

$$3x + 1 < 8$$
$$3x < 7$$
$$x < \frac{7}{3}$$

The solution set is $\left(-\infty, \frac{7}{3}\right)$. ■

EXAMPLE 3 Solve $\frac{x}{2} - \frac{x-1}{5} \geq \frac{x+2}{10} - 4$.

Solution

$$\frac{x}{2} - \frac{x-1}{5} \geq \frac{x+2}{10} - 4$$
$$10\left(\frac{x}{2} - \frac{x-1}{5}\right) \geq 10\left(\frac{x+2}{10} - 4\right)$$
$$10\left(\frac{x}{2}\right) - 10\left(\frac{x-1}{5}\right) \geq 10\left(\frac{x+2}{10}\right) - 10(4)$$
$$5x - 2(x-1) \geq x + 2 - 40$$
$$5x - 2x + 2 \geq x - 38$$
$$3x + 2 \geq x - 38$$
$$2x + 2 \geq -38$$
$$2x \geq -40$$
$$x \geq -20$$

The solution set is $[-20, \infty)$. ■

The idea of **clearing all decimals** also works with inequalities in much the same way as it does with equations. We can multiply both sides of an inequality by an appropriate power of ten and then proceed to solve in the usual way. The next two examples illustrate this procedure.

EXAMPLE 4 Solve $x \geq 1.6 + 0.2x$.

Solution

$$x \geq 1.6 + 0.2x$$
$$10(x) \geq 10(1.6 + 0.2x) \quad \text{Multiply both sides by 10.}$$
$$10x \geq 16 + 2x$$
$$8x \geq 16$$
$$x \geq 2$$

The solution set is $[2, \infty)$. ■

EXAMPLE 5 Solve $0.08x + 0.09(x + 100) \geq 43$.

Solution

$$0.08x + 0.09(x + 100) \geq 43$$

$$100(0.08x + 0.09(x + 100)) \geq 100(43) \qquad \text{Multiply both sides by 100.}$$

$$8x + 9(x + 100) \geq 4300$$

$$8x + 9x + 900 \geq 4300$$

$$17x + 900 \geq 4300$$

$$17x \geq 3400$$

$$x \geq 200$$

The solution set is $[200, \infty)$. ■

Compound Statements

We use the words *and* and *or* in mathematics to form **compound statements**. The following are examples of some compound numerical statements that use *and*. We call such statements **conjunctions**. We agree to call a conjunction true only if all of its component parts are true. Statements 1 and 2 below are true, but 3, 4, and 5 are false.

1. $3 + 4 = 7$ and $-4 < -3$. true

2. $-3 < -2$ and $-6 > -10$. true

3. $6 > 5$ and $-4 < -8$. false

4. $4 < 2$ and $0 < 10$. false

5. $-3 + 2 = 1$ and $5 + 4 = 8$. false

We call compound statements that use *or* **disjunctions**. The following are some examples of disjunctions that involve numerical statements.

6. $0.14 > 0.13$ or $0.235 < 0.237$. true

7. $\frac{3}{4} > \frac{1}{2}$ or $-4 + (-3) = 10$. true

8. $-\frac{2}{3} > \frac{1}{3}$ or $(0.4)(0.3) = 0.12$. true

9. $\frac{2}{5} < -\frac{2}{5}$ or $7 + (-9) = 16$. false

A disjunction is true if at least one of its component parts is true. In other words, disjunctions are false only if all of the component parts are false. In the statements above, 6, 7, and 8 are true, but 9 is false.

Now let's consider finding solutions for some compound statements that involve algebraic inequalities. Keep in mind that our previous agreements for labeling conjunctions and disjunctions true or false form the basis for our reasoning.

EXAMPLE 6 Graph the solution set for the conjunction $x > -1$ *and* $x < 3$.

Solution The key word is *and*, so we need to satisfy both inequalities. Thus, all numbers between -1 and 3 are solutions and we can indicate this on a number line as in Figure 2.4.

Figure 2.4

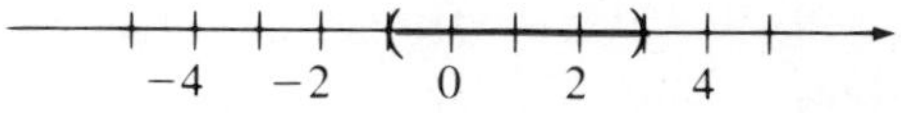

Using interval notation we can represent the interval enclosed in parentheses in Figure 2.4 by $(-1, 3)$. Using set builder notation we can express the same interval as $\{x \mid -1 < x < 3\}$, where the statement $-1 < x < 3$ is read "negative one is less than x and x is less than three." In other words, x is between -1 and 3. ■

Example 6 represents another concept pertaining to sets. The set of all elements common to two sets is called the **intersection** of the two sets. Thus, in Example 6 we found the intersection of the two sets $\{x \mid x > -1\}$ and $\{x \mid x < 3\}$ to be the set $\{x \mid -1 < x < 3\}$. In general, we define the intersection of two sets as follows.

DEFINITION 2.1

> The **intersection** of two sets A and B (written $A \cap B$) is the set of all elements that are both in A and in B. Using set builder notation we can write
>
> $$A \cap B = \{x \mid x \in A \text{ and } x \in B\}.$$

EXAMPLE 7 Solve the conjunction $3x + 1 > -5$ *and* $2x + 5 > 7$ and graph its solution set on a number line.

Solution First, let's simplify both inequalities.

$$3x + 1 > -5 \quad \text{and} \quad 2x + 5 > 7$$

$$3x > -6 \quad \text{and} \quad 2x > 2$$

$$x > -2 \quad \text{and} \quad x > 1$$

Since it is a conjunction we must satisfy both inequalities. Thus, all numbers greater than 1 are solutions and the solution set is $(1, \infty)$. We indicate the graph of the solution set in Figure 2.5.

Figure 2.5

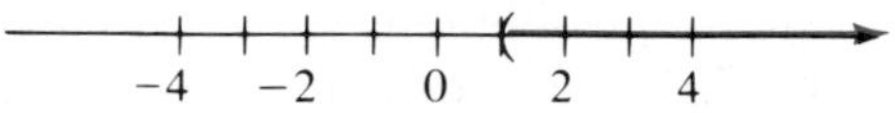

■

We can solve a conjunction such as $3x + 1 > -3$ and $3x + 1 < 7$, in which the same algebraic expression (in this case $3x + 1$) is contained in both inequalities, by using the **compact form** $-3 < 3x + 1 < 7$ as follows.

$$-3 < 3x + 1 < 7$$

$$-4 < 3x < 6 \qquad \text{Add } -1 \text{ to the left side, middle, and right side.}$$

$$-\frac{4}{3} < x < 2 \qquad \text{Multiply through by } 1/3.$$

The solution set is $\left(-\frac{4}{3}, 2\right)$.

The word *and* ties the concept of a conjunction to the set concept of intersection. In a like manner, the word *or* links the idea of a disjunction to the set concept of **union**. We define the union of two sets as follows.

DEFINITION 2.2

The **union** of two sets A and B (written $A \cup B$) is the set of all elements that are in A or in B, or in both. Using set builder notation we can write

$$A \cup B = \{x \mid x \in A \text{ or } x \in B\}.$$

EXAMPLE 8 Graph the solution set for the disjunction $x < -1$ *or* $x > 2$, and express it using interval notation.

Solution The key word is *or*, so all numbers that satisfy either inequality (or both) are solutions. Thus, all numbers less than -1, along with all numbers greater than 2, are the solutions. The graph of the solution set is shown in Figure 2.6.

Figure 2.6

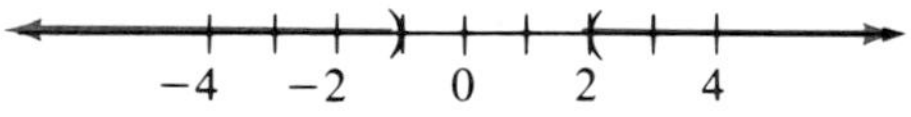

−4 −2 0 2 4

Using interval notation and the set concept of union, we can express the solution set as $(-\infty, -1) \cup (2, \infty)$. ■

Example 8 illustrates that in terms of set vocabulary, the solution set of a disjunction is the union of the solution sets of the component parts of the disjunction. Note that there is *no compact form* for writing $x < -1$ or $x > 2$ *or for any disjunction.*

EXAMPLE 9 Solve the disjunction $2x - 5 < -11$ *or* $5x + 1 \geq 6$ and graph its solution set on a number line.

Solution First, let's simplify both inequalities.

$$2x - 5 < -11 \quad \text{or} \quad 5x + 1 \geq 6$$

$$2x < -6 \quad \text{or} \quad 5x \geq 5$$

$$x < -3 \quad \text{or} \quad x \geq 1$$

Since it is a disjunction, all numbers less than -3, along with all numbers greater than or equal to 1, will satisfy it. Thus, the solution set is $(-\infty, -3) \cup [1, \infty)$ and its graph is shown in Figure 2.7.

Figure 2.7

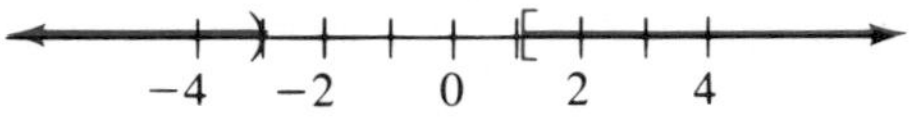

−4 −2 0 2 4

■

In summary, to solve a compound sentence involving an inequality, proceed as follows.

1. Solve separately each inequality in the compound sentence.
2. If it is a *conjunction*, the solution set is the *intersection* of the solution sets of each inequality.
3. If it is a *disjunction*, the solution set is the *union* of the solution sets of each inequality.

The following agreements on the use of interval notation should be added to the list on page 78.

Set	*Graph*	*Interval notation*
$\{x \mid a < x < b\}$	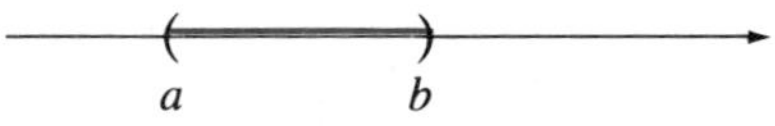 a b	(a, b)
$\{x \mid a \leq x < b\}$	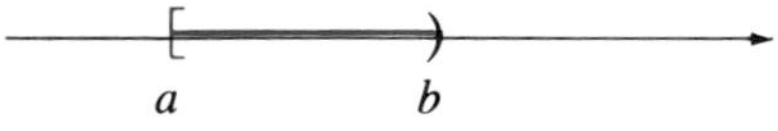 a b	$[a, b)$
$\{x \mid a < x \leq b\}$	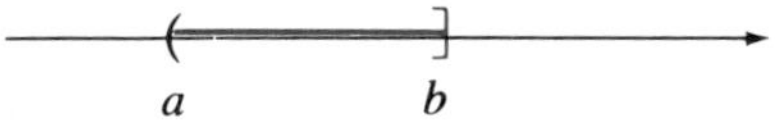 a b	$(a, b]$
$\{x \mid a \leq x \leq b\}$	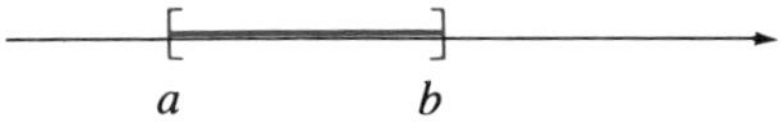 a b	$[a, b]$

Problem Solving

Let's conclude this section by considering some word problems that contain inequality statements.

PROBLEM 1 Rhonda had scores of 94, 84, 86, and 88 on her first four exams of the semester. What score must she obtain on the 5th exam to have an average of 90 or better for the five exams?

Solution Let s represent the score Rhonda needs on the 5th exam. Since the average is computed by adding all scores and dividing by the number of scores, we have the following inequality to solve.

$$\frac{94 + 84 + 86 + 88 + s}{5} \geq 90$$

Solving this inequality we obtain

$$\frac{352 + s}{5} \geq 90$$

$$5\left(\frac{352 + s}{5}\right) \geq 5(90) \qquad \text{Multiply both sides by 5.}$$

$$352 + s \geq 450$$

$$s \geq 98.$$

She must receive a score of 98 or better. ■

PROBLEM 2 An investor has \$1000 to invest. Suppose she invests \$500 at 8% interest. At what rate must she invest the other \$500 so that the two investments together yield more than \$100 of yearly interest?

Solution Let r represent the unknown rate of interest. We can use the following guideline to set up an inequality.

interest from 8% investment	+	interest from r percent investment	>	\$100
↓		↓		↓
(8%)(\$500)	+	r(\$500)	>	\$100

Solving this inequality yields

$$40 + 500r > 100$$

$$500r > 60$$

$$r > \frac{60}{500}$$

$$r > 0.12. \qquad \text{Change to a decimal.}$$

She must invest the other \$500 at a rate greater than 12%. ■

PROBLEM 3 If the temperature for a 24-hour period ranged between 41°F and 59°F, inclusive, (that is, $41 \leq F \leq 59$) what was the range in Celsius degrees?

Solution Use the formula $F = \frac{9}{5}C + 32$, to solve the following compound inequality.

$$41 \leq \frac{9}{5}C + 32 \leq 59$$

Solving this yields

$$9 \le \frac{9}{5}C \le 27 \qquad \text{Add } -32.$$

$$\frac{5}{9}(9) \le \frac{5}{9}\left(\frac{9}{5}C\right) \le \frac{5}{9}(27) \qquad \text{Multiply by } \frac{5}{9}.$$

$$5 \le C \le 15.$$

The range is between 5°C and 15°C, inclusive. ■

Problem Set 2.6

For Problems 1–18, solve each of the inequalities and express the solution sets in interval notation.

1. $\frac{2}{5}x + \frac{1}{3}x > \frac{44}{15}$ $(4, \infty)$

2. $\frac{1}{4}x - \frac{4}{3}x < -13$ $(12, \infty)$

3. $x - \frac{5}{6} < \frac{x}{2} + 3$ $\left(-\infty, \frac{23}{3}\right)$

4. $x + \frac{2}{7} > \frac{x}{2} - 5$ $\left(-\frac{74}{7}, \infty\right)$

5. $\frac{x-2}{3} + \frac{x+1}{4} \ge \frac{5}{2}$ $[5, \infty)$

6. $\frac{x-1}{3} + \frac{x+2}{5} \le \frac{3}{5}$ $(-\infty, 1]$

7. $\frac{3-x}{6} + \frac{x+2}{7} \le 1$ $[-9, \infty)$

8. $\frac{4-x}{5} + \frac{x+1}{6} \ge 2$ $(-\infty, -31]$

9. $\frac{x+3}{8} - \frac{x+5}{5} \ge \frac{3}{10}$ $\left(-\infty, -\frac{37}{3}\right]$

10. $\frac{x-4}{6} - \frac{x-2}{9} \le \frac{5}{18}$ $(-\infty, 13]$

11. $\frac{4x-3}{6} - \frac{2x-1}{12} < -2$ $\left(-\infty, -\frac{19}{6}\right)$

12. $\frac{3x+2}{9} - \frac{2x+1}{3} > -1$ $\left(-\infty, \frac{8}{3}\right)$

13. $.06x + .08(250 - x) \ge 19$ $(-\infty, 50]$

14. $.08x + .09(2x) \ge 130$ $[500, \infty)$

15. $.09x + .1(x + 200) > 77$ $(300, \infty)$

16. $.07x + .08(x + 100) > 38$ $(200, \infty)$

17. $x \ge 3.4 + .15x$ $[4, \infty)$

18. $x \ge 2.1 + .3x$ $[3, \infty)$

For Problems 19–34, graph the solution set for each compound inequality and express the solution sets in interval notation.

19. $x > -1$ and $x < 2$ $(-1, 2)$

20. $x > 1$ and $x < 4$ $(1, 4)$

21. $x \le 2$ and $x > -1$ $(-1, 2]$

22. $x \le 4$ and $x \ge -2$ $[-2, 4]$

23. $x > 2$ or $x < -1$ $(-\infty, -1) \cup (2, \infty)$

24. $x > 1$ or $x < -4$ $(-\infty, -4) \cup (1, \infty)$

25. $x \le 1$ or $x > 3$ $(-\infty, 1] \cup (3, \infty)$

26. $x < -2$ or $x \ge 1$ $(-\infty, -2) \cup [1, \infty)$

27. $x > 0$ and $x > -1$ $(0, \infty)$

28. $x > -2$ and $x > 2$ $(2, \infty)$

29. $x < 0$ and $x > 4$ $\varnothing$

30. $x > 1$ or $x < 2$ $(-\infty, \infty)$

31. $x > -2$ or $x < 3$ $(-\infty, \infty)$

32. $x > 3$ and $x < -1$ $\varnothing$

33. $x > -1$ or $x > 2$ $(-1, \infty)$

34. $x < -2$ or $x < 1$ $(-\infty, 1)$

For Problems 35–44, solve each compound inequality and graph the solution sets. Express solution sets in interval notation.

35. $x - 2 > -1$ and $x - 2 < 1$ $(1, 3)$

36. $x + 3 > -2$ and $x + 3 < 2$ $(-5, -1)$

37. $x + 2 < -3$ or $x + 2 > 3$ $(-\infty, -5) \cup (1, \infty)$

38. $x - 4 < -2$ or $x - 4 > 2$ $(-\infty, 2) \cup (6, \infty)$

39. $2x - 1 \geq 5$ and $x > 0$ $[3, \infty)$

40. $3x + 2 > 17$ and $x \geq 0$ $(5, \infty)$

41. $5x - 2 < 0$ and $3x - 1 > 0$ $(\frac{1}{3}, \frac{2}{5})$

42. $x + 1 > 0$ and $3x - 4 < 0$ $(-1, \frac{4}{3})$

43. $3x + 2 < -1$ or $3x + 2 > 1$ $(-\infty, -1) \cup (-\frac{1}{3}, \infty)$

44. $5x - 2 < -2$ or $5x - 2 > 2$ $(-\infty, 0) \cup (\frac{4}{5}, \infty)$

For Problems 45–56, solve each compound inequality using the compact form. Express the solution sets in interval notation.

45. $-3 < 2x + 1 < 5$ $(-2, 2)$

46. $-7 < 3x - 1 < 8$ $(-2, 3)$

47. $-17 \leq 3x - 2 \leq 10$ $[-5, 4]$

48. $-25 \leq 4x + 3 \leq 19$ $[-7, 4]$

49. $1 < 4x + 3 < 9$ $(-\frac{1}{2}, \frac{3}{2})$

50. $0 < 2x + 5 < 12$ $(-\frac{5}{2}, \frac{7}{2})$

51. $-6 < 4x - 5 < 6$ $(-\frac{1}{4}, \frac{11}{4})$

52. $-2 < 3x + 4 < 2$ $(-2, -\frac{2}{3})$

53. $-4 \leq \dfrac{x - 1}{3} \leq 4$ $[-11, 13]$

54. $-1 \leq \dfrac{x + 2}{4} \leq 1$ $[-6, 2]$

55. $-3 < 2 - x < 3$ $(-1, 5)$

56. $-4 < 3 - x < 4$ $(-1, 7)$

For Problems 57–67, solve each problem by setting up and solving an appropriate inequality.

57. Suppose that Lance has \$500 to invest. If he invests \$300 at 9% interest, at what rate must he invest the remaining \$200 so that the two investments yield more than \$47 in yearly interest? More than 10%

58. Mona invests \$100 at 8% yearly interest. How much does she have to invest at 9% so that the total yearly interest from the two investments exceeds \$26? More than \$200

59. The average height of the two forwards and the center of a basketball team is 6 feet and 8 inches. What must the average height of the two guards be so that the team average is at least 6 feet and 4 inches? 5 feet and 10 inches or better

60. Ron has scores of 52, 84, 65, and 74 on his first four math exams. What score must he make on the 5th exam to have an average of 70 or better for the five exams? 75 or better

61. Marsha bowled 142 and 170 in her first two games. What must she bowl in the third game to have an average of at least 160 for the three games? 168 or better

62. Debbie had scores of 95, 82, 93, and 84 on her first four exams of the semester. What score must she obtain on the fifth exam to have an average of 90 or better for the five exams? 96 or better

63. Suppose that Scott shot rounds of 82, 84, 78, and 79 on the first four days of a golf tournament. What must he shoot on the fifth day of the tournament to average 80 or less for the five days? 77 or less

64. The temperatures for a 24-hour period range between $-4°$F and $23°$F, inclusive. What was the range in Celsius degrees? $\left(\text{Use } F = \frac{9}{5}C + 32.\right)$ $-20° \leq C \leq -5°$

65. Oven temperatures for baking various foods usually range between 325°F and 425°F, inclusive. Express this range in Celsius degrees. (Round answers to the nearest degree.) $163° \leq C \leq 218°$

66. A person's intelligence quotient (I) is found by dividing mental age (M), as indicated by standard tests, by the chronological age (C), and then multiplying this ratio by 100. The formula $I = \dfrac{100M}{C}$ can be used. If the I range of a group of 11-year-olds is given by $80 \leq I \leq 140$, find the mental age range of this group. $8.8 \leq M \leq 15.4$

67. Repeat Problem 66 for an I range of 70 to 125, inclusive, for a group of 9-year-olds. $6.3 \leq M \leq 11.25$

2.7 Equations and Inequalities Involving Absolute Value

In Section 1.2 we defined the absolute value of a real number by

$$|a| = \begin{cases} a, & \text{if } a \geq 0 \\ -a, & \text{if } a < 0 \end{cases}.$$

We also interpreted the absolute value of any real number to be the distance between the number and zero on a number line. For example, $|6| = 6$ translates to 6 units between 6 and 0. Likewise, $|-8| = 8$ translates to 8 units between -8 and 0.

The interpretation of absolute value as distance on a number line provides a straightforward approach to solving a variety of equations and inequalities involving absolute value. First, let's consider some equations.

EXAMPLE 1 Solve $|x| = 2$.

Solution Thinking in terms of *distance between the number and zero*, we see that x must be 2 or -2. That is, the equation $|x| = 2$ is equivalent to

$$x = -2 \quad \text{or} \quad x = 2.$$

The solution set is $\{-2, 2\}$. ■

EXAMPLE 2 Solve $|x + 2| = 5$.

Solution The number, $x + 2$, must be -5 or 5. Thus, $|x + 2| = 5$ is equivalent to

$$x + 2 = -5 \quad \text{or} \quad x + 2 = 5.$$

Solving each equation of the disjunction yields

$$x + 2 = -5 \quad \text{or} \quad x + 2 = 5$$

$$x = -7 \quad \text{or} \quad x = 3.$$

The solution set is $\{-7, 3\}$.

CHECK

$$|x+2| = 5 \qquad\qquad |x+2| = 5$$
$$|-7+2| \stackrel{?}{=} 5 \qquad\qquad |3+2| \stackrel{?}{=} 5$$
$$|-5| \stackrel{?}{=} 5 \qquad\qquad |5| \stackrel{?}{=} 5$$
$$5 = 5 \qquad\qquad 5 = 5$$

■

The following general property should seem reasonable from the distance interpretation of absolute value.

PROPERTY 2.1

> $|ax + b| = k$ is equivalent to $ax + b = -k$ or $ax + b = k$, where k is a positive number.

Example 3 demonstrates our format for solving equations of the form $|ax + b| = k$.

EXAMPLE 3 Solve $|5x + 3| = 7$.

Solution

$$|5x + 3| = 7$$
$$5x + 3 = -7 \quad \text{or} \quad 5x + 3 = 7$$
$$5x = -10 \quad \text{or} \quad 5x = 4$$
$$x = -2 \quad \text{or} \quad x = \frac{4}{5}$$

The solution set is $\left\{-2, \frac{4}{5}\right\}$. Check these solutions! ■

The *distance interpretation* for absolute value also provides a good basis for solving some inequalities that involve absolute value. Consider the following examples.

EXAMPLE 4 Solve $|x| < 2$ and graph the solution set.

Solution The number, x, must be *less than two units away from zero*. Thus, $|x| < 2$ is equivalent to

$$x > -2 \quad \text{and} \quad x < 2.$$

The solution set is $(-2, 2)$ and its graph is shown in Figure 2.8.

Figure 2.8

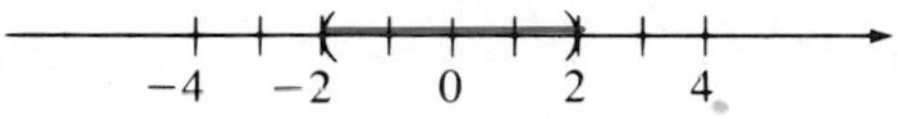

−4 −2 0 2 4

■

EXAMPLE 5 Solve and graph the solutions for $|x + 3| < 1$.

Solution Let's continue to think in terms of "distance" on a number line. The number, $x + 3$, must be *less than one unit away from zero*. Thus, $|x + 3| < 1$ is equivalent to

$$x + 3 > -1 \quad \text{and} \quad x + 3 < 1.$$

Solving this conjunction yields

$$x + 3 > -1 \quad \text{and} \quad x + 3 < 1$$
$$x > -4 \quad \text{and} \quad x < -2.$$

The solution set is $(-4, -2)$ and its graph is shown in Figure 2.9.

Figure 2.9

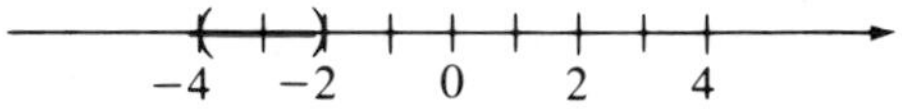

−4 −2 0 2 4

■

Take another look at Examples 4 and 5. The following general property should seem reasonable.

PROPERTY 2.2

> $|ax + b| < k$ is equivalent to $ax + b > -k$ and $ax + b < k$, where k is a positive number.

Remember that we can write a conjunction such as $ax + b > -k$ and $ax + b < k$ in the compact form $-k < ax + b < k$. The compact form provides a very convenient format for solving inequalities such as $|3x - 1| < 8$, as Example 6 illustrates.

EXAMPLE 6 Solve and graph the solutions for $|3x - 1| < 8$.

Solution

$$|3x - 1| < 8$$
$$-8 < 3x - 1 < 8$$
$$-7 < 3x < 9 \qquad \text{Add 1 to left side, middle, and right side.}$$
$$\frac{1}{3}(-7) < \frac{1}{3}(3x) < \frac{1}{3}(9) \qquad \text{Multiply through by } \frac{1}{3}.$$
$$-\frac{7}{3} < x < 3$$

The solution set is $\left(-\frac{7}{3}, 3\right)$ and its graph is shown in Figure 2.10.

Figure 2.10

$-\frac{7}{3}$

−4 −2 0 2 4

■

The distance interpretation also clarifies a property that pertains to *greater than* situations involving absolute value. Consider the following examples.

EXAMPLE 7 Solve and graph the solutions for $|x| > 1$.

Solution The number, x, must be *more than one unit away from zero*. Thus, $|x| > 1$ is equivalent to

$$x < -1 \quad \text{or} \quad x > 1.$$

The solution set is $(-\infty, -1) \cup (1, \infty)$ and its graph is shown in Figure 2.11.

Figure 2.11

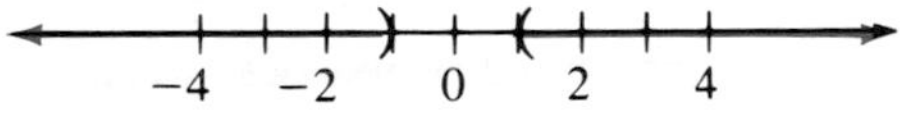

−4 −2 0 2 4

■

EXAMPLE 8 Solve and graph the solutions for $|x - 1| > 3$.

Solution The number, $x - 1$, must be *more than three units away from zero*. Thus, $|x - 1| > 3$ is equivalent to

$$x - 1 < -3 \quad \text{or} \quad x - 1 > 3.$$

Solving this disjunction yields

$$x - 1 < -3 \quad \text{or} \quad x - 1 > 3$$

$$x < -2 \quad \text{or} \quad x > 4.$$

The solution set is $(-\infty, -2) \cup (4, \infty)$ and its graph is shown in Figure 2.12.

Figure 2.12

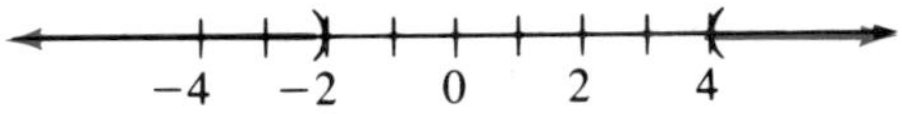

−4 −2 0 2 4

■

Examples 7 and 8 illustrate the following general property.

PROPERTY 2.3

> $|ax + b| > k$ is equivalent to $ax + b < -k$ or $ax + b > k$, where k is a positive number.

Therefore, solving inequalities of the form $|ax + b| > k$ can take on the format in Example 9.

EXAMPLE 9 Solve and graph the solutions for $|3x - 1| > 2$.

Solution

$$|3x - 1| > 2$$

$$3x - 1 < -2 \quad \text{or} \quad 3x - 1 > 2$$

$$3x < -1 \quad \text{or} \quad 3x > 3$$

$$x < -\frac{1}{3} \quad \text{or} \quad x > 1$$

The solution set is $\left(-\infty, -\frac{1}{3}\right) \cup (1, \infty)$ and its graph is shown in Figure 2.13.

Figure 2.13

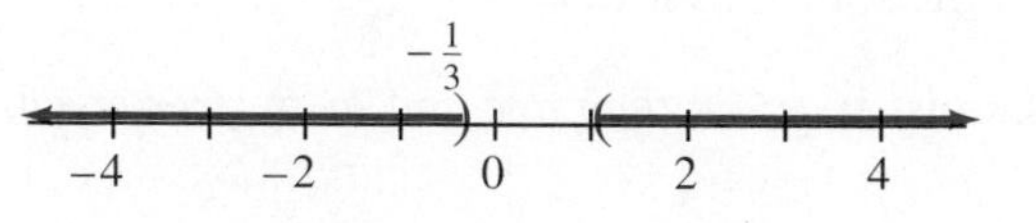

Properties 2.1, 2.2, and 2.3 provide the basis for solving a variety of equations and inequalities involving absolute value. However, if at any time you become doubtful as to what property applies, don't forget the distance interpretation. Furthermore, note that in each of the properties, k is a positive number. If k is a nonpositive number, we can determine the solution sets by inspection, as indicated by the following examples.

The solution set of $|x + 3| = 0$ is $\{-3\}$ because the number $x + 3$ has to be 0.

$|2x - 5| = -3$ has *no solutions* because the absolute value (distance) cannot be negative. (The solution set is $\varnothing$, the null set.)

$|x - 7| < -4$ has *no solutions* because we cannot obtain an absolute value less than -4. (The solution set is $\varnothing$.)

$|2x - 1| > -1$ is *satisfied by all real numbers* because the absolute value of $(2x - 1)$, regardless of what number is substituted for x, will always be greater than -1. (The solution set is the set of all real numbers, which we can express in interval notation as $(-\infty, \infty)$).

Problem Set 2.7

Solve and graph the solutions for each of the following.

1. $|x| < 5$ $(-5, 5)$

2. $|x| < 1$ $(-1, 1)$

3. $|x| \leq 2$ $[-2, 2]$

4. $|x| \leq 4$ $[-4, 4]$

5. $|x| > 2$ $(-\infty, -2) \cup (2, \infty)$

6. $|x| > 3$ $(-\infty, -3) \cup (3, \infty)$

7. $|x - 1| < 2$ $(-1, 3)$

8. $|x - 2| < 4$ $(-2, 6)$

9. $|x + 2| \leq 4$ $[-6, 2]$

10. $|x + 1| \leq 1$ $[-2, 0]$

11. $|x + 2| > 1$ $(-\infty, -3) \cup (-1, \infty)$

12. $|x + 1| > 3$ $(-\infty, -4) \cup (2, \infty)$

13. $|x - 3| \geq 2$ $(-\infty, 1] \cup [5, \infty)$

14. $|x - 2| \geq 1$ $(-\infty, 1] \cup [3, \infty)$

Solve each of the following.

15. $|x - 1| = 8$ $\{-7, 9\}$

16. $|x + 2| = 9$ $\{-11, 7\}$

17. $|x - 2| > 6$ $(-\infty, -4) \cup (8, \infty)$

18. $|x - 3| > 9$ $(-\infty, -6) \cup (12, \infty)$

19. $|x + 3| < 5$ $(-8, 2)$

20. $|x + 1| < 8$ $(-9, 7)$

21. $|2x - 4| = 6$ $\{-1, 5\}$

22. $|3x - 4| = 14$ $\{-\frac{10}{3}, 6\}$

23. $|2x - 1| \leq 9$ $[-4, 5]$

24. $|3x + 1| \leq 13$ $[-\frac{14}{3}, 4]$

25. $|4x + 2| \geq 12$ $(-\infty, -\frac{7}{2}] \cup [\frac{5}{2}, \infty)$

26. $|5x - 2| \geq 10$ $(-\infty, -\frac{8}{5}] \cup [\frac{12}{5}, \infty)$

27. $|3x + 4| = 11$ $\{-5, \frac{7}{3}\}$

28. $|5x - 7| = 14$ $\{-\frac{7}{5}, \frac{21}{5}\}$

29. $|4 - 2x| = 6$ $\{-1, 5\}$

30. $|3 - 4x| = 8$ $\{-\frac{5}{4}, \frac{11}{4}\}$

31. $|2 - x| > 4$ $(-\infty, -2) \cup (6, \infty)$

32. $|4 - x| > 3$ $(-\infty, 1) \cup (7, \infty)$

33. $|1 - 2x| < 2$ $(-\frac{1}{2}, \frac{3}{2})$

34. $|2 - 3x| < 5$ $(-1, \frac{7}{3})$

35. $|5x + 9| \leq 16$ $[-5, \frac{7}{5}]$

36. $|7x - 6| \geq 22$ $(-\infty, -\frac{16}{7}] \cup [4, \infty)$

37. $\left|x - \dfrac{3}{4}\right| = \dfrac{2}{3}$ $\left\{\dfrac{1}{12}, \dfrac{17}{12}\right\}$

38. $\left|x + \dfrac{1}{2}\right| = \dfrac{3}{5}$ $\left\{-\dfrac{11}{10}, \dfrac{1}{10}\right\}$

39. $|-2x + 7| \leq 13$ $[-3, 10]$

40. $|-3x - 4| \leq 15$ $[-\frac{19}{3}, \frac{11}{3}]$

41. $\left|\dfrac{x - 3}{4}\right| < 2$ $(-5, 11)$

42. $\left|\dfrac{x + 2}{3}\right| < 1$ $(-5, 1)$

43. $\left|\dfrac{2x + 1}{2}\right| > 1$ $\left(-\infty, -\dfrac{3}{2}\right) \cup \left(\dfrac{1}{2}, \infty\right)$

44. $\left|\dfrac{3x - 1}{4}\right| > 3$ $\left(-\infty, -\dfrac{11}{3}\right) \cup \left(\dfrac{13}{3}, \infty\right)$

45. $|2x - 3| + 2 = 5$ $\{0, 3\}$

46. $|3x - 1| - 1 = 9$ $\{-3, \frac{11}{3}\}$

47. $|x + 7| - 3 \geq 4$ $(-\infty, -14] \cup [0, \infty)$

48. $|x - 2| + 4 \geq 10$ $(-\infty, -4] \cup [8, \infty)$

49. $|2x - 1| + 1 \leq 6$ $[-2, 3]$

50. $|4x + 3| - 2 \leq 5$ $[-\frac{5}{2}, 1]$

Solve each of the following *by inspection*. Don't forget the distance interpretation for absolute value.

51. $|2x + 1| = -4$ $\varnothing$

52. $|5x - 1| = -2$ $\varnothing$

53. $|3x - 1| > -2$ $(-\infty, \infty)$

54. $|4x + 3| < -4$ $\varnothing$

55. $|5x - 2| = 0$ $\{\frac{2}{5}\}$

56. $|3x - 1| - 0$ $\{\frac{1}{3}\}$

57. $|4x - 6| < -1$ $\varnothing$

58. $|x + 9| > -6$ $(-\infty, \infty)$

59. $|x + 4| < 0$ $\varnothing$

60. $|x + 6| > 0$ $(-\infty, -6) \cup (-6, \infty)$

Miscellaneous Problems

Solve each of the following equations.

61. $|3x + 1| = |2x + 3|$ [*Hint*: $3x + 1 = 2x + 3$ or $3x + 1 = -(2x + 3)$] $\{-\frac{4}{5}, 2\}$

62. $|-2x - 3| = |x + 1|$ $\{-2, -\frac{4}{3}\}$

63. $|2x - 1| = |x - 3|$ $\{-2, \frac{4}{3}\}$

64. $|x - 2| = |x + 6|$ $\{-2\}$

65. $|x + 1| = |x - 4|$ $\{\frac{3}{2}\}$

66. $|x + 1| = |x - 1|$ $\{0\}$

67. Use the definition of absolute value to help prove Property 2.1.

68. Use the definition of absolute value to help prove Property 2.2.

69. Use the definition of absolute value to help prove Property 2.3.

Chapter 2 Summary

(2.1) **Solving an algebraic equation** refers to the process of finding the number (or numbers) that makes the algebraic equation a true numerical statement. We call such numbers the **solutions** or **roots** of the equation that **satisfy** the equation. We call the set of all solutions of an equation the **solution set**. The general procedure for solving an equation is to continue replacing the given equation with **equivalent, but simpler**, equations until we arrive at one that can be solved by inspection. Two properties of equality play an important role in the process of solving equations.

Addition Property of Equality $a = b$ if and only if $a + c = b + c$.

Multiplication Property of Equality For $c \neq 0$, $a = b$ if and only if $ac = bc$.

(2.2) To solve an equation involving fractions, first **clear the equation of all fractions**. It is usually easiest to begin by multiplying both sides of the equation by the least common multiple of all of the denominators in the equation (by the *least common denominator* or LCD).

Keep the following suggestions in mind as you solve word problems:

1. Read the problem carefully.
2. Sketch any figure, diagram, or chart that might be helpful.
3. Choose a meaningful variable.
4. Look for a guideline.
5. Form an equation or inequality.
6. Solve the equation or inequality.
7. Check your answers.

(2.3) To solve equations containing decimals, you can **clear the equation of all decimals** by multiplying both sides by an appropriate power of ten.

(2.4) **Formulas** are rules frequently stated in symbolic form using equations. We can solve a formula such as $P = 2l + 2w$ for $l\left(l = \dfrac{P - 2w}{2}\right)$ or for $w\left(w = \dfrac{P - 2l}{2}\right)$ by applying the addition and multiplication properties of equality.

We often use formulas as **guidelines** for solving word problems.

(2.5) **Solving an algebraic inequality** refers to the process of finding the numbers that make the algebraic inequality a true numerical statement. We call such numbers the **solutions** and we call the set of all solutions the **solution set**.

The general procedure for solving an inequality is to continue replacing the given inequality with **equivalent, but simpler**, inequalities until we arrive at one that we can solve by inspection. The following properties form the basis for solving algebraic inequalities.

1. $a > b$ if and only if $a + c > b + c$. (addition property)
2. **(a)** For $c > 0$, $a > b$ if and only if $ac > bc$.
   (b) For $c < 0$, $a > b$ if and only if $ac < bc$.
   (multiplication properties)

(2.6) To solve compound sentences that involve inequalities, we proceed as follows.

1. Solve separately each inequality in the compound sentence.

2. If it is a **conjunction**, the solution set is the **intersection** of the solution sets of each inequality.
3. If it is a **disjunction**, the solution set is the **union** of the solution sets of each inequality.

We define the intersection and union of two sets as follows.

Intersection $A \cap B = \{x \mid x \in A \text{ and } x \in B\}$.

Union $A \cup B = \{x \mid x \in A \text{ or } x \in B\}$.

The following are some examples of solution sets we examined in Sections 2.5 and 2.6.

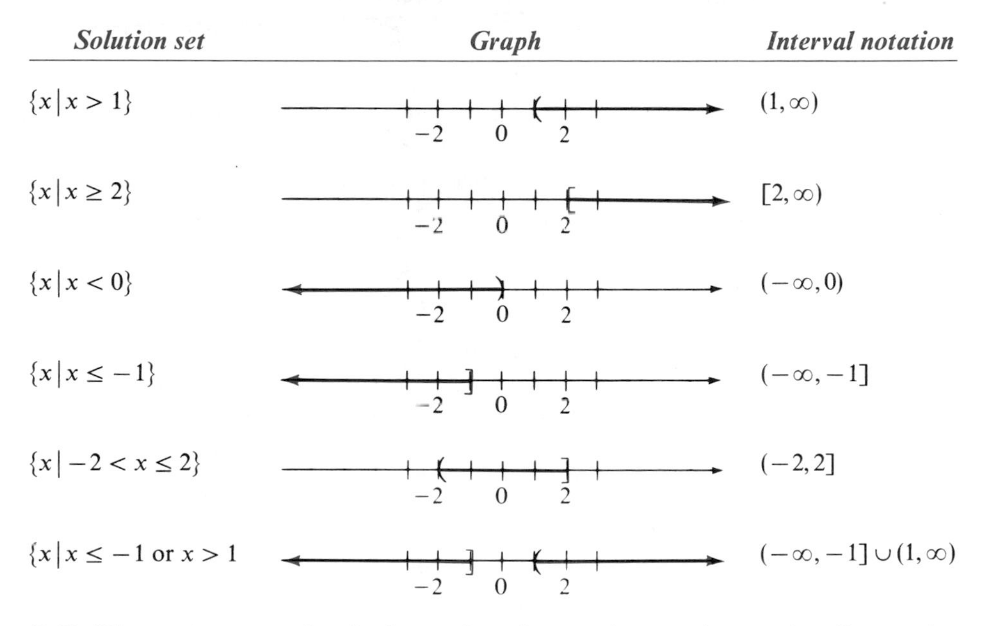

Solution set	*Graph*	*Interval notation*
$\{x \mid x > 1\}$		$(1, \infty)$
$\{x \mid x \geq 2\}$		$[2, \infty)$
$\{x \mid x < 0\}$		$(-\infty, 0)$
$\{x \mid x \leq -1\}$		$(-\infty, -1]$
$\{x \mid -2 < x \leq 2\}$		$(-2, 2]$
$\{x \mid x \leq -1 \text{ or } x > 1$		$(-\infty, -1] \cup (1, \infty)$

(2.7) We can interpret the **absolute value** of a number on the number line as the distance between that number and zero. The following properties form the basis for solving equations and inequalities involving absolute value.

1. $|ax + b| = k$ is equivalent to $ax + b = -k$ or $ax + b = k$.
2. $|ax + b| < k$ is equivalent to $-k < ax + b < k$.
3. $|ax + b| > k$ is equivalent to $ax + b < -k$ or $ax + b > k$.

$k > 0$

Chapter 2 Review Problem Set

For Problems 1–15, solve each of the equations.

1. $5(x - 6) = 3(x + 2)$ $\{18\}$

2. $2(2x + 1) - (x - 4) = 4(x + 5)$ $\{-14\}$

3. $-(2n - 1) + 3(n + 2) = 7$ $\{0\}$

4. $2(3n - 4) + 3(2n - 3) = -2(n + 5)$ $\{\frac{1}{2}\}$

5. $\frac{3t-2}{4} = \frac{2t+1}{3}$ $\{10\}$

6. $\frac{x+6}{5} + \frac{x-1}{4} = 2$ $\left\{\frac{7}{3}\right\}$

7. $1 - \frac{2x-1}{6} = \frac{3x}{8}$ $\left\{\frac{28}{17}\right\}$

8. $\frac{2x+1}{3} + \frac{3x-1}{5} = \frac{1}{10}$ $\left\{-\frac{1}{38}\right\}$

9. $\frac{3n-1}{2} - \frac{2n+3}{7} = 1$ $\left\{\frac{27}{17}\right\}$

10. $|3x-1| = 11$ $\{-\frac{10}{3}, 4\}$

11. $0.06x + 0.08(x+100) = 15$ $\{50\}$

12. $0.4(t-6) = 0.3(2t+5)$ $\{-\frac{39}{2}\}$

13. $0.1(n+300) = 0.09n + 32$ $\{200\}$

14. $0.2(x-0.5) - 0.3(x+1) = 0.4$ $\{-8\}$

15. $|2n+3| = 4$ $\{-\frac{7}{2}, \frac{1}{2}\}$

For Problems 16–20, solve each equation for x.

16. $ax - b = b + 2$ $x = \frac{2b+2}{a}$

17. $ax = bx + c$ $x = \frac{c}{a-b}$

18. $m(x+a) = p(x+b)$ $x = \frac{pb-ma}{m-p}$

19. $5x - 7y = 11$ $x = \frac{11+7y}{5}$

20. $\frac{x-a}{b} = \frac{y+1}{c}$ $x = \frac{by+b+ac}{c}$

For Problems 21–24, solve each of the formulas for the indicated variable.

21. $A = \pi r^2 + \pi rs$ for s $s = \frac{A-\pi r^2}{\pi r}$

22. $A = \frac{1}{2}h(b_1+b_2)$ for b_2 $b_2 = \frac{2A-hb_1}{h}$

23. $R = \frac{R_1R_2}{R_1+R_2}$ for R_1 $R_1 = \frac{RR_2}{R_2-R}$

24. $\frac{1}{R} = \frac{1}{R_1} + \frac{1}{R_2}$ for R $R = \frac{R_1R_2}{R_1+R_2}$

For Problems 25–36, solve each of the inequalities.

25. $5x - 2 \geq 4x - 7$ $[-5, \infty)$

26. $3 - 2x < -5$ $(4, \infty)$

27. $2(3x-1) - 3(x-3) > 0$ $(-\frac{7}{3}, \infty)$

28. $3(x+4) \leq 5(x-1)$ $[\frac{17}{2}, \infty)$

29. $\frac{5}{6}n - \frac{1}{3}n < \frac{1}{6}$ $\left(-\infty, \frac{1}{3}\right)$

30. $\frac{n-4}{5} + \frac{n-3}{6} > \frac{7}{15}$ $\left(\frac{5}{11}, \infty\right)$

31. $s \geq 4.5 + 0.25s$ $[6, \infty)$

32. $0.07x + 0.09(500 - x) \geq 43$ $(-\infty, 100]$

33. $|2x-1| < 11$ $(-5, 6)$

34. $|3x+1| > 10$ $(-\infty, -\frac{11}{3}) \cup (3, \infty)$

35. $-3(2t-1) - (t+2) > -6(t-3)$ $(-\infty, -17)$

36. $\frac{2}{3}(x-1) + \frac{1}{4}(2x+1) < \frac{5}{6}(x-2)$ $\left(-\infty, -\frac{15}{4}\right)$

For Problems 37–44, graph the solutions of each compound inequality.

37. $x > -1$ and $x < 1$ $(-1, 1)$

38. $x > 2$ or $x \leq -3$ $(-\infty, -3] \cup (2, \infty)$

39. $x > 2$ and $x > 3$ $(3, \infty)$

40. $x < 2$ or $x > -1$ $(-\infty, \infty)$

41. $2x + 1 > 3$ or $2x + 1 < -3$ $(-\infty, -2) \cup (1, \infty)$

42. $2 \leq x + 4 \leq 5$ $(-2, 1)$

43. $-1 < 4x - 3 \leq 9$ $(\frac{1}{2}, 3]$

44. $x + 1 > 3$ and $x - 3 < -5$ $\varnothing$

Solve each of the following problems by setting up and solving an appropriate equation or inequality.

45. The width of a rectangle is 2 meters more than one-third of the length. The perimeter of the rectangle is 44 meters. Find the length and width of the rectangle. 15 m by 7 m

46. A total of \$500 was invested, part of it at 7% interest and the remainder at 8%. If the total yearly interest from both investments amounted to \$38, how much was invested at each rate? \$200 at 7% and \$300 at 8%

47. Susan's average score for her first three psychology exams is 84. What must she get on the 4th exam so that her average for the four exams is 85 or better? 88 or better

48. Find three consecutive integers such that the sum of one-half of the smallest and one-third of the largest is one less than the other integer. 4, 5, and 6

49. Pat is paid time-and-a-half for each hour he works over 36 hours in a week. Last week he worked 42 hours for a total of \$472.50. What is his normal hourly rate? \$10.50

50. Louise has a collection of nickels, dimes, and quarters worth \$24.75. The number of dimes is 10 more than twice the number of nickels and the number of quarters is 25 more than the number of dimes. How many coins of each kind does she have? 20, 50, 75

51. If the complement of an angle is one-tenth of the supplement of the angle, find the measure of the angle. $80°$

52. A retailer has some sweaters that cost her \$38 each. She wants to sell them at a profit of 20% of her cost. What price should she charge for the sweaters? \$45.60

53. Nora scored 16, 22, 18, and 14 points for each of the first four basketball games. How many points does she need to score in the 5th game so that her average for the first five games is at least 20 points per game? 30 or more

54. Gladys leaves a town driving at a rate of 40 miles per hour. Two hours later, Heidi leaves from the same place traveling the same route and catches Gladys in 5 hours and 20 minutes. How fast was Heidi traveling? 55 miles per hour

55. In $1\frac{1}{4}$ hours more time, Rita, riding her bicycle at 12 miles per hour, rode 2 miles farther than Sonya, who was riding her bicycle at 16 miles per hour. How long did each girl ride? Sonya for $3\frac{1}{4}$ hours and Rita for $4\frac{1}{2}$ hours

56. How many cups of orange juice must be added to 50 cups of a punch that is 10% orange juice to obtain a punch that is 20% orange juice? $6\frac{1}{4}$ cups.

Thoughts into Words

57. Give a step-by-step description of how to solve an equation such as $3(x + 2) - (x - 3) = 4(x - 1) + 7(x - 9)$.

58. What is an algebraic equation?

59. What role does the symmetric property of equality play when solving equations?

60. Why must potential answers to word problems be checked back into the original statement of the problem?

61. Explain the difference between arithmetic and algebra.

62. Explain the difference between a conjunction and a disjunction. Give an example of each, outside of the field of mathematics.

Chapter 3

Polynomials

The main thrust of this text is to develop algebraic skills, to use skills to solve equations and inequalities, and to use equations and inequalities to solve word problems. This should be readily apparent throughout this chapter. The work will center around a class of algebraic expressions called **polynomials**.

3.1

Polynomials: Sums and Differences

Recall that algebraic expressions such as $5x$, $-6y^2$, $7xy$, $14a^2b$, and $-17ab^2c^3$ are called terms. A **term** is an indicated product and may contain any number of factors. The variables involved in a term are called **literal factors** and the numerical factor is called the **numerical coefficient**. Thus, in $7xy$ the x and y are literal factors, 7 is the numerical coefficient, and the term is *in two variables* (x and y).

Terms that contain variables with only nonnegative integers as exponents are called **monomials**. The previously listed terms, $5x$, $-6y^2$, $7xy$, $14a^2b$, and $-17ab^2c^3$, are all monomials. (We shall work with some algebraic expressions later such as $7x^{-1}y^{-1}$ and $6a^{-2}b^{-3}$, which are not monomials.)

The **degree** of a monomial is the sum of the exponents of the literal factors.

$7xy$ is of degree 2, $\quad$ $5x$ is of degree 1,

$14a^2b$ is of degree 3, $\quad$ $-6y^2$ is of degree 2,

$-17ab^2c^3$ is of degree 6

If the monomial contains only one variable, then the exponent of the variable is the degree of the monomial. The two examples on the right illustrate this point. We say that any nonzero constant term is of degree zero.

A **polynomial** is a monomial or a finite sum (or difference) of monomials. Thus,

$$4x^2, \quad 3x^2 - 2x - 4, \quad 7x^4 - 6x^3 + 4x^2 + x - 1,$$

$$3x^2y - 2xy^2, \quad \frac{1}{5}a^2 - \frac{2}{3}b^2, \quad \text{and} \quad 14$$

are examples of polynomials. In addition to calling a polynomial with one term a **monomial**, we also classify polynomials with two terms as **binomials** and those with three terms as **trinomials**.

The **degree of a polynomial** is the degree of the term with the highest degree in the polynomial. The following examples help illustrate some of this terminology.

The polynomial $4x^3y^4$ is a monomial in two variables of degree 7.

The polynomial $4x^2y - 2xy$ is a binomial in two variables of degree 3.

The polynomial $9x^2 - 7x + 1$ is a trinomial in one variable of degree 2.

Combining Similar Terms

Remember that *similar* or *like terms* are terms that have the same literal factors. In the preceding chapters we have frequently simplified algebraic expressions by combining similar terms, as the next examples illustrate.

$$
\begin{aligned}
2x + 3y + 7x + 8y &= 2x + 7x + 3y + 8y \\
&= (2 + 7)x + (3 + 8)y \\
&= 9x + 11y.
\end{aligned}
$$

The steps in the dashed boxes are usually done mentally.

$$
\begin{aligned}
4a - 7 - 9a + 10 &= 4a + (-7) + (-9a) + 10 \\
&= 4a + (-9a) + (-7) + 10 \\
&= (4 + (-9))a + (-7) + 10 \\
&= -5a + 3.
\end{aligned}
$$

Both the addition and subtraction of polynomials rely on basically the same ideas. The commutative, associative, and distributive properties provide the basis for rearranging, regrouping, and combining similar terms. Let's consider some examples.

EXAMPLE 1 Add $4x^2 + 5x + 1$ and $7x^2 - 9x + 4$.

Solution We usually use the horizontal format for such work. Thus,

$$
\begin{aligned}
(4x^2 + 5x + 1) + (7x^2 - 9x + 4) &= (4x^2 + 7x^2) + (5x - 9x) + (1 + 4) \\
&= 11x^2 - 4x + 5.
\end{aligned}
$$

■

EXAMPLE 2 Add $5x - 3$, $3x + 2$, and $8x + 6$.

Solution

$$
\begin{aligned}
(5x - 3) + (3x + 2) + (8x + 6) &= (5x + 3x + 8x) + (-3 + 2 + 6) \\
&= 16x + 5
\end{aligned}
$$

■

EXAMPLE 3 Find the indicated sum: $(-4x^2y + xy^2) + (7x^2y - 9xy^2) + (5x^2y - 4xy^2)$.

Solution

$$
\begin{aligned}
&(-4x^2y + xy^2) + (7x^2y - 9xy^2) + (5x^2y - 4xy^2) \\
&\quad = (-4x^2y + 7x^2y + 5x^2y) + (xy^2 - 9xy^2 - 4xy^2) \\
&\quad = 8x^2y - 12xy^2
\end{aligned}
$$

■

The idea of subtraction as *adding the opposite* ($a - b = a + (-b)$) extends to polynomials in general. We can form the opposite of a polynomial by taking the opposite of each term. For example, the opposite of $3x^2 - 7x + 1$ is $-3x^2 + 7x - 1$. Symbolically we express this as

$$-(3x^2 - 7x + 1) = -3x^2 + 7x - 1.$$

Now consider the following subtraction problems.

EXAMPLE 4 Subtract $3x^2 + 7x - 1$ from $7x^2 - 2x - 4$.

Solution Use the horizontal format to obtain

$$\begin{aligned}(7x^2 - 2x - 4) - (3x^2 + 7x - 1) \\ &= (7x^2 - 2x - 4) + (-3x^2 - 7x + 1) \\ &= (7x^2 - 3x^2) + (-2x - 7x) + (-4 + 1) \\ &= 4x^2 - 9x - 3.\end{aligned}$$

■

EXAMPLE 5 Subtract $-3y^2 + y - 2$ from $4y^2 + 7$.

Solution

$$\begin{aligned}(4y^2 + 7) - (-3y^2 + y - 2) &= (4y^2 + 7) + (3y^2 - y + 2) \\ &= (4y^2 + 3y^2) + (-y) + (7 + 2) \\ &= 7y^2 - y + 9.\end{aligned}$$

■

The next example demonstrates the use of the vertical format for this work.

EXAMPLE 6 Subtract $4x^2 - 7xy + 5y^2$ from $3x^2 - 2xy + y^2$.

Solution

$$\begin{array}{r}3x^2 - 2xy + y^2 \\ \underline{4x^2 - 7xy + 5y^2}\end{array}$$

Notice which polynomial goes on the bottom and how the similar terms are aligned.

Now we can *mentally form the opposite of the bottom polynomial* and add.

$$\begin{array}{r}3x^2 - 2xy + y^2 \\ \underline{4x^2 - 7xy + 5y^2} \\ -x^2 + 5xy - 4y^2\end{array}$$

The opposite of $4x^2 - 7xy + 5y^2$ is $-4x^2 + 7xy - 5y^2$.

■

The distributive property along with the properties "$a = 1(a)$" and "$-a = -1(a)$" can also be used when adding and subtracting polynomials. The next examples illustrate this approach.

EXAMPLE 7 Perform the indicated operations: $(5x - 2) + (2x - 1) - (3x + 4)$.

Solution

$$\begin{aligned}(5x - 2) + (2x - 1) - (3x + 4) &= 1(5x - 2) + 1(2x - 1) - 1(3x + 4) \\ &= 1(5x) - 1(2) + 1(2x) - 1(1) - 1(3x) - 1(4) \\ &= 5x - 2 + 2x - 1 - 3x - 4 \\ &= 5x + 2x - 3x - 2 - 1 - 4 \\ &= 4x - 7\end{aligned}$$

■

Certainly you can do some of the steps mentally and our format can be simplified as in the next two examples.

EXAMPLE 8 Perform the indicated operations: $(5a^2 - 2b) - (2a^2 + 4) + (-7b - 3)$.

Solution

$$\begin{aligned}(5a^2 - 2b) - (2a^2 + 4) + (-7b - 3) &= 5a^2 - 2b - 2a^2 - 4 - 7b - 3 \\ &= 3a^2 - 9b - 7\end{aligned}$$

■

EXAMPLE 9 Simplify $(4t^2 - 7t - 1) - (t^2 + 2t - 6)$.

Solution

$$\begin{aligned}(4t^2 - 7t - 1) - (t^2 + 2t - 6) &= 4t^2 - 7t - 1 - t^2 - 2t + 6 \\ &= 3t^2 - 9t + 5\end{aligned}$$

■

Remember that a polynomial in parentheses preceded by a negative sign can be written without the parentheses by replacing each term with its opposite. Thus, in Example 9, $-(t^2 + 2t - 6) = -t^2 - 2t + 6$. Finally, let's consider a simplification problem that contains grouping symbols within grouping symbols.

EXAMPLE 10 Simplify $7x + [3x - (2x + 7)]$.

Solution

$$\begin{aligned}7x + [3x - (2x + 7)] &= 7x + [3x - 2x - 7] \\ &= 7x + [x - 7] \\ &= 7x + x - 7 \\ &= 8x - 7\end{aligned}$$

Remove the innermost parentheses first.

■

Problem Set 3.1

For Problems 1–10, determine the degree of the given polynomials.

1. $7xy + 6y$ 2
2. $-5x^2y^2 + 6xy^2 + x$ 4
3. $-x^2y + 2xy^2 - xy$ 3
4. $5x^3y^2 - 6x^3y^3$ 6
5. $5x^2 - 7x - 2$ 2
6. $7x^3 - 2x + 4$ 3
7. $8x^6 + 9$ 6
8. $5y^6 + y^4 - 2y^2 - 8$ 6
9. -12 0
10. $7x - 2y$ 1

For Problems 11–20, add the given polynomials.

11. $3x - 7$ and $7x + 4$ $10x - 3$
12. $9x + 6$ and $5x - 3$ $14x + 3$
13. $-5t - 4$ and $-6t + 9$ $-11t + 5$
14. $-7t + 14$ and $-3t - 6$ $-10t + 8$
15. $3x^2 - 5x - 1$ and $-4x^2 + 7x - 1$ $-x^2 + 2x - 2$
16. $6x^2 + 8x + 4$ and $-7x^2 - 7x - 10$ $-x^2 + x - 6$
17. $12a^2b^2 - 9ab$ and $5a^2b^2 + 4ab$ $17a^2b^2 - 5ab$
18. $15a^2b^2 - ab$ and $-20a^2b^2 - 6ab$ $-5a^2b^2 - 7ab$
19. $2x - 4$, $-7x + 2$, and $-4x + 9$ $-9x + 7$
20. $-x^2 - x - 4$, $2x^2 - 7x + 9$, and $-3x^2 + 6x - 10$ $-2x^2 - 2x - 5$

For Problems 21–30, subtract the polynomials using the horizontal format.

21. $5x - 2$ from $3x + 4$ $-2x + 6$

22. $7x + 5$ from $2x - 1$ $-5x - 6$

23. $-4a - 5$ from $6a + 2$ $10a + 7$

24. $5a + 7$ from $-a - 4$ $-6a - 11$

25. $3x^2 - x + 2$ from $7x^2 + 9x + 8$ $4x^2 + 10x + 6$

26. $5x^2 + 4x - 7$ from $3x^2 + 2x - 9$ $-2x^2 - 2x - 2$

27. $2a^2 - 6a - 4$ from $-4a^2 + 6a + 10$ $-6a^2 + 12a + 14$

28. $-3a^2 - 6a + 3$ from $3a^2 + 6a - 11$ $6a^2 + 12a - 14$

29. $2x^3 + x^2 - 7x - 2$ from $5x^3 + 2x^2 + 6x - 13$ $3x^3 + x^2 + 13x - 11$

30. $6x^3 + x^2 + 4$ from $9x^3 - x - 2$ $3x^3 - x^2 - x - 6$

For Problems 31–40, subtract the polynomials using the vertical format.

31. $5x - 2$ from $12x + 6$ $7x + 8$

32. $3x - 7$ from $2x + 1$ $-x + 8$

33. $-4x + 7$ from $-7x - 9$ $-3x - 16$

34. $-6x - 2$ from $5x + 6$ $11x + 8$

35. $2x^2 + x + 6$ from $4x^2 - x - 2$ $2x^2 - 2x - 8$

36. $4x^2 - 3x - 7$ from $-x^2 - 6x + 9$ $-5x^2 - 3x + 16$

37. $x^3 + x^2 - x - 1$ from $-2x^3 + 6x^2 - 3x + 8$ $-3x^3 + 5x^2 - 2x + 9$

38. $2x^3 - x + 6$ from $x^3 + 4x^2 + 1$ $-x^3 + 4x^2 + x - 5$

39. $-5x^2 + 6x - 12$ from $2x - 1$ $5x^2 - 4x + 11$

40. $2x^2 - 7x - 10$ from $-x^3 - 12$ $-x^3 - 2x^2 + 7x - 2$

For Problems 41–46, perform the operations as described.

41. Subtract $2x^2 - 7x - 1$ from the sum of $x^2 + 9x - 4$ and $-5x^2 - 7x + 10$. $-6x^2 + 9x + 7$

42. Subtract $4x^2 + 6x + 9$ from the sum of $-3x^2 - 9x + 6$ and $-2x^2 + 6x - 4$. $-9x^2 - 9x - 7$

43. Subtract $-x^2 - 7x - 1$ from the sum of $4x^2 + 3$ and $-7x^2 + 2x$. $-2x^2 + 9x + 4$

44. Subtract $-4x^2 + 6x - 3$ from the sum of $-3x + 4$ and $9x^2 - 6$. $13x^2 - 9x + 1$

45. Subtract the sum of $5n^2 - 3n - 2$ and $-7n^2 + n + 2$ from $12n^2 - n + 9$. $-10n^2 + n + 9$

46. Subtract the sum of $-6n^2 + 2n - 4$ and $4n^2 - 2n + 4$ from $-n^2 - n + 1$. $n^2 - n + 1$

For Problems 47–56, perform the indicated operations.

47. $(5x + 2) + (7x - 1) + (-4x - 3)$ $8x - 2$

48. $(-3x + 1) + (6x - 2) + (9x - 4)$ $12x - 5$

49. $(12x - 9) - (-3x + 4) - (7x + 1)$ $8x - 14$

50. $(6x + 4) - (4x - 2) - (-x - 1)$ $3x + 7$

51. $(2x^2 - 7x - 1) + (-4x^2 - x + 6) + (-7x^2 - 4x - 1)$ $-9x^2 - 12x + 4$

52. $(5x^2 + x + 4) + (-x^2 + 2x + 4) + (-14x^2 - x + 6)$ $-10x^2 + 2x + 14$

53. $(7x^2 - x - 4) - (9x^2 - 10x + 8) + (12x^2 + 4x - 6)$ $10x^2 + 13x - 18$

54. $(-6x^2 + 2x + 5) - (4x^2 + 4x - 1) + (7x^2 + 4)$ $-3x^2 - 2x + 10$

55. $(n^2 - 7n - 9) - (-3n + 4) - (2n^2 - 9)$ $-n^2 - 4n - 4$

56. $(6n^2 - 4) - (5n^2 + 9) - (6n + 4)$ $n^2 - 6n - 17$

For Problems 57–70, simplify by removing the inner parentheses first and working outward.

57. $3x - [5x - (x + 6)]$ $-x + 6$

58. $7x - [2x - (-x - 4)]$ $4x - 4$

59. $2x^2 - [-3x^2 - (x^2 - 4)]$ $6x^2 - 4$

60. $4x^2 - [-x^2 - (5x^2 - 6)]$ $10x^2 - 6$

61. $-2n^2 - [n^2 - (-4n^2 + n + 6)]$ $-7n^2 + n + 6$

62. $-7n^2 - [3n^2 - (-n^2 - n + 4)]$ $-11n^2 - n + 4$

63. $[4t^2 - (2t + 1) + 3] - [3t^2 + (2t - 1) - 5]$ $t^2 - 4t + 8$

64. $-(3n^2 - 2n + 4) - [2n^2 - (n^2 + n + 3)]$ $-4n^2 + 3n - 1$

65. $[2n^2 - (2n^2 - n + 5)] + [3n^2 + (n^2 - 2n - 7)]$ $4n^2 - n - 12$

66. $3x^2 - [4x^2 - 2x - (x^2 - 2x + 6)]$ 6

67. $[7xy - (2x - 3xy + y)] - [3x - (x - 10xy - y)]$ $-4x - 2y$

68. $[9xy - (4x + xy - y)] - [4y - (2x - xy + 6y)]$ $7xy - 2x + 3y$

69. $[4x^3 - (2x^2 - x - 1)] - [5x^3 - (x^2 + 2x - 1)]$ $-x^3 - x^2 + 3x$

70. $[x^3 - (x^2 - x + 1)] - [-x^3 + (7x^2 - x + 10)]$ $2x^3 - 8x^2 + 2x - 11$

3.2 Products and Quotients of Monomials

Suppose that we want to find the product of two monomials such as $3x^2y$ and $4x^3y^2$. To proceed, use the properties of real numbers and the idea that exponents indicate repeated multiplication.

$$\begin{aligned}(3x^2y)(4x^3y^2) &= (3 \cdot x \cdot x \cdot y)(4 \cdot x \cdot x \cdot x \cdot y \cdot y)\\ &= 3 \cdot 4 \cdot x \cdot x \cdot x \cdot x \cdot x \cdot y \cdot y \cdot y\\ &= 12x^5y^3\end{aligned}$$

You can use such an approach to find the product of any two monomials. However, there are some basic properties of exponents that make the process of multiplying monomials a much easier task. Let's consider each of these properties and illustrate its use when multiplying monomials. The following examples motivate the first property.

$$x^2 \cdot x^3 = (x \cdot x)(x \cdot x \cdot x) = x^5;$$

$$a^4 \cdot a^2 = (a \cdot a \cdot a \cdot a)(a \cdot a) = a^6;$$

$$b^3 \cdot b^4 = (b \cdot b \cdot b)(b \cdot b \cdot b \cdot b) = b^7.$$

In general,

$$\begin{aligned}b^n \cdot b^m &= \underbrace{(b \cdot b \cdot b \cdot \cdots b)}_{\substack{n \text{ factors} \\ \text{of } b}}\underbrace{(b \cdot b \cdot b \cdot \cdots b)}_{\substack{m \text{ factors} \\ \text{of } b}}\\ &= \underbrace{b \cdot b \cdot b \cdot \cdots b}_{(n+m) \text{ factors of } b}\\ &= b^{n+m}.\end{aligned}$$

We can state the first property as follows.

PROPERTY 3.1

> If b is any real number and n and m are positive integers, then
> $$b^n \cdot b^m = b^{n+m}.$$

Property 3.1 states "to find the product of two positive integral powers of the same base, add the exponents and use this sum as the exponent of the common base."

$$x^7 \cdot x^8 = x^{7+8} = x^{15}, \qquad y^6 \cdot y^4 = y^{6+4} = y^{10},$$
$$2^3 \cdot 2^8 = 2^{3+8} = 2^{11}, \qquad (-3)^4 \cdot (-3)^5 = (-3)^{4+5} = (-3)^9,$$
$$\left(\frac{2}{3}\right)^7 \cdot \left(\frac{2}{3}\right)^5 = \left(\frac{2}{3}\right)^{7+5} = \left(\frac{2}{3}\right)^{12}$$

The following examples illustrate the use of Property 3.1 along with the commutative and associative properties of multiplication to form the basis for multiplying monomials. The steps enclosed in the dashed boxes might be performed mentally.

EXAMPLE 1

$$\begin{aligned}(3x^2y)(4x^3y^2) &= \boxed{3 \cdot 4 \cdot x^2 \cdot x^3 \cdot y \cdot y^2} \\ &= \boxed{12x^{2+3}y^{1+2}} \\ &= 12x^5y^3\end{aligned}$$

■

EXAMPLE 2

$$\begin{aligned}(-5a^3b^4)(7a^2b^5) &= \boxed{-5 \cdot 7 \cdot a^3 \cdot a^2 \cdot b^4 \cdot b^5} \\ &= \boxed{-35a^{3+2}b^{4+5}} \\ &= -35a^5b^9\end{aligned}$$

■

EXAMPLE 3

$$\begin{aligned}\left(\frac{3}{4}xy\right)\left(\frac{1}{2}x^5y^6\right) &= \boxed{\frac{3}{4} \cdot \frac{1}{2} \cdot x \cdot x^5 \cdot y \cdot y^6} \\ &= \boxed{\frac{3}{8}x^{1+5}y^{1+6}} \\ &= \frac{3}{8}x^6y^7\end{aligned}$$

■

EXAMPLE 4

$$\begin{aligned}(-ab^2)(-5a^2b) &= \boxed{(-1)(-5)(a)(a^2)(b^2)(b)} \\ &= \boxed{5a^{1+2}b^{2+1}} \\ &= 5a^3b^3\end{aligned}$$

■

EXAMPLE 5

$$\begin{aligned}(2x^2y^2)(3x^2y)(4y^3) &= \boxed{2 \cdot 3 \cdot 4 \cdot x^2 \cdot x^2 \cdot y^2 \cdot y \cdot y^3} \\ &= \boxed{24x^{2+2}y^{2+1+3}} \\ &= 24x^4y^6\end{aligned}$$

■

The following examples demonstrate another useful property of exponents.

$$(x^2)^3 = x^2 \cdot x^2 \cdot x^2 = x^{2+2+2} = x^6;$$

$$(a^3)^2 = a^3 \cdot a^3 = a^{3+3} = a^6;$$

$$(b^4)^3 = b^4 \cdot b^4 \cdot b^4 = b^{4+4+4} = b^{12}.$$

In general,

$$(b^n)^m = \underbrace{b^n \cdot b^n \cdot b^n \cdot \cdots b^n}_{m \text{ factors of } b^n}$$

$$= b^{\overbrace{n+n+n+\cdots+n}^{m \text{ of these}}}$$

$$= b^{mn}.$$

We can state the property as follows.

PROPERTY 3.2

> If b is any real number and m and n are positive integers, then
>
> $$(b^n)^m = b^{mn}.$$

Use Property 3.2 to find "the power of a power" as follows.

$$(x^4)^5 = x^{5(4)} = x^{20}, \qquad (y^6)^3 = y^{3(6)} = y^{18},$$

$$(2^3)^7 = 2^{7(3)} = 2^{21}$$

A third property of exponents pertains to raising a monomial to a power. Consider the following examples, which are used to introduce the property.

$$(3x)^2 = (3x)(3x) = 3 \cdot 3 \cdot x \cdot x = 3^2 \cdot x^2;$$

$$(4y^2)^3 = (4y^2)(4y^2)(4y^2) = 4 \cdot 4 \cdot 4 \cdot y^2 \cdot y^2 \cdot y^2 = (4)^3(y^2)^3;$$

$$(-2a^3b^4)^2 = (-2a^3b^4)(-2a^3b^4) = (-2)(-2)(a^3)(a^3)(b^4)(b^4)$$
$$= (-2)^2(a^3)^2(b^4)^2.$$

In general,

$$(ab)^n = \underbrace{(ab)(ab)(ab) \cdot \cdots (ab)}_{n \text{ factors of } ab}$$

$$= \underbrace{(a \cdot a \cdot a \cdot a \cdot \cdots a)}_{n \text{ factors of } a}\underbrace{(b \cdot b \cdot b \cdot \cdots b)}_{n \text{ factors of } b}$$

$$= a^n b^n.$$

We can formally state the property as follows.

PROPERTY 3.3

> If a and b are real numbers and n is a positive integer, then
>
> $(ab)^n = a^n b^n$.

Property 3.3, along with Property 3.2, form the basis for raising a monomial to a power, as in the next examples.

EXAMPLE 6

$$\begin{aligned}(x^2y^3)^4 &= (x^2)^4(y^3)^4 && \text{Use } (ab)^n = a^n b^n.\\ &= x^8y^{12} && \text{Use } (b^n)^m = b^{mn}.\end{aligned}$$ ■

EXAMPLE 7

$$\begin{aligned}(3a^5)^3 &= (3)^3(a^5)^3\\ &= 27a^{15}\end{aligned}$$ ■

EXAMPLE 8

$$\begin{aligned}(-2xy^4)^5 &= (-2)^5(x)^5(y^4)^5\\ &= -32x^5y^{20}\end{aligned}$$ ■

Dividing Monomials

To develop an effective process for dividing by a monomial we need yet another property of exponents. This property is a direct consequence of the definition of an exponent. Study the following examples.

$$\frac{x^4}{x^3} = \frac{x \cdot x \cdot x \cdot x}{x \cdot x \cdot x} = x, \qquad \frac{x^3}{x^3} = \frac{x \cdot x \cdot x}{x \cdot x \cdot x} = 1,$$

$$\frac{a^5}{a^2} = \frac{a \cdot a \cdot a \cdot a \cdot a}{a \cdot a} = a^3, \qquad \frac{y^5}{y^5} = \frac{y \cdot y \cdot y \cdot y \cdot y}{y \cdot y \cdot y \cdot y \cdot y} = 1,$$

$$\frac{y^8}{y^4} = \frac{y \cdot y \cdot y \cdot y \cdot y \cdot y \cdot y \cdot y}{y \cdot y \cdot y \cdot y} = y^4$$

We can state the general property as follows.

PROPERTY 3.4

> If b is any nonzero real number and m and n are positive integers, then
>
> **1.** $\frac{b^n}{b^m} = b^{n-m}$, when $n > m$
>
> **2.** $\frac{b^n}{b^m} = 1$, when $n = m$.

Apply Property 3.4 to the previous examples to yield

$$\frac{x^4}{x^3} = x^{4-3} = x^1 = x, \qquad \frac{x^3}{x^3} = 1,$$

$$\frac{a^5}{a^2} = a^{5-2} = a^3, \qquad \frac{y^5}{y^5} = 1,$$

$$\frac{y^8}{y^4} = y^{8-4} = y^4.$$

(We will discuss the situation when $n < m$ in a later chapter.)

Property 3.4, along with our knowledge of dividing integers, provides the basis for dividing monomials. The following examples demonstrate the process.

$$\frac{24x^5}{3x^2} = 8x^{5-2} = 8x^3, \qquad \frac{-36a^{13}}{-12a^5} = 3a^{13-5} = 3a^8,$$

$$\frac{-56x^9}{7x^4} = -8x^{9-4} = -8x^5, \qquad \frac{72b^5}{8b^5} = 9 \quad \left(\frac{b^5}{b^5} = 1\right),$$

$$\frac{48y^7}{-12y} = -4y^{7-1} = -4y^6, \qquad \frac{12x^4y^7}{2x^2y^4} = 6x^{4-2}y^{7-4} = 6x^2y^3$$

Problem Set 3.2

Find each of the following products.

1. $(4x^3)(9x)$ $36x^4$
2. $(6x^3)(7x^2)$ $42x^5$
3. $(-2x^2)(6x^3)$ $-12x^5$
4. $(2xy)(-4x^2y)$ $-8x^3y^2$
5. $(-a^2b)(-4ab^3)$ $4a^3b^4$
6. $(-8a^2b^2)(-3ab^3)$ $24a^3b^5$
7. $(x^2yz^2)(-3xyz^4)$ $-3x^3y^2z^6$
8. $(-2xy^2z^2)(-x^2y^3z)$ $2x^3y^5z^3$
9. $(5xy)(-6y^3)$ $-30xy^4$
10. $(-7xy)(4x^4)$ $-28x^5y$
11. $(3a^2b)(9a^2b^4)$ $27a^4b^5$
12. $(-8a^2b^2)(-12ab^5)$ $96a^3b^7$
13. $(m^2n)(-mn^2)$ $-m^3n^3$
14. $(-x^3y^2)(xy^3)$ $-x^4y^5$
15. $\left(\frac{2}{5}xy^2\right)\left(\frac{3}{4}x^2y^4\right)$ $\frac{3}{10}x^3y^6$
16. $\left(\frac{1}{2}x^2y^6\right)\left(\frac{2}{3}xy\right)$ $\frac{1}{3}x^3y^7$
17. $\left(-\frac{3}{4}ab\right)\left(\frac{1}{5}a^2b^3\right)$ $-\frac{3}{20}a^3b^4$
18. $\left(-\frac{2}{7}a^2\right)\left(\frac{3}{5}ab^3\right)$ $-\frac{6}{35}a^3b^3$
19. $\left(-\frac{1}{2}xy\right)\left(\frac{1}{3}x^2y^3\right)$ $-\frac{1}{6}x^3y^4$
20. $\left(\frac{3}{4}x^4y^5\right)(-x^2y)$ $-\frac{3}{4}x^6y^6$
21. $(3x)(-2x^2)(-5x^3)$ $30x^6$
22. $(-2x)(-6x^3)(x^2)$ $12x^6$
23. $(-6x^2)(3x^3)(x^4)$ $-18x^9$
24. $(-7x^2)(3x)(4x^3)$ $-84x^6$
25. $(x^2y)(-3xy^2)(x^3y^3)$ $-3x^6y^6$
26. $(xy^2)(-5xy)(x^2y^4)$ $-5x^4y^7$
27. $(-3y^2)(-2y^2)(-4y^5)$ $-24y^9$
28. $(-y^3)(-6y)(-8y^4)$ $-48y^8$

29. $(4ab)(-2a^2b)(7a)$ $-56a^4b^2$

30. $(3b)(-2ab^2)(7a)$ $-42a^2b^3$

31. $(-ab)(-3ab)(-6ab)$ $-18a^3b^3$

32. $(-3a^2b)(-ab^2)(-7a)$ $-21a^4b^3$

33. $\left(\frac{2}{3}xy\right)(-3x^2y)(5x^4y^5)$ $-10x^7y^7$

34. $\left(\frac{3}{4}x\right)(-4x^2y^2)(9y^3)$ $-27x^3y^5$

35. $(12y)(-5x)\left(-\frac{5}{6}x^4y\right)$ $50x^5y^2$

36. $(-12x)(3y)\left(-\frac{3}{4}xy^6\right)$ $27x^2y^7$

Raise each of the following monomials to the indicated power.

37. $(3xy^2)^3$ $27x^3y^6$

38. $(4x^2y^3)^3$ $64x^6y^9$

39. $(-2x^2y)^5$ $-32x^{10}y^5$

40. $(-3xy^4)^3$ $-27x^3y^{12}$

41. $(-x^4y^5)^4$ $x^{16}y^{20}$

42. $(-x^5y^2)^4$ $x^{20}y^8$

43. $(ab^2c^3)^6$ $a^6b^{12}c^{18}$

44. $(a^2b^3c^5)^5$ $a^{10}b^{15}c^{25}$

45. $(2a^2b^3)^6$ $64a^{12}b^{18}$

46. $(2a^3b^2)^6$ $64a^{18}b^{12}$

47. $(9xy^4)^2$ $81x^2y^8$

48. $(8x^2y^5)^2$ $64x^4y^{10}$

49. $(-3ab^3)^4$ $81a^4b^{12}$

50. $(-2a^2b^4)^4$ $16a^8b^{16}$

51. $-(2ab)^4$ $-16a^4b^4$

52. $-(3ab)^4$ $-81a^4b^4$

53. $-(xy^2z^3)^6$ $-x^6y^{12}z^{18}$

54. $-(xy^2z^3)^8$ $-x^8y^{16}z^{24}$

55. $(-5a^2b^2c)^3$ $-125a^6b^6c^3$

56. $(-4abc^4)^3$ $-64a^3b^3c^{12}$

57. $(-xy^4z^2)^7$ $-x^7y^{28}z^{14}$

58. $(-x^2y^4z^5)^5$ $-x^{10}y^{20}z^{25}$

Find each of the following quotients.

59. $\dfrac{9x^4y^5}{3xy^2}$ $3x^3y^3$

60. $\dfrac{12x^2y^7}{6x^2y^3}$ $2y^4$

61. $\dfrac{25x^5y^6}{-5x^2y^4}$ $-5x^3y^2$

62. $\dfrac{56x^6y^4}{-7x^2y^3}$ $-8x^4y$

63. $\dfrac{-54ab^2c^3}{-6abc}$ $9bc^2$

64. $\dfrac{-48a^3bc^5}{-6a^2c^4}$ $8abc$

65. $\dfrac{-18x^2y^2z^6}{xyz^2}$ $-18xyz^4$

66. $\dfrac{-32x^4y^5z^8}{x^2yz^3}$ $-32x^2y^4z^5$

67. $\dfrac{a^3b^4c^7}{-abc^5}$ $-a^2b^3c^2$

68. $\dfrac{-a^4b^5c}{a^2b^4c}$ $-a^2b$

69. $\dfrac{-72x^2y^4}{-8x^2y^4}$ 9

70. $\dfrac{-96x^4y^5}{12x^4y^4}$ $-8y$

71. $\dfrac{14ab^3}{-14ab}$ $-b^2$

72. $\dfrac{-12abc^2}{12bc}$ $-ac$

73. $\dfrac{-36x^3y^5}{2y^5}$ $-18x^3$

74. $\dfrac{-48xyz^2}{2xz}$ $-24yz$

Find each of the following products. Assume that the variables in the exponents represent positive integers. For example,

$$(x^{2n})(x^{3n}) = x^{2n+3n} = x^{5n}.$$

75. $(2x^n)(3x^{2n})$ $6x^{3n}$

76. $(3x^{2n})(x^{3n-1})$ $3x^{5n-1}$

77. $(a^{2n-1})(a^{3n+4})$ a^{5n+3}

78. $(a^{5n-1})(a^{5n+1})$ a^{10n}

79. $(x^{3n-2})(x^{n+2})$ x^{4n}

80. $(x^{n-1})(x^{4n+3})$ x^{5n+2}

81 $(a^{5n-2})(a^3)$ a^{5n+1}

82. $(x^{3n-4})(x^4)$ x^{3n}

83. $(2x^n)(-5x^n)$ $-10x^{2n}$

84. $(4x^{2n-1})(-3x^{n+1})$ $-12x^{3n}$

85. $(-3a^2)(-4a^{n+2})$ $12a^{n+4}$

86. $(-5x^{n-1})(-6x^{2n+4})$ $30x^{3n+3}$

87. $(x^n)(2x^{2n})(3x^2)$ $6x^{3n+2}$

88. $(2x^n)(3x^{3n-1})(-4x^{2n+5})$ $-24x^{6n+4}$

89. $(3x^{n-1})(x^{n+1})(4x^{2-n})$ $12x^{n+2}$

90. $(-5x^{n+2})(x^{n-2})(4x^{3-2n})$ $-20x^3$

3.3 Multiplying Polynomials

We usually state the distributive property as "$a(b + c) = ab + ac$"; however, we can extend it as follows.

$$a(b + c + d) = ab + ac + ad,$$

$$a(b + c + d + e) = ab + ac + ad + ae, \qquad \text{etc.}$$

The commutative and associative properties, the properties of exponents, and the distributive property work together to find the product of a monomial and a polynomial. The following examples illustrate this idea.

EXAMPLE 1

$$\begin{aligned} 3x^2(2x^2 + 5x + 3) &= 3x^2(2x^2) + 3x^2(5x) + 3x^2(3) \\ &= 6x^4 + 15x^3 + 9x^2 \end{aligned}$$

■

EXAMPLE 2

$$\begin{aligned} -2xy(3x^3 - 4x^2y - 5xy^2 + y^3) &= -2xy(3x^3) - (-2xy)(4x^2y) \\ &\quad -(-2xy)(5xy^2) + (-2xy)(y^3) \\ &= -6x^4y + 8x^3y^2 + 10x^2y^3 - 2xy^4 \end{aligned}$$

■

Now let's consider the product of two polynomials neither of which is a monomial. Consider the following examples.

EXAMPLE 3

$$\begin{aligned} (x + 2)(y + 5) &= x(y + 5) + 2(y + 5) \\ &= x(y) + x(5) + 2(y) + 2(5) \\ &= xy + 5x + 2y + 10 \end{aligned}$$

■

Notice that each term of the first polynomial is multiplied times each term of the second polynomial.

EXAMPLE 4

$$\begin{aligned} (x - 3)(y + z + 3) &= x(y + z + 3) - 3(y + z + 3) \\ &= xy + xz + 3x - 3y - 3z - 9 \end{aligned}$$

■

Frequently, multiplying polynomials will produce similar terms that can be combined to simplify the resulting polynomial.

EXAMPLE 5

$$\begin{aligned} (x + 5)(x + 7) &= x(x + 7) + 5(x + 7) \\ &= x^2 + 7x + 5x + 35 \\ &= x^2 + 12x + 35 \end{aligned}$$

■

EXAMPLE 6

$$\begin{aligned} (x - 2)(x^2 - 3x + 4) &= x(x^2 - 3x + 4) - 2(x^2 - 3x + 4) \\ &= x^3 - 3x^2 + 4x - 2x^2 + 6x - 8 \\ &= x^3 - 5x^2 + 10x - 8 \end{aligned}$$

■

EXAMPLE 7

$$\begin{aligned}(3x - 2y)(x^2 + xy - y^2) &= 3x(x^2 + xy - y^2) - 2y(x^2 + xy - y^2)\\ &= 3x^3 + 3x^2y - 3xy^2 - 2x^2y - 2xy^2 + 2y^3\\ &= 3x^3 + x^2y - 5xy^2 + 2y^3\end{aligned}$$

■

It helps to be able to find the product of two binomials without showing all of the intermediate steps. This is quite easy to do with a *three-step shortcut pattern* as demonstrated by the following examples.

EXAMPLE 8

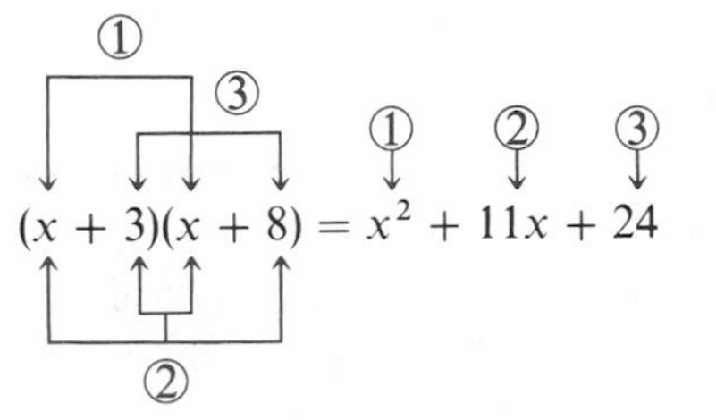

$$(x + 3)(x + 8) = x^2 + 11x + 24$$

Step①. Multiply $x \cdot x$.
Step②. Multiply $3 \cdot x$ and $8 \cdot x$ and combine.
Step③. Multiply $3 \cdot 8$. ■

EXAMPLE 9

$$(3x + 2)(2x - 1) = 6x^2 + x - 2$$

■

Now see if you can use the pattern to find the following products.

$(x + 2)(x + 6) = ?$

$(x - 3)(x + 5) = ?$

$(2x + 5)(3x + 7) = ?$

$(3x - 1)(4x - 3) = ?$

Your answers should be $x^2 + 8x + 12$, $x^2 + 2x - 15$, $6x^2 + 29x + 35$, and $12x^2 - 13x + 3$. Keep in mind that this shortcut pattern applies only to finding the product of two binomials.

We can use exponents to indicate repeated multiplication of polynomials. For example, $(x + 3)^2$ means $(x + 3)(x + 3)$; and $(x + 4)^3$ means $(x + 4)(x + 4)(x + 4)$. To square a binomial we can simply write it as the product of two equal binomials and apply the shortcut pattern. Thus,

$$(x + 3)^2 = (x + 3)(x + 3) = x^2 + 6x + 9,$$

$$(x-6)^2 = (x-6)(x-6) = x^2 - 12x + 36,$$

$$(3x-4)^2 = (3x-4)(3x-4) = 9x^2 - 24x + 16$$

When squaring binomials, be careful not to forget the middle term. That is to say, $(x+3)^2 \neq x^2 + 3^2$; instead, $(x+3)^2 = x^2 + 6x + 9$.

When multiplying binomials, there are some special patterns that you should recognize. We can use these patterns to find products and later we will use some of them when factoring polynomials.

PATTERN

$$(a+b)^2 = (a+b)(a+b) = a^2 + 2ab + b^2$$

square of first term of binomial + twice the product of the two terms of binomial + square of second term of binomial

Examples $(x+4)^2 = x^2 + 8x + 16,$

$$(2x+3y)^2 = 4x^2 + 12xy + 9y^2,$$

$$(5a+7b)^2 = 25a^2 + 70ab + 49b^2$$

PATTERN

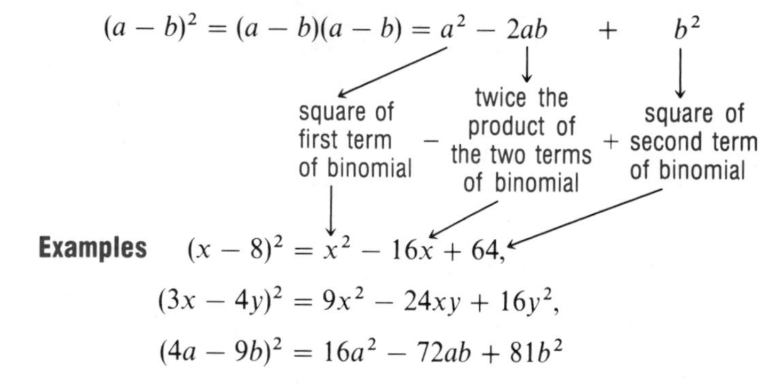

PATTERN

$$(a+b)(a-b) = a^2 - b^2.$$

square of first term of binomials − square of second term of binomials

Examples $(x+7)(x-7) = x^2 - 49,$

$$(2x+y)(2x-y) = 4x^2 - y^2,$$

$$(3a-2b)(3a+2b) = 9a^2 - 4b^2$$

Now suppose that we want to cube a binomial. One approach is as follows.

$$\begin{aligned}(x+4)^3 &= (x+4)(x+4)(x+4)\\ &= (x+4)(x^2+8x+16)\\ &= x(x^2+8x+16)+4(x^2+8x+16)\\ &= x^3+8x^2+16x+4x^2+32x+64\\ &= x^3+12x^2+48x+64\end{aligned}$$

Another approach is to cube a general binomial and then use the resulting pattern.

PATTERN

$$\begin{aligned}(a+b)^3 &= (a+b)(a+b)(a+b)\\ &= (a+b)(a^2+2ab+b^2)\\ &= a(a^2+2ab+b^2)+b(a^2+2ab+b^2)\\ &= a^3+2a^2b+ab^2+a^2b+2ab^2+b^3\\ &= a^3+3a^2b+3ab^2+b^3\end{aligned}$$

Let's use the pattern $(a+b)^3 = a^3+3a^2b+3ab^2+b^3$ to cube the binomial $x+4$.

$$\begin{aligned}(x+4)^3 &= x^3+3x^2(4)+3x(4)^2+4^3\\ &= x^3+12x^2+48x+64\end{aligned}$$

Since $a-b = a+(-b)$ we can easily develop a pattern for cubing $a-b$. ■

PATTERN

$$\begin{aligned}(a-b)^3 &= [a+(-b)]^3\\ &= a^3+3a^2(-b)+3a(-b)^2+(-b)^3\\ &= a^3-3a^2b+3ab^2-b^3\end{aligned}$$

Now let's use the pattern $(a-b)^3 = a^3-3a^2b+3ab^2-b^3$ to cube the binomial $3x-2y$.

$$\begin{aligned}(3x-2y)^3 &= (3x)^3-3(3x)^2(2y)+3(3x)(2y)^2-(2y)^3\\ &= 27x^3-54x^2y+36xy^2-8y^3\end{aligned}$$

Finally, we need to realize that if the patterns are forgotten or do not apply, then we can revert back to applying the distributive property.

$$\begin{aligned}(2x-1)(x^2-4x+6) &= 2x(x^2-4x+6)-1(x^2-4x+6)\\ &= 2x^3-8x^2+12x-x^2+4x-6\\ &= 2x^3-9x^2+16x-6\end{aligned}$$

■

Problem Set 3.3

Find the following indicated products. Remember the shortcut for multiplying binomials and the other special patterns we discussed in this section.

1. $2xy(5xy^2+3x^2y^3)$ $10x^2y^3+6x^3y^4$
2. $3x^2y(6y^2-5x^2y^4)$ $18x^2y^3-15x^4y^5$

3. $-3a^2b(4ab^2 - 5a^3)$ $-12a^3b^3 + 15a^5b$

4. $-7ab^2(2b^3 - 3a^2)$ $-14ab^5 + 21a^3b^2$

5. $8a^3b^4(3ab - 2ab^2 + 4a^2b^2)$ $24a^4b^5 - 16a^4b^6 + 32a^5b^6$

6. $9a^3b(2a - 3b + 7ab)$ $18a^4b - 27a^3b^2 + 63a^4b^2$

7. $-x^2y(6xy^2 + 3x^2y^3 - x^3y)$ $-6x^3y^3 - 3x^4y^4 + x^5y^2$

8. $-ab^2(5a + 3b - 6a^2b^3)$ $-5a^2b^2 - 3ab^3 + 6a^3b^5$

9. $(a + 2b)(x + y)$ $ax + ay + 2bx + 2by$

10. $(t - s)(x + y)$ $tx + ty - sx - sy$

11. $(a - 3b)(c + 4d)$ $ac + 4ad - 3bc - 12bd$

12. $(a - 4b)(c - d)$ $ac - ad - 4bc + 4bd$

13. $(x + 6)(x + 10)$ $x^2 + 16x + 60$

14. $(x + 2)(x + 10)$ $x^2 + 12x + 20$

15. $(y - 5)(y + 11)$ $y^2 + 6y - 55$

16. $(y - 3)(y + 9)$ $y^2 + 6y - 27$

17. $(n + 2)(n - 7)$ $n^2 - 5n - 14$

18. $(n + 3)(n - 12)$ $n^2 - 9n - 36$

19. $(x + 6)(x - 6)$ $x^2 - 36$

20. $(t + 8)(t - 8)$ $t^2 - 64$

21. $(x - 6)^2$ $x^2 - 12x + 36$

22. $(x - 2)^2$ $x^2 - 4x + 4$

23. $(x - 6)(x - 8)$ $x^2 - 14x + 48$

24. $(x - 3)(x - 13)$ $x^2 - 16x + 39$

25. $(x + 1)(x - 2)(x - 3)$ $x^3 - 4x^2 + x + 6$

26. $(x - 1)(x + 4)(x - 6)$ $x^3 - 3x^2 - 22x + 24$

27. $(x - 3)(x + 3)(x - 1)$ $x^3 - x^2 - 9x + 9$

28. $(x - 5)(x + 5)(x - 8)$ $x^3 - 8x^2 - 25x + 200$

29. $(t + 9)^2$ $t^2 + 18t + 81$

30. $(t + 13)^2$ $t^2 + 26t + 169$

31. $(y - 7)^2$ $y^2 - 14y + 49$

32. $(y - 4)^2$ $y^2 - 8y + 16$

33. $(4x + 5)(x + 7)$ $4x^2 + 33x + 35$

34. $(6x + 5)(x + 3)$ $6x^2 + 23x + 15$

35. $(3y - 1)(3y + 1)$ $9y^2 - 1$

36. $(5y - 2)(5y + 2)$ $25y^2 - 4$

37. $(7x - 2)(2x + 1)$ $14x^2 + 3x - 2$

38. $(6x - 1)(3x + 2)$ $18x^2 + 9x - 2$

39. $(1 + t)(5 - 2t)$ $5 + 3t - 2t^2$

40. $(3 - t)(2 + 4t)$ $6 + 10t - 4t^2$

41. $(3t + 7)^2$ $9t^2 + 42t + 49$

42. $(4t + 6)^2$ $16t^2 + 48t + 36$

43. $(2 - 5x)(2 + 5x)$ $4 - 25x^2$

44. $(6 - 3x)(6 + 3x)$ $36 - 9x^2$

45. $(7x - 4)^2$ $49x^2 - 56x + 16$

46. $(5x - 7)^2$ $25x^2 - 70x + 49$

47. $(6x + 7)(3x - 10)$ $18x^2 - 39x - 70$

48. $(4x - 7)(7x + 4)$ $28x^2 - 33x - 28$

49. $(2x - 5y)(x + 3y)$ $2x^2 + xy - 15y^2$

50. $(x - 4y)(3x + 7y)$ $3x^2 - 5xy - 28y^2$

51. $(5x - 2a)(5x + 2a)$ $25x^2 - 4a^2$

52. $(9x - 2y)(9x + 2y)$ $81x^2 - 4y^2$

53. $(t + 3)(t^2 - 3t - 5)$ $t^3 - 14t - 15$

54. $(t - 2)(t^2 + 7t + 2)$ $t^3 + 5t^2 - 12t - 4$

55. $(x - 4)(x^2 + 5x - 4)$ $x^3 + x^2 - 24x + 16$

56. $(x + 6)(2x^2 - x - 7)$ $2x^3 + 11x^2 - 13x - 42$

57. $(2x - 3)(x^2 + 6x + 10)$ $2x^3 + 9x^2 + 2x - 30$

58. $(3x + 4)(2x^2 - 2x - 6)$ $6x^3 + 2x^2 - 26x - 24$

59. $(4x - 1)(3x^2 - x + 6)$ $12x^3 - 7x^2 + 25x - 6$

60. $(5x - 2)(6x^2 + 2x - 1)$ $30x^3 - 2x^2 - 9x + 2$

61. $(x^2 + 2x + 1)(x^2 + 3x + 4)$ $x^4 + 5x^3 + 11x^2 + 11x + 4$

62. $(x^2 - x + 6)(x^2 - 5x - 8)$ $x^4 - 6x^3 + 3x^2 - 22x - 48$

63. $(2x^2 + 3x - 4)(x^2 - 2x - 1)$ $2x^4 - x^3 - 12x^2 + 5x + 4$

64. $(3x^2 - 2x + 1)(2x^2 + x - 2)$ $6x^4 - x^3 - 6x^2 + 5x - 2$

65. $(x + 2)^3$ $x^3 + 6x^2 + 12x + 8$

66. $(x + 1)^3$ $x^3 + 3x^2 + 3x + 1$

67. $(x - 4)^3$ $x^3 - 12x^2 + 48x - 64$

68. $(x - 5)^3$ $x^3 - 15x^2 + 75x - 125$

69. $(2x + 3)^3$ $8x^3 + 36x^2 + 54x + 27$

70. $(3x + 1)^3$ $27x^3 + 27x^2 + 9x + 1$

71. $(4x - 1)^3$ $64x^3 - 48x^2 + 12x - 1$

72. $(3x - 2)^3$ $27x^3 - 54x^2 + 36x - 8$

73. $(5x + 2)^3$ $125x^3 + 150x^2 + 60x + 8$

74. $(4x - 5)^3$ $64x^3 - 240x^2 + 300x - 125$

For Problems 75–84, find the indicated products. Assume all variables that appear as exponents represent positive integers.

75. $(x^n - 4)(x^n + 4)$ $x^{2n} - 16$

76. $(x^{3a} - 1)(x^{3a} + 1)$ $x^{6a} - 1$

77. $(x^a + 6)(x^a - 2)$ $x^{2a} + 4x^a - 12$

78. $(x^a + 4)(x^a - 9)$ $x^{2a} - 5x^a - 36$

79. $(2x^n + 5)(3x^n - 7)$ $6x^{2n} + x^n - 35$

80. $(3x^n + 5)(4x^n - 9)$ $12x^{2n} - 7x^n - 45$

81. $(x^{2a} - 7)(x^{2a} - 3)$ $x^{4a} - 10x^{2a} + 21$

82. $(x^{2a} + 6)(x^{2a} - 4)$ $x^{4a} + 2x^{2a} - 24$

83. $(2x^n + 5)^2$ $4x^{2n} + 20x^n + 25$

84. $(3x^n - 7)^2$ $9x^{2n} - 42x^n + 49$

Miscellaneous Problems

85. We have used the following two multiplication patterns.

$$(a + b)^2 = a^2 + 2ab + b^2,$$

$$(a + b)^3 = a^3 + 3a^2b + 3ab^2 + b^3$$

By multiplying, we can extend these patterns as follows.

$$(a + b)^4 = a^4 + 4a^3b + 6a^2b^2 + 4ab^3 + b^4,$$

$$(a + b)^5 = a^5 + 5a^4b + 10a^3b^2 + 10a^2b^3 + 5ab^4 + b^5$$

Based on these results, see if you can determine a pattern that will allow you to complete each of the following without using the long multiplication process.

(a) $(a + b)^6$ $a^6 + 6a^5b + 15a^4b^2 + 20a^3b^3 + 15a^2b^4 + 6ab^5 + b^6$

(b) $(a + b)^7$ $a^7 + 7a^6b + 21a^5b^2 + 35a^4b^3 + 35a^3b^4 + 21a^2b^5 + 7ab^6 + b^7$

(c) $(a + b)^8$ $a^8 + 8a^7b + 28a^6b^2 + 56a^5b^3 + 70a^4b^4 + 56a^3b^5 + 28a^2b^6 + 8ab^7 + b^8$

(d) $(a + b)^9$ $a^9 + 9a^8b + 36a^7b^2 + 84a^6b^3 + 126a^5b^4 + 126a^4b^5 + 84a^3b^6 + 36a^2b^7 + 9ab^8 + b^9$

86. Find each of the following indicated products. These patterns will be used again in Section 3.5.

(a) $(x - 1)(x^2 + x + 1)$ $x^3 - 1$

(b) $(x + 1)(x^2 - x + 1)$ $x^3 + 1$

(c) $(x + 3)(x^2 - 3x + 9)$ $x^3 + 27$

(d) $(x - 4)(x^2 + 4x + 16)$ $x^3 - 64$

(e) $(2x - 3)(4x^2 + 6x + 9)$ $8x^3 - 27$

(f) $(3x + 5)(9x^2 - 15x + 25)$ $27x^3 + 125$

3.4

Factoring: Use of the Distributive Property

Recall that 2 and 3 are said to be *factors* of 6 because the product of 2 and 3 is 6. Likewise, in an indicated product such as $7ab$, the 7, a, and b are called factors of the product. If a positive integer greater than 1 has no factors that are positive integers other than itself and 1, then it is called a **prime number**. Thus, the prime numbers less than 20 are 2, 3, 5, 7, 11, 13, 17, and 19. A positive integer greater than 1 that is not a prime number is called a **composite number**. The composite numbers less than 20 are 4, 6, 8, 9, 10, 12, 14, 15, 16, and 18. Every composite number is the product of prime numbers. Consider the following examples.

$$4 = 2 \cdot 2, \qquad 63 = 3 \cdot 3 \cdot 7,$$

$$12 = 2 \cdot 2 \cdot 3, \qquad 121 = 11 \cdot 11,$$

$$35 = 5 \cdot 7$$

The indicated product form that contains only prime factors is called the **prime factorization form** of a number. Thus, the prime factorization form of 63 is $3 \cdot 3 \cdot 7$. We also say that the number has been **completely factored** when it is in the prime factorization form.

In general, factoring is the reverse of multiplication. Previously, we have used the distributive property to find the product of a monomial and a polynomial, as in the next examples.

$$3(x + 2) = 3(x) + 3(2) = 3x + 6,$$

$$5(2x - 1) = 5(2x) - 5(1) = 10x - 5,$$

$$x(x^2 + 6x - 4) = x(x^2) + x(6x) - x(4) = x^3 + 6x^2 - 4x$$

Now we shall also use the distributive property (in the form "$ab + ac = a(b + c)$") to reverse the process, that is, to factor a given polynomial. Consider the following examples. (The steps in the dashed boxes can be done mentally.)

$$3x + 6 = \boxed{3(x) + 3(2)} = 3(x + 2),$$

$$10x - 5 = \boxed{5(2x) - 5(1)} = 5(2x - 1),$$

$$x^3 + 6x^2 - 4x = \boxed{x(x^2) + x(6x) - x(4)} = x(x^2 + 6x - 4)$$

Note that in each example a given polynomial has been factored into the product of a monomial and a polynomial. Obviously, polynomials could be factored in a variety of ways. Consider some factorizations of $3x^2 + 12x$.

$$3x^2 + 12x = 3x(x + 4), \quad \text{or} \quad 3x^2 + 12x = 3(x^2 + 4x), \quad \text{or}$$

$$3x^2 + 12x = x(3x + 12), \quad \text{or} \quad 3x^2 + 12x = \frac{1}{2}(6x^2 + 24x)$$

We are, however, primarily interested in the first of the previous factorization forms, which we refer to as the **completely factored form**. A polynomial with integral coefficients is in completely factored form if:

1. It is expressed as a product of polynomials with *integral coefficients*;
2. No polynomial, other than a monomial, within the factored form can be further factored into polynomials with integral coefficients.

Do you see why only the first of the above factored forms of $3x^2 + 12x$ is said to be in completely factored form? In each of the other three forms the polynomial inside the parentheses can be further factored. Furthermore, in the last form, $\frac{1}{2}(6x^2 + 24x)$, the condition of using only integral coefficients is violated.

The factoring process that we discuss in this section ($ab + ac = a(b + c)$) is often referred to as **factoring out the highest common monomial factor**. The key idea in this process is to recognize the monomial factor that is common to all terms. For example, observe that each term of the polynomial $2x^3 + 4x^2 + 6x$ has a factor of $2x$. Thus, we write

$$2x^3 + 4x^2 + 6x = 2x(\qquad\qquad)$$

and insert within the parentheses the appropriate polynomial factor. We determine the terms of this polynomial factor by dividing each term of the original polynomial by the factor of $2x$. The final completely factored form is

$$2x^3 + 4x^2 + 6x = 2x(x^2 + 2x + 3).$$

The following examples further demonstrate this process of factoring out the highest common monomial factor.

$$12x^3 + 16x^2 = 4x^2(3x + 4), \qquad 30x^3 + 42x^4 - 24x^5 = 6x^3(5 + 7x - 4x^2),$$

$$8ab - 18b = 2b(4a - 9), \qquad 8y^3 + 4y^2 = 4y^2(2y + 1),$$

$$6x^2y^3 + 27xy^4 = 3xy^3(2x + 9y)$$

Note that in each example the common monomial factor itself is not in a completely factored form. For example, $4x^2(3x + 4)$ is not written as $2 \cdot 2x \cdot x \cdot (3x + 4)$.

Sometimes there may be a common binomial factor rather than a common monomial factor. For example, each of the two terms of the expression $x(y + 2) + z(y + 2)$ has a binomial factor of $(y + 2)$. Thus, we can factor $(y + 2)$ from each term and our result is as follows.

$$x(y + 2) + z(y + 2) = (y + 2)(x + z)$$

Consider a few more examples that involve a common binomial factor.

$$a^2(b + 1) + 2(b + 1) = (b + 1)(a^2 + 2),$$

$$x(2y - 1) - y(2y - 1) = (2y - 1)(x - y),$$

$$x(x + 2) + 3(x + 2) = (x + 2)(x + 3)$$

It may be that the original polynomial exhibits no apparent common monomial or binomial factor, which is the case with $ab + 3a + bc + 3c$. However, by factoring a from the first two terms and c from the last two terms we get

$$ab + 3a + bc + 3c = a(b + 3) + c(b + 3).$$

Now a common binomial factor of $(b + 3)$ is obvious and we can proceed as before.

$$a(b + 3) + c(b + 3) = (b + 3)(a + c)$$

We refer to this factoring process as **factoring by grouping**. Let's consider a few more examples of this type.

$ab^2 - 4b^2 + 3a - 12 = b^2(a - 4) + 3(a - 4)$	factor b^2 from first two terms and 3 from last two terms
$= (a - 4)(b^2 + 3),$	factor common binomial from both terms
$x^2 - x + 5x - 5 = x(x - 1) + 5(x - 1)$	factor x from first two terms and 5 from last two terms
$= (x - 1)(x + 5),$	factor common binomial from both terms
$x^2 + 2x - 3x - 6 = x(x + 2) - 3(x + 2)$	factor x from first two terms and -3 from last two terms
$= (x + 2)(x - 3)$	factor common binomial factor from both terms

It may be necessary to rearrange some terms first before applying the distributive property. Terms that contain common factors need to be grouped together and this may be done in more than one way. The next example illustrates this idea.

$$\begin{aligned} 4a^2 - bc^2 - a^2b + 4c^2 &= 4a^2 - a^2b + 4c^2 - bc^2 \\ &= a^2(4 - b) + c^2(4 - b) \\ &= (4 - b)(a^2 + c^2), \qquad \text{or} \end{aligned}$$

$$\begin{aligned} 4a^2 - bc^2 - a^2b + 4c^2 &= 4a^2 + 4c^2 - bc^2 - a^2b \\ &= 4(a^2 + c^2) - b(c^2 + a^2) \\ &= 4(a^2 + c^2) - b(a^2 + c^2) \\ &= (a^2 + c^2)(4 - b) \end{aligned}$$

Equations and Problem Solving

One reason that factoring is an important algebraic skill is that it extends our techniques for solving equations. Each time that we examine a factoring technique we will then use it to help solve certain types of equations.

We need another property of equality before considering some equations for which the highest-common-factor technique is useful. Suppose that the product of two numbers is zero. Do you agree that we can conclude that at least one of the numbers must be zero? Let's state a property that formalizes this idea.

PROPERTY 3.5

> Let a and b be real numbers,
>
> $ab = 0$ if and only if $a = 0$ or $b = 0$.

Property 3.5, along with the highest-common-factor pattern, provides us with another technique for solving equations.

EXAMPLE 1 Solve $x^2 + 6x = 0$.

Solution

$$x^2 + 6x = 0$$

$$x(x + 6) = 0 \qquad \text{Factor the left side.}$$

$$x = 0 \quad \text{or} \quad x + 6 = 0 \qquad \text{Use Property 3.5.}$$

$$x = 0 \quad \text{or} \quad x = -6$$

Thus, both 0 and -6 will satisfy the original equation and the solution set is $\{-6, 0\}$. ■

EXAMPLE 2 Solve $a^2 = 11a$.

Solution

$$a^2 = 11a$$

$$a^2 - 11a = 0 \qquad \text{Add } -11a \text{ to both sides.}$$

$$a(a - 11) = 0 \qquad \text{Factor the left side.}$$

$$a = 0 \quad \text{or} \quad a - 11 = 0 \qquad \text{Use Property 3.5.}$$

$$a = 0 \quad \text{or} \quad a = 11$$

The solution set is $\{0, 11\}$. ■

EXAMPLE 3 Solve $3n^2 - 5n = 0$.

Solution

$$3n^2 - 5n = 0$$

$$n(3n - 5) = 0$$

$$n = 0 \quad \text{or} \quad 3n - 5 = 0$$

$$n = 0 \quad \text{or} \quad 3n = 5$$

$$n = 0 \quad \text{or} \quad n = \frac{5}{3}$$

The solution set is $\left\{0, \frac{5}{3}\right\}$. ■

EXAMPLE 4 Solve $3ax^2 + bx = 0$ for x.

Solution

$$3ax^2 + bx = 0$$

$$x(3ax + b) = 0$$

$$x = 0 \quad \text{or} \quad 3ax + b = 0$$

$$x = 0 \quad \text{or} \quad 3ax = -b$$

$$x = 0 \quad \text{or} \quad x = -\frac{b}{3a}$$

The solution set is $\left\{0, -\frac{b}{3a}\right\}$. ■

Many of the problems that we are going to solve in the next few sections have a geometric setting. So let's briefly review a few formulas from geometry that we can put to use in a problem-solving situation.

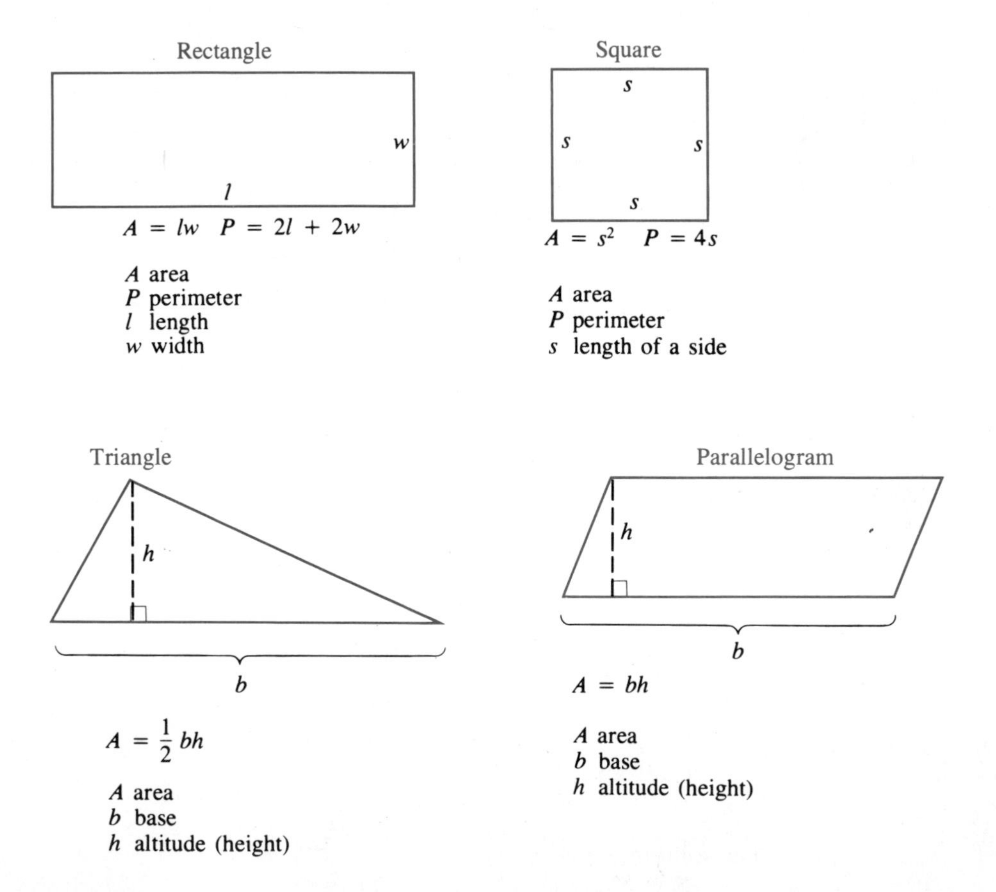

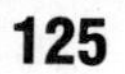

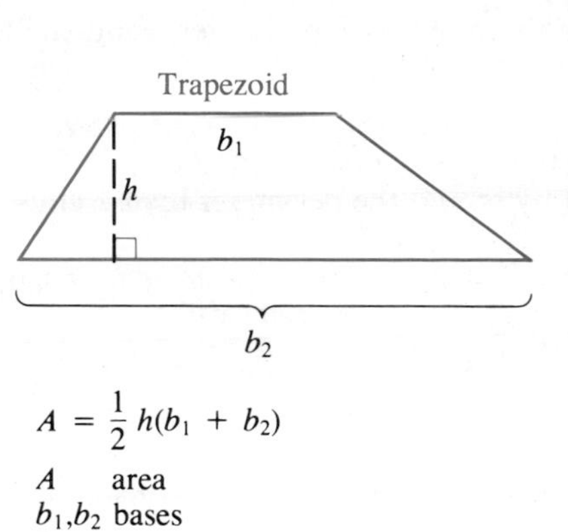

$$A = \frac{1}{2}h(b_1 + b_2)$$

A area
b_1, b_2 bases
h altitude

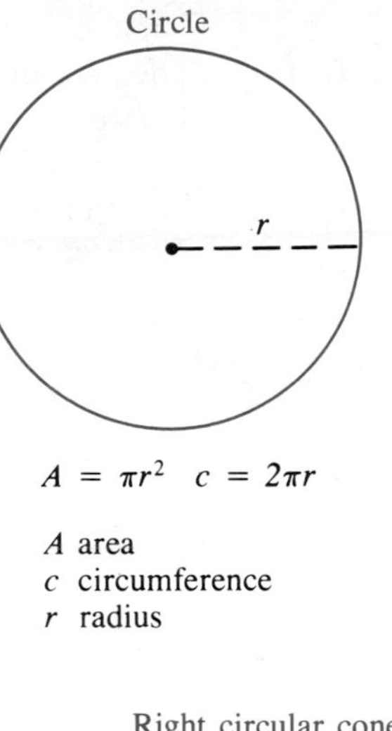

$$A = \pi r^2 \quad c = 2\pi r$$

A area
c circumference
r radius

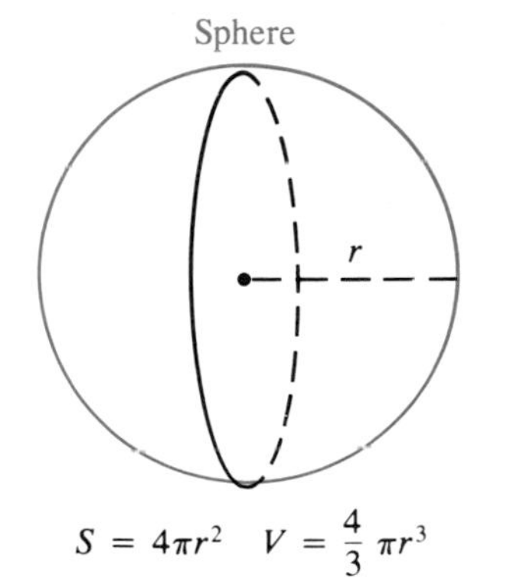

$$S = 4\pi r^2 \quad V = \frac{4}{3}\pi r^3$$

S surface area
V volume
r radius

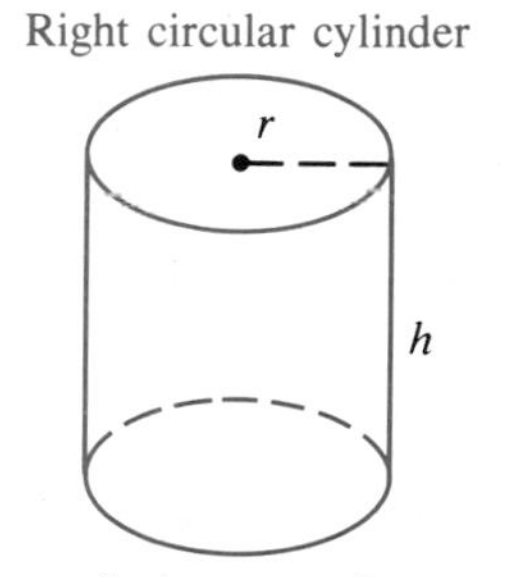

$$V = \pi r^2 h \quad S = 2\pi r^2 + 2\pi rh$$

V volume
S total surface area
r radius
h altitude (height)

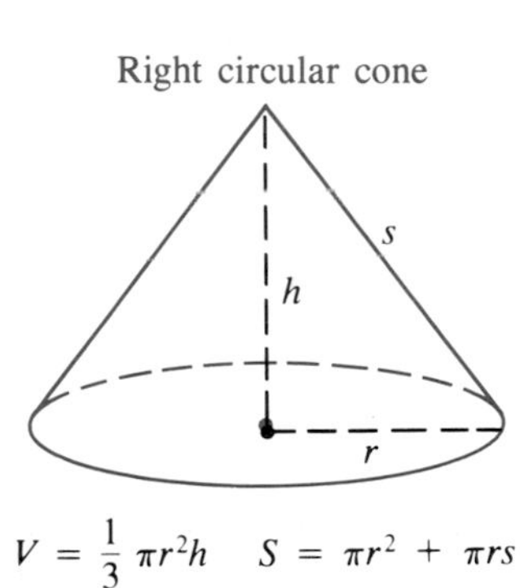

$$V = \frac{1}{3}\pi r^2 h \quad S = \pi r^2 + \pi rs$$

V volume
S total surface area
r radius
h altitude (height)
s slant height

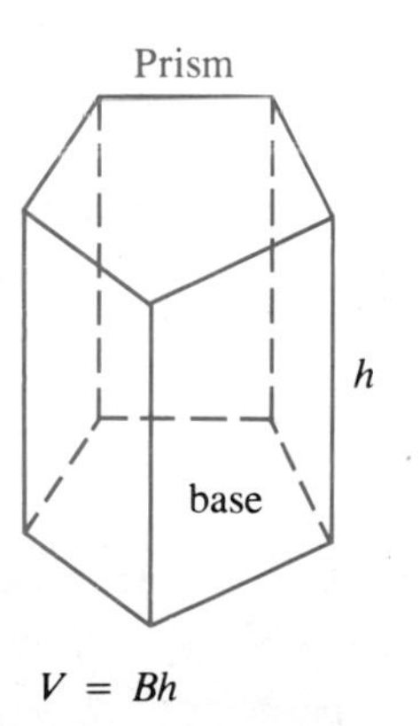

$$V = Bh$$

V volume
B area of base
h altitude (height)

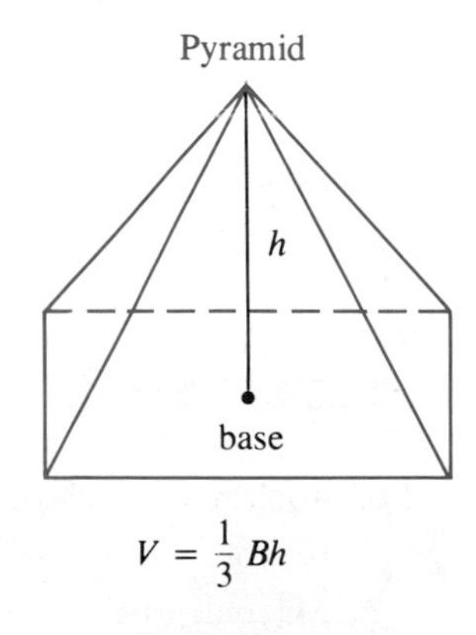

$$V = \frac{1}{3}Bh$$

V volume
B area of base
h altitude (height)

PROBLEM 1 The area of a square is three times its perimeter. Find the length of a side of the square.

Solution Let s represent the length of a side of the square.
The area is represented by s^2 and the perimeter by $4s$. Thus,

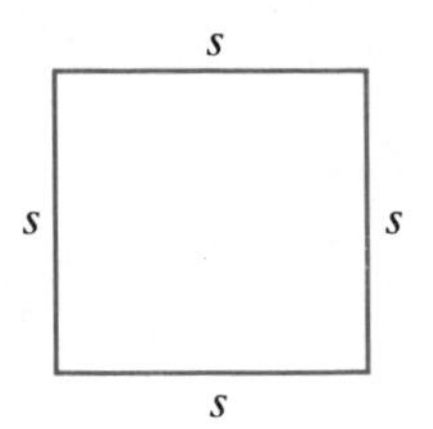

$$s^2 = 3(4s) \qquad \text{The area is to be three times the perimeter.}$$

$$s^2 = 12s$$

$$s^2 - 12s = 0$$

$$s(s - 12) = 0$$

$$s = 0 \quad \text{or} \quad s = 12.$$

Since 0 is not a reasonable solution, it must be a 12-by-12 square. (Be sure to check this answer in the *original statement of the problem*!) ■

PROBLEM 2 Suppose that the volume of a right circular cylinder is numerically equal to the total surface area of the cylinder. If the height of the cylinder is equal to the length of a radius of the base, find the height.

Solution Since $r = h$, the formula for volume $V = \pi r^2 h$ becomes $V = \pi r^3$ and the formula for the total surface area $S = 2\pi r^2 + 2\pi rh$ becomes $S = 2\pi r^2 + 2\pi r^2$ or $S = 4\pi r^2$. Therefore, we can set up and solve the following equation.

$$\pi r^3 = 4\pi r^2$$

$$\pi r^3 - 4\pi r^2 = 0$$

$$\pi r^2(r - 4) = 0$$

$$\pi r^2 = 0 \quad \text{or} \quad r - 4 = 0$$

$$r = 0 \quad \text{or} \quad r = 4$$

Since 0 is not a reasonable answer, the height must be 4 units. ■

Problem Set 3.4

For problems 1–10, classify each number as prime or composite.

1. 63 composite **2.** 81 composite **3.** 59 prime **4.** 83 prime
5. 51 composite **6.** 69 composite **7.** 91 composite **8.** 119 composite
9. 71 prime **10.** 101 prime

For Problems 11–20, factor each of the composite numbers into the product of prime numbers. For example, $30 = 2 \cdot 3 \cdot 5$.

11. 28 $2 \cdot 2 \cdot 7$

12. 39 $3 \cdot 13$

13. 44 $2 \cdot 2 \cdot 11$

14. 49 $7 \cdot 7$

15. 56 $2 \cdot 2 \cdot 2 \cdot 7$

16. 64 $2 \cdot 2 \cdot 2 \cdot 2 \cdot 2 \cdot 2$ or 2^6

17. 72 $2 \cdot 2 \cdot 2 \cdot 3 \cdot 3$

18. 84 $2 \cdot 2 \cdot 3 \cdot 7$

19. 87 $3 \cdot 29$

20. 91 $7 \cdot 13$

For Problems 21–46, factor completely each of the following.

21. $6x + 3y$ $3(2x + y)$

22. $12x + 8y$ $4(3x + 2y)$

23. $6x^2 + 14x$ $2x(3x + 7)$

24. $15x^2 + 6x$ $3x(5x + 2)$

25. $28y^2 - 4y$ $4y(7y - 1)$

26. $42y^2 - 6y$ $6y(7y - 1)$

27. $20xy - 15x$ $5x(4y - 3)$

28. $27xy - 36y$ $9y(3x - 4)$

29. $7x^3 + 10x^2$ $x^2(7x + 10)$

30. $12x^3 - 10x^2$ $2x^2(6x - 5)$

31. $18a^2b + 27ab^2$ $9ab(2a + 3b)$

32. $24a^3b^2 + 36a^2b$ $12a^2b(2ab + 3)$

33. $12x^3y^4 - 39x^4y^3$ $3x^3y^3(4y - 13x)$

34. $15x^4y^2 - 45x^5y^4$ $15x^4y^2(1 - 3xy^2)$

35. $8x^4 + 12x^3 - 24x^2$ $4x^2(2x^2 + 3x - 6)$

36. $6x^5 - 18x^3 + 24x$ $6x(x^4 - 3x^2 + 4)$

37. $5x + 7x^2 + 9x^4$ $x(5 + 7x + 9x^3)$

38. $9x^2 - 17x^4 + 21x^5$ $x^2(9 - 17x^2 + 21x^3)$

39. $15x^2y^3 + 20xy^2 + 35x^3y^4$ $5xy^2(3xy + 4 + 7x^2y^2)$

40. $8x^5y^3 - 6x^4y^5 + 12x^2y^3$ $2x^2y^3(4x^3 - 3x^2y^2 + 6)$

41. $x(y + 2) + 3(y + 2)$ $(y + 2)(x + 3)$

42. $x(y - 1) + 5(y - 1)$ $(y - 1)(x + 5)$

43. $3x(2a + b) - 2y(2a + b)$ $(2a + b)(3x - 2y)$

44. $5x(a - b) + y(a - b)$ $(a - b)(5x + y)$

45. $x(x + 2) + 5(x + 2)$ $(x + 2)(x + 5)$

46. $x(x - 1) - 3(x - 1)$ $(x - 1)(x - 3)$

For Problems 47–64, factor each of the following by grouping.

47. $ax + 4x + ay + 4y$ $(a + 4)(x + y)$

48. $ax - 2x + ay - 2y$ $(a - 2)(x + y)$

49. $ax - 2bx + ay - 2by$ $(a - 2b)(x + y)$

50. $2ax - bx + 2ay - by$ $(2a - b)(x + y)$

51. $3ax - 3bx - ay + by$ $(a - b)(3x - y)$

52. $5ax - 5bx - 2ay + 2by$ $(a - b)(5x - 2y)$

53. $2ax + 2x + ay + y$ $(a + 1)(2x + y)$

54. $3bx + 3x + by + y$ $(b + 1)(3x + y)$

55. $ax^2 - x^2 + 2a - 2$ $(a - 1)(x^2 + 2)$

56. $ax^2 - 2x^2 + 3a - 6$ $(a - 2)(x^2 + 3)$

57. $2ac + 3bd + 2bc + 3ad$ $(a + b)(2c + 3d)$

58. $2bx + cy + cx + 2by$ $(2b + c)(x + y)$

59. $ax - by + bx - ay$ $(a + b)(x - y)$

60. $2a^2 - 3bc - 2ab + 3ac$ $(a - b)(2a + 3c)$

61. $x^2 + 9x + 6x + 54$ $(x + 9)(x + 6)$

62. $x^2 - 2x + 5x - 10$ $(x - 2)(x + 5)$

63. $2x^2 + 8x + x + 4$ $(x + 4)(2x + 1)$

64. $3x^2 + 18x - 2x - 12$ $(x + 6)(3x - 2)$

For Problems 65–80, solve each of the equations.

65. $x^2 + 7x = 0$ $\{-7, 0\}$

66. $x^2 + 9x = 0$ $\{-9, 0\}$

67. $x^2 - x = 0$ $\{0, 1\}$

68. $x^2 - 14x = 0$ $\{0, 14\}$

69. $a^2 = 5a$ $\{0, 5\}$

70. $b^2 = -7b$ $\{-7, 0\}$

71. $-2y = 4y^2$ $\{-\frac{1}{2}, 0\}$

72. $-6x = 2x^2$ $\{-3, 0\}$

73. $3x^2 + 7x = 0$ $\{-\frac{7}{3}, 0\}$

74. $-4x^2 + 9x = 0$ $\{0, \frac{9}{4}\}$

75. $4x^2 = 5x$ $\{0, \frac{5}{4}\}$

76. $3x = 11x^2$ $\{0, \frac{3}{11}\}$

77. $x - 4x^2 = 0$ $\{0, \frac{1}{4}\}$

78. $x - 6x^2 = 0$ $\{0, \frac{1}{6}\}$

79. $12a = -a^2$ $\{-12, 0\}$

80. $-5a = -a^2$ $\{0, 5\}$

For Problems 81–86, solve each equation for the indicated variable.

81. $5bx^2 - 3ax = 0$ for x $\{0, \frac{3a}{5b}\}$

82. $ax^2 + bx = 0$ for x $\{-\frac{b}{a}, 0\}$

83. $2by^2 = -3ay$ for y $\{-\frac{3a}{2b}, 0\}$

84. $3ay^2 = by$ for y $\{0, \frac{b}{3a}\}$

85. $y^2 - ay + 2by - 2ab = 0$ for y $\{a, -2b\}$

86. $x^2 + ax + bx + ab = 0$ for x $\{-a, -b\}$

For Problems 87–96, set up an equation and solve each of the following problems.

87. The square of a number equals seven times the number. Find the number. 0 or 7

88. Suppose that the area of a square is six times its perimeter. Find the length of a side of the square. 24 units

89. The area of a circular region is numerically equal to three times the circumference of the circle. Find the length of a radius of the circle. 6 units

90. Find the length of a radius of a circle such that the circumference of the circle is numerically equal to the area of the circle. 2 units

91. Suppose that the area of a circle is numerically equal to the perimeter of a square and that the length of a radius of the circle is equal to the length of a side of the square. Find the length of a side of the square. Express your answer in terms of π. $\frac{4}{\pi}$ units

92. Find the length of a radius of a sphere such that the surface area of the sphere is numerically equal to the volume of the sphere. 3 units

93. Suppose that the area of a square lot is twice the area of an adjoining rectangular plot of ground. If the rectangular plot is 50 feet wide and its length is the same as the length of a side of the square lot, find the dimensions of both the square and the rectangle. The square is 100 feet by 100 feet and the rectangle is 50 feet by 100 feet.

94. The area of a square is one-fourth as large as the area of a triangle. One side of the triangle is 16 inches long and the altitude to that side is the same length as a side of the square. Find the length of a side of the square. 2 inches

95. Suppose that the volume of a sphere is numerically equal to twice the surface area of the sphere. Find the length of a radius of the sphere. 6 units

96. Suppose that a radius of a sphere is equal in length to a radius of a circle. If the volume of the sphere is numerically equal to four times the area of the circle, find the length of a radius for both the sphere and the circle. 3 units

Miscellaneous Problems

For Problems 97–102, factor each expression and assume that all variables that appear as exponents represent positive integers.

97. $2x^{2a} - 3x^a$ $x^a(2x^a - 3)$

98. $6x^{2a} + 8x^a$ $2x^a(3x^a + 4)$

99. $y^{3m} + 5y^{2m}$ $y^{2m}(y^m + 5)$

100. $3y^{5m} - y^{4m} - y^{3m}$ $y^{3m}(3y^{2m} - y^m - 1)$

101. $2x^{6a} - 3x^{5a} + 7x^{4a}$ $x^{4a}(2x^{2a} - 3x^a + 7)$

102. $6x^{3a} - 10x^{2a}$ $2x^{2a}(3x^a - 5)$

3.5 The Difference of Two Squares

In Section 3.3 we examined some special multiplication patterns. One of these patterns was the following.

$$(a + b)(a - b) = a^2 - b^2$$

This same pattern, viewed as a factoring pattern, we referred to as:

Difference of Two Squares

$$a^2 - b^2 = (a + b)(a - b).$$

Applying the pattern is a fairly simple process as these next examples demonstrate. Again the steps in dashed boxes are usually performed mentally.

$$x^2 - 16 = \boxed{(x)^2 - (4)^2} = (x + 4)(x - 4),$$

$$4x^2 - 25 = \boxed{(2x)^2 - (5)^2} = (2x + 5)(2x - 5),$$

$$16x^2 - 9y^2 = \boxed{(4x)^2 - (3y)^2} = (4x + 3y)(4x - 3y),$$

$$1 - a^2 = \boxed{(1)^2 - (a)^2} = (1 + a)(1 - a)$$

Since multiplication is commutative, the order of writing the factors is not important. For example, $(x + 4)(x - 4)$ can also be written as $(x - 4)(x + 4)$.

You must be careful not to assume an analogous factoring pattern for the *sum* of two squares; *it does not exist.* For example, $x^2 + 4 \neq (x + 2)(x + 2)$ because $(x + 2)(x + 2) = x^2 + 4x + 4$. We say that a polynomial such as $x^2 + 4$ is a **prime polynomial** or that it is *not factorable using integers.*

Sometimes the difference-of-two-squares pattern can be applied more than once as the next examples illustrate.

$$x^4 - y^4 = (x^2 + y^2)(x^2 - y^2) = (x^2 + y^2)(x + y)(x - y),$$

$$16x^4 - 81y^4 = (4x^2 + 9y^2)(4x^2 - 9y^2) = (4x^2 + 9y^2)(2x + 3y)(2x - 3y)$$

It may also be that the squares are other than simple monomial squares as in these next three examples.

$$(x + 3)^2 - y^2 = ((x + 3) + y)((x + 3) - y) = (x + 3 + y)(x + 3 - y),$$

$$\begin{aligned} 4x^2 - (2y + 1)^2 &= (2x + (2y + 1))(2x - (2y + 1)) \\ &= (2x + 2y + 1)(2x - 2y - 1), \end{aligned}$$

$$
\begin{aligned}
(x-1)^2-(x+4)^2 &= ((x-1)+(x+4))((x-1)-(x+4)) \\
&= (x-1+x+4)(x-1-x-4) \\
&= (2x+3)(-5)
\end{aligned}
$$

It is possible to apply both the technique of *factoring out a common monomial factor* and the pattern of the *difference of two squares* to the same problem. *In general, it is best to first look for a common monomial factor.* Consider the following examples.

$$
\begin{aligned}
2x^2-50 &= 2(x^2-25) \\
&= 2(x+5)(x-5),
\end{aligned}
$$

$$
\begin{aligned}
48y^3-27y &= 3y(16y^2-9) \\
&= 3y(4y+3)(4y-3),
\end{aligned}
$$

$$
\begin{aligned}
9x^2-36 &= 9(x^2-4) \\
&= 9(x+2)(x-2)
\end{aligned}
$$

WORD OF CAUTION The polynomial $9x^2-36$ can be factored as follows.

$$
\begin{aligned}
9x^2-36 &= (3x+6)(3x-6) \\
&= 3(x+2)(3)(x-2) \\
&= 9(x+2)(x-2)
\end{aligned}
$$

However, when taking this approach there seems to be a tendency to stop at the step $(3x+6)(3x-6)$. Therefore, remember the suggestion to *first look for a common monomial factor*.

The following examples should help you summarize all of our factoring techniques thus far.

$7x^2+28=7(x^2+4)$,

$4x^2y-14xy^2=2xy(2x-7y)$,

$x^2-4=(x+2)(x-2)$,

$18-2x^2=2(9-x^2)=2(3+x)(3-x)$,

y^2+9 is not factorable using integers,

$5x+13y$ is not factorable using integers,

$x^4-16=(x^2+4)(x^2-4)=(x^2+4)(x+2)(x-2)$

The Sum and Difference of Two Cubes

As we pointed out before, no sum-of-squares pattern exists analogous to the difference-of-squares factoring pattern. That is to say, a polynomial such as x^2+9 is not factorable using integers. However, there do exist patterns for both the *sum and difference of two cubes*. These patterns are as follows.

Sum and Difference of Two Cubes

$$a^3 + b^3 = (a + b)(a^2 - ab + b^2),$$
$$a^3 - b^3 = (a - b)(a^2 + ab + b^2).$$

Note how we apply these patterns in the next four examples.

$$x^3 + 27 = (x)^3 + (3)^3 = (x + 3)(x^2 - 3x + 9),$$
$$8a^3 + 125b^3 = (2a)^3 + (5b)^3 = (2a + 5b)(4a^2 - 10ab + 25b^2),$$
$$x^3 - 1 = (x)^3 - (1)^3 = (x - 1)(x^2 + x + 1),$$
$$27y^3 - 64x^3 = (3y)^3 - (4x)^3 = (3y - 4x)(9y^2 + 12xy + 16x^2)$$

Equations and Problem Solving

Remember that each time we pick up a new factoring technique we also develop more power for solving equations. Let's consider how we can use the difference-of-two-squares factoring pattern to help solve certain types of equations.

EXAMPLE 1 Solve $x^2 = 16$.

Solution

$$x^2 = 16$$
$$x^2 - 16 = 0$$
$$(x + 4)(x - 4) = 0$$
$$x + 4 = 0 \quad \text{or} \quad x - 4 = 0$$
$$x = -4 \quad \text{or} \quad x = 4$$

The solution set is $\{-4, 4\}$. (Be sure to check these solutions in the original equation!) ■

EXAMPLE 2 Solve $9x^2 = 64$.

Solution

$$9x^2 = 64$$
$$9x^2 - 64 = 0$$
$$(3x + 8)(3x - 8) = 0$$
$$3x + 8 = 0 \quad \text{or} \quad 3x - 8 = 0$$

$$3x = -8 \quad \text{or} \quad 3x = 8$$

$$x = -\frac{8}{3} \quad \text{or} \quad x = \frac{8}{3}$$

The solution set is $\left\{-\frac{8}{3}, \frac{8}{3}\right\}$. ■

EXAMPLE 3 Solve $7x^2 - 7 = 0$.

Solution

$$7x^2 - 7 = 0$$

$$7(x^2 - 1) = 0$$

$$x^2 - 1 = 0 \quad \text{Multiply both sides by } \frac{1}{7}.$$

$$(x + 1)(x - 1) = 0$$

$$x + 1 = 0 \quad \text{or} \quad x - 1 = 0$$

$$x = -1 \quad \text{or} \quad x = 1$$

The solution set is $\{-1, 1\}$. ■

In the previous examples we have been using the property "$ab = 0$ if and only if $a = 0$ or $b = 0$." This property can be extended to any number of factors whose product is zero. Thus, for three factors the property could be stated "$abc = 0$ if and only if $a = 0$ or $b = 0$ or $c = 0$." The next two examples illustrate this idea.

EXAMPLE 4 Solve $x^4 - 16 = 0$.

Solution

$$x^4 - 16 = 0$$

$$(x^2 + 4)(x^2 - 4) = 0$$

$$(x^2 + 4)(x + 2)(x - 2) = 0$$

$$x^2 + 4 = 0 \quad \text{or} \quad x + 2 = 0 \quad \text{or} \quad x - 2 = 0$$

$$x^2 = -4 \quad \text{or} \quad x = -2 \quad \text{or} \quad x = 2$$

The solution set is $\{-2, 2\}$. (Since no real numbers when squared will produce -4, the equation $x^2 = -4$ yields no additional real solutions.) ■

EXAMPLE 5 Solve $x^3 - 49x = 0$.

Solution

$$x^3 - 49x = 0$$

$$x(x^2 - 49) = 0$$

$$x(x + 7)(x - 7) = 0$$

$$x = 0 \quad \text{or} \quad x + 7 = 0 \quad \text{or} \quad x - 7 = 0$$

$$x = 0 \quad \text{or} \quad x = -7 \quad \text{or} \quad x = 7$$

The solution set is $\{-7, 0, 7\}$. ■

The more that we know about solving equations the better off we are for solving word problems.

PROBLEM 1 The combined area of two squares is 40 square centimeters. Each side of one square is three times as long as a side of the other square. Find the dimensions of each of the squares.

Solution Let s represent the length of a side of the smaller square. Then $3s$ represents the length of a side of the larger square.

$$s^2 + (3s)^2 = 40$$

$$s^2 + 9s^2 = 40$$

$$10s^2 = 40$$

$$s^2 = 4$$

$$s^2 - 4 = 0$$

$$(s + 2)(s - 2) = 0$$

$$s + 2 = 0 \quad \text{or} \quad s - 2 = 0$$

$$s = -2 \quad \text{or} \quad s = 2$$

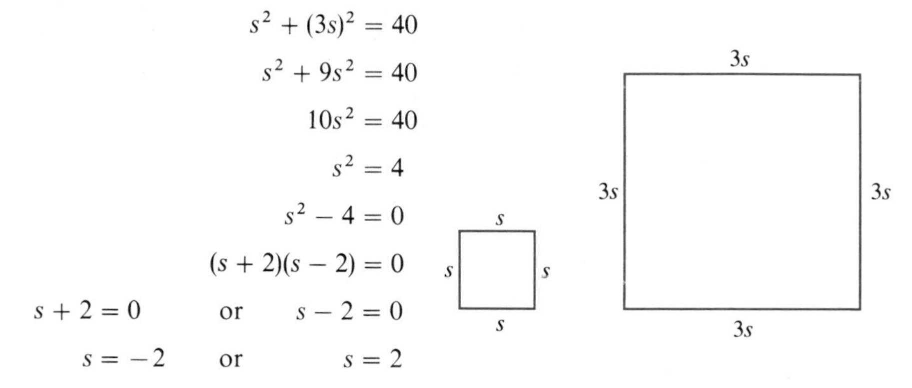

Since s represents the length of a side of a square, the solution -2 has to be disregarded. Thus, the length of a side of the small square is 2 centimeters and the large square has sides of $3(2) = 6$ centimeters. ■

Problem Set 3.5

For Problems 1–20, use the difference-of-squares pattern to factor each of the following.

1. $x^2 - 1$ $(x + 1)(x - 1)$

2. $x^2 - 9$ $(x + 3)(x - 3)$

3. $16x^2 - 25$ $(4x + 5)(4x - 5)$

4. $4x^2 - 49$ $(2x + 7)(2x - 7)$

5. $9x^2 - 25y^2$ $(3x + 5y)(3x - 5y)$

6. $x^2 - 64y^2$ $(x + 8y)(x - 8y)$

7. $25x^2y^2 - 36$ $(5xy + 6)(5xy - 6)$

8. $x^2y^2 - a^2b^2$ $(xy + ab)(xy - ab)$

9. $4x^2 - y^4$ $(2x + y^2)(2x - y^2)$

10. $x^6 - 9y^2$ $(x^3 + 3y)(x^3 - 3y)$

11. $1 - 144n^2$ $(1 + 12n)(1 - 12n)$

12. $25 - 49n^2$ $(5 + 7n)(5 - 7n)$

13. $(x + 2)^2 - y^2$ $(x + 2 + y)(x + 2 - y)$

14. $(3x + 5)^2 - y^2$ $(3x + 5 + y)(3x + 5 - y)$

15. $4x^2 - (y + 1)^2$ $(2x + y + 1)(2x - y - 1)$

16. $x^2 - (y - 5)^2$ $(x + y - 5)(x - y + 5)$

17. $9a^2-(2b+3)^2$ $(3a+2b+3)(3a-2b-3)$ **18.** $16s^2-(3t+1)^2$ $(4s+3t+1)(4s-3t-1)$

19. $(x+2)^2-(x+7)^2$ $-5(2x+9)$ **20.** $(x-1)^2-(x-8)^2$ $7(2x-9)$

For Problems 21–44, factor each of the following completely. Indicate any that are not factorable using integers. Don't forget to first look for a common monomial factor.

21. $9x^2-36$ $9(x+2)(x-2)$ **22.** $8x^2-72$ $8(x+3)(x-3)$

23. $5x^2+5$ $5(x^2+1)$ **24.** $7x^2+28$ $7(x^2+4)$

25. $8y^2-32$ $8(y+2)(y-2)$ **26.** $5y^2-80$ $5(y+4)(y-4)$

27. a^3b-9ab $ab(a+3)(a-3)$ **28.** $x^3y^2-xy^2$ $xy^2(x+1)(x-1)$

29. $16x^2+25$ Not factorable **30.** x^4-16 $(x+2)(x-2)(x^2+4)$

31. n^4-81 $(n+3)(n-3)(n^2+9)$ **32.** $4x^2+9$ Not factorable

33. $3x^3+27x$ $3x(x^2+9)$ **34.** $20x^3+45x$ $5x(4x^2+9)$

35. $4x^3y-64xy^3$ $4xy(x+4y)(x-4y)$ **36.** $12x^3-27xy^2$ $3x(2x+3y)(2x-3y)$

37. $6x-6x^3$ $6x(1+x)(1-x)$ **38.** $1-16x^4$ $(1+2x)(1-2x)(1+4x^2)$

39. $1-x^4y^4$ $(1+xy)(1-xy)(1+x^2y^2)$ **40.** $20x-5x^3$ $5x(2+x)(2-x)$

41. $4x^2-64y^2$ $4(x+4y)(x-4y)$ **42.** $9x^2-81y^2$ $9(x+3y)(x-3y)$

43. $3x^4-48$ $3(x+2)(x-2)(x^2+4)$ **44.** $2x^5-162x$ $2x(x+3)(x-3)(x^2+9)$

For Problems 45–56, use the sum or difference-of-cubes patterns to factor each of the following.

45. a^3-64 $(a-4)(a^2+4a+16)$ **46.** a^3-27 $(a-3)(a^2+3a+9)$

47. x^3+1 $(x+1)(x^2-x+1)$ **48.** x^3+8 $(x+2)(x^2-2x+4)$

49. $27x^3+64y^3$ $(3x+4y)(9x^2-12xy+16y^2)$

50. $8x^3+27y^3$ $(2x+3y)(4x^2-6xy+9y^2)$

51. $1-27a^3$ $(1-3a)(1+3a+9a^2)$ **52.** $1-8x^3$ $(1-2x)(1+2x+4x^2)$

53. x^3y^3-1 $(xy-1)(x^2y^2+xy+1)$

54. $125x^3+27y^3$ $(5x+3y)(25x^2-15xy+9y^2)$

55. x^6-y^6 $(x+y)(x-y)(x^4+x^2y^2+y^4)$

56. x^6+y^6 $(x^2+y^2)(x^4-x^2y^2+y^4)$

For Problems 57–70, find all real number solutions for each equation.

57. $x^2-25=0$ $\{-5,5\}$ **58.** $x^2-1=0$ $\{-1,1\}$ **59.** $9x^2-49=0$ $\{-\frac{7}{3},\frac{7}{3}\}$

60. $4y^2=25$ $\{-\frac{5}{2},\frac{5}{2}\}$ **61.** $8x^2-32=0$ $\{-2,2\}$ **62.** $3x^2-108=0$ $\{-6,6\}$

63. $3x^3=3x$ $\{-1,0,1\}$ **64.** $4x^3=64x$ $\{-4,0,4\}$ **65.** $20-5x^2=0$ $\{-2,2\}$

66. $54-6x^2=0$ $\{-3,3\}$ **67.** $x^4-81=0$ $\{-3,3\}$ **68.** $x^5-x=0$ $\{-1,0,1\}$

69. $6x^3+24x=0$ $\{0\}$ **70.** $4x^3+12x=0$ $\{0\}$

For Problems 71–80, set up an equation and solve each of the following problems.

71. The cube of a number equals nine times the same number. Find the number. -3, 0, or 3

72. The cube of a number equals the square of the same number. Find the number. 0 or 1

73. The combined area of two circles is 80π square centimeters. The length of a radius of one circle is twice the length of a radius of the other circle. Find the length of the radius of each circle. 4 and 8

74. The combined area of two squares is 26 square meters. The sides of the larger square are five times as long as the sides of the smaller square. Find the dimensions of each of the squares. 1 by 1and 5 by 5

75. A rectangle is twice as long as it is wide and its area is 50 square meters. Find the length and the width of the rectangle. 10 meters long and 5 meters wide

76. Suppose that the length of a rectangle is one and one-third times as long as its width. The area of the rectangle is 48 square centimeters. Find the length and width of the rectangle. 8 centimeters long and 6 centimeters wide

77. The total surface area of a right circular cylinder is 54π square inches. If the altitude of the cylinder is twice the length of a radius, find the altitude of the cylinder. 6 inches

78. The total surface area of a right circular cone is 108π square feet. If the slant height of the cone is twice the length of a radius of the base, find the length of a radius. 6 feet

79. The sum of the areas of a circle and a square is $(16\pi + 64)$ square yards. If a side of the square is twice the length of a radius of the circle, find the length of a side of the square. 8 yards

80. The length of an altitude of a triangle is one-third the length of the side to which it is drawn. If the area of the triangle is 6 square centimeters, find the length of that altitude. 2 centimeters

3.6 Factoring Trinomials

Expressing a trinomial as the product of two binomials is one of the most common types of factoring used in algebra. Like before, to develop a factoring technique we first look at some multiplication ideas. Let's consider the product $(x + a)(x + b)$ and use the distributive property to show how each term of the resulting trinomial is formed.

$$\begin{aligned}(x + a)(x + b) &= x(x + b) + a(x + b)\\ &= x(x) + x(b) + a(x) + a(b)\\ &= x^2 + (a + b)x + ab\end{aligned}$$

Notice that the coefficient of the middle term is the *sum* of a and b and the last term is the *product* of a and b. These two relationships can be used to factor trinomials. Let's consider some examples.

EXAMPLE 1 Factor $x^2 + 8x + 12$.

Solution We need to complete the following with two integers whose sum is 8 and whose product is 12.

$$x^2 + 8x + 12 = (x + \_\_\_)(x + \_\_\_)$$

The possible pairs of factors of 12 are 1(12), 2(6), and 3(4). Since $6 + 2 = 8$, we can complete the factoring as follows.

$$x^2 + 8x + 12 = (x + 6)(x + 2)$$

To *check* our answer we find the product of $(x + 6)$ and $(x + 2)$. ■

EXAMPLE 2 Factor $x^2 - 10x + 24$.

Solution We need two integers whose product is 24 and whose sum is -10. Let's use a small table to organize our thinking.

Product	*Sum*
$(-1)(-24) = 24$	$-1 + (-24) = -25$
$(-2)(-12) = 24$	$-2 + (-12) = -14$
$(-3)(-8) = 24$	$-3 + (-8) = -11$
$(-4)(-6) = 24$	$-4 + (-6) = -10$

The bottom line contains the numbers that we need. Thus,

$$x^2 - 10x + 24 = (x - 4)(x - 6).$$ ■

EXAMPLE 3 Factor $x^2 + 7x - 30$.

Solution We need two integers whose product is -30 and whose sum is 7.

Product	*Sum*
$(-1)(30) = -30$	$-1 + 30 = 29$
$1(-30) = -30$	$1 + (-30) = -29$
$2(-15) = -30$	$2 + (-15) = -13$
$-2(15) = -30$	$-2 + 15 = 13$
$-3(10) = -30$	$-3 + 10 = 7$

no need to search any further

The numbers that we need are -3 and 10 and we can complete the factoring.

$$x^2 + 7x - 30 = (x + 10)(x - 3)$$ ■

EXAMPLE 4 Factor $x^2 + 7x + 16$.

Solution We need two integers whose product is 16 and whose sum is 7.

Product	*Sum*
$1(16) = 16$	$1 + 16 = 17$
$2(8) = 16$	$2 + 8 = 10$
$4(4) = 16$	$4 + 4 = 8$

Since we have exhausted all possible pairs of factors of 16 and no two factors have a sum of 7, we conclude that $x^2 + 7x + 16$ *is not factorable using integers.* ■

The tables in Examples 2, 3, and 4 have been used to illustrate one way of organizing your thoughts for such problems. Normally you will probably do such factoring problems mentally without taking the time to formulate a table. Notice, however, that in Example 4 the table helped us to be absolutely sure that we tried all the possibilities. Whether or not you use the table, keep in mind that the key ideas are the product and sum relationships.

EXAMPLE 5 Factor $n^2 - n - 72$.

Solution Notice that the coefficient of the middle term is -1. So, we are looking for two integers whose product is -72, and since their sum is -1, the absolute value of the negative number must be one larger than the positive number. The numbers are -9 and 8, and we can complete the factoring.

$$n^2 - n - 72 = (n - 9)(n + 8)$$ ■

EXAMPLE 6 Factor $t^2 - 2t - 168$.

Solution We need two integers whose product is -168 and whose sum is 2. Since the absolute value of the constant term is rather large, it might help to look at it in prime factored form.

$$168 = 2 \cdot 2 \cdot 2 \cdot 3 \cdot 7$$

Now we can mentally form two numbers by using all of these factors in different combinations. Using two 2's and a 3 in one number, and the other 2 and the 7 in the other number produces $2 \cdot 2 \cdot 3 = 12$ and $2 \cdot 7 = 14$. Since the coefficient of the middle term of the trinomial is 2, we know that we must use 14 and -12. Thus, we obtain

$$t^2 + 2t - 168 = (t + 14)(t - 12).$$ ■

Trinomials of the Form $ax^2 + bx + c$

We have been factoring trinomials of the form $x^2 + bx + c$, that is, trinomials where the coefficient of the squared term is one. Now let's consider factoring trinomials where the coefficient of the squared term is not one. First, let's illustrate an informal trial and error technique that works quite well for certain types of trinomials. This technique is based on our knowledge of multiplication of binomials.

EXAMPLE 7 Factor $2x^2 + 11x + 5$.

Solution By looking at the first term, $2x^2$, and the positive signs of the other two terms, we know that the binomials are of the form

$$(x + \underline{\hspace{2em}})(2x + \underline{\hspace{2em}}).$$

Since the factors of the last term, 5, are 1 and 5, we have only the following two possibilities to try.

$$(x + 1)(2x + 5) \qquad \text{or} \qquad (x + 5)(2x + 1)$$

By checking the middle term formed in each of these products, we find that the second possibility yields the correct middle term of $11x$. Therefore,

$$2x^2 + 11x + 5 = (x + 5)(2x + 1).$$ ■

EXAMPLE 8 Factor $10x^2 - 17x + 3$.

Solution First, observe that $10x^2$ can be written as $x \cdot 10x$ or $2x \cdot 5x$. Secondly, since the middle term of the trinomial is negative and the last term is positive, we know that the binomials are of the form

$$(x - \underline{\hspace{2em}})(10x - \underline{\hspace{2em}}) \qquad \text{or} \qquad (2x - \underline{\hspace{2em}})(5x - \underline{\hspace{2em}}).$$

Since the factors of the last term, 3, are 1 and 3, the following possibilities exist.

$$(x - 1)(10x - 3), \qquad (2x - 1)(5x - 3),$$

$$(x - 3)(10x - 1), \qquad (2x - 3)(5x - 1)$$

By checking the middle term formed in each of these products, we find that the product $(2x - 3)(5x - 1)$ yields the desired middle term of $-17x$. Therefore,

$$10x^2 - 17x + 3 = (2x - 3)(5x - 1).$$ ■

EXAMPLE 9 Factor $4x^2 + 6x + 9$.

Solution The first term, $4x^2$, and the positive signs of the middle and last terms indicate that the binomials are of the form

$$(x + \underline{\hspace{2em}})(4x + \underline{\hspace{2em}}) \qquad \text{or} \qquad (2x + \underline{\hspace{2em}})(2x + \underline{\hspace{2em}}).$$

Since the factors of 9 are 1 and 9 or 3 and 3, we have the following five possibilities to try.

$$(x + 1)(4x + 9), \qquad (2x + 1)(2x + 9),$$

$$(x + 9)(4x + 1), \qquad (2x + 3)(2x + 3),$$

$$(x + 3)(4x + 3)$$

When we try all of these possibilities we find that none of them yields a middle term of $6x$. Therefore, $4x^2 + 6x + 9$ is *not factorable using integers.* ■

By now it is obvious that factoring trinomials of the form $ax^2 + bx + c$ can be tedious. The key idea is to organize your work so that all possibilities are considered. We have suggested one possible format in the previous three examples. As you practice such problems, you may come across a format of your own. Whatever works best for you is the right approach.

There is another more systematic technique that you may wish to use with some trinomials. It is an extension of the technique we used at the beginning of this section. To see the basis of this technique, let's look at the following product.

$$\begin{aligned}(px + r)(qx + s) &= px(qx) + px(s) + r(qx) + r(s)\\ &= (pq)x^2 + (ps + rq)x + rs\end{aligned}$$

Notice that the product of the coefficient of the x^2 term and the constant term is $pqrs$. Likewise, the product of the two coefficients of x, ps, and rq, is also $pqrs$. Therefore, when factoring the trinomial $(pq)x^2 + (ps + rq)x + rs$, the two coefficients of x must have a sum of $(ps) + (rq)$ and a product of $pqrs$. Let's see how this works in some examples.

EXAMPLE 10 Factor $4x^2 - 4x - 15$.

Solution

$4x^2 - 4x - 15$ sum of -4

product of $4(-15) = -60$

We need two integers whose sum is -4 and whose product is -60. The integers -10 and 6 satisfy these conditions. Thus, we can express the middle term, $-4x$, as $-10x + 6x$, and we can factor as follows.

$$\begin{aligned}4x^2 - 4x - 15 &= 4x^2 - 10x + 6x - 15\\ &= 2x(2x - 5) + 3(2x - 5)\\ &= (2x - 5)(2x + 3)\end{aligned}$$

■

EXAMPLE 11 Factor $20x^2 + 39x - 18$.

Solution

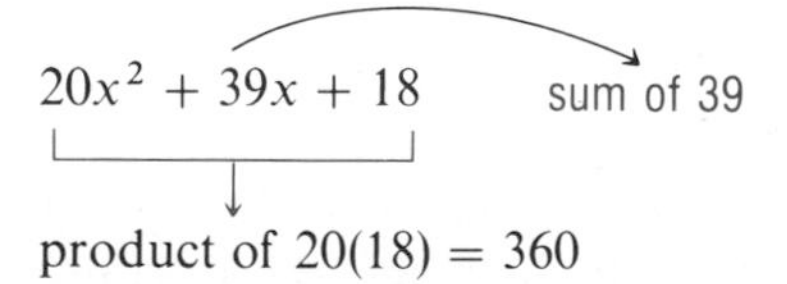

We need two integers whose sum is 39 and whose product is 360. To help find these integers, let's prime factor 360.

$$360 = 2 \cdot 2 \cdot 2 \cdot 3 \cdot 3 \cdot 5$$

Now by grouping these factors in various ways we find that $2 \cdot 2 \cdot 2 \cdot 3 = 24$ and $3 \cdot 5 = 15$ and $24 + 15 = 39$. So the numbers are 15 and 24 and the middle term, $39x$, can be written as $15x + 24x$. Thus, we can factor as follows.

$$\begin{aligned} 20x^2 + 39x + 18 &= 20x^2 + 15x + 24x + 18 \\ &= 5x(4x + 3) + 6(4x + 3) \\ &= (4x + 3)(5x + 6) \end{aligned}$$

■

Perhaps at this time it would be helpful for you to go back and take another look at Examples 7 through 11. Both of the techniques we used have their strengths and weaknesses. The more that you work with them, the more comfortable you will feel using them.

Summary of Factoring Techniques

Before we summarize our work with factoring techniques let's look at two more special factoring patterns. In Section 3.3 we used the following two patterns to square binomials.

$$(a + b)^2 = a^2 + 2ab + b^2, \qquad (a - b)^2 = a^2 - 2ab + b^2$$

These patterns can also be used for factoring purposes.

$$a^2 + 2ab + b^2 = (a + b)^2, \qquad a^2 - 2ab + b^2 = (a - b)^2$$

The trinomials on the left sides are called **perfect square trinomials**; they are the result of squaring a binomial. We can always factor perfect square trinomials using the usual techniques for factoring trinomials. However, they are easily recognized by the nature of their terms. For example, $4x^2 + 12x + 9$ is a perfect square trinomial because

1. the first term is a perfect square; $(2x)^2$
2. the last term is a perfect square; $(3)^2$
3. the middle term is twice the product of the quantities being squared in the first and last terms. $2(2x)(3)$

Likewise, $9x^2 - 30x + 25$ is a perfect square trinomial because

1. the first term is a perfect square; $(3x)^2$
2. the last term is a perfect square; $(5)^2$
3. the middle term is the negative of twice the product of the quantities being squared in the first and last terms. $-2(3x)(5)$

Once we know that we have a perfect square trinomial, then the factors follow immediately from the two basic patterns. Thus,

$$4x^2 + 12x + 9 = (2x + 3)^2, \qquad 9x^2 - 30x + 25 = (3x - 5)^2.$$

Here are some additional examples of perfect square trinomials and their factored forms.

$$x^2 + 14x + 49 = (x)^2 + 2(x)(7) + (7)^2 = (x + 7)^2,$$
$$n^2 - 16n + 64 = (n)^2 - 2(n)(8) + (8)^2 = (n - 8)^2,$$
$$36a^2 + 60ab + 25b^2 = (6a)^2 + 2(6a)(5b) + (5b)^2 = (6a + 5b)^2,$$
$$16x^2 - 8xy + y^2 = (4x)^2 - 2(4x)(y) + (y)^2 = (4x - y)^2$$

↓

Perhaps you will want to do this step mentally after you feel comfortable with the process.

As we indicated previously, factoring is an important algebraic skill. We have learned some basic factoring techniques one at a time, but you must be able to apply them when the situation presents itself. So let's review the techniques and consider a variety of examples that demonstrate their use.

In this chapter we have discussed:

1. factoring by using the distributive property to factor out a common monomial (or binomial) factor;
2. factoring by applying the difference-of-two-squares pattern;
3. factoring by applying the sum or difference-of-two-cubes patterns;
4. factoring of trinomials into the product of two binomials (the *perfect square trinomial* pattern is a special case of this technique).

As a general guideline, *always look for a common monomial factor first and then proceed with the other techniques.* Study the following examples carefully and be sure that you agree with the indicated factors.

$$2x^2 + 20x + 48 = 2(x^2 + 10x + 24) = 2(x + 4)(x + 6),$$

$$16a^2 - 64 = 16(a^2 - 4) = 16(a + 2)(a - 2),$$

$$3x^3y^3 + 27xy = 3xy(x^2y^2 + 9),$$

$$30n^2 - 31n + 5 = (5n - 1)(6n - 5),$$

$$2x^3 - 16 = 2(x^3 - 8) = 2(x - 2)(x^2 + 2x + 4),$$

$$t^4 + 3t^2 + 2 = (t^2 + 2)(t^2 + 1)$$

$x^2 + 3x - 21$ is not factorable using integers.

Problem Set 3.6

Factor completely each of the following. Indicate any that are not factorable using integers.

1. $x^2 + 9x + 20$ $(x + 5)(x + 4)$

2. $x^2 + 11x + 24$ $(x + 3)(x + 8)$

3. $x^2 - 11x + 28$ $(x - 4)(x - 7)$

4. $x^2 - 8x + 12$ $(x - 2)(x - 6)$

5. $a^2 + 5a - 36$ $(a + 9)(a - 4)$

6. $a^2 + 6a - 40$ $(a + 10)(a - 4)$

7. $y^2 + 20y + 84$ $(y + 6)(y + 14)$

8. $y^2 + 21y + 98$ $(y + 7)(y + 14)$

9. $x^2 - 5x - 14$ $(x - 7)(x + 2)$

10. $x^2 - 3x - 54$ $(x - 9)(x + 6)$

11. $x^2 + 9x + 12$ Not factorable

12. $35 - 2x - x^2$ $(7 + x)(5 - x)$

13. $6 + 5x - x^2$ $(6 - x)(1 + x)$

14. $x^2 + 8x - 24$ Not factorable

15. $x^2 + 15xy + 36y^2$ $(x + 3y)(x + 12y)$

16. $x^2 - 14xy + 40y^2$ $(x - 4y)(x - 10y)$

17. $a^2 - ab - 56b^2$ $(a - 8b)(a + 7b)$

18. $a^2 + 2ab - 63b^2$ $(a + 9b)(a - 7b)$

19. $15x^2 + 23x + 6$ $(3x + 1)(5x + 6)$

20. $9x^2 + 30x + 16$ $(3x + 2)(3x + 8)$

21. $12x^2 - x - 6$ $(4x - 3)(3x + 2)$

22. $20x^2 - 11x - 3$ $(5x + 1)(4x - 3)$

23. $8a^2 - 6a - 27$ $(2a + 3)(4a - 9)$

24. $12a^2 + 4a - 5$ $(2a - 1)(6a + 5)$

25. $12n^2 - 43n - 20$ $(n - 4)(12n + 5)$

26. $16n^2 + 43n - 15$ $(n + 3)(16n - 5)$

27. $3x^2 + 10x + 4$ Not factorable

28. $20n^2 - 47n + 21$ $(4n - 7)(5n - 3)$

29. $20n^2 - 64n - 21$ $(2n - 7)(10n + 3)$

30. $4x^2 - x + 6$ Not factorable

31. $16x^2 + 62x - 45$ $(8x - 5)(2x + 9)$

32. $12x^2 + 4x - 33$ $(2x - 3)(6x + 11)$

33. $6 - 29x - 42x^2$ $(1 - 6x)(6 + 7x)$

34. $4 + 11x - 20x^2$ $(1 + 4x)(4 - 5x)$

35. $20y^2 + 31y - 9$ $(5y + 9)(4y - 1)$

36. $8y^2 + 22y - 21$ $(4y - 3)(2y + 7)$

37. $24n^2 - 2n - 5$ $(12n + 5)(2n - 1)$

38. $12n^2 - n - 35$ $(4n - 7)(3n + 5)$

39. $5n^2 + 33n + 18$ $(5n + 3)(n + 6)$

40. $7n^2 + 31n + 12$ $(7n + 3)(n + 4)$

41. $x^2 + 25x + 150$ $(x + 10)(x + 15)$

42. $x^2 + 21x + 108$ $(x + 9)(x + 12)$

43. $n^2 - 36n + 320$ $(n - 16)(n - 20)$

44. $n^2 - 26n + 168$ $(n - 12)(n - 14)$

45. $t^2 + 3t - 180$ $(t + 15)(t - 12)$

46. $t^2 - 2t - 143$ $(t + 11)(t - 13)$

47. $t^4 - 5t^2 + 6$ $(t^2 - 3)(t^2 - 2)$

48. $t^4 + 10t^2 + 24$ $(t^2 + 4)(t^2 + 6)$

49. $10x^4 + 3x^2 - 4$ $(2x^2 - 1)(5x^2 + 4)$

50. $3x^4 + 7x^2 - 6$ $(3x^2 - 2)(x^2 + 3)$

51. $x^4 - 9x^2 + 8$ $(x + 1)(x - 1)(x^2 - 8)$

52. $x^4 - x^2 - 12$ $(x + 2)(x - 2)(x^2 + 3)$

53. $18n^4 + 25n^2 - 3$ $(3n + 1)(3n - 1)(2n^2 + 3)$

54. $4n^4 + 3n^2 - 27$ $(2n + 3)(2n - 3)(n^2 + 3)$

55. $x^4 - 17x^2 + 16$ $(x + 1)(x - 1)(x + 4)(x - 4)$

56. $x^4 - 13x^2 + 36$ $(x + 2)(x - 2)(x + 3)(x - 3)$

The following problems should help you pull together all of the factoring techniques of this chapter. Factor completely each of the following; indicate any that are not factorable using integers.

57. $2t^2 - 8$ $2(t + 2)(t - 2)$

58. $14w^2 - 29w - 15$ $(7w + 3)(2w - 5)$

59. $12x^2 + 7xy - 10y^2$ $(4x + 5y)(3x - 2y)$

60. $8x^2 + 2xy - y^2$ $(4x - y)(2x + y)$

61. $18n^3 + 39n^2 - 15n$ $3n(2n + 5)(3n - 1)$

62. $n^2 + 18n + 77$ $(n + 11)(n + 7)$

63. $n^2 - 17n + 60$ $(n - 12)(n - 5)$

64. $(x + 5)^2 - y^2$ $(x + 5 + y)(x + 5 - y)$

65. $36a^2 - 12a + 1$ $(6a - 1)^2$

66. $2n^2 - n - 5$ Not factorable

67. $6x^2 + 54$ $6(x^2 + 9)$

68. $x^5 - x$ $x(x^2 + 1)(x + 1)(x - 1)$

69. $3x^2 + x - 5$ Not factorable

70. $10x^2 + 39x - 27$ $(5x - 3)(2x + 9)$

71. $x^2 - (y - 7)^2$ $(x + y - 7)(x - y + 7)$

72. $2n^3 + 6n^2 + 10n$ $2n(n^2 + 3n + 5)$

73. $1 - 16x^4$ $(1 + 4x^2)(1 + 2x)(1 - 2x)$

74. $9a^2 - 42a + 49$ $(3a - 7)^2$

75. $12n^2 + 59n + 72$ $(4n + 9)(3n + 8)$

76. $x^3 - 9x$ $x(x + 3)(x - 3)$

77. $n^3 - 49n$ $n(n + 7)(n - 7)$

78. $4x^2 + 16$ $4(x^2 + 4)$

79. $x^2 - 7x - 8$ $(x - 8)(x + 1)$

80. $x^2 + 3x - 54$ $(x + 9)(x - 6)$

81. $3x^4 - 81x$ $3x(x - 3)(x^2 + 3x + 9)$

82. $x^3 + 125$ $(x + 5)(x^2 - 5x + 25)$

83. $x^4 + 6x^2 + 9$ $(x^2 + 3)^2$

84. $18x^2 - 12x + 2$ $2(3x - 1)^2$

85. $x^4 - 5x^2 - 36$ $(x + 3)(x - 3)(x^2 + 4)$

86. $6x^4 - 5x^2 - 21$ $(2x^2 + 3)(3x^2 - 7)$

87. $18w^2 + 9w - 35$ $(6w - 7)(3w + 5)$

88. $10x^3 + 15x^2 + 20x$ $5x(2x^2 + 3x + 4)$

89. $25n^2 + 64$ Not factorable

90. $4x^2 - 37x + 40$ $(4x - 5)(x - 8)$

91. $2n^3 + 14n^2 - 20n$ $2n(n^2 + 7n - 10)$

92. $25t^2 - 100$ $25(t + 2)(t - 2)$

93. $2xy + 6x + y + 3$ $(2x + 1)(y + 3)$

94. $3xy + 15x - 2y - 10$ $(3x - 2)(y + 5)$

Miscellaneous Problems

For Problems 95–100, factor each trinomial and assume that all variables that appear as exponents represent positive integers.

95. $x^{2a} + 2x^a - 24$ $(x^a - 4)(x^a + 6)$

96. $x^{2a} + 10x^a + 21$ $(x^a + 3)(x^a + 7)$

97. $6x^{2a} - 7x^a + 2$ $(2x^a - 1)(3x^a - 2)$

98. $4x^{2a} + 20x^a + 25$ $(2x^a + 5)^2$

99. $12x^{2n} + 7x^n - 12$ $(3x^n + 4)(4x^n - 3)$

100. $20x^{2n} + 21x^n - 5$ $(5x^n - 1)(4x^n + 5)$

Consider the following approach to factoring $(x - 2)^2 + 3(x - 2) - 10$.

$$(x - 2)^2 + 3(x - 2) - 10 = y^2 + 3y - 10 \quad \text{Replace } x - 2 \text{ with } y.$$
$$= (y + 5)(y - 2) \quad \text{Factor.}$$
$$= (x - 2 + 5)(x - 2 - 2) \quad \text{Replace } y \text{ with } x - 2.$$
$$= (x + 3)(x - 4)$$

Use this approach to factor Problems 101–106.

101. $(x - 3)^2 + 10(x - 3) + 24$ $(x + 1)(x + 3)$

102. $(x + 1)^2 - 8(x + 1) + 15$ $(x - 4)(x - 2)$

103. $(2x + 1)^2 + 3(2x + 1) - 28$ $2(x + 4)(2x - 3)$

104. $(3x - 2)^2 - 5(3x - 2) - 36$ $(3x - 11)(3x + 2)$

105. $6(x - 4)^2 + 7(x - 4) - 3$ $(2x - 5)(3x - 13)$

106. $15(x + 2)^2 - 13(x + 2) + 2$ $(3x + 4)(5x + 9)$

3.7 Equations and Problem Solving

The techniques presented in the previous section for factoring trinomials provide us with more power to solve equations. That is to say, the property "$ab = 0$ if and only if $a = 0$ or $b = 0$" continues to play an important role as we solve equations that involve trinomials that can be factored. Let's consider some examples.

EXAMPLE 1 Solve $x^2 - 11x - 12 = 0$.

Solution

$$x^2 - 11x - 12 = 0$$
$$(x - 12)(x + 1) = 0$$
$$x - 12 = 0 \quad \text{or} \quad x + 1 = 0$$
$$x = 12 \quad \text{or} \quad x = -1$$

The solution set is $\{-1, 12\}$. (Check these solutions in the original equation.) ■

EXAMPLE 2 Solve $20x^2 + 7x - 3 = 0$.

Solution

$$20x^2 + 7x - 3 = 0$$
$$(4x - 1)(5x + 3) = 0$$
$$4x - 1 = 0 \quad \text{or} \quad 5x + 3 = 0$$
$$4x = 1 \quad \text{or} \quad 5x = -3$$
$$x = \frac{1}{4} \quad \text{or} \quad x = -\frac{3}{5}$$

The solution set is $\left\{-\frac{3}{5}, \frac{1}{4}\right\}$. ■

EXAMPLE 3 Solve $-2n^2 - 10n + 12 = 0$.

Solution

$$-2n^2 - 10n + 12 = 0$$
$$-2(n^2 + 5n - 6) = 0$$
$$n^2 + 5n - 6 = 0 \qquad \text{Multiply both sides by } -\frac{1}{2}.$$
$$(n + 6)(n - 1) = 0$$
$$n + 6 = 0 \quad \text{or} \quad n - 1 = 0$$
$$n = -6 \quad \text{or} \quad n = 1$$

The solution set is $\{-6, 1\}$. ■

EXAMPLE 4 Solve $16x^2 - 56x + 49 = 0$.

Solution

$$16x^2 - 56x + 49 = 0$$
$$(4x - 7)^2 = 0$$

$$(4x - 7)(4x - 7) = 0$$

$$4x - 7 = 0 \quad \text{or} \quad 4x - 7 = 0$$

$$4x = 7 \quad \text{or} \quad 4x = 7$$

$$x = \frac{7}{4} \quad \text{or} \quad x = \frac{7}{4}$$

The only solution is $\frac{7}{4}$; thus, the solution set is $\left\{\frac{7}{4}\right\}$. ■

Sometimes the original form of an equation needs to be changed before a factoring technique becomes apparent. The next two examples demonstrate situations of this type.

EXAMPLE 5 Solve $9a(a + 1) = 4$.

Solution

$$9a(a + 1) = 4$$

$$9a^2 + 9a = 4$$

$$9a^2 + 9a - 4 = 0$$

$$(3a + 4)(3a - 1) = 0$$

$$3a + 4 = 0 \quad \text{or} \quad 3a - 1 = 0$$

$$3a = -4 \quad \text{or} \quad 3a = 1$$

$$a = -\frac{4}{3} \quad \text{or} \quad a = \frac{1}{3}$$

The solution set is $\left\{-\frac{4}{3}, \frac{1}{3}\right\}$. (Perhaps you should check these solutions in the original equation.) ■

EXAMPLE 6 Solve $(x - 1)(x + 9) = 11$.

Solution

$$(x - 1)(x + 9) = 11$$

$$x^2 + 8x - 9 = 11$$

$$x^2 + 8x - 20 = 0$$

$$(x + 10)(x - 2) = 0$$

$$x + 10 = 0 \quad \text{or} \quad x - 2 = 0$$

$$x = -10 \quad \text{or} \quad x = 2$$

The solution set is $\{-10, 2\}$. ■

Problem Solving

As you might expect, the increase in our power to solve equations broadens our base for solving problems. Now we are ready to tackle some problems that can be translated into the types of equations presented in the previous examples of this section.

PROBLEM 1 A room contains 78 chairs. The number of chairs per row is one more than twice the number of rows. Find the number of rows and the number of chairs per row.

Solution Let r represent the number of rows. Then $2r + 1$ represents the number of chairs per row.

$$r(2r + 1) = 78$$ The number of rows times the number of chairs per row yields the total number of chairs.

$$2r^2 + r = 78$$

$$2r^2 + r - 78 = 0$$

$$(2r + 13)(r - 6) = 0$$

$$2r + 13 = 0 \quad \text{or} \quad r - 6 = 0$$

$$2r = -13 \quad \text{or} \quad r = 6$$

$$r = -\frac{13}{2} \quad \text{or} \quad r = 6$$

The solution $-\frac{13}{2}$ must be disregarded, so there are 6 rows and $2r + 1$ or $2(6) + 1 = 13$ chairs per row. ■

PROBLEM 2 A strip of uniform width is to be cut off of both sides and both ends of a sheet of paper that is 8 inches by 11 inches in order to reduce the size of the paper to an area of 40 square inches. Find the width of the strip.

Solution Let x represent the width of the strip, as indicated in the following diagram.

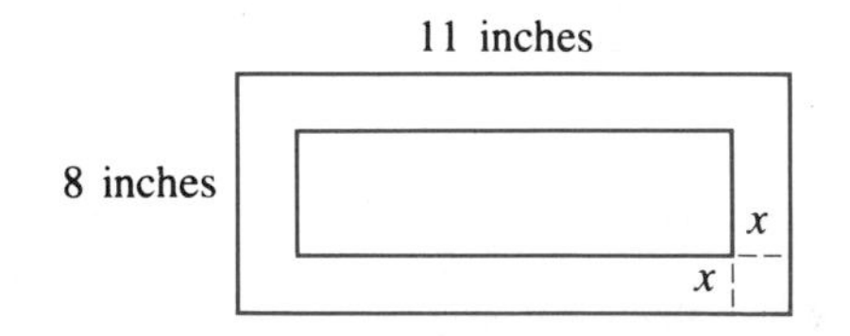

The length of the paper after the strips of width x are cut from both ends and both sides will be $11 - 2x$ and the width of the newly formed rectangle will be $8 - 2x$. Since the area ($A = lw$) is to be 40 square inches, we can set up and solve the following equation.

$$(11 - 2x)(8 - 2x) = 40$$

$$88 - 38x + 4x^2 = 40$$

$$4x^2 - 38x + 48 = 0$$

$$2x^2 - 19x + 24 = 0$$

$$(2x - 3)(x - 8) = 0$$

$$2x - 3 = 0 \quad \text{or} \quad x - 8 = 0$$

$$2x = 3 \quad \text{or} \quad x = 8$$

$$x = \frac{3}{2} \quad \text{or} \quad x = 8$$

The solution of 8 must be discarded since the width of the original sheet is only 8 inches. Therefore, the strip to be cut off of all four sides must be $1\frac{1}{2}$ inches wide. (Check this answer!) ■

The Pythagorean Theorem, an important theorem pertaining to right triangles, can sometimes serve as a guideline for solving problems that deal with right triangles. The Pythagorean Theorem states that "in any right triangle, the square of the longest side (called the hypotenuse) is equal to the sum of the squares of the other two sides (called legs)". Let's use this relationship to help solve a problem.

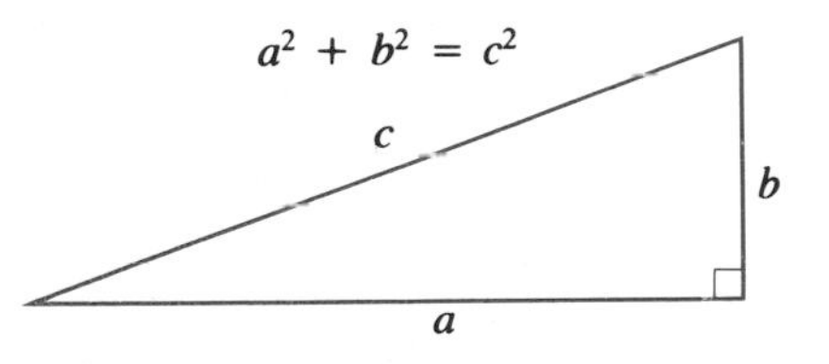

PROBLEM 3 One leg of a right triangle is 2 centimeters more than twice the length of the other leg. The hypotenuse is 1 centimeter longer than the longer of the two legs. Find the lengths of the three sides of the right triangle.

Solution Let l represent the length of the shortest leg. Then $2l + 2$ represents the length of the other leg and $2l + 3$ represents the length of the hypotenuse. Use the Pythagorean Theorem as a guideline to set up and solve the following equation.

$$l^2 + (2l + 2)^2 = (2l + 3)^2$$

$$l^2 + 4l^2 + 8l + 4 = 4l^2 + 12l + 9$$

$$l^2 - 4l - 5 = 0$$

$$(l-5)(l+1)=0$$

$$l-5=0 \quad \text{or} \quad l+1=0$$

$$l=5 \quad \text{or} \quad l=-1$$

The negative solution must be discarded, so the length of one leg is 5 centimeters; the other leg is $2(5)+2=12$ centimeters long and the hypotenuse is $2(5)+3=13$ centimeters long. ■

Problem Set 3.7

Solve each of the following equations. You will need to use factoring techniques that we discussed throughout this chapter.

1. $x^2+4x+3=0$ $\{-3,-1\}$
2. $x^2+7x+10=0$ $\{-5,-2\}$
3. $x^2+18x+72=0$ $\{-12,-6\}$
4. $n^2+20n+91=0$ $\{-13,-7\}$
5. $n^2-13n+36=0$ $\{4,9\}$
6. $n^2-10n+16=0$ $\{2,8\}$
7. $x^2+4x-12=0$ $\{-6,2\}$
8. $x^2+7x-30=0$ $\{-10,3\}$
9. $w^2-4w=5$ $\{-1,5\}$
10. $s^2-4s=21$ $\{-3,7\}$
11. $n^2+25n+156=0$ $\{-13,-12\}$
12. $n(n-24)=-128$ $\{8,16\}$
13. $3t^2+14t-5=0$ $\{-5,\frac{1}{3}\}$
14. $4t^2-19t-30=0$ $\{-\frac{5}{4},6\}$
15. $6x^2+25x+14=0$ $\{-\frac{7}{2},-\frac{2}{3}\}$
16. $25x^2+30x+8=0$ $\{-\frac{4}{5},-\frac{2}{5}\}$
17. $3t(t-4)=0$ $\{0,4\}$
18. $1-x^4=0$ $\{-1,1\}$
19. $-6n^2+13n-2=0$ $\{\frac{1}{6},2\}$
20. $(x+1)^2-4=0$ $\{-3,1\}$
21. $2n^3=72n$ $\{-6,0,6\}$
22. $a(a-1)=2$ $\{-1,2\}$
23. $(x-5)(x+3)=9$ $\{-4,6\}$
24. $3w^3-24w^2+36w=0$ $\{0,2,6\}$
25. $16-x^4=0$ $\{-2,2\}$
26. $16t^2-72t+81=0$ $\{\frac{9}{4}\}$
27. $n^2+7n-44=0$ $\{-11,4\}$
28. $2x^3=50x$ $\{-5,0,5\}$
29. $3x^2=75$ $\{-5,5\}$
30. $x^2+x-2=0$ $\{-2,1\}$
31. $15x^2+34x+15=0$ $\{-\frac{5}{3},-\frac{3}{5}\}$
32. $20x^2+41x+20=0$ $\{-\frac{5}{4},-\frac{4}{5}\}$
33. $8n^2-47n-6=0$ $\{-\frac{1}{8},6\}$
34. $7x^2+62x-9=0$ $\{-9,\frac{1}{7}\}$
35. $28n^2-47n+15=0$ $\{\frac{3}{7},\frac{5}{4}\}$
36. $24n^2-38n+15=0$ $\{\frac{3}{4},\frac{5}{6}\}$
37. $35n^2-18n-8=0$ $\{-\frac{2}{7},\frac{4}{5}\}$
38. $8n^2-6n-5=0$ $\{-\frac{1}{2},\frac{5}{4}\}$
39. $-3x^2-19x+14=0$ $\{-7,\frac{2}{3}\}$
40. $5x^2=43x-24$ $\{\frac{3}{5},8\}$
41. $n(n+2)=360$ $\{-20,18\}$
42. $n(n+1)=182$ $\{-14,13\}$
43. $9x^4-37x^2+4=0$ $\{-2,-\frac{1}{3},\frac{1}{3},2\}$
44. $4x^4-13x^2+9=0$ $\{-\frac{3}{2},-1,1,\frac{3}{2}\}$
45. $3x^4-46x^2-32=0$ $\{-4,4\}$
46. $x^4-9x^2=0$ $\{-3,0,3\}$
47. $2x^4+x^2-3=0$ $\{-1,1\}$
48. $x^4+5x^2-36=0$ $\{-2,2\}$
49. $12x^3+46x^2+40x=0$ $\{-\frac{5}{2},-\frac{4}{3},0\}$
50. $5x(3x-2)=0$ $\{0,\frac{2}{3}\}$

51. $(3x-1)^2-16=0$ $\{-1, \frac{5}{3}\}$

52. $(x+8)(x-6)=-24$ $\{-6, 4\}$

53. $4a(a+1)=3$ $\{-\frac{3}{2}, \frac{1}{2}\}$

54. $-18n^2-15n+7=0$ $\{-\frac{7}{6}, \frac{1}{3}\}$

Set up an equation and solve each of the following problems.

55. Find two consecutive integers whose product is 72. 8 and 9 or -9 and -8

56. Find two consecutive even whole numbers whose product is 224. 14 and 16

57. Find two integers whose product is 105 such that one of the integers is one more than twice the other integer. 7 and 15

58. Find two integers whose product is 104 such that one of the integers is three less than twice the other integer. 8 and 13

59. The perimeter of a rectangle is 32 inches and the area is 60 square inches. Find the length and width of the rectangle. 10 inches by 6 inches

60. Suppose that the length of a certain rectangle is two centimeters more than three times its width. If the area of the rectangle is 56 square centimeters, find its length and width. 14 centimeters by 4 centimeters

61. The sum of the squares of two consecutive integers is 85. Find the integers. -7 and -6 or 6 and 7

62. The sum of the areas of two circles is 65π square feet. The length of a radius of the larger circle is one foot less than twice the length of a radius of the smaller circle. Find the length of a radius of each circle. 4 feet and 7 feet

63. The combined area of a square and a rectangle is 64 square centimeters. The width of the rectangle is 2 centimeters more than the length of a side of the square and the length of the rectangle is 2 centimeters more than its width. Find the dimensions of the square and the rectangle. 4 centimeters by 4 centimeters and 6 centimeters by 8 centimeters

64. The Meyers have an apple orchard that contains 90 trees. The number of trees in each row is 3 more than twice the number of rows. Find the number of rows and the number of trees per row. 6 rows and 15 trees per row

65. The lengths of the three sides of a right triangle are represented by consecutive whole numbers. Find the lengths of the three sides. 3, 4, and 5 units

66. The area of the floor of a rectangular room is 175 square feet. The length of the room is $1\frac{1}{2}$ feet longer than the width. Find the length of the room. 14 feet

67. Suppose that the length of one leg of a right triangle is 3 inches more than the length of the other leg. If the length of the hypotenuse is 15 inches, find the lengths of the two legs. 9 inches and 12 inches

68. The lengths of the three sides of a right triangle are represented by consecutive even whole numbers. Find the lengths of the three sides. 6, 8, and 10 units

69. The area of a triangular sheet of paper is 28 square inches. One side of the triangle is two inches more than three times the length of the altitude to that side. Find the length of that side and the altitude to the side. An altitude of 4 inches and a side 14 inches long

70. A strip of uniform width is shaded along both sides and both ends of a rectangular poster that is 12 inches by 16 inches. How wide is the shaded strip if one-half of the poster is shaded? 2 inches

Chapter 3 Summary

(3.1) A **term** is an indicated product and may contain any number of factors. The variables involved in a term are called **literal factors** and the numerical factor is called the **numerical coefficient**. Terms that contain variables with only nonnegative integers as exponents are called **monomials**. The **degree** of a monomial is the sum of the exponents of the literal factors.

A **polynomial** is a monomial or a finite sum (or difference) of monomials. We classify polynomials as follows.

polynomial with one term ⟶ monomial

polynomial with two terms ⟶ binomial

polynomial with three terms ⟶ trinomial

Similar or **like terms** have the same literal factors. The commutative, associative, and distributive properties provide the basis for rearranging, regrouping, and combining similar terms.

(3.2) The following properties provide the basis for multiplying and dividing monomials.

1. $b^n \cdot b^m = b^{n+m}$

2. $(b^n)^m = b^{mn}$

3. $(ab)^n = a^n b^n$

4. **(a)** $\dfrac{b^n}{b^m} = b^{n-m}$ if $n > m$

(b) $\dfrac{b^n}{b^m} = 1$ if $n = m$

(3.3) The commutative and associative properties, the properties of exponents, and the distributive property work together to form a basis for multiplying polynomials. The following can be used as multiplication patterns.

$$(a + b)^2 = a^2 + 2ab + b^2,$$

$$(a - b)^2 = a^2 - 2ab + b^2,$$

$$(a + b)(a - b) = a^2 - b^2,$$

$$(a + b)^3 = a^3 + 3a^2b + 3ab^2 + b^3,$$

$$(a - b)^3 = a^3 - 3a^2b + 3ab^2 - b^3$$

(3.4) If a positive integer greater than 1 has no factors that are positive integers other than itself and 1, then it is called a **prime number**. A positive integer greater than 1 that is not a prime number is called a **composite number**.

The indicated product form that contains only prime factors is called the **prime factorization form** of a number.

An expression such as $ax + bx + ay + by$ can be factored as follows.

$$\begin{aligned} ax + bx + ay + by &= x(a + b) + y(a + b) \\ &= (a + b)(x + y) \end{aligned}$$

This is called **factoring by grouping**.

The distributive property in the form "$ab + ac = a(b + c)$" forms the basis for **factoring out the highest common monomial factor**.

Expressing polynomials in factored form and then applying the property "$ab = 0$ if and only if $a = 0$ or $b = 0$" provides us with another technique for solving equations.

(3.5) The factoring pattern

$$a^2 - b^2 = (a + b)(a - b)$$

is called the **difference of two squares**.

The difference-of-two-squares factoring pattern, along with the property "$ab - 0$ if and only if $a = 0$ or $b = 0$" provides us with another technique for solving equations. The factoring patterns

$$a^3 + b^3 = (a + b)(a^2 - ab + b^2), \qquad \text{and}$$

$$a^3 - b^3 = (a - b)(a^2 + ab + b^2)$$

are called the **sum and difference of two cubes**.

(3.6) Expressing a trinomial (for which the coefficient of the squared term is one) as a product of two binomials is based on the following relationship.

$$(x + a)(x + b) = x^2 + (a + b)x + ab$$

The coefficient of the middle term is the *sum* of a and b, and the last term is the *product* of a and b.

If the coefficient of the squared term of a trinomial does not equal one, then the following relationship holds.

$$(px + r)(qx + s) = (pq)x^2 + (ps + rq)x + rs$$

The two coefficients of x, ps and rq, must have a sum of $(ps) + (rq)$ and a product of $pqrs$. Thus, to factor something like $6x^2 + 7x - 3$, we need to find two integers whose product is $6(-3) = -18$ and whose sum is 7. The integers are 9 and -2 and we can factor as follows.

$$\begin{aligned} 6x^2 + 7x - 3 &= 6x^2 + 9x - 2x - 3 \\ &= 3x(2x + 3) - 1(2x + 3) \\ &= (2x + 3)(3x - 1) \end{aligned}$$

A **perfect square trinomial** is the result of squaring a binomial. There are two basic perfect square trinomial factoring patterns as follows.

$$a^2 + 2ab + b^2 = (a + b)^2,$$
$$a^2 - 2ab + b^2 = (a - b)^2$$

(3.7) The factoring techniques we discussed in this chapter along with the property "$ab = 0$ if and only if $a = 0$ or $b = 0$" provide the basis for expanding our equation solving processes.

The ability to solve more types of equations increases our capabilities for problem solving.

Chapter 3 Review Problem Set

Perform the indicated operations and simplify each of the following.

1. $(3x - 2) + (4x - 6) + (-2x + 5)$ $5x - 3$

2. $(8x^2 + 9x - 3) - (5x^2 - 3x - 1)$ $3x^2 + 12x - 2$

3. $(6x^2 - 2x - 1) + (4x^2 + 2x + 5) - (-2x^2 + x - 1)$ $12x^2 - x + 5$

4. $(-5x^2y^3)(4x^3y^4)$ $-20x^5y^7$

5. $(-2a^2)(3ab^2)(a^2b^3)$ $-6a^5b^5$

6. $5a^2(3a^2 - 2a - 1)$ $15a^4 - 10a^3 - 5a^2$

7. $(4x - 3y)(6x + 5y)$ $24x^2 + 2xy - 15y^2$

8. $(x + 4)(3x^2 - 5x - 1)$ $3x^3 + 7x^2 - 21x - 4$

9. $(4x^2y^3)^4$ $256x^8y^{12}$

10. $(3x - 2y)^2$ $9x^2 - 12xy + 4y^2$

11. $(-2x^2y^3z)^3$ $-8x^6y^9z^3$

12. $\dfrac{-39x^3y^4}{3xy^3}$ $-13x^2y$

13. $[3x - (2x - 3y + 1)] - [2y - (x - 1)]$ $2x + y - 2$

14. $(x^2 - 2x - 5)(x^2 + 3x - 7)$ $x^4 + x^3 - 18x^2 - x + 35$

15. $(7 - 3x)(3 + 5x)$ $21 + 26x - 15x^2$

16. $-(3ab)(2a^2b^3)^2$ $-12a^5b^7$

17. $\left(\frac{1}{2}ab\right)(8a^3b^2)(-2a^3)$ $-8a^7b^3$

18. $(7x - 9)(x + 4)$ $7x^2 + 19x - 36$

19. $(3x + 2)(2x^2 - 5x + 1)$ $6x^3 - 11x^2 - 7x + 2$

20. $(3x^{n+1})(2x^{3n-1})$ $6x^{4n}$

21. $(2x + 5y)^2$ $4x^2 + 20xy + 25y^2$

22. $(x - 2)^3$ $x^3 - 6x^2 + 12x - 8$

23. $(2x + 5)^3$ $8x^3 + 60x^2 + 150x + 125$

Factor completely each of the following. Indicate any that are not factorable using integers.

24. $x^2 + 3x - 28$ $(x + 7)(x - 4)$

25. $2t^2 - 18$ $2(t + 3)(t - 3)$

26. $4n^2 + 9$ Not factorable

27. $12n^2 - 7n + 1$ $(4n - 1)(3n - 1)$

28. $x^6 - x^2$ $x^2(x^2 + 1)(x + 1)(x - 1)$

29. $x^3 - 6x^2 - 72x$ $x(x - 12)(x + 6)$

30. $6a^3b + 4a^2b^2 - 2a^2bc$ $2a^2b(3a + 2b - c)$

31. $x^2 - (y - 1)^2$ $(x - y + 1)(x + y - 1)$

32. $8x^2 + 12$ $4(2x^2 + 3)$

33. $12x^2 + x - 35$ $(4x + 7)(3x - 5)$

34. $16n^2 - 40n + 25$ $(4n-5)^2$

35. $4n^2 - 8n$ $4n(n-2)$

36. $3w^3 + 18w^2 - 24w$ $3w(w^2+6w-8)$

37. $20x^2 + 3xy - 2y^2$ $(5x+2y)(4x-y)$

38. $16a^2 - 64a$ $16a(a-4)$

39. $3x^3 - 15x^2 - 18x$ $3x(x+1)(x-6)$

40. $n^2 - 8n - 128$ $(n+8)(n-16)$

41. $t^4 - 22t^2 - 75$ $(t+5)(t-5)(t^2+3)$

42. $35x^2 - 11x - 6$ $(5x-3)(7x+2)$

43. $15 - 14x + 3x^2$ $(3-x)(5-3x)$

44. $64n^3 - 27$ $(4n-3)(16n^2+12n+9)$

45. $16x^3 + 250$ $2(2x+5)(4x^2-10x+25)$

Solve each of the following equations.

46. $4x^2 - 36 = 0$ $\{-3, 3\}$

47. $x^2 + 5x - 6 = 0$ $\{-6, 1\}$

48. $49n^2 - 28n + 4 = 0$ $\{\frac{2}{7}\}$

49. $(3x-1)(5x+2) = 0$ $\{-\frac{2}{5}, \frac{1}{3}\}$

50. $(3x-4)^2 - 25 = 0$ $\{-\frac{1}{3}, 3\}$

51. $6a^3 = 54a$ $\{-3, 0, 3\}$

52. $x^5 = x$ $\{-1, 0, 1\}$

53. $-n^2 + 2n + 63 = 0$ $\{-7, 9\}$

54. $7n(7n+2) = 8$ $\{-\frac{4}{7}, \frac{2}{7}\}$

55. $30w^2 - w - 20 = 0$ $\{-\frac{4}{5}, \frac{5}{6}\}$

56. $5x^4 - 19x^2 - 4 = 0$ $\{-2, 2\}$

57. $9n^2 - 30n + 25 = 0$ $\{\frac{5}{3}\}$

58. $n(2n+4) = 96$ $\{-8, 6\}$

59. $7x^2 + 33x - 10 = 0$ $\{-5, \frac{2}{7}\}$

60. $(x+1)(x+2) = 42$ $\{-8, 5\}$

61. $x^2 + 12x - x - 12 = 0$ $\{-12, 1\}$

62. $2x^4 + 9x^2 + 4 = 0$ $\varnothing$

63. $30 - 19x - 5x^2 = 0$ $\{-5, \frac{6}{5}\}$

64. $3t^3 - 27t^2 + 24t = 0$ $\{0, 1, 8\}$

65. $-4n^2 - 39n + 10 = 0$ $\{-10, \frac{1}{4}\}$

Set up an equation and solve each of the following problems.

66. Find three consecutive integers such that the product of the smallest and largest is 1 less than 9 times the middle integer. 8, 9, and 10 or -1, 0, and 1

67. Find two integers whose sum is 2 and whose product is -48. -6 and 8

68. Find two consecutive odd whole numbers whose product is 195. 13 and 15

69. Two cars leave an intersection at the same time, one traveling north and the other traveling east. Some time later, they are 20 miles apart and the car going east had traveled 4 miles further than the other car. How far had each car traveled? 12 miles and 16 miles

70. The perimeter of a rectangle is 32 meters and its area is 48 square meters. Find the length and width of the rectangle. 4 meters by 12 meters

71. A room contains 144 chairs. The number of chairs per row is 2 less than twice the number of rows. Find the number of rows and the number of chairs per row. 9 rows and 16 chairs per row

72. The area of a triangle is 39 square feet. The length of one side is 1 foot more than twice the altitude to that side. Find the length of that side and the altitude to the side. The side is 13 feet long and the altitude is 6 feet.

73. A rectangular-shaped pool 20 feet by 30 feet has a sidewalk of uniform width around the pool. The area of the sidewalk is 336 square feet. Find the width of the sidewalk. 3 feet

74. The sum of the areas of two squares is 89 square centimeters. The length of a side of the larger square is 3 centimeters more than the length of a side of the smaller square. Find the dimensions of each square. 5 centimeters by 5 centimeters and 8 centimeters by 8 centimeters

75. The total surface area of a right circular cylinder is 32π square inches. If the altitude of the cylinder is three times the length of a radius, find the altitude of the cylinder. 6 inches

Cumulative Review Problem Set

For Problems 1–10, evaluate each algebraic expression for the given values of the variables. Don't forget that in some cases it may be helpful to simplify the algebraic expression before evaluating it.

1. $x^2 - 2xy + y^2$ for $x = -2$ and $y = -4$ 4

2. $-n^2 + 2n - 4$ for $n = -3$ -19

3. $2x^2 - 5x + 6$ for $x = 3$ 9

4. $3(2x - 1) - 2(x + 4) - 4(2x - 7)$ for $x = -1$ 21

5. $-(2n - 1) + 5(2n - 3) - 6(3n + 4)$ for $n = 4$ -78

6. $2(a - 4) - (a - 1) + (3a - 6)$ for $a = -5$ -33

7. $(3x^2 - 4x - 7) - (4x^2 - 7x + 8)$ for $x = -4$ -43

8. $-2(3x - 5y) - 4(x + 2y) + 3(-2x - 3y)$ for $x = 2$ and $y = -3$ -11

9. $5(-x^2 - x + 3) - (2x^2 - x + 6) - 2(x^2 + 4x - 6)$ for $x = 2$ -39

10. $3(x^2 - 4xy + 2y^2) - 2(x^2 - 6xy - y^2)$ for $x = -5$ and $y = -2$ 57

For Problems 11–18, perform the indicated operations and express your answers in simplest form.

11. $4(3x - 2) - 2(4x - 1) - (2x + 5)$ $2x - 11$

12. $(-6ab^2)(2ab)(-3b^3)$ $36a^2b^6$

13. $(5x - 7)(6x + 1)$ $30x^2 - 37x - 7$

14. $(-2x - 3)(x + 4)$ $-2x^2 - 11x - 12$

15. $(-4a^2b^3)^3$ $-64a^6b^9$

16. $(x + 2)(5x - 6)(x - 2)$ $5x^3 - 6x^2 - 20x + 24$

17. $(x - 3)(x^2 - x - 4)$ $x^3 - 4x^2 - x + 12$

18. $(x^2 + x + 4)(2x^2 - 3x - 7)$ $2x^4 - x^3 - 2x^2 - 19x - 28$

For Problems 19–28, factor each of the algebraic expressions completely.

19. $7x^2 - 7$ $7(x + 1)(x - 1)$

20. $4a^2 - 4ab + b^2$ $(2a - b)^2$

21. $3x^2 - 17x - 56$ $(3x + 7)(x - 8)$

22. $1 - x^3$ $(1 - x)(1 + x + x^2)$

23. $xy - 5x + 2y - 10$ $(y - 5)(x + 2)$

24. $3x^2 - 24x + 48$ $3(x - 4)^2$

25. $4n^4 - n^2 - 3$ $(4n^2 + 3)(n + 1)(n - 1)$

26. $32x^4 + 108x$ $4x(2x + 3)(4x^2 - 6x + 9)$

27. $4x^2 + 36$ $4(x^2 + 9)$

28. $20x^2 + x - 30$ $(5x - 6)(4x + 5)$

For Problems 29–32, solve each equation for the indicated variable.

29. $5x - 2y = 6$ for x $x = \dfrac{2y + 6}{5}$

30. $3x + 4y = 12$ for y $y = \dfrac{12 - 3x}{4}$

31. $V = 2\pi rh + 2\pi r^2$ for h $h = \dfrac{V - 2\pi r^2}{2\pi r}$

32. $\dfrac{1}{R} = \dfrac{1}{R_1} + \dfrac{1}{R_2}$ for R_1 $R_1 = \dfrac{RR_2}{R_2 - R}$

For Problems 33–50, solve each of the equations.

33. $(x - 2)(x + 5) = 8$ $\{-6, 3\}$

34. $(5n - 2)(3n + 7) = 0$ $\{-\frac{7}{3}, \frac{2}{5}\}$

35. $-2(n - 1) + 3(2n + 1) = -11$ $\{-4\}$

36. $x^2 + 7x - 18 = 0$ $\{-9, 2\}$

37. $8x^2 - 8 = 0$ $\{-1, 1\}$

38. $\dfrac{3}{4}(x - 2) - \dfrac{2}{5}(2x - 3) = \dfrac{1}{5}$ $\{-10\}$

39. $.1(x - .1) - .4(x + 2) = -5.31$ $\{15\}$

40. $\dfrac{2x - 1}{2} - \dfrac{5x + 2}{3} = 3$ $\left\{-\dfrac{25}{4}\right\}$

41. $|3n - 2| = 7$ $\{-\frac{5}{3}, 3\}$

42. $|2x - 1| = |x + 4|$ $\{-1, 5\}$

43. $.08(x + 200) = .07x + 20$ $\{400\}$

44. $2x^2 - 12x - 80 = 0$ $\{-4, 10\}$

45. $x^3 = 16x$ $\{-4, 0, 4\}$

46. $x(x + 2) - 3(x + 2) = 0$ $\{-2, 3\}$

47. $-12n^2 - 29n + 8 = 0$ $\{-\frac{8}{3}, \frac{1}{4}\}$

48. $3y(y + 1) = 90$ $\{-6, 5\}$

49. $2x^3 + 6x^2 - 20x = 0$ $\{-5, 0, 2\}$

50. $(3n - 1)(2n + 3) = (n + 4)(6n - 5)$ $\{\frac{17}{12}\}$

For Problems 51–58, solve each of the inequalities.

51. $-5(3n + 4) < -2(7n - 1)$ $(-22, \infty)$

52. $7(x + 1) - 8(x - 2) < 0$ $(23, \infty)$

53. $|2x - 1| > 7$ $(-\infty, -3) \cup (4, \infty)$

54. $|3x + 7| < 14$ $(-7, \frac{7}{3})$

55. $.09x + .1(x + 200) > 77$ $(300, \infty)$

56. $\dfrac{2x - 1}{4} - \dfrac{x - 2}{6} \leq \dfrac{3}{8}$ $\left(-\infty, \dfrac{7}{8}\right]$

57. $-(x - 1) + 2(3x - 1) \geq 2(x + 4) - (x - 1)$ $[\frac{5}{2}, \infty)$

58. $\dfrac{1}{4}(x - 2) + \dfrac{3}{7}(2x - 1) < \dfrac{3}{14}$ $\left(-\infty, \dfrac{32}{31}\right)$

For Problems 59–72, solve each problem by setting up and solving an appropriate equation or inequality.

59. Find three consecutive odd integers such that three times the first minus the second is one more than the third. 7, 9, and 11

60. Audrey has a collection of 48 coins consisting of nickels, dimes, and quarters. The number of dimes is one less than twice the number of nickels, and the number of quarters is ten larger than the number of dimes. How many coins of each denomination are there in the collection? 8 nickels, 15 dimes, and 25 quarters

61. The sum of the present ages of Joey and his mother is 46 years. In four years, Joey will be 3 years less than one-half as old as his mother at that time. Find the present ages of Joey and his mother. 12 and 34

62. The difference of the measures of two supplementary angles is 56°. Find the measure of each angle. 62° and 118°

63. Norm invested a certain amount of money at 8% interest and $200 more than that amount at 9%. His total yearly interest was $86. How much did he invest at each rate? $400 at 8% and $600 at 9%

64. Sanchez has a collection of pennies, nickels, and dimes worth $9.35. He has 5 more nickels than pennies and twice as many dimes as pennies. How many coins of each kind does he have? 35 pennies, 40 nickels, and 70 dimes

65. Sandy starts with her bicycle at 8 miles per hour. Fifty minutes later, Billie starts riding along the same route at 12 miles per hour. How long will it take Billie to overtake Sandy? 1 hour and 40 minutes

66. How many milliliters of pure acid must be added to 150 milliliters of a 30% solution of acid to obtain a 40% solution? 25 milliliters

67. Doreen bowled 152 and 174 in her first two games. What must she bowl in the third game to have an average of at least 160 for the three games? 154 or better

68. Brad had scores of 88, 92, 93, and 89 on his first four algebra tests. What score must he obtain on the fifth test to have an average better than 90 for the five tests? Better than 88

69. Suppose that the area of a square is one-half the area of a triangle. One side of the triangle is 16 inches long and the altitude to that side is the same length as a side of the square. Find the length of a side of the square. 4 inches

70. A rectangle is twice as long as it is wide and its area is 98 square meters. Find the length and width of the rectangle. 7 meters by 14 meters

71. A room contains 96 chairs. The number of chairs per row is 4 more than the number of rows. Find the number of rows and the number of chairs per row. 8 and 12

72. One leg of a right triangle is 3 feet longer than the other leg. The hypotenuse is 3 feet longer than the longer leg. Find the lengths of the three sides of the right triangle. 9 feet, 12 feet, and 15 feet

Thoughts into Words

73. Explain in your own words how to multiply two binomials.

74. Illustrate as many uses of the distributive property as you can.

75. What does the phrase "factor completely" mean to you?

76. Explain your thought process for factoring $30x^2 + 13x - 56$.

77. What does the expression "not factorable using integers" mean to you?

78. Discuss the role that factoring plays in solving equations.

Chapter 4

Rational Expressions

Rational expressions are to algebra what rational numbers are to arithmetic. Most of the work that we will do with rational expressions in this chapter parallels the work you have previously done with arithmetic fractions. The same basic properties used to explain reducing, adding, subtracting, multiplying, and dividing arithmetic fractions will serve as a basis for our work with rational expressions. The techniques of factoring we studied in the previous chapter will also play an important role in this chapter. We will conclude the chapter by working with some fractional equations that contain rational expressions.

4.1 Simplifying Rational Expressions

Any number that can be written in the form $\frac{a}{b}$, where a and b are integers and b is not zero, is called a **rational number**. The following are examples of rational numbers.

$$\frac{1}{2}, \quad \frac{3}{4}, \quad \frac{15}{7}, \quad \frac{-5}{6}, \quad \frac{7}{-8}, \quad \frac{-12}{-17}$$

Numbers such as 6, -4, 0, $4\frac{1}{2}$, 0.7 and 0.21 are also rational because we can express them as the indicated quotient of two integers. For example,

$$6 = \frac{6}{1} = \frac{12}{2} = \frac{18}{3} \text{ and so on,} \qquad 4\frac{1}{2} = \frac{9}{2},$$

$$-4 = \frac{4}{-1} = \frac{-4}{1} = \frac{8}{-2} \text{ and so on,} \qquad 0.7 = \frac{7}{10},$$

$$0 = \frac{0}{1} = \frac{0}{2} = \frac{0}{3} \text{ and so on,} \qquad 0.21 = \frac{21}{100}.$$

Our work with division of integers helps with the next examples.

$$\frac{8}{-2} = \frac{-8}{2} = -\frac{8}{2} = -4, \qquad \frac{12}{3} = \frac{-12}{-3} = 4$$

Observe the following general properties.

PROPERTY 4.1

1. $\dfrac{-a}{b} = \dfrac{a}{-b} = -\dfrac{a}{b}$
2. $\dfrac{-a}{-b} = \dfrac{a}{b}$

Therefore, a rational number such as $\frac{-2}{5}$ can also be written as $\frac{2}{-5}$ or $-\frac{2}{5}$.

We use the following property, often referred to as the **fundamental principle of fractions**, to reduce fractions to lowest terms or express fractions in simplest or reduced form.

PROPERTY 4.2

> If b and k are nonzero integers and a is any integer, then
> $$\frac{a \cdot k}{b \cdot k} = \frac{a}{b}.$$

Let's apply Properties 4.1 and 4.2 to the following examples.

EXAMPLE 1 Reduce $\frac{18}{24}$ to lowest terms.

Solution

$$\frac{18}{24} = \frac{3 \cdot 6}{4 \cdot 6} = \frac{3}{4}$$ ■

EXAMPLE 2 Change $\frac{40}{48}$ to simplest form.

Solution

$$\frac{\overset{5}{\cancel{40}}}{\underset{6}{\cancel{48}}} = \frac{5}{6}$$

A common factor of 8 has been divided out of both numerator and denominator. ■

EXAMPLE 3 Express $\frac{-36}{63}$ in reduced form.

Solution

$$\frac{-36}{63} = -\frac{36}{63} = -\frac{4 \cdot 9}{7 \cdot 9} = -\frac{4}{7}$$ ■

EXAMPLE 4 Reduce $\frac{72}{-90}$ to simplest form.

Solution

$$\frac{72}{-90} = -\frac{72}{90} = \frac{\cancel{2} \cdot 2 \cdot 2 \cdot \cancel{3} \cdot \cancel{3}}{\cancel{2} \cdot \cancel{3} \cdot \cancel{3} \cdot 5} = -\frac{4}{5}$$ ■

In Example 4, notice the use of the prime factored forms to help identify the common factors of the numerator and denominator.

Rational Expressions

A **rational expression** is the indicated quotient of two polynomials. The following are examples of rational expressions.

$$\frac{3x^2}{5},\quad \frac{x-2}{x+3},\quad \frac{x^2+5x-1}{x^2-9},\quad \frac{xy^2+x^2y}{xy},\quad \frac{a^3-3a^2-5a-1}{a^4+a^3+6}$$

Because we must avoid division by zero, no values that create a denominator of zero can be assigned to variables. Thus, the rational expression $\frac{x-2}{x+3}$ is meaningful for all values of x except for $x = -3$. Rather than making restrictions for each individual expression, we will merely assume that all denominators represent nonzero real numbers.

Property 4.2 $\left(\frac{a \cdot k}{b \cdot k} = \frac{a}{b}\right)$ serves as the basis for simplifying rational expressions, as the next examples illustrate.

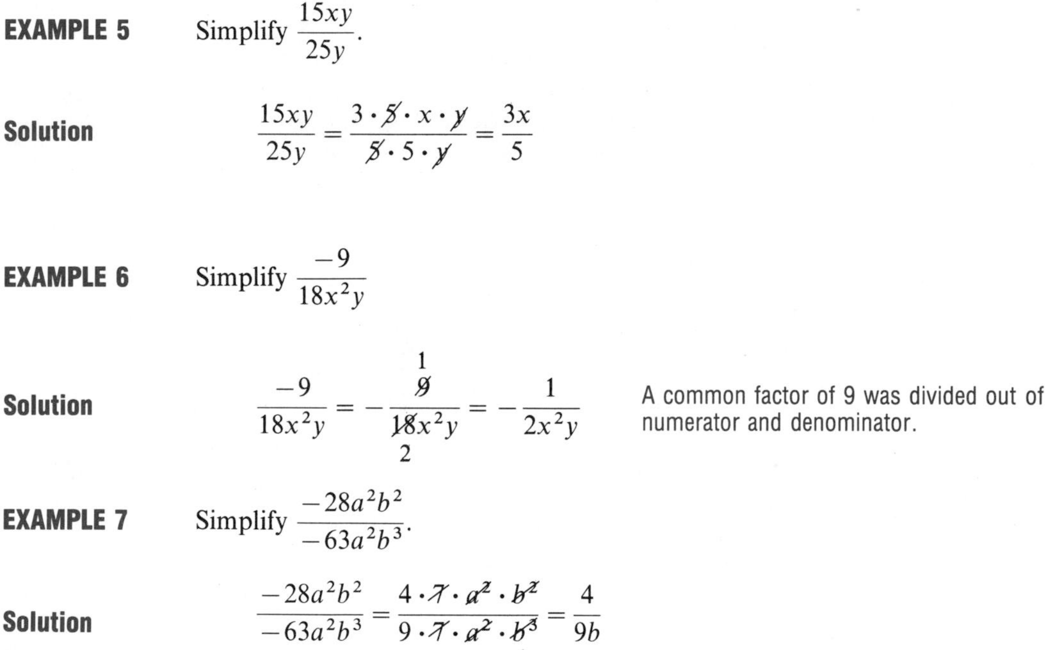

EXAMPLE 5 Simplify $\frac{15xy}{25y}$.

Solution

$$\frac{15xy}{25y} = \frac{3 \cdot \cancel{5} \cdot x \cdot \cancel{y}}{\cancel{5} \cdot 5 \cdot \cancel{y}} = \frac{3x}{5}$$ ■

EXAMPLE 6 Simplify $\frac{-9}{18x^2y}$

Solution

$$\frac{-9}{18x^2y} = -\frac{\overset{1}{\cancel{9}}}{\underset{2}{\cancel{18}}x^2y} = -\frac{1}{2x^2y}$$

A common factor of 9 was divided out of numerator and denominator. ■

EXAMPLE 7 Simplify $\frac{-28a^2b^2}{-63a^2b^3}$.

Solution

$$\frac{-28a^2b^2}{-63a^2b^3} = \frac{4 \cdot \cancel{7} \cdot \cancel{a^2} \cdot \cancel{b^2}}{9 \cdot \cancel{7} \cdot \cancel{a^2} \cdot \underset{b}{\cancel{b^3}}} = \frac{4}{9b}$$ ■

The factoring techniques from Chapter 3 can be used to factor numerators and/or denominators so that we can apply the property $\frac{a \cdot k}{b \cdot k} = \frac{a}{b}$. Examples 8–12 should clarify this process.

EXAMPLE 8 Simplify $\dfrac{x^2 + 4x}{x^2 - 16}$.

Solution

$$\frac{x^2 + 4x}{x^2 - 16} = \frac{x\cancel{(x + 4)}}{(x - 4)\cancel{(x + 4)}} = \frac{x}{x - 4}$$ ■

EXAMPLE 9 Simplify $\dfrac{4a^2 + 12a + 9}{2a + 3}$.

Solution

$$\frac{4a^2 + 12a + 9}{2a + 3} = \frac{\cancel{(2a + 3)}(2a + 3)}{1\cancel{(2a + 3)}} = \frac{2a + 3}{1} = 2a + 3$$ ■

EXAMPLE 10 Simplify $\dfrac{5n^2 + 6n - 8}{10n^2 - 3n - 4}$.

Solution

$$\frac{5n^2 + 6n - 8}{10n^2 - 3n - 4} = \frac{\cancel{(5n - 4)}(n + 2)}{\cancel{(5n - 4)}(2n + 1)} = \frac{n + 2}{2n + 1}$$ ■

EXAMPLE 11 Simplify $\dfrac{6x^3y - 6xy}{x^3 + 5x^2 + 4x}$.

Solution

$$\frac{6x^3y - 6xy}{x^3 + 5x^2 + 4x} = \frac{6xy(x^2 - 1)}{x(x^2 + 5x + 4)} = \frac{6\cancel{x}y\cancel{(x + 1)}(x - 1)}{\cancel{x}\cancel{(x + 1)}(x + 4)} = \frac{6y(x - 1)}{x + 4}$$ ■

Note that in Example 11 we left the numerator of the final fraction in factored form. This is often done if expressions other than monomials are involved. Either $\dfrac{6y(x - 1)}{x + 4}$ or $\dfrac{6xy - 6y}{x + 4}$ is an acceptable answer.

Remember that the quotient of any nonzero real number and its opposite is -1. For example, $\dfrac{6}{-6} = -1$ and $\dfrac{-8}{8} = -1$. Likewise, the indicated quotient of any polynomial and its opposite is equal to -1. For example,

$\dfrac{a}{-a} = -1$ because a and $-a$ are opposites;

$\dfrac{a - b}{b - a} = -1$ because $a - b$ and $b - a$ are opposites;

$\dfrac{x^2 - 4}{4 - x^2} = -1$ because $x^2 - 4$ and $4 - x^2$ are opposites.

The final example of this section illustrates the use of this idea when simplifying rational expressions.

EXAMPLE 12 Simplify $\dfrac{6a^2 - 7a + 2}{10a - 15a^2}$.

Solution

$$\frac{6a^2 - 7a + 2}{10a - 15a^2} = \frac{(2a-1)(3a-2)}{5a(2-3a)}$$

$$= -\frac{2a-1}{5a} \quad \text{or} \quad \frac{1-2a}{5a}$$

■

Problem Set 4.1

Express each of the following rational numbers in reduced form.

1. $\dfrac{27}{36}$
2. $\dfrac{14}{21}$
3. $\dfrac{45}{54}$
4. $\dfrac{-14}{42}$
5. $\dfrac{24}{-60}$
6. $\dfrac{45}{-75}$
7. $\dfrac{-16}{-56}$
8. $\dfrac{-30}{-42}$

Simplify each of the following.

9. $\dfrac{12xy}{42y}$
10. $\dfrac{21xy}{35x}$
11. $\dfrac{18a^2}{45ab}$
12. $\dfrac{48ab}{84b^2}$
13. $\dfrac{-14y^3}{56xy^2}$
14. $\dfrac{-14x^2y^3}{63xy^2}$
15. $\dfrac{54c^2d}{-78cd^2}$
16. $\dfrac{60x^3z}{-64xyz^2}$
17. $\dfrac{-40x^3y}{-24xy^4}$
18. $\dfrac{-30x^2y^2z^2}{-35xz^3}$
19. $\dfrac{x^2 - 4}{x^2 + 2x}$
20. $\dfrac{xy + y^2}{x^2 - y^2}$
21. $\dfrac{18x + 12}{12x - 6}$
22. $\dfrac{20x + 50}{15x - 30}$
23. $\dfrac{a^2 + 7a + 10}{a^2 - 7a - 18}$
24. $\dfrac{a^2 + 4a - 32}{3a^2 + 26a + 16}$
25. $\dfrac{2n^2 + n - 21}{10n^2 + 33n - 7}$
26. $\dfrac{4n^2 - 15n - 4}{7n^2 - 30n + 8}$
27. $\dfrac{5x^2 + 7}{10x}$
28. $\dfrac{12x^2 + 11x - 15}{20x^2 - 23x + 6}$
29. $\dfrac{6x^2 + x - 15}{8x^2 - 10x - 3}$
30. $\dfrac{4x^2 + 8x}{x^3 + 8}$
31. $\dfrac{3x^2 - 12x}{x^3 - 64}$
32. $\dfrac{x^2 - 14x + 49}{6x^2 - 37x - 35}$

33. $\frac{3x^2+17x-6}{9x^2-6x+1}$ $\frac{x+6}{3x-1}$

34. $\frac{9y^2-1}{3y^2+11y-4}$ $\frac{3y+1}{y+4}$

35. $\frac{2x^3+3x^2-14x}{x^2y+7xy-18y}$ $\frac{x(2x+7)}{y(x+9)}$

36. $\frac{3x^3+12x}{9x^2+18x}$ $\frac{x^2+4}{3(x+2)}$

37. $\frac{5y^2+22y+8}{25y^2-4}$ $\frac{y+4}{5y-2}$

38. $\frac{16x^3y+24x^2y^2-16xy^3}{24x^2y+12xy^2-12y^3}$ $\frac{2x(x+2y)}{3(x+y)}$

39. $\frac{15x^3-15x^2}{5x^3+5x}$ $\frac{3x(x-1)}{x^2+1}$

40. $\frac{5n^2+18n-8}{3n^2+13n+4}$ $\frac{5n-2}{3n+1}$

41. $\frac{4x^2y+8xy^2-12y^3}{18x^3y-12x^2y^2-6xy^3}$ $\frac{2(x+3y)}{3x(3x+y)}$

42. $\frac{3+x-2x^2}{2+x-x^2}$ $\frac{3-2x}{2-x}$

43. $\frac{3n^2+14n-24}{7n^2+44n+12}$ $\frac{3n-4}{7n+2}$

44. $\frac{x^4-2x^2-15}{2x^4+9x^2+9}$ $\frac{x^2-5}{2x^2+3}$

45. $\frac{8+18x-5x^2}{10+31x+15x^2}$ $\frac{4-x}{5+3x}$

46. $\frac{6x^4-11x^2+4}{2x^4+17x^2-9}$ $\frac{3x^2-4}{x^2+9}$

47. $\frac{27x^4-x}{6x^3+10x^2-4x}$ $\frac{9x^2+3x+1}{2(x+2)}$

48. $\frac{64x^4+27x}{12x^3-27x^2-27x}$ $\frac{16x^2-12x+9}{3(x-3)}$

49. $\frac{-40x^3+24x^2+16x}{20x^3+28x^2+8x}$ $\frac{-2(x-1)}{x+1}$

50. $\frac{-6x^3-21x^2+12x}{-18x^3-42x^2+120x}$ $\frac{2x-1}{2(3x-5)}$

Simplify each of the following. You will need to use factoring by grouping.

51. $\frac{xy+ay+bx+ab}{xy+ay+cx+ac}$ $\frac{y+b}{y+c}$

52. $\frac{xy+2y+3x+6}{xy+2y+4x+8}$ $\frac{y+3}{y+4}$

53. $\frac{ax-3x+2ay-6y}{2ax-6x+ay-3y}$ $\frac{x+2y}{2x+y}$

54. $\frac{x^2-2x+ax-2a}{x^2-2x+3ax-6a}$ $\frac{x+a}{x+3a}$

55. $\frac{5x^2+5x+3x+3}{5x^2+3x-30x-18}$ $\frac{x+1}{x-6}$

56. $\frac{x^2+3x+4x+12}{2x^2+6x-x-3}$ $\frac{x+4}{2x-1}$

57. $\frac{2st-30-12s+5t}{3st-6-18s+t}$ $\frac{2s+5}{3s+1}$

58. $\frac{nr-6-3n+2r}{nr+10+2r+5n}$ $\frac{r-3}{r+5}$

Simplify each of the following. You may want to refer to Example 12 in this section.

59. $\frac{5x-7}{7-5x}$ -1

60. $\frac{4a-9}{9-4a}$ -1

61. $\frac{n^2-49}{7-n}$ $-n-7$

62. $\frac{9-y}{y^2-81}$ $-\frac{1}{y+9}$

63. $\frac{2y-2xy}{x^2y-y}$ $-\frac{2}{x+1}$

64. $\frac{3x-x^2}{x^2-9}$ $-\frac{x}{x+3}$

65. $\frac{2x^3-8x}{4x-x^3}$ -2

66. $\frac{x^2-(y-1)^2}{(y-1)^2-x^2}$ -1

67. $\frac{n^2-5n-24}{40+3n-n^2}$ $-\frac{n+3}{n+5}$

68. $\frac{x^2+2x-24}{20-x-x^2}$ $-\frac{x+6}{x+5}$

4.2 Multiplying and Dividing Rational Expressions

We define multiplication of rational numbers in common fraction form as follows.

DEFINITION 4.1

If a, b, c, and d are integers with b and d not equal to zero, then

$$\frac{a}{b} \cdot \frac{c}{d} = \frac{a \cdot c}{b \cdot d} = \frac{ac}{bd}.$$

To multiply rational numbers in common fraction form we merely **multiply numerators and multiply denominators** as the next examples demonstrate. (The steps in the dashed boxes are usually done mentally.)

$$\frac{2}{3} \cdot \frac{4}{5} = \boxed{\frac{2 \cdot 4}{3 \cdot 5}} = \frac{8}{15},$$

$$\frac{-3}{4} \cdot \frac{5}{7} = \boxed{\frac{-3 \cdot 5}{4 \cdot 7} = \frac{-15}{28}} = -\frac{15}{28},$$

$$-\frac{5}{6} \cdot \frac{13}{3} = \boxed{\frac{-5}{6} \cdot \frac{13}{3} = \frac{-5 \cdot 13}{6 \cdot 3} = \frac{-65}{18}} = -\frac{65}{18}$$

We also agree, when multiplying rational numbers, to express the final product in reduced form. The following examples show some different formats to *multiply and simplify* rational numbers.

$$\frac{3}{4} \cdot \frac{4}{7} = \frac{3 \cdot \cancel{4}}{\cancel{4} \cdot 7} = \frac{3}{7};$$

$$\frac{\overset{1}{\cancel{8}}}{\underset{1}{\cancel{9}}} \cdot \frac{\overset{3}{\cancel{27}}}{\underset{4}{\cancel{32}}} = \frac{3}{4};$$

A common factor of 9 has been divided out of 9 and 27, and a common factor of 8 has been divided out of 8 and 32.

$$\left(-\frac{28}{25}\right)\left(-\frac{65}{78}\right) = \frac{\cancel{2} \cdot 2 \cdot 7 \cdot \cancel{5} \cdot \cancel{13}}{\cancel{5} \cdot 5 \cdot \cancel{2} \cdot 3 \cdot \cancel{13}} = \frac{14}{15}.$$

We should recognize that a *negative times a negative is positive.* Also, notice the use of prime factors to help recognize common factors.

Multiplication of rational expressions follows the same basic pattern as multiplication of rational numbers in common fraction form. That is to say, "we multiply numerators and multiply denominators and express the final product in simplified or reduced form." Let's consider some examples.

$$\frac{3x}{4y} \cdot \frac{8y^2}{9x} = \frac{\cancel{3} \cdot \overset{2}{\cancel{8}} \cdot \cancel{x} \cdot \overset{y}{\cancel{y^2}}}{\cancel{4} \cdot \underset{3}{\cancel{9}} \cdot \cancel{x} \cdot \cancel{y}} = \frac{2y}{3},$$

Notice the use of the commutative property of multiplication to rearrange factors in a more convenient form for identifying common factors of the numerator and denominator.

$$\frac{-4a}{6a^2b^2} \cdot \frac{9ab}{12a^2} = -\frac{\cancel{4} \cdot \overset{\cancel{3}}{\cancel{9}} \cdot \cancel{a^2} \cdot \cancel{b}}{\underset{2}{\cancel{6}} \cdot \underset{\cancel{3}}{\cancel{12}} \cdot \underset{a^2}{\cancel{a^4}} \cdot \underset{b}{\cancel{b^2}}} = -\frac{1}{2a^2b},$$

$$\frac{12x^2y}{-18xy} \cdot \frac{-24xy^2}{56y^3} = \frac{\overset{2}{\cancel{12}} \cdot \overset{\cancel{3}}{\cancel{24}} \cdot \overset{x^2}{\cancel{x^3}} \cdot \cancel{y^3}}{\underset{\cancel{3}}{\cancel{18}} \cdot \underset{7}{\cancel{56}} \cdot \cancel{x} \cdot \underset{y}{\cancel{y^4}}} = \frac{2x^2}{7y}$$

You should recognize that the first fraction is equivalent to $-\frac{12x^2y}{18xy}$ and the second to $-\frac{24xy^2}{56y^3}$; thus, the product is positive.

If the rational expressions contain polynomials (other than monomials) that are factorable, then our work may take on the following format.

EXAMPLE 1 Multiply and simplify $\frac{y}{x^2 - 4} \cdot \frac{x + 2}{y^2}$.

Solution

$$\frac{y}{x^2 - 4} \cdot \frac{x + 2}{y^2} = \frac{\cancel{y}\cancel{(x + 2)}}{\underset{y}{\cancel{y^2}}\cancel{(x + 2)}(x - 2)} = \frac{1}{y(x - 2)}$$ ■

In Example 1, notice that we combined the steps of multiplying numerators and denominators and factoring the polynomials. Also, notice that we left the final answer in factored form. Either $\frac{1}{y(x - 2)}$ or $\frac{1}{xy - 2y}$ would be an acceptable answer.

EXAMPLE 2 Multiply and simplify $\frac{x^2 - x}{x + 5} \cdot \frac{x^2 + 5x + 4}{x^4 - x^2}$.

Solution

$$\frac{x^2 - x}{x + 5} \cdot \frac{x^2 + 5x + 4}{x^4 - x^2} = \frac{\cancel{x}\cancel{(x - 1)}\cancel{(x + 1)}(x + 4)}{(x + 5)\underset{x}{\cancel{(x^2)}}\cancel{(x - 1)}\cancel{(x + 1)}} = \frac{x + 4}{x(x + 5)}$$ ■

EXAMPLE 3 Multiply and simplify $\frac{6n^2 + 7n - 5}{n^2 + 2n - 24} \cdot \frac{4n^2 + 21n - 18}{12n^2 + 11n - 15}$.

Solution

$$\frac{6n^2 + 7n - 5}{n^2 + 2n - 24} \cdot \frac{4n^2 + 21n - 18}{12n^2 + 11n - 15}$$

$$= \frac{\cancel{(3n+5)}(2n-1)\cancel{(4n-3)}\cancel{(n+6)}}{\cancel{(n+6)}(n-4)\cancel{(3n+5)}\cancel{(4n-3)}} = \frac{2n-1}{n-4}$$ ■

Dividing Rational Expressions

We define division of rational numbers in common fraction form as follows.

DEFINITION 4.2

> If a, b, c, and d are integers with b, c, and d not equal to zero, then
>
> $$\frac{a}{b} \div \frac{c}{d} = \frac{a}{b} \cdot \frac{d}{c} = \frac{ad}{bc}.$$

Definition 4.2 states that to divide two rational numbers in fractional form we **invert the divisor and multiply**. We call the numbers $\frac{c}{d}$ and $\frac{d}{c}$ **reciprocals** or **multiplicative inverses** of each other because their product is one. Thus, we can describe division as to **divide by a fraction, multiply by its reciprocal**. The following examples demonstrate the use of Definition 4.2.

$$\frac{7}{8} \div \frac{5}{6} = \frac{7}{\underset{4}{\cancel{8}}} \cdot \frac{\overset{3}{\cancel{6}}}{5} = \frac{21}{20}, \qquad \frac{-5}{9} \div \frac{15}{18} = -\frac{\cancel{5}}{\cancel{9}} \cdot \frac{\overset{2}{\cancel{18}}}{\underset{3}{\cancel{15}}} = -\frac{2}{3},$$

$$\frac{14}{-19} \div \frac{21}{-38} = \left(-\frac{14}{19}\right) \div \left(-\frac{21}{38}\right) = \left(-\frac{\overset{2}{\cancel{14}}}{\cancel{19}}\right)\left(-\frac{\overset{2}{\cancel{38}}}{\underset{3}{\cancel{21}}}\right) = \frac{4}{3}$$

We define division of algebraic rational expressions in the same way as division of rational numbers. That is, the quotient of two rational expressions is the product of the first expression times the reciprocal of the second. Consider the following examples.

EXAMPLE 4 Divide and simplify $\frac{16x^2y}{24xy^3} \div \frac{9xy}{8x^2y^2}$.

Solution

$$\frac{16x^2y}{24xy^3} \div \frac{9xy}{8x^2y^2} = \frac{16x^2y}{24xy^3} \cdot \frac{8x^2y^2}{9xy} = \frac{16 \cdot \cancel{8} \cdot \overset{x^2}{\cancel{x^4}} \cdot \cancel{y^3}}{\underset{3}{\cancel{24}} \cdot 9 \cdot \cancel{x^2} \cdot \underset{y}{\cancel{y^4}}} = \frac{16x^2}{27y}$$ ■

EXAMPLE 5 Divide and simplify $\dfrac{3a^2 + 12}{3a^2 - 15a} \div \dfrac{a^4 - 16}{a^2 - 3a - 10}$.

Solution

$$\frac{3a^2 + 12}{3a^2 - 15a} \div \frac{a^4 - 16}{a^2 - 3a - 10} = \frac{3a^2 + 12}{3a^2 - 15a} \cdot \frac{a^2 - 3a - 10}{a^4 - 16}$$

$$= \frac{\cancel{3}\cancel{(a^2 + 4)}\cancel{(a - 5)}\cancel{(a + 2)}}{\cancel{3}a\cancel{(a - 5)}\cancel{(a^2 + 4)}\cancel{(a + 2)}(a - 2)}$$

$$= \frac{1}{a(a - 2)}$$ ■

EXAMPLE 6 Divide and simplify $\dfrac{28t^3 - 51t^2 - 27t}{49t^2 + 42t + 9} \div (4t - 9)$.

Solution

$$\frac{28t^3 - 51t^2 - 27t}{49t^2 + 42t + 9} \div \frac{4t - 9}{1} = \frac{28t^3 - 51t^2 - 27t}{49t^2 + 42t + 9} \cdot \frac{1}{4t - 9}$$

$$= \frac{t\cancel{(7t + 3)}\cancel{(4t - 9)}}{\cancel{(7t + 3)}(7t + 3)\cancel{(4t - 9)}}$$

$$= \frac{t}{7t + 3}$$ ■

In a problem such as Example 6, it may be helpful to write the divisor with a denominator of 1. Thus, we write $4t - 9$ as $\dfrac{4t - 9}{1}$; then its reciprocal is obviously $\dfrac{1}{4t - 9}$.

Let's consider one final example that involves both multiplication and division.

EXAMPLE 7 Perform the indicated operations and simplify.

$$\frac{x^2 + 5x}{3x^2 - 4x - 20} \cdot \frac{x^2y + y}{2x^2 + 11x + 5} \div \frac{xy^2}{6x^2 - 17x - 10}$$

Solution

$$\frac{x^2 + 5x}{3x^2 - 4x - 20} \cdot \frac{x^2y + y}{2x^2 + 11x + 5} \div \frac{xy^2}{6x^2 - 17x - 10}$$

$$= \frac{x^2 + 5x}{3x^2 - 4x - 20} \cdot \frac{x^2y + y}{2x^2 + 11x + 5} \cdot \frac{6x^2 - 17x - 10}{xy^2}$$

$$= \frac{\cancel{x}\cancel{(x + 5)}(\cancel{y})(x^2 + 1)\cancel{(2x + 1)}\cancel{(3x - 10)}}{\cancel{(3x - 10)}(x + 2)\cancel{(2x + 1)}\cancel{(x + 5)}(\cancel{x})(\underset{y}{\cancel{y^2}})} = \frac{x^2 + 1}{y(x + 2)}$$ ■

Problem Set 4.2

Perform the following indicated operations involving rational numbers. Express final answers in reduced form.

1. $\frac{7}{12} \cdot \frac{6}{35}$ $\frac{1}{10}$

2. $\frac{5}{8} \cdot \frac{12}{20}$ $\frac{3}{8}$

3. $\frac{-4}{9} \cdot \frac{18}{30}$ $-\frac{4}{15}$

4. $\frac{-6}{9} \cdot \frac{36}{48}$ $-\frac{1}{2}$

5. $\frac{3}{-8} \cdot \frac{-6}{12}$ $\frac{3}{16}$

6. $\frac{-12}{16} \cdot \frac{18}{-32}$ $\frac{27}{64}$

7. $\left(-\frac{5}{7}\right) \div \frac{6}{7}$ $-\frac{5}{6}$

8. $\left(-\frac{5}{9}\right) \div \frac{10}{3}$ $-\frac{1}{6}$

9. $\frac{-9}{5} \div \frac{27}{10}$ $= \frac{2}{3}$

10. $\frac{4}{7} \div \frac{16}{-21}$ $-\frac{3}{4}$

11. $\frac{4}{9} \cdot \frac{6}{11} \div \frac{4}{15}$ $\frac{10}{11}$

12. $\frac{2}{3} \cdot \frac{6}{7} \div \frac{8}{3}$ $\frac{3}{14}$

Perform the following indicated operations involving rational expressions. Express final answers in simplest form.

13. $\frac{6xy}{9y^4} \cdot \frac{30x^3y}{-48x}$ $-\frac{5x^3}{12y^2}$

14. $\frac{-14xy^4}{18y^2} \cdot \frac{24x^2y^3}{35y^2}$ $-\frac{8x^3y^3}{15}$

15. $\frac{5a^2b^2}{11ab} \cdot \frac{22a^3}{15ab^2}$ $\frac{2a^3}{3b}$

16. $\frac{10a^2}{5b^2} \cdot \frac{15b^3}{2a^4}$ $\frac{15b}{a^2}$

17. $\frac{5xy}{8y^2} \cdot \frac{18x^2y}{15}$ $\frac{3x^3}{4}$

18. $\frac{4x^2}{5y^2} \cdot \frac{15xy}{24x^2y^2}$ $\frac{x}{2y^3}$

19. $\frac{5x^4}{12x^2y^3} \div \frac{9}{5xy}$ $\frac{25x^3}{108y^2}$

20. $\frac{7x^2y}{9xy^3} \div \frac{3x^4}{2x^2y^2}$ $\frac{14}{27x}$

21. $\frac{9a^2c}{12bc^2} \div \frac{21ab}{14c^3}$ $\frac{ac^2}{2b^2}$

22. $\frac{3ab^3}{4c} \div \frac{21ac}{12bc^3}$ $\frac{3b^4c}{7}$

23. $\frac{9x^2y^3}{14x} \cdot \frac{21y}{15xy^2} \cdot \frac{10x}{12y^3}$ $\frac{3x}{4y}$

24. $\frac{5xy}{7a} \cdot \frac{14a^2}{15x} \cdot \frac{3a}{8y}$ $\frac{a^2}{4}$

25. $\frac{3x+6}{5y} \cdot \frac{x^2+4}{x^2+10x+16}$ $\frac{3(x^2+4)}{5y(x+8)}$

26. $\frac{5xy}{x+6} \cdot \frac{x^2-36}{x^2-6x}$ $5y$

27. $\frac{5a^2+20a}{a^3-2a^2} \cdot \frac{a^2-a-12}{a^2-16}$ $\frac{5(a+3)}{a(a-2)}$

28. $\frac{2a^2+6}{a^2-a} \cdot \frac{a^3-a^2}{8a-4}$ $\frac{a(a^2+3)}{2(2a-1)}$

29. $\frac{3n^2+15n-18}{3n^2+10n-48} \cdot \frac{12n^2-17n-40}{8n^2+2n-10}$ $\frac{3}{2}$

30. $\frac{10n^2+21n-10}{5n^2+33n-14} \cdot \frac{2n^2+6n-56}{2n^2-3n-20}$ 2

31. $\frac{9y^2}{x^2+12x+36} \div \frac{12y}{x^2+6x}$ $\frac{3xy}{4(x+6)}$

32. $\frac{7xy}{x^2-4x+4} \div \frac{14y}{x^2-4}$ $\frac{x(x+2)}{2(x-2)}$

33. $\frac{x^2-4xy+4y^2}{7xy^2} \div \frac{4x^2-3xy-10y^2}{20x^2y+25xy^2}$ $\frac{5(x-2y)}{7y}$

34. $\frac{x^2+5xy-6y^2}{xy^2-y^3} \cdot \frac{2x^2+15xy+18y^2}{xy+4y^2}$ $\frac{(x+6y)^2(2x+3y)}{y^3(x+4y)}$

35. $\frac{5-14n-3n^2}{1-2n-3n^2} \cdot \frac{9+7n-2n^2}{27-15n+2n^2}$ $\frac{5+n}{3-n}$

36. $\frac{6-n-2n^2}{12-11n+2n^2} \cdot \frac{24-26n+5n^2}{2+3n+n^2}$ $\frac{6-5n}{1+n}$

37. $\frac{3x^4+2x^2-1}{3x^4+14x^2-5} \cdot \frac{x^4-2x^2-35}{x^4-17x^2+70}$ $\frac{x^2+1}{x^2-10}$

38. $\dfrac{2x^4 + x^2 - 3}{2x^4 + 5x^2 + 2} \cdot \dfrac{3x^4 + 10x^2 + 8}{3x^4 + x^2 - 4}$ $\dfrac{2x^2 + 3}{2x^2 + 1}$

39. $\dfrac{6x^2 - 35x + 25}{4x^2 - 11x - 45} \div \dfrac{18x^2 + 9x - 20}{24x^2 + 74x + 45}$ $\dfrac{6x + 5}{3x + 4}$

40. $\dfrac{21t^2 + 22t - 8}{5t^2 - 43t - 18} \div \dfrac{12t^2 + 7t - 12}{20t^2 - 7t - 6}$ $\dfrac{7t - 2}{t - 9}$

41. $\dfrac{10t^3 + 25t}{20t + 10} \cdot \dfrac{2t^2 - t - 1}{t^5 - t}$ $\dfrac{2t^2 + 5}{2(t^2 + 1)(t + 1)}$

42. $\dfrac{t^4 - 81}{t^2 - 6t + 9} \cdot \dfrac{6t^2 - 11t - 21}{5t^2 + 8t - 21}$ $\dfrac{(t^2 + 9)(6t + 7)}{5t - 7}$

43. $\dfrac{4t^2 + t - 5}{t^3 - t^2} \cdot \dfrac{t^4 + 6t^3}{16t^2 + 40t + 25}$ $\dfrac{t(t + 6)}{4t + 5}$

44. $\dfrac{9n^2 - 12n + 4}{n^2 - 4n - 32} \cdot \dfrac{n^2 + 4n}{3n^3 - 2n^2}$ $\dfrac{3n - 2}{n(n - 8)}$

45. $\dfrac{nr + 3n + 2r + 6}{nr + 3n - 3r - 9} \cdot \dfrac{n^2 - 9}{n^3 - 4n}$ $\dfrac{n + 3}{n(n - 2)}$

46. $\dfrac{xy + xc + ay + ac}{xy - 2xc + ay - 2ac} \cdot \dfrac{2x^3 - 8x}{12x^3 + 20x^2 - 8x}$ $\dfrac{(y + c)(x - 2)}{2(y - 2c)(3x - 1)}$

47. $\dfrac{x^2 - x}{4y} \cdot \dfrac{10xy^2}{2x - 2} \div \dfrac{3x^2 + 3x}{15x^2y^2}$ $\dfrac{25x^3y^3}{4(x + 1)}$

48. $\dfrac{4xy^2}{7x} \cdot \dfrac{14x^3y}{12y} \div \dfrac{7y}{9x^3}$ $\dfrac{6x^6y}{7}$

49. $\dfrac{a^2 - 4ab + 4b^2}{6a^2 - 4ab} \cdot \dfrac{3a^2 + 5ab - 2b^2}{6a^2 + ab - b^2} \div \dfrac{a^2 - 4b^2}{8a + 4b}$ $\dfrac{2(a - 2b)}{a(3a - 2b)}$

50. $\dfrac{2x^2 + 3x}{2x^3 - 10x^2} \cdot \dfrac{x^2 - 8x + 15}{3x^3 - 27x} \div \dfrac{14x + 21}{x^2 - 6x - 27}$ $\dfrac{x - 9}{42x^2}$

4.3 Adding and Subtracting Rational Expressions

We can define addition and subtraction of rational numbers as follows.

DEFINITION 4.3

If a, b, and c are integers and b is not zero, then

$$\frac{a}{b} + \frac{c}{b} = \frac{a + c}{b}, \quad \text{addition}$$

$$\frac{a}{b} - \frac{c}{b} = \frac{a - c}{b} \quad \text{subtraction}$$

We **can add or subtract rational numbers with a common denominator by adding or subtracting the numerators and placing the result over the common denominator**. The following examples illustrate Definition 4.3.

$$\frac{2}{9}+\frac{3}{9}=\frac{2+3}{9}=\frac{5}{9},$$

$$\frac{7}{8}-\frac{3}{8}=\frac{7-3}{8}=\frac{4}{8}=\frac{1}{2},$$ Don't forget to reduce!

$$\frac{4}{6}+\frac{-5}{6}=\frac{4+(-5)}{6}=\frac{-1}{6}=-\frac{1}{6},$$

$$\frac{7}{10}+\frac{4}{-10}=\frac{7}{10}+\frac{-4}{10}=\frac{7+(-4)}{10}=\frac{3}{10}$$

We use this same *common denominator* approach when adding or subtracting rational expressions, as in these next examples.

$$\frac{3}{x}+\frac{9}{x}=\frac{3+9}{x}=\frac{12}{x},$$

$$\frac{8}{x-2}-\frac{3}{x-2}=\frac{8-3}{x-2}=\frac{5}{x-2},$$

$$\frac{9}{4y}+\frac{5}{4y}=\frac{9+5}{4y}=\frac{14}{4y}=\frac{7}{2y},$$ Don't forget to simplify the final answer!

$$\frac{n^2}{n-1}-\frac{1}{n-1}=\frac{n^2-1}{n-1}=\frac{(n+1)\cancel{(n-1)}}{\cancel{n-1}}=n+1,$$

$$\frac{6a^2}{2a+1}+\frac{13a+5}{2a+1}=\frac{6a^2+13a+5}{2a+1}=\frac{\cancel{(2a+1)}(3a+5)}{\cancel{2a+1}}=3a+5$$

Technically, in each of the previous examples that involve rational expressions, we should restrict the variables to exclude division by zero. For example, $\frac{3}{x}+\frac{9}{x}=\frac{12}{x}$ is true for all real number values for x, *except* $x=0$. Likewise, $\frac{8}{x-2}-\frac{3}{x-2}=\frac{5}{x-2}$ as long as x does not equal 2. Rather than taking the time and space to write down restrictions for each problem we will merely assume that such restrictions exist.

If rational numbers that do not have a common denominator are to be added or subtracted, then we apply the fundamental principle of fractions $\left(\frac{a}{b}=\frac{ak}{bk}\right)$ to obtain equivalent fractions with a common denominator. Equivalent fractions are fractions such as $\frac{1}{2}$ and $\frac{2}{4}$ that name the same number. Consider the following example.

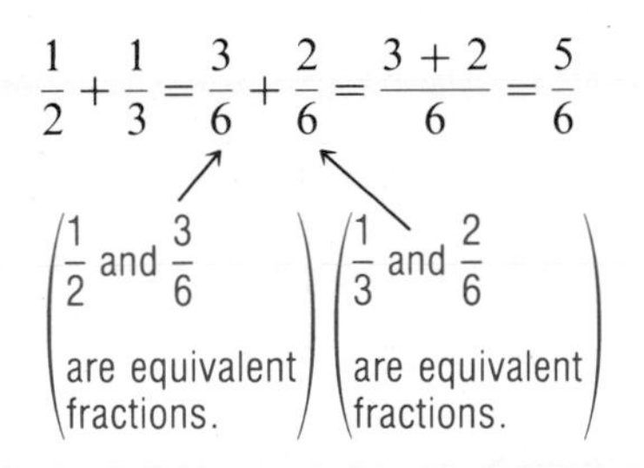

Notice that we chose 6 as our common denominator and 6 is the **least common multiple** of the original denominators 2 and 3. (The least common multiple of a set of whole numbers is the smallest nonzero whole number divisible by each of the numbers.) In general, we use the least common multiple of the denominators of the fractions to be added or subtracted as a **least common denominator** (LCD).

A least common denominator may be found by inspection or by using the prime factored forms of the numbers. Let's consider some examples using each of these techniques.

EXAMPLE 1 Subtract $\frac{5}{6} - \frac{3}{8}$.

Solution By inspection we can see that the LCD is 24. Thus, both fractions can be changed to equivalent fractions, each with a denominator of 24.

$$\frac{5}{6} - \frac{3}{8} = \left(\frac{5}{6}\right)\underbrace{\left(\frac{4}{4}\right)}_{\text{form of 1}} - \left(\frac{3}{8}\right)\underbrace{\left(\frac{3}{3}\right)}_{\text{form of 1}} = \frac{20}{24} - \frac{9}{24} = \frac{11}{24}$$ ■

In Example 1, notice that the fundamental principle of fractions, $\frac{a}{b} = \frac{a \cdot k}{b \cdot k}$, can be written as $\frac{a}{b} = \left(\frac{a}{b}\right)\left(\frac{k}{k}\right)$. This latter form emphasizes the fact that one is the multiplication identity element.

EXAMPLE 2 Perform the indicated operations $\frac{3}{5} + \frac{1}{6} - \frac{13}{15}$.

Solution Again by inspection we can determine that the LCD is 30. Thus, we can proceed as follows.

$$\frac{3}{5}+\frac{1}{6}-\frac{13}{15}=\left(\frac{3}{5}\right)\left(\frac{6}{6}\right)+\left(\frac{1}{6}\right)\left(\frac{5}{5}\right)-\left(\frac{13}{15}\right)\left(\frac{2}{2}\right)$$

$$=\frac{18}{30}+\frac{5}{30}-\frac{26}{30}=\frac{18+5-26}{30}$$

$$=\frac{-3}{30}=-\frac{1}{10}.$$ Don't forget to reduce! ■

EXAMPLE 3 Add $\frac{7}{18}+\frac{11}{24}$.

Solution Let's use the prime factored forms of the denominators to help find the LCD.

$$18 = 2 \cdot 3 \cdot 3, \qquad 24 = 2 \cdot 2 \cdot 2 \cdot 3$$

The LCD must contain three factors of 2 since 24 contains three 2s. The LCD must also contain two factors of 3 since 18 has two 3s. Thus, the

$$\text{LCD} = 2 \cdot 2 \cdot 2 \cdot 3 \cdot 3 = 72.$$

Now we can proceed as usual.

$$\frac{7}{18}+\frac{11}{24}=\left(\frac{7}{18}\right)\left(\frac{4}{4}\right)+\left(\frac{11}{24}\right)\left(\frac{3}{3}\right)=\frac{28}{72}+\frac{33}{72}=\frac{61}{72}$$ ■

Adding and subtracting rational expressions with different denominators follows the same basic routine as adding or subtracting rational numbers with different denominators. Study the following examples carefully and notice the similarity to our previous work with rational numbers.

EXAMPLE 4 Add $\frac{x+2}{4}+\frac{3x+1}{3}$.

Solution By inspection we see that the LCD is 12.

$$\frac{x+2}{4}+\frac{3x+1}{3}=\left(\frac{x+2}{4}\right)\left(\frac{3}{3}\right)+\left(\frac{3x+1}{3}\right)\left(\frac{4}{4}\right)$$

$$=\frac{3(x+2)}{12}+\frac{4(3x+1)}{12}$$

$$=\frac{3x+6+12x+4}{12}$$

$$=\frac{15x+10}{12}$$ ■

Notice the final result in Example 4. The numerator, $15x + 10$, could be factored as $5(3x + 2)$. However, since this produces no common factors with the denominator, the fraction cannot be simplified. Thus, the final answer can be left as $\frac{15x + 10}{12}$, or it would also be acceptable to express it as $\frac{5(3x + 2)}{12}$.

EXAMPLE 5 Subtract $\frac{a - 2}{2} - \frac{a - 6}{6}$.

Solution By inspection we see that the LCD is 6.

$$\frac{a-2}{2} - \frac{a-6}{6} = \left(\frac{a-2}{2}\right)\left(\frac{3}{3}\right) - \frac{a-6}{6}$$

$$= \frac{3(a-2) - (a-6)}{6}$$

Be careful with this sign as you move to the next step!

$$= \frac{3a - 6 - a + 6}{6}$$

$$= \frac{2a}{6} = \frac{a}{3}$$

Don't forget to simplify. ■

EXAMPLE 6 Perform the indicated operations $\frac{x + 3}{10} + \frac{2x + 1}{15} - \frac{x - 2}{18}$.

Solution If you cannot determine the LCD by inspection, then use the prime factored forms of the denominators.

$$10 = 2 \cdot 5, \qquad 15 = 3 \cdot 5, \qquad 18 = 2 \cdot 3 \cdot 3$$

The LCD must contain one factor of 2, two factors of 3, and one factor of 5. Thus, the LCD is $2 \cdot 3 \cdot 3 \cdot 5 = 90$.

$$\frac{x+3}{10} + \frac{2x+1}{15} - \frac{x-2}{18} = \left(\frac{x+3}{10}\right)\left(\frac{9}{9}\right) + \left(\frac{2x+1}{15}\right)\left(\frac{6}{6}\right) - \left(\frac{x-2}{18}\right)\left(\frac{5}{5}\right)$$

$$= \frac{9(x+3) + 6(2x+1) - 5(x-2)}{90}$$

$$= \frac{9x + 27 + 12x + 6 - 5x + 10}{90}$$

$$= \frac{16x + 43}{90}$$

■

A denominator that contains variables does not create any serious difficulties; our approach remains basically the same.

EXAMPLE 7 Add $\frac{3}{2x} + \frac{5}{3y}$.

Solution Using an LCD of $6xy$ we can proceed as follows.

$$\frac{3}{2x} + \frac{5}{3y} = \left(\frac{3}{2x}\right)\left(\frac{3y}{3y}\right) + \left(\frac{5}{3y}\right)\left(\frac{2x}{2x}\right)$$

$$= \frac{9y}{6xy} + \frac{10x}{6xy}$$

$$= \frac{9y + 10x}{6xy}$$

■

EXAMPLE 8 Subtract $\frac{7}{12ab} - \frac{11}{15a^2}$.

Solution We can prime factor the numerical coefficients of the denominators to help find the LCD.

$$\left.\begin{aligned} 12ab &= 2 \cdot 2 \cdot 3 \cdot a \cdot b \\ 15a^2 &= 3 \cdot 5 \cdot a^2 \end{aligned}\right\} \longrightarrow \begin{aligned} \text{LCD} &= 2 \cdot 2 \cdot 3 \cdot 5 \cdot a^2 \cdot b \\ &= 60a^2b \end{aligned}$$

$$\frac{7}{12ab} - \frac{11}{15a^2} = \left(\frac{7}{12ab}\right)\left(\frac{5a}{5a}\right) - \left(\frac{11}{15a^2}\right)\left(\frac{4b}{4b}\right)$$

$$= \frac{35a}{60a^2b} - \frac{44b}{60a^2b}$$

$$= \frac{35a - 44b}{60a^2b}$$

■

EXAMPLE 9 Add $\frac{x}{x-3} + \frac{4}{x}$.

Solution By inspection the LCD is $x(x-3)$.

$$\frac{x}{x-3} + \frac{4}{x} = \left(\frac{x}{x-3}\right)\left(\frac{x}{x}\right) + \left(\frac{4}{x}\right)\left(\frac{x-3}{x-3}\right)$$

$$= \frac{x^2}{x(x-3)} + \frac{4(x-3)}{x(x-3)}$$

$$= \frac{x^2 + 4x - 12}{x(x-3)} \quad \text{or} \quad \frac{(x+6)(x-2)}{x(x-3)}$$

■

EXAMPLE 10 Subtract $\frac{2x}{x+1} - 3$.

Solution

$$\frac{2x}{x+1} - 3 = \frac{2x}{x+1} - 3\left(\frac{x+1}{x+1}\right)$$
$$= \frac{2x}{x+1} - \frac{3(x+1)}{x+1}$$
$$= \frac{2x - 3x - 3}{x+1}$$
$$= \frac{-x-3}{x+1}$$

■

Problem Set 4.3

Perform the following indicated operations involving rational numbers. Be sure to express answers in reduced form.

1. $\frac{1}{4}+\frac{5}{6}$ $\frac{13}{12}$
2. $\frac{3}{5}+\frac{1}{6}$ $\frac{23}{30}$
3. $\frac{7}{8}-\frac{3}{5}$ $\frac{11}{40}$
4. $\frac{7}{9}-\frac{1}{6}$ $\frac{11}{18}$
5. $\frac{6}{5}+\frac{1}{-4}$ $\frac{19}{20}$
6. $\frac{7}{8}+\frac{5}{-12}$ $\frac{11}{24}$
7. $\frac{8}{15}+\frac{3}{25}$ $\frac{49}{75}$
8. $\frac{5}{9}-\frac{11}{12}$ $-\frac{13}{36}$
9. $\frac{1}{5}+\frac{5}{6}-\frac{7}{15}$ $\frac{17}{30}$
10. $\frac{2}{3}-\frac{7}{8}+\frac{1}{4}$ $\frac{1}{24}$
11. $\frac{1}{3}-\frac{1}{4}-\frac{3}{14}$ $-\frac{11}{84}$
12. $\frac{5}{6}-\frac{7}{9}-\frac{3}{10}$ $-\frac{11}{45}$

Add or subtract the following rational expressions as indicated. Be sure to express answers in simplest form.

13. $\frac{2x}{x-1}+\frac{4}{x-1}$ $\frac{2x+4}{x-1}$
14. $\frac{3x}{2x+1}-\frac{5}{2x+1}$ $\frac{3x-5}{2x+1}$
15. $\frac{4a}{a+2}+\frac{8}{a+2}$ 4
16. $\frac{6a}{a-3}-\frac{18}{a-3}$ 6
17. $\frac{3(y-2)}{7y}+\frac{4(y-1)}{7y}$ $\frac{7y-10}{7y}$
18. $\frac{2x-1}{4x^2}+\frac{3(x-2)}{4x^2}$ $\frac{5x-7}{4x^2}$
19. $\frac{x-1}{2}+\frac{x+3}{3}$ $\frac{5x+3}{6}$
20. $\frac{x-2}{4}+\frac{x+6}{5}$ $\frac{9x+14}{20}$
21. $\frac{2a-1}{4}+\frac{3a+2}{6}$ $\frac{12a+1}{12}$
22. $\frac{a-4}{6}+\frac{4a-1}{8}$ $\frac{16a-19}{24}$
23. $\frac{n+2}{6}-\frac{n-4}{9}$ $\frac{n+14}{18}$
24. $\frac{2n+1}{9}-\frac{n+3}{12}$ $\frac{5n-5}{36}$

25. $\dfrac{3x-1}{3} - \dfrac{5x+2}{5} - \dfrac{11}{15}$

27. $\dfrac{x-2}{5} - \dfrac{x+3}{6} + \dfrac{x+1}{15} \ \dfrac{3x-25}{30}$

29. $\dfrac{3}{8x} + \dfrac{7}{10x} \ \dfrac{43}{40x}$

31. $\dfrac{5}{7x} - \dfrac{11}{4y} \ \dfrac{20y-77x}{28xy}$

33. $\dfrac{4}{3x} + \dfrac{5}{4y} - 1 \ \dfrac{16y+15x-12xy}{12xy}$

35. $\dfrac{7}{10x^2} + \dfrac{11}{15x} \ \dfrac{21+22x}{30x^2}$

37. $\dfrac{10}{7n} - \dfrac{12}{4n^2} \ \dfrac{10n-21}{7n^2}$

39. $\dfrac{3}{n^2} - \dfrac{2}{5n} + \dfrac{4}{3} \ \dfrac{45-6n+20n^2}{15n^2}$

41. $\dfrac{3}{x} - \dfrac{5}{3x^2} - \dfrac{7}{6x} \ \dfrac{11x-10}{6x^2}$

43. $\dfrac{6}{5t^2} - \dfrac{4}{7t^3} + \dfrac{9}{5t^3} \ \dfrac{42t+43}{35t^3}$

45. $\dfrac{5b}{24a^2} - \dfrac{11a}{32b} \ \dfrac{20b^2-33a^3}{96a^2b}$

47. $\dfrac{7}{9xy^3} - \dfrac{4}{3x} + \dfrac{5}{2y^2} \ \dfrac{14-24y^3+45xy}{18xy^3}$

49. $\dfrac{2x}{x-1} + \dfrac{3}{x} \ \dfrac{2x^2+3x-3}{x(x-1)}$

51. $\dfrac{a-2}{a} - \dfrac{3}{a+4} \ \dfrac{a^2-a-8}{a(a+4)}$

53. $\dfrac{-3}{4n+5} - \dfrac{8}{3n+5} \ \dfrac{-41n-55}{(4n+5)(3n+5)}$

55. $\dfrac{-1}{x+4} + \dfrac{4}{7x-1} \ \dfrac{-3x+17}{(x+4)(7x-1)}$

57. $\dfrac{7}{3x-5} - \dfrac{5}{2x+7} \ \dfrac{-x+74}{(3x-5)(2x+7)}$

59. $\dfrac{5}{3x-2} + \dfrac{6}{4x+5} \ \dfrac{38x+13}{(3x-2)(4x+5)}$

61. $\dfrac{3x}{2x+5} + 1 \ \dfrac{5x+5}{2x+5}$

63. $\dfrac{4x}{x-5} - 3 \ \dfrac{x+15}{x-5}$

26. $\dfrac{4x-3}{6} - \dfrac{8x-2}{12} - \dfrac{1}{3}$

28. $\dfrac{x+1}{4} + \dfrac{x-3}{6} - \dfrac{x-2}{8} \ \dfrac{7x}{24}$

30. $\dfrac{5}{6x} - \dfrac{3}{10x} \ \dfrac{8}{15x}$

32. $\dfrac{5}{12x} - \dfrac{9}{8y} \ \dfrac{10y-27x}{24xy}$

34. $\dfrac{7}{3x} - \dfrac{8}{7y} - 2 \ \dfrac{49y-24x-42xy}{21xy}$

36. $\dfrac{7}{12a^2} - \dfrac{5}{16a} \ \dfrac{28-15a}{48a^2}$

38. $\dfrac{6}{8n^2} - \dfrac{3}{5n} \ \dfrac{3(5-4n)}{20n^2}$

40. $\dfrac{1}{n^2} + \dfrac{3}{4n} - \dfrac{5}{6} \ \dfrac{12+9n-10n^2}{12n^2}$

42. $\dfrac{7}{3x^2} - \dfrac{9}{4x} - \dfrac{5}{2x} \ \dfrac{28-57x}{12x^2}$

44. $\dfrac{5}{7t} + \dfrac{3}{4t^2} + \dfrac{1}{14t} \ \dfrac{22t+21}{28t^2}$

46. $\dfrac{9}{14x^2y} - \dfrac{4x}{7y^2} \ \dfrac{9y-8x^3}{14x^2y^2}$

48. $\dfrac{7}{16a^2b} + \dfrac{3a}{20b^2} \ \dfrac{35b+12a^3}{80a^2b^2}$

50. $\dfrac{3x}{x-4} - \dfrac{2}{x} \ \dfrac{3x^2-2x+8}{x(x-4)}$

52. $\dfrac{a+1}{a} - \dfrac{2}{a+1} \ \dfrac{a^2+1}{a(a+1)}$

54. $\dfrac{-2}{n-6} - \dfrac{6}{2n+3} \ \dfrac{-10n+30}{(n-6)(2n+3)}$

56. $\dfrac{-3}{4x+3} + \dfrac{5}{2x-5} \ \dfrac{14x+30}{(4x+3)(2x-5)}$

58. $\dfrac{5}{x-1} - \dfrac{3}{2x-3} \ \dfrac{7x-12}{(x-1)(2x-3)}$

60. $\dfrac{3}{2x+1} + \dfrac{2}{3x+4} \ \dfrac{13x+14}{(2x+1)(3x+14)}$

62. $2 + \dfrac{4x}{3x-1} \ \dfrac{10x-2}{3x-1}$

64. $\dfrac{7x}{x+4} - 2 \ \dfrac{5x-8}{x+4}$

65. $-1 - \frac{3}{2x+1} \quad \frac{-2x-4}{2x+1}$

66. $-2 - \frac{5}{4x-3} \quad \frac{-8x+1}{4x-3}$

67. Recall that the indicated quotient of a polynomial and its opposite is -1. For example, $\frac{x-2}{2-x}$ simplifies to -1. Keep this idea in mind as you add or subtract the following rational expressions.

(a) $\frac{1}{x-1} - \frac{x}{x-1} \quad -1$

(b) $\frac{3}{2x-3} - \frac{2x}{2x-3} \quad -1$

(c) $\frac{4}{x-4} - \frac{x}{x-4} + 1 \quad 0$

(d) $-1 + \frac{2}{x-2} - \frac{x}{x-2} \quad -2$

68. Consider the addition problem $\frac{8}{x-2} + \frac{5}{2-x}$. Notice that the denominators are opposites of each other. If the property $\frac{a}{-b} = -\frac{a}{b}$ is applied to the second fraction, we have $\frac{5}{2-x} = -\frac{5}{x-2}$. Thus, we proceed as follows.

$$\frac{8}{x-2} + \frac{5}{2-x} = \frac{8}{x-2} - \frac{5}{x-2} = \frac{8-5}{x-2} = \frac{3}{x-2}$$

Use this approach to do the following problems.

(a) $\frac{7}{x-1} + \frac{2}{1-x} \quad \frac{5}{x-1}$

(b) $\frac{5}{2x-1} + \frac{8}{1-2x} \quad -\frac{3}{2x-1}$

(c) $\frac{4}{a-3} - \frac{1}{3-a} \quad \frac{5}{a-3}$

(d) $\frac{10}{a-9} - \frac{5}{9-a} \quad \frac{15}{a-9}$

(e) $\frac{x^2}{x-1} - \frac{2x-3}{1-x} \quad x+3$

(f) $\frac{x^2}{x-4} - \frac{3x-28}{4-x} \quad x+7$

4.4 More on Rational Expressions and Complex Fractions

In this section we shall expand our work with adding and subtracting rational expressions and we will also discuss the process of simplifying complex fractions. Before we begin, however, this seems like an appropriate time to offer a bit of advice regarding your study of algebra. Success in algebra depends upon having a good understanding of the concepts as well as the ability to perform the various computations. As for the computational work, you should adopt a carefully organized format that shows as many steps as *you need* in order to minimize the chances of making careless errors. Don't be eager to find shortcuts for certain computations before you have a thorough understanding of the steps involved in the process. This is especially appropriate advice at the beginning of this section.

Study the following examples very carefully. Notice the same basic procedure for each problem of: (1) finding the LCD, (2) changing each fraction to an equivalent fraction that has the LCD as its denominator, (3) adding or subtracting numerators and placing this result over the LCD, and (4) looking for possibilities to simplify the resulting fraction.

EXAMPLE 1 Add $\frac{8}{x^2 - 4x} + \frac{2}{x}$.

Solution

$$\left.\begin{aligned} x^2 - 4x &= x(x-4) \\ x &= x \end{aligned}\right\} \longrightarrow \text{LCD is } x(x-4)$$

$$\begin{aligned}
\frac{8}{x^2 - 4x} + \frac{2}{x} &= \frac{8}{x(x-4)} + \frac{2}{x} \\
&= \frac{8}{x(x-4)} + \left(\frac{2}{x}\right)\left(\frac{x-4}{x-4}\right) \\
&= \frac{8}{x(x-4)} + \frac{2(x-4)}{x(x-4)} \\
&= \frac{8 + 2x - 8}{x(x-4)} \\
&= \frac{2x}{x(x-4)} = \frac{2}{x-4}
\end{aligned}$$

■

EXAMPLE 2 Subtract $\frac{a}{a^2 - 4} - \frac{3}{a+2}$.

Solution

$$\left.\begin{aligned} a^2 - 4 &= (a+2)(a-2) \\ a + 2 &= a + 2 \end{aligned}\right\} \longrightarrow \text{LCD is } (a+2)(a-2)$$

$$\begin{aligned}
\frac{a}{a^2 - 4} - \frac{3}{a+2} &= \frac{a}{(a+2)(a-2)} - \frac{3}{a+2} \\
&= \frac{a}{(a+2)(a-2)} - \left(\frac{3}{a+2}\right)\left(\frac{a-2}{a-2}\right) \\
&= \frac{a}{(a+2)(a-2)} - \frac{3(a-2)}{(a+2)(a-2)} \\
&= \frac{a - 3a + 6}{(a+2)(a-2)} \\
&= \frac{-2a + 6}{(a+2)(a-2)} \quad \text{or} \quad \frac{-2(a-3)}{(a+2)(a-2)}
\end{aligned}$$

■

EXAMPLE 3 Add $\dfrac{3n}{n^2 + 6n + 5} + \dfrac{4}{n^2 - 7n - 8}$.

Solution

$$\left.\begin{aligned} n^2 + 6n + 5 &= (n + 5)(n + 1) \\ n^2 - 7n - 8 &= (n - 8)(n + 1) \end{aligned}\right\} \longrightarrow \text{LCD is } (n+1)(n+5)(n-8)$$

$$\begin{aligned}
\frac{3n}{n^2 + 6n + 5} + \frac{4}{n^2 - 7n - 8} &= \frac{3n}{(n + 5)(n + 1)} + \frac{4}{(n - 8)(n + 1)} \\
&= \left(\frac{3n}{(n + 5)(n + 1)}\right)\left(\frac{n - 8}{n - 8}\right) \\
&\quad + \left(\frac{4}{(n - 8)(n + 1)}\right)\left(\frac{n + 5}{n + 5}\right) \\
&= \frac{3n(n - 8)}{(n + 5)(n + 1)(n - 8)} + \frac{4(n + 5)}{(n + 5)(n + 1)(n - 8)} \\
&= \frac{3n^2 - 24n + 4n + 20}{(n + 5)(n + 1)(n - 8)} \\
&= \frac{3n^2 - 20n + 20}{(n + 5)(n + 1)(n - 8)}
\end{aligned}$$

■

EXAMPLE 4 Perform the indicated operations.

$$\frac{2x^2}{x^4 - 1} + \frac{x}{x^2 - 1} - \frac{1}{x - 1}.$$

Solution

$$\left.\begin{aligned} x^4 - 1 &= (x^2 + 1)(x + 1)(x - 1) \\ x^2 - 1 &= (x + 1)(x - 1) \\ x - 1 &= x - 1 \end{aligned}\right\} \longrightarrow \text{LCD is } (x^2 + 1)(x + 1)(x - 1)$$

$$\begin{aligned}
\frac{2x^2}{x^4 - 1} + \frac{x}{x^2 - 1} - \frac{1}{x - 1} &= \frac{2x^2}{(x^2 + 1)(x + 1)(x - 1)} + \frac{x}{(x + 1)(x - 1)} - \frac{1}{x - 1} \\
&= \frac{2x^2}{(x^2 + 1)(x + 1)(x - 1)} + \left(\frac{x}{(x + 1)(x - 1)}\right)\left(\frac{x^2 + 1}{x^2 + 1}\right) \\
&\quad - \left(\frac{1}{x - 1}\right)\left(\frac{(x^2 + 1)(x + 1)}{(x^2 + 1)(x + 1)}\right) \\
&= \frac{2x^2}{(x^2 + 1)(x + 1)(x - 1)} + \frac{x(x^2 + 1)}{(x^2 + 1)(x + 1)(x - 1)} \\
&\quad - \frac{(x^2 + 1)(x + 1)}{(x^2 + 1)(x + 1)(x - 1)}
\end{aligned}$$

$$= \frac{2x^2 + x^3 + x - x^3 - x^2 - x - 1}{(x^2 + 1)(x + 1)(x - 1)}$$

$$= \frac{x^2 - 1}{(x^2 + 1)(x + 1)(x - 1)}$$

$$= \frac{\cancel{(x + 1)}\cancel{(x - 1)}}{(x^2 + 1)\cancel{(x + 1)}\cancel{(x - 1)}}$$

$$= \frac{1}{x^2 + 1}$$ ■

Complex Fractions

Fractional forms that contain rational numbers or rational expressions in the numerators and/or denominators are called **complex fractions**. The following are examples of complex fractions.

$$\frac{\frac{3}{5}}{\frac{7}{8}}, \quad \frac{\frac{4}{x}}{\frac{2}{xy}}, \quad \frac{\frac{1}{2} + \frac{3}{4}}{\frac{5}{6} - \frac{3}{8}}, \quad \frac{\frac{3}{x} + \frac{2}{y}}{\frac{5}{x} - \frac{6}{y^2}}$$

It is often necessary to *simplify* a complex fraction. Let's examine some techniques for simplifying complex fractions with the four previous examples.

EXAMPLE 5 Simplify $\dfrac{\frac{3}{5}}{\frac{7}{8}}$.

Solution This type of problem creates nothing new since it is merely a division problem. Thus,

$$\frac{\frac{3}{5}}{\frac{7}{8}} = \frac{3}{5} \div \frac{7}{8} = \frac{3}{5} \cdot \frac{8}{7} = \frac{24}{35}.$$ ■

EXAMPLE 6 Simplify $\dfrac{\frac{4}{x}}{\frac{2}{xy}}$.

Solution

$$\frac{\frac{4}{x}}{\frac{2}{xy}} = \frac{4}{x} \div \frac{2}{xy} = \frac{\overset{2}{\cancel{4}}}{\cancel{x}} \cdot \frac{\cancel{x}y}{\cancel{2}} = 2y$$

■

EXAMPLE 7 Simplify $\dfrac{\frac{1}{2}+\frac{3}{4}}{\frac{5}{6}-\frac{3}{8}}$.

Let's look at two possible "attacks" for such a problem.

Solution A

$$\frac{\frac{1}{2}+\frac{3}{4}}{\frac{5}{6}-\frac{3}{8}} = \frac{\frac{2}{4}+\frac{3}{4}}{\frac{20}{24}-\frac{9}{24}} = \frac{\frac{5}{4}}{\frac{11}{24}} = \frac{5}{\cancel{4}} \cdot \frac{\overset{6}{\cancel{24}}}{11} = \frac{30}{11}$$

Solution B The LCD of all four denominators (2, 4, 6, and 8) is 24. Multiply the entire complex fraction by a form of 1, namely, $\frac{24}{24}$.

$$\frac{\frac{1}{2}-\frac{3}{4}}{\frac{5}{6}-\frac{3}{8}} = \left(\frac{24}{24}\right)\frac{\frac{1}{2}+\frac{3}{4}}{\frac{5}{6}-\frac{3}{8}}$$

$$= \frac{24\left(\frac{1}{2}+\frac{3}{4}\right)}{24\left(\frac{5}{6}-\frac{3}{8}\right)}$$

$$= \frac{24\left(\frac{1}{2}\right)+24\left(\frac{3}{4}\right)}{24\left(\frac{5}{6}\right)-24\left(\frac{3}{8}\right)}$$

$$= \frac{12+18}{20-9} = \frac{30}{11}$$

■

EXAMPLE 8 Simplify $\dfrac{\frac{3}{x}+\frac{2}{y}}{\frac{5}{x}-\frac{6}{y^2}}$.

Solution A

$$\frac{\frac{3}{x}+\frac{2}{y}}{\frac{5}{x}-\frac{6}{y^2}}=\frac{\left(\frac{3}{x}\right)\left(\frac{y}{y}\right)+\left(\frac{2}{y}\right)\left(\frac{x}{x}\right)}{\left(\frac{5}{x}\right)\left(\frac{y^2}{y^2}\right)-\left(\frac{6}{y^2}\right)\left(\frac{x}{x}\right)}$$

$$=\frac{\frac{3y}{xy}+\frac{2x}{xy}}{\frac{5y^2}{xy^2}-\frac{6x}{xy^2}}$$

$$=\frac{\frac{3y+2x}{xy}}{\frac{5y^2-6x}{xy^2}}$$

$$=\frac{3y+2x}{xy}\div\frac{5y^2-6x}{xy^2}$$

$$=\frac{3y+2x}{\cancel{xy}}\cdot\frac{\overset{y}{\cancel{xy^2}}}{5y^2-6x}$$

$$=\frac{y(3y+2x)}{5y^2-6x}$$

Solution B The LCD of all four denominators ($x, y, x,$ and y^2) is xy^2. Multiplify the entire complex fraction by a form of 1, namely, $\frac{xy^2}{xy^2}$.

$$\frac{\frac{3}{x}+\frac{2}{y}}{\frac{5}{x}-\frac{6}{y^2}}=\left(\frac{xy^2}{xy^2}\right)\frac{\frac{3}{x}+\frac{2}{y}}{\frac{5}{x}-\frac{6}{y^2}}$$

$$=\frac{xy^2\left(\frac{3}{x}+\frac{2}{y}\right)}{xy^2\left(\frac{5}{x}-\frac{6}{y^2}\right)}$$

$$=\frac{xy^2\left(\frac{3}{x}\right)+xy^2\left(\frac{2}{y}\right)}{xy^2\left(\frac{5}{x}\right)-xy^2\left(\frac{6}{y^2}\right)}$$

$$=\frac{3y^2+2xy}{5y^2-6x} \quad \text{or} \quad \frac{y(3y+2x)}{5y^2-6x}$$

■

Certainly either approach (Solution A or Solution B) will work with problems such as Examples 7 and 8. Examine Solution B in both examples carefully. This approach works very effectively with complex fractions where the LCD of all the denominators is easy to find. (Don't be misled by the length of Solution B for Example 7. We were especially careful to show every step.) Let's conclude this section with two more examples that involve algebraic complex fractions.

EXAMPLE 9 Simplify $\dfrac{\dfrac{1}{a}+\dfrac{1}{a-1}}{\dfrac{1}{a-1}-\dfrac{1}{a}}$.

Solution Multiply the entire complex fraction by a form of 1, namely, $\dfrac{a(a-1)}{a(a-1)}$.

$$\frac{\dfrac{1}{a}+\dfrac{1}{a-1}}{\dfrac{1}{a-1}-\dfrac{1}{a}} = \left(\frac{a(a-1)}{a(a-1)}\right)\frac{\dfrac{1}{a}+\dfrac{1}{a-1}}{\dfrac{1}{a-1}-\dfrac{1}{a}}$$

$$= \frac{a(a-1)\left(\dfrac{1}{a}+\dfrac{1}{a-1}\right)}{a(a-1)\left(\dfrac{1}{a-1}-\dfrac{1}{1}\right)}$$

$$= \frac{a(a-1)\left(\dfrac{1}{a}\right)+a(a-1)\left(\dfrac{1}{a-1}\right)}{a(a-1)\left(\dfrac{1}{a-1}\right)-a(a-1)\left(\dfrac{1}{a}\right)}$$

$$= \frac{a-1+a}{a-(a-1)} = \frac{2a-1}{a-a+1}$$

$$= \frac{2a-1}{1} \quad \text{or} \quad 2a-1$$

■

EXAMPLE 10 Simplify $1-\dfrac{n}{1-\dfrac{1}{n}}$.

Solution First simplify the complex fraction $\dfrac{n}{1-\dfrac{1}{n}}$ by multiplying by $\dfrac{n}{n}$.

$$\left(\frac{n}{1-\frac{1}{n}}\right)\left(\frac{n}{n}\right)=\frac{n^2}{n-1}$$

Now we can perform the subtraction.

$$1-\frac{n^2}{n-1}=\left(\frac{n-1}{n-1}\right)\left(\frac{1}{1}\right)-\frac{n^2}{n-1}$$

$$=\frac{n-1}{n-1}-\frac{n^2}{n-1}$$

$$=\frac{n-1-n^2}{n-1} \quad \text{or} \quad \frac{-n^2+n-1}{n-1}$$

■

Problem Set 4.4

Perform the indicated operations and express your answers in simplest form.

1. $\dfrac{2x}{x^2+4x}+\dfrac{5}{x}$ $\dfrac{7x+20}{x(x+4)}$

2. $\dfrac{3x}{x^2-6x}+\dfrac{4}{x}$ $\dfrac{7x-24}{x(x-6)}$

3. $\dfrac{4}{x^2+7x}-\dfrac{1}{x}$ $\dfrac{-x-3}{x(x+7)}$

4. $\dfrac{-10}{x^2-9x}-\dfrac{2}{x}$ $\dfrac{-2x+8}{x(x-9)}$

5. $\dfrac{x}{x^2-1}+\dfrac{5}{x+1}$ $\dfrac{6x-5}{(x+1)(x-1)}$

6. $\dfrac{2x}{x^2-16}+\dfrac{7}{x-4}$ $\dfrac{9x+28}{(x+4)(x-4)}$

7. $\dfrac{6a+4}{a^2-1}-\dfrac{5}{a-1}$ $\dfrac{1}{a+1}$

8. $\dfrac{4a-4}{a^2-4}-\dfrac{3}{a+2}$ $\dfrac{1}{a-2}$

9. $\dfrac{2n}{n^2-25}-\dfrac{3}{4n+20}$ $\dfrac{5n+15}{4(n+5)(n-5)}$

10. $\dfrac{3n}{n^2-36}-\dfrac{2}{5n+30}$ $\dfrac{13n+12}{5(n+6)(n-6)}$

11. $\dfrac{5}{x}-\dfrac{5x-30}{x^2+6x}+\dfrac{x}{x+6}$ $\dfrac{x^2+60}{x(x+6)}$

12. $\dfrac{3}{x+1}+\dfrac{x+5}{x^2-1}-\dfrac{3}{x-1}$ $\dfrac{1}{x+1}$

13. $\dfrac{3}{x^2+9x+14}+\dfrac{5}{2x^2+15x+7}$ $\dfrac{11x+13}{(x+2)(x+7)(2x+1)}$

14. $\dfrac{6}{x^2+11x+24}+\dfrac{4}{3x^2+13x+12}$ $\dfrac{22x+56}{(x+3)(x+8)(3x+4)}$

15. $\dfrac{1}{a^2-3a-10}-\dfrac{4}{a^2+4a-45}$ $\dfrac{-3a+1}{(a-5)(a+2)(a+9)}$

16. $\dfrac{6}{a^2-3a-54}-\dfrac{10}{a^2+5a-6}$ $\dfrac{-4a+84}{(a+6)(a-9)(a-1)}$

17. $\dfrac{3a}{20a^2-11a-3}+\dfrac{1}{12a^2+7a-12}$ $\dfrac{9a^2+17a+1}{(5a+1)(4a-3)(3a+4)}$

18. $\dfrac{2a}{6a^2+11a-10}+\dfrac{a}{2a^2-3a-20}$ $\dfrac{5a^2-10a}{(2a+5)(3a-2)(a-4)}$

19. $\dfrac{5}{x^2+3} - \dfrac{2}{x^2+4x-21} \quad \dfrac{3x^2+20x-111}{(x^2+3)(x+7)(x-3)}$

20. $\dfrac{7}{x^2+1} - \dfrac{3}{x^2+7x-60} \quad \dfrac{4x^2+49x-423}{(x^2+1)(x+12)(x-5)}$

21. $\dfrac{2}{y^2+6y-16} - \dfrac{4}{y+8} - \dfrac{3}{y-2} \quad \dfrac{-7y-14}{(y+8)(y-2)}$

22. $\dfrac{7}{y-6} - \dfrac{10}{y+12} + \dfrac{4}{y^2+6y-72} \quad \dfrac{-3y+148}{(y-6)(y+12)}$

23. $x - \dfrac{x^2}{x-2} + \dfrac{3}{x^2-4} \quad \dfrac{-2x^2-4x+3}{(x+2)(x-2)}$

24. $x + \dfrac{5}{x^2-25} - \dfrac{x^2}{x+5} \quad \dfrac{5x^2-25x+5}{(x+5)(x-5)}$

25. $\dfrac{x+3}{x+10} + \dfrac{4x-3}{x^2+8x-20} + \dfrac{x-1}{x-2} \quad \dfrac{2x^2+14x-19}{(x+10)(x-2)}$

26. $\dfrac{2x-1}{x+3} + \dfrac{x+4}{x-6} + \dfrac{3x-1}{x^2-3x-18} \quad \dfrac{3x^2-3x+17}{(x-6)(x+3)}$

27. $\dfrac{n}{n-6} + \dfrac{n+3}{n+8} + \dfrac{12n+26}{n^2+2n-48} \quad \dfrac{2n+1}{n-6}$

28. $\dfrac{n-1}{n+4} + \dfrac{n}{n+6} + \dfrac{2n+18}{n^2+10n+24} \quad \dfrac{2n+3}{n+6}$

29. $\dfrac{4x-3}{2x^2+x-1} - \dfrac{2x+7}{3x^2+x-2} - \dfrac{3}{3x-2} \quad \dfrac{2x^2-32x+16}{(x+1)(2x-1)(3x-2)}$

30. $\dfrac{2x+5}{x^2+3x-18} - \dfrac{3x-1}{x^2+4x-12} + \dfrac{5}{x-2} \quad \dfrac{4x^2+26x-103}{(x-3)(x+6)(x-2)}$

31. $\dfrac{n}{n^2+1} + \dfrac{n^2+3n}{n^4-1} - \dfrac{1}{n-1} \quad \dfrac{1}{(n^2+1)(n+1)}$

32. $\dfrac{2n^2}{n^4-16} - \dfrac{n}{n^2-4} + \dfrac{1}{n+2} \quad \dfrac{-8}{(n^2+4)(n+2)(n-2)}$

33. $\dfrac{15x^2-10}{5x^2-7x+2} - \dfrac{3x+4}{x-1} - \dfrac{2}{5x-2} \quad \dfrac{-16x}{(5x-2)(x-1)}$

34. $\dfrac{32x+9}{12x^2+x-6} - \dfrac{3}{4x+3} - \dfrac{x+5}{3x-2} \quad \dfrac{-4x^2}{(4x+3)(3x-2)}$

35. $\dfrac{t+3}{3t-1} + \dfrac{8t^2+8t+2}{3t^2-7t+2} - \dfrac{2t+3}{t-2} \quad \dfrac{t+1}{t-2}$

36. $\dfrac{t-3}{2t+1} + \dfrac{2t^2+19t-46}{2t^2-9t-5} - \dfrac{t+4}{t-5} \quad \dfrac{t+7}{2t+1}$

Simplify each of the following complex fractions.

37. $\dfrac{\dfrac{1}{2}-\dfrac{1}{4}}{\dfrac{5}{8}+\dfrac{3}{4}} \quad \dfrac{2}{11}$

38. $\dfrac{\dfrac{3}{8}+\dfrac{3}{4}}{\dfrac{5}{8}-\dfrac{7}{12}} \quad 27$

39. $\dfrac{\dfrac{3}{28}-\dfrac{5}{14}}{\dfrac{5}{7}+\dfrac{1}{4}} \quad -\dfrac{7}{27}$

40. $\dfrac{\dfrac{5}{9}+\dfrac{7}{36}}{\dfrac{3}{18}-\dfrac{5}{12}}$ $\quad -3$

41. $\dfrac{\dfrac{5}{6y}}{\dfrac{10}{3xy}}$ $\quad \dfrac{x}{4}$

42. $\dfrac{\dfrac{9}{8xy^2}}{\dfrac{5}{4x^2}}$ $\quad \dfrac{9x}{10y^2}$

43. $\dfrac{\dfrac{3}{x}-\dfrac{2}{y}}{\dfrac{4}{y}-\dfrac{7}{xy}}$ $\quad \dfrac{3y-2x}{4x-7}$

44. $\dfrac{\dfrac{9}{x}+\dfrac{7}{x^2}}{\dfrac{5}{y}+\dfrac{3}{y^2}}$ $\quad \dfrac{9xy^2-7y^2}{5x^2y+3x^2}$

45. $\dfrac{\dfrac{6}{a}-\dfrac{5}{b^2}}{\dfrac{12}{a^2}+\dfrac{2}{b}}$ $\quad \dfrac{6ab^2-5a^2}{12b^2+2a^2b}$

46. $\dfrac{\dfrac{4}{ab}-\dfrac{3}{b^2}}{\dfrac{1}{a}+\dfrac{3}{b}}$ $\quad \dfrac{4b-3a}{b^2+3ab}$

47. $\dfrac{\dfrac{2}{x}-3}{\dfrac{3}{y}+4}$ $\quad \dfrac{2y-3xy}{3x+4xy}$

48. $\dfrac{1+\dfrac{3}{x}}{1-\dfrac{6}{x}}$ $\quad \dfrac{x+3}{x-6}$

49. $\dfrac{3+\dfrac{2}{n+4}}{5-\dfrac{1}{n+4}}$ $\quad \dfrac{3n+14}{5n+19}$

50. $\dfrac{4+\dfrac{6}{n-1}}{7-\dfrac{4}{n-1}}$ $\quad \dfrac{4n+2}{7n-11}$

51. $\dfrac{5-\dfrac{2}{n-3}}{4-\dfrac{1}{n-3}}$ $\quad \dfrac{5n-17}{4n-13}$

52. $\dfrac{\dfrac{3}{n-5}-2}{1-\dfrac{4}{n-5}}$ $\quad \dfrac{-2n+13}{n-9}$

53. $\dfrac{\dfrac{-1}{y-2}+\dfrac{5}{x}}{\dfrac{3}{x}-\dfrac{4}{xy-2x}}$ $\quad \dfrac{-x+5y-10}{3y-10}$

54. $\dfrac{\dfrac{-2}{x}-\dfrac{4}{x+2}}{\dfrac{3}{x^2+2x}+\dfrac{3}{x}}$ $\quad \dfrac{-6x-4}{3x+9}$

55. $\dfrac{\dfrac{2}{x-3}-\dfrac{3}{x+3}}{\dfrac{5}{x^2-9}-\dfrac{2}{x-3}}$ $\quad \dfrac{-x+15}{-2x-1}$

56. $\dfrac{\dfrac{2}{x-y}+\dfrac{3}{x+y}}{\dfrac{5}{x+y}-\dfrac{1}{x^2-y^2}}$ $\quad \dfrac{5x-y}{5x-5y-1}$

57. $\dfrac{3a}{2-\dfrac{1}{a}}-1$ $\quad \dfrac{3a^2-2a+1}{2a-1}$

58. $\dfrac{a}{\dfrac{1}{a}+4}+1$ $\quad \dfrac{a^2+4a+1}{4a+1}$

59. $2-\dfrac{x}{3-\dfrac{2}{x}}$ $\quad \dfrac{-x^2+6x-4}{3x-2}$

60. $1+\dfrac{x}{1+\dfrac{1}{x}}$ $\quad \dfrac{x^2+x+1}{x+1}$

4.5 Dividing Polynomials

In Chapter 3 we saw how the property $\dfrac{b^n}{b^m}=b^{n-m}$ along with our knowledge of dividing integers was used to divide monomials. For example,

$$\frac{12x^3}{3x}=4x^2, \qquad \frac{-36x^4y^5}{4xy^2}=-9x^3y^3$$

In Section 4.3, we used $\frac{a}{b} + \frac{c}{b} = \frac{a+c}{b}$ and $\frac{a}{b} - \frac{c}{b} = \frac{a-c}{b}$ as the basis for adding and subtracting rational expressions. These same equalities, viewed as $\frac{a+b}{c} = \frac{a}{c} + \frac{b}{c}$ and $\frac{a-c}{b} = \frac{a}{b} - \frac{c}{b}$, along with our knowledge of dividing monomials, provide the basis for dividing polynomials by monomials. Consider the following examples.

$$\frac{18x^3 + 24x^2}{6x} = \frac{18x^3}{6x} + \frac{24x^2}{6x} = 3x^2 + 4x,$$

$$\frac{35x^2y^3 - 55x^3y^4}{5xy^2} = \frac{35x^2y^3}{5xy^2} - \frac{55x^3y^4}{5xy^2} = 7xy - 11x^2y^2$$

To divide a polynomial by a monomial we divide each term of the polynomial by the monomial. As with many skills, once you feel comfortable with the process you may then want to perform some of the steps mentally. Your work could take on the following format.

$$\frac{40x^4y^5 + 72x^5y^7}{8x^2y} = 5x^2y^4 + 9x^3y^6, \qquad \frac{36a^3b^4 - 45a^4b^6}{-9a^2b^3} = -4ab + 5a^2b^3$$

In Section 4.1 we saw that a fraction like $\frac{3x^2 + 11x - 4}{x + 4}$ can be simplified as follows.

$$\frac{3x^2 + 11x - 4}{x + 4} = \frac{(3x - 1)\cancel{(x + 4)}}{\cancel{x + 4}} = 3x - 1$$

We can obtain the same result by using a dividing process similar to long division in arithmetic. The process is as follows.

Step 1. Use the conventional long division format and arrange both the dividend and the divisor in descending powers of the variable.

$$x + 4\overline{)3x^2 + 11x - 4}$$

Step 2. Find the first term of the quotient by dividing the first term of the dividend by the first term of the divisor.

$$\begin{array}{r} 3x \\ x + 4\overline{)3x^2 + 11x - 4} \end{array}$$

Step 3. Multiply the entire divisor by the term of the quotient found in Step 2 and position this product to be subtracted from the dividend.

$$\begin{array}{r} 3x \\ x + 4\overline{)3x^2 + 11x - 4} \\ \underline{3x^2 + 12x} \end{array}$$

Step 4. Subtract

$$\begin{array}{r} 3x \\ x + 4\overline{)3x^2 + 11x - 4} \\ \underline{3x^2 + 12x } \\ -x - 4 \end{array}$$

Remember to add the opposite! → $3x^2 + 12x$

$(3x^2 + 11x - 4) - (3x^2 + 12x) = -x - 4$ → $-x - 4$

Step 5. Repeat the process beginning with Step 2; use the polynomial that resulted from the subtraction in Step 4 as a new dividend.

$$\begin{array}{r}3x \quad - 1\\ x+4\overline{)3x^2 + 11x - 4}\\ \underline{3x^2 + 12x}\\ -x - 4\\ \underline{-x - 4}\end{array}$$

In the next example let's *think* in terms of the previous step-by-step procedure but arrange our work in a more compact form.

EXAMPLE 1 Divide $5x^2 + 6x - 8$ by $x + 2$.

Solution

$$\begin{array}{r}5x \quad - 4\\ x+2\overline{)5x^2 + 6x - 8}\\ \underline{5x^2 + 10x}\\ -4x - 8\\ \underline{-4x - 8}\\ 0\end{array}$$

Think Steps

1. $\dfrac{5x^2}{x} = 5x.$

2. $5x(x + 2) = 5x^2 + 10x.$

3. $(5x^2 + 6x - 8) - (5x^2 + 10x) = -4x - 8.$

4. $\dfrac{-4x}{x} = -4.$

5. $-4(x + 2) = -4x - 8.$ ■

Recall that to check a division problem we can multiply the divisor times the quotient and add the remainder. In other words,

$$\text{dividend} = (\text{divisor})(\text{quotient}) + \text{remainder}.$$

Sometimes the remainder is expressed as a fractional part of the divisor. The relationship then becomes

$$\frac{\text{dividend}}{\text{divisor}} = \text{quotient} + \frac{\text{remainder}}{\text{divisor}}.$$

EXAMPLE 2 Divide $2x^2 - 3x + 1$ by $x - 5$.

Solution

$$\begin{array}{r}2x \quad + 7\\ x-5\overline{)2x^2 - 3x + 1}\\ \underline{2x^2 - 10x}\\ 7x + 1\\ \underline{7x - 35}\\ 36 \leftarrow \text{remainder}\end{array}$$

Thus,

$$\frac{2x^2 - 3x + 1}{x - 5} = 2x + 7 + \frac{36}{x - 5}, \qquad x \neq 5.$$

CHECK $(x-5)(2x+7)+36 \stackrel{?}{=} 2x^2-3x+1$

$$2x^2-3x-35+36 \stackrel{?}{=} 2x^2-3x+1$$

$$2x^2-3x+1 = 2x^2-3x+1$$

■

Each of the next two examples illustrates another point regarding the division process. Study them carefully and then you should be ready to work the exercises in the next problem set.

EXAMPLE 3 Divide t^3-8 by $t-2$.

Solution

$$\begin{array}{r} t^2+2t\ \ +4 \\ t-2\overline{)t^3+0t^2+0t-8} \\ \underline{t^3-2t^2} \qquad\qquad \\ 2t^2+0t-8 \\ \underline{2t^2-4t} \quad\ \ \\ 4t-8 \\ \underline{4t-8} \\ 0 \end{array}$$

← Notice the insertion of a "t squared" and a "t-term" with zero coefficients

Check this result! ■

EXAMPLE 4 Divide y^3+3y^2-2y-1 by y^2+2y.

Solution

$$\begin{array}{r} y+1 \qquad\qquad \\ y^2+2y\overline{)y^3+3y^2-2y-1} \\ \underline{y^3+2y^2} \qquad\qquad \\ y^2-2y-1 \\ \underline{y^2+2y} \quad\ \ \\ -4y-1 \end{array}$$

← a remainder of $-4y-1$

(The division process is complete when the degree of the remainder is less than the degree of the divisor.) Thus,

$$\frac{y^3+3y^2-2y-1}{y^2+2y} = y+1+\frac{-4y-1}{y^2+2y}.$$

■

REMARK If the divisor is of the form $x-k$, where the coefficient of the x-term is one, then the format of the division process described in this section can be simplified by a procedure called **synthetic division**. This procedure is outlined later in the text.

Problem Set 4.5

Perform the following divisions of polynomials by monomials.

1. $\dfrac{9x^4 + 18x^3}{3x}$ $3x^3 + 6x^2$

2. $\dfrac{12x^3 - 24x^2}{6x^2}$ $2x - 4$

3. $\dfrac{-24x^6 + 36x^8}{4x^2}$ $-6x^4 + 9x^6$

4. $\dfrac{-35x^5 - 42x^3}{-7x^2}$ $5x^3 + 6x$

5. $\dfrac{15a^3 - 25a^2 - 40a}{5a}$ $3a^2 - 5a - 8$

6. $\dfrac{-16a^4 + 32a^3 - 56a^2}{-8a}$ $2a^3 - 4a^2 + 7a$

7. $\dfrac{13x^3 - 17x^2 + 28x}{-x}$ $-13x^2 + 17x - 28$

8. $\dfrac{14xy - 16x^2y^2 - 20x^3y^4}{-xy}$ $-14 + 16xy + 20x^2y^3$

9. $\dfrac{-18x^2y^2 + 24x^3y^2 - 48x^2y^3}{6xy}$ $-3xy + 4x^2y - 8xy^2$

10. $\dfrac{-27a^3b^4 - 36a^2b^3 + 72a^2b^5}{9a^2b^2}$ $-3ab^2 - 4b + 8b^3$

Perform the following divisions.

11. $\dfrac{x^2 - 7x - 78}{x + 6}$ $x - 13$

12. $\dfrac{x^2 + 11x - 60}{x - 4}$ $x + 15$

13. $(x^2 + 12x - 160) \div (x - 8)$ $x + 20$

14. $(x^2 - 18x - 175) \div (x + 7)$ $x - 25$

15. $\dfrac{2x^2 - x - 4}{x - 1}$ $2x + 1 - \dfrac{3}{x - 1}$

16. $\dfrac{3x^2 - 2x - 7}{x + 2}$ $3x - 8 + \dfrac{9}{x + 2}$

17. $\dfrac{15x^2 + 22x - 5}{3x + 5}$ $5x - 1$

18. $\dfrac{12x^2 - 32x - 35}{2x - 7}$ $6x + 5$

19. $\dfrac{3x^3 + 7x^2 - 13x - 21}{x + 3}$ $3x^2 - 2x - 7$

20. $\dfrac{4x^3 - 21x^2 + 3x + 10}{x - 5}$ $4x^2 - x - 2$

21. $(2x^3 + 9x^2 - 17x + 6) \div (2x - 1)$ $x^2 + 5x - 6$

22. $(3x^3 - 5x^2 - 23x - 7) \div (3x + 1)$ $x^2 - 2x - 7$

23. $(4x^3 - x^2 - 2x + 6) \div (x - 2)$ $4x^2 + 7x + 12 + \dfrac{30}{x - 2}$

24. $(6x^3 - 2x^2 + 4x - 3) \div (x + 1)$ $6x^2 - 8x + 12 - \dfrac{15}{x + 1}$

25. $(x^4 - 10x^3 + 19x^2 + 33x - 18) \div (x - 6)$ $x^3 - 4x^2 - 5x + 3$

26. $(x^4 + 2x^3 - 16x^2 + x + 6) \div (x - 3)$ $x^3 + 5x^2 - x - 2$

27. $\dfrac{x^3 - 125}{x - 5}$ $x^2 + 5x + 25$

28. $\dfrac{x^3 + 64}{x + 4}$ $x^2 - 4x + 16$

29. $(x^3 + 64) \div (x + 1)$ $x^2 - x + 1 + \dfrac{63}{x + 1}$

30. $(x^3 - 8) \div (x - 4)$ $x^2 + 4x + 16 + \dfrac{56}{x - 4}$

31. $(2x^3 - x - 6) \div (x + 2)$ $2x^2 - 4x + 7 - \frac{20}{x+2}$

32. $(5x^3 + 2x - 3) \div (x - 2)$ $5x^2 + 10x + 22 + \frac{41}{x-2}$

33. $\frac{4a^2 - 8ab + 4b^2}{a - b}$ $4a - 4b$

34. $\frac{3x^2 - 2xy - 8y^2}{x - 2y}$ $3x + 4y$

35. $\frac{4x^3 - 5x^2 + 2x - 6}{x^2 - 3x}$ $4x + 7 + \frac{23x - 6}{x^2 - 3x}$

36. $\frac{3x^3 + 2x^2 - 5x - 1}{x^2 + 2x}$ $3x - 4 + \frac{3x - 1}{x^2 + 2x}$

37. $\frac{8y^3 - y^2 - y + 5}{y^2 + y}$ $8y - 9 + \frac{8y + 5}{y^2 + y}$

38. $\frac{5y^3 - 6y^2 - 7y - 2}{y^2 - y}$ $5y - 1 + \frac{-8y - 2}{y^2 - y}$

39. $(2x^3 + x^2 - 3x + 1) \div (x^2 + x - 1)$ $2x - 1$

40. $(3x^3 - 4x^2 + 8x + 8) \div (x^2 - 2x + 4)$ $3x + 2$

41. $(4x^3 - 13x^2 + 8x - 15) \div (4x^2 - x + 5)$ $x - 3$

42. $(5x^3 + 8x^2 - 5x - 2) \div (5x^2 - 2x - 1)$ $x + 2$

43. $(5a^3 + 7a^2 - 2a - 9) \div (a^2 + 3a - 4)$ $5a - 8 + \frac{42a - 41}{a^2 + 3a - 4}$

44. $(4a^3 - 2a^2 + 7a - 1) \div (a^2 - 2a + 3)$ $4a + 6 + \frac{7a - 19}{a^2 - 2a + 3}$

45. $(2n^4 + 3n^3 - 2n^2 + 3n - 4) \div (n^2 + 1)$ $2n^2 + 3n - 4$

46. $(3n^4 + n^3 - 7n^2 - 2n + 2) \div (n^2 - 2)$ $3n^2 + n - 1$

47. $(x^5 - 1) \div (x - 1)$ $x^4 + x^3 + x^2 + x + 1$

48. $(x^5 + 1) \div (x + 1)$ $x^4 - x^3 + x^2 - x + 1$

49. $(x^4 - 1) \div (x + 1)$ $x^3 - x^2 + x - 1$

50. $(x^4 - 1) \div (x - 1)$ $x^3 + x^2 + x + 1$

51. $(3x^4 + x^3 - 2x^2 - x + 6) \div (x^2 - 1)$ $3x^2 + x + 1 + \frac{7}{x^2 - 1}$

52. $(4x^3 - 2x^2 + 7x - 5) \div (x^2 + 2)$ $4x - 2 - \frac{x + 1}{x^2 + 2}$

4.6 Fractional Equations

The fractional equations used in this text are of two basic types. One type has only constants as denominators and the other type contains variables in the denominators.

In Chapter 2 we considered fractional equations that involve only constants in the denominators. Let's briefly review our approach to solving such equations as we will be using that same basic technique to solve any type of fractional equation.

EXAMPLE 1 Solve $\frac{x-2}{3}+\frac{x+1}{4}=\frac{1}{6}$.

Solution

$$\frac{x-2}{3}+\frac{x+1}{4}=\frac{1}{6}$$

$$12\left(\frac{x-2}{3}+\frac{x+1}{4}\right)=12\left(\frac{1}{6}\right)$$

Multiply both sides by 12, which is the LCD of all of the denominators.

$$4(x-2)+3(x+1)=2$$

$$4x-8+3x+3=2$$

$$7x-5=2$$

$$7x=7$$

$$x=1$$

The solution set is $\{1\}$. Check it! ■

If an equation contains a variable (or variables) in one or more denominators, then we proceed in essentially the same way as in Example 1 above, **except we must avoid any value of the variable that makes a denominator zero.** Consider the following examples.

EXAMPLE 2 Solve $\frac{5}{n}+\frac{1}{2}=\frac{9}{n}$.

Solution First, we need to realize that ***n* cannot equal zero.** (Let's indicate this restriction so that it is not forgotten!) Then we can proceed as follows.

$$\frac{5}{n}+\frac{1}{2}=\frac{9}{n}, \qquad n \neq 0$$

$$2n\left(\frac{5}{n}+\frac{1}{2}\right)=2n\left(\frac{9}{n}\right)$$

Multiply both sides by the LCD, which is $2n$.

$$10+n=18$$

$$n=8$$

The solution set is $\{8\}$. Check it! ■

EXAMPLE 3 Solve $\frac{35-x}{x}=7+\frac{3}{x}$.

Solution

$$\frac{35-x}{x}=7+\frac{3}{x}, \qquad x \neq 0$$

$$x\left(\frac{35-x}{x}\right) = x\left(7+\frac{3}{x}\right) \qquad \text{Multiply both sides by } x.$$

$$35 - x = 7x + 3$$

$$32 = 8x$$

$$4 = x$$

The solution set is $\{4\}$. ■

EXAMPLE 4 Solve $\dfrac{3}{a-2} = \dfrac{4}{a+1}$.

Solution

$$\frac{3}{a-2} = \frac{4}{a+1}, \qquad a \neq 2 \text{ and } a \neq -1$$

$$(a-2)(a+1)\left(\frac{3}{a-2}\right) = (a-2)(a+1)\left(\frac{4}{a+1}\right) \qquad \text{Multiply both sides by } (a-2)(a+1).$$

$$3(a+1) = 4(a-2)$$

$$3a + 3 = 4a - 8$$

$$11 = a$$

The solution set is $\{11\}$. ■

Keep in mind that listing the restrictions at the beginning of a problem does not replace *checking* the potential solutions. In Example 4, 11 needs to be checked in the original equation.

EXAMPLE 5 Solve $\dfrac{a}{a-2} + \dfrac{2}{3} = \dfrac{2}{a-2}$.

Solution

$$\frac{a}{a-2} + \frac{2}{3} = \frac{2}{a-2}, \qquad a \neq 2$$

$$3(a-2)\left(\frac{a}{a-2} + \frac{2}{3}\right) = 3(a-2)\left(\frac{2}{a-2}\right) \qquad \text{Multiply both sides by } 3(a-2).$$

$$3a + 2(a-2) = 6$$

$$3a + 2a - 4 = 6$$

$$5a = 10$$

$$a = 2$$

Because our initial restriction was $a \neq 2$, we conclude that this equation *has no solution*. Thus, the solution set is $\varnothing$. ■

Example 5 demonstrates the importance of recognizing the restrictions that must be made to exclude division by zero.

Ratio and Proportion

A **ratio** is the comparison of two numbers by division. The fractional form is frequently used to express ratios. For example, the ratio of a and b can be written as $\frac{a}{b}$. A statement of equality between two ratios is called a **proportion**. Thus, if $\frac{a}{b}$ and $\frac{c}{d}$ are two equal ratios, the proportion $\frac{a}{b} = \frac{c}{d}$ $(b \neq 0$ and $d \neq 0)$ can be formed.

We can deduce an important property of proportions as follows.

$$\frac{a}{b} = \frac{c}{d}, \qquad b \neq 0 \text{ and } d \neq 0$$

$$bd\left(\frac{a}{b}\right) = bd\left(\frac{c}{d}\right) \qquad \text{Multiply both sides by } bd.$$

$$ad = bc$$

We sometimes refer to this as:

> **Cross-Multiplication Property of Proportions**
>
> If $\frac{a}{b} = \frac{c}{d}$ $(b \neq 0$ and $d \neq 0)$, then $ad = bc$.

We can treat some fractional equations as proportions and solve them by using the cross-multiplication idea, as in the next examples.

EXAMPLE 6 Solve $\frac{5}{x+6} = \frac{7}{x-5}$.

Solution

$$\frac{5}{x+6} = \frac{7}{x-5},$$

$$5(x-5) = 7(x+6) \qquad \text{Apply the cross-multiplication property.}$$

$$5x - 25 = 7x + 42$$

$$-67 = 2x$$

$$-\frac{67}{2} = x$$

The solution set is $\left\{-\frac{67}{2}\right\}$. ■

EXAMPLE 7 Solve $\frac{x}{7} = \frac{4}{x+3}$.

Solution

$$\frac{x}{7} = \frac{4}{x+3}, \qquad x \neq -3$$

$$x(x+3) = 7(4) \qquad \text{cross-multiplication property}$$

$$x^2 + 3x = 28$$

$$x^2 + 3x - 28 = 0$$

$$(x+7)(x-4) = 0$$

$$x + 7 = 0 \quad \text{or} \quad x - 4 = 0$$

$$x = -7 \quad \text{or} \quad x = 4$$

The solution set is $\{-7, 4\}$. Check these solutions in the original equation. ■

Problem Solving

The ability to solve fractional equations broadens our base for solving word problems. We are now ready to tackle some word problems that translate into fractional equations.

PROBLEM 1 The sum of a number and its reciprocal is $\frac{10}{3}$. Find the number.

Solution Let n represent the number. Then $\frac{1}{n}$ represents its reciprocal.

$$n + \frac{1}{n} = \frac{10}{3}, \qquad n \neq 0$$

$$3n\left(n + \frac{1}{n}\right) = 3n\left(\frac{10}{3}\right)$$

$$3n^2 + 3 = 10n$$

$$3n^2 - 10n + 3 = 0$$

$$(3n-1)(n-3) = 0$$

$$3n - 1 = 0 \quad \text{or} \quad n - 3 = 0$$

$$3n = 1 \quad \text{or} \quad n = 3$$

$$n = \frac{1}{3} \quad \text{or} \quad n = 3$$

If the number is $\frac{1}{3}$, then its reciprocal is $\frac{1}{\frac{1}{3}} = 3$. If the number is 3, then its reciprocal is $\frac{1}{3}$. ■

Now let's consider a problem where we can use the relationship

$$\frac{\text{dividend}}{\text{divisor}} = \text{quotient} + \frac{\text{remainder}}{\text{divisor}}$$

as a guideline.

PROBLEM 2 The sum of two numbers is 52. If the larger is divided by the smaller, the quotient is 9 and the remainder is 2. Find the numbers.

Solution Let n represent the smaller number. Then $52 - n$ represents the larger number. Let's use the relationship we discussed above as a guideline and proceed as follows.

$$\frac{\text{dividend}}{\text{divisor}} = \text{quotient} + \frac{\text{remainder}}{\text{divisor}}$$

$$\frac{52 - n}{n} = 9 + \frac{2}{n}, \qquad n \neq 0$$

$$n\left(\frac{52 - n}{n}\right) = n\left(9 + \frac{2}{n}\right)$$

$$52 - n = 9n + 2$$

$$50 = 10n$$

$$5 = n$$

If $n = 5$, then $52 - n$ equals 47. The numbers are 5 and 47. ■

We can very conveniently set up some problems and solve them using the concepts of ratio and proportion. Let's conclude this section with two such examples.

PROBLEM 3 On a certain map, $1\frac{1}{2}$ inches represents 25 miles. If two cities are $5\frac{1}{4}$ inches apart on the map, find the number of miles between the cities.

Solution Let m represent the number of miles between the two cities. The following proportion can be set up and solved.

$$\frac{1\frac{1}{2}}{25} = \frac{5\frac{1}{4}}{m}, \qquad m \neq 0$$

$$\frac{\frac{3}{2}}{25} = \frac{\frac{21}{4}}{m}$$

$$\frac{3}{2}m = 25\left(\frac{21}{4}\right) \qquad \text{cross-multiplication property}$$

$$\frac{2}{3}\left(\frac{3}{2}m\right) = \frac{\cancel{2}}{\cancel{3}}(25)\left(\frac{\overset{7}{\cancel{21}}}{\underset{2}{\cancel{4}}}\right) \qquad \text{Multiply both sides by } \frac{2}{3}.$$

$$m = \frac{175}{2} = 87\frac{1}{2}$$

The distance between the two cities is $87\frac{1}{2}$ miles. ■

PROBLEM 4 A sum of \$750 is to be divided between two people in the ratio of 2 to 3. How much does each person receive?

Solution Let d represent the amount of money that one person receives. Then $750 - d$ represents the amount for the other person.

$$\frac{d}{750 - d} = \frac{2}{3}, \qquad d \neq 750$$

$$3d = 2(750 - d)$$

$$3d = 1500 - 2d$$

$$5d = 1500$$

$$d = 300$$

If $d = 300$, then $750 - d$ equals 450. Therefore, one person receives \$300 and the other person receives \$450. ■

Problem Set 4.6

Solve each of the following equations.

1. $\frac{x+1}{4} + \frac{x-2}{6} = \frac{3}{4}$ $\{2\}$

2. $\frac{x+2}{5} + \frac{x-1}{6} = \frac{3}{5}$ $\{1\}$

3. $\dfrac{x+3}{2} - \dfrac{x-4}{7} = 1$ $\{-3\}$

4. $\dfrac{x+4}{3} - \dfrac{x-5}{9} = 1$ $\{-4\}$

5. $\dfrac{5}{n} + \dfrac{1}{3} = \dfrac{7}{n}$ $\{6\}$

6. $\dfrac{3}{n} + \dfrac{1}{6} = \dfrac{11}{3n}$ $\{4\}$

7. $\dfrac{7}{2x} + \dfrac{3}{5} = \dfrac{2}{3x}$ $\left\{-\dfrac{85}{18}\right\}$

8. $\dfrac{9}{4x} + \dfrac{1}{3} = \dfrac{5}{2x}$ $\left\{\dfrac{3}{4}\right\}$

9. $\dfrac{3}{4x} + \dfrac{5}{6} = \dfrac{4}{3x}$ $\left\{\dfrac{7}{10}\right\}$

10. $\dfrac{5}{7x} - \dfrac{5}{6} = \dfrac{1}{6x}$ $\left\{\dfrac{23}{35}\right\}$

11. $\dfrac{47-n}{n} = 8 + \dfrac{2}{n}$ $\{5\}$

12. $\dfrac{45-n}{n} = 6 + \dfrac{3}{n}$ $\{6\}$

13. $\dfrac{n}{65-n} = 8 + \dfrac{2}{65-n}$ $\{58\}$

14. $\dfrac{n}{70-n} = 7 + \dfrac{6}{70-n}$ $\{62\}$

15. $n + \dfrac{1}{n} = \dfrac{17}{4}$ $\left\{\dfrac{1}{4}, 4\right\}$

16. $n + \dfrac{1}{n} = \dfrac{37}{6}$ $\left\{\dfrac{1}{6}, 6\right\}$

17. $n - \dfrac{2}{n} = \dfrac{23}{5}$ $\left\{-\dfrac{2}{5}, 5\right\}$

18. $n - \dfrac{3}{n} = \dfrac{26}{3}$ $\left\{-\dfrac{1}{3}, 9\right\}$

19. $\dfrac{5}{7x-3} = \dfrac{3}{4x-5}$ $\{-16\}$

20. $\dfrac{3}{2x-1} = \dfrac{5}{3x+2}$ $\{11\}$

21. $\dfrac{-2}{x-5} = \dfrac{1}{x+9}$ $\left\{-\dfrac{13}{3}\right\}$

22. $\dfrac{5}{2a-1} = \dfrac{-6}{3a+2}$ $\left\{-\dfrac{4}{27}\right\}$

23. $\dfrac{x}{x+1} - 2 = \dfrac{3}{x-3}$ $\{-3, 1\}$

24. $\dfrac{x}{x-2} + 1 = \dfrac{8}{x-1}$ $\{3\}$

25. $\dfrac{a}{a+5} - 2 = \dfrac{3a}{a+5}$ $\left\{-\dfrac{5}{2}\right\}$

26. $\dfrac{a}{a-3} - \dfrac{3}{2} = \dfrac{3}{a-3}$ $\varnothing$

27. $\dfrac{5}{x+6} = \dfrac{6}{x-3}$ $\{-51\}$

28. $\dfrac{3}{x-1} = \dfrac{4}{x+2}$ $\{10\}$

29. $\dfrac{3x-7}{10} = \dfrac{2}{x}$ $\left\{-\dfrac{5}{3}, 4\right\}$

30. $\dfrac{x}{-4} = \dfrac{3}{12x-25}$ $\left\{\dfrac{3}{4}, \dfrac{4}{3}\right\}$

31. $\dfrac{x}{x-6} - 3 = \dfrac{6}{x-6}$ $\varnothing$

32. $\dfrac{x}{x+1} + 3 = \dfrac{4}{x+1}$ $\left\{\dfrac{1}{4}\right\}$

33. $\dfrac{3s}{s+2} + 1 = \dfrac{35}{2(3s+1)}$ $\left\{-\dfrac{11}{8}, 2\right\}$

34. $\dfrac{s}{2s-1} - 3 = \dfrac{-32}{3(s+5)}$ $\left\{-1, \dfrac{13}{15}\right\}$

35. $2 - \dfrac{3x}{x-4} = \dfrac{14}{x+7}$ $\{-29, 0\}$

36. $-1 + \dfrac{2x}{x+3} = \dfrac{-4}{x+4}$ $\{-5, 0\}$

37. $\dfrac{n+6}{27} = \dfrac{1}{n}$ $\{-9, 3\}$

38. $\dfrac{n}{5} = \dfrac{10}{n-5}$ $\{-5, 10\}$

39. $\dfrac{3n}{n-1} - \dfrac{1}{3} = \dfrac{-40}{3n-18}$ $\left\{-2, \dfrac{23}{8}\right\}$

40. $\dfrac{n}{n+1} + \dfrac{1}{2} = \dfrac{-2}{n+2}$ $\left\{-3, -\dfrac{2}{3}\right\}$

41. $\dfrac{-3}{4x+5} = \dfrac{2}{5x-7}$ $\left\{\dfrac{11}{23}\right\}$

42. $\dfrac{7}{x+4} = \dfrac{3}{x-8}$ $\{17\}$

43. $\dfrac{2x}{x-2} + \dfrac{15}{x^2 - 7x + 10} = \dfrac{3}{x-5}$ $\left\{3, \dfrac{7}{2}\right\}$

44. $\dfrac{x}{x-4} - \dfrac{2}{x+3} = \dfrac{20}{x^2 - x - 12}$ $\{-4, 3\}$

Set up an algebraic equation and solve each of the following problems.

45. A sum of \$1750 is to be divided between two people in the ratio of 3 to 4. How much does each person receive? \$750 and \$1000

46. A blueprint has a scale of 1 inch represents 5 feet. Find the dimensions of a rectangular room that measures $3\frac{1}{2}$ inches by $5\frac{3}{4}$ inches on the blueprint. $17\frac{1}{2}$ feet by $28\frac{3}{4}$ feet

47. One angle of a triangle has a measure of 60° and the measures of the other two angles are in a ratio of 2 to 3. Find the measures of the other two angles. 48° and 72°

48. The ratio of the complement of an angle to its supplement is 1 to 4. Find the measure of the angle. 60°

49. The sum of a number and its reciprocal is $\frac{53}{14}$. Find the number. $\frac{2}{7}$ or $\frac{7}{2}$

50. The sum of two numbers is 80. If the larger is divided by the smaller, the quotient is 7 and the remainder is 8. Find the numbers. 9 and 71

51. If a home valued at \$50,000 is assessed \$900 in real estate taxes, at the same rate how much are the taxes on a home valued at \$60,000? \$1080

52. The ratio of male students to female students at a certain university is 5 to 7. If there is a total of 16,200 students, find the number of male students and the number of female students. 6750 males and 9450 females

53. Suppose that, together, Laura and Tammy sold \$120.75 worth of candy for the annual school fair. If the ratio of Tammy's sales to Laura's sales was 4 to 3, how much did each sell? \$69 for Tammy and \$51.75 for Laura

54. The total value of a house and a lot is \$68,000. If the ratio of the value of the house to the value of the lot is 7 to 1, find the value of the house. \$59,500

55. The sum of two numbers is 90. If the larger is divided by the smaller, the quotient is 10 and the remainder is 2. Find the numbers. 8 and 82

56. What number must be added to the numerator and denominator of $\frac{2}{5}$ to produce a rational number that is equivalent to $\frac{7}{8}$? 19

57. A 20-foot board is to be cut into two pieces whose lengths are in the ratio of 7 to 3. Find the lengths of the two pieces. 14 feet and 6 feet

58. An inheritance of \$300,000 is to be divided between a son and the local heart fund in the ratio of 3 to 1. How much money will the son receive? \$225,000

59. Suppose that in a certain precinct, 1150 people voted in the presidential last election. If the ratio of female voters to male voters was 3 to 2, how many females and how many males voted? 690 females and 460 males

60. The perimeter of a rectangle is 114 centimeters. If the ratio of its width to its length is 7 to 12, find the dimensions of the rectangle. 21 centimeters by 36 centimeters

4.7

More Fractional Equations and Applications

Let's begin this section by considering a few more fractional equations. We will continue to solve them using the same basic techniques as in the previous section. That is, we will multiply both sides of the equation by the least common denominator of all of the denominators in the equation, with the necessary restrictions to avoid division by zero. Some of the denominators in these problems will require factoring before we can determine a least common denominator.

EXAMPLE 1 Solve $\dfrac{x}{2x-8}+\dfrac{16}{x^2-16}=\dfrac{1}{2}$.

Solution

$$\frac{x}{2x-8}+\frac{16}{x^2-16}=\frac{1}{2}$$

$$\frac{x}{2(x-4)}+\frac{16}{(x+4)(x-4)}=\frac{1}{2}, \qquad x \neq 4 \text{ and } x \neq -4$$

$$2(x-4)(x+4)\left(\frac{x}{2(x-4)}+\frac{16}{(x+4)(x-4)}\right)=2(x+4)(x-4)\left(\frac{1}{2}\right)$$

$$x(x+4)+2(16)=(x+4)(x-4)$$

$$x^2+4x+32=x^2-16$$

$$4x=-48$$

$$x=-12$$

The solution set is $\{-12\}$. Perhaps you should check it! ■

In Example 1, notice that the restrictions were not indicated until the denominators were expressed in factored form. It is usually easier to determine the necessary restrictions at this step.

EXAMPLE 2 Solve $\dfrac{3}{n-5}-\dfrac{2}{2n+1}=\dfrac{n+3}{2n^2-9n-5}$.

Solution

$$\frac{3}{n-5}-\frac{2}{2n-1}=\frac{n+3}{2n^2-9n-5}$$

$$\frac{3}{n-5}-\frac{2}{2n+1}=\frac{n+3}{(2n+1)(n-5)}, \qquad n \neq -\frac{1}{2} \text{ and } n \neq 5$$

$$(2n+1)(n-5)\left(\frac{3}{n-5}-\frac{2}{2n+1}\right)=(2n+1)(n-5)\left(\frac{n+3}{(2n+1)(n-5)}\right)$$

$$3(2n+1)-2(n-5)=n+3$$

$$6n+3-2n+10=n+3$$

$$4n+13=n+3$$

$$3n=-10$$

$$n=-\frac{10}{3}$$

The solution set is $\left\{-\frac{10}{3}\right\}$. ■

EXAMPLE 3 Solve $2+\frac{4}{x-2}=\frac{8}{x^2-2x}$.

Solution

$$2+\frac{4}{x-2}=\frac{8}{x^2-2x}$$

$$2+\frac{4}{x-2}=\frac{8}{x(x-2)}, \qquad x \neq 0 \text{ and } x \neq 2$$

$$x(x-2)\left(2+\frac{4}{x-2}\right)=x(x-2)\left(\frac{8}{x(x-2)}\right)$$

$$2x(x-2)+4x=8$$

$$2x^2-4x+4x=8$$

$$2x^2=8$$

$$x^2=4$$

$$x^2-4=0$$

$$(x+2)(x-2)=0$$

$$x+2=0 \quad \text{or} \quad x-2=0$$

$$x=-2 \quad \text{or} \quad x=2$$

Since our initial restriction indicated that $x \neq 2$, then the *only solution* is -2. Thus, the solution set is $\{-2\}$. ■

In Section 2.4, we discussed using the properties of equality to change the form of various formulas. For example, we considered the simple interest formula "$A = P + Prt$" and changed its form by solving for P as follows.

$$A = P + Prt$$

$$A = P(1 + rt)$$

$$\frac{A}{1 + rt} = P \qquad \text{Multiply both sides by } \frac{1}{1 + rt}$$

If the formula is in the form of a fractional equation, then the techniques of these last two sections are applicable. Consider the following example.

EXAMPLE 4 If the original cost of some business property is C dollars and it is depreciated linearly over N years, its value, V, at the end of T years is given by

$$V = C\left(1 - \frac{T}{N}\right).$$

Solve this formula for N in terms of V, C, and T.

Solution

$$V = C\left(1 - \frac{T}{N}\right)$$

$$V = C - \frac{CT}{N}$$

$$N(V) = N\left(C - \frac{CT}{N}\right) \qquad \text{Multiply both sides by } N.$$

$$NV = NC - CT$$

$$NV - NC = -CT$$

$$N(V - C) = -CT$$

$$N = \frac{-CT}{V - C}$$

$$N = -\frac{CT}{V - C}.$$

■

Problem Solving

In Section 2.4 we solved some uniform motion problems. The formula $d = rt$ was used in the analysis of the problems and we used guidelines that involve distance relationships. Now let's consider some uniform motion problems for which guidelines that involve either times or rates are appropriate. These problems will generate fractional equations to solve.

PROBLEM 1 An airplane travels 2050 miles in the same time that a car travels 260 miles. If the rate of the plane is 358 miles per hour greater than the rate of the car, find the rate of each.

Solution Let r represent the rate of the car. Then $r + 358$ represents the rate of the plane. The fact that the times are equal can be a guideline.

time of plane equals time of car

$$\frac{\text{distance of plane}}{\text{rate of plane}} = \frac{\text{distance of car}}{\text{rate of car}}$$

$$\frac{2050}{r + 358} = \frac{260}{r}$$

$$2050r = 260(r + 358)$$

$$2050r = 260r + 93{,}080$$

$$1790r = 93{,}080$$

$$r = 52$$

If $r = 52$, then $r + 358$ equals 410. Thus, the rate of the car is 52 miles per hour and the rate of the plane is 410 miles per hour. ■

PROBLEM 2 It takes a freight train 2 hours longer to travel 300 miles than it takes an express train to travel 280 miles. The rate of the express train is 20 miles per hour greater than the rate of the freight train. Find the times and rates of both trains.

Solution Let t represent the time of the express train. Then $t + 2$ represents the time of the freight train. Let's record the information of this problem in a table as follows.

	Distance	*Time*	$r = \frac{d}{t}$
Express train	280	t	$\frac{280}{t}$
Freight train	300	$t + 2$	$\frac{300}{t + 2}$

The fact that the rate of the express train is 20 miles per hour greater than the rate of the freight train can be a guideline.

rate of express equals rate of freight train plus 20

$$\frac{280}{t} = \frac{300}{t + 2} + 20$$

$$t(t + 2)\left(\frac{280}{t}\right) = t(t + 2)\left(\frac{300}{t + 2} + 20\right)$$

$$280(t + 2) = 300t + 20t(t + 2)$$

$$280t + 560 = 300t + 20t^2 + 40t$$
$$280t + 560 = 340t + 20t^2$$
$$0 = 20t^2 + 60t - 560$$
$$0 = t^2 + 3t - 28$$
$$0 = (t + 7)(t - 4)$$
$$t + 7 = 0 \quad \text{or} \quad t - 4 = 0$$
$$t = -7 \quad \text{or} \quad t = 4$$

The negative solution must be discarded, so the time of the express train (t) is 4 hours and the time of the freight train ($t + 2$) is 6 hours. The rate of the express train $\left(\frac{280}{t}\right)$ is $\frac{280}{4} = 70$ miles per hour and the rate of the freight train $\left(\frac{300}{t+2}\right)$ is $\frac{300}{6} = 50$ miles per hour. ■

REMARK Note that to solve Problem 1 we went directly to a guideline without the use of a table, but for Problem 2 we used a table. Again, remember that this is a personal preference; we are merely exposing you to a variety of techniques.

Uniform motion problems are a special case of a larger group of problems we refer to as **rate-time** problems. For example, if a certain machine can produce 150 items in 10 minutes, then we say that the machine is producing at a rate of $\frac{150}{10} = 15$ items per minute. Likewise, if a person can do a certain job in 3 hours, then, assuming a constant rate of work we say that the person is working at a rate of $\frac{1}{3}$ of the job per hour. In general, if Q is the quantity of something done in t units of time, then the rate, r, is given by $r = \frac{Q}{t}$. We state the rate in terms of *so much quantity per unit of time*. (In uniform motion problems the "quantity" is distance.) Let's consider some examples of rate-time problems.

PROBLEM 3 If Jim can mow a lawn in 50 minutes and his son, Todd, can mow the same lawn in 40 minutes, how long will it take them to mow the lawn if they work together?

Solution Jim's rate is $\frac{1}{50}$ of the lawn per minute and Todd's rate is $\frac{1}{40}$ of the lawn per minute. If we let m represent the number of minutes that they work together, then $\frac{1}{m}$ represents their rate when working together. Therefore, since the sum of the individual rates must equal the rate when working together, we can set up and solve the following equation.

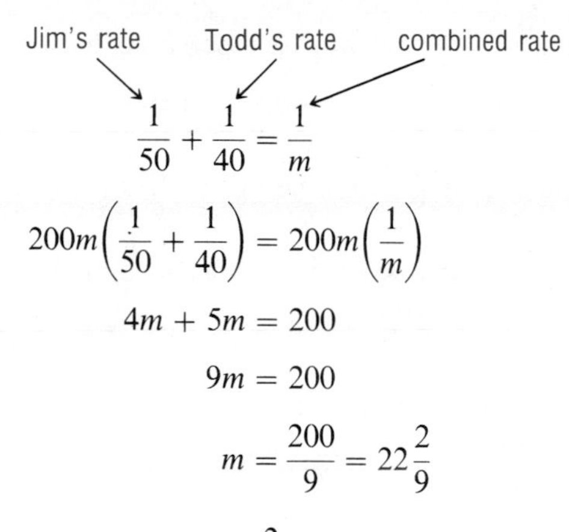

$$\frac{1}{50} + \frac{1}{40} = \frac{1}{m}$$

$$200m\left(\frac{1}{50} + \frac{1}{40}\right) = 200m\left(\frac{1}{m}\right)$$

$$4m + 5m = 200$$

$$9m = 200$$

$$m = \frac{200}{9} = 22\frac{2}{9}$$

It should take them $22\frac{2}{9}$ minutes. ■

PROBLEM 4 Working together, Linda and Kathy can type a term paper in $3\frac{3}{5}$ hours. Linda can type the paper by herself in 6 hours. How long would it take Kathy to type the paper by herself?

Solution Their rate working together is $\dfrac{1}{3\frac{3}{5}} = \dfrac{1}{\frac{18}{5}} = \dfrac{5}{18}$ of the job per hour and Linda's rate is $\frac{1}{6}$ of the job per hour. If we let h represent the number of hours that it would take Kathy by herself, then her rate is $\frac{1}{h}$ of the job per hour. Thus, we have

$$\underset{\text{Linda's rate}}{\frac{1}{6}} + \underset{\text{Kathy's rate}}{\frac{1}{h}} = \underset{\text{combined rate}}{\frac{5}{18}}.$$

Solving this equation yields

$$18h\left(\frac{1}{6} + \frac{1}{h}\right) = 18h\left(\frac{5}{18}\right)$$

$$3h + 18 = 5h$$

$$18 = 2h$$

$$9 = h.$$

It would take Kathy 9 hours to type the paper by herself. ■

One final example of this section illustrates another approach that some people find meaningful for rate-time problems. For this approach, think in terms of fractional parts of the job. For example, if a person can do a certain job in 5 hours, then at the end of 2 hours he or she has done $\frac{2}{5}$ of the job. (Again, assume a constant rate of work.) At the end of 4 hours, he or she has finished $\frac{4}{5}$ of the job and, in general, at the end of h hours, he or she has done $\frac{h}{5}$ of the job. Let's see how this works in a problem.

PROBLEM 5 It takes Pat 12 hours to complete a task. After he had been working for 3 hours, he was joined by his brother, Mike, and together they finished the task in 5 hours. How long would it take Mike to do the job by himself?

Solution Let h represent the number of hours that it would take Mike by himself. Since Pat has been working for 3 hours, he has done $\frac{3}{12} = \frac{1}{4}$ of the job before Mike joins him. Thus, there is $\frac{3}{4}$ of the original job to be done while working together. We can set up and solve the following equation.

fractional part of the remaining $\frac{3}{4}$ of the job that Pat does ↓ + fractional part of the remaining $\frac{3}{4}$ of the job that Mike does ↓

$$\frac{5}{12} + \frac{5}{h} = \frac{3}{4}$$

$$12h\left(\frac{5}{12} + \frac{5}{h}\right) = 12h\left(\frac{3}{4}\right)$$

$$5h + 60 = 9h$$

$$60 = 4h$$

$$15 = h$$

It would take Mike 15 hours to do the entire job by himself. ■

Problem Set 4.7

Solve each of the following equations.

1. $\frac{x}{4x-4} + \frac{5}{x^2-1} = \frac{1}{4}$ $\{-21\}$

2. $\frac{x}{3x-6} + \frac{4}{x^2-4} = \frac{1}{3}$ $\{-8\}$

3. $3 + \frac{6}{t-3} = \frac{6}{t^2-3t}$ $\{-1, 2\}$

4. $2 + \frac{4}{t-1} = \frac{4}{t^2-t}$ $\{-2\}$

5. $\dfrac{3}{n-5}+\dfrac{4}{n+7}=\dfrac{2n+11}{n^2+2n-35}$ $\{2\}$

6. $\dfrac{2}{n+3}+\dfrac{3}{n-4}=\dfrac{2n-1}{n^2-n-12}$ $\left\{-\dfrac{2}{3}\right\}$

7. $\dfrac{5x}{2x+6}-\dfrac{4}{x^2-9}=\dfrac{5}{2}$ $\left\{\dfrac{37}{15}\right\}$

8. $\dfrac{3x}{5x+5}-\dfrac{2}{x^2-1}=\dfrac{3}{5}$ $\left\{-\dfrac{7}{3}\right\}$

9. $1+\dfrac{1}{n-1}=\dfrac{1}{n^2-n}$ $\{-1\}$

10. $3+\dfrac{9}{n-3}=\dfrac{27}{n^2-3n}$ $\{-3\}$

11. $\dfrac{2}{n-2}-\dfrac{n}{n+5}=\dfrac{10n+15}{n^2+3n-10}$ $\{-1\}$

12. $\dfrac{n}{n+3}+\dfrac{1}{n-4}=\dfrac{11-n}{n^2-n-12}$ $\{-2\}$

13. $\dfrac{2}{2x-3}-\dfrac{2}{10x^2-13x-3}=\dfrac{x}{5x+1}$ $\left\{0,\dfrac{13}{2}\right\}$

14. $\dfrac{1}{3x+4}+\dfrac{6}{6x^2+5x-4}=\dfrac{x}{2x-1}$ $\left\{-\dfrac{5}{3},1\right\}$

15. $\dfrac{2x}{x+3}-\dfrac{3}{x-6}=\dfrac{29}{x^2-3x-18}$ $\left\{-2,\dfrac{19}{2}\right\}$

16. $\dfrac{x}{x-4}-\dfrac{2}{x+8}=\dfrac{63}{x^2+4x-32}$ $\{-11,5\}$

17. $\dfrac{a}{a-5}+\dfrac{2}{a-6}=\dfrac{2}{a^2-11a+30}$ $\{-2\}$

18. $\dfrac{a}{a+2}+\dfrac{3}{a+4}=\dfrac{14}{a^2+6a+8}$ $\{-8,1\}$

19. $\dfrac{-1}{2x-5}+\dfrac{2x-4}{4x^2-25}=\dfrac{5}{6x+15}$ $\left\{-\dfrac{1}{5}\right\}$

20. $\dfrac{-2}{3x+2}+\dfrac{x-1}{9x^2-4}=\dfrac{3}{12x-8}$ $\left\{\dfrac{6}{29}\right\}$

21. $\dfrac{7y+2}{12y^2+11y-15}-\dfrac{1}{3y+5}=\dfrac{2}{4y-3}$ $\varnothing$

22. $\dfrac{5y-4}{6y^2+y-12}-\dfrac{2}{2y+3}=\dfrac{5}{3y-4}$ $\{-1\}$

23. $\dfrac{2n}{6n^2+7n-3}-\dfrac{n-3}{3n^2+11n-4}=\dfrac{5}{2n^2+11n+12}$ $\left\{\dfrac{7}{2}\right\}$

24. $\dfrac{x+1}{2x^2+7x-4}-\dfrac{x}{2x^2-7x+3}=\dfrac{1}{x^2+x-12}$ $\left\{-\dfrac{1}{4}\right\}$

25. $\dfrac{1}{2x^2-x-1}+\dfrac{3}{2x^2+x}=\dfrac{2}{x^2-1}$ $\{-3\}$

26. $\dfrac{2}{n^2+4n}+\dfrac{3}{n^2-3n-28}=\dfrac{5}{n^2-6n-7}$ $\left\{-\dfrac{14}{29}\right\}$

27. $\dfrac{x+1}{x^3-9x}-\dfrac{1}{2x^2+x-21}=\dfrac{1}{2x^2+13x+21}$ $\left\{-\dfrac{7}{9}\right\}$

28. $\dfrac{x}{2x^2+5x}-\dfrac{x}{2x^2+7x+5}=\dfrac{2}{x^2+x}$ $\left\{-\dfrac{10}{3}\right\}$

29. $\dfrac{4t}{4t^2-t-3}+\dfrac{2-3t}{3t^2-t-2}=\dfrac{1}{12t^2+17t+6}$ $\left\{-\dfrac{7}{6}\right\}$

30. $\dfrac{2t}{2t^2+9t+10}+\dfrac{1-3t}{3t^2+4t-4}=\dfrac{4}{6t^2+11t-10}$ $\left\{-\dfrac{1}{7}\right\}$

For Problems 31–44, solve each equation for the indicated variable.

31. $y=\dfrac{5}{6}x+\dfrac{2}{9}$ for x $x=\dfrac{18y-4}{15}$

32. $y=\dfrac{3}{4}x-\dfrac{2}{3}$ for x $x=\dfrac{12y+8}{9}$

33. $\dfrac{-2}{x-4}=\dfrac{5}{y-1}$ for y $y=\dfrac{-5x+22}{2}$

34. $\dfrac{7}{y-3}=\dfrac{3}{x+1}$ for y $y=\dfrac{7x+16}{3}$

35. $I = \dfrac{100M}{C}$ for M $M = \dfrac{IC}{100}$

36. $V = C\left(1 - \dfrac{T}{N}\right)$ for T $T = \dfrac{N(C - V)}{C}$

37. $\dfrac{R}{S} = \dfrac{T}{S + T}$ for R $R = \dfrac{ST}{S + T}$

38. $\dfrac{1}{R} = \dfrac{1}{S} + \dfrac{1}{T}$ for R $R = \dfrac{ST}{T + S}$

39. $\dfrac{y - 1}{x - 3} = \dfrac{b - 1}{a - 3}$ for y $y = \dfrac{bx - x - 3b + a}{a - 3}$

40. $y = -\dfrac{a}{b}x + \dfrac{c}{d}$ for x $x = \dfrac{bc - bdy}{ad}$

41. $\dfrac{x}{a} + \dfrac{y}{b} = 1$ for y $y = \dfrac{ab - bx}{a}$

42. $\dfrac{y - b}{x} = m$ for y $y = mx + b$

43. $\dfrac{y - 1}{x + 6} = \dfrac{-2}{3}$ for y $y = \dfrac{-2x - 9}{3}$

44. $\dfrac{y + 5}{x - 2} = \dfrac{3}{7}$ for y $y = \dfrac{3x - 41}{7}$

Set up an equation and solve each of the following problems.

45. Kent drives his Mazda 270 miles in the same time that Dave drives his Datsun 250 miles. If Kent averages 4 miles per hour faster than Dave, find their rates. 50 mph and 54 mph

46. Suppose that Wendy rides her bicycle 30 miles in the same time that it takes Kim to ride her bicycle 20 miles. If Wendy rides 5 miles per hour faster than Kim, find the rate of each. 10 miles per hour for Kim and 15 miles per hour for Wendy

47. An inlet pipe can fill a tank in 10 minutes. A drain can empty the tank in 12 minutes. If the tank is empty and both the pipe and drain are open, how long will it take before the tank overflows? 60 minutes.

48. Barry can do a certain job in 3 hours, while it takes Roy 5 hours to do the same job. How long would it take them to do the job working together? $1\frac{7}{8}$ hours

49. Connie can type 600 words in 5 minutes less than it takes Katie to type 600 words. If Connie types at a rate of 20 words per minute faster than Katie types, find the typing rate of each woman. 60 words per minute for Connie and 40 words per minute for Katie

50. Walt can mow a lawn in 1 hour, while his son, Mike, can mow the same lawn in 50 minutes. One day Mike started mowing the lawn by himself and worked for 30 minutes. Then Walt joined him and they finished the lawn. How long did it take them to finish mowing the lawn after Walt started to help? $10\frac{10}{11}$ minutes.

51. Plane A can travel 1400 miles in one hour less time than it takes plane B to travel 2000 miles. The rate of plane *B* is 50 miles per hour faster than the rate of plane *A*. Find the times and rates of both planes. Plane B could travel at 400 mph for 5 hours and plane A at 350 mph for 4 hours or plane B could travel at 250 mph for 8 hours and plane A at 200 mph for 7 hours

52. To travel 60 miles, it takes Sue, riding a moped, 2 hours less time than it takes Ann to travel 50 miles riding a bicycle. Sue travels 10 miles per hour faster than Ann. Find the times and rates of both girls. 10 mph for 5 hrs and 20 mph for 3 hrs

53. It takes Amy twice as long to deliver papers as it does Nancy. How long would it take each if they can deliver the papers together in 40 minutes? 60 minutes and 120 minutes

54. If two inlet pipes are both open, they can fill a pool in 1 hour and 12 minutes. One of the pipes can fill the pool by itself in 2 hours. How long would it take the other pipe to fill the pool by itself? 3 hours

55. Dan agreed to mow a vacant lot for $12. It took him an hour longer than what he had anticipated, so he earned $1 per hour less than he originally calculated. How long had he anticipated that it would take him to mow the lot? 3 hours

56. Last week Al bought some golf balls for \$20. The next day they were on sale for \$0.50 per ball less and he bought \$22.50 worth of balls. If he purchased 5 more balls on the second day than he did on the first day, how many did he buy each day and at what price per ball? 10 golf balls at \$2 each on the first day and 15 balls at \$1.50 each on the 2nd day

57. Debbie rode her bicycle out into the country for a distance of 24 miles. On the way back, she took a much shorter route of 12 miles and made the return trip in one-half hour less time. If her rate out into the country was 4 miles per hour faster than her rate on the return trip, find both rates. 16 mph and 12 mph or 12 mph and 8 mph

58. Nick jogs for 10 miles and then walks another 10 miles. He jogs $2\frac{1}{2}$ miles per hour faster than he walks and the entire distance of 20 miles takes 6 hours. Find the rate that he walks and the rate that he jogs. Walks at $2\frac{1}{2}$ miles per hour and jogs at 5 miles per hour

Chapter 4 Summary

(4.1) Any number that can be written in the form $\frac{a}{b}$, where a and b are integers and $b \neq 0$, is called a **rational number**.

A **rational expression** is defined as the indicated quotient of two polynomials. The following properties pertain to rational numbers and rational expressions.

1. $\frac{-a}{b} = \frac{a}{-b} = -\frac{a}{b}$
2. $\frac{-a}{-b} = \frac{a}{b}$
3. $\frac{a \cdot k}{b \cdot k} = \frac{a}{b}$ fundamental principle of fractions

(4.2) Multiplication and division of rational expressions are based upon the following

1. $\frac{a}{b} \cdot \frac{c}{d} = \frac{ac}{bd}$ multiplication
2. $\frac{a}{b} \div \frac{c}{d} = \frac{a}{b} \cdot \frac{d}{c} = \frac{ad}{bc}$ division

(4.3) Addition and subtraction of rational expressions are based upon the following.

1. $\frac{a}{b} + \frac{c}{b} = \frac{a+c}{b}$ addition

2. $\dfrac{a}{b} - \dfrac{c}{b} = \dfrac{a-c}{b}$ subtraction

(4.4) The following basic procedure is used to add or subtract rational expressions.

1. Find the LCD of all denominators.
2. Change each fraction to an equivalent fraction that has the LCD as its denominator.
3. Add or subtract numerators and place this result over the LCD.
4. Look for possibilities to simplify the resulting fraction.

Fractional forms that contain rational numbers or rational expressions in the numerators and/or denominators are called **complex fractions**. The fundamental principle of fractions serves as a basis for simplifying complex fractions.

(4.5) To divide a polynomial by a monomial we divide each term of the polynomial by the monomial.

The procedure for dividing a polynomial by a polynomial, rather than a monomial, resembles the long division process in arithmetic. (See the examples in Section 4.5.)

(4.6) **To solve a fractional equation**, it is often easiest to begin by multiplying both sides of the equation by the LCD of all the denominators in the equation. If an equation contains a variable in one or more denominators, then we must be careful to avoid any value of the variable that makes the denominator zero.

A **ratio** is the comparison of two numbers by division. A statement of equality between two ratios is a **proportion**.

We can treat some fractional equations as proportions and we can solve them by applying the following property. This property is often called the **cross-multiplication** property.

If $\dfrac{a}{b} = \dfrac{c}{d}$, then $ad = bc$.

(4.7) The techniques we use to solve fractional equations we can also use to change the form of formulas that contain rational expressions and to use them in solving problems.

Chapter 4 Review Problem Set

For Problems 1–6, simplify each of the rational expressions.

1. $\dfrac{26x^2y^3}{39x^4y^2} \quad \dfrac{2y}{3x^2}$

2. $\dfrac{a^2-9}{a^2+3a} \quad \dfrac{a-3}{a}$

3. $\dfrac{n^2-3n-10}{n^2+n-2} \quad \dfrac{n-5}{n-1}$

4. $\dfrac{x^4-1}{x^3-x} \quad \dfrac{x^2+1}{x}$

5. $\dfrac{8x^3-2x^2-3x}{12x^2-9x} \quad \dfrac{2x+1}{3}$

6. $\dfrac{x^4-7x^2-30}{2x^4+7x^2+3} \quad \dfrac{x^2-10}{2x^2+1}$

For Problems 7–10, simplify each complex fraction.

7. $\dfrac{\frac{5}{8}-\frac{1}{2}}{\frac{1}{6}+\frac{3}{4}}$ $\quad \frac{3}{22}$

8. $\dfrac{\frac{3}{2x}+\frac{5}{3y}}{\frac{4}{x}-\frac{3}{4y}}$ $\quad \frac{18y+20x}{48y-9x}$

9. $\dfrac{\frac{3}{x-2}-\frac{4}{x^2-4}}{\frac{2}{x+2}+\frac{1}{x-2}}$ $\quad \frac{3x+2}{3x-2}$

10. $1-\dfrac{1}{2-\frac{1}{x}}$ $\quad \frac{x-1}{2x-1}$

For Problems 11–22, perform the indicated operations and express answers in simplest form.

11. $\frac{6xy^2}{7y^3} \div \frac{15x^2y}{5x^2}$ $\quad \frac{2x}{7y^2}$

12. $\frac{9ab}{3a+6} \cdot \frac{a^2-4a-12}{a^2-6a}$ $\quad 3b$

13. $\frac{n^2+10n+25}{n^2-n} \cdot \frac{5n^3-3n^2}{5n^2+22n-15}$ $\quad \frac{n(n+5)}{n-1}$

14. $\frac{x^2-2xy-3y^2}{x^2+9y^2} \div \frac{2x^2+xy-y^2}{2x^2-xy}$ $\quad \frac{x(x-3y)}{x^2+9y^2}$

15. $\frac{2x+1}{5}+\frac{3x-2}{4}$ $\quad \frac{23x-6}{20}$

16. $\frac{3}{2n}+\frac{5}{3n}-\frac{1}{9}$ $\quad \frac{57-2n}{18n}$

17. $\frac{3x}{x+7}-\frac{2}{x}$ $\quad \frac{3x^2-2x-14}{x(x+7)}$

18. $\frac{10}{x^2-5x}+\frac{2}{x}$ $\quad \frac{2}{x-5}$

19. $\frac{3}{n^2-5n-36}+\frac{2}{n^2+3n-4}$ $\quad \frac{5n-21}{(n-9)(n+4)(n-1)}$

20. $\frac{3}{2y+3}+\frac{5y-2}{2y^2-9y-18}-\frac{1}{y-6}$ $\quad \frac{6y-23}{(2y+3)(y-6)}$

21. $(18x^2+9x-2) \div (3x+2)$ $\quad 6x-1$

22. $(3x^3+5x^2-6x-2) \div (x+4)$ $\quad 3x^2-7x+22-\frac{90}{x+4}$

For Problems 23–32, solve each equation.

23. $\frac{4x+5}{3}+\frac{2x-1}{5}=2$ $\quad \left\{\frac{4}{13}\right\}$

24. $\frac{3}{4x}+\frac{4}{5}=\frac{9}{10x}$ $\quad \left\{\frac{3}{16}\right\}$

25. $\frac{a}{a-2}-\frac{3}{2}=\frac{2}{a-2}$ $\quad \varnothing$

26. $\frac{4}{5y-3}=\frac{2}{3y+7}$ $\quad \{-17\}$

27. $n+\frac{1}{n}=\frac{53}{14}$ $\quad \left\{\frac{2}{7},\frac{7}{2}\right\}$

28. $\frac{1}{2x-7}+\frac{x-5}{4x^2-49}=\frac{4}{6x-21}$ $\quad \{22\}$

29. $\frac{x}{2x+1}-1=\frac{-4}{7(x-2)}$ $\quad \left\{-\frac{6}{7},3\right\}$

30. $\frac{2x}{-5}=\frac{3}{4x-13}$ $\quad \left\{\frac{3}{4},\frac{5}{2}\right\}$

31. $\frac{2n}{2n^2+11n-21}-\frac{n}{n^2+5n-14}=\frac{3}{n^2+5n-14}$ $\quad \left\{\frac{9}{7}\right\}$

32. $\frac{2}{t^2-t-6}+\frac{t+1}{t^2+t-12}=\frac{t}{t^2+6t+8}$ $\quad \left\{-\frac{5}{4}\right\}$

33. Solve $\frac{y-6}{x+1} = \frac{3}{4}$ for y. $y = \frac{3x+27}{4}$

34. Solve $\frac{x}{a} - \frac{y}{b} = 1$ for y. $y = \frac{bx-ab}{a}$

For Problems 35–40, set up an equation and solve the problem.

35. A sum of \$1400 is to be divided between two people in the ratio of $\frac{3}{5}$. How much does each person receive? \$525 and \$875

36. Working together, Dan and Don can mow a lawn in 12 minutes. Don can mow the lawn by himself in 10 minutes less time than it takes Dan by himself. How long does it take each of them to mow the lawn alone? 20 minutes for Don and 30 minutes for Dan

37. Suppose that car A can travel 250 miles in 3 hours less time than it takes car B to travel 440 miles. The rate of car B is 5 miles per hour faster than car A. Find the rates of both cars. 50 miles per hour and 55 miles per hour

38. Mark can overhaul an engine in 20 hours and Phil can do the same job by himself in 30 hours. If they both work together for a time and then Mark finishes the job by himself in 5 hours, how long did they work together? 9 hours

39. Kelly contracted to paint a house for \$640. It took him 20 hours longer than he had anticipated, so he earned \$1.60 per hour less than he had calculated. How long had he anticipated that it would take him to paint the house? 80 hours

40. Kent rode his bicycle 66 miles in $4\frac{1}{2}$ hours. For the first 40 miles he averaged a certain rate and then for the last 26 miles he reduced his rate by 3 miles per hour. Find his rate for the last 26 miles. 13 miles per hour

Thoughts into Words

41. Compare the concept of a rational number in arithmetic to the concept of a rational expression in algebra.

42. Explain in your own words how to divide two rational expressions.

43. Explain how to add two rational expressions such as $\frac{3x+4}{8} + \frac{5x-2}{12}$.

44. What is the difference between the concept of "least common multiple" and the concept of "least common denominator"?

45. Why is it important to sometimes consider more than one way to do a problem?

46. Describe the process of long division of polynomials.

Chapter 5

Exponents and Radicals

It is not uncommon in mathematics to find two separately developed concepts that are closely related to each other. In this chapter we will first develop the concepts of exponent and root individually and then show how they merge to become even more functional as a unified idea.

5.1

Using Integers as Exponents

Thus far in the text we have used only positive integers as exponents. In Chapter 1 the expression b^n, where b is any real number and n is a positive integer, was defined by

$$b^n = b \cdot b \cdot b \cdot \cdots \cdot b \qquad n \text{ factors of } b.$$

Then in Chapter 3 some of the parts of the following property served as a basis for manipulation with polynomials.

PROPERTY 5.1

If m and n are positive integers and a and b are real numbers (except $b \neq 0$ whenever it appears in a denominator), then

1. $b^n \cdot b^m = b^{n+m}$ **2.** $(b^n)^m = b^{mn}$

3. $(ab)^n = a^n b^n$ **4.** $\left(\frac{a}{b}\right)^n = \frac{a^n}{b^n}$

5. $\frac{b^n}{b^m} = b^{n-m}$ when $n > m$

$\frac{b^n}{b^m} = 1$ when $n = m$

$\frac{b^n}{b^m} = \frac{1}{b^{m-n}}$ when $n < m$

We are now ready to extend the concept of an exponent to include the use of zero and the negative integers as exponents.

First, let's consider the use of zero as an exponent. We want to use zero in such a way that the previously listed properties continue to hold. If "$b^n \cdot b^m = b^{n+m}$" is to hold, then $x^4 \cdot x^0 = x^{4+0} = x^4$. In other words, x^0 *acts like* 1 because $x^4 \cdot x^0 = x^4$. This line of reasoning suggests the following definition.

DEFINITION 5.1

If b is a nonzero real number, then

$$b^0 = 1.$$

According to Definition 5.1 the following statements are all true.

$5^0 = 1,$ $(-413)^0 = 1,$

$\left(\frac{3}{11}\right)^0 = 1,$ $n^0 = 1, \quad n \neq 0,$

$(x^3y^4)^0 = 1, \quad x \neq 0, y \neq 0$

We can use a similar line of reasoning to motivate a definition for the use of negative integers as exponents. Consider the example $x^4 \cdot x^{-4}$. If "$b^n \cdot b^m = b^{n+m}$" is to hold, then $x^4 \cdot x^{-4} = x^{4+(-4)} = x^0 = 1$. Thus, x^{-4} must be the reciprocal of x^4, since their product is 1. That is to say,

$$x^{-4} = \frac{1}{x^4}.$$

This suggests the following general definition.

DEFINITION 5.2

> If n is a positive integer and b is a nonzero real number, then
>
> $$b^{-n} = \frac{1}{b^n}.$$

According to Definition 5.2 the following statements are true.

$$x^{-5} = \frac{1}{x^5}, \qquad 2^{-4} = \frac{1}{2^4} = \frac{1}{16},$$

$$10^{-2} = \frac{1}{10^2} = \frac{1}{100} \text{ or } 0.01, \qquad \frac{2}{x^{-3}} = \frac{2}{\frac{1}{x^3}} = (2)\left(\frac{x^3}{1}\right) = 2x^3,$$

$$\left(\frac{3}{4}\right)^{-2} = \frac{1}{\left(\frac{3}{4}\right)^2} = \frac{1}{\frac{9}{16}} = \frac{16}{9}$$

It can be verified (although it is beyond the scope of this text) that all of the parts of Property 5.1 hold for *all integers*. In fact, the following equality can replace the three separate statements for part (5).

$$\frac{b^n}{b^m} = b^{n-m} \quad \text{for all integers } n \text{ and } m$$

Let's restate Property 5.1 as it holds for all integers and include, at the right, a "name tag" for easy reference.

PROPERTY 5.2

> If m and n are integers and a and b are real numbers, except $b \neq 0$ whenever it appears in a denominator, then
>
> 1. $b^n \cdot b^m = b^{n+m}$ — product of two powers
> 2. $(b^n)^m = b^{mn}$ — power of a power
> 3. $(ab)^n = a^n b^n$ — power of a product
> 4. $\left(\frac{a}{b}\right)^n = \frac{a^n}{b^n}$ — power of a quotient
> 5. $\frac{b^n}{b^m} = b^{n-m}$ — quotient of two powers

Having the use of all integers as exponents allows us to work with a large variety of numerical and algebraic expressions. Let's consider some examples that illustrate the use of the various parts of Property 5.2.

EXAMPLE 1 Simplify each of the following numerical expressions.

(a) $10^{-3} \cdot 10^{2}$ **(b)** $(2^{-3})^{-2}$ **(c)** $(2^{-1} \cdot 3^{2})^{-1}$

(d) $\left(\dfrac{2^{-3}}{3^{-2}}\right)^{-1}$ **(e)** $\dfrac{10^{-2}}{10^{-4}}$

Solution

(a) $10^{-3} \cdot 10^{2} = 10^{-3+2}$ product of two powers

$= 10^{-1}$

$= \dfrac{1}{10^{1}} = \dfrac{1}{10}$

(b) $(2^{-3})^{-2} = 2^{(-2)(-3)}$ power of a power

$= 2^{6} = 64$

(c) $(2^{-1} \cdot 3^{2})^{-1} = (2^{-1})^{-1}(3^{2})^{-1}$ power of a product

$= 2^{1} \cdot 3^{-2}$

$= \dfrac{2^{1}}{3^{2}} = \dfrac{2}{9}$

(d) $\left(\dfrac{2^{-3}}{3^{-2}}\right)^{-1} = \dfrac{(2^{-3})^{-1}}{(3^{-2})^{-1}}$ power of a quotient

$= \dfrac{2^{3}}{3^{2}} = \dfrac{8}{9}$

(e) $\dfrac{10^{-2}}{10^{-4}} = 10^{-2-(-4)}$ quotient of two powers

$= 10^{2} = 100$ ■

EXAMPLE 2 Simplify each of the following; express final results without using zero or negative integers as exponents.

(a) $x^{2} \cdot x^{-5}$ **(b)** $(x^{-2})^{4}$ **(c)** $(x^{2}y^{-3})^{-4}$

(d) $\left(\dfrac{a^{3}}{b^{-5}}\right)^{-2}$ **(e)** $\dfrac{x^{-4}}{x^{-2}}$

Solution

(a) $x^{2} \cdot x^{-5} = x^{2+(-5)}$ product of two powers

$= x^{-3}$

$= \dfrac{1}{x^{3}}$

(b) $(x^{-2})^4 = x^{4(-2)}$ power of a power

$= x^{-8}$

$= \dfrac{1}{x^8}$

(c) $(x^2y^{-3})^{-4} = (x^2)^{-4}(y^{-3})^{-4}$ power of a product

$= x^{-4(2)}y^{-4(-3)}$

$= x^{-8}y^{12}$

$= \dfrac{y^{12}}{x^8}$

(d) $\left(\dfrac{a^3}{b^{-5}}\right)^{-2} = \dfrac{(a^3)^{-2}}{(b^{-5})^{-2}}$ power of a quotient

$= \dfrac{a^{-6}}{b^{10}}$

$= \dfrac{1}{a^6b^{10}}$

(e) $\dfrac{x^{-4}}{x^{-2}} = x^{-4-(-2)}$ quotient of two powers

$= x^{-2}$

$= \dfrac{1}{x^2}$ ■

EXAMPLE 3 Find the indicated products and quotients; express your results using positive integral exponents only.

(a) $(3x^2y^{-4})(4x^{-3}y)$ **(b)** $\dfrac{12a^3b^2}{-3a^{-1}b^5}$ **(c)** $\left(\dfrac{15x^{-1}y^2}{5xy^{-4}}\right)^{-1}$

Solution

(a) $(3x^2y^{-4})(4x^{-3}y) = 12x^{2+(-3)}y^{-4+1}$

$= 12x^{-1}y^{-3}$

$= \dfrac{12}{xy^3}$

(b) $\dfrac{12a^3b^2}{-3a^{-1}b^5} = -4a^{3-(-1)}b^{2-5}$

$= -4a^4b^{-3}$

$= -\dfrac{4a^4}{b^3}$

(c) $\left(\dfrac{15x^{-1}y^{2}}{5xy^{-4}}\right)^{-1} = (3x^{-1-1}y^{2-(-4)})^{-1}$ Notice that we are first simplifying inside the parentheses.

$$= (3x^{-2}y^{6})^{-1}$$
$$= 3^{-1}x^{2}y^{-6}$$
$$= \frac{x^2}{3y^6}.$$

The final examples of this section show the simplification of numerical and algebraic expressions that involve sums and differences. In such cases, we use Definition 5.2 to change from negative to positive exponents so that we can proceed in the usual way.

EXAMPLE 4 Simplify $2^{-3} + 3^{-1}$.

Solution

$$2^{-3} + 3^{-1} = \frac{1}{2^3} + \frac{1}{3^1}$$
$$= \frac{1}{8} + \frac{1}{3}$$
$$= \frac{3}{24} + \frac{8}{24}$$
$$= \frac{11}{24}$$

EXAMPLE 5 Simplify $(4^{-1} - 3^{-2})^{-1}$.

Solution

$$(4^{-1} - 3^{-2})^{-1} = \left(\frac{1}{4^1} - \frac{1}{3^2}\right)^{-1}$$ Apply $b^{-n} = \dfrac{1}{b^n}$ to 4^{-1} and to 3^{-2}.

$$= \left(\frac{1}{4} - \frac{1}{9}\right)^{-1}$$
$$= \left(\frac{9}{36} - \frac{4}{36}\right)^{-1}$$
$$= \left(\frac{5}{36}\right)^{-1}$$
$$= \frac{1}{\left(\frac{5}{36}\right)^{1}}$$ Apply $b^{-n} = \dfrac{1}{b^n}$.

$$= \frac{1}{\frac{5}{36}} = \frac{36}{5}$$

EXAMPLE 6 Express $a^{-1} + b^{-2}$ as a single fraction involving positive exponents only.

Solution

$$a^{-1} + b^{-2} = \frac{1}{a^1} + \frac{1}{b^2}$$

$$= \left(\frac{1}{a}\right)\left(\frac{b^2}{b^2}\right) + \left(\frac{1}{b^2}\right)\left(\frac{a}{a}\right) \quad \text{Use } ab^2 \text{ as the LCD.}$$

$$= \frac{b^2}{ab^2} + \frac{a}{ab^2}$$

$$= \frac{b^2 + a}{ab^2}$$

■

Problem Set 5.1

Simplify each of the following numerical expressions.

1. 3^{-3} $\frac{1}{27}$
2. 2^{-4} $\frac{1}{16}$
3. -10^{-2} $-\frac{1}{100}$
4. 10^{-3} $\frac{1}{1000}$
5. $\frac{1}{3^{-4}}$ 81
6. $\frac{1}{2^{-6}}$ 64
7. $-\left(\frac{1}{3}\right)^{-3}$ -27
8. $\left(\frac{1}{2}\right)^{-3}$ 8
9. $\left(-\frac{1}{2}\right)^{-3}$ -8
10. $\left(\frac{2}{7}\right)^{-2}$ $\frac{49}{4}$
11. $\left(-\frac{3}{4}\right)^{0}$ 1
12. $\frac{1}{\left(\frac{4}{5}\right)^{-2}}$ $\frac{16}{25}$
13. $\frac{1}{\left(\frac{3}{7}\right)^{-2}}$ $\frac{9}{49}$
14. $-\left(\frac{5}{6}\right)^{0}$ -1
15. $2^7 \cdot 2^{-3}$ 16
16. $3^{-4} \cdot 3^6$ 9
17. $10^{-5} \cdot 10^2$ $\frac{1}{1000}$
18. $10^4 \cdot 10^{-6}$ $\frac{1}{100}$
19. $10^{-1} \cdot 10^{-2}$ $\frac{1}{1000}$
20. $10^{-2} \cdot 10^{-2}$ $\frac{1}{10000}$
21. $(3^{-1})^{-3}$ 27
22. $(2^{-2})^{-4}$ 256
23. $(5^3)^{-1}$ $\frac{1}{125}$
24. $(3^{-1})^3$ $\frac{1}{27}$
25. $(2^3 \cdot 3^{-2})^{-1}$ $\frac{9}{8}$
26. $(2^{-2} \cdot 3^{-1})^{-3}$ 1728
27. $(4^2 \cdot 5^{-1})^2$ $\frac{256}{25}$
28. $(2^{-3} \cdot 4^{-1})^{-1}$ 32
29. $\left(\frac{2^{-1}}{5^{-2}}\right)^{-1}$ $\frac{2}{25}$
30. $\left(\frac{2^{-4}}{3^{-2}}\right)^{-2}$ $\frac{256}{81}$
31. $\left(\frac{2^{-1}}{3^{-2}}\right)^{2}$ $\frac{81}{4}$
32. $\left(\frac{3^2}{5^{-1}}\right)^{-1}$ $\frac{1}{45}$
33. $\frac{3^3}{3^{-1}}$ 81
34. $\frac{2^{-2}}{2^3}$ $\frac{1}{32}$
35. $\frac{10^{-2}}{10^2}$ $\frac{1}{10{,}000}$
36. $\frac{10^{-2}}{10^{-5}}$ 1000
37. $2^{-2} + 3^{-2}$ $\frac{13}{36}$
38. $2^{-4} + 5^{-1}$ $\frac{21}{80}$
39. $\left(\frac{1}{3}\right)^{-1} - \left(\frac{2}{5}\right)^{-1}$ $\frac{1}{2}$
40. $\left(\frac{3}{2}\right)^{-1} - \left(\frac{1}{4}\right)^{-1}$ $-\frac{10}{3}$
41. $(2^{-3} + 3^{-2})^{-1}$ $\frac{72}{17}$
42. $(5^{-1} - 2^{-3})^{-1}$ $\frac{40}{3}$

Simplify each of the following; express final results without using zero or negative integers as exponents.

43. $x^2 \cdot x^{-8}$ $\frac{1}{x^6}$
44. $x^{-3} \cdot x^{-4}$ $\frac{1}{x^7}$
45. $a^3 \cdot a^{-5} \cdot a^{-1}$ $\frac{1}{a^3}$
46. $b^{-2} \cdot b^3 \cdot b^{-6}$ $\frac{1}{b^5}$
47. $(a^{-4})^2$ $\frac{1}{a^8}$
48. $(b^4)^{-3}$ $\frac{1}{b^{12}}$
49. $(x^2y^{-6})^{-1}$ $\frac{y^6}{x^2}$
50. $(x^5y^{-1})^{-3}$ $\frac{y^3}{x^{15}}$
51. $(ab^3c^{-2})^{-4}$ $\frac{c^8}{a^4b^{12}}$
52. $(a^3b^{-3}c^{-2})^{-5}$ $\frac{b^{15}c^{10}}{a^{15}}$
53. $(2x^3y^{-4})^{-3}$ $\frac{y^{12}}{8x^9}$
54. $(4x^5y^{-2})^{-2}$ $\frac{y^4}{16x^{10}}$
55. $\left(\frac{x^{-1}}{y^{-4}}\right)^{-3}$ $\frac{x^3}{y^{12}}$
56. $\left(\frac{y^3}{x^{-4}}\right)^{-2}$ $\frac{1}{x^8y^6}$
57. $\left(\frac{3a^{-2}}{2b^{-1}}\right)^{-2}$ $\frac{4a^4}{9b^2}$
58. $\left(\frac{2xy^2}{5a^{-1}b^{-2}}\right)^{-1}$ $\frac{5}{2ab^2xy^2}$
59. $\frac{x^{-6}}{x^{-4}}$ $\frac{1}{x^2}$
60. $\frac{a^{-2}}{a^2}$ $\frac{1}{a^4}$
61. $\frac{a^3b^{-2}}{a^{-2}b^{-4}}$ a^5b^2
62. $\frac{x^{-3}y^{-4}}{x^2y^{-1}}$ $\frac{1}{x^5y^3}$

Find the indicated products and quotients; express results using positive integral exponents only.

63. $(2xy^{-1})(3x^{-2}y^4)$ $\frac{6y^3}{x}$
64. $(-4x^{-1}y^2)(6x^3y^{-4})$ $-\frac{24x^2}{y^2}$
65. $(-7a^2b^{-5})(-a^{-2}b^7)$ $7b^2$
66. $(-9a^{-3}b^{-6})(-12a^{-1}b^4)$ $\frac{108}{a^4b^2}$
67. $\frac{28x^{-2}y^{-3}}{4x^{-3}y^{-1}}$ $\frac{7x}{y^2}$
68. $\frac{63x^2y^{-4}}{7xy^{-4}}$ $9x$
69. $\frac{-72a^2b^{-4}}{6a^3b^{-7}}$ $-\frac{12b^3}{a}$
70. $\frac{108a^{-5}b^{-4}}{9a^{-2}b}$ $\frac{12}{a^3b^5}$
71. $\left(\frac{35x^{-1}y^{-2}}{7x^4y^3}\right)^{-1}$ $\frac{x^5y^5}{5}$
72. $\left(\frac{-48ab^2}{-6a^3b^5}\right)^{-2}$ $\frac{a^4b^6}{64}$
73. $\left(\frac{-36a^{-1}b^{-6}}{4a^{-1}b^4}\right)^{-2}$ $\frac{b^{20}}{81}$
74. $\left(\frac{8xy^3}{-4x^4y}\right)^{-3}$ $-\frac{x^9}{8y^6}$

Express each of the following as a single fraction involving positive exponents only.

75. $x^{-2} + x^{-3}$ $\frac{x+1}{x^3}$
76. $x^{-1} + x^{-5}$ $\frac{x^4+1}{x^5}$
77. $x^{-3} - y^{-1}$ $\frac{y-x^3}{x^3y}$
78. $2x^{-1} - 3y^{-2}$ $\frac{2y^2-3x}{xy^2}$
79. $3a^{-2} + 4b^{-1}$ $\frac{3b+4a^2}{a^2b}$
80. $a^{-1} + a^{-1}b^{-3}$ $\frac{b^3+1}{ab^3}$
81. $x^{-1}y^{-2} - xy^{-1}$ $\frac{1-x^2y}{xy^2}$
82. $x^2y^{-2} - x^{-1}y^{-3}$ $\frac{x^3y-1}{xy^3}$
83. $2x^{-1} - 3x^{-2}$ $\frac{2x-3}{x^2}$
84. $5x^{-2}y + 6x^{-1}y^{-2}$ $\frac{5y^3+6x}{x^2y^2}$

5.2 Roots and Radicals

To **square a number** means to raise it to the second power, that is, to use the number as a factor twice.

$4^2 = 4 \cdot 4 = 16,$ read "four squared equals sixteen"

$10^2 = 10 \cdot 10 = 100,$

$$\left(\frac{1}{2}\right)^2 = \frac{1}{2} \cdot \frac{1}{2} = \frac{1}{4},$$

$$(-3)^2 = (-3)(-3) = 9$$

A **square root of a number** is one of its two equal factors. Thus, 4 is a square root of 16 because $4 \cdot 4 = 16$. Likewise, -4 is also a square root of 16 because $(-4)(-4) = 16$. In general, a is a square root of b if $a^2 = b$. The following generalizations are a direct consequence of the previous statement.

1. Every positive real number has two square roots; one is positive and the other is negative. They are opposites of each other.
2. Negative real numbers have no real number square roots because any nonzero real number is positive when squared.
3. The square root of 0 is 0.

The symbol $\sqrt{\ }$, called a **radical sign**, is used to designate the nonnegative square root. The number under the radical sign is called the **radicand**. The entire expression, such as $\sqrt{16}$, is called a **radical**.

$\sqrt{16} = 4,$ — $\sqrt{16}$ indicates the *nonnegative* or **principal square root** of 16.

$-\sqrt{16} = -4,$ — $-\sqrt{16}$ indicates the negative square root of 16.

$\sqrt{0} = 0,$ — Zero has only one square root. Technically, we could write $-\sqrt{0} = -0 = 0$.

$\sqrt{-4}$ is not a real number,

$-\sqrt{-4}$ is not a real number

In general, the following definition is useful.

DEFINITION 5.3

If $a \geq 0$ and $b \geq 0$, then $\sqrt{b} = a$ if and only if $a^2 = b$; a is called the **principal square root of *b*.**

To **cube a number** means to raise it to the third power, that is, to use the number as a factor three times.

$2^3 = 2 \cdot 2 \cdot 2 = 8,$ read "two cubed equals eight"

$4^3 = 4 \cdot 4 \cdot 4 = 64,$

$$\left(\frac{2}{3}\right)^3 = \frac{2}{3} \cdot \frac{2}{3} \cdot \frac{2}{3} = \frac{8}{27},$$

$$(-2)^3 = (-2)(-2)(-2) = -8$$

A **cube root of a number** is one of its three equal factors. Thus, 2 is a cube root of 8 because $2 \cdot 2 \cdot 2 = 8$. (In fact, 2 is the only real number that is a cube root of 8.)

Furthermore, -2 is a cube root of -8 because $(-2)(-2)(-2) = -8$. (In fact, -2 is the only real number that is a cube root of -8.)

In general, a is a cube root of b if $a^3 = b$. The following generalizations are a direct consequence of the previous statement.

1. Every positive real number has one positive real number cube root.
2. Every negative real number has one negative real number cube root.
3. The cube root of 0 is 0.

REMARK Technically, every nonzero real number has three cube roots, but only one of them is a real number. The other two roots are classified as complex numbers. We are restricting our work at this time to the set of real numbers.

The symbol $\sqrt[3]{}$ designates the cube root of a number. Thus, we can write

$$\sqrt[3]{8} = 2, \qquad \sqrt[3]{\frac{1}{27}} = \frac{1}{3},$$

$$\sqrt[3]{-8} = -2, \qquad \sqrt[3]{-\frac{1}{27}} = -\frac{1}{3}$$

In general, the following definition is useful.

DEFINITION 5.4

$$\sqrt[3]{b} = a \quad \text{if and only if } a^3 = b.$$

In Definition 5.4, if $b \geq 0$ then $a \geq 0$ whereas if $b < 0$ then $a < 0$. The number a is called **the principal cube root of b** or simply **the cube root of b**.

The concept of root can be extended to fourth roots, fifth roots, sixth roots, and, in general, nth roots. We can make the following generalizations.

If n is an even positive integer, then the following statements are true.

1. Every positive real number has exactly two real nth roots—one positive and one negative. For example, the real fourth roots of 16 are 2 and -2.
2. Negative real numbers do not have real nth roots. For example, there are no real fourth roots of -16.

If n is an odd positive integer greater than one, then the following statements are true.

1. Every real number has exactly one real nth root.
2. The real nth root of a positive number is positive. For example, the fifth root of 32 is 2.
3. The real nth root of a negative number is negative. For example, the fifth root of -32 is -2.

In general, the following definition is useful.

DEFINITION 5.5

> $\sqrt[n]{b} = a$ if and only if $a^n = b$.

In Definition 5.5, if n is an even positive integer, then a and b are both nonnegative. If n is an odd positive integer greater than one, then a and b are both nonnegative or both negative. The symbol $\sqrt[n]{\ }$ designates the **principal *n*th root**. Consider the following examples.

$$\sqrt[4]{81} = 3 \quad \text{because } 3^4 = 81,$$

$$\sqrt[5]{32} = 2 \quad \text{because } 2^5 = 32,$$

$$\sqrt[5]{-32} = -2 \quad \text{because } (-2)^5 = -32$$

To complete our terminology, the n in the radical $\sqrt[n]{b}$ is called the **index** of the radical. If $n = 2$, we commonly write $\sqrt{b}$ instead of $\sqrt[2]{b}$. In the future as we use symbols such as $\sqrt[n]{b}$, $\sqrt[m]{y}$, and $\sqrt[r]{x}$, we will assume the previous agreements relative to the existence of real roots (without listing the various restrictions) unless a special restriction is necessary.

The following property is a direct consequence of Definition 5.5.

PROPERTY 5.3

> **1.** $(\sqrt[n]{b})^n = b$ — n is any positive integer greater than one.
>
> **2.** $\sqrt[n]{b^n} = b$ — n is any positive integer greater than one if $b \geq 0$; n is an odd positive integer greater than one if $b < 0$.

The following examples demonstrate the use of Property 5.3.

$$\sqrt{16^2} = (\sqrt{16})^2 = 4^2 = 16,$$

$$\sqrt[3]{64^3} = (\sqrt[3]{64})^3 = 4^3 = 64,$$

$$\sqrt[3]{(-8)^3} = (\sqrt[3]{-8})^3 = (-2)^3 = -8,$$

but $\sqrt{(-16)^2} \neq (\sqrt{-16})^2$ because $\sqrt{-16}$ is not a real number.

Let's use some examples to lead into the next very useful property of radicals.

$$\sqrt{4 \cdot 9} = \sqrt{36} = 6 \quad \text{and} \quad \sqrt{4} \cdot \sqrt{9} = 2 \cdot 3 = 6,$$

$$\sqrt{16 \cdot 25} = \sqrt{400} = 20 \quad \text{and} \quad \sqrt{16} \cdot \sqrt{25} = 4 \cdot 5 = 20,$$

$$\sqrt[3]{8 \cdot 27} = \sqrt[3]{216} = 6 \quad \sqrt[3]{8} \cdot \sqrt[3]{27} = 2 \cdot 3 = 6,$$

$$\sqrt[3]{(-8)(27)} = \sqrt[3]{-216} = -6 \quad \text{and} \quad \sqrt[3]{-8} \cdot \sqrt[3]{27} = (-2)(3) = -6$$

In general, we can state the following property.

PROPERTY 5.4

> $\sqrt[n]{bc} = \sqrt[n]{b}\sqrt[n]{c}$ — $\sqrt[n]{b}$ and $\sqrt[n]{c}$ are real numbers

Property 5.4 states that **the nth root of a product is equal to the product of the nth roots.**

Simplest Radical Form

The definition of nth root, along with Property 5.4, provides the basis for changing radicals to simplest radical form. The concept of **simplest radical form** takes on additional meaning as we encounter more complicated expressions, but for now it simply means that the radicand is not to contain any perfect powers of the index. Let's consider some examples to clarify this idea.

EXAMPLE 1 Express each of the following in simplest radical form.

(a) $\sqrt{8}$ **(b)** $\sqrt{45}$ **(c)** $\sqrt[3]{24}$ **(d)** $\sqrt[3]{54}$

Solution

(a) $\sqrt{8} = \sqrt{4 \cdot 2} = \sqrt{4}\sqrt{2} = 2\sqrt{2}$

↑ 4 is a perfect square.

(b) $\sqrt{45} = \sqrt{9 \cdot 5} = \sqrt{9}\sqrt{5} = 3\sqrt{5}$

↑ 9 is a perfect square.

(c) $\sqrt[3]{24} = \sqrt[3]{8 \cdot 3} = \sqrt[3]{8}\sqrt[3]{3} = 2\sqrt[3]{3}$

↑ 8 is a perfect cube.

(d) $\sqrt[3]{54} = \sqrt[3]{27 \cdot 2} = \sqrt[3]{27}\sqrt[3]{2} = 3\sqrt[3]{2}$

↑ 27 is a perfect cube. ■

The first step in each example is to express the radicand of the given radical as the product of two factors, one of which must be a perfect nth power other than 1. Also, observe the radicands of the final radicals. In each case, the radicand *cannot* be the product of two factors; one of which must be a perfect nth power other than 1. We say that the final radicals $2\sqrt{2}$, $3\sqrt{5}$, $2\sqrt[3]{3}$, and $3\sqrt[3]{2}$ are in **simplest radical form.**

You may vary the steps somewhat in changing to simplest radical form, but the final result should be the same. Consider some different approaches to change

$\sqrt{72}$ to simplest form.

$$\sqrt{72} = \sqrt{9}\sqrt{8} = 3\sqrt{8} = 3\sqrt{4}\sqrt{2} = 3 \cdot 2\sqrt{2} = 6\sqrt{2}, \quad \text{or}$$
$$\sqrt{72} = \sqrt{4}\sqrt{18} = 2\sqrt{18} = 2\sqrt{9}\sqrt{2} = 2 \cdot 3\sqrt{2} = 6\sqrt{2}, \quad \text{or}$$
$$\sqrt{72} = \sqrt{36}\sqrt{2} = 6\sqrt{2}$$

Another variation of the technique for changing radicals to simplest form is to prime factor the radicand and then to look for perfect nth powers in exponential form. The following example illustrates the use of this technique.

EXAMPLE 2 Express each of the following in simplest radical form.

(a) $\sqrt{50}$ **(b)** $3\sqrt{80}$ **(c)** $\sqrt[3]{108}$

Solution

(a) $\sqrt{50} = \sqrt{2 \cdot 5 \cdot 5} = \sqrt{5^2}\sqrt{2} = 5\sqrt{2}$

(b) $3\sqrt{80} = 3\sqrt{2 \cdot 2 \cdot 2 \cdot 2 \cdot 5} = 3\sqrt{2^4}\sqrt{5} = 3 \cdot 2^2\sqrt{5} = 12\sqrt{5}$

(c) $\sqrt[3]{108} = \sqrt[3]{2 \cdot 2 \cdot 3 \cdot 3 \cdot 3} = \sqrt[3]{3^3}\sqrt[3]{4} = 3\sqrt[3]{4}$ ■

Another property of nth roots is demonstrated by the following examples.

$$\sqrt{\frac{36}{9}} = \sqrt{4} = 2 \quad \text{and} \quad \frac{\sqrt{36}}{\sqrt{9}} = \frac{6}{3} = 2,$$

$$\sqrt[3]{\frac{64}{8}} = \sqrt[3]{8} = 2 \quad \text{and} \quad \frac{\sqrt[3]{64}}{\sqrt[3]{8}} = \frac{4}{2} = 2,$$

$$\sqrt[3]{\frac{-8}{64}} = \sqrt[3]{-\frac{1}{8}} = -\frac{1}{2} \quad \text{and} \quad \frac{\sqrt[3]{-8}}{\sqrt[3]{64}} = \frac{-2}{4} = -\frac{1}{2}$$

In general, we can state the following property.

PROPERTY 5.5

$$\sqrt[n]{\frac{b}{c}} = \frac{\sqrt[n]{b}}{\sqrt[n]{c}}$$ $\sqrt[n]{b}$ and $\sqrt[n]{c}$ are real numbers and $c \neq 0$.

Property 5.5 states that **the nth root of a quotient is equal to the quotient of the nth roots**.

To evaluate radicals such as $\sqrt{\frac{4}{25}}$ and $\sqrt[3]{\frac{27}{8}}$, for which the numerator and denominator of the fractional radicand are perfect nth powers, you may use Property 5.5 or merely rely on the definition of nth root.

$$\sqrt{\frac{4}{25}} = \frac{\sqrt{4}}{\sqrt{25}} = \frac{2}{5} \quad \text{or} \quad \sqrt{\frac{4}{25}} = \frac{2}{5} \quad \text{because } \frac{2}{5}\cdot\frac{2}{5} = \frac{4}{25},$$

Property 5.5 (first equality in each line); definition of nth root (the "because" statements)

$$\sqrt[3]{\frac{27}{8}} = \frac{\sqrt[3]{27}}{\sqrt[3]{8}} = \frac{3}{2} \quad \text{or} \quad \sqrt[3]{\frac{27}{8}} = \frac{3}{2} \quad \text{because } \frac{3}{2}\cdot\frac{3}{2}\cdot\frac{3}{2} = \frac{27}{8}$$

Radicals such as $\sqrt{\frac{28}{9}}$ and $\sqrt[3]{\frac{24}{27}}$ in which only the denominators of the radicand are perfect nth powers can be simplified as follows.

$$\sqrt{\frac{28}{9}} = \frac{\sqrt{28}}{\sqrt{9}} = \frac{\sqrt{28}}{3} = \frac{\sqrt{4}\sqrt{7}}{3} = \frac{2\sqrt{7}}{3},$$

$$\sqrt[3]{\frac{24}{27}} = \frac{\sqrt[3]{24}}{\sqrt[3]{27}} = \frac{\sqrt[3]{24}}{3} = \frac{\sqrt[3]{8}\sqrt[3]{3}}{3} = \frac{2\sqrt[3]{3}}{3}$$

Before considering more examples, let's summarize some ideas that pertain to the simplifying of radicals. A radical is said to be in **simplest radical form** if the following conditions are satisfied.

1. No fraction appears with a radical sign. $\sqrt{\frac{3}{4}}$ violates this condition.
2. No radical appears in the denominator. $\frac{\sqrt{2}}{\sqrt{3}}$ violates this condition.
3. No radicand when expressed in prime factored form contains a factor raised to a power equal to or greater than the index. $\sqrt{2^3 \cdot 5}$ violates this condition.

Now let's consider an example in which neither the numerator nor the denominator of the radicand is a perfect nth power.

EXAMPLE 3 Simplify $\sqrt{\frac{2}{3}}$.

Solution

$$\sqrt{\frac{2}{3}} = \frac{\sqrt{2}}{\sqrt{3}} = \frac{\sqrt{2}}{\sqrt{3}} \cdot \frac{\sqrt{3}}{\sqrt{3}} = \frac{\sqrt{6}}{3}$$

form of 1 (refers to $\frac{\sqrt{3}}{\sqrt{3}}$)

■

We refer to the process we used to simplify the radical in Example 3 as **rationalizing the denominator**. Notice that the denominator becomes a rational number. The process of rationalizing the denominator can often be accomplished in more than one way, as we will see in the next example.

EXAMPLE 4 Simplify $\dfrac{\sqrt{5}}{\sqrt{8}}$.

Solution A

$$\frac{\sqrt{5}}{\sqrt{8}} = \frac{\sqrt{5}}{\sqrt{8}} \cdot \frac{\sqrt{8}}{\sqrt{8}} = \frac{\sqrt{40}}{8} = \frac{\sqrt{4}\sqrt{10}}{8} = \frac{2\sqrt{10}}{8} = \frac{\sqrt{10}}{4}$$

Solution B

$$\frac{\sqrt{5}}{\sqrt{8}} = \frac{\sqrt{5}}{\sqrt{8}} \cdot \frac{\sqrt{2}}{\sqrt{2}} = \frac{\sqrt{10}}{\sqrt{16}} = \frac{\sqrt{10}}{4}$$

Solution C

$$\frac{\sqrt{5}}{\sqrt{8}} = \frac{\sqrt{5}}{\sqrt{4}\sqrt{2}} = \frac{\sqrt{5}}{2\sqrt{2}} = \frac{\sqrt{5}}{2\sqrt{2}} \cdot \frac{\sqrt{2}}{\sqrt{2}} = \frac{\sqrt{10}}{4}$$ ■

The three approaches to Example 4 again illustrate the need to think first and then push the pencil. You may find one approach easier than another.

To conclude this section, study the following examples and check the final radicals according to the three conditions previously listed for **simplest radical form**.

EXAMPLE 5 Simplify each of the following.

(a) $\dfrac{3\sqrt{2}}{5\sqrt{3}}$ **(b)** $\dfrac{3\sqrt{7}}{2\sqrt{18}}$ **(c)** $\sqrt[3]{\dfrac{5}{9}}$ **(d)** $\dfrac{\sqrt[3]{5}}{\sqrt[3]{16}}$

Solution

(a) $$\frac{3\sqrt{2}}{5\sqrt{3}} = \frac{3\sqrt{2}}{5\sqrt{3}} \cdot \underbrace{\frac{\sqrt{3}}{\sqrt{3}}}_{\text{form of 1}} = \frac{3\sqrt{6}}{5\sqrt{9}} = \frac{3\sqrt{6}}{15} = \frac{\sqrt{6}}{5}$$

(b) $$\frac{3\sqrt{7}}{2\sqrt{18}} = \frac{3\sqrt{7}}{2\sqrt{18}} \cdot \underbrace{\frac{\sqrt{2}}{\sqrt{2}}}_{\text{form of 1}} = \frac{3\sqrt{14}}{2\sqrt{36}} = \frac{3\sqrt{14}}{12} = \frac{\sqrt{14}}{4}$$

(c) $$\sqrt[3]{\frac{5}{9}} = \frac{\sqrt[3]{5}}{\sqrt[3]{9}} = \frac{\sqrt[3]{5}}{\sqrt[3]{9}} \cdot \underbrace{\frac{\sqrt[3]{3}}{\sqrt[3]{3}}}_{\text{form of 1}} = \frac{\sqrt[3]{15}}{\sqrt[3]{27}} = \frac{\sqrt[3]{15}}{3}$$

(d) $$\frac{\sqrt[3]{5}}{\sqrt[3]{16}} = \frac{\sqrt[3]{5}}{\sqrt[3]{16}} \cdot \underbrace{\frac{\sqrt[3]{4}}{\sqrt[3]{4}}}_{\text{form of 1}} = \frac{\sqrt[3]{20}}{\sqrt[3]{64}} = \frac{\sqrt[3]{20}}{4}$$ ■

Problem Set 5.2

Evaluate each of the following; for example, $\sqrt{25} = 5$.

1. $\sqrt{64}$ 8
2. $\sqrt{49}$ 7
3. $-\sqrt{100}$ -10
4. $-\sqrt{81}$ -9
5. $\sqrt[3]{27}$ 3
6. $\sqrt[3]{216}$ 6
7. $\sqrt[3]{-64}$ -4
8. $\sqrt[3]{-125}$ -5
9. $\sqrt[4]{81}$ 3
10. $-\sqrt[4]{16}$ -2
11. $\sqrt{\frac{16}{25}}$ $\frac{4}{5}$
12. $\sqrt{\frac{25}{64}}$ $\frac{5}{8}$
13. $-\sqrt{\frac{36}{49}}$ $-\frac{6}{7}$
14. $\sqrt{\frac{16}{64}}$ $\frac{1}{2}$
15. $\sqrt{\frac{9}{36}}$ $\frac{1}{2}$
16. $\sqrt{\frac{144}{36}}$ 2
17. $\sqrt[3]{\frac{27}{64}}$ $\frac{3}{4}$
18. $\sqrt[3]{-\frac{8}{27}}$ $-\frac{2}{3}$
19. $\sqrt[3]{8^3}$ 8
20. $\sqrt[4]{16^4}$ 16

Change each of the following radicals to simplest radical form.

21. $\sqrt{27}$ $3\sqrt{3}$
22. $\sqrt{48}$ $4\sqrt{3}$
23. $\sqrt{32}$ $4\sqrt{2}$
24. $\sqrt{98}$ $7\sqrt{2}$
25. $\sqrt{80}$ $4\sqrt{5}$
26. $\sqrt{125}$ $5\sqrt{5}$
27. $\sqrt{160}$ $4\sqrt{10}$
28. $\sqrt{112}$ $4\sqrt{7}$
29. $4\sqrt{18}$ $12\sqrt{2}$
30. $5\sqrt{32}$ $20\sqrt{2}$
31. $-6\sqrt{20}$ $-12\sqrt{5}$
32. $-4\sqrt{54}$ $-12\sqrt{6}$
33. $\frac{2}{5}\sqrt{75}$ $2\sqrt{3}$
34. $\frac{1}{3}\sqrt{90}$ $\sqrt{10}$
35. $\frac{3}{2}\sqrt{24}$ $3\sqrt{6}$
36. $\frac{3}{4}\sqrt{45}$ $\frac{9}{4}\sqrt{5}$
37. $-\frac{5}{6}\sqrt{28}$ $-\frac{5}{3}\sqrt{7}$
38. $-\frac{2}{3}\sqrt{96}$ $-\frac{8}{3}\sqrt{6}$
39. $\sqrt{\frac{19}{4}}$ $\frac{\sqrt{19}}{2}$
40. $\sqrt{\frac{22}{9}}$ $\frac{\sqrt{22}}{3}$
41. $\sqrt{\frac{27}{16}}$ $\frac{3\sqrt{3}}{4}$
42. $\sqrt{\frac{8}{25}}$ $\frac{2\sqrt{2}}{5}$
43. $\sqrt{\frac{75}{81}}$ $\frac{5\sqrt{3}}{9}$
44. $\sqrt{\frac{24}{49}}$ $\frac{2\sqrt{6}}{7}$
45. $\sqrt{\frac{2}{7}}$ $\frac{\sqrt{14}}{7}$
46. $\sqrt{\frac{3}{8}}$ $\frac{\sqrt{6}}{4}$
47. $\sqrt{\frac{2}{3}}$ $\frac{\sqrt{6}}{3}$
48. $\sqrt{\frac{7}{12}}$ $\frac{\sqrt{21}}{6}$
49. $\frac{\sqrt{5}}{\sqrt{12}}$ $\frac{\sqrt{15}}{6}$
50. $\frac{\sqrt{3}}{\sqrt{7}}$ $\frac{\sqrt{21}}{7}$
51. $\frac{\sqrt{11}}{\sqrt{24}}$ $\frac{\sqrt{66}}{12}$
52. $\frac{\sqrt{5}}{\sqrt{48}}$ $\frac{\sqrt{15}}{12}$
53. $\frac{\sqrt{18}}{\sqrt{27}}$ $\frac{\sqrt{6}}{3}$
54. $\frac{\sqrt{10}}{\sqrt{20}}$ $\frac{\sqrt{2}}{2}$
55. $\frac{\sqrt{35}}{\sqrt{7}}$ $\sqrt{5}$
56. $\frac{\sqrt{42}}{\sqrt{6}}$ $\sqrt{7}$
57. $\frac{2\sqrt{3}}{\sqrt{7}}$ $\frac{2\sqrt{21}}{7}$
58. $\frac{3\sqrt{2}}{\sqrt{6}}$ $\sqrt{3}$
59. $-\frac{4\sqrt{12}}{\sqrt{5}}$ $-\frac{8\sqrt{15}}{5}$

60. $\frac{-6\sqrt{5}}{\sqrt{18}}$ $-\sqrt{10}$
61. $\frac{3\sqrt{2}}{4\sqrt{3}}$ $\frac{\sqrt{6}}{4}$
62. $\frac{6\sqrt{5}}{5\sqrt{12}}$ $\frac{\sqrt{15}}{5}$
63. $\frac{-8\sqrt{18}}{10\sqrt{50}}$ $-\frac{12}{25}$
64. $\frac{4\sqrt{45}}{-6\sqrt{20}}$ -1
65. $\sqrt[3]{16}$ $2\sqrt[3]{2}$
66. $\sqrt[3]{40}$ $2\sqrt[3]{5}$
67. $2\sqrt[3]{81}$ $6\sqrt[3]{3}$
68. $-3\sqrt[3]{54}$ $-9\sqrt[3]{2}$
69. $\frac{2}{\sqrt[3]{9}}$ $\frac{2\sqrt[3]{3}}{3}$
70. $\frac{3}{\sqrt[3]{3}}$ $\sqrt[3]{9}$
71. $\frac{\sqrt[3]{27}}{\sqrt[3]{4}}$ $\frac{3\sqrt[3]{2}}{2}$
72. $\frac{\sqrt[3]{8}}{\sqrt[3]{16}}$ $\frac{\sqrt[3]{4}}{2}$
73. $\frac{\sqrt[3]{6}}{\sqrt[3]{4}}$ $\frac{\sqrt[3]{12}}{2}$
74. $\frac{\sqrt[3]{4}}{\sqrt[3]{2}}$ $\sqrt[3]{2}$

75. Note the key $\boxed{\sqrt{\ }}$ on your calculator. It calculates the square root of the displayed number. Use your calculator to find an approximation, to the nearest thousandth, for each of the following.

(a) $\sqrt{2}$ 1.414 (b) $\sqrt{75}$ 8.660 (c) $\sqrt{156}$ 12.490
(d) $\sqrt{691}$ 26.287 (e) $\sqrt{3249}$ 57.000 (f) $\sqrt{45123}$ 212.422
(g) $\sqrt{0.14}$ .374 (h) $\sqrt{0.023}$ .152 (i) $\sqrt{0.8649}$ .930

76. Your calculator may have a key labeled $\boxed{\sqrt[x]{y}}$. It calculates cube roots, fourth roots, fifth roots, and so on. Be sure you can use the $\boxed{\sqrt[x]{y}}$ key on your calculator by evaluating each of the following. (If your calculator does not have a $\boxed{\sqrt[x]{y}}$ key, wait until we study Section 5.6!)

(a) $\sqrt[3]{729}$ 9 (b) $\sqrt[3]{2744}$ 14 (c) $\sqrt[4]{4096}$ 8
(d) $\sqrt[4]{234256}$ 22 (e) $\sqrt[5]{7776}$ 6 (f) $\sqrt[5]{371293}$ 13

77. See how your calculator reacts to each of the following.

(a) $\sqrt{-4}$ (b) $\sqrt[3]{-8}$ (c) $\sqrt[4]{-16}$

5.3 Combining Radicals and Simplifying Radicals That Contain Variables

Recall our use of the distributive property as the basis for combining similar terms. For example,

$$3x + 2x = (3 + 2)x = 5x,$$

$$8y - 5y = (8 - 5)y = 3y,$$

$$\frac{2}{3}a^2 + \frac{3}{4}a^2 = \left(\frac{2}{3} + \frac{3}{4}\right)a^2 = \left(\frac{8}{12} + \frac{9}{12}\right)a^2 = \frac{17}{12}a^2.$$

In a like manner, expressions that contain radicals can often be simplified by using the distributive property as follows.

$$3\sqrt{2} + 5\sqrt{2} = (3 + 5)\sqrt{2} = 8\sqrt{2},$$
$$7\sqrt[3]{5} - 3\sqrt[3]{5} = (7 - 3)\sqrt[3]{5} = 4\sqrt[3]{5},$$
$$4\sqrt{7} + 5\sqrt{7} + 6\sqrt{11} - 2\sqrt{11} = (4 + 5)\sqrt{7} + (6 - 2)\sqrt{11} = 9\sqrt{7} + 4\sqrt{11}$$

Notice that *to add or subtract radicals they must have the same index and the same radicand.* Thus, we cannot simplify an expression such as $5\sqrt{2} + 7\sqrt{11}$.

Simplifying by combining radicals sometimes requires that you first express the given radicals in simplest form and then apply the distributive property. The following examples illustrate this idea.

EXAMPLE 1 Simplify $3\sqrt{8} + 2\sqrt{18} - 4\sqrt{2}$.

Solution

$$\begin{aligned} 3\sqrt{8} + 2\sqrt{18} - 4\sqrt{2} &= 3\sqrt{4}\sqrt{2} + 2\sqrt{9}\sqrt{2} - 4\sqrt{2} \\ &= 6\sqrt{2} + 6\sqrt{2} - 4\sqrt{2} \\ &= (6 + 6 - 4)\sqrt{2} = 8\sqrt{2} \end{aligned}$$

■

EXAMPLE 2 Simplify $\frac{1}{4}\sqrt{45} + \frac{1}{3}\sqrt{20}$.

Solution

$$\begin{aligned} \frac{1}{4}\sqrt{45} + \frac{1}{3}\sqrt{20} &= \frac{1}{4}\sqrt{9}\sqrt{5} + \frac{1}{3}\sqrt{4}\sqrt{5} \\ &= \frac{1}{4}\cdot 3\cdot\sqrt{5} + \frac{1}{3}\cdot 2\cdot\sqrt{5} \\ &= \frac{3}{4}\sqrt{5} + \frac{2}{3}\sqrt{5} = \left(\frac{3}{4} + \frac{2}{3}\right)\sqrt{5} \\ &= \left(\frac{9}{12} + \frac{8}{12}\right)\sqrt{5} = \frac{17}{12}\sqrt{5} \end{aligned}$$

■

EXAMPLE 3 Simplify $5\sqrt[3]{2} - 2\sqrt[3]{16} - 6\sqrt[3]{54}$.

Solution

$$\begin{aligned} 5\sqrt[3]{2} - 2\sqrt[3]{16} - 6\sqrt[3]{54} &= 5\sqrt[3]{2} - 2\sqrt[3]{8}\sqrt[3]{2} - 6\sqrt[3]{27}\sqrt[3]{2} \\ &= 5\sqrt[3]{2} - 2\cdot 2\cdot\sqrt[3]{2} - 6\cdot 3\cdot\sqrt[3]{2} \\ &= 5\sqrt[3]{2} - 4\sqrt[3]{2} - 18\sqrt[3]{2} \\ &= (5 - 4 - 18)\sqrt[3]{2} \\ &= -17\sqrt[3]{2} \end{aligned}$$

■

Radicals That Contain Variables

Before we discuss the process of simplifying *radicals that contain variables*, there is one technicality that we should call to your attention. Let's look at some examples to clarify the point. Consider the radical $\sqrt{x^2}$ as follows.

Let $x = 3$, then $\sqrt{x^2} = \sqrt{3^2} = \sqrt{9} = 3$.

Let $x = -3$, then $\sqrt{x^2} = \sqrt{(-3)^2} = \sqrt{9} = 3$.

Thus, if $x \geq 0$, then $\sqrt{x^2} = x$, *but* if $x < 0$, then $\sqrt{x^2} = -x$. Using the concept of absolute value we can state that *for all real numbers*, $\sqrt{x^2} = |x|$.

Now consider the radical $\sqrt{x^3}$. Since x^3 is negative when x is negative, we need to restrict x to the nonnegative reals when working with $\sqrt{x^3}$. Thus, we can write if $x \geq 0$, then $\sqrt{x^3} = \sqrt{x^2}\sqrt{x} = x\sqrt{x}$, and no absolute value sign is necessary. Finally, let's consider the radical $\sqrt[3]{x^3}$.

Let $x = 2$, then $\sqrt[3]{x^3} = \sqrt[3]{2^3} = \sqrt[3]{8} = 2$.

Let $x = -2$, then $\sqrt[3]{x^3} = \sqrt[3]{(-2)^3} = \sqrt[3]{-8} = -2$.

Thus, it is correct to write: $\sqrt[3]{x^3} = x$ for all real numbers, and again no absolute value sign is necessary.

The previous discussion indicates that technically every radical expression involving variables in the radicand needs to be analyzed individually as to the necessary restrictions imposed on the variables. However, to avoid considering such restrictions on a problem-to-problem basis we shall merely *assume that all variables represent positive real numbers.*

Let's consider the process of simplifying radicals that contain variables in the radicand. Study the following examples and notice that it is the same basic approach we used in Section 5.2.

EXAMPLE 4 Simplify each of the following.

(a) $\sqrt{8x^3}$ **(b)** $\sqrt{45x^3y^7}$ **(c)** $\sqrt{180a^4b^3}$ **(d)** $\sqrt[3]{40x^4y^8}$

Solution

(a) $\sqrt{8x^3} = \sqrt{4x^2}\sqrt{2x} = 2x\sqrt{2x}$

$4x^2$ is a perfect square.

(b) $\sqrt{45x^3y^7} = \sqrt{9x^2y^6}\sqrt{5xy} = 3xy^3\sqrt{5xy}$

$9x^2y^6$ is a perfect square.

(c) If the numerical coefficient of the radicand is quite large, you may want to look at it in prime factored form.

$$\begin{aligned}\sqrt{180a^4b^3} &= \sqrt{2\cdot 2\cdot 3\cdot 3\cdot 5\cdot a^4\cdot b^3}\\ &= \sqrt{36\cdot 5\cdot a^4\cdot b^3}\\ &= \sqrt{36a^4b^2}\sqrt{5b}\\ &= 6a^2b\sqrt{5b}\end{aligned}$$

(d) $\sqrt[3]{40x^4y^8} = \sqrt[3]{8x^3y^6}\sqrt[3]{5xy^2} = 2xy^2\sqrt[3]{5xy^2}$ ■

($8x^3y^6$ is a perfect cube.)

Before we consider more examples, let's restate (so as to include radicands containing variables) the conditions necessary for a radical to be in *simplest radical form.*

1. A radicand contains no polynomial factor raised to a power equal to or greater than the index of the radical. $\sqrt{x^3}$ violates this condition.

2. No fraction appears within a radical sign. $\sqrt{\frac{2x}{3y}}$ violates this condition.

3. No radical appears in the denominator. $\frac{3}{\sqrt[3]{4x}}$ violates this condition.

EXAMPLE 5 Express each of the following in simplest radical form.

(a) $\sqrt{\frac{2x}{3y}}$ **(b)** $\frac{\sqrt{5}}{\sqrt{12a^3}}$ **(c)** $\frac{\sqrt{8x^2}}{\sqrt{27y^5}}$

(d) $\frac{3}{\sqrt[3]{4x}}$ **(e)** $\frac{\sqrt[3]{16x^2}}{\sqrt[3]{9y^5}}$

Solution

(a) $\sqrt{\frac{2x}{3y}} = \frac{\sqrt{2x}}{\sqrt{3y}} = \frac{\sqrt{2x}}{\sqrt{3y}}\cdot\frac{\sqrt{3y}}{\sqrt{3y}} = \frac{\sqrt{6xy}}{3y}$

(form of 1)

(b) $\frac{\sqrt{5}}{\sqrt{12a^3}} = \frac{\sqrt{5}}{\sqrt{12a^3}}\cdot\frac{\sqrt{3a}}{\sqrt{3a}} = \frac{\sqrt{15a}}{\sqrt{36a^4}} = \frac{\sqrt{15a}}{6a^2}$

(form of 1)

(c) $\frac{\sqrt{8x^2}}{\sqrt{27y^5}} = \frac{\sqrt{4x^2}\sqrt{2}}{\sqrt{9y^4}\sqrt{3y}} = \frac{2x\sqrt{2}}{3y^2\sqrt{3y}} = \frac{2x\sqrt{2}}{3y^2\sqrt{3y}}\cdot\frac{\sqrt{3y}}{\sqrt{3y}}$

$= \frac{2x\sqrt{6y}}{(3y^2)(3y)} = \frac{2x\sqrt{6y}}{9y^3}$

(d) $\frac{3}{\sqrt[3]{4x}} = \frac{3}{\sqrt[3]{4x}} \cdot \frac{\sqrt[3]{2x^2}}{\sqrt[3]{2x^2}} = \frac{3\sqrt[3]{2x^2}}{\sqrt[3]{8x^3}} = \frac{3\sqrt[3]{2x^2}}{2x}$

(e) $\frac{\sqrt[3]{16x^2}}{\sqrt[3]{9y^5}} = \frac{\sqrt[3]{16x^2}}{\sqrt[3]{9y^5}} \cdot \frac{\sqrt[3]{3y}}{\sqrt[3]{3y}} = \frac{\sqrt[3]{48x^2y}}{\sqrt[3]{27y^6}} = \frac{\sqrt[3]{8}\sqrt[3]{6x^2y}}{3y^2} = \frac{2\sqrt[3]{6x^2y}}{3y^2}$ ■

Notice that in **(c)** we did some simplifying first before rationalizing the denominator, whereas in **(b)** we proceeded immediately to rationalize the denominator. This is an individual choice and you should probably do it both ways a few times to help determine your preference.

Problem Set 5.3

Use the distributive property to help simplify each of the following. For example,

$$\begin{aligned} 3\sqrt{8} - \sqrt{32} &= 3\sqrt{4}\sqrt{2} - \sqrt{16}\sqrt{2} \\ &= 3(2)\sqrt{2} - 4\sqrt{2} \\ &= 6\sqrt{2} - 4\sqrt{2} \\ &= (6-4)\sqrt{2} = 2\sqrt{2}. \end{aligned}$$

1. $5\sqrt{18} - 2\sqrt{2}$ $13\sqrt{2}$

2. $7\sqrt{12} + 4\sqrt{3}$ $18\sqrt{3}$

3. $7\sqrt{12} + 10\sqrt{48}$ $54\sqrt{3}$

4. $6\sqrt{8} - 5\sqrt{18}$ $-3\sqrt{2}$

5. $-2\sqrt{50} - 5\sqrt{32}$ $-30\sqrt{2}$

6. $-2\sqrt{20} - 7\sqrt{45}$ $-25\sqrt{5}$

7. $3\sqrt{20} - \sqrt{5} - 2\sqrt{45}$ $-\sqrt{5}$

8. $6\sqrt{12} + \sqrt{3} - 2\sqrt{48}$ $5\sqrt{3}$

9. $-9\sqrt{24} + 3\sqrt{54} - 12\sqrt{6}$ $-21\sqrt{6}$

10. $13\sqrt{28} - 2\sqrt{63} - 7\sqrt{7}$ $13\sqrt{7}$

11. $\frac{3}{4}\sqrt{7} - \frac{2}{3}\sqrt{28}$ $-\frac{7\sqrt{7}}{12}$

12. $\frac{3}{5}\sqrt{5} - \frac{1}{4}\sqrt{80}$ $-\frac{2\sqrt{5}}{5}$

13. $\frac{3}{5}\sqrt{40} + \frac{5}{6}\sqrt{90}$ $\frac{37\sqrt{10}}{10}$

14. $\frac{3}{8}\sqrt{96} - \frac{2}{3}\sqrt{54}$ $-\frac{\sqrt{6}}{2}$

15. $\frac{3\sqrt{18}}{5} - \frac{5\sqrt{72}}{6} + \frac{3\sqrt{98}}{4}$ $\frac{41\sqrt{2}}{20}$

16. $\frac{-2\sqrt{20}}{3} + \frac{3\sqrt{45}}{4} - \frac{5\sqrt{80}}{6}$ $-\frac{29\sqrt{5}}{12}$

17. $5\sqrt[3]{3} + 2\sqrt[3]{24} - 6\sqrt[3]{81}$ $-9\sqrt[3]{3}$

18. $-3\sqrt[3]{2} - 2\sqrt[3]{16} + \sqrt[3]{54}$ $-4\sqrt[3]{2}$

19. $-\sqrt[3]{16} + 7\sqrt[3]{54} - 9\sqrt[3]{2}$ $10\sqrt[3]{2}$

20. $4\sqrt[3]{24} - 6\sqrt[3]{3} + 13\sqrt[3]{81}$ $41\sqrt[3]{3}$

Express each of the following in simplest radical form. All variables represent positive real numbers.

21. $\sqrt{32x}$ $4\sqrt{2x}$

22. $\sqrt{50y}$ $5\sqrt{2y}$

23. $\sqrt{75x^2}$ $5x\sqrt{3}$

24. $\sqrt{108y^2}$ $6y\sqrt{3}$

25. $\sqrt{20x^2y}$ $2x\sqrt{5y}$

26. $\sqrt{80xy^2}$ $4y\sqrt{5x}$

27. $\sqrt{64x^3y^7}$ $8xy^3\sqrt{xy}$

28. $\sqrt{36x^5y^6}$ $6x^2y^3\sqrt{x}$

29. $\sqrt{54a^4b^3}$ $3a^2b\sqrt{6b}$

30. $\sqrt{96a^7b^8}$ $4a^3b^4\sqrt{6a}$

31. $\sqrt{63x^6y^8}$ $3x^3y^4\sqrt{7}$

32. $\sqrt{28x^4y^{12}}$ $2x^2y^6\sqrt{7}$

33. $2\sqrt{40a^3}$ $4a\sqrt{10a}$ **34.** $4\sqrt{90a^5}$ $12a^2\sqrt{10a}$ **35.** $\frac{2}{3}\sqrt{96xy^3}$ $\frac{8y}{3}\sqrt{6xy}$

36. $\frac{4}{5}\sqrt{125x^4y}$ $4x^2\sqrt{5y}$ **37.** $\sqrt{\frac{2x}{5y}}$ $\frac{\sqrt{10xy}}{5y}$ **38.** $\sqrt{\frac{3x}{2y}}$ $\frac{\sqrt{6xy}}{2y}$

39. $\sqrt{\frac{5}{12x^4}}$ $\frac{\sqrt{15}}{6x^2}$ **40.** $\sqrt{\frac{7}{8x^2}}$ $\frac{\sqrt{14}}{4x}$ **41.** $\frac{5}{\sqrt{18y}}$ $\frac{5\sqrt{2y}}{6y}$

42. $\frac{3}{\sqrt{12x}}$ $\frac{\sqrt{3x}}{2x}$ **43.** $\frac{\sqrt{7x}}{\sqrt{8y^5}}$ $\frac{\sqrt{14xy}}{4y^3}$ **44.** $\frac{\sqrt{5y}}{\sqrt{18x^3}}$ $\frac{\sqrt{10xy}}{6x^2}$

45. $\frac{\sqrt{18y^3}}{\sqrt{16x}}$ $\frac{3y\sqrt{2xy}}{4x}$ **46.** $\frac{\sqrt{2x^3}}{\sqrt{9y}}$ $\frac{x\sqrt{2xy}}{3y}$ **47.** $\frac{\sqrt{24a^2b^3}}{\sqrt{7ab^6}}$ $\frac{2\sqrt{42ab}}{7b^2}$

48. $\frac{\sqrt{12a^2b}}{\sqrt{5a^3b^3}}$ $\frac{2\sqrt{15a}}{5ab}$ **49.** $\sqrt[3]{24y}$ $2\sqrt[3]{3y}$ **50.** $\sqrt[3]{16x^2}$ $2\sqrt[3]{2x^2}$

51. $\sqrt[3]{16x^4}$ $2x\sqrt[3]{2x}$ **52.** $\sqrt[3]{54x^3}$ $3x\sqrt[3]{2}$ **53.** $\sqrt[3]{56x^6y^8}$ $2x^2y^2\sqrt[3]{7y^2}$

54. $\sqrt[3]{81x^5y^6}$ $3xy^2\sqrt[3]{3x^2}$ **55.** $\sqrt[3]{\frac{7}{9x^2}}$ $\frac{\sqrt[3]{21x}}{3x}$ **56.** $\sqrt[3]{\frac{5}{2x}}$ $\frac{\sqrt[3]{20x^2}}{2x}$

57. $\frac{\sqrt[3]{3y}}{\sqrt[3]{16x^4}}$ $\frac{\sqrt[3]{12x^2y}}{4x^2}$ **58.** $\frac{\sqrt[3]{2y}}{\sqrt[3]{3x}}$ $\frac{\sqrt[3]{18x^2y}}{3x}$

59. $\frac{\sqrt[3]{12xy}}{\sqrt[3]{3x^2y^5}}$ $\frac{\sqrt[3]{4x^2y^2}}{xy^2}$ **60.** $\frac{5}{\sqrt[3]{9xy^2}}$ $\frac{5\sqrt[3]{3x^2y}}{3xy}$

61. $\sqrt{8x + 12y}$ [*Hint*: $\sqrt{8x + 12y} = \sqrt{4(2x + 3y)}$] $2\sqrt{2x + 3y}$

62. $\sqrt{4x + 4y}$ $2\sqrt{x + y}$ **63.** $\sqrt{16x + 48y}$ $4\sqrt{x + 3y}$ **64.** $\sqrt{27x + 18y}$ $3\sqrt{3x + 2y}$

Use the distributive property to help simplify each of the following. All variables represent positive real numbers.

65. $-3\sqrt{4x} + 5\sqrt{9x} + 6\sqrt{16x}$ $33\sqrt{x}$ **66.** $-2\sqrt{25x} - 4\sqrt{36x} + 7\sqrt{64x}$ $22\sqrt{x}$

67. $2\sqrt{18x} - 3\sqrt{8x} - 6\sqrt{50x}$ $-30\sqrt{2x}$ **68.** $4\sqrt{20x} + 5\sqrt{45x} - 10\sqrt{80x}$ $-17\sqrt{5x}$

69. $5\sqrt{27n} - \sqrt{12n} - 6\sqrt{3n}$ $7\sqrt{3n}$ **70.** $4\sqrt{8n} + 3\sqrt{18n} - 2\sqrt{72n}$ $5\sqrt{2n}$

71. $7\sqrt{4ab} - \sqrt{16ab} - 10\sqrt{25ab}$ $-40\sqrt{ab}$ **72.** $4\sqrt{ab} - 9\sqrt{36ab} + 6\sqrt{49ab}$ $-8\sqrt{ab}$

73. $-3\sqrt{2x^3} + 4\sqrt{8x^3} - 3\sqrt{32x^3}$ $-7x\sqrt{2x}$

74. $2\sqrt{40x^5} - 3\sqrt{90x^5} + 5\sqrt{160x^5}$ $15x^2\sqrt{10x}$

Miscellaneous Problems

75. **(a)** Use 1.414 as a rational approximation for $\sqrt{2}$ and find the value of $\frac{1}{\sqrt{2}}$ to the nearest one-thousandth. .707

(b) Change $\frac{1}{\sqrt{2}}$ to simplest radical form. $\frac{\sqrt{2}}{2}$

(c) Use 1.414 as a rational approximation for $\sqrt{2}$ and find the value of $\frac{\sqrt{2}}{2}$ to the nearest one-thousandth. .707

(d) Use 1.414 as a rational approximation for $\sqrt{2}$ and find the value of the following to the nearest one-thousandth.

(1) $\frac{5}{\sqrt{2}}$ 3.536 **(2)** $\frac{3}{4\sqrt{2}}$ .530

(3) $\sqrt{8}+\sqrt{18}$ 7.070 **(4)** $3\sqrt{32}-4\sqrt{50}$ -11.312

(5) $\frac{3}{\sqrt{8}}$ 1.061 **(6)** $\frac{-5}{\sqrt{18}}$ -1.178

5.4

Products and Quotients Involving Radicals

As we have seen, Property 5.4 ($\sqrt[n]{bc}=\sqrt[n]{b}\sqrt[n]{c}$) is used to express one radical as the product of two radicals and also to express the product of two radicals as one radical. In fact, we have used the property for both purposes within the framework of simplifying radicals. For example,

$$\frac{\sqrt{3}}{\sqrt{32}}=\frac{\sqrt{3}}{\sqrt{16}\sqrt{2}}=\frac{\sqrt{3}}{4\sqrt{2}}=\frac{\sqrt{3}}{4\sqrt{2}}\cdot\frac{\sqrt{2}}{\sqrt{2}}=\frac{\sqrt{6}}{8}.$$

$\uparrow$ $\sqrt[n]{bc}=\sqrt[n]{b}\sqrt[n]{c}$ $\uparrow$ $\sqrt[n]{b}\sqrt[n]{c}=\sqrt[n]{bc}$

The following examples demonstrate the use of Property 5.4 to multiply radicals and to express the product in simplest form.

EXAMPLE 1 Multiply and simplify where possible.

(a) $(2\sqrt{3})(3\sqrt{5})$ **(b)** $(3\sqrt{8})(5\sqrt{2})$

(c) $(7\sqrt{6})(3\sqrt{8})$ **(d)** $(2\sqrt[3]{6})(5\sqrt[3]{4})$

Solution

(a) $(2\sqrt{3})(3\sqrt{5})=2\cdot3\cdot\sqrt{3}\cdot\sqrt{5}=6\sqrt{15}.$

(b) $(3\sqrt{8})(5\sqrt{2})=3\cdot5\cdot\sqrt{8}\cdot\sqrt{2}=15\sqrt{16}=15\cdot4=60.$

(c) $(7\sqrt{6})(3\sqrt{8})=7\cdot3\cdot\sqrt{6}\cdot\sqrt{8}=21\sqrt{48}=21\sqrt{16}\sqrt{3}=21\cdot4\cdot\sqrt{3}$
$=84\sqrt{3}$

(d) $(2\sqrt[3]{6})(5\sqrt[3]{4})=2\cdot5\cdot\sqrt[3]{6}\cdot\sqrt[3]{4}=10\sqrt[3]{24}$
$=10\sqrt[3]{8}\sqrt[3]{3}$
$=10\cdot2\cdot\sqrt[3]{3}$
$=20\sqrt[3]{3}$ ■

Recall the use of the distributive property when finding the product of a monomial and a polynomial. For example, $3x^2(2x + 7) = 3x^2(2x) + 3x^2(7) = 6x^3 + 21x^2$. In a similar manner, the distributive property, along with Property 5.4, provide the basis for finding certain special products that involve radicals. The following examples illustrate this idea.

EXAMPLE 2 Multiply and simplify where possible.

(a) $\sqrt{3}(\sqrt{6} + \sqrt{12})$
(b) $2\sqrt{2}(4\sqrt{3} - 5\sqrt{6})$
(c) $\sqrt{6x}(\sqrt{8x} + \sqrt{12xy})$
(d) $\sqrt[3]{2}(5\sqrt[3]{4} - 3\sqrt[3]{16})$

Solution

(a)
$$\begin{aligned}\sqrt{3}(\sqrt{6} + \sqrt{12}) &= \sqrt{3}\sqrt{6} + \sqrt{3}\sqrt{12}\\ &= \sqrt{18} + \sqrt{36}\\ &= \sqrt{9}\sqrt{2} + 6\\ &= 3\sqrt{2} + 6\end{aligned}$$

(b)
$$\begin{aligned}2\sqrt{2}(4\sqrt{3} - 5\sqrt{6}) &= (2\sqrt{2})(4\sqrt{3}) - (2\sqrt{2})(5\sqrt{6})\\ &= 8\sqrt{6} - 10\sqrt{12}\\ &= 8\sqrt{6} - 10\sqrt{4}\sqrt{3}\\ &= 8\sqrt{6} - 20\sqrt{3}\end{aligned}$$

(c)
$$\begin{aligned}\sqrt{6x}(\sqrt{8x} + \sqrt{12xy}) &= (\sqrt{6x})(\sqrt{8x}) + (\sqrt{6x})(\sqrt{12xy})\\ &= \sqrt{48x^2} + \sqrt{72x^2y}\\ &= \sqrt{16x^2}\sqrt{3} + \sqrt{36x^2}\sqrt{2y}\\ &= 4x\sqrt{3} + 6x\sqrt{2y}\end{aligned}$$

(d)
$$\begin{aligned}\sqrt[3]{2}(5\sqrt[3]{4} - 3\sqrt[3]{16}) &= (\sqrt[3]{2})(5\sqrt[3]{4}) - (\sqrt[3]{2})(3\sqrt[3]{16})\\ &= 5\sqrt[3]{8} - 3\sqrt[3]{32}\\ &= 5 \cdot 2 - 3\sqrt[3]{8}\sqrt[3]{4}\\ &= 10 - 6\sqrt[3]{4}\end{aligned}$$

■

The distributive property also plays a central role in determining the product of two binomials. For example, $(x + 2)(x + 3) = x(x + 3) + 2(x + 3) = x^2 + 3x + 2x + 6 = x^2 + 5x + 6$. Finding the product of two binomial expressions that involve radicals can be handled in a similar fashion, as in the next examples.

EXAMPLE 3 Find the following products and simplify.

(a) $(\sqrt{3} + \sqrt{5})(\sqrt{2} + \sqrt{6})$
(b) $(2\sqrt{2} - \sqrt{7})(3\sqrt{2} + 5\sqrt{7})$
(c) $(\sqrt{8} + \sqrt{6})(\sqrt{8} - \sqrt{6})$
(d) $(\sqrt{x} + \sqrt{y})(\sqrt{x} - \sqrt{y})$

Solution

(**a**) $(\sqrt{3} + \sqrt{5})(\sqrt{2} + \sqrt{6}) = \sqrt{3}(\sqrt{2} + \sqrt{6}) + \sqrt{5}(\sqrt{2} + \sqrt{6})$

$$= \sqrt{3}\sqrt{2} + \sqrt{3}\sqrt{6} + \sqrt{5}\sqrt{2} + \sqrt{5}\sqrt{6}$$
$$= \sqrt{6} + \sqrt{18} + \sqrt{10} + \sqrt{30}$$
$$= \sqrt{6} + 3\sqrt{2} + \sqrt{10} + \sqrt{30}$$

(**b**) $(2\sqrt{2} - \sqrt{7})(3\sqrt{2} + 5\sqrt{7}) = 2\sqrt{2}(3\sqrt{2} + 5\sqrt{7}) - \sqrt{7}(3\sqrt{2} + 5\sqrt{7})$

$$= (2\sqrt{2})(3\sqrt{2}) + (2\sqrt{2})(5\sqrt{7})$$
$$-(\sqrt{7})(3\sqrt{2}) - (\sqrt{7})(5\sqrt{7})$$
$$= 12 + 10\sqrt{14} - 3\sqrt{14} - 35$$
$$= -23 + 7\sqrt{14}$$

(**c**) $(\sqrt{8} + \sqrt{6})(\sqrt{8} - \sqrt{6}) = \sqrt{8}(\sqrt{8} - \sqrt{6}) + \sqrt{6}(\sqrt{8} - \sqrt{6})$

$$= \sqrt{8}\sqrt{8} - \sqrt{8}\sqrt{6} + \sqrt{6}\sqrt{8} - \sqrt{6}\sqrt{6}$$
$$= 8 - \sqrt{48} + \sqrt{48} - 6$$
$$- 2$$

(**d**) $(\sqrt{x} + \sqrt{y})(\sqrt{x} - \sqrt{y}) = \sqrt{x}(\sqrt{x} - \sqrt{y}) + \sqrt{y}(\sqrt{x} - \sqrt{y})$

$$= \sqrt{x}\sqrt{x} - \sqrt{x}\sqrt{y} + \sqrt{y}\sqrt{x} - \sqrt{y}\sqrt{y}$$
$$= x - \sqrt{xy} + \sqrt{xy} - y$$
$$= x - y$$

■

Notice parts (**c**) and (**d**) of Example 3. They fit the special product pattern $(a + b)(a - b) = a^2 - b^2$. Furthermore, in each case the final product is in rational form. This suggests a way of rationalizing the denominator of an expression that contains a binomial denominator with radicals. Consider the following example.

EXAMPLE 4 Simplify $\dfrac{4}{\sqrt{5} + \sqrt{2}}$ by rationalizing the denominator.

Solution

$$\frac{4}{\sqrt{5} + \sqrt{2}} = \frac{4}{\sqrt{5} + \sqrt{2}} \cdot \left(\frac{\sqrt{5} - \sqrt{2}}{\sqrt{5} - \sqrt{2}}\right)$$

a form of 1

$$= \frac{4(\sqrt{5} - \sqrt{2})}{(\sqrt{5} + \sqrt{2})(\sqrt{5} - \sqrt{2})} = \frac{4(\sqrt{5} - \sqrt{2})}{5 - 2}$$

$$= \frac{4(\sqrt{5} - \sqrt{2})}{3} \quad \text{or} \quad \frac{4\sqrt{5} - 4\sqrt{2}}{3}$$

either answer is acceptable

■

The next examples further illustrate the process of rationalizing and simplifying expressions that contain binomial denominators.

EXAMPLE 5 For each of the following, rationalize the denominator and simplify.

(a) $\dfrac{\sqrt{3}}{\sqrt{6}-9}$ (b) $\dfrac{7}{3\sqrt{5}+2\sqrt{3}}$

(c) $\dfrac{\sqrt{x}+2}{\sqrt{x}-3}$ (d) $\dfrac{2\sqrt{x}-3\sqrt{y}}{\sqrt{x}+\sqrt{y}}$

Solution

(a)
$$\frac{\sqrt{3}}{\sqrt{6}-9}=\frac{\sqrt{3}}{\sqrt{6}-9}\cdot\frac{\sqrt{6}+9}{\sqrt{6}+9}$$
$$=\frac{\sqrt{3}(\sqrt{6}+9)}{(\sqrt{6}-9)(\sqrt{6}+9)}$$
$$=\frac{\sqrt{18}+9\sqrt{3}}{6-81}$$
$$=\frac{3\sqrt{2}+9\sqrt{3}}{-75}$$
$$=\frac{3(\sqrt{2}+3\sqrt{3})}{(-3)(25)}$$
$$=-\frac{\sqrt{2}+3\sqrt{3}}{25}\quad\text{or}\quad\frac{-\sqrt{2}-3\sqrt{3}}{25}$$

(b)
$$\frac{7}{3\sqrt{5}+2\sqrt{3}}=\frac{7}{3\sqrt{5}+2\sqrt{3}}\cdot\frac{3\sqrt{5}-2\sqrt{3}}{3\sqrt{5}-2\sqrt{3}}$$
$$=\frac{7(3\sqrt{5}-2\sqrt{3})}{(3\sqrt{5}+2\sqrt{3})(3\sqrt{5}-2\sqrt{3})}$$
$$=\frac{7(3\sqrt{5}-2\sqrt{3})}{45-12}$$
$$=\frac{7(3\sqrt{5}-2\sqrt{3})}{33}\quad\text{or}\quad\frac{21\sqrt{5}-14\sqrt{3}}{33}$$

(c)
$$\frac{\sqrt{x}+2}{\sqrt{x}-3}=\frac{\sqrt{x}+2}{\sqrt{x}-3}\cdot\frac{\sqrt{x}+3}{\sqrt{x}+3}=\frac{(\sqrt{x}+2)(\sqrt{x}+3)}{(\sqrt{x}-3)(\sqrt{x}+3)}$$
$$=\frac{x+3\sqrt{x}+2\sqrt{x}+6}{x-9}$$
$$=\frac{x+5\sqrt{x}+6}{x-9}$$

(d)
$$\frac{2\sqrt{x}-3\sqrt{y}}{\sqrt{x}+\sqrt{y}} = \frac{2\sqrt{x}-3\sqrt{y}}{\sqrt{x}+\sqrt{y}} \cdot \frac{\sqrt{x}-\sqrt{y}}{\sqrt{x}-\sqrt{y}}$$
$$= \frac{(2\sqrt{x}-3\sqrt{y})(\sqrt{x}-\sqrt{y})}{(\sqrt{x}+\sqrt{y})(\sqrt{x}-\sqrt{y})}$$
$$= \frac{2x-2\sqrt{xy}-3\sqrt{xy}+3y}{x-y}$$
$$= \frac{2x-5\sqrt{xy}+3y}{x-y}$$ ■

Problem Set 5.4

Multiply and simplify where possible.

1. $\sqrt{6}\sqrt{12}$ $6\sqrt{2}$
2. $\sqrt{8}\sqrt{6}$ $4\sqrt{3}$
3. $(3\sqrt{3})(2\sqrt{6})$ $18\sqrt{2}$
4. $(5\sqrt{2})(3\sqrt{12})$ $30\sqrt{6}$
5. $(4\sqrt{2})(-6\sqrt{5})$ $-24\sqrt{10}$
6. $(-7\sqrt{3})(2\sqrt{5})$ $-14\sqrt{15}$
7. $(-3\sqrt{3})(-4\sqrt{8})$ $24\sqrt{6}$
8. $(-5\sqrt{8})(-6\sqrt{7})$ $60\sqrt{14}$
9. $(5\sqrt{6})(4\sqrt{6})$ 120
10. $(3\sqrt{7})(2\sqrt{7})$ 42
11. $(2\sqrt[3]{4})(6\sqrt[3]{2})$ 24
12. $(4\sqrt[3]{3})(5\sqrt[3]{9})$ 60
13. $(4\sqrt[3]{6})(7\sqrt[3]{4})$ $56\sqrt[3]{3}$
14. $(9\sqrt[3]{6})(2\sqrt[3]{9})$ $54\sqrt[3]{2}$

Find the following products and express answers in simplest radical form. All variables represent nonnegative real numbers.

15. $\sqrt{2}(\sqrt{3}+\sqrt{5})$ $\sqrt{6}+\sqrt{10}$
16. $\sqrt{3}(\sqrt{7}+\sqrt{10})$ $\sqrt{21}+\sqrt{30}$
17. $3\sqrt{5}(2\sqrt{2}-\sqrt{7})$ $6\sqrt{10}-3\sqrt{35}$
18. $5\sqrt{6}(2\sqrt{5}-3\sqrt{11})$ $10\sqrt{30}-15\sqrt{66}$
19. $2\sqrt{6}(3\sqrt{8}-5\sqrt{12})$ $24\sqrt{3}-60\sqrt{2}$
20. $4\sqrt{2}(3\sqrt{12}+7\sqrt{6})$ $24\sqrt{6}+56\sqrt{3}$
21. $-4\sqrt{5}(2\sqrt{5}+4\sqrt{12})$ $-40-32\sqrt{15}$
22. $-5\sqrt{3}(3\sqrt{12}-9\sqrt{8})$ $-90+90\sqrt{6}$
23. $3\sqrt{x}(5\sqrt{2}+\sqrt{y})$ $15\sqrt{2x}+3\sqrt{xy}$
24. $\sqrt{2x}(3\sqrt{y}-7\sqrt{5})$ $3\sqrt{2xy}-7\sqrt{10x}$
25. $\sqrt{xy}(5\sqrt{xy}-6\sqrt{x})$ $5xy-6x\sqrt{y}$
26. $4\sqrt{x}(2\sqrt{xy}+2\sqrt{x})$ $8x\sqrt{y}+8x$
27. $\sqrt{5y}(\sqrt{8x}+\sqrt{12y^2})$ $2\sqrt{10xy}+2y\sqrt{15y}$
28. $\sqrt{2x}(\sqrt{12xy}-\sqrt{8y})$ $2x\sqrt{6y}-4\sqrt{xy}$
29. $5\sqrt{3}(2\sqrt{8}-3\sqrt{18})$ $-25\sqrt{6}$
30. $2\sqrt{2}(3\sqrt{12}-\sqrt{27})$ $6\sqrt{6}$
31. $(\sqrt{3}+4)(\sqrt{3}-7)$ $-25-3\sqrt{3}$
32. $(\sqrt{2}+6)(\sqrt{2}-2)$ $-10+4\sqrt{2}$
33. $(\sqrt{5}-6)(\sqrt{5}-3)$ $23-9\sqrt{5}$
34. $(\sqrt{7}-2)(\sqrt{7}-8)$ $23-10\sqrt{7}$
35. $(3\sqrt{5}-2\sqrt{3})(2\sqrt{7}+\sqrt{2})$ $6\sqrt{35}+3\sqrt{10}-4\sqrt{21}-2\sqrt{6}$
36. $(\sqrt{2}+\sqrt{3})(\sqrt{5}-\sqrt{7})$ $\sqrt{10}-\sqrt{14}+\sqrt{15}-\sqrt{21}$
37. $(2\sqrt{6}+3\sqrt{5})(\sqrt{8}-3\sqrt{12})$ $8\sqrt{3}-36\sqrt{2}+6\sqrt{10}-18\sqrt{15}$
38. $(5\sqrt{2}-4\sqrt{6})(2\sqrt{8}+\sqrt{6})$ $16-22\sqrt{3}$
39. $(2\sqrt{6}+5\sqrt{5})(3\sqrt{6}-\sqrt{5})$ $11+13\sqrt{30}$

40. $(7\sqrt{3} - \sqrt{7})(2\sqrt{3} + 4\sqrt{7})$ $14 + 26\sqrt{21}$

41. $(3\sqrt{2} - 5\sqrt{3})(6\sqrt{2} - 7\sqrt{3})$ $141 - 51\sqrt{6}$

42. $(\sqrt{8} - 3\sqrt{10})(2\sqrt{8} - 6\sqrt{10})$ $196 - 48\sqrt{5}$

43. $(\sqrt{6} + 4)(\sqrt{6} - 4)$ -10

44. $(\sqrt{7} - 2)(\sqrt{7} + 2)$ 3

45. $(\sqrt{2} + \sqrt{10})(\sqrt{2} - \sqrt{10})$ -8

46. $(2\sqrt{3} + \sqrt{11})(2\sqrt{3} - \sqrt{11})$ 1

47. $(\sqrt{2x} + \sqrt{3y})(\sqrt{2x} - \sqrt{3y})$ $2x - 3y$

48. $(2\sqrt{x} - 5\sqrt{y})(2\sqrt{x} + 5\sqrt{y})$ $4x - 25y$

49. $2\sqrt[3]{3}(5\sqrt[3]{4} + \sqrt[3]{6})$ $10\sqrt[3]{12} + 2\sqrt[3]{18}$

50. $2\sqrt[3]{2}(3\sqrt[3]{6} - 4\sqrt[3]{5})$ $6\sqrt[3]{12} - 8\sqrt[3]{10}$

51. $3\sqrt[3]{4}(2\sqrt[3]{2} - 6\sqrt[3]{4})$ $12 - 36\sqrt[3]{2}$

52. $3\sqrt[3]{3}(4\sqrt[3]{9} + 5\sqrt[3]{7})$ $36 + 15\sqrt[3]{21}$

For each of the following, rationalize the denominator and simplify. All variables represent positive real numbers.

53. $\dfrac{2}{\sqrt{7} + 1}$ $\dfrac{\sqrt{7} - 1}{3}$

54. $\dfrac{6}{\sqrt{5} + 2}$ $6\sqrt{5} - 12$

55. $\dfrac{3}{\sqrt{2} - 5}$ $\dfrac{-3\sqrt{2} - 15}{23}$

56. $\dfrac{-4}{\sqrt{6} - 3}$ $\dfrac{4\sqrt{6} + 12}{3}$

57. $\dfrac{1}{\sqrt{2} + \sqrt{7}}$ $\dfrac{\sqrt{7} - \sqrt{2}}{5}$

58. $\dfrac{3}{\sqrt{3} + \sqrt{10}}$ $\dfrac{-3\sqrt{3} + 3\sqrt{10}}{7}$

59. $\dfrac{\sqrt{2}}{\sqrt{10} - \sqrt{3}}$ $\dfrac{2\sqrt{5} + \sqrt{6}}{7}$

60. $\dfrac{\sqrt{3}}{\sqrt{7} - \sqrt{2}}$ $\dfrac{\sqrt{21} + \sqrt{6}}{5}$

61. $\dfrac{\sqrt{3}}{2\sqrt{5} + 4}$ $\dfrac{\sqrt{15} - 2\sqrt{3}}{2}$

62. $\dfrac{\sqrt{7}}{3\sqrt{2} - 5}$ $\dfrac{-3\sqrt{14} - 5\sqrt{7}}{7}$

63. $\dfrac{6}{3\sqrt{7} - 2\sqrt{6}}$ $\dfrac{6\sqrt{7} + 4\sqrt{6}}{13}$

64. $\dfrac{5}{2\sqrt{5} + 3\sqrt{7}}$ $\dfrac{-10\sqrt{5} + 15\sqrt{7}}{43}$

65. $\dfrac{\sqrt{6}}{3\sqrt{2} + 2\sqrt{3}}$ $\sqrt{3} - \sqrt{2}$

66. $\dfrac{3\sqrt{6}}{5\sqrt{3} - 4\sqrt{2}}$ $\dfrac{45\sqrt{2} + 24\sqrt{3}}{43}$

67. $\dfrac{2}{\sqrt{x} + 4}$ $\dfrac{2\sqrt{x} - 8}{x - 16}$

68. $\dfrac{3}{\sqrt{x} + 7}$ $\dfrac{3\sqrt{x} - 21}{x - 49}$

69. $\dfrac{\sqrt{x}}{\sqrt{x} - 5}$ $\dfrac{x + 5\sqrt{x}}{x - 25}$

70. $\dfrac{\sqrt{x}}{\sqrt{x} - 1}$ $\dfrac{x + \sqrt{x}}{x - 1}$

71. $\dfrac{\sqrt{x} - 2}{\sqrt{x} + 6}$ $\dfrac{x - 8\sqrt{x} + 12}{x - 36}$

72. $\dfrac{\sqrt{x} + 1}{\sqrt{x} - 10}$ $\dfrac{x + 11\sqrt{x} + 10}{x - 100}$

73. $\dfrac{\sqrt{x}}{\sqrt{x} + 2\sqrt{y}}$ $\dfrac{x - 2\sqrt{xy}}{x - 4y}$

74. $\dfrac{\sqrt{y}}{2\sqrt{x} - \sqrt{y}}$ $\dfrac{2\sqrt{xy} + y}{4x - y}$

75. $\dfrac{3\sqrt{y}}{2\sqrt{x} - 3\sqrt{y}}$ $\dfrac{6\sqrt{xy} + 9y}{4x - 9y}$

76. $\dfrac{2\sqrt{x}}{3\sqrt{x} + 5\sqrt{y}}$ $\dfrac{6x - 10\sqrt{xy}}{9x - 25y}$

5.5 Equations Involving Radicals

We often refer to equations that contain radicals with variables in a radicand as **radical equations**. In this section we shall discuss techniques for solving such equations that contain one or more radicals. To solve radical equations we need the following property of equality.

PROPERTY 5.6

> Let a and b be real numbers and n be a positive integer.
>
> If $a = b$, then $a^n = b^n$.

Property 5.6 states that we can *raise both sides of an equation to a positive integral power*. However, raising both sides of an equation to a positive integral power sometimes produces results that do not satisfy the original equation. Let's consider two examples to illustrate this point.

EXAMPLE 1 Solve $\sqrt{2x - 5} = 7$.

Solution

$$\sqrt{2x-5} = 7$$

$$(\sqrt{2x-5})^2 = 7^2 \qquad \text{Square both sides.}$$

$$2x - 5 = 49$$

$$2x = 54$$

$$x = 27$$

CHECK $\sqrt{2x-5} = 7$

$$\sqrt{2(27)-5} \stackrel{?}{=} 7$$

$$\sqrt{49} \stackrel{?}{=} 7$$

$$7 = 7$$

The solution set for $\sqrt{2x-5} = 7$ is $\{27\}$. ■

EXAMPLE 2 Solve $\sqrt{3a + 4} = -4$.

Solution

$$\sqrt{3a+4} = -4$$

$$(\sqrt{3a+4})^2 = (-4)^2 \qquad \text{Square both sides.}$$

$$3a + 4 = 16$$

$$3a = 12$$

$$a = 4$$

CHECK $\sqrt{3a+4} = -4$

$$\sqrt{3(4)+4} \stackrel{?}{=} -4$$

$$\sqrt{16} \stackrel{?}{=} -4$$

$$4 \neq -4$$

Since 4 does not check, the original equation *has no real number solution.* Thus, the solution set is $\varnothing$. ■

In general, raising both sides of an equation to a positive integral power produces an equation that has all of the solutions of the original equation; it may also have some extra solutions that will not satisfy the original equation. Such extra solutions are called **extraneous solutions**. Therefore, when using Property 5.6 you *must* check each potential solution in the original equation.

Let's consider some examples to illustrate different situations that arise when solving radical equations.

EXAMPLE 3 Solve $\sqrt{2t-4} = t-2$.

Solution

$$\sqrt{2t-4} = t-2$$

$$(\sqrt{2t-4})^2 = (t-2)^2 \qquad \text{Square both sides.}$$

$$2t-4 = t^2-4t+4$$

$$0 = t^2-6t+8$$

$$0 = (t-2)(t-4) \qquad \text{Factor the right side.}$$

$$t-2=0 \quad \text{or} \quad t-4=0 \qquad \text{Apply: } ab=0 \text{ if and only if } a=0 \text{ or } b=0.$$

$$t=2 \quad \text{or} \quad t=4$$

CHECK $\sqrt{2t-4} = t-2$ or $\sqrt{2t-4} = t-2$

$$\sqrt{2(2)-4} \stackrel{?}{=} 2-2 \quad \text{or} \quad \sqrt{2(4)-4} \stackrel{?}{=} 4-2$$

$$\sqrt{0} \stackrel{?}{=} 0 \qquad \sqrt{4} \stackrel{?}{=} 2$$

$$0 = 0 \qquad 2 = 2$$

The solution set is $\{2, 4\}$. ■

EXAMPLE 4 Solve $\sqrt{y} + 6 = y$.

Solution

$$\sqrt{y} + 6 = y$$

$$\sqrt{y} = y-6$$

$$(\sqrt{y})^2 = (y-6)^2 \qquad \text{Square both sides.}$$

$$y = y^2 - 12y + 36$$

$$0 = y^2 - 13y + 36$$

$$0 = (y - 4)(y - 9) \qquad \text{Factor the right side.}$$

$$y - 4 = 0 \quad \text{or} \quad y - 9 = 0 \qquad \text{Apply: } ab = 0 \text{ if and only if } a = 0 \text{ or } b = 0.$$

$$y = 4 \quad \text{or} \quad y = 9$$

CHECK

$$\sqrt{y} + 6 = y \qquad \sqrt{y} + 6 = y$$

$$\sqrt{4} + 6 \stackrel{?}{=} 4 \quad \text{or} \quad \sqrt{9} + 6 \stackrel{?}{=} 9$$

$$2 + 6 \stackrel{?}{=} 4 \qquad 3 + 6 \stackrel{?}{=} 9$$

$$8 \neq 4 \qquad 9 = 9$$

The only solution is 9; the solution set is $\{9\}$. ■

In Example 4, note that we changed the form of the original equation $\sqrt{y} + 6 = y$ to $\sqrt{y} = y - 6$ before squaring both sides. Squaring both sides of $\sqrt{y} + 6 = y$ produces $y + 12\sqrt{y} + 36 = y^2$, a more complex equation that still contains a radical. So, again it pays to think ahead a few steps before carrying out the details. Now let's consider an example involving a cube root.

EXAMPLE 5 Solve $\sqrt[3]{n^2 - 1} = 2$.

Solution

$$\sqrt[3]{n^2 - 1} = 2$$

$$(\sqrt[3]{n^2 - 1})^3 = 2^3 \qquad \text{Cube both sides.}$$

$$n^2 - 1 = 8$$

$$n^2 - 9 = 0$$

$$(n + 3)(n - 3) = 0$$

$$n + 3 = 0 \quad \text{or} \quad n - 3 = 0$$

$$n = -3 \quad \text{or} \quad n = 3$$

CHECK

$$\sqrt[3]{n^2 - 1} = 2 \qquad \sqrt[3]{n^2 - 1} = 2$$

$$\sqrt[3]{(-3)^2 - 1} \stackrel{?}{=} 2 \quad \text{or} \quad \sqrt[3]{3^2 - 1} \stackrel{?}{=} 2$$

$$\sqrt[3]{8} \stackrel{?}{=} 2 \qquad \sqrt[3]{8} \stackrel{?}{=} 2$$

$$2 = 2 \qquad 2 = 2$$

The solution set is $\{-3, 3\}$. ■

It may be necessary to square both sides of an equation, simplify the resulting equation, and then square both sides again. Our final example of this section illustrates this type of problem.

EXAMPLE 6 Solve $\sqrt{x+2} = 7 - \sqrt{x+9}$.

Solution

$$\sqrt{x+2} = 7 - \sqrt{x+9}$$

$$(\sqrt{x+2})^2 = (7 - \sqrt{x+9})^2 \quad \text{Square both sides.}$$

$$x + 2 = 49 - 14\sqrt{x+9} + x + 9$$

$$x + 2 = x + 58 - 14\sqrt{x+9}$$

$$-56 = -14\sqrt{x+9}$$

$$4 = \sqrt{x+9}$$

$$(4)^2 = (\sqrt{x+9})^2 \quad \text{Square both sides.}$$

$$16 = x + 9$$

$$7 = x$$

CHECK

$$\sqrt{x+2} = 7 - \sqrt{x+9}$$

$$\sqrt{7+2} \stackrel{?}{=} 7 - \sqrt{7+9}$$

$$\sqrt{9} \stackrel{?}{=} 7 - \sqrt{16}$$

$$3 \stackrel{?}{=} 7 - 4$$

$$3 = 3$$

The solution set is $\{7\}$. ■

Problem Set 5.5

Solve each of the following equations. Don't forget to check each of your potential solutions.

1. $\sqrt{5x} = 10$ $\{20\}$
2. $\sqrt{3x} = 9$ $\{27\}$
3. $\sqrt{2x} + 4 = 0$ $\varnothing$
4. $\sqrt{4x} + 5 = 0$ $\varnothing$
5. $2\sqrt{n} = 5$ $\{\frac{25}{4}\}$
6. $5\sqrt{n} = 3$ $\{\frac{9}{25}\}$
7. $3\sqrt{n} - 2 = 0$ $\{\frac{4}{9}\}$
8. $2\sqrt{n} - 7 = 0$ $\{\frac{49}{4}\}$
9. $\sqrt{3y+1} = 4$ $\{5\}$
10. $\sqrt{2y-3} = 5$ $\{14\}$
11. $\sqrt{4y-3} - 6 = 0$ $\{\frac{39}{4}\}$
12. $\sqrt{3y+5} - 2 = 0$ $\{-\frac{1}{3}\}$
13. $\sqrt{2x-5} = -1$ $\varnothing$
14. $\sqrt{4x-3} = -4$ $\varnothing$
15. $\sqrt{5x+2} = \sqrt{6x+1}$ $\{1\}$
16. $\sqrt{4x+2} = \sqrt{3x+4}$ $\{2\}$
17. $\sqrt{3x+1} = \sqrt{7x-5}$ $\{\frac{3}{2}\}$
18. $\sqrt{6x+5} = \sqrt{2x+10}$ $\{\frac{5}{4}\}$
19. $\sqrt{3x-2} - \sqrt{x+4} = 0$ $\{3\}$
20. $\sqrt{7x-6} - \sqrt{5x+2} = 0$ $\{4\}$

21. $5\sqrt{t-1} = 6$ $\{\frac{61}{25}\}$
22. $4\sqrt{t+3} = 6$ $\{-\frac{3}{4}\}$
23. $\sqrt{x^2+7} = 4$ $\{-3, 3\}$
24. $\sqrt{x^2+3} - 2 = 0$ $\{-1, 1\}$
25. $\sqrt{x^2+13x+37} = 1$ $\{-9, -4\}$
26. $\sqrt{x^2+5x-20} = 2$ $\{-8, 3\}$
27. $\sqrt{x^2-x+1} = x+1$ $\{0\}$
28. $\sqrt{n^2-2n-4} = n$ $\varnothing$
29. $\sqrt{x^2+3x+7} = x+2$ $\{3\}$
30. $\sqrt{x^2+2x+1} = x+3$ $\{-2\}$
31. $\sqrt{-4x+17} = x-3$ $\{4\}$
32. $\sqrt{2x-1} = x-2$ $\{5\}$
33. $\sqrt{n+4} = n+4$ $\{-4, -3\}$
34. $\sqrt{n+6} = n+6$ $\{-6, -5\}$
35. $\sqrt{3y} = y-6$ $\{12\}$
36. $2\sqrt{n} = n-3$ $\{9\}$
37. $4\sqrt{x} + 5 = x$ $\{25\}$
38. $\sqrt{-x} - 6 = x$ $\{-4\}$
39. $\sqrt[3]{x-2} = 3$ $\{29\}$
40. $\sqrt[3]{x+1} = 4$ $\{63\}$
41. $\sqrt[3]{2x+3} = -3$ $\{-15\}$
42. $\sqrt[3]{3x-1} = -4$ $\{-21\}$
43. $\sqrt[3]{2x+5} = \sqrt[3]{4-x}$ $\{-\frac{1}{3}\}$
44. $\sqrt[3]{3x-1} = \sqrt[3]{2-5x}$ $\{\frac{3}{8}\}$
45. $\sqrt{x+19} - \sqrt{x+28} = -1$ $\{-3\}$
46. $\sqrt{x+4} = \sqrt{x-1} + 1$ $\{5\}$
47. $\sqrt{3x+1} + \sqrt{2x+4} = 3$ $\{0\}$
48. $\sqrt{2x-1} - \sqrt{x+3} = 1$ $\{13\}$
49. $\sqrt{n-4} + \sqrt{n+4} = 2\sqrt{n-1}$ $\{5\}$
50. $\sqrt{n-3} + \sqrt{n+5} = 2\sqrt{n}$ $\{4\}$
51. $\sqrt{t+3} - \sqrt{t-2} = \sqrt{7-t}$ $\{2, 6\}$
52. $\sqrt{t+7} - 2\sqrt{t-8} = \sqrt{t-5}$ $\{9\}$

5.6 The Merging of Exponents and Roots

Recall that the basic properties of positive integral exponents led to a definition for the use of negative integers as exponents. In this section, the properties of integral exponents are used to form definitions for the use of rational numbers as exponents. These definitions will tie together the concepts of *exponent* and *root*.

Let's consider the following comparisons.

From our study of radicals we know that

$$(\sqrt{5})^2 = 5,$$
$$(\sqrt[3]{8})^3 = 8,$$
$$(\sqrt[4]{21})^4 = 21.$$

If $(b^n)^m = b^{mn}$ is to hold when n equals a rational number of the form $\frac{1}{p}$, where p is a positive integer greater than one, then

$$\left(5^{\frac{1}{2}}\right)^2 = 5^{2\left(\frac{1}{2}\right)} = 5^1 = 5,$$
$$\left(8^{\frac{1}{3}}\right)^3 = 8^{3\left(\frac{1}{3}\right)} = 8^1 = 8,$$
$$\left(21^{\frac{1}{4}}\right)^4 = 21^{4\left(\frac{1}{4}\right)} = 21^1 = 21.$$

It would seem reasonable to make the following definition.

DEFINITION 5.6

> If b is a real number, n is a positive integer greater than one, and $\sqrt[n]{b}$ exists, then
>
> $$b^{\frac{1}{n}} = \sqrt[n]{b}.$$

Definition 5.6 states that $b^{\frac{1}{n}}$ *means the nth root of* b. We shall assume that b and n are chosen so that $\sqrt[n]{b}$ exists. For example, $(-25)^{\frac{1}{2}}$ is not meaningful at this time because $\sqrt{-25}$ is not a real number.

Consider the following examples that demonstrate the use of Definition 5.6.

$$25^{\frac{1}{2}} = \sqrt{25} = 5, \qquad 16^{\frac{1}{4}} = \sqrt[4]{16} = 2,$$

$$8^{\frac{1}{3}} = \sqrt[3]{8} = 2, \qquad \left(\frac{36}{49}\right)^{\frac{1}{2}} = \sqrt{\frac{36}{49}} = \frac{6}{7},$$

$$(-27)^{\frac{1}{3}} = \sqrt[3]{-27} = -3$$

The following definition provides the basis for the use of *all* rational numbers as exponents.

DEFINITION 5.7

> If $\frac{m}{n}$ is a rational number, where n is a positive integer greater than one, and b is a real number such that $\sqrt[n]{b}$ exists, then
>
> $$b^{\frac{m}{n}} = \sqrt[n]{b^m} = (\sqrt[n]{b})^m.$$

In Definition 5.7, notice that the denominator of the exponent is the index of the radical; and the numerator of the exponent is either the exponent of the radicand or the exponent of the root.

Whether we use the form $\sqrt[n]{b^m}$ or $(\sqrt[n]{b})^m$ for computational purposes depends somewhat on the magnitude of the problem. Let's use both forms on two problems to illustrate this point.

$$\begin{aligned} 8^{\frac{2}{3}} &= \sqrt[3]{8^2} & \text{or} \quad 8^{\frac{2}{3}} &= (\sqrt[3]{8})^2 \\ &= \sqrt[3]{64} & &= 2^2 \\ &= 4 & &= 4 \end{aligned}$$

$$\begin{aligned} 27^{\frac{2}{3}} &= \sqrt[3]{27^2} & \text{or} \quad 27^{\frac{2}{3}} &= (\sqrt[3]{27})^2 \\ &= \sqrt[3]{729} & &= 3^2 \\ &= 9 & &= 9 \end{aligned}$$

To compute $8^{\frac{2}{3}}$, either form seems to work about as well as the other one. However, to compute $27^{\frac{2}{3}}$, it should be obvious that $(\sqrt[3]{27})^2$ is much easier to handle than $\sqrt[3]{27^2}$.

EXAMPLE 1 Simplify each of the following numerical expressions.

(a) $25^{\frac{3}{2}}$ (b) $16^{\frac{3}{4}}$ (c) $(32)^{-\frac{2}{5}}$

(d) $(-64)^{\frac{2}{3}}$ (e) $-8^{\frac{1}{3}}$

Solution

(a) $25^{\frac{3}{2}} = (\sqrt{25})^3 = 5^3 = 125$

(b) $16^{\frac{3}{4}} = (\sqrt[4]{16})^3 = 2^3 = 8$

(c) $(32)^{-\frac{2}{5}} = \dfrac{1}{(32)^{\frac{2}{5}}} = \dfrac{1}{(\sqrt[5]{32})^2} = \dfrac{1}{2^2} = \dfrac{1}{4}$

(d) $(-64)^{\frac{2}{3}} = (\sqrt[3]{-64})^2 = (-4)^2 = 16$

(e) $-8^{\frac{1}{3}} = -\sqrt[3]{8} = -2$ ■

The basic laws of exponents stated in Property 5.2 are true for all rational exponents. Therefore, without restating Property 5.2 we shall henceforth use it for rational as well as integral exponents.

Some problems can be handled better in exponential form and others in radical form. Thus, we must be able to switch forms with a certain amount of ease. Let's consider some examples where we switch from one form to the other.

EXAMPLE 2 Write each of the following expressions in radical form.

(a) $x^{\frac{3}{4}}$ (b) $3y^{\frac{2}{5}}$

(c) $x^{\frac{1}{4}}y^{\frac{3}{4}}$ (d) $(x + y)^{\frac{2}{3}}$

Solution

(a) $x^{\frac{3}{4}} = \sqrt[4]{x^3}$ (b) $3y^{\frac{2}{5}} = 3\sqrt[5]{y^2}$

(c) $x^{\frac{1}{4}}y^{\frac{3}{4}} = (xy^3)^{\frac{1}{4}} = \sqrt[4]{xy^3}$ (d) $(x + y)^{\frac{2}{3}} = \sqrt[3]{(x + y)^2}$ ■

EXAMPLE 3 Write each of the following using positive rational exponents.

(a) $\sqrt{xy}$ (b) $\sqrt[4]{a^3b}$ (c) $4\sqrt[3]{x^2}$ (d) $\sqrt[5]{(x + y)^4}$

Solution

(a) $\sqrt{xy} = (xy)^{\frac{1}{2}} = x^{\frac{1}{2}}y^{\frac{1}{2}}$ (b) $\sqrt[4]{a^3b} = (a^3b)^{\frac{1}{4}} = a^{\frac{3}{4}}b^{\frac{1}{4}}$

(c) $4\sqrt[3]{x^2} = 4x^{\frac{2}{3}}$ (d) $\sqrt[5]{(x + y)^4} = (x + y)^{\frac{4}{5}}$ ■

The basic properties of exponents provide the basis for simplifying algebraic expressions that contain rational exponents, as these next examples illustrate.

EXAMPLE 4 Simplify each of the following. Express final results using positive exponents only.

(a) $\left(3x^{\frac{1}{2}}\right)\left(4x^{\frac{2}{3}}\right)$ **(b)** $\left(5a^{\frac{1}{3}}b^{\frac{1}{2}}\right)^2$ **(c)** $\dfrac{12y^{\frac{1}{3}}}{6y^{\frac{1}{2}}}$ **(d)** $\left(\dfrac{3x^{\frac{2}{5}}}{2y^{\frac{2}{3}}}\right)^4$

Solution

(a)
$$\left(3x^{\frac{1}{2}}\right)\left(4x^{\frac{2}{3}}\right) = 3 \cdot 4 \cdot x^{\frac{1}{2}} \cdot x^{\frac{2}{3}}$$
$$= 12x^{\frac{1}{2}+\frac{2}{3}} \qquad b^n \cdot b^m = b^{n+m}$$
$$= 12x^{\frac{3}{6}+\frac{4}{6}}$$
$$= 12x^{\frac{7}{6}}$$

(b)
$$\left(5a^{\frac{1}{3}}b^{\frac{1}{2}}\right)^2 = 5^2 \cdot \left(a^{\frac{1}{3}}\right)^2 \cdot \left(b^{\frac{1}{2}}\right)^2 \qquad (ab)^n = a^n b^n$$
$$= 25a^{\frac{2}{3}}b \qquad (b^n)^m = b^{mn}$$

(c)
$$\frac{12y^{\frac{1}{3}}}{6y^{\frac{1}{2}}} = 2y^{\frac{1}{3}-\frac{1}{2}} \qquad \frac{b^n}{b^m} = b^{n-m}$$
$$= 2y^{\frac{2}{6}-\frac{3}{6}}$$
$$= 2y^{-\frac{1}{6}}$$
$$= \frac{2}{y^{\frac{1}{6}}}$$

(d)
$$\left(\frac{3x^{\frac{2}{5}}}{2y^{\frac{2}{3}}}\right)^4 = \frac{\left(3x^{\frac{2}{5}}\right)^4}{\left(2y^{\frac{2}{3}}\right)^4} \qquad \left(\frac{a}{b}\right)^n = \frac{a^n}{b^n}$$
$$= \frac{3^4 \cdot \left(x^{\frac{2}{5}}\right)^4}{2^4 \cdot \left(y^{\frac{2}{3}}\right)^4} \qquad (ab)^n = a^n b^n$$
$$= \frac{81x^{\frac{8}{5}}}{16y^{\frac{8}{3}}} \qquad (b^n)^m = b^{mn}$$

■

The link between exponents and roots also provides a basis for multiplying and dividing some radicals even if they have a different index. The general procedure is to change from radical form to exponential form, apply the properties of exponents, and then change back to radical form. Let's consider three examples to illustrate this process.

EXAMPLE 5 Perform the indicated operations and express the answer in simplest radical form.

(a) $\sqrt{2}\sqrt[3]{2}$ (b) $\dfrac{\sqrt{5}}{\sqrt[3]{5}}$ (c) $\dfrac{\sqrt{4}}{\sqrt[3]{2}}$

Solution

(a)
$$\begin{aligned}\sqrt{2}\sqrt[3]{2} &= 2^{\frac{1}{2}} \cdot 2^{\frac{1}{3}} \\ &= 2^{\frac{1}{2}+\frac{1}{3}} \\ &= 2^{\frac{3}{6}+\frac{2}{6}} \\ &= 2^{\frac{5}{6}} \\ &= \sqrt[6]{2^5} = \sqrt[6]{32}\end{aligned}$$

(b)
$$\begin{aligned}\frac{\sqrt{5}}{\sqrt[3]{5}} &= \frac{5^{\frac{1}{2}}}{5^{\frac{1}{3}}} \\ &= 5^{\frac{1}{2}-\frac{1}{3}} \\ &= 5^{\frac{3}{6}-\frac{2}{6}} \\ &= 5^{\frac{1}{6}} = \sqrt[6]{5}\end{aligned}$$

(c)
$$\begin{aligned}\frac{\sqrt{4}}{\sqrt[3]{2}} &= \frac{4^{\frac{1}{2}}}{2^{\frac{1}{3}}} \\ &= \frac{(2^2)^{\frac{1}{2}}}{2^{\frac{1}{3}}} \\ &= \frac{2^1}{2^{\frac{1}{3}}} \\ &= 2^{1-\frac{1}{3}} \\ &= 2^{\frac{2}{3}} = \sqrt[3]{2^2} = \sqrt[3]{4}\end{aligned}$$

■

Problem Set 5.6

Simplify each of the following numerical expressions.

1. $81^{\frac{1}{2}}$ 9
2. $64^{\frac{1}{2}}$ 8
3. $27^{\frac{1}{3}}$ 3
4. $(-32)^{\frac{1}{5}}$ -2
5. $(-8)^{\frac{1}{3}}$ -2
6. $\left(-\frac{27}{8}\right)^{\frac{1}{3}}$ $-\frac{3}{2}$
7. $-25^{\frac{1}{2}}$ -5
8. $-64^{\frac{1}{3}}$ -4
9. $36^{-\frac{1}{2}}$ $\frac{1}{6}$
10. $81^{-\frac{1}{2}}$ $\frac{1}{9}$
11. $\left(\frac{1}{27}\right)^{-\frac{1}{3}}$ 3
12. $\left(-\frac{8}{27}\right)^{-\frac{1}{3}}$ $-\frac{3}{2}$
13. $4^{\frac{3}{2}}$ 8
14. $64^{\frac{2}{3}}$ 16
15. $27^{\frac{4}{3}}$ 81

16. $4^{\frac{7}{2}}$ 128

17. $(-1)^{\frac{7}{3}}$ -1

18. $(-8)^{\frac{4}{3}}$ 16

19. $-4^{\frac{5}{2}}$ -32

20. $-16^{\frac{3}{2}}$ -64

21. $\left(\frac{27}{8}\right)^{\frac{4}{3}}$ $\frac{81}{16}$

22. $\left(\frac{8}{125}\right)^{\frac{2}{3}}$ $\frac{4}{25}$

23. $\left(\frac{1}{8}\right)^{-\frac{2}{3}}$ 4

24. $\left(-\frac{1}{27}\right)^{-\frac{2}{3}}$ 9

25. $64^{-\frac{7}{6}}$ $\frac{1}{128}$

26. $32^{-\frac{4}{5}}$ $\frac{1}{16}$

27. $-25^{\frac{3}{2}}$ -125

28. $-16^{\frac{3}{4}}$ -8

29. $125^{\frac{4}{3}}$ 625

30. $81^{\frac{5}{4}}$ 243

Write each of the following in radical form. For example, $3x^{\frac{2}{3}} = 3\sqrt[3]{x^2}$.

31. $x^{\frac{4}{3}}$ $\sqrt[3]{x^4}$

32. $x^{\frac{2}{5}}$ $\sqrt[5]{x^2}$

33. $3x^{\frac{1}{2}}$ $3\sqrt{x}$

34. $5x^{\frac{1}{4}}$ $5\sqrt[4]{x}$

35. $(2y)^{\frac{1}{3}}$ $\sqrt[3]{2y}$

36. $(3xy)^{\frac{1}{2}}$ $\sqrt{3xy}$

37. $(2x - 3y)^{\frac{1}{2}}$ $\sqrt{2x - 3y}$

38. $(5x + y)^{\frac{1}{3}}$ $\sqrt[3]{5x + y}$

39. $(2a - 3b)^{\frac{2}{3}}$ $\sqrt[3]{(2a - 3b)^2}$

40. $(5a + 7b)^{\frac{3}{5}}$ $\sqrt[5]{(5a + 7b)^3}$

41. $x^{\frac{2}{3}}y^{\frac{1}{3}}$ $\sqrt[3]{x^2y}$

42. $x^{\frac{3}{7}}y^{\frac{5}{7}}$ $\sqrt[7]{x^3y^5}$

43. $-3x^{\frac{1}{5}}y^{\frac{2}{5}}$ $-3\sqrt[5]{xy^2}$

44. $-4x^{\frac{3}{4}}y^{\frac{1}{4}}$ $-4\sqrt[4]{x^3y}$

Write each of the following using positive rational exponents. For example,

$$\sqrt{ab} = (ab)^{\frac{1}{2}} = a^{\frac{1}{2}}b^{\frac{1}{2}}.$$

45. $\sqrt{5y}$ $5^{\frac{1}{2}}y^{\frac{1}{2}}$

46. $\sqrt{2xy}$ $2^{\frac{1}{2}}x^{\frac{1}{2}}y^{\frac{1}{2}}$

47. $3\sqrt{y}$ $3y^{\frac{1}{2}}$

48. $5\sqrt{ab}$ $5a^{\frac{1}{2}}b^{\frac{1}{2}}$

49. $\sqrt[3]{xy^2}$ $x^{\frac{1}{3}}y^{\frac{2}{3}}$

50. $\sqrt[5]{x^2y^4}$ $x^{\frac{2}{5}}y^{\frac{4}{5}}$

51. $\sqrt[4]{a^2b^3}$ $a^{\frac{1}{2}}b^{\frac{3}{4}}$

52. $\sqrt[6]{ab^5}$ $a^{\frac{1}{6}}b^{\frac{5}{6}}$

53. $\sqrt[5]{(2x - y)^3}$ $(2x - y)^{\frac{3}{5}}$

54. $\sqrt[7]{(3x - y)^4}$ $(3x - y)^{\frac{4}{7}}$

55. $5x\sqrt{y}$ $5xy^{\frac{1}{2}}$

56. $4y\sqrt[3]{x}$ $4yx^{\frac{1}{3}}$

57. $-\sqrt[3]{x + y}$ $-(x + y)^{\frac{1}{3}}$

58. $-\sqrt[5]{(x - y)^2}$ $-(x - y)^{\frac{2}{5}}$

Simplify each of the following. Express final results using positive exponents only. For example,

$$\left(2x^{\frac{1}{2}}\right)\left(3x^{\frac{1}{3}}\right) = 6x^{\frac{5}{6}}.$$

59. $\left(2x^{\frac{2}{5}}\right)\left(6x^{\frac{1}{4}}\right)$ $12x^{\frac{13}{20}}$

60. $\left(3x^{\frac{1}{4}}\right)\left(5x^{\frac{1}{3}}\right)$ $15x^{\frac{7}{12}}$

61. $\left(y^{\frac{2}{3}}\right)\left(y^{-\frac{1}{4}}\right)$ $y^{\frac{5}{12}}$

62. $\left(y^{\frac{3}{4}}\right)\left(y^{-\frac{1}{2}}\right)$ $y^{\frac{1}{4}}$

63. $\left(x^{\frac{2}{5}}\right)\left(4x^{-\frac{1}{2}}\right)$ $4/x^{\frac{1}{10}}$

64. $\left(2x^{\frac{1}{3}}\right)\left(x^{-\frac{1}{2}}\right)$ $2/x^{\frac{1}{6}}$

65. $\left(4x^{\frac{1}{2}}y\right)^2$ $16xy^2$

66. $\left(3x^{\frac{1}{4}}y^{\frac{1}{5}}\right)^3$ $27x^{\frac{3}{4}}y^{\frac{3}{5}}$

67. $(8x^6y^3)^{\frac{1}{3}}$ $2x^2y$

68. $(9x^2y^4)^{\frac{1}{2}}$ $3xy^2$

69. $\dfrac{24x^{\frac{3}{5}}}{6x^{\frac{1}{3}}}$ $4x^{\frac{4}{15}}$

70. $\dfrac{18x^{\frac{1}{2}}}{9x^{\frac{1}{3}}}$ $2x^{\frac{1}{6}}$

71. $\dfrac{48b^{\frac{1}{3}}}{12b^{\frac{3}{4}}}$ $\dfrac{4}{b^{\frac{5}{12}}}$

72. $\dfrac{56a^{\frac{1}{6}}}{8a^{\frac{1}{4}}}$ $\dfrac{7}{a^{\frac{1}{12}}}$

73. $\left(\dfrac{6x^{\frac{2}{5}}}{7y^{\frac{2}{3}}}\right)^2$ $\dfrac{36x^{\frac{4}{5}}}{49y^{\frac{4}{3}}}$

74. $\left(\dfrac{2x^{\frac{1}{3}}}{3y^{\frac{1}{4}}}\right)^4$ $\dfrac{16x^{\frac{4}{3}}}{81y}$

75. $\left(\dfrac{x^2}{y^3}\right)^{-\frac{1}{2}}$ $\dfrac{y^{\frac{3}{2}}}{x}$

76. $\left(\dfrac{a^3}{b^{-2}}\right)^{-\frac{1}{3}}$ $\dfrac{1}{ab^{\frac{2}{3}}}$

77. $\left(\dfrac{18x^{\frac{1}{3}}}{9x^{\frac{1}{4}}}\right)^2$ $4x^{\frac{1}{6}}$

78. $\left(\dfrac{72x^{\frac{3}{4}}}{6x^{\frac{1}{2}}}\right)^2$ $144x^{\frac{1}{2}}$

79. $\left(\dfrac{60a^{\frac{1}{5}}}{15a^{\frac{3}{4}}}\right)^2$ $\dfrac{16}{a^{\frac{11}{10}}}$

80. $\left(\dfrac{64a^{\frac{1}{3}}}{16a^{\frac{5}{9}}}\right)^3$ $\dfrac{64}{a^{\frac{2}{3}}}$

Perform the indicated operations and express answers in simplest radical form. (See Example 5.)

81. $\sqrt[3]{3}\sqrt{3}$ $\sqrt[6]{243}$

82. $\sqrt{2}\sqrt[4]{2}$ $\sqrt[4]{8}$

83. $\sqrt[4]{6}\sqrt{6}$ $\sqrt[4]{216}$

84. $\sqrt[3]{5}\sqrt{5}$ $\sqrt[6]{3125}$

85. $\dfrac{\sqrt[3]{3}}{\sqrt[4]{3}}$ $\sqrt[12]{3}$

86. $\dfrac{\sqrt{2}}{\sqrt[3]{2}}$ $\sqrt[6]{2}$

87. $\dfrac{\sqrt[3]{8}}{\sqrt[4]{4}}$ $\sqrt{2}$

88. $\dfrac{\sqrt{9}}{\sqrt[3]{3}}$ $\sqrt[3]{9}$

89. $\dfrac{\sqrt[4]{27}}{\sqrt{3}}$ $\sqrt[4]{3}$

90. $\dfrac{\sqrt[3]{16}}{\sqrt[6]{4}}$ 2

Miscellaneous Problems

91. If your calculator has $\boxed{y^x}$ and $\boxed{1/x}$ keys, then you can use them to evaluate cube roots, fourth roots, fifth roots, and so on. For example, some calculators evaluate $\sqrt[3]{4913}$ as follows:

$$4913 \;\boxed{y^x}\; 3 \;\boxed{1/x} = 17.$$

Use your calculator to evalute each of the following.

(a) $\sqrt[3]{1728}$ 12 **(b)** $\sqrt[3]{5832}$ 18 **(c)** $\sqrt[4]{2401}$ 7

(d) $\sqrt[4]{65536}$ 16 **(e)** $\sqrt[5]{161051}$ 11 **(f)** $\sqrt[5]{6436343}$ 23

92. Definition 5.7 stated that

$$b^{\frac{m}{n}} = \sqrt[n]{b^m} = (\sqrt[n]{b})^m.$$

Use your calculator to verify each of the following.

(a) $\sqrt[3]{27^2} = (\sqrt[3]{27})^2$ **(b)** $\sqrt[3]{8^5} = (\sqrt[3]{8})^5$ **(c)** $\sqrt[4]{16^3} = (\sqrt[4]{16})^3$

(d) $\sqrt[3]{16^2} = (\sqrt[3]{16})^2$ **(e)** $\sqrt[5]{9^4} = (\sqrt[5]{9})^4$ **(f)** $\sqrt[3]{12^4} = (\sqrt[3]{12})^4$

93. Use your calculator to evaluate each of the following.

(a) $16^{\frac{5}{2}}$ 1024 **(b)** $25^{\frac{7}{2}}$ 78125 **(c)** $16^{\frac{9}{4}}$ 512

(d) $27^{\frac{5}{3}}$ 243 **(e)** $343^{\frac{2}{3}}$ 49 **(f)** $512^{\frac{4}{3}}$ 4096

94. Use your calculator to estimate each of the following to the nearest one-thousandth.

(a) $7^{\frac{4}{3}}$ 13.391 **(b)** $10^{\frac{4}{5}}$ 6.310 **(c)** $12^{\frac{2}{5}}$ 2.702

(d) $19^{\frac{2}{5}}$ 3.247 **(e)** $7^{\frac{3}{4}}$ 4.304 **(f)** $10^{\frac{5}{4}}$ 17.783

95. **(a)** Since $\frac{4}{5} = 0.8$, we can evaluate $10^{\frac{4}{5}}$ by evaluating $10^{0.8}$, which involves a shorter sequence of "calculator steps." Evaluate parts (b), (c), (d), (e), and (f) of Problem 94 and take advantage of decimal exponents.

(b) What problem is created when we try to evaluate $7^{\frac{4}{3}}$ by changing the exponent to decimal form?

5.7 Scientific Notation

Many applications of mathematics involve the use of very large and very small numbers. For example,

1. A light year—the distance that a ray of light travels in one year—is approximately 5,900,000,000,000 miles;
2. In 1982 the national debt was approximately 950,000,000,000 dollars;
3. In the metric system, a millimicron equals 0.000000001 of a meter;
4. The weight of an oxygen molecule is approximately 0.000000000000000000000053 of a gram.

Working with numbers of this type in standard form is quite cumbersome. It is much more convenient to represent very small and very large numbers in **scientific notation**, sometimes called **scientific form.** We express a number in scientific notation when we write it as a product of a number between 1 and 10 and an integral power of 10. Symbolically, a number written in scientific notation has the form

$$(N)(10)^k$$

where N is a number between 1 and 10, written in decimal form, and k is an integer. Consider the following examples that show a comparison between ordinary notation and scientific notation.

Ordinary notation	*Scientific notation*
2.14	$(2.14)(10)^0$
31.78	$(3.178)(10)^1$
412.9	$(4.129)(10)^2$
8,000,000	$(8)(10)^6$
0.14	$(1.4)(10)^{-1}$
0.0379	$(3.79)(10)^{-2}$
0.00000049	$(4.9)(10)^{-7}$

To switch from ordinary notation to scientific notation, you can use the following procedure.

> Write the given number as the product of a number between 1 and 10 and a power of 10. Determine the exponent of 10 by counting the number of places that the decimal point was moved when going from the original number to the number between 1 and 10. This exponent is (a) negative if the original number is less than 1, (b) positive if the original number is greater than 10, and (c) 0 if the original number itself is between 1 and 10.

Thus, we can write

$$0.00467 = (4.67)(10)^{-3},$$

$$87{,}000 = (8.7)(10)^{4},$$

$$3.1416 = (3.1416)(10)^{0}.$$

To switch from scientific notation to ordinary notation you can use the following procedure.

> Move the decimal point the number of places indicated by the exponent of 10. The decimal point is moved to the right if the exponent is positive and to the left if it is negative.

Thus, we can write

$$(4.78)(10)^{4} = 47{,}800,$$

$$(8.4)(10)^{-3} = 0.0084.$$

Scientific notation can frequently be used to simplify numerical calculations. We merely change the numbers to scientific notation and use the appropriate properties of exponents. Consider the following examples.

EXAMPLE 1 Perform the indicated operations.

(a) $(0.00024)(20{,}000)$

(b) $\dfrac{7{,}800{,}000}{0.0039}$

(c) $\dfrac{(0.00069)(0.0034)}{(0.0000017)(0.023)}$

(d) $\sqrt{0.000004}$

Solution

(a)
$$\begin{aligned}(0.00024)(20{,}000) &= (2.4)(10)^{-4}(2)(10)^{4}\\ &= (2.4)(2)(10)^{-4}(10)^{4}\\ &= (4.8)(10)^{0}\\ &= (4.8)(1)\\ &= 4.8\end{aligned}$$

$$\textbf{(b)}\quad \frac{7{,}800{,}000}{0.0039} = \frac{(7.8)(10)^6}{(3.9)(10)^{-3}}$$
$$= (2)(10)^9$$
$$= 2{,}000{,}000{,}000$$

$$\textbf{(c)}\quad \frac{(0.00069)(0.0034)}{(0.0000017)(0.023)} = \frac{(6.9)(10)^{-4}(3.4)(10)^{-3}}{(1.7)(10)^{-6}(2.3)(10)^{-2}}$$
$$= \frac{(\overset{3}{\cancel{6.9}})(\overset{2}{\cancel{3.4}})(10)^{-7}}{(\cancel{1.7})(\cancel{2.3})(10)^{-8}}$$
$$= (6)(10)^1$$
$$= 60$$

$$\textbf{(d)}\quad \sqrt{0.000004} = \sqrt{(4)(10)^{-6}}$$
$$= ((4)(10)^{-6})^{\frac{1}{2}}$$
$$= 4^{\frac{1}{2}}\,((10)^{-6})^{\frac{1}{2}}$$
$$= (2)(10)^{-3}$$
$$= 0.002$$

■

Many calculators are equipped to display numbers in scientific notation. The display panel shows the number between 1 and 10 and the appropriate exponent of 10. For example, evaluating $(3{,}800{,}000)^2$ yields

1.444 13

Thus, $(3{,}800{,}000)^2 = (1.444)(10)^{13} = 14{,}440{,}000{,}000{,}000$.

Similarly, the answer for $(0.000168)^2$ is displayed as

2.8224 −08

Thus, $(0.000168)^2 = (2.8224)(10)^{-8} = 0.000000028224$.

Calculators vary as to the number of digits displayed in the number between 1 and 10 when using scientific notation. For example, we used two different calculators to estimate $(6729)^6$ and obtained the following results.

9.2833 22

9.283316768 22

.

Obviously, you need to know the capabilities of your calculator when working with problems in scientific notation.

Many calculators also allow the entry of a number in scientific notation. Such calculators are equipped with an enter-the-exponent key (often labeled as [EE] or [E EX]). Thus, a number such as $(3.14)(10)^8$ might be entered as follows.

Enter	*Press*	*Display*
3.14	EE	3.14 00
8		3.14 08

Furthermore, it may be that your calculator will perform the switch from ordinary notation to scientific notation with a routine such as the following.

Enter	*Press*	*Display*
4721	EE	4721 00
	=	4.721 03

Be sure that you know the routine that will accomplish this switch on your calculator.

It should be evident from this brief discussion that even when using a calculator, you need to have a thorough understanding of scientific notation.

Problem Set 5.7

Write each of the following in scientific notation. For example, $27{,}800 = (2.78)(10)^4$.

1. 89 $(8.9)(10)^1$
2. 117 $(1.17)(10)^2$
3. 4290 $(4.29)(10)^3$
4. 812,000 $(8.12)(10)^5$
5. 6,120,000 $(6.12)(10)^6$
6. 72,400,000 $(7.24)(10)^7$
7. 40,000,000 $(4)(10)^7$
8. 500,000,000 $(5)(10)^8$
9. 376.4 $(3.764)(10)^2$
10. 9126.21 $(9.12621)(10)^3$
11. 0.347 $(3.47)(10)^{-1}$
12. 0.2165 $(2.165)(10)^{-1}$
13. 0.0214 $(2.14)(10)^{-2}$
14. 0.0037 $(3.7)(10)^{-3}$
15. 0.00005 $(5)(10)^{-5}$
16. 0.00000082 $(8.2)(10)^{-7}$
17. 0.00000000194 $(1.94)(10)^{-9}$
18. 0.00000000003 $(3)(10)^{-11}$

Write each of the following in ordinary notation. For example $(3.18)(10)^2 = 318$.

19. $(2.3)(10)^1$ 23
20. $(1.62)(10)^2$ 162
21. $(4.19)(10)^3$ 4190
22. $(7.631)(10)^4$ 76310
23. $(5)(10)^8$ 500,000,000
24. $(7)(10)^9$ 7,000,000,000
25. $(3.14)(10)^{10}$ 31,400,000,000
26. $(2.04)(10)^{12}$ 2,040,000,000,000
27. $(4.3)(10)^{-1}$ .43
28. $(5.2)(10)^{-2}$ .052
29. $(9.14)(10)^{-4}$ .000914
30. $(8.76)(10)^{-5}$ .0000876
31. $(5.123)(10)^{-8}$ .00000005123
32. $(6)(10)^{-9}$ .000000006

Use scientific notation and the properties of exponents to help perform the following operations.

33. $(0.0037)(0.00002)$ .000000074
34. $(0.00003)(0.00025)$ .0000000075
35. $(0.00007)(11{,}000)$ .77
36. $(0.000004)(120{,}000)$ .48
37. $\dfrac{360{,}000{,}000}{0.0012}$ 300,000,000,000
38. $\dfrac{66{,}000{,}000{,}000}{0.022}$ 3,000,000,000,000
39. $\dfrac{0.000064}{16{,}000}$ .000000004
40. $\dfrac{0.00072}{0.0000024}$ 300

41. $\dfrac{(60{,}000)(0.006)}{(0.0009)(400)}$ 1000

42. $\dfrac{(0.00063)(960{,}000)}{(3{,}200)(0.0000021)}$ 90,000

43. $\dfrac{(0.0045)(60000)}{(1800)(0.00015)}$ 1000

44. $\dfrac{(0.00016)(300)(0.028)}{0.064}$ .021

45. $\sqrt{9{,}000{,}000}$ 3000

46. $\sqrt{0.00000009}$ .0003

47. $\sqrt[3]{8000}$ 20

48. $\sqrt[3]{0.001}$ .1

49. $(90{,}000)^{\frac{3}{2}}$ 27,000,000

50. $(8000)^{\frac{2}{3}}$ 400

51. Sometimes it is more convenient to express a number as a product of a power of 10 and a number that is not between 1 and 10. For example, suppose that we want to calculate $\sqrt{640{,}000}$. We can proceed as follows.

$$\begin{aligned}\sqrt{640{,}000} &= \sqrt{(64)(10)^4}\\ &= ((64)(10)^4)^{\frac{1}{2}}\\ &= (64)^{\frac{1}{2}}(10^4)^{\frac{1}{2}}\\ &= (8)(10)^2\\ &= 8(100) = 800\end{aligned}$$

Compute each of the following.

(a) $\sqrt{49{,}000{,}000}$ 7000 (b) $\sqrt{0.0025}$ .05 (c) $\sqrt{14400}$ 120

(d) $\sqrt{0.000121}$ .011 (e) $\sqrt[3]{27000}$ 30 (f) $\sqrt[3]{0.000064}$ .04

52. Use your calculator to evaluate each of the following. Express final answers in ordinary notation.

(a) $(27{,}000)^2$ 729,000,000 (b) $(450{,}000)^2$ 202,500,000,000 (c) $(14{,}800)^2$ 219,040,000

(d) $(1700)^3$ 4,913,000,000 (e) $(900)^4$ 656,100,000,000 (f) $(60)^5$ 777,600,000

(g) $(0.0213)^2$ .00045369 (h) $(0.000213)^2$ .000000045369

(i) $(0.000198)^2$ .000000039204

(j) $(0.000009)^3$ .000000000000000729

53. Use your calculator to estimate each of the following. Express final answers in scientific notation with the number between 1 and 10 rounded to the nearest one-thousandth.

(a) $(4576)^4$ $(4.385)(10)^{14}$ (b) $(719)^{10}$ $(3.692)(10)^{28}$ (c) $(28)^{12}$ $(2.322)(10)^{17}$

(d) $(8619)^6$ $(4.100)(10)^{23}$ (e) $(314)^5$ $(3.052)(10)^{12}$ (f) $(145{,}723)^2$ $(2.124)(10)^{10}$

54. Use your calculator to estimate each of the following. Express final answers in ordinary notation rounded to the nearest one-thousandth.

(a) $(1.09)^5$ 1.539 (b) $(1.08)^{10}$ 2.159 (c) $(1.14)^7$ 2.502

(d) $(1.12)^{20}$ 9.646 (e) $(0.785)^4$ .380 (f) $(0.492)^5$ .029

Chapter 5 Summary

(5.1) The following properties form the basis for manipulating with exponents.

1. $b^n \cdot b^m = b^{n+m}$ product of two powers

2. $(b^n)^m = b^{mn}$ power of a power

3. $(ab)^n = a^n b^n$ power of a product

4. $\left(\frac{a}{b}\right)^n = \frac{a^n}{b^n}$ power of a quotient

5. $\frac{b^n}{b^m} = b^{n-m}$ quotient of two powers

(5.2) and (5.3) The **principal *n*th root of *b*** is designated by $\sqrt[n]{b}$, where n is the **index** and b is the **radicand**.

A radical expression is in **simplest radical form** if:

1. A radicand contains no polynomial factor raised to a power equal to or greater than the index of the radical;
2. No fraction appears within a radical sign;
3. No radical appears in the denominator.

The following properties are used to express radicals in simplest form.

$$\sqrt[n]{bc} = \sqrt[n]{b}\sqrt[n]{c}, \qquad \sqrt[n]{\frac{b}{c}} = \frac{\sqrt[n]{b}}{\sqrt[n]{c}}$$

Simplifying by combining radicals sometimes requires that you first express the given radicals in simplest form and then apply the distributive property.

(5.4) The distributive property and the property $\sqrt[n]{b}\sqrt[n]{c} = \sqrt[n]{bc}$ are used to find products of expressions that involve radicals.

The special product pattern $(a + b)(a - b) = a^2 - b^2$ suggests a procedure for **rationalizing the denominator** of an expression that contains a binomial denominator with radicals.

(5.5) Equations that contain radicals with variables in a radicand are called **radical equations**. The property, if $a = b$, then $a^n = b^n$, forms the basis for solving radical equations. Raising both sides of an equation to a positive integral power may produce **extraneous solutions**, that is, solutions that do not satisfy the original equation. Therefore, you **must check** each potential solution.

(5.6) If b is a real number, n is a positive integer greater than one, and $\sqrt[n]{b}$ exists, then

$$b^{\frac{1}{n}} = \sqrt[n]{b}.$$

Thus, $b^{\frac{1}{n}}$ means **the *n*th root of *b***.

If $\frac{m}{n}$ is a rational number, when n is a positive integer greater than one, and b is a real number such that $\sqrt[n]{b}$ exists, then

$$b^{\frac{m}{n}} = \sqrt[n]{b^m} = (\sqrt[n]{b})^m.$$

Both $\sqrt[n]{b^m}$ and $(\sqrt[n]{n})^m$ can be used for computational purposes.

We need to be able to switch back and forth between **exponential** and **radical form**. The link between exponents and roots provides a basis for multiplying and dividing some radicals even if they have different indices.

(5.7) The **scientific form** of a number is expressed as

$$(N)(10)^k$$

where N is a number between 1 and 10, written in decimal form, and k is an integer. Scientific notation is often convenient to use with very small and very large numbers. For example, .000046 can be expressed as $(4.6)(10^{-5})$ and 92,000,000 can be written as $(9.2)(10)^7$.

Scientific notation can often be used to simplify numerical calculations. For example,

$$\begin{aligned}(.000016)(30000) &= (1.6)(10)^{-5}(3)(10)^4\\ &= (4.8)(10)^{-1} = .48.\end{aligned}$$

Chapter 5 Review Problem Set

Evaluate each of the following numerical expressions.

1. 4^{-3} $\frac{1}{64}$ **2.** $\left(\frac{2}{3}\right)^{-2}$ $\frac{9}{4}$ **3.** $(3^2 \cdot 3^{-3})^{-1}$ 3 **4.** $\sqrt[3]{-8}$ -2

5. $\sqrt[4]{\frac{16}{81}}$ $\frac{2}{3}$ **6.** $4^{\frac{5}{2}}$ 32 **7.** $(-1)^{-\frac{2}{3}}$ 1 **8.** $\left(\frac{8}{27}\right)^{\frac{2}{3}}$ $\frac{4}{9}$

9. $-16^{\frac{3}{2}}$ -64 **10.** $\frac{2^3}{2^{-2}}$ 32 **11.** $(4^{-2} \cdot 4^2)^{-1}$ 1 **12.** $\left(\frac{3^{-1}}{3^2}\right)^{-1}$ 27

Express each of the following radicals in simplest radical form.

13. $\sqrt{54}$ $3\sqrt{6}$ **14.** $\sqrt{48x^3y}$ $4x\sqrt{3xy}$ **15.** $\frac{4\sqrt{3}}{\sqrt{6}}$ $2\sqrt{2}$

16. $\sqrt{\frac{5}{12x^3}}$ $\frac{\sqrt{15x}}{6x^2}$ **17.** $\sqrt[3]{56}$ $2\sqrt[3]{7}$ **18.** $\frac{\sqrt[3]{2}}{\sqrt[3]{9}}$ $\frac{\sqrt[3]{6}}{3}$

19. $\sqrt{\frac{9}{5}}$ $\frac{3\sqrt{5}}{5}$ **20.** $\sqrt{\frac{3x^3}{7}}$ $\frac{x\sqrt{21x}}{7}$ **21.** $\sqrt[3]{108x^4y^8}$ $3xy^2\sqrt[3]{4xy^2}$

22. $\frac{3}{4}\sqrt{150}$ $\frac{15\sqrt{6}}{4}$ **23.** $\frac{2}{3}\sqrt{45xy^3}$ $2y\sqrt{5xy}$ **24.** $\frac{\sqrt{8x^2}}{\sqrt{2x}}$ $2\sqrt{x}$

Multiply and simplify.

25. $(3\sqrt{8})(4\sqrt{5})$ $24\sqrt{10}$ **26.** $(5\sqrt[3]{2})(6\sqrt[3]{4})$ 60

27. $3\sqrt{2}(4\sqrt{6} - 2\sqrt{7})$ $24\sqrt{3} - 6\sqrt{14}$ **28.** $(\sqrt{x} + 3)(\sqrt{x} - 5)$ $x - 2\sqrt{x} - 15$

29. $(2\sqrt{5} - \sqrt{3})(2\sqrt{5} + \sqrt{3})$ 17 **30.** $(3\sqrt{2} + \sqrt{6})(5\sqrt{2} - 3\sqrt{6})$ $12 - 8\sqrt{3}$

31. $(2\sqrt{a} + \sqrt{b})(3\sqrt{a} - 4\sqrt{b})$ $6a - 5\sqrt{ab} - 4b$

32. $(4\sqrt{8} - \sqrt{2})(\sqrt{8} + 3\sqrt{2})$ 70

Rationalize the denominator and simplify.

33. $\dfrac{4}{\sqrt{7}-1}$ $\dfrac{2(\sqrt{7}+1)}{3}$

34. $\dfrac{\sqrt{3}}{\sqrt{8}+\sqrt{5}}$ $\dfrac{2\sqrt{6}-\sqrt{15}}{3}$

35. $\dfrac{3}{2\sqrt{3}+3\sqrt{5}}$ $\dfrac{3\sqrt{5}-2\sqrt{3}}{11}$

36. $\dfrac{3\sqrt{2}}{2\sqrt{6}-\sqrt{10}}$ $\dfrac{6\sqrt{3}+3\sqrt{5}}{7}$

Simplify each of the following and express the final results using positive exponents.

37. $(x^{-3}y^4)^{-2}$ $\frac{x^6}{y^8}$

38. $\left(\dfrac{2a^{-1}}{3b^4}\right)^{-3}$ $\dfrac{27a^3b^{12}}{8}$

39. $\left(4x^{\frac{1}{2}}\right)\left(5x^{\frac{1}{5}}\right)$ $20x^{\frac{7}{10}}$

40. $\dfrac{42a^{\frac{3}{4}}}{6a^{\frac{1}{3}}}$ $7a^{\frac{5}{12}}$

41. $\left(\dfrac{x^3}{y^4}\right)^{-\frac{1}{3}}$ $\dfrac{y^{\frac{4}{3}}}{x}$

42. $\left(\dfrac{6x^{-2}}{2x^4}\right)^{-2}$ $\dfrac{x^{12}}{9}$

Use the distributive property to help simplify each of the following.

43. $3\sqrt{45}-2\sqrt{20}-\sqrt{80}$ $\sqrt{5}$

44. $4\sqrt[3]{24}+3\sqrt[3]{3}-2\sqrt[3]{81}$ $5\sqrt[3]{3}$

45. $3\sqrt{24}-\dfrac{2\sqrt{54}}{5}+\dfrac{\sqrt{96}}{4}$ $\dfrac{29\sqrt{6}}{5}$

46. $-2\sqrt{12x}+3\sqrt{27x}-5\sqrt{48x}$ $-15\sqrt{3x}$

Express each of the following as a single fraction involving positive exponents only.

47. $x^{-2}+y^{-1}$ $\dfrac{y+x^2}{x^2y}$

48. $a^{-2}-2a^{-1}b^{-1}$ $\dfrac{b-2a}{a^2b}$

Solve each of the following equations.

49. $\sqrt{7x-3}=4$ $\{\frac{19}{7}\}$

50. $\sqrt{2y+1}=\sqrt{5y-11}$ $\{4\}$

51. $\sqrt{2x}=x-4$ $\{8\}$

52. $\sqrt{n^2-4n-4}=n$ $\varnothing$

53. $\sqrt[3]{2x-1}=3$ $\{14\}$

54. $\sqrt{t^2+9t-1}=3$ $\{-10, 1\}$

55. $\sqrt{x^2+3x-6}=x$ $\{2\}$

56. $\sqrt{x+1}-\sqrt{2x}=-1$ $\{8\}$

Use scientific notation and the properties of exponents to help perform the following calculations.

57. (.00002)(.0003) .000000006

58. (120,000)(300,000) 36,000,000,000

59. (.000015)(400,000) 6

60. $\dfrac{.000045}{.0003}$ .15

61. $\dfrac{(.00042)(.0004)}{.006}$ .000028

62. $\sqrt{.000004}$ .002

63. $\sqrt[3]{.000000008}$ .002

64. $(4000000)^{\frac{3}{2}}$. 8,000,000,000

Thoughts into Words

65. Explain how to simplify $(2^{-1}\cdot 3^{-2})^{-1}$ and how to simplify $(2^{-1}+3^{-2})^{-1}$.

66. Why do we say that 25 has two square roots (5 and -5), but we write $\sqrt{25}=5$?

67. How is the multiplication property of one used when simplifying radicals?

68. Explain the concept of "extraneous solutions."

69. Explain the importance of scientific notation.

Cumulative Review Problem Set

For Problems 1–5, evaluate each algebraic expression for the given values of the variables.

1. $\dfrac{4a^2b^3}{12a^3b}$ for $a = 5$ and $b = -8$ $\quad \dfrac{64}{15}$

2. $\dfrac{\dfrac{1}{x} + \dfrac{1}{y}}{\dfrac{1}{x} - \dfrac{1}{y}}$ for $x = 4$ and $y = 7$ $\quad \dfrac{11}{3}$

3. $\dfrac{3}{n} + \dfrac{5}{2n} - \dfrac{4}{3n}$ for $n = 25$ $\quad \dfrac{1}{6}$

4. $\dfrac{4}{x-1} - \dfrac{2}{x+2}$ for $x = \dfrac{1}{2}$ $\quad -\dfrac{44}{5}$

5. $2\sqrt{2x+y} - 5\sqrt{3x-y}$ for $x = 5$ and $y = 6$ $\quad -7$

For Problems 6–17, perform the indicated operations and express answers in simplified form.

6. $(3a^2b)(-2ab)(4ab^3)$ $\quad -24a^4b^5$

7. $(x+3)(2x^2 - x - 4)$ $\quad 2x^3 + 5x^2 - 7x - 12$

8. $\dfrac{6xy^2}{14y} \cdot \dfrac{7x^2y}{8x}$ $\quad \dfrac{3x^2y^2}{8}$

9. $\dfrac{a^2 + 6a - 40}{a^2 - 4a} \div \dfrac{2a^2 + 19a - 10}{a^3 + a^2}$ $\quad \dfrac{a(a+1)}{2a-1}$

10. $\dfrac{3x+4}{6} - \dfrac{5x-1}{9}$ $\quad \dfrac{-x+14}{18}$

11. $\dfrac{4}{x^2+3x} + \dfrac{5}{x}$ $\quad \dfrac{5x+19}{x(x+3)}$

12. $\dfrac{3n^2 + n}{n^2 + 10n + 16} \cdot \dfrac{2n^2 - 8}{3n^3 - 5n^2 - 2n}$ $\quad \dfrac{2}{n+8}$

13. $\dfrac{3}{5x^2 + 3x - 2} - \dfrac{2}{5x^2 - 22x + 8}$ $\quad \dfrac{x-14}{(5x-2)(x+1)(x-4)}$

14. $\dfrac{y^3 - 7y^2 + 16y - 12}{y-2}$ $\quad y^2 - 5y + 6$

15. $(4x^3 - 17x^2 + 7x + 10) \div (4x - 5)$ $\quad x^2 - 3x - 2$

16. $(3\sqrt{2} + 2\sqrt{5})(5\sqrt{2} - \sqrt{5})$ $\quad 20 + 7\sqrt{10}$

17. $(\sqrt{x} - 3\sqrt{y})(2\sqrt{x} + 4\sqrt{y})$ $\quad 2x - 2\sqrt{xy} - 12y$

For Problems 18–25, evaluate each of the numerical expressions.

18. $-\sqrt{\dfrac{9}{64}}$ $\quad -\dfrac{3}{8}$

19. $\sqrt[3]{-\dfrac{8}{27}}$ $\quad -\dfrac{2}{3}$

20. $\sqrt[3]{.008}$ $\quad .2$

21. $32^{-\frac{1}{5}}$ $\quad \frac{1}{2}$

22. $3^0 + 3^{-1} + 3^{-2}$ $\quad \frac{13}{9}$

23. $-9^{\frac{3}{2}}$ $\quad -27$

24. $\left(\dfrac{3}{4}\right)^{-2}$ $\quad \dfrac{16}{9}$

25. $\dfrac{1}{\left(\dfrac{2}{3}\right)^{-3}}$ $\quad \dfrac{8}{27}$

For Problems 26–31, factor each of the algebraic expressions completely.

26. $3x^4 + 81x$ $\quad 3x(x+3)(x^2 - 3x + 9)$

27. $6x^2 + 19x - 20$ $\quad (6x-5)(x+4)$

28. $12 + 13x - 14x^2$ $(4 + 7x)(3 - 2x)$

29. $9x^4 + 68x^2 - 32$ $(3x + 2)(3x - 2)(x^2 + 8)$

30. $2ax - ay - 2bx + by$ $(2x - y)(a - b)$

31. $27x^3 - 8y^3$ $(3x - 2y)(9x^2 + 6xy + 4y^2)$

For Problems 32–49, solve each of the equations.

32. $3(x - 2) - 2(3x + 5) = 4(x - 1)$ $\{-\frac{12}{7}\}$

33. $.06n + .08(n + 50) = 25$ $\{150\}$

34. $4\sqrt{x} + 5 = x$ $\{25\}$

35. $\sqrt[3]{n^2 - 1} = -1$ $\{0\}$

36. $6x^2 - 24 = 0$ $\{-2, 2\}$

37. $a^2 + 14a + 49 = 0$ $\{-7\}$

38. $3n^2 + 14n - 24 = 0$ $\{-6, \frac{4}{3}\}$

39. $\dfrac{2}{5x - 2} = \dfrac{4}{6x + 1}$ $\left\{\dfrac{5}{4}\right\}$

40. $\sqrt{2x - 1} - \sqrt{x + 2} = 0$ $\{3\}$

41. $5x - 4 = \sqrt{5x - 4}$ $\{\frac{4}{5}, 1\}$

42. $|3x - 1| = 11$ $\{-\frac{10}{3}, 4\}$

43. $(3x - 2)(4x - 1) = 0$ $\{\frac{1}{4}, \frac{2}{3}\}$

44. $(2x + 1)(x - 2) = 7$ $\{-\frac{3}{2}, 3\}$

45. $\dfrac{5}{6x} - \dfrac{2}{3} = \dfrac{7}{10x}$ $\left\{\dfrac{1}{5}\right\}$

46. $\dfrac{3}{y + 4} + \dfrac{2y - 1}{y^2 - 16} = \dfrac{-2}{y - 4}$ $\left\{\dfrac{5}{7}\right\}$

47. $6x^4 - 23x^2 - 4 = 0$ $\{-2, 2\}$

48. $3n^3 + 3n = 0$ $\{0\}$

49. $n^2 - 13n - 114 = 0$ $\{-6, 19\}$

For Problems 50–55, solve each of the inequalities.

50. $6 - 2x \geq 10$ $(-\infty, -2]$

51. $4(2x - 1) < 3(x + 5)$ $(-\infty, \frac{19}{5})$

52. $\dfrac{n + 1}{4} + \dfrac{n - 2}{12} > \dfrac{1}{6}$ $\left(\dfrac{1}{4}, \infty\right)$

53. $|2x - 1| < 5$ $(-2, 3)$

54. $|3x + 2| > 11$ $(-\infty, -\frac{13}{3}) \cup (3, \infty)$

55. $\dfrac{1}{2}(3x - 1) - \dfrac{2}{3}(x + 4) \leq \dfrac{3}{4}(x - 1)$ $(-\infty, 29]$

For Problems 56–61, solve each problem by setting up and solving an appropriate equation.

56. How many liters of a 60% acid solution must be added to 14 liters of a 10% acid solution to produce a 25% acid solution? 6 liters

57. A sum of \$2250 is to be divided between two people in the ratio of 2 to 3. How much does each person receive? \$900 and \$1350

58. The length of a picture without its border is 7 inches less than twice its width. If the border is 1 inch wide and its area is 62 square inches, what are the dimensions of the picture alone? 12 inches by 17 inches

59. Lolita and Doug working together can paint a shed in 3 hours and 20 minutes. If Doug can paint the shed by himself in 10 hours, how long would it take Lolita to paint the shed by herself? 5 hours

60. Angie bought some golf balls for \$14. If each ball had cost \$.25 less, she could have purchased one more ball for the same amount of money. How many golf balls did Angie buy? 7 golf balls.

61. A jogger who can run an 8-minute mile starts a half mile ahead of a jogger who can run a 6-minute mile. How long will it take the faster jogger to catch the slower jogger? 12 minutes

Chapter 6

Quadratic Equations and Inequalities

Solving equations is one of the central themes of this text. Let's pause for a moment and reflect back upon the different types of equations that we have solved.

Type of equation	*Examples*
First-degree equations in one variable	$3x + 2x = x - 4$; $5(x + 4) = 12$; $\frac{x+2}{3} + \frac{x-1}{4} = 2$.
Second-degree equations in one variable *that are factorable*	$x^2 + 5x = 0$; $x^2 + 5x + 6 = 0$; $x^2 - 9 = 0$; $x^2 - 10x + 25 = 0$.
Fractional equations	$\frac{2}{x} + \frac{3}{x} = 4$; $\frac{5}{a-1} = \frac{6}{a-2}$; $\frac{2}{x^2-9} + \frac{3}{x+3} = \frac{4}{x-3}$.
Radical equations	$\sqrt{x} = 2$; $\sqrt{3x-2} = 5$; $\sqrt{5y+1} = \sqrt{3y+4}$.

As the above chart shows, we have solved second-degree equations in one variable, but only those that are factorable. In this chapter we will expand our work to include more general types of second-degree equations as well as inequalities in one variable.

6.1 Complex Numbers

Since the square of any real number is nonnegative, a simple equation such as $x^2 = -4$ has no solutions in the set of real numbers. However, we can handle this situation by expanding the set of real numbers into a larger set called the **complex numbers**. This section is devoted to introducing and learning to manipulate complex numbers.

To provide a solution for the equation $x^2 + 1 = 0$, we use the number i, such that

$$i^2 = -1.$$

The number i is not a real number and is often called the **imaginary unit**, but the number i^2 is the real number -1. The imaginary unit i is used to define a complex number as follows.

DEFINITION 6.1

A **complex number** is any number that can be expressed in the form

$$a + bi,$$

where a and b are real numbers.

The form $a + bi$ is called the **standard form** of a complex number. The real number a is called the **real part** of the complex number and b is called the **imaginary part**. (Note that b is a real number even though it is called the imaginary part.) The following exemplify this terminology.

1. The number $7 + 5i$ is a complex number that has a real part of 7 and an imaginary part of 5.
2. The number $\frac{2}{3} + i\sqrt{2}$ is a complex number that has a real part of $\frac{2}{3}$ and an imaginary part of $\sqrt{2}$. (It is easy to mistake $\sqrt{2i}$ for $\sqrt{2}\,i$. Thus, it is common to write $i\sqrt{2}$ instead of $\sqrt{2}\,i$ to avoid any difficulties with the radical sign.)
3. The number $-4 - 3i$ can be written in the standard form $-4 + (-3i)$ and therefore is a complex number that has a real part of -4 and an imaginary part of -3. (The form $-4 - 3i$ is often used knowing that it means $-4 + (-3i)$.)
4. The number $-9i$ can be written as $0 + (-9i)$; thus, it is a complex number that has a real part of 0 and an imaginary part of -9. (Complex numbers, such as $-9i$, for which $a = 0$ and $b \neq 0$ are called **pure imaginary numbers**.)
5. The real number 4 can be written as $4 + 0i$ and is thus a complex number that has a real part of 4 and an imaginary part of 0.

From number 5 we see that the set of real numbers is a subset of the set of complex numbers. The following diagram indicates the organizational format of the complex numbers.

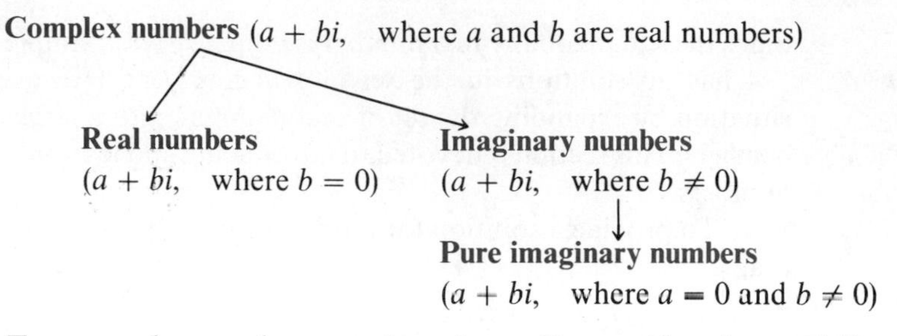

Two complex numbers $a + bi$ and $c + di$ are said to be **equal** if and only if $a = c$ and $b = d$.

Adding and Subtracting Complex Numbers

To **add complex numbers**, we simply add their real parts and add their imaginary parts. Thus,

$$(a + bi) + (c + di) = (a + c) + (b + d)i.$$

The following examples show addition of two complex numbers.

1. $(4 + 3i) + (5 + 9i) = (4 + 5) + (3 + 9)i = 9 + 12i.$

2. $(-6 + 4i) + (8 - 7i) = (-6 + 8) + (4 - 7)i$
$= 2 - 3i.$

3. $\left(\frac{1}{2} + \frac{3}{4}i\right) + \left(\frac{2}{3} + \frac{1}{5}i\right) = \left(\frac{1}{2} + \frac{2}{3}\right) + \left(\frac{3}{4} + \frac{1}{5}\right)i$
$= \left(\frac{3}{6} + \frac{4}{6}\right) + \left(\frac{15}{20} + \frac{4}{20}\right)i$
$= \frac{7}{6} + \frac{19}{20}i.$

The set of complex numbers is closed with respect to addition; that is, the sum of two complex numbers is a complex number. Furthermore, the commutative and associative properties of addition hold for all complex numbers. The addition identity element is $0 + 0i$ (or simply the real number 0). The additive inverse of $a + bi$ is $-a - bi$, because

$$(a + bi) + (-a - bi) = 0.$$

To **subtract complex numbers**, $c + di$ from $a + bi$, add the additive inverse of $c + di$. Thus,

$$\begin{aligned}(a + bi) - (c + di) &= (a + bi) + (-c - di)\\ &= (a - c) + (b - d)i.\end{aligned}$$

In other words, we subtract the real parts and subtract the imaginary parts, as in the next examples.

1. $(9 + 8i) - (5 + 3i) = (9 - 5) + (8 - 3)i$
$= 4 + 5i.$

2. $(3 - 2i) - (4 - 10i) = (3 - 4) + (-2 - (-10))i$
$= -1 + 8i.$

Products and Quotients of Complex Numbers

Since $i^2 = -1$, i is a square root of negative one; so we let $i = \sqrt{-1}$. It should also be evident that $-i$ is a square root of negative one since

$$(-i)^2 = (-i)(-i) = i^2 = -1.$$

Thus, in the set of complex numbers, -1 has two square roots, i and $-i$. We express these symbolically as

$$\sqrt{-1} = i \quad \text{and} \quad -\sqrt{-1} = -i.$$

Let us extend our definition so that in the set of complex numbers every negative real number has two square roots. We simply define $\sqrt{-b}$, where b is a positive real number, to be the number whose square is $-b$. Thus,

$$(\sqrt{-b})^2 = -b, \quad \text{for } b > 0.$$

Furthermore, since $(i\sqrt{b})(i\sqrt{b}) = i^2(b) = -1(b) = -b$ we see that

$$\sqrt{-b} = i\sqrt{b}.$$

In other words, a square root of any negative real number can be represented as the product of a real number and the imaginary unit i. Consider the following examples.

$$\sqrt{-4} = i\sqrt{4} = 2i,$$
$$\sqrt{-17} = i\sqrt{17},$$
$$\sqrt{-24} = i\sqrt{24} = i\sqrt{4}\sqrt{6} = 2i\sqrt{6}$$

Notice that we simplified the radical $\sqrt{24}$ to $2\sqrt{6}$.

We should also observe that $-\sqrt{-b}$, where $b > 0$, is a square root of $-b$ since

$$(-\sqrt{-b})^2 = (-i\sqrt{b})^2 = i^2(b) = -1(b) = -b.$$

Thus, in the set of complex numbers, $-b$ (where $b > 0$) has two square roots, $i\sqrt{b}$ and $-i\sqrt{b}$. We express these symbolically as

$$\sqrt{-b} = i\sqrt{b} \quad \text{and} \quad -\sqrt{-b} = -i\sqrt{b}.$$

We must be very careful with the use of the symbol $\sqrt{-b}$, where $b > 0$. Some relationships that are true in the set of real numbers that involve the square root symbol do not hold if the square root symbol does not represent a real number. For example, $\sqrt{a}\sqrt{b} = \sqrt{ab}$ *does not hold* if a and b are both negative numbers.

Correct $\sqrt{-4}\sqrt{-9} = (2i)(3i) = 6i^2 = 6(-1) = -6.$

Incorrect $\sqrt{-4}\sqrt{-9} = \sqrt{(-4)(-9)} = \sqrt{36} = 6.$

To avoid difficulty with this idea, you should rewrite all expressions of the form $\sqrt{-b}$, where $b > 0$, in the form $i\sqrt{b}$ before doing *any computations*. The following examples further demonstrate this point.

1. $\sqrt{-5}\sqrt{-7} = (i\sqrt{5})(i\sqrt{7}) = i^2\sqrt{35} = (-1)\sqrt{35} = -\sqrt{35}.$

2. $\sqrt{-2}\sqrt{-8} = (i\sqrt{2})(i\sqrt{8}) = i^2\sqrt{16} = (-1)(4) = -4.$

3. $\sqrt{-6}\sqrt{-8} = (i\sqrt{6})(i\sqrt{8}) = i^2\sqrt{48} = (-1)\sqrt{16}\sqrt{3} = -4\sqrt{3}.$

4. $\dfrac{\sqrt{-75}}{\sqrt{-3}} = \dfrac{i\sqrt{75}}{i\sqrt{3}} = \dfrac{\sqrt{75}}{\sqrt{3}} = \sqrt{\dfrac{75}{3}} = \sqrt{25} = 5.$

5. $\dfrac{\sqrt{-48}}{\sqrt{12}} = \dfrac{i\sqrt{48}}{\sqrt{12}} = i\sqrt{\dfrac{48}{12}} = i\sqrt{4} = 2i.$

Since complex numbers have a *binomial form*, we find the *product* of two complex numbers in the same way that we find the product of two binomials. Then by replacing i^2 with -1, we are able to simplify and express the final result in standard form. Consider the following examples.

6.
$$\begin{aligned}(2 + 3i)(4 + 5i) &= 2(4 + 5i) + 3i(4 + 5i)\\ &= 8 + 10i + 12i + 15i^2\\ &= 8 + 22i + 15i^2\\ &= 8 + 22i + 15(-1) = -7 + 22i.\end{aligned}$$

7.
$$\begin{aligned}(-3 + 6i)(2 - 4i) &= -3(2 - 4i) + 6i(2 - 4i)\\ &= -6 + 12i + 12i - 24i^2\\ &= -6 + 24i - 24(-1)\\ &= -6 + 24i + 24 = 18 + 24i.\end{aligned}$$

8.
$$\begin{aligned}(1 - 7i)^2 &= (1 - 7i)(1 - 7i)\\ &= 1(1 - 7i) - 7i(1 - 7i)\\ &= 1 - 7i - 7i + 49i^2\\ &= 1 - 14i + 49(-1)\\ &= 1 - 14i - 49\\ &= -48 - 14i.\end{aligned}$$

9.
$$\begin{aligned}(2 + 3i)(2 - 3i) &= 2(2 - 3i) + 3i(2 - 3i)\\ &= 4 - 6i + 6i - 9i^2\\ &= 4 - 9(-1)\\ &= 4 + 9\\ &= 13.\end{aligned}$$

Example 9 illustrates an important situation. The complex numbers $2 + 3i$ and $2 - 3i$ are called conjugates of each other; in general, the two complex numbers $a + bi$ and $a - bi$ are called **conjugates** of each other. *The product of a complex number and its conjugate is always a real number.* This can be shown as follows.

$$\begin{aligned}(a + bi)(a - bi) &= a(a - bi) + bi(a - bi)\\ &= a^2 - abi + abi - b^2i^2\\ &= a^2 - b^2(-1)\\ &= a^2 + b^2\end{aligned}$$

We use conjugates to *simplify expressions* such as $\dfrac{3i}{5 + 2i}$ that *indicate the quotient* of two complex numbers. To eliminate i in the denominator and change the indicated quotient to the standard form of a complex number, we can multiply both the numerator and the denominator by the conjugate of the denominator as follows.

$$\begin{aligned}\frac{3i}{5 + 2i} &= \frac{3i(5 - 2i)}{(5 + 2i)(5 - 2i)}\\ &= \frac{15i - 6i^2}{25 - 4i^2}\\ &= \frac{15i - 6(-1)}{25 - 4(-1)}\\ &= \frac{15i + 6}{29}\\ &= \frac{6}{29} + \frac{15}{29}i\end{aligned}$$

The following examples further clarify the process of *dividing* complex numbers.

10. $$\begin{aligned}\frac{2 - 3i}{4 - 7i} &= \frac{(2 - 3i)(4 + 7i)}{(4 - 7i)(4 + 7i)}\\ &= \frac{8 + 14i - 12i - 21i^2}{16 - 49i^2}\\ &= \frac{8 + 2i - 21(-1)}{16 - 49(-1)}\\ &= \frac{8 + 2i + 21}{16 + 49}\\ &= \frac{29 + 2i}{65}\\ &= \frac{29}{65} + \frac{2}{65}i.\end{aligned}$$

$4 + 7i$ is the conjugate of $4 - 7i$.

11. $\dfrac{4-5i}{2i} = \dfrac{(4-5i)(-2i)}{(2i)(-2i)}$ $\qquad -2i$ is the conjugate of $2i$.

$$= \frac{-8i+10i^2}{-4i^2}$$

$$= \frac{-8i+10(-1)}{-4(-1)}$$

$$= \frac{-8i-10}{4}$$

$$= -\frac{5}{2} - 2i.$$

For a problem such as number 11 in which the denominator is a pure imaginary number, we can change to standard form by choosing a multiplier other than the conjugate. Consider the following alternate approach for number 11.

$$\frac{4-5i}{2i} = \frac{(4-5i)(i)}{(2i)(i)}$$

$$= \frac{4i-5i^2}{2i^2}$$

$$= \frac{4i-5(-1)}{2(-1)}$$

$$= \frac{4i+5}{-2}$$

$$= -\frac{5}{2} - 2i$$

Problem Set 6.1

Label each of the following statements true or false.

1. Every complex number is a real number. False
2. Every real number is a complex number. True
3. The real part of the complex number $6i$ is 0. True
4. Every complex number is a pure imaginary number. False
5. The sum of two complex numbers is always a complex number. True
6. The imaginary part of the complex number 7 is 0. True
7. The sum of two complex numbers is sometimes a real number. True
8. The sum of two pure imaginary numbers is always a pure imaginary number. False

Add or subtract as indicated.

9. $(6+3i)+(4+5i)$ $10+8i$
10. $(5+2i)+(7+10i)$ $12+12i$
11. $(-8+4i)+(2+6i)$ $-6+10i$
12. $(5-8i)+(-7+2i)$ $-2-6i$

13. $(3 + 2i) - (5 + 7i)$ $-2 - 5i$

14. $(1 + 3i) - (4 + 9i)$ $-3 - 6i$

15. $(-7 + 3i) - (5 - 2i)$ $-12 + 5i$

16. $(-8 + 4i) - (9 - 4i)$ $-17 + 8i$

17. $(-3 - 10i) + (2 - 13i)$ $-1 - 23i$

18. $(-4 - 12i) + (-3 + 16i)$ $-7 + 4i$

19. $(4 - 8i) - (8 - 3i)$ $-4 - 5i$

20. $(12 - 9i) - (14 - 6i)$ $-2 - 3i$

21. $(-1 - i) - (-2 - 4i)$ $1 + 3i$

22. $(-2 - 3i) - (-4 - 14i)$ $2 + 11i$

23. $\left(\frac{3}{2} + \frac{1}{3}i\right) + \left(\frac{1}{6} - \frac{3}{4}i\right)$ $\frac{5}{3} - \frac{5}{12}i$

24. $\left(\frac{2}{3} - \frac{1}{5}i\right) + \left(\frac{3}{5} - \frac{3}{4}i\right)$ $\frac{19}{15} - \frac{19}{20}i$

25. $\left(-\frac{5}{9} + \frac{3}{5}i\right) - \left(\frac{4}{3} - \frac{1}{6}i\right)$ $-\frac{17}{9} + \frac{23}{30}i$

26. $\left(\frac{3}{8} - \frac{5}{2}i\right) - \left(\frac{5}{6} + \frac{1}{7}i\right)$ $-\frac{11}{24} - \frac{37}{14}i$

Write each of the following in terms of i and simplify. For example, $\sqrt{-20} = i\sqrt{20} = i\sqrt{4}\sqrt{5} = 2i\sqrt{5}$.

27. $\sqrt{-81}$ $9i$

28. $\sqrt{-49}$ $7i$

29. $\sqrt{-14}$ $i\sqrt{14}$

30. $\sqrt{-33}$ $i\sqrt{33}$

31. $\sqrt{-\frac{16}{25}}$ $\frac{4}{5}i$

32. $\sqrt{-\frac{64}{36}}$ $\frac{4}{3}i$

33. $\sqrt{-18}$ $3i\sqrt{2}$

34. $\sqrt{-84}$ $2i\sqrt{21}$

35. $\sqrt{-75}$ $5i\sqrt{3}$

36. $\sqrt{-63}$ $3i\sqrt{7}$

37. $3\sqrt{-28}$ $6i\sqrt{7}$

38. $5\sqrt{-72}$ $30i\sqrt{2}$

39. $-2\sqrt{-80}$ $-8i\sqrt{5}$

40. $-6\sqrt{-27}$ $-18i\sqrt{3}$

41. $12\sqrt{-90}$ $36i\sqrt{10}$

42. $9\sqrt{-40}$ $18i\sqrt{10}$

Write each of the following in terms of i, perform the indicated operations, and simplify. For example,

$$\begin{aligned}\sqrt{-3}\sqrt{-8} &= (i\sqrt{3})(i\sqrt{8})\\ &= i^2\sqrt{24}\\ &= (-1)\sqrt{4}\sqrt{6}\\ &= -2\sqrt{6}.\end{aligned}$$

43. $\sqrt{-4}\sqrt{-16}$ -8

44. $\sqrt{-81}\sqrt{-25}$ -45

45. $\sqrt{-3}\sqrt{-5}$ $-\sqrt{15}$

46. $\sqrt{-7}\sqrt{-10}$ $-\sqrt{70}$

47. $\sqrt{-9}\sqrt{-6}$ $-3\sqrt{6}$

48. $\sqrt{-8}\sqrt{-16}$ $-8\sqrt{2}$

49. $\sqrt{-15}\sqrt{-5}$ $-5\sqrt{3}$

50. $\sqrt{-2}\sqrt{-20}$ $-2\sqrt{10}$

51. $\sqrt{-2}\sqrt{-27}$ $-3\sqrt{6}$

52. $\sqrt{-3}\sqrt{-15}$ $-3\sqrt{5}$

53. $\sqrt{6}\sqrt{-8}$ $4i\sqrt{3}$

54. $\sqrt{-75}\sqrt{3}$ $15i$

55. $\frac{\sqrt{-25}}{\sqrt{-4}}$ $\frac{5}{2}$

56. $\frac{\sqrt{-81}}{\sqrt{-9}}$ 3

57. $\frac{\sqrt{-56}}{\sqrt{-7}}$ $2\sqrt{2}$

58. $\frac{\sqrt{-72}}{\sqrt{-6}}$ $2\sqrt{3}$

59. $\frac{\sqrt{-24}}{\sqrt{6}}$ $2i$

60. $\frac{\sqrt{-96}}{\sqrt{2}}$ $4i\sqrt{3}$

Find each of the following products and express answers in the standard form of a complex number.

61. $(5i)(4i)$ $-20 + 0i$

62. $(-6i)(9i)$ $54 + 0i$

63. $(7i)(-6i)$ $42 + 0i$

64. $(-5i)(-12i)$ $-60 + 0i$

65. $(3i)(2 - 5i)$ $15 + 6i$

66. $(7i)(-9 + 3i)$ $-21 - 63i$

67. $(-6i)(-2 - 7i)$ $-42 + 12i$

68. $(-9i)(-4 - 5i)$ $-45 + 36i$

69. $(3 + 2i)(5 + 4i)$ $7 + 22i$

70. $(4 + 3i)(6 + i)$ $21 + 22i$

71. $(6 - 2i)(7 - i)$ $40 - 20i$

72. $(8 - 4i)(7 - 2i)$ $48 - 44i$

73. $(-3 - 2i)(5 + 6i)$ $-3 - 28i$

74. $(-5 - 3i)(2 - 4i)$ $-22 + 14i$

75. $(9 + 6i)(-1 - i)$ $-3 - 15i$

76. $(10 + 2i)(-2 - i)$ $-18 - 14i$

77. $(4 + 5i)^2$ $-9 + 40i$

78. $(5 - 3i)^2$ $16 - 30i$

79. $(-2 - 4i)^2$ $-12 + 16i$

80. $(-3 - 6i)^2$ $-27 + 36i$

81. $(6 + 7i)(6 - 7i)$ $85 + 0i$

82. $(5 - 7i)(5 + 7i)$ $74 + 0i$

83. $(-1 + 2i)(-1 - 2i)$ $5 + 0i$

84. $(-2 - 4i)(-2 + 4i)$ $20 + 0i$

Find each of the following quotients and express answers in the standard form of a complex number.

85. $\frac{3i}{2 + 4i}$ $\frac{3}{5} + \frac{3}{10}i$

86. $\frac{4i}{5 + 2i}$ $\frac{8}{29} + \frac{20}{29}i$

87. $\frac{-2i}{3 - 5i}$ $\frac{5}{17} - \frac{3}{17}i$

88. $\frac{-5i}{2 - 4i}$ $1 - \frac{1}{2}i$

89. $\frac{-2 + 6i}{3i}$ $2 + \frac{2}{3}i$

90. $\frac{-4 - 7i}{6i}$ $-\frac{7}{6} + \frac{2}{3}i$

91. $\frac{2}{7i}$ $0 - \frac{2}{7}i$

92. $\frac{3}{10i}$ $0 - \frac{3}{10}i$

93. $\frac{2 + 6i}{1 + 7i}$ $\frac{22}{25} - \frac{4}{25}i$

94. $\frac{5 + i}{2 + 9i}$ $\frac{19}{85} - \frac{43}{85}i$

95. $\frac{3 + 6i}{4 - 5i}$ $-\frac{18}{41} + \frac{39}{41}i$

96. $\frac{7 - 3i}{4 - 3i}$ $\frac{37}{25} + \frac{9}{25}i$

97. $\frac{-2 + 7i}{-1 + i}$ $\frac{9}{2} - \frac{5}{2}i$

98. $\frac{-3 + 8i}{-2 + i}$ $\frac{14}{5} - \frac{13}{5}i$

99. $\frac{-1 - 3i}{-2 - 10i}$ $\frac{4}{13} - \frac{1}{26}i$

100. $\frac{-3 - 4i}{-4 - 11i}$ $\frac{56}{137} - \frac{17}{137}i$

6.2 Quadratic Equations

A second-degree equation in one variable contains the variable with an exponent of two, but no higher power. Such equations are also called **quadratic equations**. The following are examples of quadratic equations.

$$x^2 = 36, \qquad y^2 + 4y = 0, \qquad x^2 + 5x - 2 = 0,$$

$$3n^2 + 2n - 1 = 0, \qquad 5x^2 + x + 2 = 3x^2 - 2x - 1$$

A quadratic equation in the variable x can also be defined as any equation that can be written in the form

$$ax^2 + bx + c = 0,$$

where a, b, and c are real numbers and $a \neq 0$. The form $ax^2 + bx + c = 0$ is called the **standard form** of a quadratic equation.

In previous chapters you solved quadratic equations (the term *quadratic* was not used at that time) by factoring and applying the property, $ab = 0$ if and only if $a = 0$ or $b = 0$. Let's review a few such examples.

EXAMPLE 1 Solve $3n^2 + 14n - 5 = 0$.

Solution

$$3n^2 + 14n - 5 = 0$$

$$(3n - 1)(n + 5) = 0 \qquad \text{Factor the left side.}$$

$$3n - 1 = 0 \quad \text{or} \quad n + 5 = 0 \qquad \text{Apply: } ab = 0 \text{ if and only if } a = 0 \text{ or } b = 0.$$

$$3n = 1 \quad \text{or} \quad n = -5$$

$$n = \frac{1}{3} \quad \text{or} \quad n = -5$$

The solution set is $\left\{-5, \frac{1}{3}\right\}$. ■

EXAMPLE 2 Solve $x^2 + 3kx - 10k^2 = 0$ for x.

Solution

$$x^2 + 3kx - 10k^2 = 0$$

$$(x + 5k)(x - 2k) = 0 \qquad \text{Factor the left side.}$$

$$x + 5k = 0 \quad \text{or} \quad x - 2k = 0 \qquad \text{Apply: } ab = 0 \text{ if and only if } a = 0 \text{ or } b = 0.$$

$$x = -5k \quad \text{or} \quad x = 2k$$

The solution set is $\{-5k, 2k\}$. ■

EXAMPLE 3 Solve $2\sqrt{x} = x - 8$.

Solution

$$2\sqrt{x} = x - 8$$

$$(2\sqrt{x})^2 = (x - 8)^2 \qquad \text{Square both sides.}$$

$$4x = x^2 - 16x + 64$$

$$0 = x^2 - 20x + 64$$

$$0 = (x - 16)(x - 4) \qquad \text{Factor the right side.}$$

$$x - 16 = 0 \quad \text{or} \quad x - 4 = 0 \qquad \text{Apply: } ab = 0 \text{ if and only if } a = 0 \text{ or } b = 0.$$

$$x = 16 \quad \text{or} \quad x = 4$$

CHECK

$$2\sqrt{x} = x - 8 \qquad\qquad 2\sqrt{x} = x - 8$$
$$2\sqrt{16} \stackrel{?}{=} 16 - 8 \quad \text{or} \quad 2\sqrt{4} \stackrel{?}{=} 4 - 8$$
$$2(4) \stackrel{?}{=} 8 \qquad\qquad 2(2) \stackrel{?}{=} -4$$
$$8 = 8 \qquad\qquad 4 \neq -4$$

The solution set is $\{16\}$. ■

We should make two comments about Example 3. First, remember that applying the property, if $a = b$, then $a^n = b^n$, might produce extraneous solutions. Therefore, we *must* check all potential solutions. Secondly, the equation $2\sqrt{x} = x - 8$ is said to be of **quadratic form** because it can be written as $2x^{\frac{1}{2}} = \left(x^{\frac{1}{2}}\right)^2 - 8$. More will be said about the phrase, quadratic form, later.

Let's consider quadratic equations of the form $x^2 = a$, where x is the variable and a is any real number. We can solve $x^2 = a$ as follows.

$$x^2 = a$$
$$x^2 - a = 0$$
$$x^2 - (\sqrt{a})^2 = 0 \qquad a = (\sqrt{a})^2$$
$$(x - \sqrt{a})(x + \sqrt{a}) = 0 \qquad \text{Factor the left side.}$$
$$x - \sqrt{a} = 0 \quad \text{or} \quad x + \sqrt{a} = 0 \qquad \text{Apply: } ab = 0 \text{ if and only if } a = 0 \text{ or } b = 0.$$
$$x = \sqrt{a} \quad \text{or} \quad x = -\sqrt{a}.$$

The solutions are $\sqrt{a}$ and $-\sqrt{a}$.

We can state the previous result as a general property and use it to solve certain types of quadratic equations.

PROPERTY 6.1

> For any real number a,
>
> $x^2 = a$ if and only if $x = \sqrt{a}$ or $x = -\sqrt{a}$.
>
> (The statement $x = \sqrt{a}$ or $x = -\sqrt{a}$ can be written as $x = \pm\sqrt{a}$.)

Property 6.1, along with our knowledge of square roots, makes it very easy to solve quadratic equations of the form $x^2 = a$.

EXAMPLE 4 Solve $x^2 = 45$.

Solution

$$x^2 = 45$$
$$x = \pm\sqrt{45}$$
$$x = \pm 3\sqrt{5} \qquad \sqrt{45} = \sqrt{9}\sqrt{5} = 3\sqrt{5}$$

The solution set is $\{\pm 3\sqrt{5}\}$. ■

EXAMPLE 5 Solve $x^2 = -9$.

Solution

$$x^2 = -9$$
$$x = \pm\sqrt{-9}$$
$$x = \pm 3i. \qquad \sqrt{-9} = i\sqrt{9} = 3i$$

Thus, the solution set is $\{\pm 3i\}$. ■

EXAMPLE 6 Solve $7n^2 = 12$.

Solution

$$7n^2 = 12$$
$$n^2 = \frac{12}{7}$$
$$n = \pm\sqrt{\frac{12}{7}}$$
$$n = \pm\frac{2\sqrt{21}}{7}. \qquad \sqrt{\frac{12}{7}} = \frac{\sqrt{12}}{\sqrt{7}}\cdot\frac{\sqrt{7}}{\sqrt{7}} = \frac{\sqrt{84}}{7} = \frac{\sqrt{4}\sqrt{21}}{7} = \frac{2\sqrt{21}}{7}$$

The solution set is $\left\{\pm\frac{2\sqrt{21}}{7}\right\}$. ■

EXAMPLE 7 Solve $(3n + 1)^2 = 25$.

Solution

$$(3n + 1)^2 = 25$$
$$(3n + 1) = \pm\sqrt{25}$$
$$3n + 1 = \pm 5$$
$$3n + 1 = 5 \quad \text{or} \quad 3n + 1 = -5$$
$$3n = 4 \quad \text{or} \quad 3n = -6$$
$$n = \frac{4}{3} \quad \text{or} \quad n = -2$$

The solution set is $\left\{-2, \frac{4}{3}\right\}$. ■

EXAMPLE 8 Solve $(x - 3)^2 = -10$.

Solution

$$(x - 3)^2 = -10$$
$$x - 3 = \pm\sqrt{-10}$$
$$x - 3 = \pm i\sqrt{10}$$
$$x = 3 \pm i\sqrt{10}$$

Thus, the solution set is $\{3 \pm i\sqrt{10}\}$. ■

Sometimes it may be necessary to change the form before we can apply Property 6.1. Let's consider one example to illustrate this idea.

EXAMPLE 9 Solve $3(2x - 3)^2 + 8 = 44$.

Solution

$$3(2x - 3)^2 + 8 = 44$$
$$3(2x - 3)^2 = 36$$
$$(2x - 3)^2 = 12$$
$$2x - 3 = \pm\sqrt{12}$$
$$2x - 3 = \pm 2\sqrt{3}$$
$$2x = 3 \pm 2\sqrt{3}$$
$$x = \frac{3 \pm 2\sqrt{3}}{2}$$

The solution set is $\left\{\frac{3 \pm 2\sqrt{3}}{2}\right\}$. ■

Back to the Pythagorean Theorem

Our work with radicals, Property 6.1, and the Pythagorean Theorem form a basis for solving a variety of problems that pertain to right triangles.

PROBLEM 1 Find c in the accompanying figure.

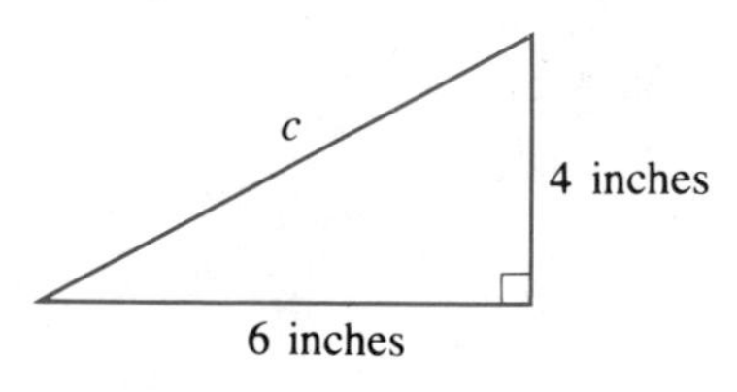

Solution Apply the Pythagorean Theorem and we obtain

$$c^2 = 6^2 + 4^2$$
$$= 36 + 16$$
$$= 52.$$

Therefore,

$$c = \pm\sqrt{52} = \pm 2\sqrt{13}.$$

Disregard the negative solution and we find that $c = 2\sqrt{13}$ inches. ■

There are two special kinds of right triangles that we use extensively in later mathematics courses. An **isosceles right triangle** is a right triangle that has both legs of the same length. Let's consider a problem that involves an isosceles right triangle.

PROBLEM 2 Find the length of each leg of an isosceles right triangle that has a hypotenuse of length 5 meters.

Solution Let's sketch an isosceles right triangle and let x represent the length of each leg. Then we can apply the Pythagorean Theorem as follows.

$$x^2 + x^2 = 5^2$$
$$2x^2 = 25$$
$$x^2 = \frac{25}{2}$$
$$x = \pm\sqrt{\frac{25}{2}} = \pm\frac{5}{\sqrt{2}} = \pm\frac{5\sqrt{2}}{2}$$

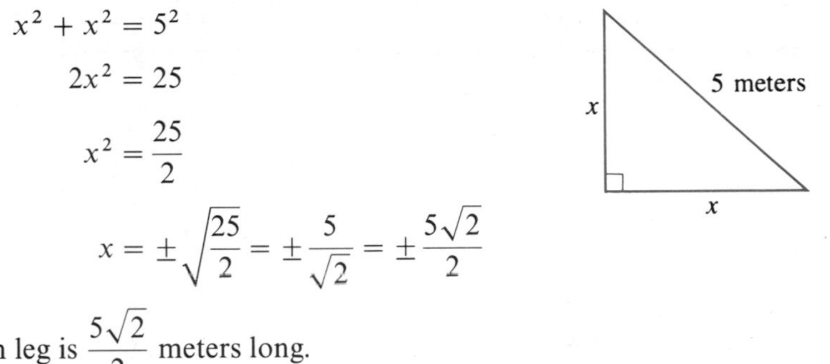

Each leg is $\frac{5\sqrt{2}}{2}$ meters long. ■

The other special kind of right triangle that we use frequently is one that contains acute angles of 30° and 60°. In such a right triangle, which we refer to as a **30°–60° right triangle**, *the side opposite the 30° angle is equal in length to one-half of the length of the hypotenuse.* This relationship, along with the Pythagorean Theorem, provides us with another problem solving technique.

PROBLEM 3 Suppose that the side opposite the 60° angle in a 30°–60° right triangle is 9 feet long. Find the length of the other leg and the length of the hypotenuse.

Solution Let x represent the length of the side opposite the 30° angle. Then $2x$ represents the length of the hypotenuse, as indicated in the accompanying diagram. Thus, we can apply the Pythagorean Theorem as follows.

$$x^2 + 9^2 = (2x)^2$$
$$x^2 + 81 = 4x^2$$
$$81 = 3x^2$$
$$27 = x^2$$
$$\pm\sqrt{27} = x$$
$$\pm 3\sqrt{3} = x$$

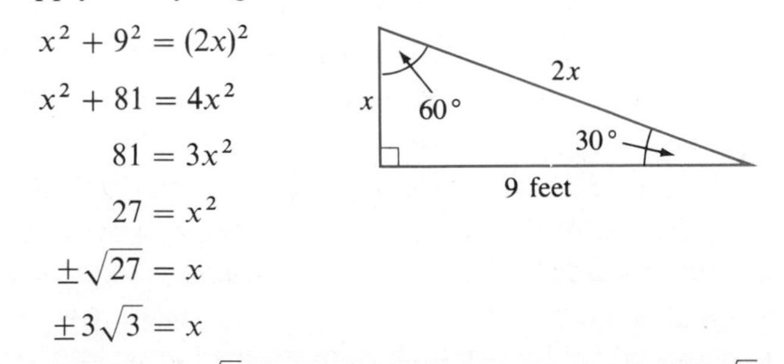

The other leg is $3\sqrt{3}$ feet long and the hypotenuse is $2(3\sqrt{3}) = 6\sqrt{3}$ feet long. ■

Problem Set 6.2

Solve each of the following quadratic equations by factoring and applying the property, $ab = 0$ if and only if $a = 0$ or $b = 0$. If necessary, return to Chapter 3 and review the factoring techniques that we have studied.

1. $x^2 - 9x = 0$ $\{0, 9\}$
2. $x^2 + 5x = 0$ $\{-5, 0\}$
3. $x^2 = -3x$ $\{-3, 0\}$
4. $x^2 = 15x$ $\{0, 15\}$
5. $3y^2 + 12y = 0$ $\{-4, 0\}$
6. $6y^2 - 24y = 0$ $\{0, 4\}$
7. $5n^2 - 9n = 0$ $\{0, \frac{9}{5}\}$
8. $4n^2 + 13n = 0$ $\{-\frac{13}{4}, 0\}$
9. $x^2 + x - 30 = 0$ $\{-6, 5\}$
10. $x^2 - 8x - 48 = 0$ $\{-4, 12\}$
11. $x^2 - 19x + 84 = 0$ $\{7, 12\}$
12. $x^2 - 21x + 104 = 0$ $\{8, 13\}$
13. $2x^2 + 19x + 24 = 0$ $\{-8, -\frac{3}{2}\}$
14. $4x^2 + 29x + 30 = 0$ $\{-6, -\frac{5}{4}\}$
15. $15x^2 + 29x - 14 = 0$ $\{-\frac{7}{3}, \frac{2}{5}\}$
16. $24x^2 + x - 10 = 0$ $\{-\frac{2}{3}, \frac{5}{8}\}$
17. $25x^2 - 30x + 9 = 0$ $\{\frac{3}{5}\}$
18. $16x^2 - 8x + 1 = 0$ $\{\frac{1}{4}\}$
19. $6x^2 - 5x - 21 = 0$ $\{-\frac{3}{2}, \frac{7}{3}\}$
20. $12x^2 - 4x - 5 = 0$ $\{-\frac{1}{2}, \frac{5}{6}\}$

Solve each of the following radical equations. Don't forget, you *must check* potential solutions.

21. $3\sqrt{x} = x + 2$ $\{1, 4\}$
22. $3\sqrt{2x} = x + 4$ $\{2, 8\}$
23. $\sqrt{2x} = x - 4$ $\{8\}$
24. $\sqrt{x} = x - 2$ $\{4\}$
25. $\sqrt{3x} + 6 = x$ $\{12\}$
26. $\sqrt{5x} + 10 = x$ $\{20\}$

Solve each of the following equations for x by factoring and applying the property, $ab = 0$ if and only if $a = 0$ or $b = 0$.

27. $x^2 - 5kx = 0$ $\{0, 5k\}$
28. $x^2 + 7kx = 0$ $\{0, -7k\}$
29. $x^2 = 16k^2x$ $\{0, 16k^2\}$
30. $x^2 = 25k^2x$ $\{0, 25k^2\}$
31. $x^2 - 12kx + 35k^2 = 0$ $\{5k, 7k\}$
32. $x^2 - 3kx - 18k^2 = 0$ $\{-3k, 6k\}$
33. $2x^2 + 5kx - 3k^2 = 0$ $\{\frac{k}{2}, -3k\}$
34. $3x^2 - 20kx - 7k^2 = 0$ $\{-\frac{k}{3}, 7k\}$

Use Property 6.1 to help solve each of the following quadratic equations.

35. $x^2 = 1$ $\{\pm 1\}$
36. $x^2 = 81$ $\{\pm 9\}$
37. $x^2 = -36$ $\{\pm 6i\}$
38. $x^2 = -49$ $\{\pm 7i\}$
39. $x^2 = 14$ $\{\pm\sqrt{14}\}$
40. $x^2 = 22$ $\{\pm\sqrt{22}\}$
41. $n^2 - 28 = 0$ $\{\pm 2\sqrt{7}\}$
42. $n^2 - 54 = 0$ $\{\pm 3\sqrt{6}\}$
43. $3t^2 = 54$ $\{\pm 3\sqrt{2}\}$
44. $4t^2 = 108$ $\{\pm 3\sqrt{3}\}$
45. $2t^2 = 7$ $\{\pm\frac{\sqrt{14}}{2}\}$
46. $3t^2 = 8$ $\{\pm\frac{2\sqrt{6}}{3}\}$
47. $15y^2 = 20$ $\{\pm\frac{2\sqrt{3}}{3}\}$
48. $14y^2 = 80$ $\{\pm\frac{2\sqrt{70}}{7}\}$
49. $10x^2 + 48 = 0$ $\{\pm\frac{2i\sqrt{30}}{5}\}$
50. $12x^2 + 50 = 0$ $\{\pm\frac{5i\sqrt{6}}{6}\}$
51. $24x^2 = 36$ $\{\pm\frac{\sqrt{6}}{2}\}$
52. $12x^2 = 49$ $\{\pm\frac{7\sqrt{3}}{6}\}$
53. $(x - 2)^2 = 9$ $\{-1, 5\}$
54. $(x + 1)^2 = 16$ $\{-5, 3\}$
55. $(x + 3)^2 = 25$ $\{-8, 2\}$
56. $(x - 2)^2 = 49$ $\{-5, 9\}$
57. $(x + 6)^2 = -4$ $\{-6 \pm 2i\}$
58. $(3x + 1)^2 = 9$ $\{-\frac{4}{3}, \frac{2}{3}\}$
59. $(2x - 3)^2 = 1$ $\{1, 2\}$
60. $(2x + 5)^2 = -4$ $\{\frac{-5 \pm 2i}{2}\}$
61. $(n - 4)^2 = 5$ $\{4 \pm \sqrt{5}\}$
62. $(n - 7)^2 = 6$ $\{7 \pm \sqrt{6}\}$
63. $(t + 5)^2 = 12$ $\{-5 \pm 2\sqrt{3}\}$
64. $(t - 1)^2 = 18$ $\{1 \pm 3\sqrt{2}\}$
65. $(3y - 2)^2 = -27$ $\{\frac{2 \pm 3i\sqrt{3}}{3}\}$
66. $(4y + 5)^2 = 80$ $\{\frac{-5 \pm 4\sqrt{5}}{4}\}$
67. $3(x + 7)^2 + 4 = 79$ $\{-12, -2\}$
68. $2(x + 6)^2 - 9 = 63$ $\{-12, 0\}$
69. $2(5x - 2)^2 + 5 = 25$ $\{\frac{2 \pm \sqrt{10}}{5}\}$
70. $3(4x - 1)^2 + 1 = -17$ $\{\frac{1 \pm i\sqrt{6}}{4}\}$

For Problems 71–76, a and b represent the lengths of the legs of a right triangle, and c represents the length of the hypotenuse. Express answers in simplest radical form.

71. Find c if $a = 4$ centimeters and $b = 6$ centimeters. $2\sqrt{13}$ centimeters

72. Find c if $a = 3$ meters and $b = 7$ meters. $\sqrt{58}$ meters
73. Find a if $c = 12$ inches and $b = 8$ inches. $4\sqrt{5}$ inches
74. Find a if $c = 8$ feet and $b = 6$ feet. $2\sqrt{7}$ feet
75. Find b if $c = 17$ yards and $a = 15$ yards. 8 yards
76. Find b if $c = 14$ meters and $a = 12$ meters. $2\sqrt{13}$ meters

For Problems 77–80, use the following isosceles right triangle. Express answers in simplest radical form.

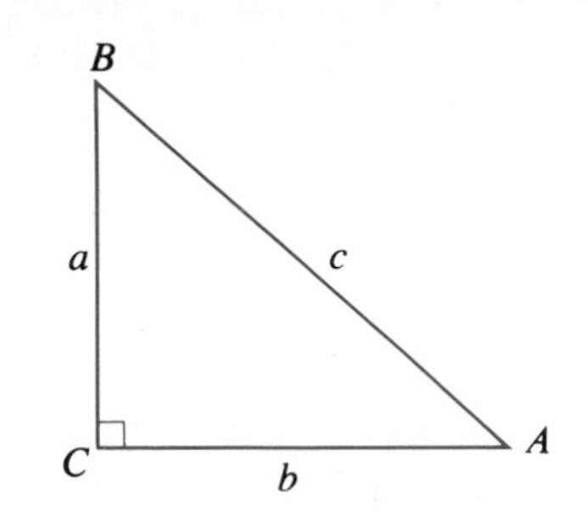

77. If $b = 6$ inches, find c. $6\sqrt{2}$ inches
78. If $a = 7$ centimeters, find c. $7\sqrt{2}$ centimeters
79. If $c = 8$ meters, find a and b. $a = b = 4\sqrt{2}$ meters
80. If $c = 9$ feet, find a and b. $a = b = \frac{9\sqrt{2}}{2}$ feet

For Problems 81–86, use the accompanying figure. Express answers in simplest radical form.

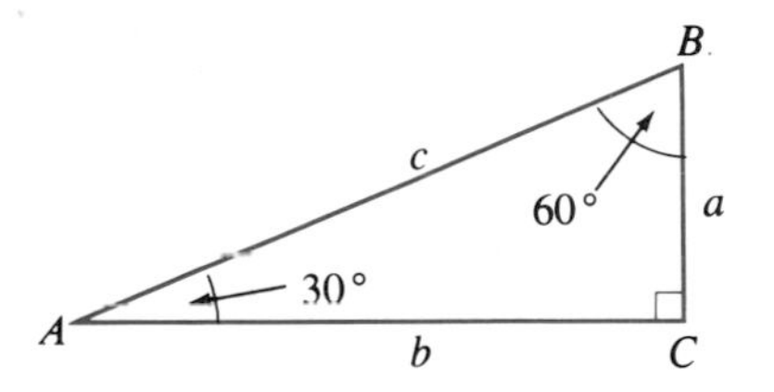

81. If $a = 3$ inches, find b and c. $b = 3\sqrt{3}$ inches and $c = 6$ inches
82. If $a = 6$ feet, find b and c. $b = 6\sqrt{3}$ feet and $c = 12$ feet
83. If $c = 14$ centimeters, find a and b. $a = 7$ centimeters and $b = 7\sqrt{3}$ centimeters
84. If $c = 9$ centimeters, find a and b. $a = 4.5$ centimeters and $b = \frac{9\sqrt{3}}{2}$ centimeters
85. If $b = 10$ feet, find a and c. $a = \frac{10\sqrt{3}}{3}$ feet and $c = \frac{20\sqrt{3}}{3}$ feet
86. If $b = 8$ meters, find a and c. $a = \frac{8\sqrt{3}}{3}$ meters and $c = \frac{16\sqrt{3}}{3}$ meters

Miscellaneous Problems

87. Find the length of the hypotenuse (h) of an isosceles right triangle if each leg is s units long. Then use this relationship and redo Problems 77–80.
88. Suppose that the side opposite the 30° angle in a 30°–60° right triangle is s units long. Express the length of the hypotenuse and the length of the other leg in terms of s. Then use these relationships and redo Problems 81–86.

6.3

Completing the Square

Thus far we have solved quadratic equations by factoring and applying the property $ab = 0$ if and only if $a = 0$ or $b = 0$, or by applying the property, $x^2 = a$ if and only if $x = \pm\sqrt{a}$. In this section we shall examine another method called **completing the square**, which will give us the power to solve *any* quadratic equation.

A factoring technique we studied in Chapter 3 relied on recognizing **perfect square trinomials**. In each of the following, the perfect square trinomial on the right side is the result of squaring the binomial on the left side.

$$(x + 4)^2 = x^2 + 8x + 16, \qquad (x - 6)^2 = x^2 - 12x + 36,$$

$$(x + 7)^2 = x^2 + 14x + 49, \qquad (x - 9)^2 = x^2 - 18x + 81$$

Notice that in each of the square trinomials *the constant term is equal to the square of one-half of the coefficient of the x-term.* This relationship allows us to form a perfect square trinomial by adding a proper constant term. For example, suppose that we want to form a perfect square trinomial from $x^2 + 10x$. Since $\frac{1}{2}(10) = 5$ and $5^2 = 25$, the perfect square trinomial $x^2 + 10x + 25$ can be formed. Let's use the previous ideas to help solve some quadratic equations.

EXAMPLE 1 Solve $x^2 + 10x - 2 = 0$.

Solution

$$x^2 + 10x - 2 = 0$$

$$x^2 + 10x = 2$$

$$x^2 + 10x + 25 = 2 + 25$$

We add 25 to the left side to form a perfect square trinomial; 25 must also be added to the right side.

$$(x + 5)^2 = 27$$

(Now we can proceed as in the last section.)

$$x + 5 = \pm\sqrt{27}$$

$$x + 5 = \pm 3\sqrt{3}$$

$$x = -5 \pm 3\sqrt{3}$$

The solution set is $\{-5 \pm 3\sqrt{3}\}$. ■

Notice from Example 1 that the method of completing the square to solve a quadratic equation is merely what the name implies. A perfect square trinomial is formed, then the equation can be changed to the necessary form for applying the property, $x^2 = a$ if and only if $x = \pm\sqrt{a}$. Let's consider another example.

EXAMPLE 2 Solve $x^2 + 4x + 7 = 0$.

Solution

$$x^2 + 4x + 7 = 0$$
$$x^2 + 4x = -7$$
$$x^2 + 4x + 4 = -7 + 4$$
$$(x + 2)^2 = -3$$
$$x + 2 = \pm\sqrt{-3}$$
$$x + 2 = \pm i\sqrt{3}$$
$$x = -2 \pm i\sqrt{3}$$

The solution set is $\{-2 \pm i\sqrt{3}\}$. ■

REMARK Even though we do not show a check for each problem, it is probably a good idea for you to check one now and then. Example 2 is a good one to check since it involves some manipulation with complex numbers.

EXAMPLE 3 Solve $x^2 - 3x + 1 = 0$.

Solution

$$x^2 - 3x + 1 = 0$$
$$x^2 - 3x = -1$$
$$x^2 - 3x + \frac{9}{4} = -1 + \frac{9}{4} \qquad \frac{1}{2}(3) = \frac{3}{2} \text{ and } \left(\frac{3}{2}\right)^2 = \frac{9}{4}$$
$$\left(x - \frac{3}{2}\right)^2 = \frac{5}{4}$$
$$x - \frac{3}{2} = \pm\sqrt{\frac{5}{4}}$$
$$x - \frac{3}{2} = \pm\frac{\sqrt{5}}{2}$$
$$x = \frac{3}{2} \pm \frac{\sqrt{5}}{2}$$
$$x = \frac{3 \pm \sqrt{5}}{2}$$

The solution set is $\left\{\frac{3 \pm \sqrt{5}}{2}\right\}$. ■

In Example 3 notice that since the coefficient of the x-term is odd, we are forced into the realm of fractions. The use of common fractions rather than decimals allows us to apply our previous work with radicals.

The relationship for a perfect square trinomial that states that *the constant term is equal to the square of one-half of the coefficient of the x-term* holds only if the coefficient of x^2 is 1. Thus, an adjustment needs to be made when solving quadratic equations that have a coefficient of x^2 other than 1. The next example shows how to make this adjustment.

EXAMPLE 4 Solve $2x^2 + 12x - 5 = 0$.

Solution

$$2x^2 + 12x - 5 = 0$$

$$2x^2 + 12x = 5$$

$$x^2 + 6x = \frac{5}{2} \qquad \text{Multiply both sides by } \frac{1}{2}.$$

$$x^2 + 6x + 9 = \frac{5}{2} + 9$$

$$x^2 + 6x + 9 = \frac{23}{2}$$

$$(x + 3)^2 = \frac{23}{2}$$

$$x + 3 = \pm\sqrt{\frac{23}{2}}$$

$$x + 3 = \pm\frac{\sqrt{46}}{2} \qquad \sqrt{\frac{23}{2}} = \frac{\sqrt{23}}{\sqrt{2}} \cdot \frac{\sqrt{2}}{\sqrt{2}} = \frac{\sqrt{46}}{2}$$

$$x = -3 \pm \frac{\sqrt{46}}{2}$$

$$x = \frac{-6 \pm \sqrt{46}}{2}$$

The solution set is $\left\{\frac{-6 \pm \sqrt{46}}{2}\right\}$. ■

As we mentioned earlier, we can use the method of completing the square to solve *any* quadratic equation. To illustrate, let's use it to solve an equation that could also be solved by factoring.

EXAMPLE 5 Solve $x^2 - 2x - 8 = 0$ by completing the square.

Solution

$$x^2 - 2x - 8 = 0$$

$$x^2 - 2x = 8$$

$$x^2 - 2x + 1 = 8 + 1$$

$$(x - 1)^2 = 9$$

$$x - 1 = \pm 3$$

$$x - 1 = 3 \quad \text{or} \quad x - 1 = -3$$

$$x = 4 \quad \text{or} \quad x = -2$$

The solution set is $\{-2, 4\}$. ■

We make no claim that using the method of completing the square with an equation such as the one in Example 5 is easier than the factoring technique. However, you should recognize that the method of completing the square will work with any quadratic equation.

Problem Set 6.3

Solve each of the following quadratic equations by using (a) the factoring method and (b) the method of completing the square.

1. $x^2 - 4x - 60 = 0$ $\{-6, 10\}$

2. $x^2 + 6x - 16 = 0$ $\{-8, 2\}$

3. $x^2 - 14x = -40$ $\{4, 10\}$

4. $x^2 - 18x = -72$ $\{6, 12\}$

5. $x^2 - 5x - 50 = 0$ $\{-5, 10\}$

6. $x^2 + 3x - 18 = 0$ $\{-6, 3\}$

7. $x(x + 7) = 8$ $\{-8, 1\}$

8. $x(x - 1) = 30$ $\{-5, 6\}$

9. $2n^2 - n - 15 = 0$ $\{-\frac{5}{2}, 3\}$

10. $3n^2 + n - 14 = 0$ $\{-\frac{7}{3}, 2\}$

11. $3n^2 + 7n - 6 = 0$ $\{-3, \frac{2}{3}\}$

12. $2n^2 + 7n - 4 = 0$ $\{-4, \frac{1}{2}\}$

13. $n(n + 6) = 160$ $\{-16, 10\}$

14. $n(n - 6) = 216$ $\{-12, 18\}$

Use the method of completing the square to solve each of the following quadratic equations.

15. $x^2 + 4x - 2 = 0$ $\{-2 \pm \sqrt{6}\}$

16. $x^2 + 2x - 1 = 0$ $\{-1 + \sqrt{2}\}$

17. $x^2 + 6x - 3 = 0$ $\{-3 \pm 2\sqrt{3}\}$

18. $x^2 + 8x - 4 = 0$ $\{-4 \pm 2\sqrt{5}\}$

19. $y^2 - 10y = 1$ $\{5 \pm \sqrt{26}\}$

20. $y^2 - 6y = -10$ $\{3 \pm i\}$

21. $n^2 - 8n + 17 = 0$ $\{4 \pm i\}$

22. $n^2 - 4n + 2 = 0$ $\{2 \pm \sqrt{2}\}$

23. $n(n + 12) = -9$ $\{-6 \pm 3\sqrt{3}\}$

24. $n(n + 14) = -4$ $\{-7 \pm 3\sqrt{5}\}$

25. $n^2 + 2n + 6 = 0$ $\{-1 \pm i\sqrt{5}\}$

26. $n^2 + n - 1 = 0$ $\{\frac{-1 \pm \sqrt{5}}{2}\}$

27. $x^2 + 3x - 2 = 0$ $\{\frac{-3 \pm \sqrt{17}}{2}\}$

28. $x^2 + 5x - 3 = 0$ $\{\frac{-5 \pm \sqrt{37}}{2}\}$

29. $x^2 + 5x + 1 = 0$ $\{\frac{-5 \pm \sqrt{21}}{2}\}$

30. $x^2 + 7x + 2 = 0$ $\{\frac{-7 \pm \sqrt{41}}{2}\}$

31. $y^2 - 7y + 3 = 0$ $\{\frac{7 \pm \sqrt{37}}{2}\}$

32. $y^2 - 9y + 30 = 0$ $\{\frac{9 \pm i\sqrt{39}}{2}\}$

33. $2x^2 + 4x - 3 = 0$ $\{\frac{-2 \pm \sqrt{10}}{2}\}$

34. $2t^2 - 4t + 1 = 0$ $\{\frac{2 \pm \sqrt{2}}{2}\}$

35. $3n^2 - 6n + 5 = 0$ $\{\frac{3 \pm i\sqrt{6}}{3}\}$

36. $3x^2 + 12x - 2 = 0$ $\{\frac{-6 \pm \sqrt{42}}{3}\}$

37. $3x^2 + 5x - 1 = 0$ $\{\frac{-5 \pm \sqrt{37}}{6}\}$

38. $2x^2 + 7x - 3 = 0$ $\{\frac{-7 \pm \sqrt{73}}{4}\}$

Solve each of the following quadratic equations and use whatever method seems most appropriate.

39. $x^2 + 8x - 48 = 0$ $\{-12, 4\}$

40. $x^2 + 5x - 14 = 0$ $\{-7, 2\}$

41. $2n^2 - 8n = -3$ $\left\{\frac{4 \pm \sqrt{10}}{2}\right\}$

42. $3x^2 + 6x = 1$ $\left\{\frac{-3 \pm 2\sqrt{3}}{3}\right\}$

43. $(3x - 1)(2x + 9) = 0$ $\left\{-\frac{9}{2}, \frac{1}{3}\right\}$

44. $(5x + 2)(x - 4) = 0$ $\left\{-\frac{2}{5}, 4\right\}$

45. $(x + 2)(x - 7) = 10$ $\{-3, 8\}$

46. $(x - 3)(x + 5) = -7$ $\{-4, 2\}$

47. $(x - 3)^2 = 12$ $\{3 \pm 2\sqrt{3}\}$

48. $x^2 = 16x$ $\{0, 16\}$

49. $3n^2 - 6n + 4 = 0$ $\left\{\frac{3 \pm i\sqrt{3}}{3}\right\}$

50. $2n^2 - 2n - 1 = 0$ $\left\{\frac{1 \pm \sqrt{3}}{2}\right\}$

51. $n(n + 8) = 240$ $\{-20, 12\}$

52. $t(t - 26) = -160$ $\{10, 16\}$

53. $3x^2 + 29x = -66$ $\left\{-6, -\frac{11}{3}\right\}$

54. $6x^2 - 13x = 28$ $\left\{-\frac{4}{3}, \frac{7}{2}\right\}$

55. $6n^2 + 23n + 21 = 0$ $\left\{-\frac{7}{3}, -\frac{3}{2}\right\}$

56. $6n^2 + n - 2 = 0$ $\left\{-\frac{2}{3}, \frac{1}{2}\right\}$

57. $x^2 + 12x = 4$ $\{-6 \pm 2\sqrt{10}\}$

58. $x^2 + 6x = -11$ $\{-3 \pm i\sqrt{2}\}$

59. $12n^2 - 7n + 1 = 0$ $\left\{\frac{1}{4}, \frac{1}{3}\right\}$

60. $5(x + 2)^2 + 1 = 16$ $\{-2 \pm \sqrt{3}\}$

61. Use the method of completing the square to solve $ax^2 + bx + c = 0$ for x, where a, b, and c are real numbers and $a \neq 0$. $\left\{\frac{-b \pm \sqrt{b^2 - 4ac}}{2a}\right\}$

Miscellaneous Problems

Solve each of the following for the indicated variable. Assume that all letters represent positive numbers.

62. $\frac{x^2}{a^2} - \frac{y^2}{b^2} = 1$ for y $y = \frac{b\sqrt{x^2 - a^2}}{a}$

63. $\frac{x^2}{a^2} + \frac{y^2}{b^2} = 1$ for x $x = \frac{a\sqrt{b^2 - y^2}}{b}$

64. $s = \frac{1}{2}gt^2$ for t $t = \frac{\sqrt{2gs}}{g}$

65. $A = \pi r^2$ for r $r = \frac{\sqrt{A\pi}}{\pi}$

Solve each of the following equations for x.

66. $x^2 + 8ax + 15a^2 = 0$ $\{-3a, -5a\}$

67. $x^2 - 5ax + 6a^2 = 0$ $\{2a, 3a\}$

68. $10x^2 - 31ax - 14a^2 = 0$ $\left\{-\frac{2a}{5}, \frac{7a}{2}\right\}$

69. $6x^2 + ax - 2a^2 = 0$ $\left\{\frac{a}{2}, -\frac{2a}{3}\right\}$

70. $4x^2 + 4bx + b^2 = 0$ $\left\{-\frac{b}{2}\right\}$

71. $9x^2 - 12bx + 4b^2 = 0$ $\left\{\frac{2b}{3}\right\}$

6.4 The Quadratic Formula

As we saw in the last section, the method of completing the square can be used to solve *any* quadratic equation. Furthermore, the equation $ax^2 + bx + c = 0$, where a, b, and c are real numbers with $a \neq 0$, can be used to represent *any* quadratic equation. These two ideas merge to produce the **quadratic formula**, a formula that can be used to solve *any* quadratic equation. The merger is accomplished by using the method of completing the square to solve the equation $ax^2 + bx + c = 0$ as follows.

$$ax^2 + bx + c = 0$$

$$ax^2 + bx = -c$$

$$x^2 + \frac{b}{a}x = -\frac{c}{a} \qquad \text{Multiply both sides by } \frac{1}{a}.$$

$$x^2 + \frac{b}{a}x + \frac{b^2}{4a^2} = -\frac{c}{a} + \frac{b^2}{4a^2} \qquad \textit{Complete the square} \text{ by adding } \frac{b^2}{4a^2} \text{ to both sides.}$$

$$\left(x + \frac{b}{2a}\right)^2 = \frac{b^2 - 4ac}{4a^2} \qquad \text{The right side is combined into a single fraction.}$$

$$x + \frac{b}{2a} = \pm\sqrt{\frac{b^2 - 4ac}{4a^2}}$$

$$x + \frac{b}{2a} = \pm\frac{\sqrt{b^2 - 4ac}}{\sqrt{4a^2}}$$

$$x + \frac{b}{2a} = \pm\frac{\sqrt{b^2 - 4ac}}{2a}$$

$$x + \frac{b}{2a} = \frac{\sqrt{b^2 - 4ac}}{2a} \quad \text{or} \quad x + \frac{b}{2a} = -\frac{\sqrt{b^2 - 4ac}}{2a}$$

$$x = -\frac{b}{2a} + \frac{\sqrt{b^2 - 4ac}}{2a} \quad \text{or} \quad x = -\frac{b}{2a} - \frac{\sqrt{b^2 - 4ac}}{2a}$$

$$x = \frac{-b + \sqrt{b^2 - 4ac}}{2a} \quad \text{or} \quad x = \frac{-b - \sqrt{b^2 - 4ac}}{2a}$$

The quadratic formula is usually stated as follows.

Quadratic Formula

$$x = \frac{-b \pm \sqrt{b^2 - 4ac}}{2a}, \qquad a \neq 0$$

We can use this formula to solve any quadratic equation by expressing the equation in the standard form, $ax^2 + bx + c = 0$, and substituting the values for a, b, and c into the formula. Let's consider some examples.

EXAMPLE 1 Solve $x^2 + 5x + 2 = 0$.

Solution The given equation is in standard form, so $a = 1, b = 5$, and $c = 2$. Substituting these values into the formula and simplifying, we obtain

$$x = \frac{-b \pm \sqrt{b^2 - 4ac}}{2a}$$

$$x = \frac{-5 \pm \sqrt{5^2 - 4(1)(2)}}{2(1)}$$

$$x = \frac{-5 \pm \sqrt{25 - 8}}{2}$$

$$x = \frac{-5 \pm \sqrt{17}}{2}.$$

The solution set is $\left\{\frac{-5 \pm \sqrt{17}}{2}\right\}$. ■

EXAMPLE 2 Solve $x^2 - 2x - 4 = 0$.

Solution We need to think of $x^2 - 2x - 4 = 0$ as $x^2 + (-2x) + (-4) = 0$ to determine the values $a = 1$, $b = -2$, $c = -4$. Substitute these values into the quadratic formula and simplify, to obtain

$$x = \frac{-b \pm \sqrt{b^2 - 4ac}}{2a}$$

$$x = \frac{-(-2) \pm \sqrt{(-2)^2 - 4(1)(-4)}}{2(1)}$$

$$x = \frac{2 \pm \sqrt{4 + 16}}{2}$$

$$x = \frac{2 \pm \sqrt{20}}{2}$$

$$x = \frac{2 \pm 2\sqrt{5}}{2}$$

$$= \frac{\cancel{2}(1 \pm \sqrt{5})}{\cancel{2}}.$$

The solution set is $\{1 \pm \sqrt{5}\}$. ■

EXAMPLE 3 Solve $x^2 - 2x + 19 = 0$.

Solution

$$x^2 - 2x + 19 = 0$$

$$x = \frac{-(-2) \pm \sqrt{(-2)^2 - 4(1)(19)}}{2(1)}$$

$$x = \frac{2 \pm \sqrt{4 - 76}}{2}$$

$$x = \frac{2 \pm \sqrt{-72}}{2}$$

$$x = \frac{2 \pm 6i\sqrt{2}}{2} \qquad \sqrt{-72} = i\sqrt{72} = i\sqrt{36}\,\sqrt{2} = 6i\sqrt{2}$$

$$x = 1 \pm 3i\sqrt{2}.$$

The solution set is $\{1 \pm 3i\sqrt{2}\}$. ■

EXAMPLE 4 Solve $2x^2 + 4x - 3 = 0$.

Solution Substitute $a = 2$, $b = 4$, and $c = -3$ into the quadratic formula and simplify to obtain

$$x = \frac{-b \pm \sqrt{b^2 - 4ac}}{2a}$$

$$x = \frac{-4 \pm \sqrt{4^2 - 4(2)(-3)}}{2(2)}$$

$$x = \frac{-4 \pm \sqrt{16 + 24}}{4}$$

$$x = \frac{-4 \pm \sqrt{40}}{4}$$

$$x = \frac{-4 \pm 2\sqrt{10}}{4}$$

$$= \frac{-2 \pm \sqrt{10}}{2}.$$

The solution set is $\left\{\frac{-2 \pm \sqrt{10}}{2}\right\}$. ■

EXAMPLE 5 Solve $n(3n - 10) = 25$.

Solution First, we need to change the equation to the standard form $an^2 + bn + c = 0$.

$$n(3n - 10) = 25$$
$$3n^2 - 10n = 25$$
$$3n^2 - 10n - 25 = 0.$$

Now substituting $a = 3$, $b = -10$, and $c = -25$ into the quadratic formula we obtain

$$n = \frac{-b \pm \sqrt{b^2 - 4ac}}{2a}$$

$$n = \frac{-(-10) \pm \sqrt{(-10)^2 - 4(3)(-25)}}{2(3)}$$

$$n = \frac{10 \pm \sqrt{100 + 300}}{2(3)}$$

$$n = \frac{10 \pm \sqrt{400}}{6}$$

$$n = \frac{10 \pm 20}{6}$$

$$n = \frac{10 + 20}{6} \quad \text{or} \quad n = \frac{10 - 20}{6}$$

$$n = 5 \quad \text{or} \quad n = -\frac{5}{3}.$$

The solution set is $\left\{-\frac{5}{3}, 5\right\}$. ■

In Example 5, notice that we used the variable n. The quadratic formula is usually stated in terms of x, but it certainly can be applied to quadratic equations in other variables. Also note in Example 5 that the polynomial $3n^2 - 10n - 25$ can be factored as $(3n + 5)(n - 5)$. Therefore, we could also solve the equation $3n^2 - 10n - 25 = 0$ using the factoring approach. We will give some guidance in the next section as to which approach to use on a particular equation.

Nature of Roots

The quadratic formula makes it easy to determine the nature of the roots of a quadratic equation without completely solving the equation. The number

$$b^2 - 4ac,$$

which appears under the radical sign in the quadratic formula, is called the **discriminant** of the quadratic equation. It is the indicator as to the kind of roots of the equation. For example, suppose that we start to solve the equation $x^2 - 4x + 7 = 0$ as follows.

$$x = \frac{-b \pm \sqrt{b^2 - 4ac}}{2a}$$

$$x = \frac{-(-4) \pm \sqrt{(-4)^2 - 4(1)(7)}}{2(1)}$$

$$x = \frac{4 \pm \sqrt{16 - 28}}{2}$$

$$x = \frac{4 \pm \sqrt{-12}}{2}$$

At this stage we should be able to look ahead and realize that we will obtain two complex solutions for the equation. (By the way, observe that these solutions are complex conjugates.) In other words, the discriminant, -12, is the indicator as to the type of roots that we will obtain.

We make the following general statements relative to the roots of a quadratic equation of the form $ax^2 + bx + c = 0$.

1. If $b^2 - 4ac < 0$, then the equation has two nonreal complex solutions.
2. If $b^2 - 4ac = 0$, then the equation has one real solution.
3. If $b^2 - 4ac > 0$, then the equation has two real solutions.

The following examples illustrate each of these situations. (You may want to solve the equations completely to verify the conclusions.)

Equation	*Discriminant*	*Nature of roots*
$x^2 - 3x + 7 = 0$	$b^2 - 4ac = (-3)^2 - 4(1)(7)$ $= 9 - 28$ $= -19$	two nonreal complex solutions
$9x^2 - 12x + 4 = 0$	$b^2 - 4ac = (-12)^2 - 4(9)(4)$ $= 144 - 144$ $= 0$	one real solution
$2x^2 + 5x - 3 = 0$	$b^2 - 4ac = (5)^2 - 4(2)(-3)$ $= 25 + 24$ $= 49$	two real solutions

There is another very useful relationship that involves the roots of a quadratic equation and the numbers a, b, and c of the general form $ax^2 + bx + c = 0$. Suppose that we let x_1 and x_2 be the two roots generated by the quadratic formula. Thus, we have

$$x_1 = \frac{-b + \sqrt{b^2 - 4ac}}{2a} \quad \text{and} \quad x_2 = \frac{-b - \sqrt{b^2 - 4ac}}{2a}.$$

REMARK A point of clarification should be made at this time. Previously, we made the statement that if $b^2 - 4ac = 0$, then the equation has one real solution. Technically, such an equation has two solutions but they are equal. For example, each factor of $(x - 2)(x - 2) = 0$ produces a solution but both solutions are the number 2. This is sometimes referred to as one real solution with a *multiplicity of two.* Using the idea of multiplicity of roots, we can say that every quadratic equation has two roots.

Now let's consider the sum and product of the two roots.

Sum $x_1 + x_2 = \dfrac{-b + \sqrt{b^2 - 4ac}}{2a} + \dfrac{-b - \sqrt{b^2 - 4ac}}{2a} = \dfrac{-2b}{2a} = \boxed{-\dfrac{b}{a}}.$

Product $(x_1)(x_2) = \left(\dfrac{-b + \sqrt{b^2 - 4ac}}{2a}\right)\left(\dfrac{-b - \sqrt{b^2 - 4ac}}{2a}\right) = \dfrac{b^2 - (b^2 - 4ac)}{4a^2}$

$$= \frac{b^2 - b^2 + 4ac}{4a^2}$$

$$= \frac{4ac}{4a^2} = \boxed{\frac{c}{a}}.$$

These relationships provide another way of checking potential solutions when solving quadratic equations. For example, back in Example 3 we solved the equation $x^2 - 2x + 19 = 0$ and obtained solutions of $1 + 3i\sqrt{2}$ and $1 - 3i\sqrt{2}$. Let's check these solutions by using the sum and product relationships.

Check for Example 3

Sum of roots $(1 + 3i\sqrt{2}) + (1 - 3i\sqrt{2}) = 2$ and $-\dfrac{b}{a} = -\dfrac{-2}{1} = 2$

Product of roots $(1 + 3i\sqrt{2})(1 - 3i\sqrt{2}) = 1 - 18i^2 = 1 + 18 = 19$
and $\dfrac{c}{a} = \dfrac{19}{1} = 19$

Likewise, a check for Example 4 is as follows.

Check for Example 4

Sum of roots $\left(\dfrac{-2 + \sqrt{10}}{2}\right) + \left(\dfrac{-2 - \sqrt{10}}{2}\right) = -\dfrac{4}{2} = -2$ and $-\dfrac{b}{a} = -\dfrac{4}{2} = -2$

Product of roots $\left(\dfrac{-2 + \sqrt{10}}{2}\right)\left(\dfrac{-2 - \sqrt{10}}{2}\right) = -\dfrac{6}{4} = -\dfrac{3}{2}$ and $\dfrac{c}{a} = \dfrac{-3}{2} = -\dfrac{3}{2}$

Notice that for both Examples 3 and 4, it was much easier to check by using the sum and product relationships than it would have been by substituting back into the original equation. Don't forget that the values for a, b, and c come from a quadratic equation of the form $ax^2 + bx + c = 0$. Therefore, in Example 5 we must be certain that no errors were made when changing the given equation $n(3n - 10) = 25$ to the form $3n^2 - 10n - 25 = 0$ if we are going to check the potential solutions by using the sum and product relationships.

Problem Set 6.4

For each quadratic equation in Problems 1–10, first use the discriminant to determine whether the equation has two nonreal complex solutions, one real solution with a multiplicity of two, or two real solutions, and then solve the equation.

1. $x^2 + 4x - 21 = 0$ Two real solutions; $\{-7, 3\}$

2. $x^2 - 3x - 54 = 0$ Two real solutions; $\{-6, 9\}$

3. $9x^2 - 6x + 1 = 0$ One real solution; $\{\frac{1}{3}\}$

4. $4x^2 + 20x + 25 = 0$ One real solution; $\{-\frac{5}{2}\}$

5. $x^2 - 7x + 13 = 0$ Two complex solutions; $\{\frac{7 \pm i\sqrt{3}}{2}\}$

6. $2x^2 - x + 5 = 0$ Two complex solutions; $\{\frac{1 \pm i\sqrt{39}}{4}\}$

7. $15x^2 + 17x - 4 = 0$ Two real solutions; $\{-\frac{4}{3}, \frac{1}{5}\}$

8. $8x^2 + 18x - 5 = 0$ Two real solutions; $\{-\frac{5}{2}, \frac{1}{4}\}$

9. $3x^2 + 4x = 2$ Two real solutions; $\{\frac{-2 \pm \sqrt{10}}{3}\}$

10. $2x^2 - 6x = -1$ Two real solutions; $\{\frac{3 \pm \sqrt{7}}{2}\}$

For Problems 11–50, use the quadratic formula to solve each of the quadratic equations. Check your solutions by using the *sum and product relationships.*

11. $x^2 + 2x - 1 = 0$ $\{-1 \pm \sqrt{2}\}$

12. $x^2 + 4x - 1 = 0$ $\{-2 \pm \sqrt{5}\}$

13. $n^2 + 5n - 3 = 0$ $\{\frac{-5 \pm \sqrt{37}}{2}\}$

14. $n^2 + 3n - 2 = 0$ $\{\frac{-3 \pm \sqrt{17}}{2}\}$

15. $a^2 - 8a = 4$ $\{4 \pm 2\sqrt{5}\}$

16. $a^2 - 6a = 2$ $\{3 \pm \sqrt{11}\}$

17. $n^2 + 5n + 8 = 0$ $\{\frac{-5 \pm i\sqrt{7}}{2}\}$

18. $2n^2 - 3n + 5 = 0$ $\{\frac{3 \pm i\sqrt{31}}{4}\}$

19. $x^2 - 18x + 80 = 0$ $\{8, 10\}$

20. $x^2 + 19x + 70 = 0$ $\{-14, -5\}$

21. $-y^2 = -9y + 5$ $\{\frac{9 \pm \sqrt{61}}{2}\}$

22. $-y^2 + 7y = 4$ $\{\frac{7 \pm \sqrt{33}}{2}\}$

23. $2x^2 + x - 4 = 0$ $\{\frac{-1 \pm \sqrt{33}}{4}\}$

24. $2x^2 + 5x - 2 = 0$ $\{\frac{-5 \pm \sqrt{41}}{4}\}$

25. $4x^2 + 2x + 1 = 0$ $\{\frac{-1 \pm i\sqrt{3}}{4}\}$

26. $3x^2 - 2x + 5 = 0$ $\{\frac{1 \pm i\sqrt{14}}{3}\}$

27. $3a^2 - 8a + 2 = 0$ $\{\frac{4 \pm \sqrt{10}}{3}\}$

28. $2a^2 - 6a + 1 = 0$ $\{\frac{3 \pm \sqrt{7}}{2}\}$

29. $-2n^2 + 3n + 5 = 0$ $\{-1, \frac{5}{2}\}$

30. $-3n^2 - 11n + 4 = 0$ $\{-4, \frac{1}{3}\}$

31. $3x^2 + 19x + 20 = 0$ $\{-5, -\frac{4}{3}\}$

32. $2x^2 - 17x + 30 = 0$ $\{\frac{5}{2}, 6\}$

33. $36n^2 - 60n + 25 = 0$ $\{\frac{5}{6}\}$

34. $9n^2 + 42n + 49 = 0$ $\{-\frac{7}{3}\}$

35. $4x^2 - 2x = 3$ $\{\frac{1 \pm \sqrt{13}}{4}\}$

36. $6x^2 - 4x = 3$ $\{\frac{2 \pm \sqrt{22}}{6}\}$

37. $5x^2 - 13x = 0$ $\{0, \frac{13}{5}\}$

38. $7x^2 + 12x = 0$ $\{-\frac{12}{7}, 0\}$

39. $3x^2 = 5$ $\{\pm\frac{\sqrt{15}}{3}\}$

40. $4x^2 - 3$ $\{\pm\frac{\sqrt{3}}{2}\}$

41. $6t^2 + t - 3 = 0$ $\left\{\frac{-1 \pm \sqrt{73}}{12}\right\}$

42. $2t^2 + 6t - 3 = 0$ $\left\{\frac{-3 \pm \sqrt{15}}{2}\right\}$

43. $n^2 + 32n + 252 = 0$ $\{-18, -14\}$

44. $n^2 - 4n - 192 = 0$ $\{-12, 16\}$

45. $12x^2 - 73x + 110 = 0$ $\left\{\frac{11}{4}, \frac{10}{3}\right\}$

46. $6x^2 + 11x - 255 = 0$ $\left\{-\frac{15}{2}, \frac{17}{3}\right\}$

47. $-2x^2 + 4x - 3 = 0$ $\left\{\frac{2 \pm i\sqrt{2}}{2}\right\}$

48. $-2x^2 + 6x - 5 = 0$ $\left\{\frac{3 \pm i}{2}\right\}$

49. $-6x^2 + 2x + 1 = 0$ $\left\{\frac{1 \pm \sqrt{7}}{6}\right\}$

50. $-2x^2 + 4x + 1 = 0$ $\left\{\frac{2 \pm \sqrt{6}}{2}\right\}$

Miscellaneous Problems

The solution set for $x^2 - 4x - 37 = 0$ is $\{2 \pm \sqrt{41}\}$. With a calculator, we found a rational approximation, to the nearest one-thousandth, for each of these solutions.

$$2 - \sqrt{41} = -4.403 \quad \text{and} \quad 2 + \sqrt{41} = 8.403.$$

Thus, the solution set is $\{-4.403, 8.403\}$, with answers rounded to the nearest one-thousandth.

Solve each of the following equations and express solutions to the nearest one-thousandth.

51. $x^2 - 6x - 10 = 0$ $\{-1.359, 7.359\}$

52. $x^2 - 16x - 24 = 0$ $\{-1.381, 17.381\}$

53. $x^2 + 6x - 44 = 0$ $\{-10.280, 4.280\}$

54. $x^2 + 10x - 46 = 0$ $\{-13.426, 3.426\}$

55. $x^2 + 8x + 2 = 0$ $\{-.259, 7.742\}$

56. $x^2 + 9x + 3 = 0$ $\{-.347, -8.653\}$

57. $4x^2 - 6x + 1 = 0$ $\{.191, 1.309\}$

58. $5x^2 - 9x + 1 = 0$ $\{.119, 1.681\}$

59. $2x^2 - 11x - 5 = 0$ $\{-.422, 5.922\}$

60. $3x^2 - 12x - 10 = 0$ $\{-.708, 4.708\}$

For Problems 61–63, use the discriminant to help solve each problem.

61. Determine k so that the solutions of $x^2 - 2x + k = 0$ are complex but nonreal. $k > 1$

62. Determine k so that $4x^2 - kx + 1 = 0$ has two equal real solutions. $k = 4$ or $k = -4$

63. Determine k so that $3x^2 - kx - 2 = 0$ has real solutions. any real number value for k

6.5

More Quadratic Equations and Applications

Which method should be used to solve a particular quadratic equation? There is no definite answer to that question; it depends upon the *type* of equation and your personal preference. In the following examples we will state reasons for choosing a specific technique. However, keep in mind that usually this is a decision *you* must make as the need arises. So become familiar with the strengths and weaknesses of each method.

EXAMPLE 1 Solve $2x^2 - 3x - 1 = 0$.

Solution Because of the leading coefficient of 2 and the constant term of -1, there are very few factoring possibilities to consider. Therefore, with such problems, first try the factoring approach. Unfortunately, this particular polynomial is not factorable using integers. Thus, let's use the quadratic formula to solve the equation.

$$x = \frac{-b \pm \sqrt{b^2 - 4ac}}{2a}$$

$$x = \frac{-(-3) \pm \sqrt{(-3)^2 - 4(2)(-1)}}{2(2)}$$

$$x = \frac{3 \pm \sqrt{9 + 8}}{4}$$

$$x = \frac{3 \pm \sqrt{17}}{4}$$

CHECK We can use the *sum and product-of-roots* relationships for our checking purposes.

Sum of roots $\frac{3 + \sqrt{17}}{4} + \frac{3 - \sqrt{17}}{4} = \frac{6}{4} = \frac{3}{2}$ and $-\frac{b}{a} = -\frac{-3}{2} = \frac{3}{2}$.

Product of roots

$$\left(\frac{3 + \sqrt{17}}{4}\right)\left(\frac{3 - \sqrt{17}}{4}\right) = \frac{9 - 17}{16} = -\frac{8}{16} = -\frac{1}{2} \quad \text{and} \quad \frac{c}{a} = \frac{-1}{2} = -\frac{1}{2}.$$

The solution set is $\left\{\frac{3 \pm \sqrt{17}}{4}\right\}$. ■

EXAMPLE 2 Solve $\frac{3}{n} + \frac{10}{n + 6} = 1$.

Solution

$$\frac{3}{n} + \frac{10}{n + 6} = 1, \qquad n \neq 0 \text{ and } n \neq -6.$$

$$n(n + 6)\left(\frac{3}{n} + \frac{10}{n + 6}\right) = 1(n)(n + 6)$$

Multiply both sides by $n(n + 6)$, which is the LCD.

$$3(n + 6) + 10n = n(n + 6)$$

$$3n + 18 + 10n = n^2 + 6n$$

$$13n + 18 = n^2 + 6n$$

$$0 = n^2 - 7n - 18$$

This is an easy one to consider the possibilities for factoring, and it factors as follows.

$$0 = (n - 9)(n + 2)$$

$$n - 9 = 0 \quad \text{or} \quad n + 2 = 0$$

$$n = 9 \quad \text{or} \quad n = -2$$

CHECK Substituting 9 and -2 back into the original equation, we obtain

$$\frac{3}{n} + \frac{10}{n+6} = 1 \qquad\qquad \frac{3}{n} + \frac{10}{n+6} = 1$$

$$\frac{3}{9} + \frac{10}{9+6} \stackrel{?}{=} 1 \qquad\qquad \frac{3}{-2} + \frac{10}{-2+6} \stackrel{?}{=} 1$$

$$\frac{1}{3} + \frac{10}{15} \stackrel{?}{=} 1 \qquad \text{or} \qquad -\frac{3}{2} + \frac{10}{4} \stackrel{?}{=} 1$$

$$\frac{1}{3} + \frac{2}{3} \stackrel{?}{=} 1 \qquad\qquad -\frac{3}{2} + \frac{5}{2} \stackrel{?}{=} 1$$

$$1 = 1 \qquad\qquad \frac{2}{2} = 1$$

The solution set is $\{-2, 9\}$. ■

We should make two comments about Example 2. First, notice the indication of the initial restrictions $n \neq 0$ and $n \neq -6$. Remember that we need to do this when solving fractional equations. Secondly, the *sum and product-of-roots* relationships were not used for checking purposes in this problem. Those relationships would only check the validity of our work from the step $0 = n^2 - 7n - 18$ to the finish. In other words, an error made in changing the original equation to quadratic form would not be detected by checking the sum and product of potential roots. Thus, with such a problem the only *absolute check* is to substitute the potential solutions back into the *original equation.*

EXAMPLE 3 Solve $x^2 + 22x + 112 = 0$.

Solution The size of the constant term makes the factoring approach a little cumbersome for this problem. Furthermore, since the leading coefficient is 1 and the coefficient of the x-term is even, the method of completing the square will work rather effectively as follows.

$$x^2 + 22x + 112 = 0$$

$$x^2 + 22x = -112$$

$$x^2 + 22x + 121 = -112 + 121$$

$$(x + 11)^2 = 9$$

$$x + 11 = \pm\sqrt{9}$$

$$x + 11 = \pm 3$$

$$x + 11 = 3 \qquad \text{or} \qquad x + 11 = -3$$

$$x = -8 \qquad \text{or} \qquad x = -14$$

CHECK

Sum of roots $-8 + (-14) = -22$ and $-\frac{b}{a} = -22.$

Product of roots $(-8)(-14) = 112$ and $\frac{c}{a} = 112.$

The solution set is $\{-14, -8\}$. ■

EXAMPLE 4 Solve $x^4 - 4x^2 - 96 = 0$.

Solution An equation such as $x^4 - 4x^2 - 96 = 0$ is not a quadratic equation, but we can solve it using the techniques that we use on quadratic equations. That is to say, we can factor the polynomial and apply the property, $ab = 0$ if and only if $a = 0$ or $b = 0$, as follows.

$$x^4 - 4x^2 - 96 = 0$$

$$(x^2 - 12)(x^2 + 8) = 0$$

$$x^2 - 12 = 0 \quad \text{or} \quad x^2 + 8 = 0$$

$$x^2 = 12 \quad \text{or} \quad x^2 = -8$$

$$x = \pm\sqrt{12} \quad \text{or} \quad x = \pm\sqrt{-8}$$

$$x = \pm 2\sqrt{3} \quad \text{or} \quad x = \pm 2i\sqrt{2}$$

The solution set is $\{\pm 2\sqrt{3}, \pm 2i\sqrt{2}\}$. (We will leave the check for this problem for you to do!) ■

REMARK Another approach to Example 4 would be to substitute y for x^2 and y^2 for x^4. The equation $x^4 - 4x^2 - 96 = 0$ becomes the quadratic equation $y^2 - 4y - 96 = 0$. Thus, we say that $x^4 - 4x^2 - 96 = 0$ is of *quadratic form.* Then we could solve the quadratic equation $y^2 - 4y - 96 = 0$ and use the equation $y = x^2$ to determine the solutions for x.

Applications

Before we conclude this section with some word problems that can be solved using quadratic equations, let's restate the suggestions for solving word problems we made in an earlier chapter.

Suggestions for Solving Word Problems

1. Read the problem carefully and make certain that you understand the meanings of all the words. Be especially alert for any technical terms used in the statement of the problem.
2. Read the problem a second time (perhaps even a third time) to get an

overview of the situation being described and to determine the known facts, as well as what is to be found.

3. Sketch any figure, diagram, or chart that might be helpful in analyzing the problem.
4. Choose a meaningful variable to represent an unknown quantity in the problem (perhaps l, if the length of a rectangle is an unknown quantity) and represent any other unknowns in terms of that variable.
5. Look for a *guideline* that you can use to set up an equation. A guideline might be a formula such as "$A = lw$" or a relationship such as "the fractional part of a job done by Bill plus the fractional part of the job done by Mary equals the total job."
6. Form an equation that contains the variable that translates the conditions of the guideline from English to algebra.
7. Solve the equation and use the solutions to determine all facts requested in the problem.
8. Check all answers back into the **original statement of the problem**.

Keep these suggestions in mind, as we now consider some word problems.

PROBLEM 1 A page for a magazine contains 70 square inches of type. The height of a page is twice the width. If the margin around the type is to be 2 inches uniformly, what are the dimensions of a page?

Solution Let x represent the width of a page. Then $2x$ represents the height of a page. Now let's draw and label a model of a page.

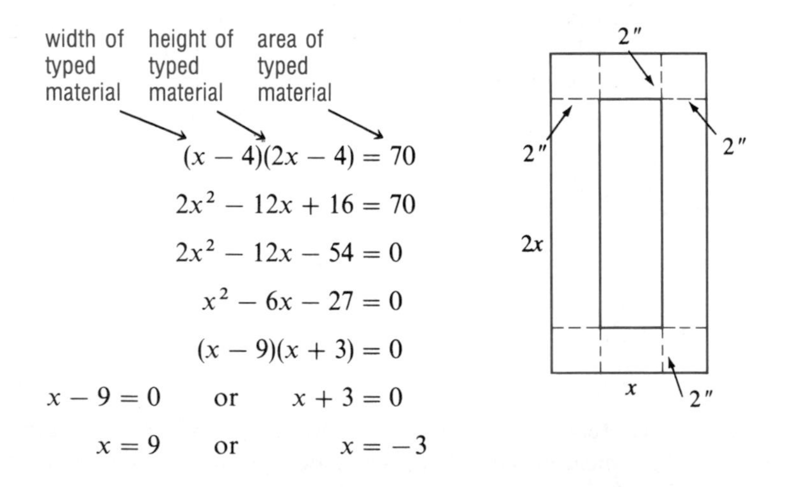

$$(x - 4)(2x - 4) = 70$$
$$2x^2 - 12x + 16 = 70$$
$$2x^2 - 12x - 54 = 0$$
$$x^2 - 6x - 27 = 0$$
$$(x - 9)(x + 3) = 0$$
$$x - 9 = 0 \quad \text{or} \quad x + 3 = 0$$
$$x = 9 \quad \text{or} \quad x = -3$$

Disregard the negative solution; the page must be 9 inches wide and its height is $2(9) = 18$ inches. ■

Let's use our knowledge of quadratic equations to analyze some applications of the business world. For example, if P dollars is invested at r rate of interest compounded annually for t years, then the amount of money, A, accumulated at the end of t years is given by the formula

$$A = P(1 + r)^t.$$

This compound interest formula serves as a guideline for the next problem.

PROBLEM 2 Suppose that $100 is invested at a certain rate of interest compounded annually for 2 years. If the accumulated value at the end of 2 years is $121, find the rate of interest.

Solution Let r represent the rate of interest. Substitute the known values into the compound interest formula to yield

$$A = P(1 + r)^t$$

$$121 = 100(1 + r)^2.$$

Solving this equation, we obtain

$$\frac{121}{100} = (1 + r)^2$$

$$\pm\sqrt{\frac{121}{100}} = (1 + r)$$

$$\pm\frac{11}{10} = 1 + r$$

$$1 + r = \frac{11}{10} \quad \text{or} \quad 1 + r = -\frac{11}{10}$$

$$r = -1 + \frac{11}{10} \quad \text{or} \quad r = -1 - \frac{11}{10}$$

$$r = \frac{1}{10} \quad \text{or} \quad r = -\frac{21}{10}.$$

We must disregard the negative solution, so $r = \frac{1}{10}$ is the only solution. Change $\frac{1}{10}$ to a percent and the rate of interest is 10%. ■

PROBLEM 3 A businesswoman bought a parcel of land on speculation for $120,000. She subdivided the land into lots and when she had sold all but 18 lots at a profit of $6000 per lot, she regained the entire cost of the land. How many lots were sold and at what price per lot?

Solution Let x represent the number of lots sold. Then $x + 18$ represents the total number of lots. Therefore, $\frac{120{,}000}{x}$ represents the selling price per lot and $\frac{120{,}000}{x+18}$ represents the cost per lot. The following equation represents the situation.

selling price per lot equals cost per lot plus \$6000

$$\frac{120{,}000}{x} = \frac{120{,}000}{x+18} + 6000$$

Solving this equation, we obtain

$$x(x+18)\left(\frac{120{,}000}{x}\right) = \left(\frac{120{,}000}{x+18} + 6000\right)(x)(x+18)$$

$$120{,}000(x+18) = 120{,}000x + 6000x(x+18)$$

$$120{,}000x + 2{,}160{,}000 = 120{,}000x + 6000x^2 + 108{,}000x$$

$$0 = 6000x^2 + 108{,}000x - 2{,}160{,}000$$

$$0 = x^2 + 18x - 360.$$

The method of completing the square works very well with this equation.

$$x^2 + 18x = 360$$

$$x^2 + 18x + 81 = 441$$

$$(x+9)^2 = 441$$

$$x + 9 = \pm\sqrt{441}$$

$$x + 9 = \pm 21$$

$$x + 9 = 21 \quad \text{or} \quad x + 9 = -21$$

$$x = 12 \quad \text{or} \quad x = -30$$

We discard the negative solution; thus, 12 lots were sold at $\frac{120{,}000}{x} = \frac{120{,}000}{12} =$ \$10,000 per lot. ■

PROBLEM 4 Barry bought a number of shares of stock for \$600. A week later the value of the stock increased \$3 per share and he sold all but 10 shares and regained his original investment of \$600. How many shares did he sell and at what price per share?

Solution Let s represent the number of shares Barry sold. Then $s + 10$ represents the number of shares purchased. Therefore $\frac{600}{s}$ represents the selling price per share and $\frac{600}{s+10}$ represents the cost per share.

$$\underset{\substack{\text{selling price} \\ \text{per share}}}{\frac{600}{s}} = \underset{\substack{\text{cost per} \\ \text{share}}}{\frac{600}{s + 10}} + 3$$

Solving this equation yields

$$s(s + 10)\left(\frac{600}{s}\right) = \left(\frac{600}{s + 10} + 3\right)(s)(s + 10)$$

$$600(s + 10) = 600s + 3s(s + 10)$$

$$600s + 6000 = 600s + 3s^2 + 30s$$

$$0 = 3s^2 + 30s - 6000$$

$$0 = s^2 + 10s - 2000.$$

Use the quadratic formula to obtain

$$s = \frac{-10 \pm \sqrt{10^2 - 4(1)(-2000)}}{2(1)}$$

$$s = \frac{-10 \pm \sqrt{100 + 8000}}{2}$$

$$s = \frac{-10 \pm \sqrt{8100}}{2}$$

$$s = \frac{-10 \pm 90}{2}$$

$$s = \frac{-10 + 90}{2} \quad \text{or} \quad s = \frac{-10 - 90}{2}$$

$$s = 40 \quad \text{or} \quad s = -50.$$

We discard the negative solution and we know that 40 shares were sold at $\frac{600}{s} = \frac{600}{40} = \$15$ per share. ■

This next problem set contains a large variety of word problems. Not only are there some business applications similar to those we discussed in this section, but there are also more problems of the types we discussed back in Chapters 3 and 4. Try to give them your best shot without referring back to examples in earlier chapters.

Problem Set 6.5

Solve each of the following quadratic equations and use the method that seems most appropriate to you.

1. $x^2 - 4x - 6 = 0$ $\{2 \pm \sqrt{10}\}$
2. $x^2 - 8x - 4 = 0$ $\{4 \pm 2\sqrt{5}\}$
3. $3x^2 + 23x - 36 = 0$ $\{-9, \frac{4}{3}\}$
4. $n^2 + 22n + 105 = 0$ $\{-15, -7\}$
5. $x^2 - 18x = 9$ $\{9 \pm 3\sqrt{10}\}$
6. $x^2 + 20x = 25$ $\{-10 \pm 5\sqrt{5}\}$
7. $2x^2 - 3x + 4 = 0$ $\{\frac{3 \pm i\sqrt{23}}{4}\}$
8. $3y^2 - 2y + 1 = 0$ $\{\frac{1 \pm i\sqrt{2}}{3}\}$
9. $135 + 24n + n^2 = 0$ $\{-15, -9\}$
10. $28 - x - 2x^2 = 0$ $\{-4, \frac{7}{2}\}$
11. $(x - 2)(x + 9) = -10$ $\{-8, 1\}$
12. $(x + 3)(2x + 1) = -3$ $\{-2, -\frac{3}{2}\}$
13. $2x^2 - 4x + 7 = 0$ $\{\frac{2 \pm i\sqrt{10}}{2}\}$
14. $3x^2 - 2x + 8 = 0$ $\{\frac{1 \pm i\sqrt{23}}{3}\}$
15. $x^2 - 18x + 15 = 0$ $\{9 \pm \sqrt{66}\}$
16. $x^2 - 16x + 14 = 0$ $\{8 \pm 5\sqrt{2}\}$
17. $20y^2 + 17y - 10 = 0$ $\{-\frac{5}{4}, \frac{2}{5}\}$
18. $12x^2 + 23x - 9 = 0$ $\{-\frac{9}{4}, \frac{1}{3}\}$
19. $4t^2 + 4t - 1 = 0$ $\{\frac{-1 \pm \sqrt{2}}{2}\}$
20. $5t^2 + 5t - 1 = 0$ $\{\frac{-5 \pm 3\sqrt{5}}{10}\}$

Solve each of the following equations.

21. $n + \frac{3}{n} = \frac{19}{4}$ $\left\{\frac{3}{4}, 4\right\}$
22. $n - \frac{2}{n} = -\frac{7}{3}$ $\left\{-3, \frac{2}{3}\right\}$
23. $\frac{3}{x} + \frac{7}{x - 1} = 1$ $\left\{\frac{11 \pm \sqrt{109}}{2}\right\}$
24. $\frac{2}{x} + \frac{5}{x + 2} = 1$ $\left\{\frac{5 \pm \sqrt{41}}{2}\right\}$
25. $\frac{12}{x - 3} + \frac{8}{x} = 14$ $\left\{\frac{3}{7}, 4\right\}$
26. $\frac{16}{x + 5} - \frac{12}{x} = -2$ $\{-10, 3\}$
27. $\frac{3}{x - 1} - \frac{2}{x} = \frac{5}{2}$ $\left\{\frac{7 \pm \sqrt{129}}{10}\right\}$
28. $\frac{4}{x + 1} + \frac{2}{x} = \frac{5}{3}$ $\left\{-\frac{2}{5}, 3\right\}$
29. $\frac{6}{x} + \frac{40}{x + 5} = 7$ $\left\{-\frac{10}{7}, 3\right\}$
30. $\frac{12}{t} + \frac{18}{t + 8} = \frac{9}{2}$ $\left\{-\frac{16}{3}, 4\right\}$
31. $\frac{5}{n - 3} - \frac{3}{n + 3} = 1$ $\{1 \pm \sqrt{34}\}$
32. $\frac{3}{t + 2} + \frac{4}{t - 2} = 2$ $\left\{\frac{7 \pm \sqrt{129}}{4}\right\}$
33. $x^4 - 18x^2 + 72 = 0$ $\{\pm\sqrt{6}, \pm 2\sqrt{3}\}$
34. $x^4 - 21x^2 + 54 = 0$ $\{\pm 3, \pm 3\sqrt{2}\}$
35. $3x^4 - 35x^2 + 72 = 0$ $\{\pm 3, \pm \frac{2\sqrt{6}}{3}\}$
36. $5x^4 - 32x^2 + 48 = 0$ $\{\pm 2, \pm \frac{2\sqrt{15}}{5}\}$
37. $3x^4 + 17x^2 + 20 = 0$ $\{\pm \frac{i\sqrt{15}}{3}, \pm 2i\}$
38. $4x^4 + 11x^2 - 45 = 0$ $\{\pm \frac{3}{2}, \pm i\sqrt{5}\}$
39. $6x^4 - 29x^2 + 28 = 0$ $\{\pm \frac{\sqrt{14}}{2}, \pm \frac{2\sqrt{3}}{3}\}$
40. $6x^4 - 31x^2 + 18 = 0$ $\{\pm \frac{\sqrt{6}}{3}, \pm \frac{3\sqrt{2}}{2}\}$

Set up an equation and solve each of the following problems.

41. Find two consecutive whole numbers such that the sum of their squares is 145. 8 and 9
42. Find two consecutive odd whole numbers such that the sum of their squares is 74. 5 and 7
43. Two positive integers differ by 3, and their product is 108. Find the numbers. 9 and 12
44. Suppose that the sum of two numbers is 20 and the sum of their squares is 232. Find the numbers. 6 and 14
45. Find two numbers such that their sum is 10 and their product is 22. $5 + \sqrt{3}$ and $5 - \sqrt{3}$

46. Find two numbers such that their sum is 6 and their product is 7. $3 + \sqrt{2}$ and $3 - \sqrt{2}$

47. Suppose that the sum of two whole numbers is 9 and the sum of their reciprocals is $\frac{1}{2}$. Find the numbers. 3 and 6

48. The difference between two whole numbers is 8 and the difference between their reciprocals is $\frac{1}{6}$. Find the two numbers. 4 and 12

49. The sum of the lengths of the two legs of a right triangle is 21 inches. If the length of the hypotenuse is 15 inches, find the length of each leg. 9 inches and 12 inches

50. The length of a rectangular floor is 1 meter less than twice its width. If a diagonal of the rectangle is 17 meters, find the length and width of the floor. 15 meters and 8 meters

51. A rectangular plot of ground measuring 12 meters by 20 meters is surrounded by a sidewalk of uniform width. The area of the sidewalk is 68 square meters. Find the width of the walk. 1 meter

52. A 5-inch-by-7-inch picture is surrounded by a frame of uniform width. The area of the picture and frame together is 80 square inches. Find the width of the frame. $1\frac{1}{2}$ inches

53. The perimeter of a rectangle is 44 inches and its area is 112 square inches. Find the length and width of the rectangle. 8 inches by 14 inches

54. A rectangular piece of cardboard is 2 units longer than it is wide. From each of its corners a square piece 2 units on a side is cut out. The flaps are then turned up to form an open box that has a volume of 70 cubic units. Find the length and width of the original piece of cardboard. 9 units by 11 units

55. Charlotte traveled 250 miles in one hour more time than it took Lorraine to travel 180 miles. Charlotte drove 5 miles per hour faster than Lorraine. How fast did each one travel? 20 mph for L and 25 mph for C or 45 mph for L and 50 mph for C

56. Larry drove 156 miles in one hour more than it took Mike to drive 108 miles. Mike drove at an average rate of 2 miles per hour faster than Larry. How fast did each one travel? 54 mph for Mike and 52 mph for Larry

57. On a 570-mile trip, Andy averaged 5 miles per hour faster for the last 240 miles than he did for the first 330 miles. The entire trip took 10 hours. How fast did he travel for the first 330 miles? 55 mph

58. On a 135-mile bicycle excursion, Maria averaged 5 miles per hour faster for the first 60 miles than she did for the last 75 miles. The entire trip took 8 hours. Find her rate for the first 60 miles. 20 mph

59. It takes Terry 2 hours longer to do a certain job than it takes Tom. They worked together for 3 hours; then Tom left and Terry finished the job in 1 hour. How long would it take each of them to do the job alone? 6 hrs for Tom and 8 hrs for Terry

60. Suppose that Arlene can mow the entire lawn in 40 minutes less time with the power mower than she can with the push mower. One day the power mower broke down after she had been mowing for 30 minutes. She finished the lawn with the push mower in 20 minutes. How long does it take Arlene to mow the entire lawn with the power mower? 40 min and 80 min

61. A man did a job for \$360. It took him 6 hours longer than he expected and therefore he earned \$2 per hour less than he anticipated. How long did he expect that it would take to do the job? 30 hrs

62. A group of students agreed to each chip in the same amount to pay for a party that would cost \$100. Then they found 5 more students interested in the party and in sharing the expenses. This decreased the amount each had to pay by \$1. How many students were involved in the party and how much did they each have to pay? 25 people at \$4 each

63. A group of customers agreed to each contribute the same amount to buy their favorite waitress a \$100 birthday gift. At the last minute, 2 of the people decided not to chip in. This increased the amount that the remaining people had to pay by \$2.50 per person. How many people actually contributed to the gift? 8 people

64. A retailer bought a number of special mugs for \$48. Two of the mugs were broken in the store, but by selling each of the other mugs \$3 above the original cost per mug she made a total profit of \$22. How many mugs did she buy and at what price per mug did she sell them? She bought 12 mugs and sold 10 of them at \$7 per mug

65. My friend Tony bought a number of shares of stock for \$720. A month later the value of the stock increased by \$8 per share and he sold all but 20 shares and regained his original investment plus a profit of \$80. How many shares did he sell and at what price per share? 40 shares at \$20 per share

66. The formula $D = \frac{n(n-3)}{2}$ yields the number of diagonals, D, in a polygon of n sides. Find the number of sides of a polygon that has 54 diagonals. 12 sides

67. The formula $S = \frac{n(n+1)}{2}$ yields the sum, S, of the first n natural numbers 1, 2, 3, 4, ... How many consecutive natural numbers starting with 1 will give a sum of 1275? 50

68. At a point 16 yards from the base of a tower, the distance to the top of the tower is 4 yards more than the height of the tower. Find the height of the tower. 30 yards

69. Suppose that \$500 is invested at a certain rate of interest compounded annually for 2 years. If the accumulated value at the end of 2 years is \$594.05, find the rate of interest. 9%

70. Suppose that \$10,000 is invested at a certain rate of interest compounded annually for 2 years. If the accumulated value at the end of 2 years is \$12,544, find the rate of interest. 12%

Miscellaneous Problems

Solve each of the following equations.

71. $x - 9\sqrt{x} + 18 = 0$ [*Hint*: Let $y = \sqrt{x}$.] $\{9, 36\}$

72. $x - 4\sqrt{x} + 3 = 0$ $\{1, 9\}$

73. $x + \sqrt{x} - 2 = 0$ $\{1\}$

74. $x^{\frac{2}{3}} + x^{\frac{1}{3}} - 6 = 0$ [*Hint*: Let $y = x^{\frac{1}{3}}$.] $\{-27, 8\}$

75. $6x^{\frac{2}{3}} - 5x^{\frac{1}{3}} - 6 = 0$ $\{-\frac{8}{27}, \frac{27}{8}\}$

76. $x^{-2} + 4x^{-1} - 12 = 0$ $\{-\frac{1}{6}, \frac{1}{2}\}$

77. $12x^{-2} - 17x^{-1} - 5 = 0$ $\{-4, \frac{3}{5}\}$

6.6

Quadratic Inequalities

We refer to the equation $ax^2 + bx + c = 0$ as the standard form of a quadratic equation in one variable. Similarly, the following forms express **quadratic inequalities** in one variable.

$$ax^2 + bx + c > 0,$$
$$ax^2 + bx + c \geq 0,$$
$$ax^2 + bx + c < 0,$$
$$ax^2 + bx + c \leq 0$$

We can use the number line very effectively to help solve quadratic inequalities where the quadratic polynomial is factorable. Let's consider some examples to illustrate the procedure.

EXAMPLE 1 Solve and graph the solutions for $x^2 + 2x - 8 > 0$.

Solution First, let's factor the polynomial.

$$x^2 + 2x - 8 > 0$$
$$(x + 4)(x - 2) > 0$$

On a number line, we shall now indicate that at $x = 2$ and $x = -4$ the product $(x + 4)(x - 2)$ equals zero.

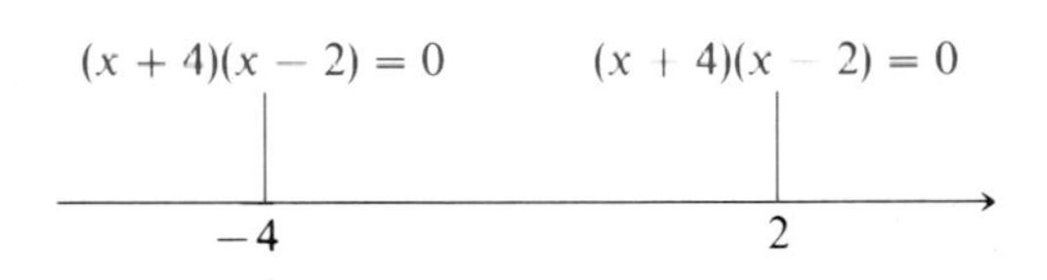

The numbers -4 and 2 divide the number line into three intervals: (1) the numbers less than -4, (2) the numbers between -4 and 2, and (3) the numbers greater than 2. We can choose a **test number** from each of these intervals and see how it affects the signs of the factors $x + 4$ and $x - 2$, and consequently, the sign of the product of these factors. For example, if $x < -4$ (try $x = -5$) then $x + 4$ is negative and $x - 2$ is negative; so their product is positive. If $-4 < x < 2$ (try $x = 0$), then $x + 4$ is positive and $x - 2$ is negative; so their product is negative. If $x > 2$ (try $x = 3$), then $x + 4$ is positive and $x - 2$ is positive; so their product is positive. This information can be conveniently arranged using a number line as follows. Note the open circles at -4 and 2 to indicate that they are not included in the solution set.

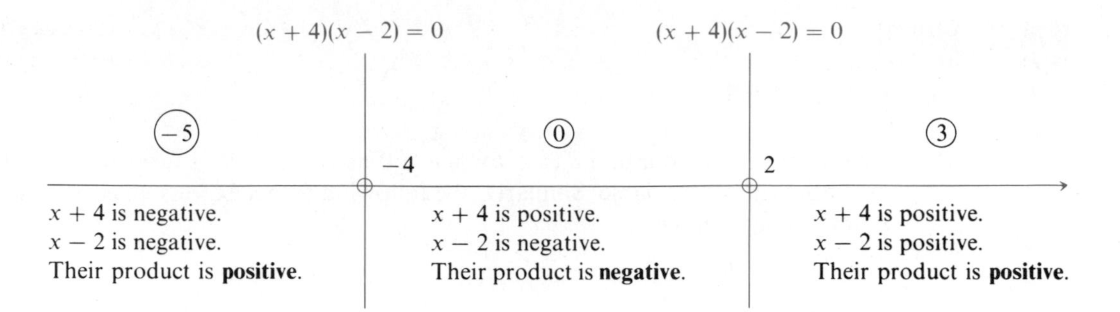

Therefore, the given inequality, $x^2 + 2x - 8 > 0$ is satisfied by numbers less than -4 along with numbers greater than 2. Using interval notation the solution set is $(-\infty, -4) \cup (2, \infty)$. These solutions can be shown on a number line as follows.

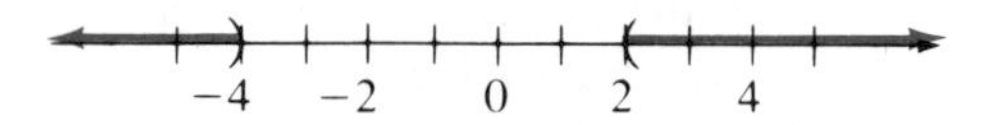

■

We refer to numbers such as -4 and 2 in the preceding example, where the given polynomial or algebraic expression equals zero or is undefined, as **critical numbers**. Let's consider some additional examples that make use of critical numbers and test numbers.

EXAMPLE 2 Solve and graph the solutions for $x^2 + 2x - 3 \le 0$.

Solution First, factor the polynomial.

$$x^2 + 2x - 3 \le 0$$

$$(x + 3)(x - 1) \le 0$$

Secondly, locate the values for which $(x + 3)(x - 1)$ equals zero. We put dots at -3 and 1 to remind ourselves that these two numbers are to be included in the solution set since the given statement includes equality. Now let's choose a test number from each of the three intervals and record the sign behavior of the factors $(x + 3)$ and $(x - 1)$.

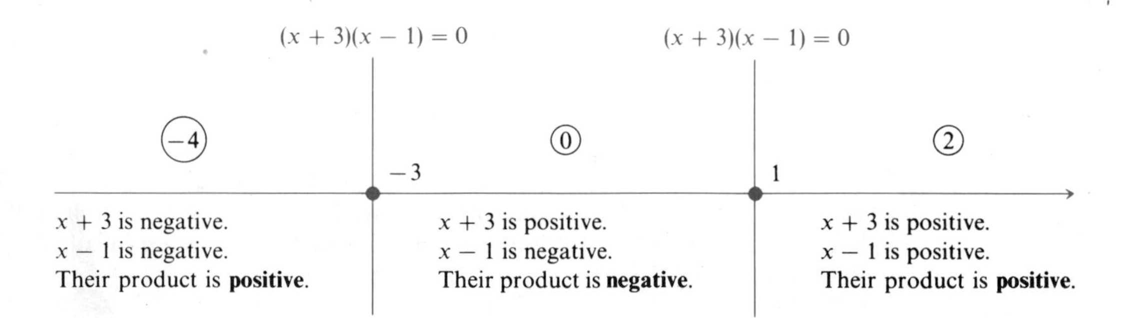

Therefore, the solution set is $[-3, 1]$ and it can be graphed as follows. ■

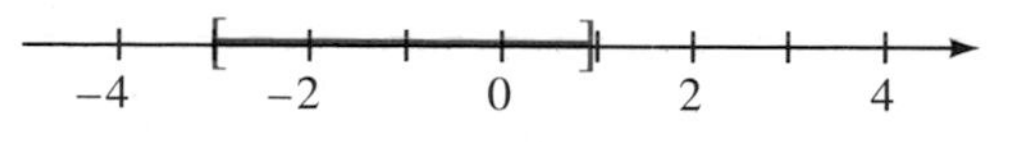

Examples 1 and 2 have indicated a systematic approach for solving quadratic inequalities where the polynomial is factorable. This same type of number line analysis can also be used to solve indicated quotients such as $\frac{x+1}{x-5} > 0$.

EXAMPLE 3 Solve and graph the solutions for $\frac{x+1}{x-5} > 0$.

Solution First, indicate that at $x = -1$ the given quotient equals zero, and at $x = 5$ the quotient is undefined. Second, choose test numbers from each of the three intervals and record the sign behavior of $(x + 1)$ and $(x - 5)$.

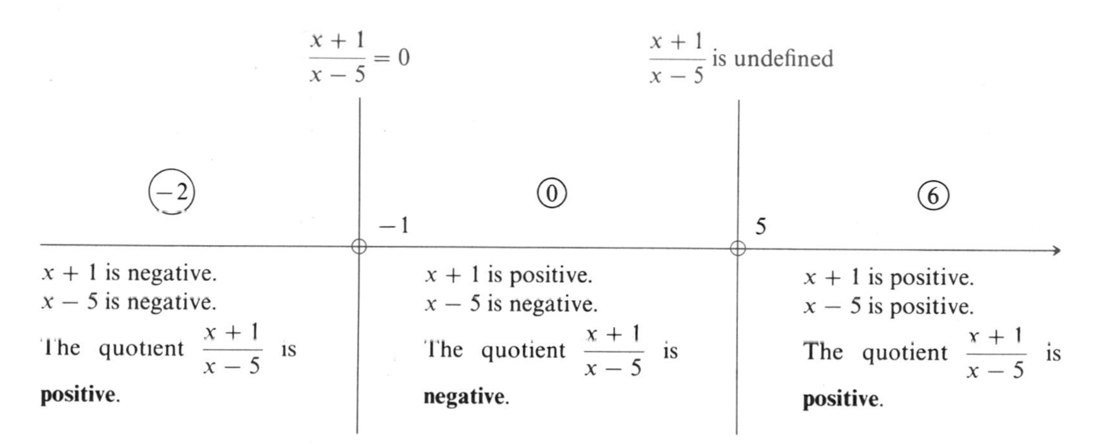

Therefore, the solution set is $(-\infty, -1) \cup (5, \infty)$ and its graph is as follows.

−4 −2 0 2 4

■

EXAMPLE 4 Solve $\frac{x+2}{x+4} \leq 0$.

Solution The indicated quotient equals zero at $x = -2$ and is undefined at $x = -4$. (Note that -2 is to be included in the solution set, but -4 is not to be included.) Now let's choose some test numbers and record the sign behavior of $(x + 2)$ and $(x + 4)$.

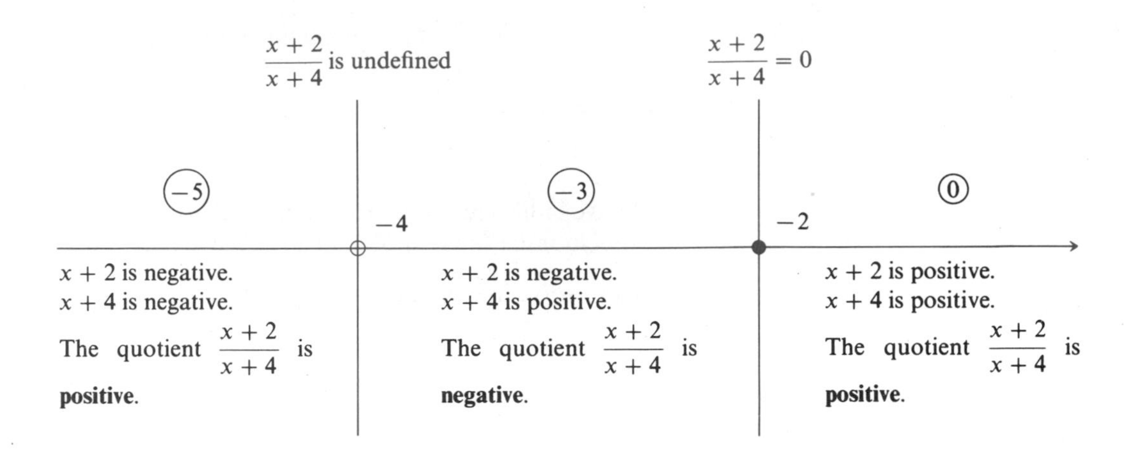

Therefore, the solution set is $(-4, -2]$. ■

The final example illustrates that sometimes we need to change the form of the given inequality before we use the number line analysis.

EXAMPLE 5 Solve $\frac{x}{x+2} \geq 3$.

Solution First, let's change the form of the given inequality as follows.

$$\frac{x}{x+2} \geq 3$$

$$\frac{x}{x+2} - 3 \geq 0 \qquad \text{Add } -3 \text{ to both sides.}$$

$$\frac{x - 3(x+2)}{x+2} \geq 0 \qquad \text{Express the left side over a common denominator.}$$

$$\frac{x - 3x - 6}{x+2} \geq 0$$

$$\frac{-2x - 6}{x+2} \geq 0.$$

Now we can proceed as we did with the previous examples. If $x = -3$, then $\frac{-2x-6}{x+2}$ equals zero; and if $x = -2$, then $\frac{-2x-6}{x+2}$ is undefined. Then choosing test numbers we can record the sign behavior of $(-2x - 6)$ and $(x + 2)$.

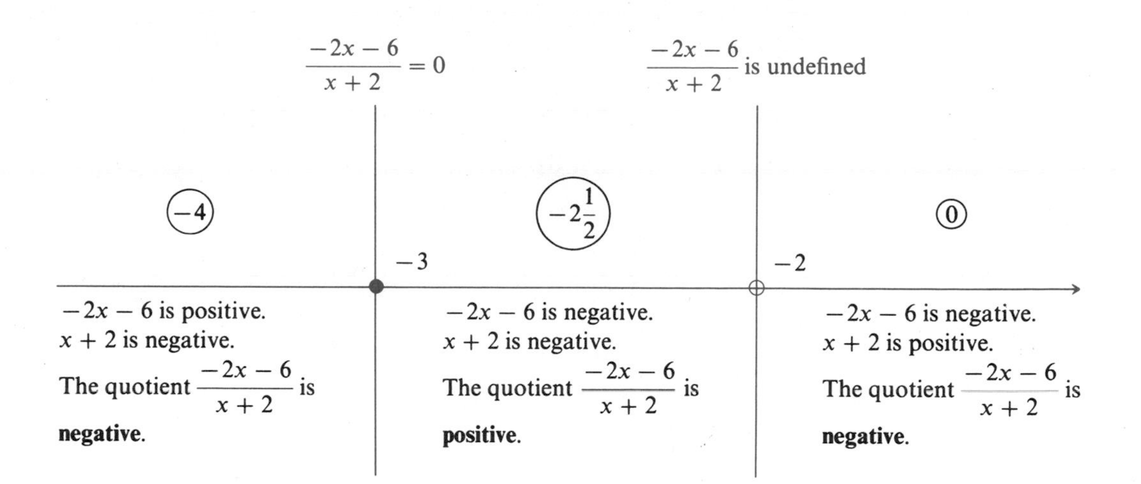

Therefore, the solution set is $[-3, -2)$. (Perhaps you should check a few numbers from this solution set back into the original inequality!) ■

Problem Set 6.6

Solve each of the following inequalities and graph each solution set on a number line.

1. $(x+2)(x-1) > 0$ $(-\infty, -2) \cup (1, \infty)$
2. $(x-2)(x+3) > 0$ $(-\infty, -3) \cup (2, \infty)$
3. $(x+1)(x+4) < 0$ $(-4, -1)$
4. $(x-3)(x-1) < 0$ $(1, 3)$
5. $(2x-1)(3x+7) \geq 0$ $(-\infty, -\frac{7}{3}] \cup [\frac{1}{2}, \infty)$
6. $(3x+2)(2x-3) \geq 0$ $(-\infty, -\frac{2}{3}] \cup [\frac{3}{2}, \infty)$
7. $(x+2)(4x-3) \leq 0$ $[-2, \frac{3}{4}]$
8. $(x-1)(2x-7) \leq 0$ $[1, \frac{7}{2}]$
9. $(x+1)(x-1)(x-3) > 0$ $(-1, 1) \cup (3, \infty)$
10. $(x+2)(x+1)(x-2) > 0$ $(-2, -1) \cup (2, \infty)$
11. $x(x+2)(x-4) \leq 0$ $(-\infty, -2] \cup [0, 4]$
12. $x(x+3)(x-3) \leq 0$ $(-\infty, -3] \cup [0, 3]$
13. $\frac{x+1}{x-2} > 0$ $(-\infty, -1) \cup (2, \infty)$
14. $\frac{x-1}{x+2} > 0$ $(-\infty, -2) \cup (1, \infty)$
15. $\frac{x-3}{x+2} < 0$ $(-2, 3)$
16. $\frac{x+2}{x-4} < 0$ $(-2, 4)$
17. $\frac{2x-1}{x} \geq 0$ $(-\infty, 0) \cup \left[\frac{1}{2}, \infty\right)$
18. $\frac{x}{3x+7} \geq 0$ $\left(-\infty, -\frac{7}{3}\right) \cup [0, \infty)$
19. $\frac{-x+2}{x-1} \leq 0$ $(-\infty, 1) \cup [2, \infty)$
20. $\frac{3-x}{x+4} \leq 0$ $(-\infty, -4) \cup [3, \infty)$

Solve each of the following inequalities.

21. $x^2 + 2x - 35 < 0$ $(-7, 5)$
22. $x^2 + 3x - 54 < 0$ $(-9, 6)$
23. $x^2 - 11x + 28 > 0$ $(-\infty, 4) \cup (7, \infty)$
24. $x^2 + 11x + 18 > 0$ $(-\infty, -9) \cup (-2, \infty)$
25. $3x^2 + 13x - 10 \leq 0$ $[-5, \frac{2}{3}]$
26. $4x^2 - x - 14 \leq 0$ $[-\frac{7}{4}, 2]$
27. $8x^2 + 22x + 5 \geq 0$ $(-\infty, -\frac{5}{2}] \cup [-\frac{1}{4}, \infty)$
28. $12x^2 - 20x + 3 \geq 0$ $(-\infty, \frac{1}{6}] \cup [\frac{3}{2}, \infty)$

29. $x(5x - 36) > 32$ $(-\infty, -\frac{4}{5}) \cup (8, \infty)$

30. $x(7x + 40) < 12$ $(-6, \frac{2}{7})$

31. $x^2 - 14x + 49 \geq 0$ $(-\infty, \infty)$

32. $(x + 9)^2 \geq 0$ $(-\infty, \infty)$

33. $4x^2 + 20x + 25 \leq 0$ $\{-\frac{5}{2}\}$

34. $9x^2 - 6x + 1 \leq 0$ $\{\frac{1}{3}\}$

35. $(x + 1)(x - 3)^2 > 0$ $(-1, 3) \cup (3, \infty)$

36. $(x - 4)^2(x - 1) \leq 0$ $(-\infty, 1] \cup \{4\}$

37. $\dfrac{2x}{x + 3} > 4$ $(-6 -3)$

38. $\dfrac{x}{x - 1} > 2$ $(1, 2)$

39. $\dfrac{x - 1}{x - 5} \leq 2$ $(-\infty, 5) \cup [9, \infty)$

40. $\dfrac{x + 2}{x + 4} \leq 3$ $(-\infty, -5] \cup (-4, \infty)$

41. $\dfrac{x + 2}{x - 3} > -2$ $\left(-\infty, \dfrac{4}{3}\right) \cup (3, \infty)$

42. $\dfrac{x - 1}{x - 2} < -1$ $\left(\dfrac{3}{2}, 2\right)$

43. $\dfrac{3x + 2}{x + 4} \leq 2$ $(-4, 6]$

44. $\dfrac{2x - 1}{x + 2} \geq -1$ $(-\infty, 2) \cup \left[-\dfrac{1}{3}, \infty\right)$

45. $\dfrac{x + 1}{x - 2} < 1$ $(-\infty, 2)$

46. $\dfrac{x + 3}{x - 4} \geq 1$ $(4, \infty)$

Miscellaneous Problems

47. The product $(x - 2)(x + 3)$ is positive if both factors are negative *or* if both factors are positive. Therefore, we can solve $(x - 2)(x + 3) > 0$ as follows.

$$(x - 2 < 0 \text{ and } x + 3 < 0) \quad \text{or} \quad (x - 2 > 0 \text{ and } x + 3 > 0)$$
$$(x < 2 \text{ and } x < -3) \quad \text{or} \quad (x > 2 \text{ and } x > -3)$$
$$x < -3 \quad \text{or} \quad x > 2$$

The solution set is $(-\infty, -3) \cup (2, \infty)$. Use this type of analysis to solve each of the following.

(a) $(x - 2)(x + 7) > 0$ $(-\infty, -7) \cup (2, \infty)$

(b) $(x - 3)(x + 9) \geq 0$ $(-\infty, -9] \cup [3, \infty)$

(c) $(x + 1)(x - 6) \leq 0$ $[-1, 6]$

(d) $(x + 4)(x - 8) < 0$ $(-4, 8)$

(e) $\dfrac{x + 4}{x - 7} > 0$ $(-\infty, -4) \cup (7, \infty)$

(f) $\dfrac{x - 5}{x + 8} \leq 0$ $(-8, 5]$

Chapter 6 Summary

(6.1) A number of the form $a + bi$, where a and b are real numbers and i is the imaginary unit defined by $i = \sqrt{-1}$, is a **complex number**.

Two complex numbers $a + bi$ and $c + di$ are said to be *equal* if and only if $a = c$ and $b = d$.

We describe addition and subtraction of complex numbers as follows.

$$(a + bi) + (c + di) = (a + c) + (b + d)i,$$
$$(a + bi) - (c + di) = (a - c) + (b - d)i$$

We can represent a square root of any negative real number as the product of a real number and the imaginary unit i. That is,

$$\sqrt{-b} = i\sqrt{b}, \quad \text{where } b \text{ is a positive real number.}$$

The product of two complex numbers conform with the product of two binomials. The **conjugate** of $a + bi$ is $a - bi$. The product of a complex number and its conjugate is a real number. Therefore, conjugates are used to simplify expressions, such as $\frac{4 + 3i}{5 - 2i}$, which indicate the quotient of two complex numbers.

(6.2) The **standard form for a quadratic equation** in one variable is

$$ax^2 + bx + c = 0,$$

where a, b, and c are real numbers and $a \neq 0$.

Some quadratic equations can be solved by *factoring* and applying the property, $ab = 0$ if and only if $a = 0$ or $b = 0$.

Don't forget that applying the property, if $a = b$, then $a^n = b^n$, might produce extraneous solutions. Therefore, we *must check* all potential solutions.

We can solve some quadratic equations by applying the property, $x^2 = a$ if and only if $x = \pm\sqrt{a}$.

(6.3) To solve a quadratic equation of the form $x^2 + bx = k$ by **completing the square**, we (1) add $\left(\frac{b}{2}\right)^2$ to both sides, (2) factor the left side, and (3) apply the property, $x^2 = a$ if and only if $x = \pm\sqrt{a}$.

(6.4) We can solve any quadratic equation of the form $ax^2 + bx + c = 0$ by the **quadratic formula**, which we usually state as

$$x = \frac{-b \pm \sqrt{b^2 - 4ac}}{2a}.$$

The **discriminant**, $b^2 - 4ac$, can be used to determine the nature of the roots of a quadratic equation as follows.

1. If $b^2 - 4ac < 0$, then the equation has two nonreal complex solutions.
2. If $b^2 - 4ac = 0$, then the equation has two equal real solutions.
3. If $b^2 - 4ac > 0$, then the equation has two unequal real solutions.

If x_1 and x_2 are roots of a quadratic equation, then the following relationships exist.

$$x_1 + x_2 = -\frac{b}{a} \quad \text{and} \quad (x_1)(x_2) = \frac{c}{a}$$

These **sum-and-product relationships** can be used to check potential solutions of quadratic equations.

(6.5) To review the strengths and weaknesses of the three basic methods for solving a quadratic equation (factoring, completing the square, the quadratic formula), go back over the examples in this section.

Keep the following suggestions in mind as you solve word problems.

1. Read the problem carefully.
2. Sketch any figure, diagram, or chart that might help organize and analyze the problem.
3. Choose a meaningful variable.
4. Look for a guideline that can be used to set up an equation.
5. Form an equation that translates the guideline from English to algebra.
6. Solve the equation and use the solutions to determine all facts requested in the problem.
7. Check all answers back into the original statement of the problem.

(6.6) The number line, along with **critical numbers** and **test numbers**, provides a good basis for solving **quadratic inequalities** where the polynomial is factorable. We can use this same basic approach to solve inequalities, such as $\frac{3x+1}{x-4} > 0$, which indicate quotients.

Chapter 6 Review Problem Set

For Problems 1–8, perform the indicated operations and express the answers in the standard form of a complex number.

1. $(-7+3i)+(9-5i)$ $2-2i$

2. $(4-10i)-(7-9i)$ $-3-i$

3. $5i(3-6i)$ $30+15i$

4. $(5-7i)(6+8i)$ $86-2i$

5. $(-2-3i)(4-8i)$ $-32+4i$

6. $(4-3i)(4+3i)$ 25

7. $\frac{4+3i}{6-2i}$ $\frac{9}{20}+\frac{13}{20}i$

8. $\frac{-1-i}{-2+5i}$ $-\frac{3}{29}+\frac{7}{29}i$

For Problems 9–12, find the discriminant of each equation and determine whether the equation has (1) two nonreal complex solutions, (2) one real solution with a multiplicity of two, (3) two real solutions. Do not solve the equations.

9. $4x^2-20x+25=0$ Two equal real solutions

10. $5x^2-7x+31=0$ Two nonreal complex solutions

11. $7x^2-2x-14=0$ Two unequal real solutions

12. $5x^2-2x=4$ Two unequal real solutions

For Problems 13–31, solve each equation.

13. $x^2-17x=0$ $\{0, 17\}$

14. $(x-2)^2=36$ $\{-4, 8\}$

15. $(2x-1)^2=-64$ $\left\{\frac{1\pm 8i}{2}\right\}$

16. $x^2-4x-21=0$ $\{-3, 7\}$

17. $x^2+2x-9=0$ $\{-1\pm\sqrt{10}\}$

18. $x^2-6x=-34$ $\{3\pm 5i\}$

19. $4\sqrt{x}=x-5$ $\{25\}$

20. $3n^2+10n-8=0$ $\left\{-4, \frac{2}{3}\right\}$

21. $n^2-10n=200$ $\{-10, 20\}$

22. $3a^2+a-5=0$ $\left\{\frac{-1\pm\sqrt{61}}{6}\right\}$

23. $x^2-x+3=0$ $\left\{\frac{1\pm i\sqrt{11}}{2}\right\}$

24. $2x^2-5x+6=0$ $\left\{\frac{5\pm i\sqrt{23}}{4}\right\}$

25. $2a^2+4a-5=0$ $\left\{\frac{-2\pm\sqrt{14}}{2}\right\}$

26. $t(t+5)=36$ $\{-9, 4\}$

27. $x^2 + 4x + 9 = 0$ $\{-2 \pm i\sqrt{5}\}$

28. $(x - 4)(x - 2) = 80$ $\{-6, 12\}$

29. $\dfrac{3}{x} + \dfrac{2}{x+3} = 1$ $\{1 \pm \sqrt{10}\}$

30. $2x^4 - 23x^2 + 56 = 0$ $\{\pm\frac{\sqrt{14}}{2}, \pm 2\sqrt{2}\}$

31. $\dfrac{3}{n-2} = \dfrac{n+5}{4}$ $\left\{\dfrac{-3 \pm \sqrt{97}}{2}\right\}$

For Problems 32–35, solve each inequality and indicate the solution set on a number line graph.

32. $x^2 + 3x - 10 > 0$ $(-\infty, -5) \cup (2, \infty)$

33. $2x^2 + x - 21 \le 0$ $[-\frac{7}{2}, 3]$

34. $\dfrac{x-4}{x+6} \ge 0$ $(-\infty, -6) \cup [4, \infty)$

35. $\dfrac{2x-1}{x+1} > 4$ $\left(-\dfrac{5}{2}, -1\right)$

For Problems 36–43, set up an equation and solve each problem.

36. Find two numbers whose sum is 6 and whose product is 2. $3 + \sqrt{7}$ and $3 - \sqrt{7}$

37. Sherry bought a number of shares of stock for \$250. Six months later the value of the stock increased by \$5 per share and she sold all but 5 shares and regained her original investment plus a profit of \$50. How many shares did she sell and at what price per share? 20 shares at \$15 per share

38. Dave traveled 270 miles in one hour more time than it took Sandy to travel 260 miles. Sandy drove 7 miles per hour faster than Dave. How fast did each one travel? 45, 52

39. The area of a square is numerically equal to twice its perimeter. Find the length of a side of the square. 8 units

40. Find two consecutive even whole numbers such that the sum of their squares is 164. 8 and 10

41. The perimeter of a rectangle is 38 inches and its area is 84 square inches. Find the length and width of the rectangle. 7 in. by 12 in.

42. It takes Billy 2 hours longer to do a certain job than it takes Janet. They worked together for 2 hours; then Janet left and Billy finished the job in 1 hour. How long would it take each of them to do the job alone? 4 hrs for Janet and 6 hrs for Billy

43. A company has a rectangular parking lot 40 meters wide and 60 meters long. They plan to increase the area of the lot by 1100 square meters by adding a strip of equal width to one side and one end. Find the width of the strip to be added. 10 meters

Thoughts into Words

44. Explain the process of "completing the square" to solve a quadratic equation.

45. Explain how to use the quadratic formula to solve $4x = 3x^2 - 2$.

46. Explain how you would solve $(x - 2)(x - 7) = 0$ and also how you would solve $(x - 2)(x - 7) = 4$.

47. One of our problem solving suggestions is to "look for a *guideline* that can be used to set up an equation." What does this suggestion mean to you?

48. Explain how to solve the inequality $(x + 1)(x - 2)(x - 3) < 0$.

49. Why is the set of real numbers a subset of the set of complex numbers?

50. How would you solve the equation $x^2 - 4x = 252$? Explain your choice of the method that you would use.

Chapter 7

Equations in Two Variables and Their Graphs

We frequently use real number lines to graph the solutions of equations and inequalities that involve one variable. For example, the solutions of $x > 2$ or $x \leq -1$ can be shown on a real number line as follows.

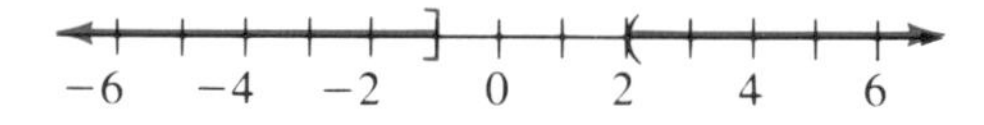

On a real number line there is a one-to-one correspondence between the real numbers and the points on a line. To each point on a line there corresponds a real number and to each real number there corresponds a point on the line. The idea of associating points with numbers can be expanded so that a correspondence is established between pairs of real numbers and the points in a plane. This will allow us to graph the solutions of equations and inequalities that contain two variables.

7.1
The Rectangular Coordinate System

Consider two number lines, one vertical and one horizontal, perpendicular to each other at the point we associate with zero on both lines (Figure 7.1). We refer to these number lines as the **horizontal and vertical axes**, or together as the **coordinate axes**. They partition the plane into four regions called **quadrants**. The quadrants are numbered counterclockwise from I through IV as indicated in Figure 7.1. The point of intersection of the two axes is called the **origin**.

Figure 7.1

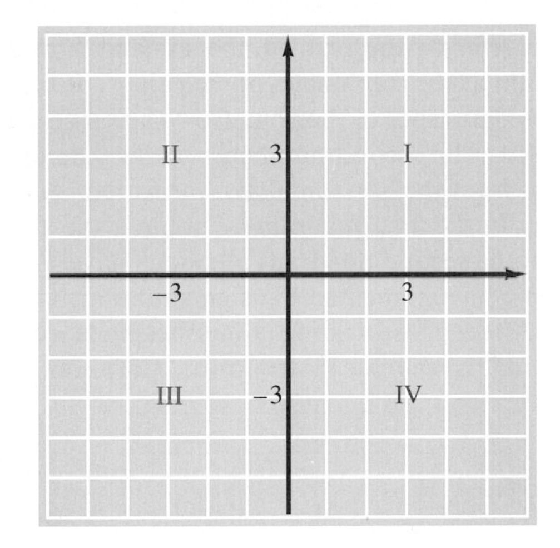

It is now possible to set up a one-to-one correspondence between **ordered pairs** of real numbers and the points in a plane. To each ordered pair of real numbers there corresponds a unique point in the plane and to each point in the plane there corresponds a unique ordered pair of real numbers. A part of this correspondence is illustrated in Figure 7.2. The ordered pair $(3, 2)$ means that the point A is located

Figure 7.2

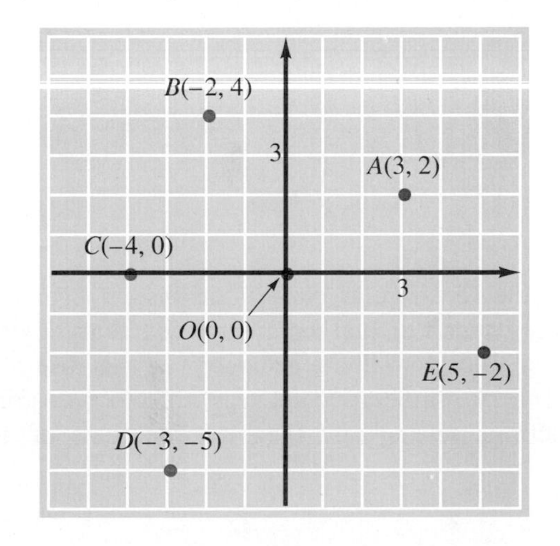

3 units to the right of and two units up from the origin. (The ordered pair $(0, 0)$ is associated with the origin O.) The ordered pair $(-3, -5)$ means that the point D is located three units to the left and five units down from the origin.

REMARK The notation $(-2, 4)$ was used earlier in this text to indicate an interval of the real number line. Now we are using the same notation to indicate an ordered pair of real numbers. This double meaning should not be confusing since the context of the material will definitely indicate the meaning of the notation being used at a particular time. Throughout this chapter we will be using the ordered-pair interpretation.

In general we refer to the real numbers a and b in an ordered pair (a, b) associated with a point as the **coordinates of the point**. The first number, a, called the **abscissa**, is the directed distance of the point from the vertical axis measured parallel to the horizontal axis. The second number, b, called the **ordinate**, is the directed distance of the point from the horizontal axis measured parallel to the vertical axis (Figure 7.3(a)). Thus, in the first quadrant all points have a positive abscissa and a positive ordinate. In the second quadrant all points have a negative abscissa and a positive ordinate. We have indicated the sign situations for all four quadrants in Figure 7.3 (b). This system of associating points in a plane with pairs of real numbers is called the **rectangular coordinate system** or the **Cartesian coordinate system**.

Figure 7.3

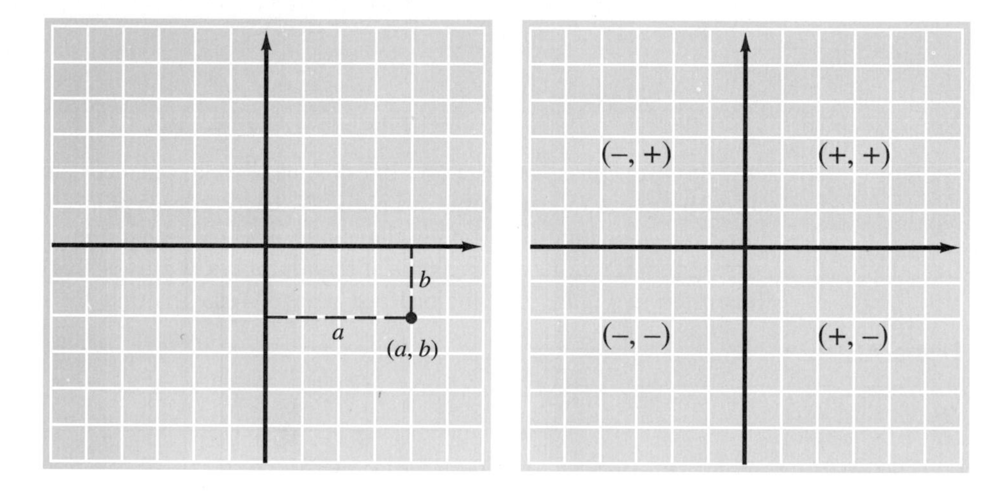

Historically, the rectangular coordinate system provided the basis for the development of the branch of mathematics called **analytic geometry**, or what we presently often refer to as **coordinate geometry**. In this discipline, René Descartes, a French 17th century mathematician, was able to transform geometric problems into an algebraic setting and then use the tools of algebra to solve the problems.

Basically, there are two kinds of problems to solve in coordinate geometry:

1. Given an algebraic equation, find its geometric graph;
2. Given a set of conditions pertaining to a geometric figure, find its algebraic equation.

In this chapter we will discuss problems of both types.

Let's begin by considering the solutions for the equation $y = x + 2$. A **solution** of an equation in two variables is an ordered pair of real numbers that satisfies the equation. When using the variables x and y, we agree that the first number of an ordered pair is a value of x and the second number is a value of y. We see that $(1, 3)$ is a solution for $y = x + 2$ because if x is replaced by 1 and y by 3, the true numerical statement $3 = 1 + 2$ is obtained. Likewise, $(-2, 0)$ is a solution since $0 = -2 + 2$ is a true statement. Infinitely many pairs of real numbers that satisfy $y = x + 2$ can be found by arbitrarily choosing values for x, and for each value of x chosen, determining a corresponding value for y. Let's use a table to record some of the solutions for $y = x + 2$.

Choose x	*Determine y from y = x + 2*	*Solutions for y = x + 2*
0	2	(0, 2)
1	3	(1, 3)
3	5	(3, 5)
5	7	(5, 7)
−2	0	(−2, 0)
−4	−2	(−4, −2)
−6	−4	(−6, −4)

Figure 7.4

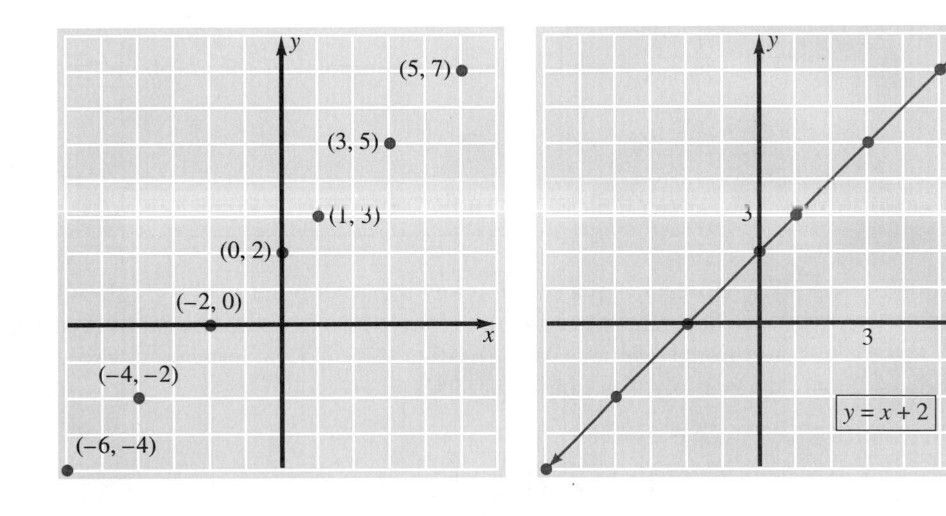

We can plot the ordered pairs as points in a coordinate plane and use the horizontal axis as the x-axis and the vertical axis as the y-axis as in Figure 7.4(a). The straight line that contains the points in Figure 7.4(b) is called the **graph of the equation** $y = x + 2$.

REMARK It is important to recognize that all points on the x-axis have ordered pairs of the form $(a, 0)$ associated with them. That is to say, the second number in the ordered pair is 0. Likewise, all points on the y-axis have ordered pairs of the form $(0, b)$ associated with them.

Graphing Techniques

For the remainder of this section we will use examples to introduce some basic graphing techniques. Then towards the end of the section we will summarize these techniques for you.

EXAMPLE 1 Graph $2x + 3y = 6$.

Solution First, let's find the points of this graph that fall on the coordinate axes. Let $x = 0$; then

$$\begin{aligned} 2(0) + 3y &= 6 \\ 3y &= 6 \\ y &= 2. \end{aligned}$$

Thus, $(0, 2)$ is a solution and locates a point of the graph on the y-axis. Let $y = 0$; then

$$\begin{aligned} 2x + 3(0) &= 6 \\ 2x &= 6 \\ x &= 3. \end{aligned}$$

Therefore, $(3, 0)$ is a solution and locates a point of the graph on the x-axis.

Second, let's change the form of the equation to make it easier to find some additional solutions. We can either solve for x in terms of y or solve for y in terms of x. Let's solve for y in terms of x.

$$\begin{aligned} 2x + 3y &= 6 \\ 3y &= 6 - 2x \\ y &= \frac{6 - 2x}{3} \end{aligned}$$

Third, a table of values can be formed that includes the two points we found previously.

x	y
0	2
3	0
6	-2
-3	4
-6	6

Plotting these points we see that they lie in a straight line and we obtain Figure 7.5.

Figure 7.5

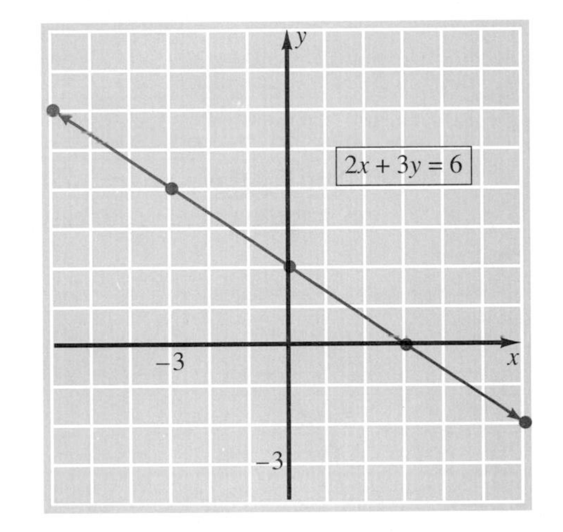

■

REMARK Look again at the table of values in Example 1. Notice that values of x were chosen so that integers were obtained for y. This is not necessary, but it does make things easier from a computational viewpoint.

The points $(3, 0)$ and $(0, 2)$ in Figure 7.5 are special points. They are the points of the graph that are on the coordinate axes. That is to say, they yield the x-intercept and the y-intercept of the graph. Let's define in general the *intercepts* of a graph.

> The x-coordinates of the points that a graph has in common with the x-axis are called the **x-intercepts** of the graph. (To compute the x-intercepts, let $y = 0$ and solve for x.)
>
> The y-coordinates of the points that a graph has in common with the y-axis are called the **y-intercepts of** the graph. (To compute the y-intercepts, let $x = 0$ and solve for y.)

EXAMPLE 2 Graph $y = (x + 2)(x - 2)$.

Solution Let's begin by finding the intercepts. If $x = 0$, then

$$y = (0 + 2)(0 - 2)$$
$$y = -4.$$

The point $(0, -4)$ is on the graph. If $y = 0$, then

$$(x + 2)(x - 2) = 0$$
$$x + 2 = 0 \quad \text{or} \quad x - 2 = 0$$
$$x = -2 \quad \text{or} \quad x = 2.$$

The points $(-2, 0)$ and $(2, 0)$ are on the graph. The given equation is in a convenient form for setting up a table of values.

x	y	
0	−4	intercepts
−2	0	
2	0	
1	−3	other points
−1	−3	
3	5	
−3	5	

Plotting these points and connecting them with a smooth curve produces Figure 7.6.

Figure 7.6

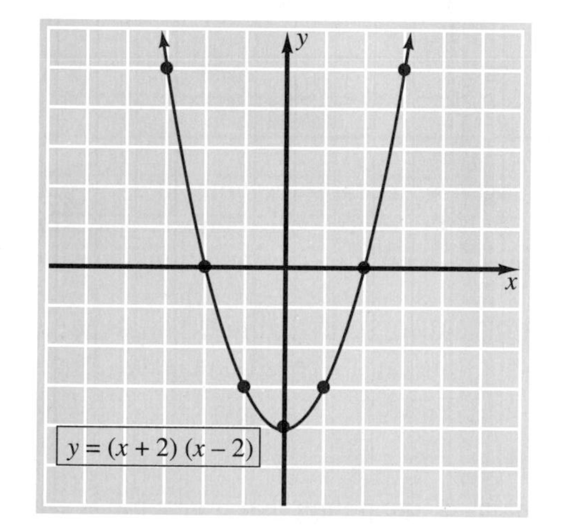

■

The curve in Figure 7.6 is called a **parabola** and we will study them in more detail in a later section. However, at this time we do want to emphasize that the parabola in Figure 7.6 is said to be *symmetric with respect to the y-axis.* In other words, the y-axis is a line of symmetry. Each half of the curve is a mirror image of the other half through the y-axis. Notice in the table of values, that for each ordered pair (x, y), the ordered pair $(-x, y)$ is also a solution. A general test for y-axis symmetry can be stated as follows.

> **y-Axis Symmetry**
>
> The graph of an equation is symmetric with respect to the y-axis if replacing x with $-x$ results in an equivalent equation.

The equation $y = x^2 - 4$ exhibits y-axis symmetry because replacing x with $-x$ produces $y = (-x)^2 - 4 = x^2 - 4$. Likewise, the equations $y = -x^2 + 2$, $y = 2x^2 + 5$, and $y = x^4 - x^2$ exhibit y-axis symmetry.

EXAMPLE 3 Graph $x = y^2$.

Solution First, we see that $(0, 0)$ is on the graph and determines both intercepts. Second, the given equation is in a convenient form for setting up a table of values.

x	y	
0	0	intercepts
1	1	other points
1	-1	
4	2	
4	-2	

Plotting these points and connecting them with a smooth curve produces Figure 7.7.

Figure 7.7

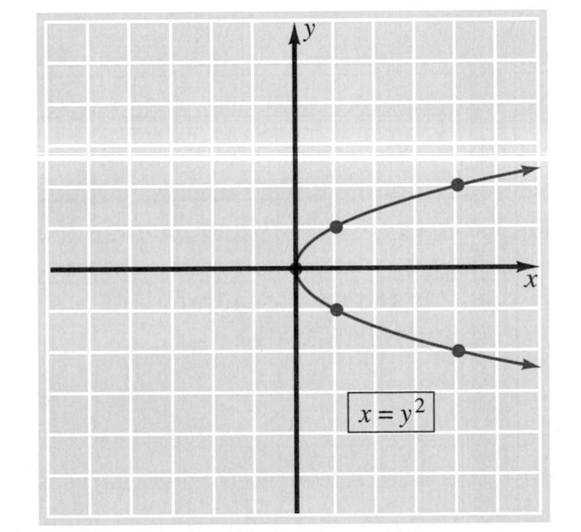

■

The parabola in Figure 7.7 is said to be *symmetric with respect to the x-axis.* Each half of the curve is a mirror image of the other half through the x-axis. Also notice in the table of values, that for each ordered pair (x, y), the ordered pair $(x, -y)$ is a solution. A general test for x-axis symmetry can be stated as follows.

x-Axis Symmetry

The graph of an equation is symmetric with respect to the x-axis if replacing y with $-y$ results in an equivalent equation.

The equation $x = y^2$ exhibits x-axis symmetry because replacing y with $-y$ produces $x = (-y)^2 = y^2$. Likewise, the equations $x + y^2 = 5$, $x = y^2 - 5$, and $x = 2y^4 - 3y^2$ exhibit x-axis symmetry.

EXAMPLE 4 Graph $y = \frac{1}{x}$.

Solution First, let's find the intercepts. Let $x = 0$; then $y = \frac{1}{x}$ becomes $y = \frac{1}{0}$ and $\frac{1}{0}$ is undefined. Thus, there is no y-intercept. Let $y = 0$; then $y = \frac{1}{x}$ becomes $0 = \frac{1}{x}$ and there are no values of x that will satisfy this equation. In other words, this graph has no points on either the x-axis or the y-axis.

Second, let's set up a table of values and keep in mind that neither x nor y can equal zero.

x	y
$\frac{1}{2}$	2
1	1
2	$\frac{1}{2}$
3	$\frac{1}{3}$
$-\frac{1}{2}$	-2
-1	-1
-2	$-\frac{1}{2}$
-3	$-\frac{1}{3}$

In Figure 7.8(a) we have plotted the points associated with the solutions from the table. Since the graph does not intersect either axis, it must consist of two branches. Thus, connecting the points in the first quadrant with a smooth curve and then connecting the points in the third quadrant with a smooth curve, we obtain the graph in Figure 7.8(b).

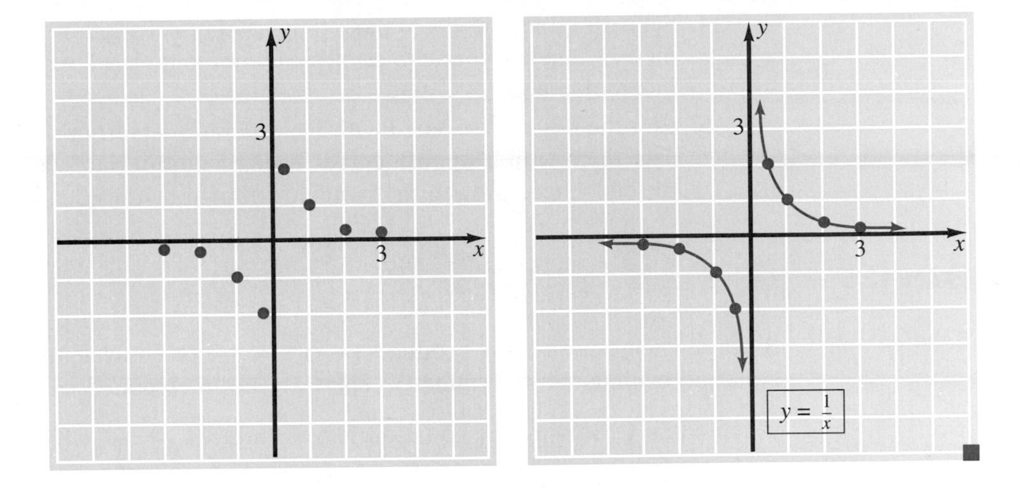

The curve in Figure 7.8 is said to be *symmetric with respect to the origin.* Each half of the curve is a mirror image of the other half through the origin. Notice in the table of values, that for each ordered pair (x, y), the ordered pair $(-x, -y)$ is also a solution. A general test for origin symmetry can be stated as follows.

> **Origin Symmetry**
>
> The graph of an equation is symmetric with respect to the origin if replacing x with $-x$ and y with $-y$ results in an equivalent equation.

The equation $y = \frac{1}{x}$ exhibits origin symmetry because replacing x with $-x$ and y with $-y$ produces $-y = \frac{1}{-x}$ which is equivalent to $y = \frac{1}{x}$. $\left(\text{We can multiply both sides of } y = \frac{1}{-x} \text{ by } -1 \text{ and obtain } y = \frac{1}{x}.\right)$ Likewise, the equations $xy = 4$, $y = x^3$, and $y = x^5$ exhibit origin symmetry.

Let's pause for a moment and pull together the graphing techniques that have been introduced thus far. Following is a list of graphing suggestions. The order of the suggestions indicates the order in which we usually attack a new graphing problem.

1. Determine the type of symmetry that the equation exhibits.
2. Find the intercepts.
3. Solve the equation for y in terms of x or for x in terms of y if it is not already in such a form.

4. Set up a table of ordered pairs that satisfy the equation. The type of symmetry will affect your choice of values in the table. (This will be illustrated in a moment.)
5. Plot the points associated with the ordered pairs from the table, and connect them with a smooth curve. Then, if appropriate, reflect this part of the curve according to the symmetry shown by the equation.

EXAMPLE 5 Graph $x^2y = -2$.

Solution Since replacing x with $-x$ produces $(-x)^2y = -2$ or equivalently $x^2y = -2$, the equation exhibits y-axis symmetry. There are no intercepts because neither x nor y can equal 0. Solving the equation for y produces $y = \frac{-2}{x^2}$. Since the equation exhibits y-axis symmetry, let's use only positive values for x and then we can reflect across the y-axis.

x	y
1	-2
2	$-\frac{1}{2}$
3	$-\frac{2}{9}$
4	$-\frac{1}{8}$
$\frac{1}{2}$	-8

Let's plot the points determined by the table, connect them with a smooth curve, and reflect this portion of the curve across the y-axis. Figure 7.9 is the result of this process.

Figure 7.9

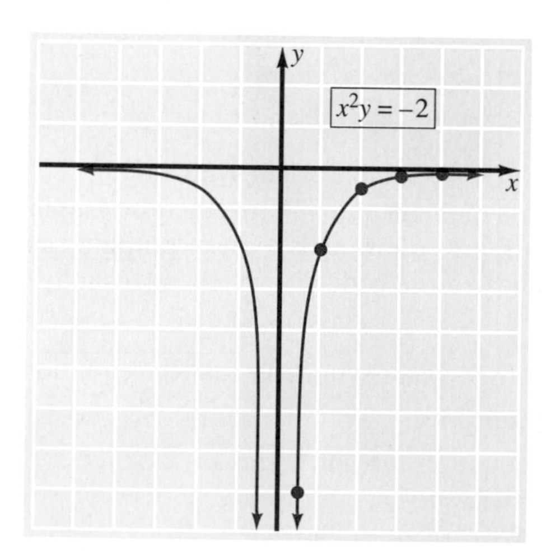

■

EXAMPLE 6 Graph $x = y^3$.

Solution Since replacing x with $-x$ and y with $-y$ produces $-x = (-y)^3 = -y^3$, which is equivalent to $x = y^3$, the given equation exhibits origin symmetry. If $x = 0$, then $y = 0$; so the origin is a point of the graph. The given equation is in an easy form for deriving a table of values.

x	y
0	0
8	2
$\frac{1}{8}$	$\frac{1}{2}$
$\frac{27}{64}$	$\frac{3}{4}$

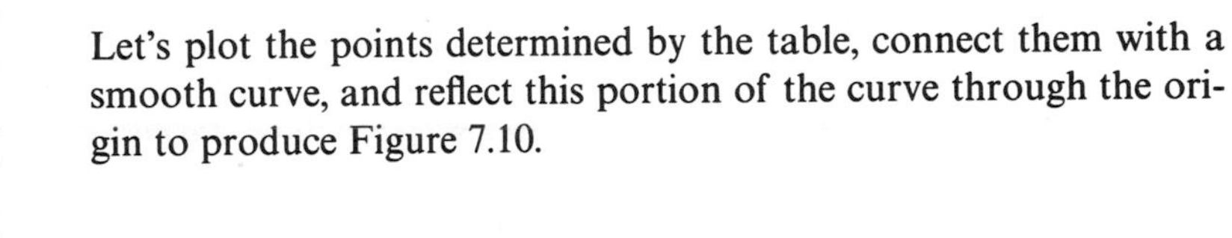

Let's plot the points determined by the table, connect them with a smooth curve, and reflect this portion of the curve through the origin to produce Figure 7.10.

Figure 7.10

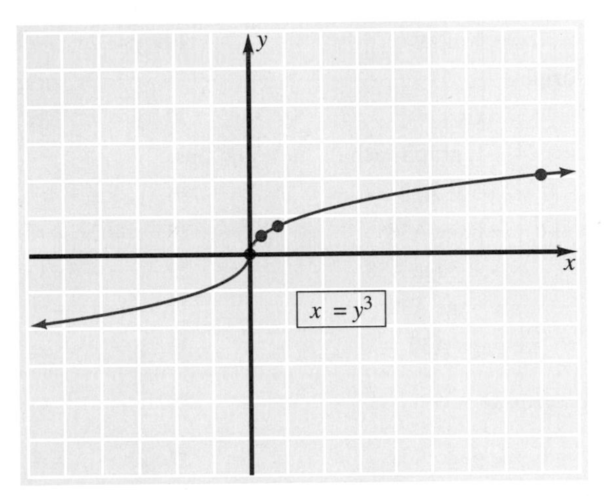

■

Problem Set 7.1

1. Indicate the quadrant that contains each of the points represented by the following ordered pairs.
   (a) $(-3, -4)$ III **(b)** $(5, -6)$ IV **(c)** $(-6, 8)$ II
   (d) $(-1, -6)$ III **(e)** $(4, 10)$ I **(f)** $(9, -1)$ IV
2. **(a)** What is the first coordinate of any point on the vertical axis in a rectangular coordinate system? 0
   (b) What is the second coordinate of any point on the horizontal axis in a rectangular coordinate system? 0

3. (a) In which quadrants do the coordinates of a point have the same sign? I and III
   (b) In which quadrants do the coordinates of a point have opposite signs? II and IV
   (c) In which quadrants is the abscissa negative? II and III
   (d) In which quadrants is the ordinate negative? III and IV

For each of the following points, determine the points that are symmetric with respect to the (a) x-axis, (b) y-axis, and (c) origin (Problems 4–9).

4. $(2, 6)$ $(2, -6); (-2, 6); (-2, -6)$
5. $(-3, 1)$ $(-3, -1); (3, 1); (3, -1)$
6. $(-2, -4)$ $(-2, 4); (2, -4); (2, 4)$
7. $(7, -2)$ $(7, 2); (-7, -2); (-7, 2)$
8. $(0, -4)$ $(0, 4); (0, -4); (0, 4)$
9. $(5, 0)$ $(5, 0); (-5, 0); (-5, 0)$

For Problems 10–23, determine the type of symmetry (x-axis, y-axis, origin) possessed by each of the graphs of the following equations. *Do not* sketch the graph.

10. $x^2 + 2y = 4$ y-axis
11. $-3x + 2y^2 = -4$ x-axis
12. $x = -y^2 + 5$ x-axis
13. $y = 4x^2 + 13$ y-axis
14. $xy = -6$ origin
15. $2x^2y^2 = 5$ x-axis, y-axis, and origin
16. $2x^2 + 3y^2 = 9$ x-axis, y-axis, and origin
17. $x^2 - 2x - y^2 = 4$ x-axis
18. $y = x^2 - 6x - 4$ none
19. $y = 2x^2 - 7x - 3$ none
20. $y = x$ origin
21. $y = 2x$ origin
22. $y = x^4 + 4$ y-axis
23. $y = x^4 - x^2 + 2$ y-axis

For Problems 24–51, graph each of the equations.

24. $y = x + 1$ see page A38
25. $y = x - 4$ see page A38
26. $y = 3x - 6$ see page A38
27. $y = 2x + 4$ see page A38
28. $y = -2x +$ see page A38
29. $y = -3x - 1$ see page A38
30. $y = x^2 - 1$ see page A39
31. $y = x^2 + 2$ see page A39
32. $y = -x^3$ see page A39
33. $y = x^3$ see page A39
34. $y = \dfrac{2}{x^2}$ see page A39
35. $y = \dfrac{-1}{x^2}$ see page A39
36. $2x + y = 6$ see page A39
37. $2x - y = 4$ see page A39
38. $y = 2x^2$ see page A39
39. $y = -3x^2$ see page A40
40. $xy = -3$ see page A40
41. $xy = 2$ see page A40
42. $x^2y = 4$ see page A40
43. $xy^2 = -4$ see page A40
44. $y^3 = x^2$ see page A40
45. $y^2 = x^3$ see page A40
46. $y = \dfrac{-2}{x^2 + 1}$ see page A40
47. $y = \dfrac{4}{x^2 + 1}$ see page A40
48. $x = -y^3$ see page A40
49. $y = x^4$ see page A40
50. $y = -x^4$ see page A40
51. $x = -y^3 + 2$ see page A41

52. The equation $F = \dfrac{9}{5}C + 32$ can be used to convert from degrees in Celsius to degrees in Fahrenheit. Complete the following table and graph the solutions on the set of axes below.

C	0	5	10	15	20	−5	−10	−15	−20	−25
F										

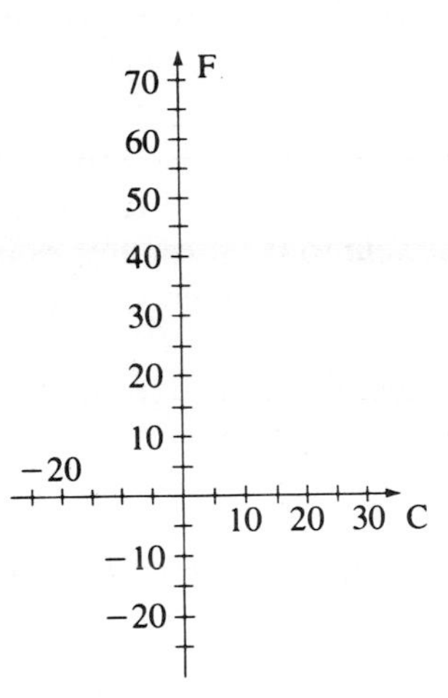

53. At $0.06 per kilowatt-hour, the equation $A = 0.06t$ determines the amount, A, of an electric bill for t hours. Complete the following table of values. $41.76; $43.20; $44.40; $46.50; $46.92

Hours	t	696	720	740	775	782
Dollars and cents	A					

54. Suppose that a used car dealer determines the selling price of his cars by using a markup of 60% of the cost. Let s represent the selling price and c the cost, and use the following equation.

$$s = c + 0.6c = 1.6c.$$

Complete the following table using the equation $s = 1.6c$. $400; $520; $920; $1432; $1750

Dollars	c	250	325	575	895	1095
Dollars	s					

7.2
Linear Equations in Two Variables

One very efficient graphing technique is the ability to recognize the kind of graph that a certain type of equation produces. For example, if we recognize that the graph of $3x + 2y = 12$ is a straight line, then it becomes a simple matter to find two points and sketch the line. Let's pursue the graphing of straight lines in a little more detail.

In general, any equation of the form $Ax + By = C$, where A, B, and C are constant (A and B not both zero) and x and y are variables, is a *linear equation* and its graph is a *straight line.* Two points of clarification about this description of a linear equation should be made. First, the choice of x and y for variables is arbitrary. Any two letters could be used to represent the variables. For example, an equation such as $3r + 2s = 9$ can be considered a linear equation in two variables. So that we are not constantly changing the labeling of the coordinate axes when graphing equations, however, it is much easier to use the same two variables in all equations. Thus, we will go along with convention and use x and y as variables. Second, the statement *any equation of the form* $Ax + By = C$ technically means any equation of the form $Ax + By = C$ or *equivalent* to the form. For example, the equation $y = 2x - 1$ is equivalent to $-2x + y = -1$ and thus is linear and produces a straight line graph.

The knowledge that any equation of the form $Ax + By = C$ produces a straight line graph, along with the fact that two points determine a straight line, makes graphing linear equations a simple process. We merely find two solutions, such as the intercepts, plot the corresponding points, and connect the points with a straight line. It is usually wise to find a third point as a check point. Let's consider an example.

EXAMPLE 1 Graph $3x - 2y = 12$.

Solution First, let's find the intercepts.

$$\text{Let } x = 0;\text{ then } 3(0) - 2y = 12$$
$$-2y = 12$$
$$y = -6$$

Thus, $(0, -6)$ is a solution.

$$\text{Let } y = 0;\text{ then } 3x - 2(0) = 12$$
$$3x = 12$$
$$x = 4.$$

Thus, $(4, 0)$ is a solution. Now let's find a third point to serve as a check point.

$$\text{Let } x = 2;\text{ then } 3(2) - 2y = 12$$
$$6 - 2y = 12$$
$$-2y = 6$$
$$y = -3.$$

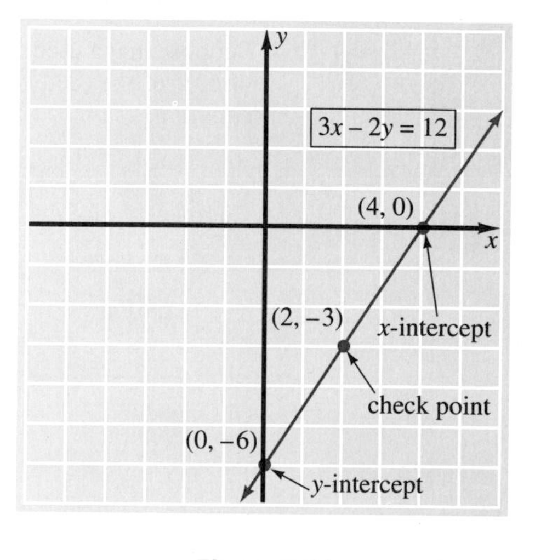

Figure 7.11

Thus, $(2, -3)$ is a solution. Plot the points associated with these three solutions and connect them with a straight line to produce the graph of $3x - 2y = 12$ in Figure 7.11. ■

Let's review our approach to Example 1. Notice that we did not solve the equation for y in terms of x or for x in terms of y. Since we know it is a straight line, there is no need for any extensive table of values; thus, there is no need to change the form of the original equation. Furthermore, the solution $(2, -3)$ served as a check point. If it had not been on the line determined by the two intercepts, then we would have known that an error had been made.

EXAMPLE 2 Graph $2x + 3y = 7$.

Solution Without showing all of our work, the following table indicates the intercepts and a check point.

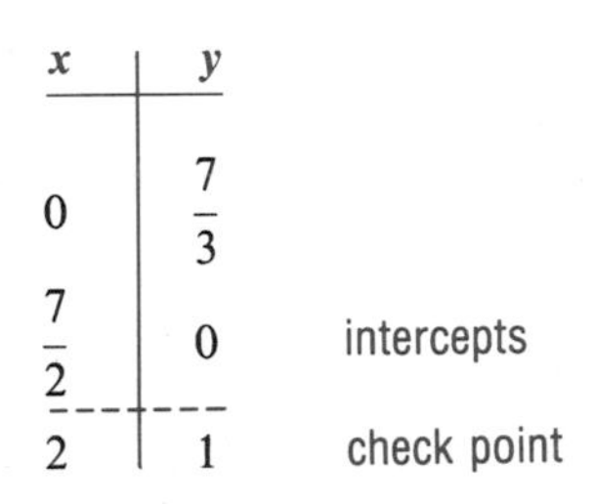

x	y	
0	$\frac{7}{3}$	
$\frac{7}{2}$	0	intercepts
2	1	check point

The points from the table are plotted and the graph of $2x + 3y = 7$ is shown in Figure 7.12.

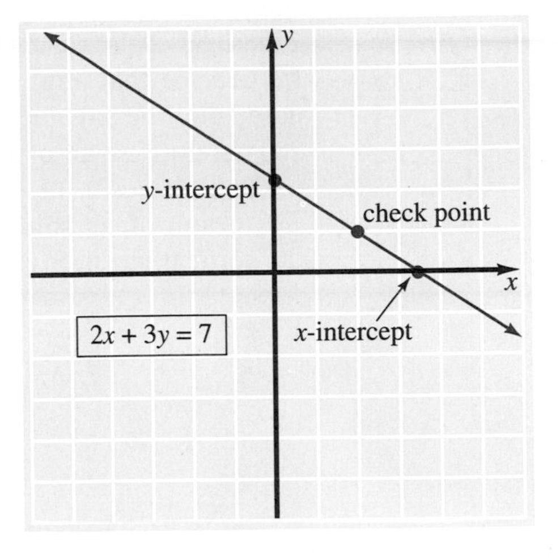

Figure 7.12 ■

It is helpful to recognize some *special* straight lines. For example, the graph of any equation of the form $Ax + By = C$ where $C = 0$ (the constant term is zero) is a straight line that contains the origin. Let's consider an example.

EXAMPLE 3 Graph $y = 2x$.

Solution Obviously $(0, 0)$ is a solution. (Also, notice that $y = 2x$ is equivalent to $-2x + y = 0$; thus, it fits the condition $Ax + By = C$, where $C = 0$.) Since both the x-intercept and y-intercept are determined by the point $(0, 0)$, another point is necessary to determine the line. Then a third point should be found as a check point.

x	y	
0	0	intercepts
2	4	additional point
-1	-2	check point

The graph of $y = 2x$ is shown in Figure 7.13.

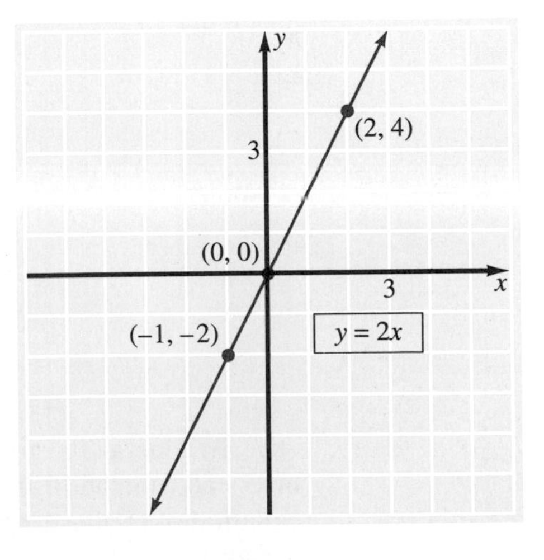

Figure 7.13 ■

EXAMPLE 4 Graph $x = 2$.

Solution Since we are considering linear equations in *two variables*, the equation $x = 2$ is equivalent to $x + 0(y) = 2$. Now we can see that any value of y can be used, but the x-value must always be 2. Therefore, some of the solutions are $(2, 0)$, $(2, 1)$, $(2, 2)$, $(2, -1)$, and $(2, -2)$. The graph of all solutions of $x = 2$ is the vertical line in Figure 7.14.

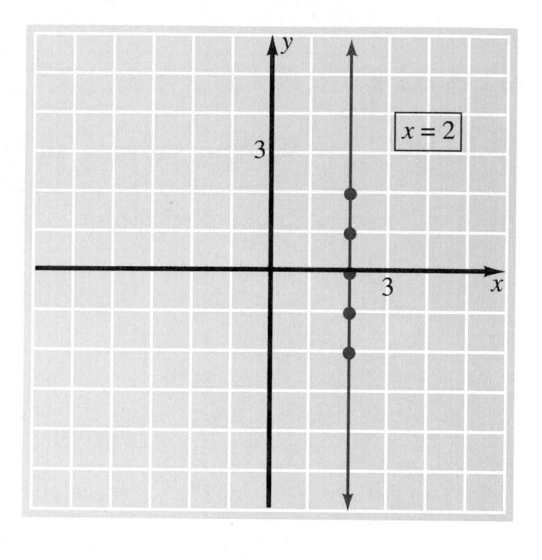

Figure 7.14 ■

EXAMPLE 5 Graph $y = -3$.

Solution The equation $y = -3$ is equivalent to $0(x) + y = -3$. Thus, any value of x can be used, but the value of y must be -3. Some solutions are $(0, -3)$, $(1, -3)$, $(2, -3)$, $(-1, -3)$, and $(-2, -3)$. The graph of $y = -3$ is the horizontal line in Figure 7.15.

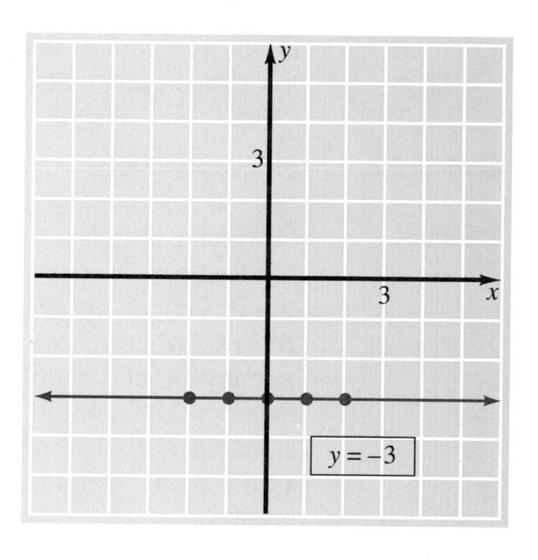

Figure 7.15 ■

In general, the graph of any equation of the form $Ax + By = C$ where $A = 0$ or $B = 0$ (not both) is a line parallel to one of the axes. More specifically, any equation of the form $x = a$, where a is a constant, is a line parallel to the y-axis that has an x-intercept of a. Any equation of the form $y = b$, where b is a constant, is a line parallel to the x-axis that has a y-intercept of b.

REMARK Even though it is not particularly helpful for graphing straight lines, you should recognize that the equations in Examples 3, 4, and 5 do exhibit symmetry. In Example 3, the equation $y = 2x$ exhibits origin symmetry. Likewise, in Example 4 the equation $x = 2$ exhibits x-axis symmetry and in Example 5 the equation $y = -3$ exhibits y-axis symmetry.

Problem Set 7.2

Graph each of the following linear equations.

1. $x + 2y = 4$ see page A41

2. $2x + y = 6$ see page A41

3. $2x - y = 2$ see page A41

4. $3x - y = 3$ see page A42

5. $3x + 2y = 6$ see page A42

6. $2x + 3y = 6$ see page A42

7. $5x - 4y = 20$ see page A42

8. $4x - 3y = -12$ see page A42

9. $x + 4y = -6$ see page A42

10. $5x + y = -2$ see page A42

11. $-x - 2y = 3$ see page A42

12. $-3x - 2y = 12$ see page A42

13. $y = x + 3$ see page A42

14. $y = x - 1$ see page A42

15. $y = -2x - 1$ see page A42

16. $y = 4x + 3$ see page A43

17. $y = \frac{1}{2}x + \frac{2}{3}$ see page A43

18. $y = \frac{2}{3}x - \frac{3}{4}$ see page A43

19. $y = -x$ see page A43

20. $y = x$ see page A43

21. $y = 3x$ see page A43

22. $y = -4x$ see page A43

23. $x = 2y - 1$ see page A43

24. $x = -3y + 2$ see page A43

25. $y = -\frac{1}{4}x + \frac{1}{6}$ see page A43

26. $y = -\frac{1}{2}x - \frac{1}{2}$ see page A43

27. $2x - 3y = 0$ see page A43

28. $3x + 4y = 0$ see page A44

29. $x = 0$ see page A44

30. $y = 0$ see page A44

31. $y = 2$ see page A44

32. $x = -3$ see page A44

33. $-3y = -x + 3$ see page A44

34. The equation $C = \frac{5}{9}(F - 32)$ can be used to convert degrees in Fahrenheit to degrees in Celsius. Let the horizontal axis represent F and the vertical axis, C, and graph the equation $C = \frac{5}{9}(F - 32)$. see page A44

35. Suppose that the daily profit from a small pizza stand is given by the equation $p = 2n - 4$ where n represents the number of pizzas sold in a day and p represents the number of dollars of profit. Label the horizontal axis n and the vertical axis p and graph the equation $p = 2n - 4$ for nonnegative values of n. see page A44

36. The cost (c) of producing n plastic toys per day is given by the equation $c = 3n + 5$. Label the horizontal axis n and the vertical axis c and graph the equation for nonnegative values of n. see page A44

Miscellaneous Problems

From our work with absolute value we know that $|x + y| = 1$ is equivalent to $x + y = 1$ or $x + y = -1$. Therefore, the graph of $|x + y| = 1$ consists of the two lines $x + y = 1$ and $x + y = -1$. Graph each of the following.

37. $|x + y| = 1$ see page A44

38. $|x - y| = 4$ see page A44

39. $|2x - y| = 4$ see page A44

40. $|3x + 2y| = 6$ see page A45

7.3 Linear Inequalities

Linear inequalities in two variables are of the form $Ax + By > C$ or $Ax + By < C$, where A, B, and C are real numbers. (Combined linear equality and inequality statements are of the form $Ax + By \geq C$ or $Ax + By \leq C$.)

Graphing linear inequalities is almost as easy as graphing linear equations. The following discussion leads into a simple, step-by-step process. Let's consider the following equation and related inequalities.

$$x + y = 2,$$

$$x + y > 2,$$

$$x + y < 2$$

The graph of $x + y = 2$ is shown in Figure 7.16. The line divides the plane into two half-planes, one above the line and one below the line. In Figure 7.17(a) we have indicated several points in the half-plane above the line. Notice that for each point, the ordered pair of real numbers satisfies the inequality $x + y > 2$. This is true for *all points* in the half-plane above the line. Therefore, the graph of $x + y > 2$ is the half-plane above the line as indicated by the shaded portion in Figure 7.17(b). We use a dashed line to indicate that points on the line do not satisfy $x + y > 2$. We would use a solid line if we were graphing $x + y \geq 2$.

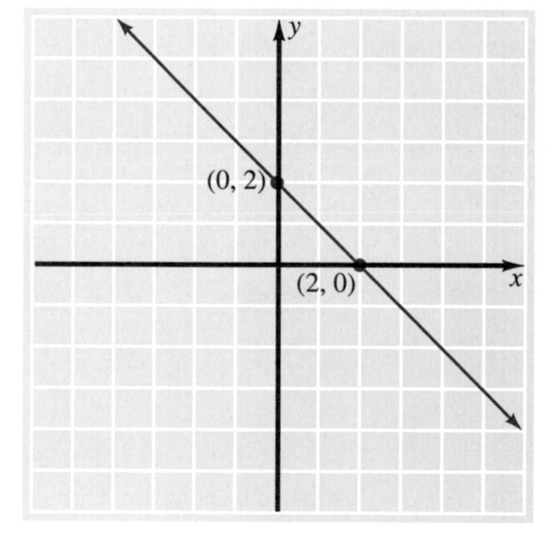

Figure 7.16

In Figure 7.18(a) several points have been indicated in the half-plane below the line $x + y = 2$. Notice that for each point the ordered pair of real numbers satisfies the inequality $x + y < 2$. This is true for *all points* in the half-plane below the line. Thus, the graph of $x + y < 2$ is the half-plane below the line as indicated in Figure 7.18(b).

Figure 7.17

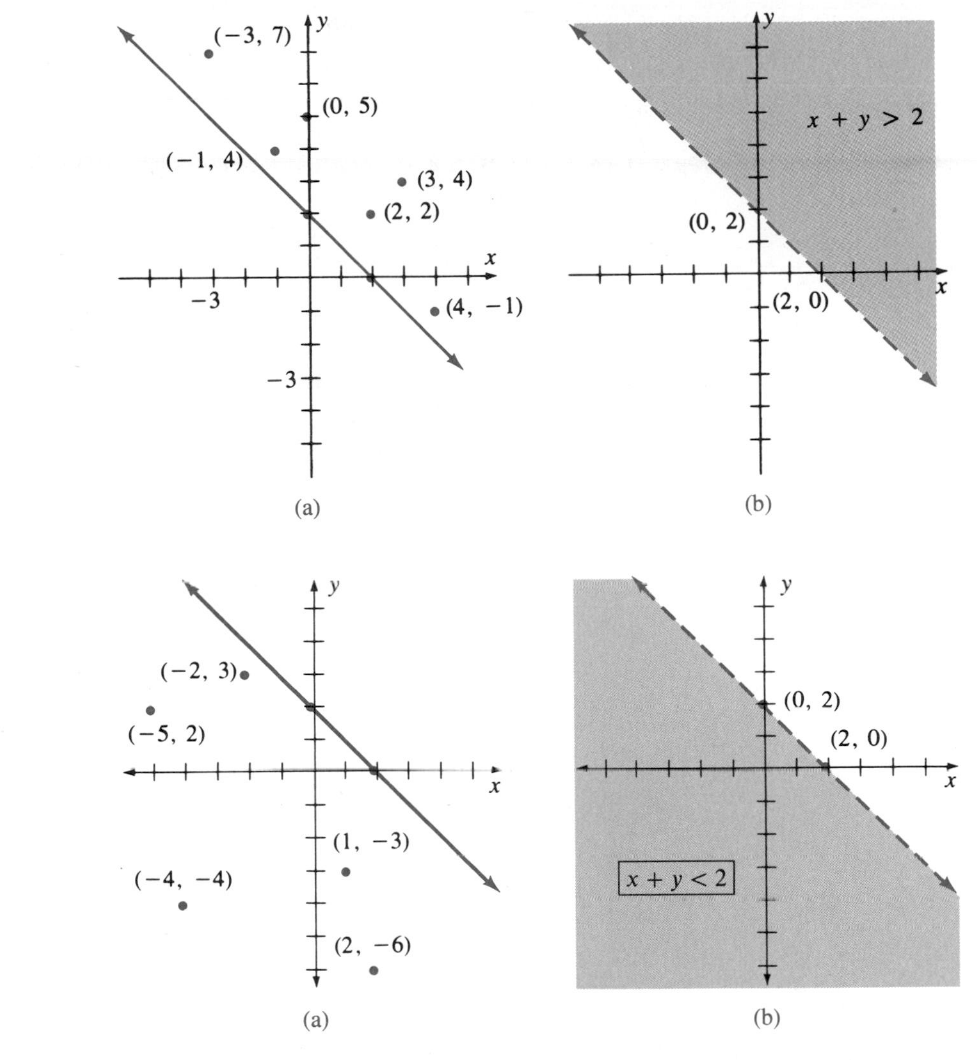

Figure 7.18

To graph a linear inequality we suggest the following steps.

1. First, graph the corresponding equality. Use a solid line if equality is included in the original statement and a dashed line if equality is not included.
2. Choose a "test point" not on the line and substitute its coordinates into the inequality. (The origin is a convenient point to use if it is not on the line.)
3. The graph of the original inequality is
   (a) the half-plane that contains the test point if the inequality is satisfied by that point, or,
   (b) the half-plane that does not contain the test point if the inequality is not satisfied by the point.

Let's apply these steps to some examples.

EXAMPLE 1 Graph $x - 2y > 4$.

Solution

Step 1. Graph $x - 2y = 4$ as a dashed line since equality is not included in $x - 2y > 4$ (Figure 7.19(a)).

Step 2. Choose the origin as a test point and substitute its coordinates into the inequality.

$$x - 2y > 4 \quad \text{becomes } 0 - 2(0) > 4, \text{ which is false.}$$

Step 3. Since the test point did not satisfy the given inequality, the graph is the half-plane that does not contain the test point. Thus, the graph of $x - 2y > 4$ is the half-plane below the line as indicated in Figure 7.19(b).

Figure 7.19

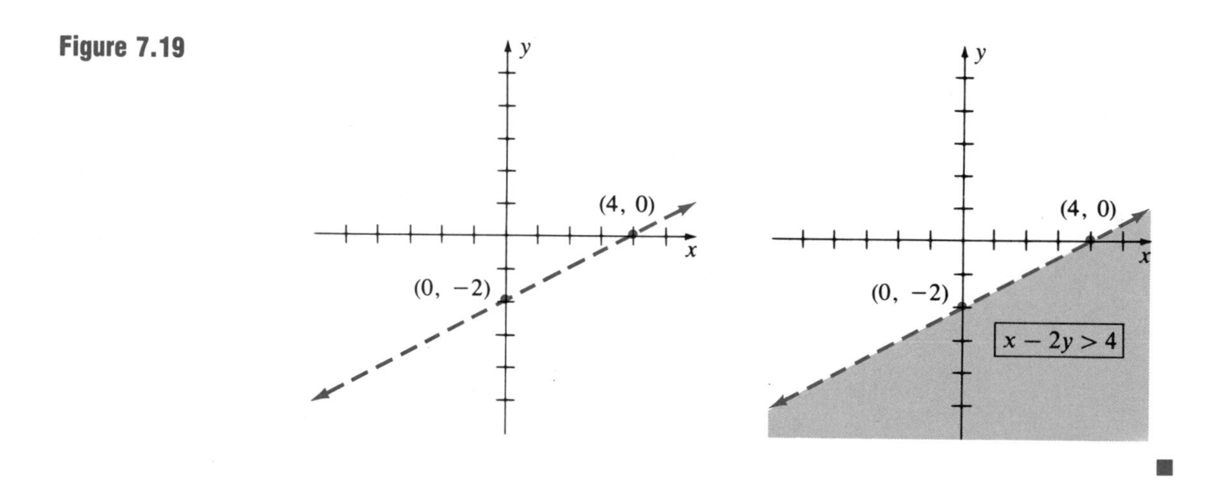

■

EXAMPLE 2 Graph $3x + 2y \leq 6$.

Solution

Step 1. Graph $3x + 2y = 6$ as a solid line since equality is included in $3x + 2y \leq 6$ (Figure 7.20(a)).

Step 2. Choose the origin as a test point and substitute its coordinates into the given statement.

$$3x + 2y \leq 6 \quad \text{becomes } 3(0) + 2(0) \leq 6, \text{ which is true.}$$

Step 3. Since the test point satisfies the given statement, all points in the same half-plane as the test point satisfy the statement. Thus, the graph of $3x + 2y \leq 6$ is the line and the half-plane below the line (Figure 7.20(b)).

Figure 7.20

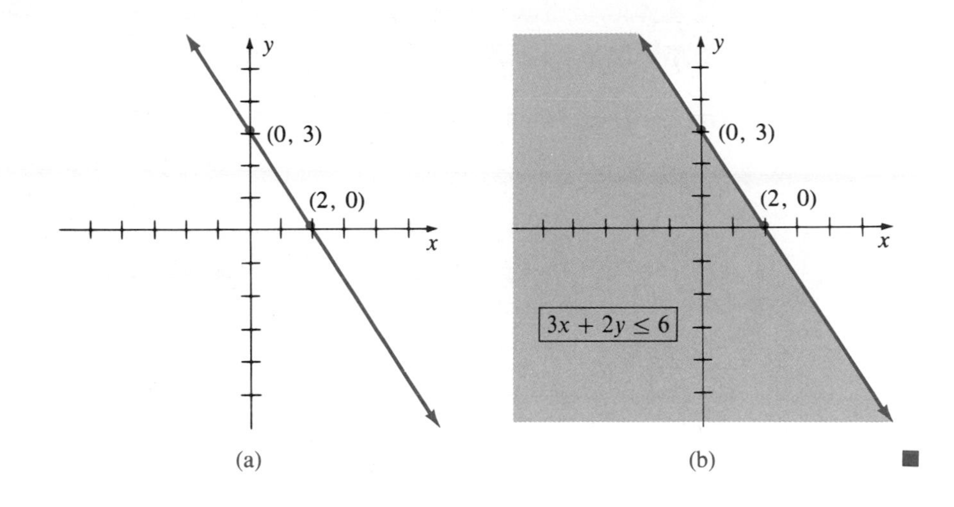

EXAMPLE 3 Graph $y \leq 3x$.

Solution

Step 1. Graph $y = 3x$ as a solid line since equality is included in the statement $y \leq 3x$ (Figure 7.21(a)).

Step 2. Since the origin is on the line we must choose some other point as a test point. Let's try (2, 1).

$y \leq 3x$ becomes $1 \leq 3(2)$, which is a true statement.

Step 3. Since the test point satisfies the given inequality, the graph is the half-plane that contains the test point. Thus, the graph of $y \leq 3x$ is the line, along with the half-plane below the line, as indicated in Figure 7.21(b).

Figure 7.21

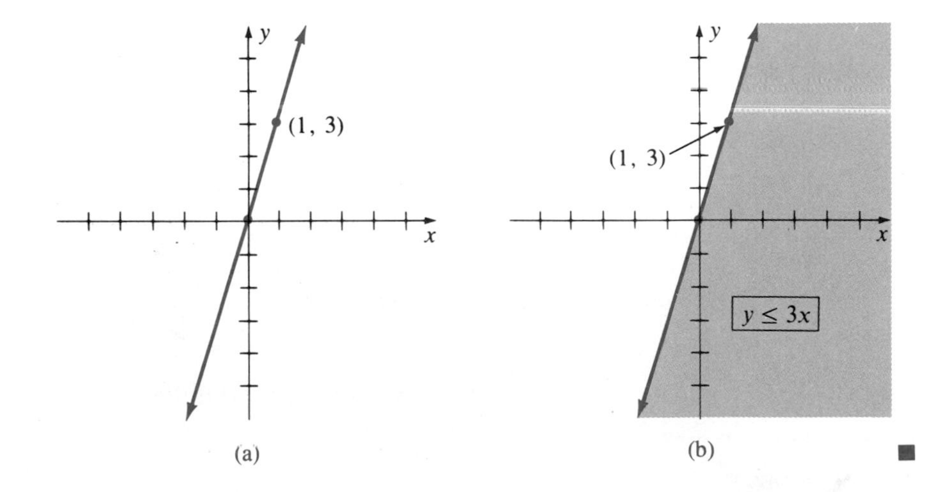

Problem Set 7.3

Graph each of the following inequalities.

1. $x - y > 2$ see page A45

2. $x + y > 4$ see page A45

3. $x + 3y < 3$ see page A45

4. $2x - y > 6$ see page A45

5. $2x + 5y \geq 10$ see page A45

6. $3x + 2y \leq 4$ see page A45

7. $y \leq -x + 2$ see page A45

8. $y \geq -2x - 1$ see page A45

9. $y > -x$ see page A45

10. $y < x$ see page A46

11. $2x - y \geq 0$ see page A46

12. $x + 2y \geq 0$ see page A46

13. $-x + 4y - 4 \leq 0$ see page A46

14. $-2x + y - 3 \leq 0$ see page A46

15. $3x + 2y > -6$ see page A46

16. $2x + 5y > -4$ see page A46

17. $y < -\frac{1}{2}x + 2$ see page A46

18. $y < -\frac{1}{3}x + 1$ see page A46

19. $x \leq 3$ see page A46

20. $y \geq -2$ see page A46

21. $x > 1$ and $y < 3$ see page A46

22. $x > -2$ and $y > -1$ see page A47

23. $x \leq -1$ and $y < 1$ see page A47

24. $x < 2$ and $y \geq -2$ see page A47

Miscellaneous Problems

25. Graph $|x| < 2$. [*Hint:* Remember that $|x| < 2$ is equivalent to $-2 < x < 2$.] see page A47

26. Graph $|y| > 1$. see page A47

27. Graph $|x + y| < 1$. see page A47

28. Graph $|x - y| > 2$. see page A47

7.4 Distance and Slope

As we work with the rectangular coordinate system it is sometimes necessary to express the length of certain line segments. In other words, we need to be able to find the *distance between* two points. Let's first consider two specific examples and then develop the general distance formula.

EXAMPLE 1 Find the distance between the points $A(2, 2)$ and $B(5, 2)$ and also between the points $C(-2, 5)$ and $D(-2, -4)$.

Solution Let's plot the points and draw $\overline{AB}$ as in Figure 7.22. Since $\overline{AB}$ is parallel to the x-axis, its length can be expressed as $|5-2|$ or $|2-5|$. (The absolute value symbol is used to ensure a nonnegative value.) Thus, the length of $\overline{AB}$ is 3 units. Likewise the length of $\overline{CD}$ is $|5-(-4)| = |-4-5| = 9$ units.

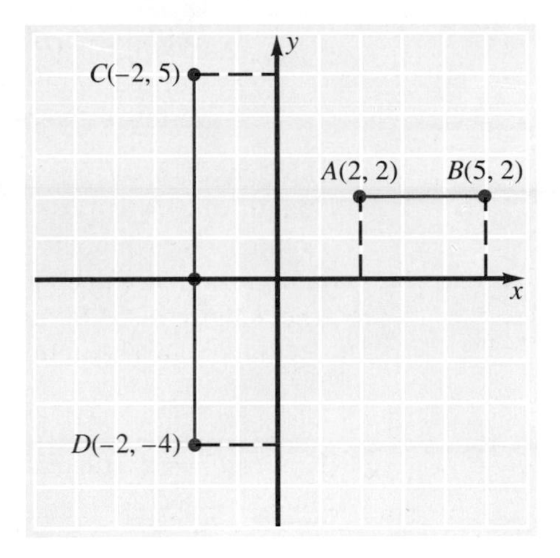

Figure 7.22

EXAMPLE 2 Find the distance between the points $A(2, 3)$ and $B(5, 7)$.

Solution Let's plot the points and form a right triangle as indicated in Figure 7.23. Notice that the coordinates of point C are $(5, 3)$. Since $\overline{AC}$ is parallel to the horizontal axis, its length is easily determined to be 3 units. Likewise, $\overline{CB}$ is parallel to the vertical axis and its length is 4 units. Let d represent the length of $\overline{AB}$ and apply the Pythagorean Theorem to obtain

$$d^2 = 3^2 + 4^2$$

$$d^2 = 9 + 16$$

$$d^2 = 25$$

$$d = \pm\sqrt{25} = \pm 5.$$

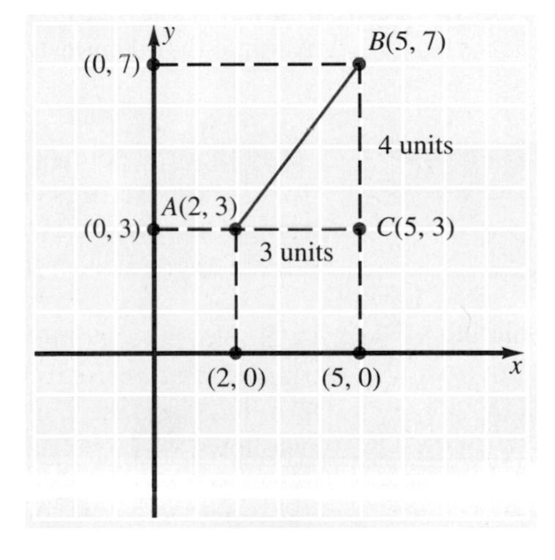

Figure 7.23

Since *distance between* is a nonnegative value, the length of $\overline{AB}$ is 5 units.

We can use the approach in Example 2 to develop a general distance formula for finding the distance between any two points in a coordinate plane. The development proceeds as follows.

1. Let $P_1(x_1, y_1)$ and $P_2(x_2, y_2)$ represent any two points in a coordinate plane;
2. Form a right triangle as indicated in Figure 7.24. The coordinates of the vertex of the right angle, point R, are (x_2, y_1).

The length of $\overline{P_1R}$ is $|x_2 - x_1|$ and the length of $\overline{RP_2}$ is $|y_2 - y_1|$. (The absolute value symbol is used to ensure a nonnegative value.) Let d represent the length of P_1P_2 and apply the Pythagorean Theorem to obtain

$$d^2 = |x_2 - x_1|^2 + |y_2 - y_1|^2.$$

Since $|a|^2 = a^2$, the **distance formula** can be stated as

$$d = \sqrt{(x_2 - x_1)^2 + (y_2 - y_1)^2}.$$

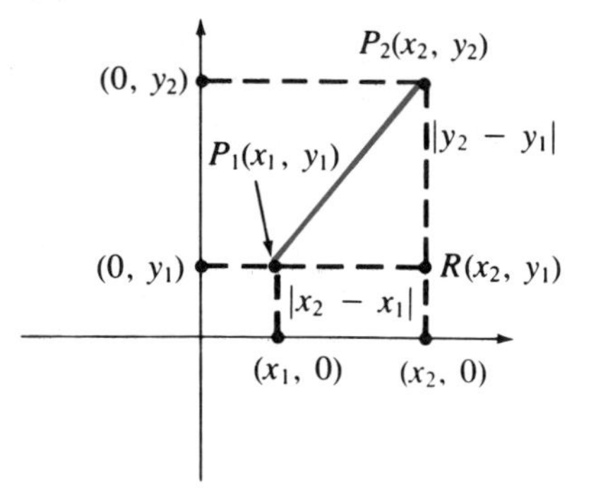

Figure 7.24

It makes no difference which point you call P_1 or P_2 when using the distance formula. Remember, if you forget the formula, don't panic, but merely form a right triangle and apply the Pythagorean Theorem as we did in Example 2. Let's consider an example that demonstrates the use of the distance formula.

EXAMPLE 3 Verify that the points $(-3, 6)$, $(3, 4)$, and $(1, -2)$ are vertices of an isosceles triangle. (An isosceles triangle has two sides of the same length.)

Solution Let's plot the points and draw the triangle (Figure 7.25). Use the distance formula to find the lengths d_1, d_2, and d_3 as follows.

$$d_1 = \sqrt{(3-1)^2 + (4-(-2))^2}$$
$$= \sqrt{2^2 + 6^2} = \sqrt{40} = 2\sqrt{10}$$
$$d_2 = \sqrt{(-3-3)^2 + (6-4)^2}$$
$$= \sqrt{(-6)^2 + 2^2} = \sqrt{40}$$
$$= 2\sqrt{10}$$
$$d_3 = \sqrt{(-3-1)^2 + (6-(-2))^2}$$
$$= \sqrt{(-4)^2 + 8^2} = \sqrt{80} = 4\sqrt{5}$$

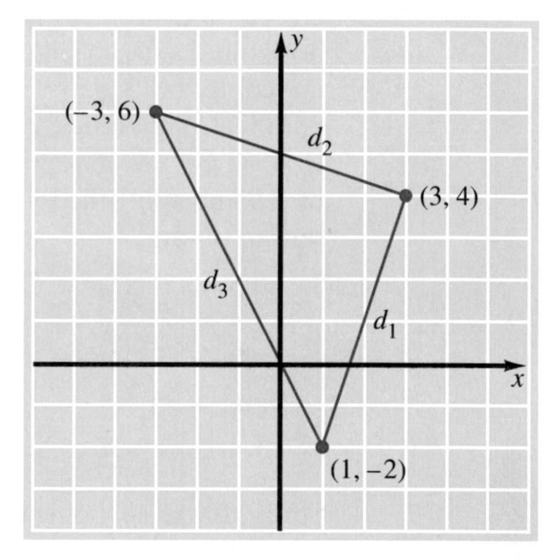

Figure 7.25

Since $d_1 = d_2$, we know that it is an isosceles triangle. ■

Slope of a Line

In coordinate geometry, the concept of **slope** is used to describe the "steepness" of lines. The slope of a line is the ratio of the vertical change compared to the horizontal change as we move from one point on a line to another point. This is illustrated in Figure 7.26 with points P_1 and P_2.

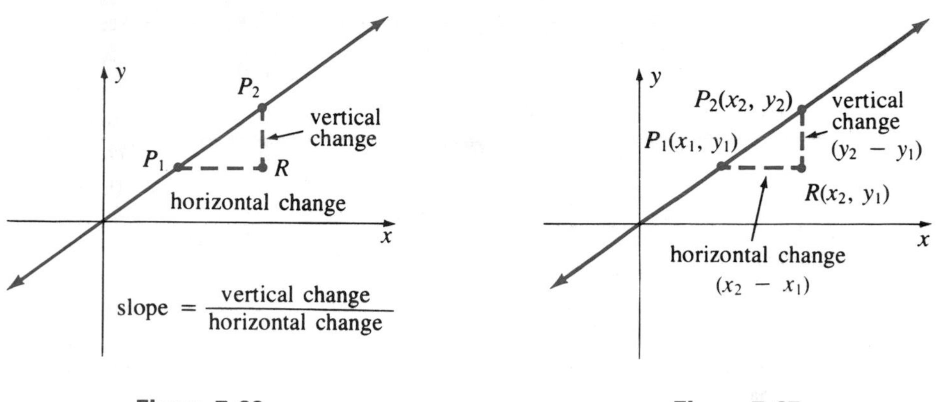

Figure 7.26 **Figure 7.27**

A precise definition for slope can be given by considering the coordinates of the points P_1, P_2, and R as indicated in Figure 7.27. The horizontal change as we move from P_1 to P_2 is $x_2 - x_1$ and the vertical change is $y_2 - y_1$. Thus, the following definition for slope is given.

DEFINITION 7.1

If points P_1 and P_2 with coordinates (x_1, y_1) and (x_2, y_2), respectively, are any two different points on a line, then the slope of the line (denoted by m) is

$$m = \frac{y_2 - y_1}{x_2 - x_1}, \qquad x_2 \neq x_1.$$

Since $\frac{y_2 - y_1}{x_2 - x_1} = \frac{y_1 - y_2}{x_1 - x_2}$, how we designate P_1 and P_2 is not important. Let's use Definition 7.1 to find the slopes of some lines.

EXAMPLE 4 Find the slope of the line determined by each of the following pairs of points and graph the lines.

(a) $(-1, 1)$ and $(3, 2)$ **(b)** $(4, -2)$ and $(-1, 5)$

(c) $(2, -3)$ and $(-3, -3)$

Solution

(a) Let $(-1, 1)$ be P_1 and $(3, 2)$ be P_2 (Figure 7.28).

$$m = \frac{y_2 - y_1}{x_2 - x_1} = \frac{2 - 1}{3 - (-1)} = \frac{1}{4}$$

Figure 7.28

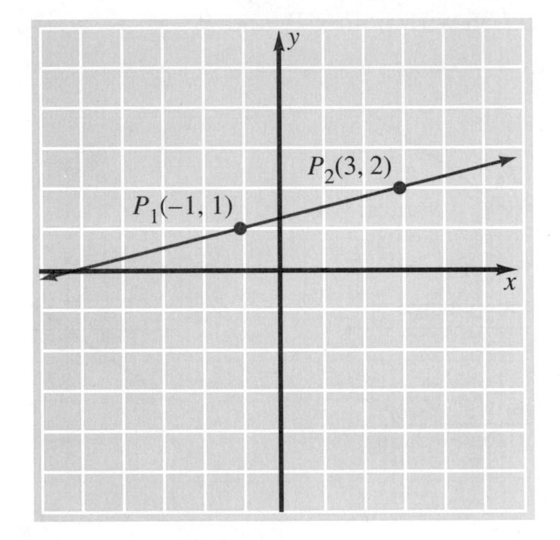

(b) Let $(4, -2)$ be P_1 and $(-1, 5)$ be P_2 (Figure 7.29).

$$m = \frac{y_2 - y_1}{x_2 - x_1} = \frac{5 - (-2)}{-1 - 4} = \frac{7}{-5} = -\frac{7}{5}$$

Figure 7.29

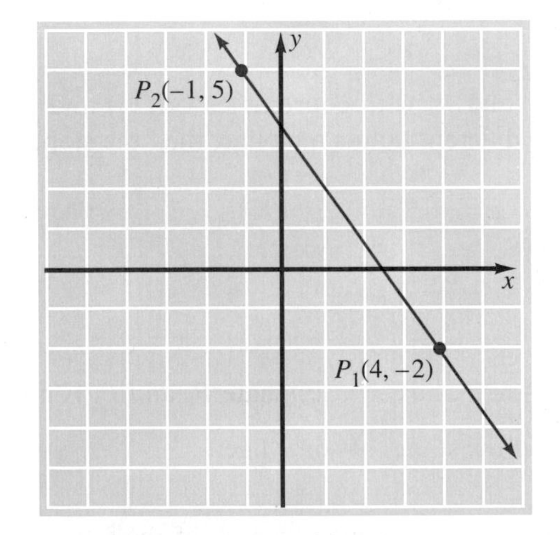

(c) Let $(2, -3)$ be P_1 and $(-3, -3)$ be P_2 (Figure 7.30).

$$m = \frac{y_2 - y_1}{x_2 - x_1} = \frac{-3 - (-3)}{-3 - 2} = \frac{0}{-5} = 0$$

Figure 7.30

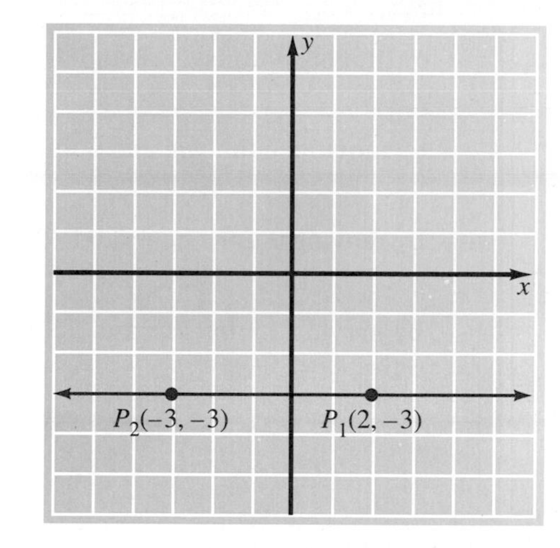

The three parts of Example 4 represent the three basic possibilities for slope; that is, the slope of a line can be positive, negative, or zero. A line that has a positive slope rises as we move from left to right as in Figure 7.28. A line that has a negative slope falls as we move from left to right as in Figure 7.29. A horizontal line, as in Figure 7.30, has a slope of zero. Finally, we need to realize that *the concept of slope is undefined for vertical lines.* This is due to the fact that for any vertical line the horizontal change as we move from one point on the line to another is zero. Thus, the ratio $\frac{y_2 - y_1}{x_2 - x_1}$ will have a denominator of zero and be undefined. So in Definition 7.1 the restriction $x_2 \neq x_1$ is made.

One final idea pertaining to the concept of slope needs to be emphasized. The slope of a line is a **ratio**, the ratio of vertical change compared to horizontal change. A slope of $\frac{2}{3}$ means that for every 2 units of vertical change there must be a corresponding 3 units of horizontal change. Thus, starting at some point on a line that has a slope of $\frac{2}{3}$ we could locate other points on the line as follows.

$\frac{2}{3} = \frac{4}{6}$ $\longrightarrow$ by moving 4 units *up* and 6 units to the *right*,

$\frac{2}{3} = \frac{8}{12}$ $\longrightarrow$ by moving 8 units *up* and 12 units to the *right*,

$\frac{2}{3} = \frac{-2}{-3}$ $\longrightarrow$ by moving 2 units *down* and 3 units to the *left*

Likewise, if a line has a slope of $-\frac{3}{4}$, then by starting at some point on the line we could locate other points on the line as follows.

$-\frac{3}{4} = \frac{-3}{4}$ $\longrightarrow$ by moving 3 units *down* and 4 units to the *right*,

$-\frac{3}{4} = \frac{3}{-4}$ $\longrightarrow$ by moving 3 units *up* and 4 units to the *left*,

$-\frac{3}{4} = \frac{-9}{12}$ $\longrightarrow$ by moving 9 units *down* and 12 units to the *right*,

$-\frac{3}{4} = \frac{15}{-20}$ $\longrightarrow$ by moving 15 units *up* and 20 units to the *left*

Problem Set 7.4

For Problems 1–12, find the distance between each of the pairs of points. Express answers in simplest radical form.

1. $(-2, -1), (7, 11)$ 15
2. $(2, 1), (10, 7)$ 10
3. $(1, -1), (3, -4)$ $\sqrt{13}$
4. $(-1, 3), (2, -2)$ $\sqrt{34}$
5. $(6, -4), (9, -7)$ $3\sqrt{2}$
6. $(-5, 2), (-1, 6)$ $4\sqrt{2}$
7. $(-3, 3), (0, -3)$ $3\sqrt{5}$
8. $(-2, -4), (4, 0)$ $2\sqrt{13}$
9. $(1, -6), (-5, -6)$ 6
10. $(-2, 3), (-2, -7)$ 10
11. $(1, 7), (4, -2)$ $3\sqrt{10}$
12. $(6, 4), (-4, -8)$ $2\sqrt{61}$

13. Verify that the points $(-3, 1)$, $(5, 7)$, and $(8, 3)$ are vertices of a right triangle. [*Hint*: if $a^2 + b^2 = c^2$, then it is a right triangle with the right angle opposite side c.]

14. Verify that the points $(0, 3)$, $(2, -3)$, and $(-4, -5)$ are vertices of an isosceles triangle.

15. Verify that the points $(7, 12)$ and $(11, 18)$ divide the line segment joining $(3, 6)$ and $(15, 24)$ into three segments of equal length.

16. Verify that $(3, 1)$ is the midpoint of the line segment joining $(-2, 6)$ and $(8, -4)$.

For Problems 17–28, graph the line determined by the two points and find the slope of the line.

17. $(1, 2), (4, 6)$ $\frac{4}{3}$
18. $(3, 1), (-2, -2)$ $\frac{3}{5}$
19. $(-4, 5), (-1, -2)$ $-\frac{7}{3}$
20. $(-2, 5), (3, -1)$ $-\frac{6}{5}$
21. $(2, 6), (6, -2)$ -2
22. $(-2, -1), (2, -5)$ -1
23. $(-6, 1), (-1, 4)$ $\frac{3}{5}$
24. $(-3, 3), (2, 3)$ 0
25. $(-2, -4), (2, -4)$ 0
26. $(1, -5), (4, -1)$ $\frac{4}{3}$
27. $(0, -2), (4, 0)$ $\frac{1}{2}$
28. $(-4, 0), (0, -6)$ $-\frac{3}{2}$

29. Find x if the line through $(-2, 4)$ and $(x, 6)$ has a slope of $\frac{2}{9}$. 7

30. Find y if the line through $(1, y)$ and $(4, 2)$ has a slope of $\frac{5}{3}$. -3

31. Find x if the line through $(x, 4)$ and $(2, -5)$ has a slope of $-\frac{9}{4}$. -2

32. Find y if the line through $(5, 2)$ and $(-3, y)$ has a slope of $-\frac{7}{8}$. 9

For Problems 33–40, you are given one point on a line and the slope of the line. Find the coordinates of three other points on the line.

33. $(2, 5)$, $m = \frac{1}{2}$ Answers will vary

34. $(3, 4)$, $m = \frac{5}{6}$ Answers will vary

35. $(-3, 4)$, $m = 3$ Answers will vary

36. $(-3, -6)$, $m = 1$ Answers will vary

37. $(5, -2)$, $m = -\frac{2}{3}$ Answers will vary

38. $(4, -1)$, $m = -\frac{3}{4}$ Answers will vary

39. $(-2, -4)$, $m = -2$ Answers will vary

40. $(-5, 3)$, $m = -3$ Answers will vary

For Problems 41–50, find the coordinates of two points on the given line and then use those coordinates to find the slope of the line.

41. $2x + 3y = 6$ $-\frac{2}{3}$

42. $4x + 5y = 20$ $-\frac{4}{5}$

43. $x - 2y = 4$ $\frac{1}{2}$

44. $3x - y = 12$ 3

45. $4x - 7y = 12$ $\frac{4}{7}$

46. $2x + 7y = 11$ $-\frac{2}{7}$

47. $y = 4$ 0

48. $x = 3$ undefined

49. $y = -5x$ -5

50. $y - 6x = 0$ 6

Miscellaneous Problems

51. The concept of slope is used for highway construction. The "grade" of a highway expressed as a percent means the number of feet that the highway changes in elevation for each 100 feet of horizontal change.

 (a) A certain highway has a 2% grade. How many feet does it rise in a horizontal distance of 1 mile? (1 mile = 5280 feet.) 105.6 feet

 (b) The grade of a highway up a hill is 30%. How much change in horizontal distance is there if the vertical height of the hill is 75 feet? 250 feet

52. Slope is often expressed as the ratio "rise to run" in construction of steps.

 (a) If the ratio "rise to run" is to be $\frac{3}{5}$ for some steps and the "rise" is 19 centimeters, find the measure of the "run" to the nearest centimeter. 32 centimeters

 (b) If the ratio "rise to run" is to be $\frac{2}{3}$ for some steps and the "run" is 28 centimeters, find the "rise" to the nearest centimeter. 19 centimeters

53. Suppose that county ordinance requires a $2\frac{1}{4}\%$ "fall" for a sewage pipe from the house to the main pipe at the street. How much vertical drop must there be for a horizontal distance of 45 feet? Express the answer to the nearest tenth of a foot. 1.0 feet

7.5

Determining the Equation of a Line

To review, there are basically two types of problems to solve in coordinate geometry:

1. Given an algebraic equation, find its geometric graph;
2. Given a set of conditions pertaining to a geometric figure, find its algebraic equation.

Problems of type 1 have been our primary concern thus far in this chapter. Now let's analyze some problems of type 2 that deal specifically with straight lines. In other words, given certain facts about a line we need to be able to determine its algebraic equation. Let's consider some examples.

EXAMPLE 1 Find the equation of the line that has a slope of $\frac{2}{3}$ and contains the point $(1, 2)$.

Solution First, let's draw the line and record the given information. Then choose a point (x, y) that represents any point on the line other than the given point $(1, 2)$. (See Figure 7.31.) The slope determined by $(1, 2)$ and (x, y) is $\frac{2}{3}$. Thus,

$$\frac{y-2}{x-1} = \frac{2}{3}$$

$$2(x-1) = 3(y-2)$$

$$2x - 2 = 3y - 6$$

$$2x - 3y = -4.$$

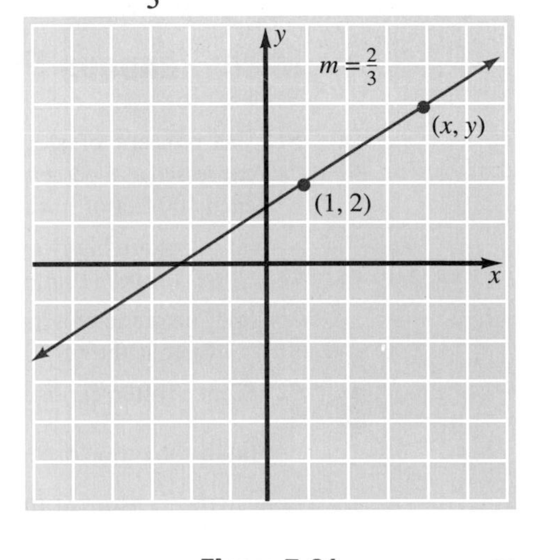

Figure 7.31 ■

EXAMPLE 2 Find the equation of the line that contains $(3, 2)$ and $(-2, 5)$.

Solution First, let's draw the line determined by the given points (Figure 7.32). Since if we know two points, we can find the slope.

$$m = \frac{y_2 - y_1}{x_2 - x_1} = \frac{3}{-5} = -\frac{3}{5}$$

Now we can use the same approach as in Example 1. Form an equation using a variable point (x, y), one of the two given points, and the slope of $-\frac{3}{5}$.

$$\frac{y-5}{x+2} = \frac{3}{-5} \qquad \left(-\frac{3}{5} = \frac{3}{-5}\right)$$

$$3(x+2) = -5(y-5)$$

$$3x + 6 = -5y + 25$$

$$3x + 5y = 19$$

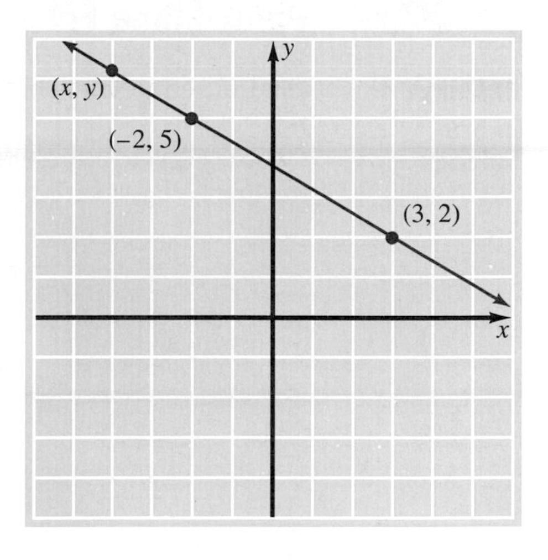

Figure 7.32 ■

EXAMPLE 3 Find the equation of the line that has a slope of $\frac{1}{4}$ and a y-intercept of 2.

Solution A y-intercept of 2 means that the point $(0, 2)$ is on the line (Figure 7.33). Choose a variable point (x, y) and proceed as in the previous examples.

$$\frac{y-2}{x-0} = \frac{1}{4}$$

$$1(x-0) = 4(y-2)$$

$$x = 4y - 8$$

$$x - 4y = -8$$

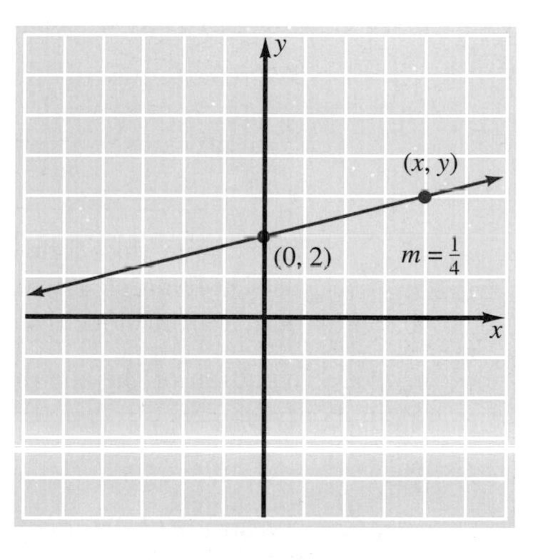

Figure 7.33 ■

Perhaps it would be helpful to pause a moment and look back over Examples 1, 2, and 3. Notice that we used the same basic approach in all three situations; that is, choose a variable point, (x, y), and use it to determine the equation that satisfies the conditions given in the problem. We should also recognize that the approach we took in the previous examples can be generalized to produce some special forms of equations of straight lines.

Point-Slope Form

EXAMPLE 4 Find the equation of the line that has a slope of m and contains the point (x_1, y_1).

Solution Choose (x, y) to represent any other point on the line (Figure 7.34) and the slope of the line is therefore given by

$$m = \frac{y - y_1}{x - x_1}, \qquad x \neq x_1$$

from which

$$y - y_1 = m(x - x_1).$$

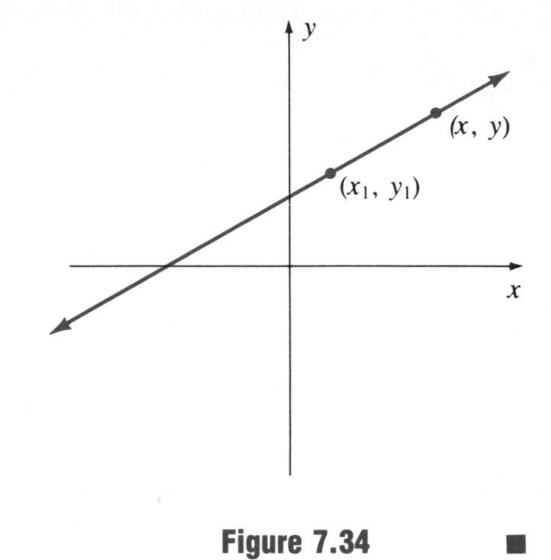

Figure 7.34 ■

We refer to the equation

$$\boxed{y - y_1 = m(x - x_1)}$$

as the **point-slope form** of the equation of a straight line. Instead of the approach we used in Example 1, we could use the point-slope form to write the equation of a line with a given slope that contains a given point. For example, we can determine the equation of the line that has a slope of $\frac{3}{5}$ and contains the point $(2, 4)$ as follows.

$$y - y_1 = m(x - x_1)$$

Substitute $(2, 4)$ for (x_1, y_1) and $\frac{3}{5}$ for m.

$$y - 4 = \frac{3}{5}(x - 2)$$

$$5(y - 4) = 3(x - 2)$$

$$5y - 20 = 3x - 6$$

$$-14 = 3x - 5y$$

Slope-Intercept Form

EXAMPLE 5 Find the equation of the line that has a slope of m and a y-intercept of b.

Solution A y-intercept of b means that the line contains the point $(0, b)$, as in Figure 7.35. Therefore, we can use the point-slope form as follows.

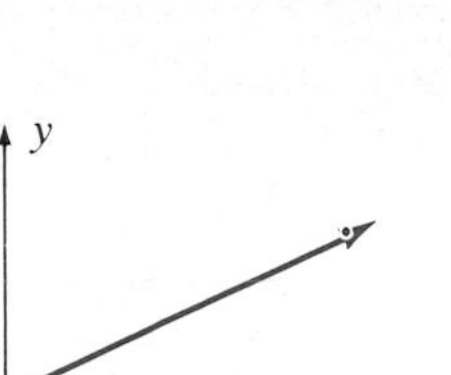

Figure 7.35

$$y - y_1 = m(x - x_1)$$
$$y - b = m(x - 0)$$
$$y - b = mx$$
$$y = mx + b$$

■

We refer to the equation

$$\boxed{y = mx + b}$$

as the **slope-intercept form** of the equation of a straight line. We use it for two primary purposes, as the next two examples illustrate.

EXAMPLE 6 Find the equation of the line that has a slope of $\frac{1}{4}$ and a y-intercept of 2.

Solution This is a restatement of Example 3, but this time we will use the slope-intercept form ($y = mx + b$) of a line to write its equation. Since $m = \frac{1}{4}$ and $b = 2$, we can substitute these values into $y = mx + b$.

$$y = mx + b$$
$$y = \frac{1}{4}x + 2$$
$$4y = x + 8 \qquad \text{Multiply both sides by 4.}$$
$$x - 4y = -8 \qquad \text{same result as in Example 3}$$

■

EXAMPLE 7 Find the slope of the line when the equation is $3x + 2y = 6$.

Solution We can solve the equation for y in terms of x and then compare it to the slope-intercept form to determine its slope. Thus,

$$3x + 2y = 6$$
$$2y = -3x + 6$$
$$y = -\frac{3}{2}x + 3.$$

$$y = -\frac{3}{2}x + 3 \qquad y = mx + b.$$

The slope of the line is $-\frac{3}{2}$. Furthermore, the y-intercept is 3. ■

In general, if the equation of a nonvertical line is written in slope-intercept form ($y = mx + b$), the coefficient of x is the slope of the line and the constant term is the y-intercept. (Remember that the concept of slope is not defined for a vertical line.)

Parallel and Perpendicular Lines

We can use two important relationships between lines and their slopes to solve certain kinds of problems. It can be shown that nonvertical parallel lines have the same slope, and that two nonvertical lines are perpendicular if the product of their slopes is -1. (Details for verifying these facts are left to another course.) In other words, if two lines have slopes m_1 and m_2, respectively, then:

1. The two lines are parallel if and only if $m_1 = m_2$;
2. The two lines are perpendicular if and only if $(m_1)(m_2) = -1$.

The following examples demonstrate the use of these properties.

EXAMPLE 8

(a) Verify that the graphs of $2x + 3y = 7$ and $4x + 6y = 11$ are parallel lines.

(b) Verify that the graphs of $8x - 12y = 3$ and $3x + 2y = 2$ are perpendicular lines.

Solution

(a) Let's change each equation to slope-intercept form.

$$2x + 3y = 7 \longrightarrow 3y = -2x + 7$$
$$y = -\frac{2}{3}x + \frac{7}{3}$$

$$4x + 6y = 11 \longrightarrow 6y = -4x + 11$$

$$y = -\frac{4}{6}x + \frac{11}{6}$$

$$y = -\frac{2}{3}x + \frac{11}{6}$$

Both lines have a slope of $-\frac{2}{3}$ and different y-intercepts. Therefore, the two lines are parallel.

(b) Solving each equation for y in terms of x, we obtain

$$8x - 12y = 3 \longrightarrow -12y = -8x + 3$$

$$y = \frac{8}{12}x - \frac{3}{12}$$

$$y = \frac{2}{3}x - \frac{1}{4}$$

$$3x + 2y = 2 \longrightarrow 2y = -3x + 2$$

$$y = -\frac{3}{2}x + 1.$$

Because $\left(\frac{2}{3}\right)\left(-\frac{3}{2}\right) = -1$ (the product of the two slopes is -1), the lines are perpendicular. ■

REMARK The statement "the product of two slopes is -1" is equivalent to saying that the two slopes are negative reciprocals of each other; that is, $m_1 = -\frac{1}{m_2}$.

EXAMPLE 9 Find the equation of the line that contains the point $(1, 4)$ and is parallel to the line determined by $x + 2y = 5$.

Solution First, let's draw a figure to help in our analysis of the problem (Figure 7.36). Since the line through $(1, 4)$ is to be parallel to the line determined by $x + 2y = 5$, it must have the same slope. So let's find the slope by changing $x + 2y = 5$ to the slope-intercept form.

$$x + 2y = 5$$

$$2y = -x + 5$$

$$y = -\frac{1}{2}x + \frac{5}{2}$$

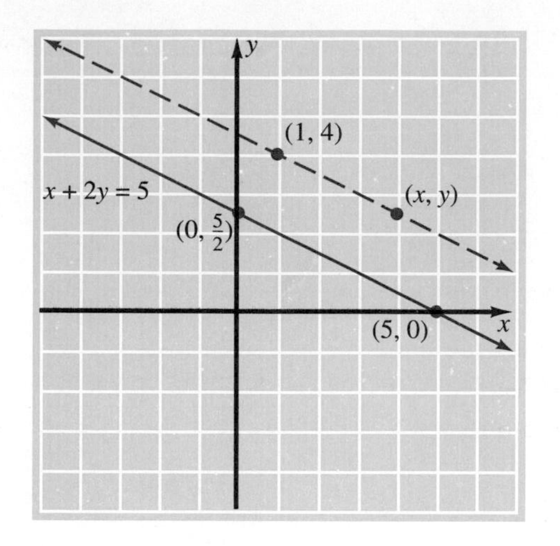

Figure 7.36

The slope of both lines is $-\frac{1}{2}$. Now we can choose a variable point (x, y) on the line through $(1, 4)$ and proceed as we did in earlier examples.

$$\frac{y-4}{x-1} = \frac{1}{-2}$$

$$1(x-1) = -2(y-4)$$

$$x - 1 = -2y + 8$$

$$x + 2y = 9$$

EXAMPLE 10 Find the equation of the line that contains the point $(-1, -2)$ and is perpendicular to the line determined by $2x - y = 6$.

Solution First, let's draw a figure to help in our analysis of the problem (Figure 7.37). Because the line through $(-1, -2)$ is to be perpendicular to the line determined by $2x - y = 6$, its slope must be the negative reciprocal of the slope of $2x - y = 6$. So let's find the slope of $2x - y = 6$ by changing it to the slope-intercept form.

$$2x - y = 6$$

$$-y = -2x + 6$$

$$y = 2x - 6 \qquad \text{The slope is 2.}$$

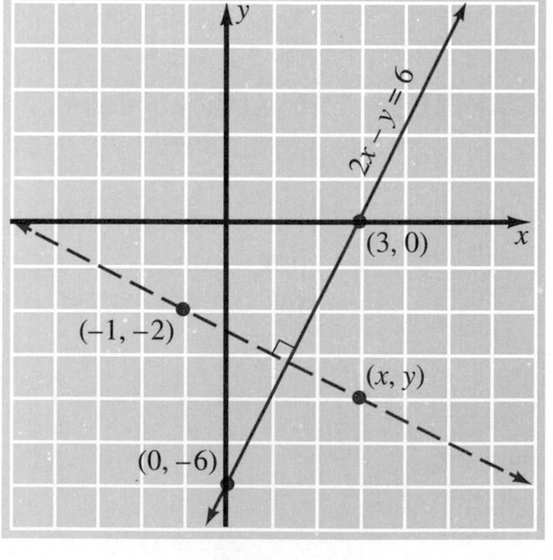

Figure 7.37

The slope of the desired line is $-\frac{1}{2}$ (the negative reciprocal of 2) and we can proceed as before by using a variable point (x, y).

$$\frac{y+2}{x+1} = \frac{1}{-2}$$

$$1(x+1) = -2(y+2)$$

$$x+1 = -2y-4$$

$$x+2y = -5$$

■

We use two forms of equations of straight lines extensively. They are the **standard form** and the **slope-intercept form** and we describe them as follows.

Standard Form $Ax + By = C$, where B and C are integers and A is a nonnegative integer (A and B not both zero).

Slope-Intercept Form $y = mx + b$ where m is a real number representing the slope and b is a real number representing the y-intercept.

Problem Set 7.5

For Problems 1–8, write the equation of each line that has the indicated slope and contains the indicated point. Express final equations in standard form.

1. $m = \frac{1}{2}$, $(3, 5)$ $x - 2y = -7$

2. $m = \frac{1}{3}$, $(2, 3)$ $x - 3y = -7$

3. $m = 3$, $(-2, 4)$ $3x - y = -10$

4. $m = -2$, $(-1, 6)$ $2x + y = 4$

5. $m = -\frac{3}{4}$, $(-1, -3)$ $3x + 4y = -15$

6. $m = -\frac{3}{5}$, $(-2, -4)$ $3x + 5y = -26$

7. $m = \frac{5}{4}$, $(4, -2)$ $5x - 4y = 28$

8. $m = \frac{3}{2}$, $(8, -2)$ $3x - 2y = 28$

For Problems 9–18, write the equation of each line that contains the indicated pair of points. Express final equations in standard form.

9. $(2, 1), (6, 5)$ $x - y = 1$

10. $(-1, 2), (2, 5)$ $x - y = -3$

11. $(-2, -3), (2, 7)$ $5x - 2y = -4$

12. $(-3, -4), (1, 2)$ $3x - 2y = -1$

13. $(-3, 2), (4, 1)$ $x + 7y = 11$

14. $(-2, 5), (3, -3)$ $8x + 5y = 9$

15. $(-1, -4), (3, -6)$ $x + 2y = -9$

16. $(3, 8), (7, 2)$ $3x + 2y = 25$

17. $(0, 0), (5, 7)$ $7x - 5y = 0$

18. $(0, 0), (-5, 9)$ $9x + 5y = 0$

For Problems 19–26, write the equation of each line that has the indicated slope (m) and y-intercept (b). Express final equations in slope-intercept form.

19. $m = \frac{3}{7}$, $b = 4$ $y = \frac{3}{7}x + 4$

20. $m = \frac{2}{9}$, $b = 6$ $y = \frac{2}{9}x + 6$

21. $m = 2$, $b = -3$ $y = 2x - 3$

22. $m = -3$, $b = -1$ $y = -3x - 1$

23. $m = -\frac{2}{5}, \quad b = 1$ $y = -\frac{2}{5}x + 1$

24. $m = -\frac{3}{7}, \quad b = 4$ $y = -\frac{3}{7}x + 4$

25. $m = 0, \quad b = -4$ $y = 0(x) - 4$

26. $m = \frac{1}{5}, \quad b = 0$ $y = \frac{1}{5}x + 0$

For Problems 27–42, write the equation of each line that satisfies the given conditions. Express final equations in standard form.

27. x-intercept of 2 and y-intercept of -4 $2x - y = 4$

28. x-intercept of -1 and y-intercept of -3 $3x + y = -3$

29. x-intercept of -3 and slope of $-\frac{5}{8}$ $5x + 8y = -15$

30. x-intercept of 5 and slope of $-\frac{3}{10}$ $3x + 10y = 15$

31. contains the point $(2, -4)$ and is parallel to the y-axis $x + 0(y) = 2$

32. contains the point $(-3, -7)$ and is parallel to the x-axis $0(x) + y = -7$

33. contains the point $(5, 6)$ and is perpendicular to the y-axis $0(x) + y = 6$

34. contains the point $(-4, 7)$ and is perpendicular to the x-axis $x + 0(y) = -4$

35. contains the point $(1, 3)$ and is parallel to the line $x + 5y = 9$ $x + 5y = 16$

36. contains the point $(-1, 4)$ and is parallel to the line $x - 2y = 6$ $x - 2y = -9$

37. contains the origin and is parallel to the line $4x - 7y = 3$ $4x - 7y = 0$

38. contains the origin and is parallel to the line $-2x - 9y = 4$ $2x + 9y = 0$

39. contains the point $(-1, 3)$ and is perpendicular to the line $2x - y = 4$ $x + 2y = 5$

40. contains the point $(-2, -3)$ and is perpendicular to the line $x + 4y = 6$ $4x - y = -5$

41. contains the origin and is perpendicular to the line $-2x + 3y = 8$ $3x + 2y = 0$

42. contains the origin and is perpendicular to the line $y = -5x$ $x - 5y = 0$

43. We can also use the slope-intercept form of a line for graphing purposes. Suppose that we want to graph the equation $y = \frac{1}{4}x + 2$. Since the y-intercept is 2, the point $(0, 2)$ is on the line. Furthermore, since the slope is $\frac{1}{4}$ we can locate another point by moving 1 unit *up* and 4 units to the *right*. Thus, the point $(4,3)$ is also on the line. The two points $(0, 2)$ and $(4, 3)$ determine the line.

Use the slope-intercept form to help graph the following lines.

(a) $y = \frac{3}{4}x + 1$ see page A48

(b) $y = \frac{2}{3}x - 2$ see page A48

(c) $y = -\frac{2}{5}x - 1$ see page A48

(d) $y = -\frac{1}{2}x + 3$ see page A48

(e) $x + 2y = 5$ see page A48

(f) $2x - y = 7$ see page A48

(g) $-y = -4x + 7$ see page A48

(h) $3x = 2y$ see page A48

(i) $7y = -2x$ see page A48

(j) $y = -3$ see page A49

(k) $x = 2$ see page A49

44. Some real-world situations can be described by the use of linear equations in two variables. If two pairs of values are known, then we can determine the equation by using the approach we used in Example 2 of this section. For each of the following, assume that the relationship can be expressed as a linear equation in two variables, and use the given information to determine the equation. Express the equation in standard form.

(a) A company produces 10 fiberglass shower stalls for \$2015 and 15 stalls for \$3015. Let y be the cost and x the number of stalls. $200x - y = -15$

(b) A company can produce 6 boxes of candy for \$8 and 10 boxes of candy for \$13. Let y represent the cost and x the number of boxes of candy. $5x - 4y = -2$

(c) Two banks on opposite corners of a town square had signs that displayed the current temperature. One bank displayed the temperature in Celsius degrees and the other in Fahrenheit. A temperature of 10°C was displayed at the same time as a temperature of 50°F. On another day, a temperature of -5°C was displayed at the same time as a temperature of 23°F. Let y represent the temperature in Fahrenheit and x the temperature in Celsius. $9x - 5y = -160$

Miscellaneous Problems

45. The equation of a line that contains the two points (x_1, y_1) and (x_2, y_2) is $\frac{y - y_1}{x - x_1} = \frac{y_2 - y_1}{x_2 - x_1}$. We often refer to this as the **two-point form** of the equation of a straight line. Use the two-point form and write the equation of each of the lines that contains the indicated pair of points. Express final equations in standard form.

(a) $(1, 1)$ and $(5, 2)$ $x - 4y = -3$

(b) $(2, 4)$ and $(-2, -1)$ $5x - 4y = -6$

(c) $(-3, 5)$ and $(3, 1)$ $2x + 3y = 9$

(d) $(-5, 1)$ and $(2, -7)$ $8x + 7y = -33$

46. Let $Ax + By = C$ and $A'x + B'y = C'$ represent two lines. Change each of these equations to slope-intercept form and then verify each of the following properties.

(a) If $\frac{A}{A'} = \frac{B}{B'} \neq \frac{C}{C'}$, then the lines are parallel.

(b) If $AA' = -BB'$, then the lines are perpendicular.

47. The properties in Problem 46 provide us with another way to write the equation of a line parallel or perpendicular to a given line that contains a given point not on the line. For example, suppose that we want the equation of the line perpendicular to $3x + 4y = 6$ that contains the point $(1, 2)$. The form $4x - 3y = k$, where k is a constant, represents a family of lines perpendicular to $3x + 4y = 6$ because we have satisfied the condition $AA' = -BB'$. Therefore, to find the specific line of the family that contains $(1, 2)$, we substitute 1 for x and 2 for y to determine k.

$$4x - 3y = k$$

$$4(1) - 3(2) = k$$

$$-2 = k$$

Thus, the equation of the desired line is $4x - 3y = -2$.

Use the properties from Problem 46 to help write the equation of each of the following lines.

(a) contains $(1, 8)$ and is parallel to $2x + 3y = 6$ $2x + 3y = 26$

(b) contains $(-1, 4)$ and is parallel to $x - 2y = 4$ $x - 2y = -9$

(c) contains $(2, -7)$ and is perpendicular to $3x - 5y = 10$ $5x + 3y = -11$

(d) contains $(-1, -4)$ and is perpendicular to $2x + 5y = 12$ $5x - 2y = 3$

7.6 Graphing Parabolas

Let's begin this section by using the concepts of intercepts and symmetry to help sketch the graph of the equation $y = x^2$.

EXAMPLE 1 Graph $y = x^2$.

Solution If we replace x with $-x$, the given equation becomes $y = (-x)^2 = x^2$; therefore, we have y-axis symmetry. The origin, (0, 0), is a point of the graph. Now we can set up a table of values that uses nonnegative values for x.

x	y
0	0
1	1
2	4
3	9
$\frac{1}{2}$	$\frac{1}{4}$

Plot the points determined by the table, connect them with a smooth curve, and reflect that portion of the curve across the y-axis to produce Figure 7.38.

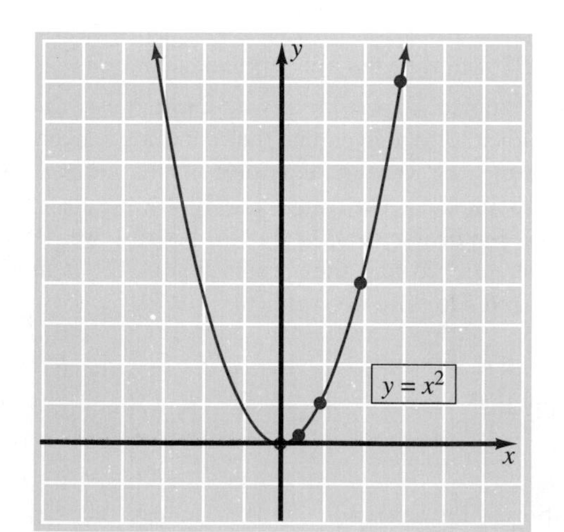

Figure 7.38

■

The curve in Figure 7.38 is called a **parabola**. The graph of any equation of the form $y = ax^2 + bx + c$, where a, b, and c are real numbers and $a \neq 0$, is a parabola. As we work with parabolas, we will use the vocabulary indicated in Figure 7.39.

One way to graph parabolas relies on the ability to find the vertex, to determine whether the parabola opens upward or downward, and to locate two points on opposite sides of the line of symmetry. For some of this information we can compare the parabolas produced by various types of equations such as $y = x^2 + k$, $y = ax^2$, $y = (x - h)^2$, and $y = a(x - h)^2 + k$ to the **basic parabola** produced by the equation $y = x^2$. First, let's consider some equations of the form $y = x^2 + k$, where k is a constant.

Figure 7.39

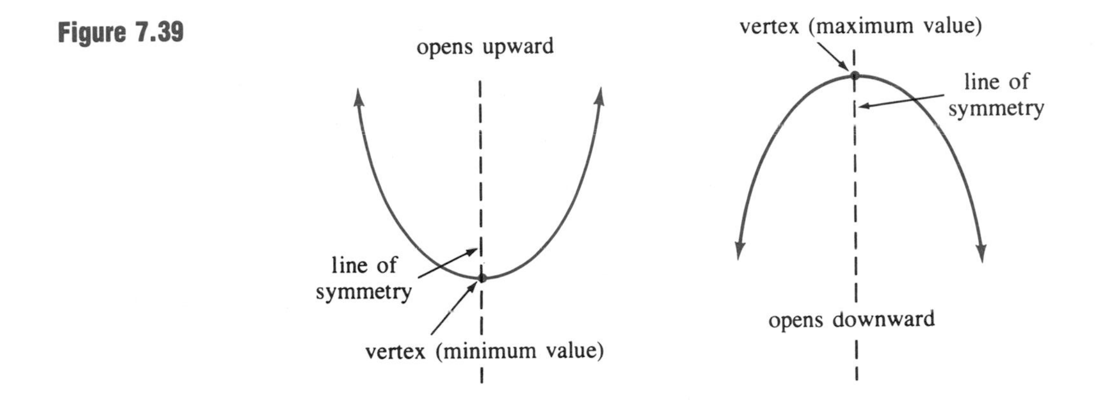

EXAMPLE 2 Graph $y = x^2 + 1$.

Solution Let's set up a table of values to compare y-values for $y = x^2 + 1$ to corresponding y-values for $y = x^2$

x	$y = x^2$	$y = x^2 + 1$
0	0	1
1	1	2
2	4	5
−1	1	2
−2	4	5

It should be evident that y-values for $y = x^2 + 1$ are *one larger* than corresponding y-values for $y = x^2$. For example, if $x = 2$ then $y = 4$ for the equation $y = x^2$, but if

$x = 2$ then $y = 5$ for the equation $y = x^2 + 1$. Thus, the graph of $y = x^2 + 1$ is the same as the graph of $y = x^2$ but *moved up* 1 *unit* (Figure 7.40).

Figure 7.40

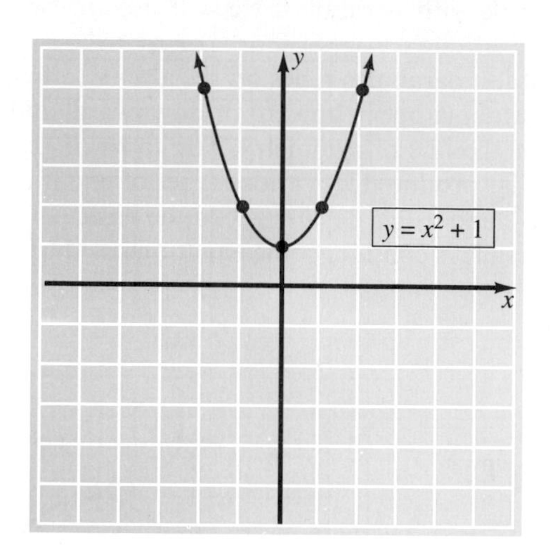

■

EXAMPLE 3 Graph $y = x^2 - 2$.

Solution The y-values for $y = x^2 - 2$ are 2 *less than* the corresponding y-values for $y = x^2$ as indicated in the following table.

x	$y = x^2$	$y = x^2 - 2$
0	0	−2
1	1	−1
2	4	2
−1	1	−1
−2	4	2

Thus, the graph of $y = x^2 - 2$ is the same as the graph of $y = x^2$, but *moved down* 2 *units* (Figure 7.41).

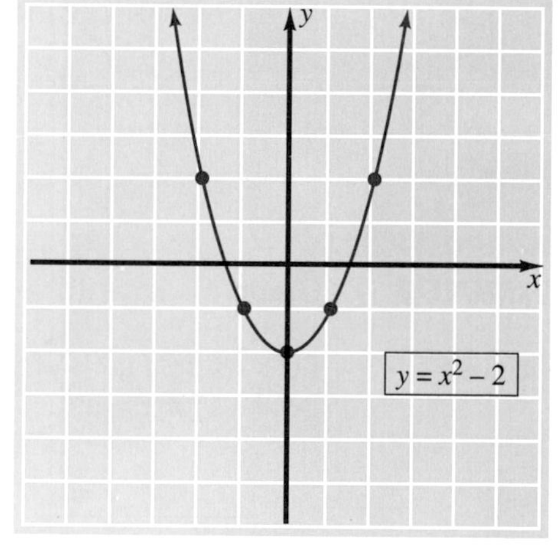

Figure 7.41 ■

In general, the graph of a quadratic equation of the form $y = x^2 + k$ is the same as the graph of $y = x^2$ but moved up or down k-units depending on whether k is positive or negative.

Now, let's consider some quadratic equations of the form $y = ax^2$, where a is a nonzero constant.

EXAMPLE 4 Graph $y = 2x^2$.

Solution Let's again use a table to make some comparisons of y-values.

x	$y = x^2$	$y = 2x^2$
0	0	0
1	1	2
2	4	8
−1	1	2
−2	4	8

Obviously, the y-values for $y = 2x^2$ are *twice* the corresponding y-values for $y = x^2$. Thus, the parabola associated with $y = 2x^2$ has the same vertex (the origin) as the graph of $y = x^2$, but it is narrower (Figure 7.42).

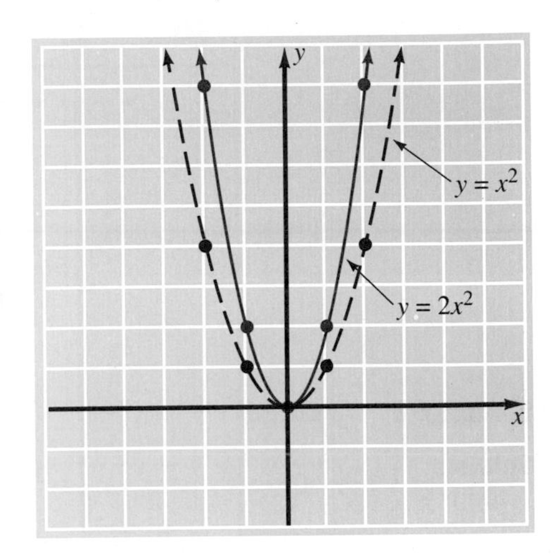

Figure 7.42

EXAMPLE 5 Graph $y = \frac{1}{2}x^2$.

Solution The following table indicates some comparisons of y-values.

x	$y = x^2$	$y = \frac{1}{2}x^2$
0	0	0
1	1	$\frac{1}{2}$
2	4	2
−1	1	$\frac{1}{2}$
−2	4	2

The y-values for $y = \frac{1}{2}x^2$ are *one-half* of the corresponding y-values for $y = x^2$.

Therefore, the graph of $y = \frac{1}{2}x^2$ is wider than the basic parabola (Figure 7.43).

Figure 7.43

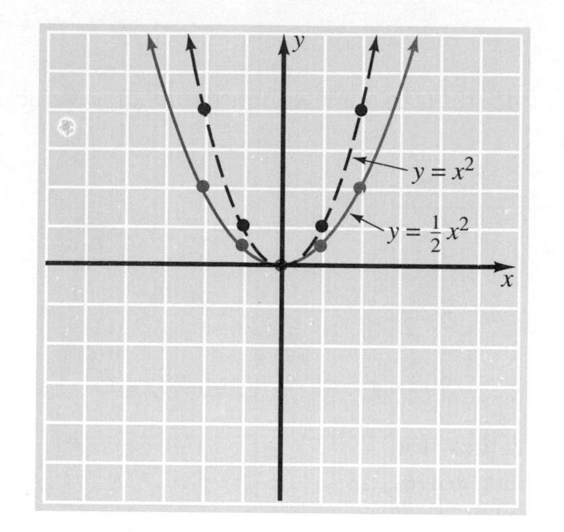

EXAMPLE 6 Graph $y = -x^2$

Solution

x	$y = x^2$	$y = -x^2$
0	0	0
1	1	−1
2	4	−4
−1	1	−1
−2	4	−4

The y-values for $y = -x^2$ are the *opposites* of the corresponding y-values for $y = x^2$. Thus, the graph of $y = -x^2$ is a reflection across the x-axis of the basic parabola (Figure 7.44).

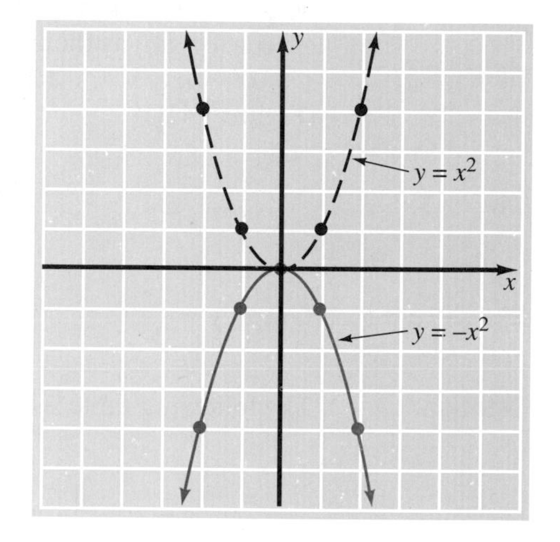

Figure 7.44

In general, the graph of a quadratic equation of the form $y = ax^2$ has its vertex at the origin and opens upward if a is positive and downward if a is negative. The parabola is narrower than the basic parabola if $|a| > 1$ and wider if $|a| < 1$.

Let's continue our investigation of quadratic equations by considering those of the form $y = (x - h)^2$, where h is a nonzero constant.

EXAMPLE 7 Graph $y = (x - 2)^2$.

Solution A fairly extensive table of values illustrates a pattern.

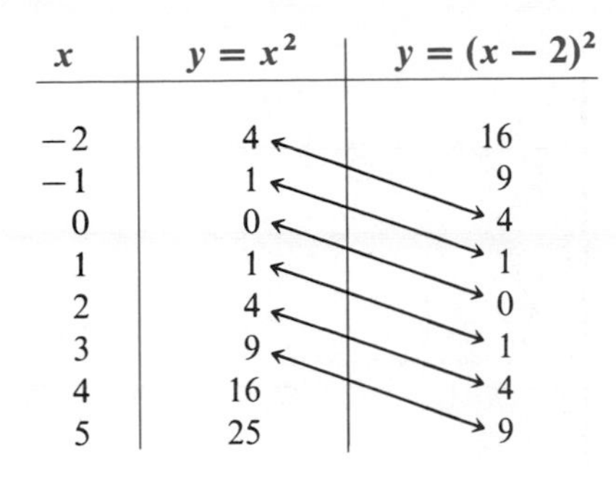

x	$y = x^2$	$y = (x - 2)^2$
-2	4	16
-1	1	9
0	0	4
1	1	1
2	4	0
3	9	1
4	16	4
5	25	9

Notice that $y = (x - 2)^2$ and $y = x^2$ take on the same y-values, *but* for different values of x. More specifically, if $y = x^2$ achieves a certain y-value at x equals a constant, then $y = (x - 2)^2$ achieves the same y-value at x equals the *constant plus two*. In other words, the graph of $y = (x - 2)^2$ is the same as the graph of $y = x^2$ *but moved two units to the right* (Figure 7.45).

Figure 7.45

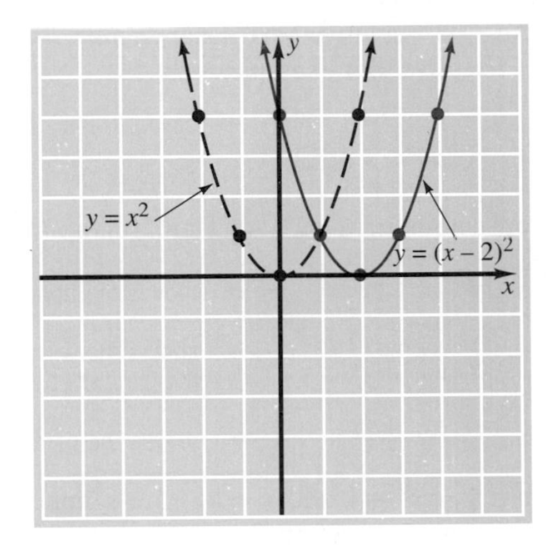

■

EXAMPLE 8 Graph $y = (x + 3)^2$.

Solution

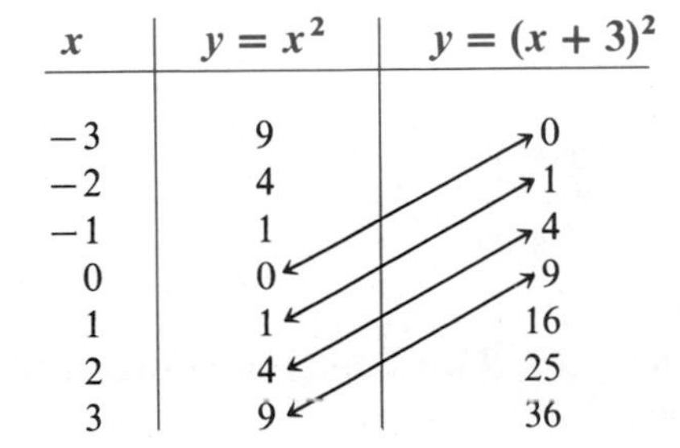

x	$y = x^2$	$y = (x + 3)^2$
-3	9	0
-2	4	1
-1	1	4
0	0	9
1	1	16
2	4	25
3	9	36

If $y = x^2$ achieves a certain y-value at x equals a constant, then $y = (x + 3)^2$ achieves that same y-value at x equals that *constant minus three*. Therefore, the graph of $y = (x + 3)^2$ is the same as the graph of $y = x^2$ *but moved three units to the left* (Figure 7.46).

Figure 7.46

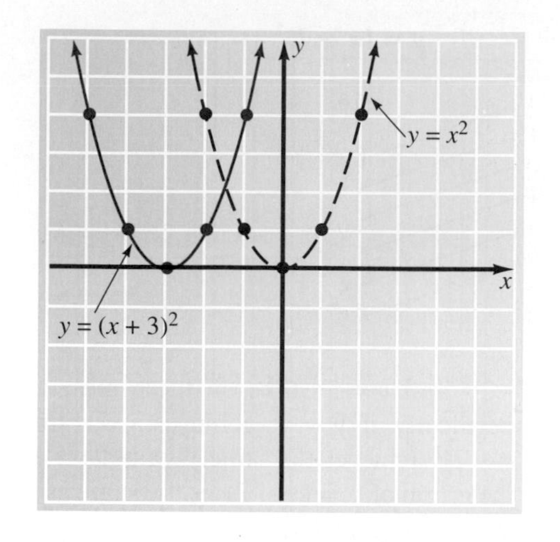

■

In general, the graph of a quadratic equation of the form $y = (x - h)^2$ is the same as the graph of $y = x^2$ but moved to the right h units if h is positive or moved to the left h units if h is negative. For example,

$y = (x - 4)^2$ $\longrightarrow$ moved to the *right* 4 units

$y = (x + 2)^2 = (x - (-2))^2$ $\longrightarrow$ moved to the *left* 2 units

The following diagram summarizes our work thus far graphing quadratic equations.

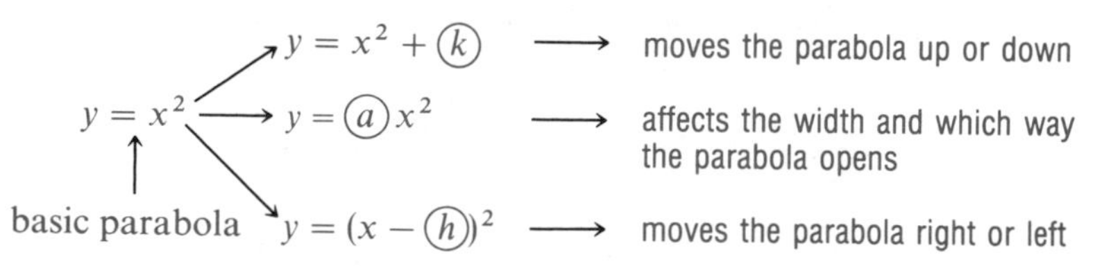

Equations of the form $y = x^2 + k$ and $y = ax^2$ are symmetrical about the y-axis. The final two examples of this section illustrate putting these ideas together to graph a quadratic equation of the form $y = a(x - h)^2 + k$.

EXAMPLE 9 Graph $y = 2(x - 3)^2 + 1$.

Solution

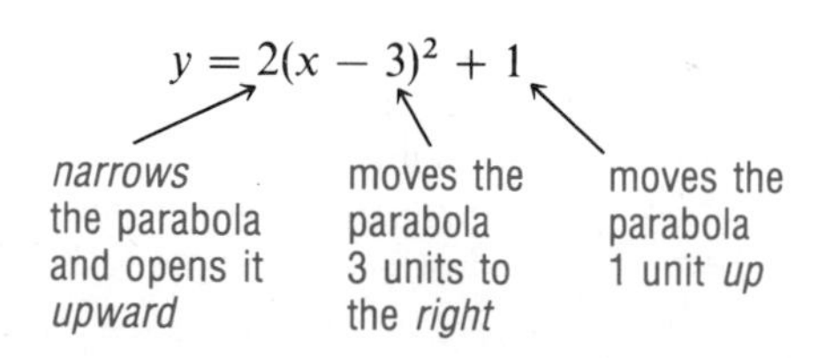

The parabola is drawn in Figure 7.47. Two points in addition to the vertex are located to determine the parabola.

Figure 7.47

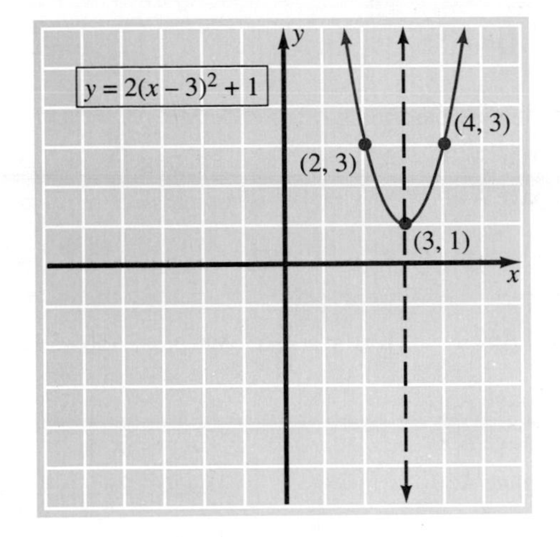

■

EXAMPLE 10 Graph $y = -\frac{1}{2}(x + 1)^2 - 2$.

Solution

$$y = -\frac{1}{2}(x + 1)^2 - 2$$

- $-\frac{1}{2}$: *widens* the parabola and opens it *downward*
- $+1$: moves the parabola 1 unit to the *left*
- -2: moves the parabola 2 units *down*

The parabola is drawn in Figure 7.48.

Figure 7.48

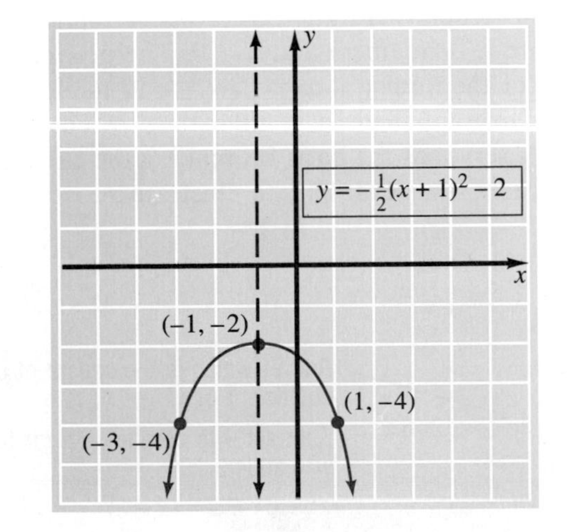

■

Problem Set 7.6

Graph each of the following parabolas.

1. $y = x^2 + 2$ see page A49
2. $y = x^2 + 3$ see page A49
3. $y = x^2 - 1$ see page A49
4. $y = x^2 - 5$ see page A49
5. $y = 4x^2$ see page A49
6. $y = 3x^2$ see page A49
7. $y = -3x^2$ see page A49
8. $y = -4x^2$ see page A49
9. $y = \frac{1}{3}x^2$ see page A49
10. $y = \frac{1}{4}x^2$ see page A50
11. $y = -\frac{1}{2}x^2$ see page A50
12. $y = -\frac{2}{3}x^2$ see page A50
13. $y = (x - 1)^2$ see page A50
14. $y = (x - 3)^2$ see page A50
15. $y = (x + 4)^2$ see page A50
16. $y = (x + 2)^2$ see page A50
17. $y = 3x^2 + 2$ see page A50
18. $y = 2x^2 + 3$ see page A50
19. $y = -2x^2 - 2$ see page A50
20. $y = \frac{1}{2}x^2 - 2$ see page A50
21. $y = (x - 1)^2 - 2$ see page A50
22. $y = (x - 2)^2 + 3$ see page A51
23. $y = (x + 2)^2 + 1$ see page A51
24. $y = (x + 1)^2 - 4$ see page A51
25. $y = 3(x - 2)^2 - 4$ see page A51
26. $y = 2(x + 3)^2 - 1$ see page A51
27. $y = -(x + 4)^2 + 1$ see page A51
28. $y = -(x - 1)^2 + 1$ see page A51
29. $y = -\frac{1}{2}(x + 1)^2 - 2$ see page A51
30. $y = -3(x - 4)^2 - 2$ see page A51

7.7

More Parabolas and Some Circles

We are now ready to graph quadratic equations of the form $y = ax^2 + bx + c$, where a, b, and c are real numbers and $a \neq 0$. The general approach is one of changing equations of the form $y = ax^2 + bx + c$ to the form $y = a(x - h)^2 + k$. Then we can proceed to graph them as we did in the previous section. The process of *completing the square* serves as the basis for making the change in the form of the equations. Let's consider some examples to illustrate the details.

EXAMPLE 1 Graph $y = x^2 + 6x + 8$.

Solution

$y = x^2 + 6x + 8$ — Add 9, which is the square of one-half the coefficient of x.

$y = x^2 + 6x + 9 + 8 - 9$ ← Subtract 9 to compensate for the 9 that was added.

$y = (x + 3)^2 - 1$ — $x^2 + 6x + 9 = (x + 3)^2$

The graph of $y = (x + 3)^2 - 1$ is the basic parabola moved three units to the left and one unit down (Figure 7.49).

Figure 7.49

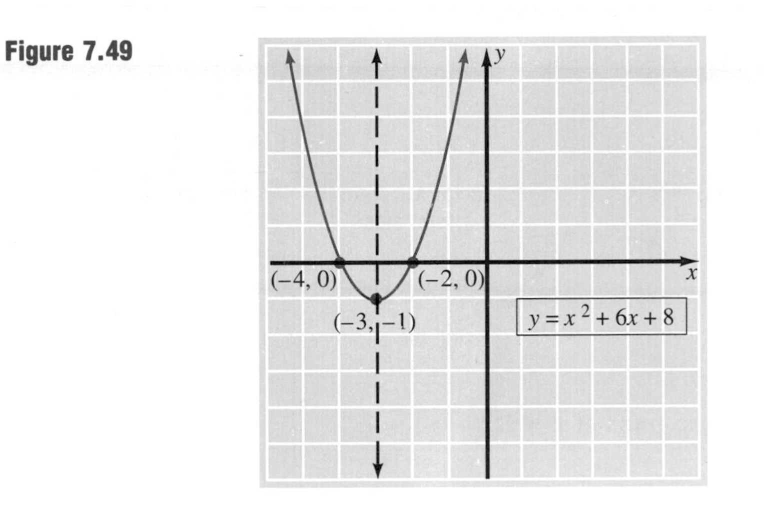

■

EXAMPLE 2 Graph $y = x^2 - 3x - 1$.

Solution

$$y = x^2 - 3x - 1$$

$$y = x^2 - 3x + \frac{9}{4} - 1 - \frac{9}{4}$$ Add and subtract $\frac{9}{4}$, which is the square of one-half the coefficient of x.

$$y = \left(x - \frac{3}{2}\right)^2 - \frac{13}{4}$$

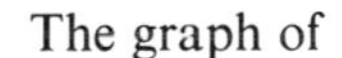

The graph of

$$y = \left(x - \frac{3}{2}\right)^2 - \frac{13}{4}$$

is the basic parabola moved $1\frac{1}{2}$ units to the right and $3\frac{1}{4}$ units down (Figure 7.50).

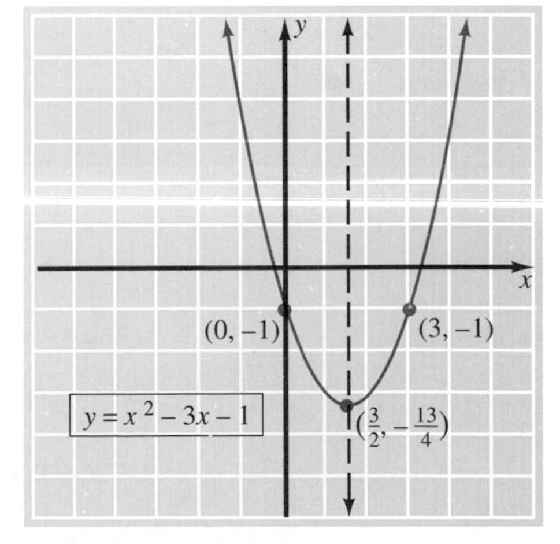

Figure 7.50

■

If the coefficient of x^2 is not 1, then a slight adjustment has to be made before we apply the process of completing the square. The next two examples illustrate this situation.

EXAMPLE 3 Graph $y = 2x^2 + 8x + 9$.

Solution

$$y = 2x^2 + 8x + 9$$

$$y = 2(x^2 + 4x) + 9$$ Factor a 2 from the first two terms on the right side.

$$y = 2(x^2 + 4x + 4) + 9 - 8$$ Add 4 inside the parentheses, which is the square of one-half the coefficient of x. Subtract 8 to compensate for the 4 added inside the parentheses times the factor 2.

$$y = 2(x + 2)^2 + 1$$

See Figure 7.51 for the graph of $y = 2(x + 2)^2 + 1$.

Figure 7.51

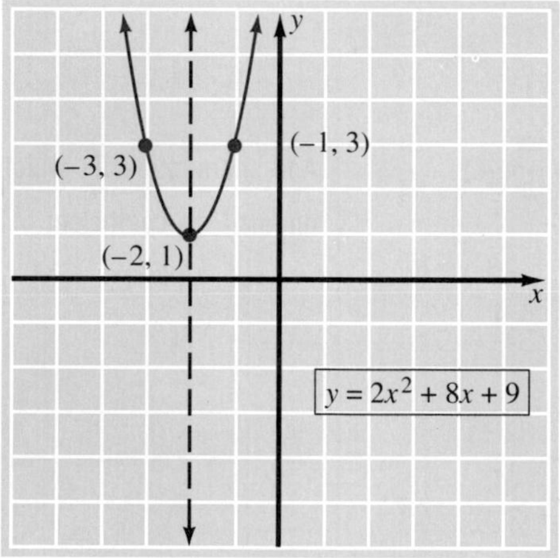

■

EXAMPLE 4 Graph $y = -3x^2 + 6x - 5$.

Solution

$$y = -3x^2 + 6x - 5$$

$$y = -3(x^2 - 2x) - 5$$ Factor −3 from the first two terms on the right side.

$$y = -3(x^2 - 2x + 1) - 5 + 3$$ Add 1 inside the parentheses to complete the square. Add 3 to compensate for the 1 added inside the parentheses times the factor of −3.

$$y = -3(x - 1)^2 - 2$$

The graph of $y = -3(x - 1)^2 - 2$ is drawn in Figure 7.52.

Figure 7.52

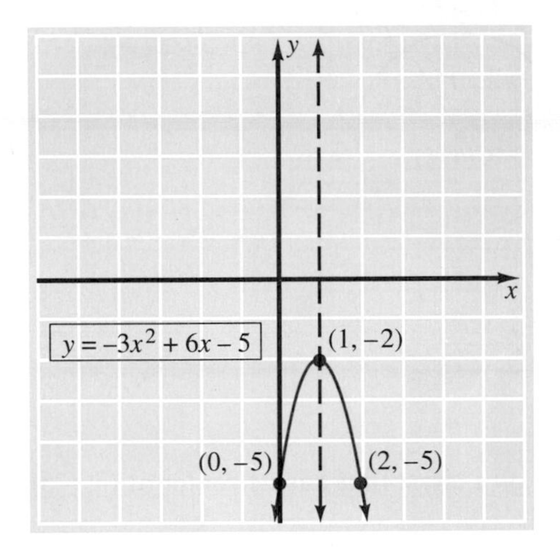

Circles

The distance formula, $d = \sqrt{(x_2 - x_1)^2 + (y_2 - y_1)^2}$ (developed in Section 7.4), when it applies to the definition of a circle produces what is known as the **standard equation of a circle**. We start with a precise definition of a circle.

DEFINITION 7.2

A **circle** is the set of all points in a plane equidistant from a given fixed point called the **center**. A line segment determined by the center and any point on the circle is called a **radius**.

Let's consider a circle that has a radius of length r and a center at (h, k) on a coordinate system (Figure 7.53). For any point P on the circle with coordinates (x, y), the length of a radius (denoted by r) can be expressed as

$$r = \sqrt{(x - h)^2 + (y - k)^2}.$$

Thus, squaring both sides of the equation, we obtain the **standard form of the equation of a circle**:

$$(x - h)^2 + (y - k)^2 = r^2.$$

We can use the standard form of the equation of a circle to solve two basic kinds of circle problems, namely, (1) given the coordinates of the center and the length of a radius of a circle, find its equation, and (2) given the equation of a circle, find its center and the length of a radius. Let's look at some examples of such problems.

Figure 7.53

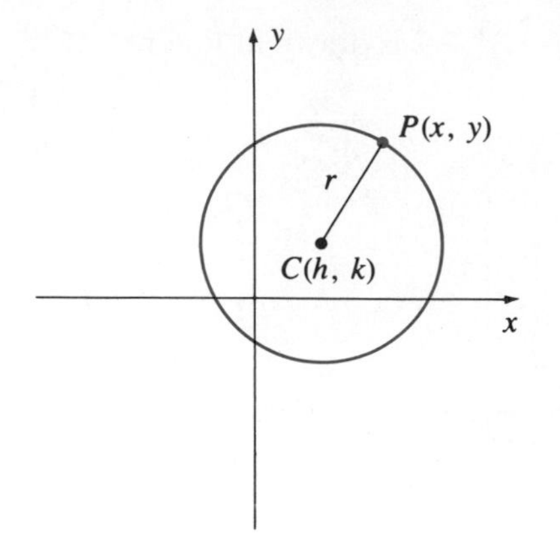

EXAMPLE 5 Write the equation of a circle that has its center at $(3, -5)$ and a radius of length 6 units.

Solution Let's substitute 3 for h, -5 for k, and 6 for r into the standard form $(x-h)^2 + (y-k)^2 = r^2$ that becomes $(x-3)^2 + (y+5)^2 = 6^2$, which we can simplify as follows.

$$(x-3)^2 + (y+5)^2 = 6^2$$

$$x^2 - 6x + 9 + y^2 + 10y + 25 = 36$$

$$x^2 + y^2 - 6x + 10y - 2 = 0$$

■

Notice in Example 5 that we simplified the equation to the form $x^2 + y^2 + Dx + Ey + F = 0$, where D, E, and F are integers. This is another commonly used form when working with circles.

EXAMPLE 6 Graph $x^2 + y^2 + 4x - 6y + 9 = 0$.

Solution This equation is of the form $x^2 + y^2 + Dx + Ey + F = 0$, so its graph is a circle. We can change the given equation into the form $(x-h)^2 + (y-k)^2 = r^2$ by completing the square on x and on y as follows.

$$x^2 + y^2 + 4x - 6y + 9 = 0$$

$$(x^2 + 4x + \_\_\_) + (y^2 - 6y + \_\_\_) = -9$$

$$(x^2 + 4x + 4) + (y^2 - 6y + 9) = -9 + 4 + 9$$

added 4 to complete the square on x | added 9 to complete the square on y | added 4 and 9 to compensate for the 4 and 9 added on the left side

$$(x + 2)^2 + (y - 3)^2 = 4$$

$$(x - (-2))^2 + (y - 3)^2 = 2^2$$

$$\uparrow \quad h \qquad \uparrow \quad k \qquad \uparrow \quad r$$

The center of the circle is at $(-2, 3)$ and the length of a radius is 2 (Figure 7.54).

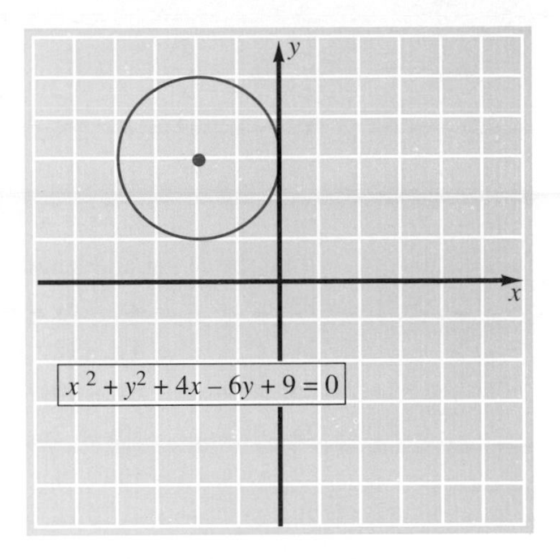

Figure 7.54 ■

As demonstrated by Examples 5 and 6, both forms, $(x - h)^2 + (y - k)^2 = r^2$ and $x^2 + y^2 + Dx + Ey + F = 0$, play an important role when solving problems that deal with circles.

Finally, we need to recognize that the standard form of a circle that has its center at the origin is $x^2 + y^2 = r^2$. This is merely the result of letting $h = 0$ and $k = 0$ in the general standard form.

$$(x - h)^2 + (y - k)^2 = r^2$$

$$(x - 0)^2 + (y - 0)^2 = r^2$$

$$x^2 + y^2 = r^2$$

Thus, by inspection we can recognize that $x^2 + y^2 = 9$ is a circle with its center at the origin; the length of a radius is 3 units. Likewise, the equation of a circle that has its center at the origin and a radius of length 6 units is $x^2 + y^2 = 36$.

Problem Set 7.7

Graph each of the following parabolas.

1. $y = x^2 - 6x + 13$ see page A51

2. $y = x^2 - 4x + 7$ see page A51

3. $y = x^2 + 2x + 6$ see page A51

4. $y = x^2 + 8x + 14$ see page A52

5. $y = x^2 - 5x + 3$ see page A52

6. $y = x^2 + 3x + 1$ see page A52

7. $y = x^2 + 7x + 14$ see page A52

8. $y = x^2 - x - 1$ see page A52

9. $y = 3x^2 - 6x + 5$ see page A52

10. $y = 2x^2 + 4x + 7$ see page A52

11. $y = 4x^2 - 24x + 32$ see page A52

12. $y = 3x^2 + 24x + 49$ see page A52

13. $y = -2x^2 - 4x - 5$ see page A52

14. $y = -2x^2 + 8x - 5$ see page A52

15. $y = -x^2 + 8x - 21$ see page A52

16. $y = -x^2 - 6x - 7$ see page A53

17. $y = 2x^2 - x + 2$ see page A53

18. $y = 2x^2 + 3x + 1$ see page A53

19. $y = 3x^2 + 2x + 1$ see page A53

20. $y = 3x^2 - x - 1$ see page A53

21. $y = -3x^2 - 7x - 2$ see page A53

22. $y = -2x^2 + x - 2$ see page A53

Find the center and length of a radius of each of the following circles.

23. $x^2 + y^2 - 2x - 6y - 6 = 0$ $(1, 3)$, $r = 4$

24. $x^2 + y^2 + 4x - 12y + 39 = 0$ $(-2, 6)$, $r = 1$

25. $x^2 + y^2 + 6x + 10y + 18 = 0$ $(-3, -5)$, $r = 4$

26. $x^2 + y^2 - 10x + 2y + 1 = 0$ $(5, -1)$, $r = 5$

27. $x^2 + y^2 = 10$ $(0, 0)$, $r = \sqrt{10}$

28. $x^2 + y^2 + 4x + 14y + 50 = 0$ $(-2, -7)$, $r = \sqrt{3}$

29. $x^2 + y^2 - 16x + 6y + 71 = 0$ $(8, -3)$, $r = \sqrt{2}$

30. $x^2 + y^2 = 12$ $(0, 0)$, $r = 2\sqrt{3}$

31. $x^2 + y^2 + 6x - 8y = 0$ $(-3, 4)$, $r = 5$

32. $x^2 + y^2 - 16x + 30y = 0$ $(8, -15)$, $r = 17$

33. $4x^2 + 4y^2 + 4x - 32y + 33 = 0$ $(-\frac{1}{2}, 4)$, $r = 2\sqrt{2}$

34. $9x^2 + 9y^2 - 6x - 12y - 40 = 0$ $(\frac{1}{3}, \frac{2}{3})$, $r = \sqrt{5}$

Write the equation of each of the following circles. Express the final equation in the form $x^2 + y^2 + Dx + Ey + F = 0$.

35. center at $(3, 5)$ and $r = 5$ $x^2 + y^2 - 6x - 10y + 9 = 0$

36. center at $(2, 6)$ and $r = 7$ $x^2 + y^2 - 4x - 12y - 9 = 0$

37. center at $(-4, 1)$ and $r = 8$ $x^2 + y^2 + 8x - 2y - 47 = 0$

38. center at $(-3, 7)$ and $r = 6$ $x^2 + y^2 + 6x - 14y + 22 = 0$

39. center at $(-2, -6)$ and $r = 3\sqrt{2}$ $x^2 + y^2 + 4x + 12y + 22 = 0$

40. center at $(-4, -5)$ and $r = 2\sqrt{3}$ $x^2 + y^2 + 8x + 10y + 29 = 0$

41. center at $(0, 0)$ and $r = 2\sqrt{5}$ $x^2 + y^2 = 20$

42. center at $(0, 0)$ and $r = \sqrt{7}$ $x^2 + y^2 = 7$

43. center at $(5, -8)$ and $r = 4\sqrt{6}$ $x^2 + y^2 - 10x + 16y - 7 = 0$

44. center at $(4, -10)$ and $r = 8\sqrt{2}$ $x^2 + y^2 - 8x + 20y - 12 = 0$

45. Find the equation of the circle that passes through the origin and has its center at $(0, 4)$. $x^2 + y^2 - 8y = 0$

46. Find the equation of the circle that passes through the origin and has its center at $(-6, 0)$. $x^2 + y^2 + 12x = 0$

47. Find the equation of the circle that passes through the origin and has its center at $(-4, 3)$. $x^2 + y^2 + 8x - 6y = 0$

48. Find the equation of the circle that passes through the origin and has its center at $(8, -15)$. $x^2 + y^2 - 16x + 30y = 0$

Miscellaneous Problems

49. The points (x, y) and (y, x) are mirror images of each other across the line $y = x$. Therefore, by interchanging x and y in the equation $y = ax^2 + bx + c$, we obtain the equation of its mirror image across the line $y = x$, namely, $x = ay^2 + by + c$. Thus to graph $x = y^2 + 2$, we can first graph $y = x^2 + 2$ and then reflect it across the line $y = x$ as indicated in the following figure.

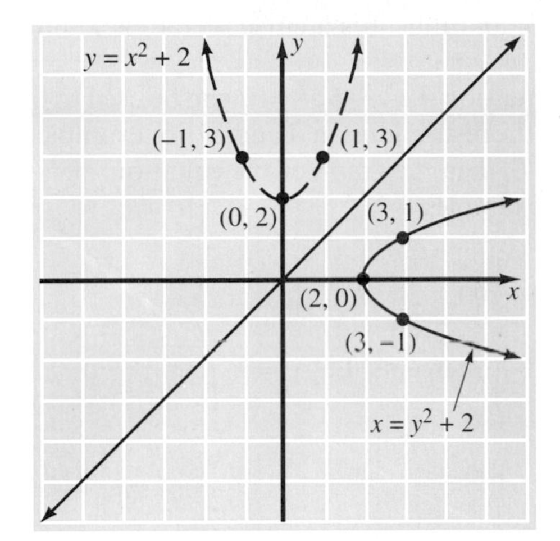

Graph each of the following parabolas.

(a) $x = y^2$ see page A53
(b) $x = -y^2$ see page A53
(c) $x = y^2 - 1$ see page A54
(d) $x = -y^2 + 3$ see page A54
(e) $x = -2y^2$ see page A54
(f) $x = 3y^2$ see page A54
(g) $x = y^2 + 4y + 7$ see page A54
(h) $x = y^2 - 2y - 3$ see page A54

50. By expanding $(x - h)^2 + (y - k)^2 = r^2$, we obtain $x^2 - 2hx + h^2 + y^2 - 2ky + k^2 - r^2 = 0$. Compare this result to the form $x^2 + y^2 + Dx + Ey + F = 0$, and we see that $D = -2h$, $E = -2k$, and $F = h^2 + k^2 - r^2$. Therefore, the center and length of a radius of a circle can be found by using $h = \dfrac{D}{-2}$, $k = \dfrac{E}{-2}$, and $r = \sqrt{h^2 + k^2 - F}$. Use these relationships to find the center and length of a radius of each of the following circles.

(a) $x^2 + y^2 - 2x - 8y + 8 = 0$ $(1, 4)$, $r = 3$
(b) $x^2 + y^2 + 4x - 14y + 49 = 0$ $(-2, 7)$, $r = 2$
(c) $x^2 + y^2 + 12x + 8y - 12 = 0$ $(-6, -4)$, $r = 8$
(d) $x^2 + y^2 - 16x + 20y + 115 = 0$ $(8, -10)$, $r = 7$
(e) $x^2 + y^2 - 12y - 45 = 0$ $(0, 6)$, $r = 9$
(f) $x^2 + y^2 + 14x = 0$ $(-7, 0)$, $r = 7$

7.8

Ellipses and Hyperbolas–Conic Sections

In the previous section we found that the graph of the equation $x^2 + y^2 = 36$ is a circle of radius 6 units with its center at the origin. More generally, it is true that any equation of the form $Ax^2 + By^2 = C$, where $A = B$ and A, B, and C are nonzero constants that have the same sign, is a circle with the center at the origin. For example, $3x^2 + 3y^2 = 12$ is equivalent to $x^2 + y^2 = 4$ (divide both sides of the given equation by 3) and thus it is a circle of radius 2 units with its center at the origin.

The general equation $Ax^2 + By^2 = C$ can be used to describe other geometric figures by changing the restrictions on A and B. For example, if A, B, and C are of the same sign, but $A \neq B$, then the graph of the equation $Ax^2 + By^2 = C$ is an **ellipse**. Let's consider two examples.

EXAMPLE 1 Graph $4x^2 + 25y^2 = 100$.

Solution Let's find the x- and y-intercepts. Let $x = 0$; then

$$4(0)^2 + 25y^2 = 100$$
$$25y^2 = 100$$
$$y^2 = 4$$
$$y = \pm 2.$$

Thus, the points $(0, 2)$ and $(0, -2)$ are on the graph. Let $y = 0$; then

$$4x^2 + 25(0)^2 = 100$$
$$4x^2 = 100$$
$$x^2 = 25$$
$$x = \pm 5.$$

Thus, the points $(5, 0)$ and $(-5, 0)$ are also on the graph. Plot the four points we have and knowing that it is an ellipse gives us a pretty good sketch of the figure (Figure 7.55).

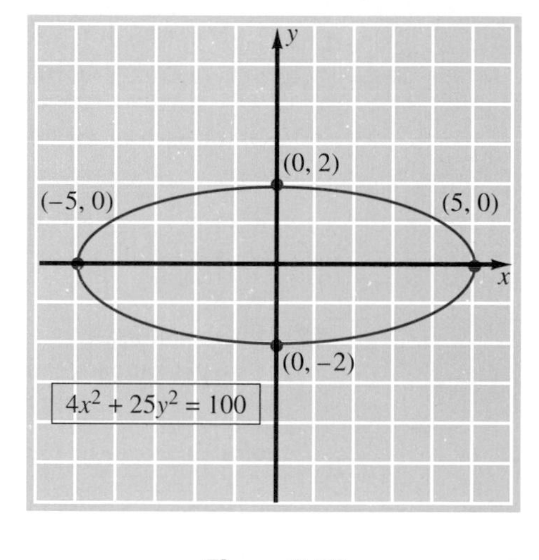

Figure 7.55 ■

In Figure 7.55 the line segment with endpoints at $(-5, 0)$ and $(5, 0)$ is called the **major axis** of the ellipse. The shorter line segment with endpoints at $(0, -2)$ and $(0, 2)$ is called the **minor axis**. Establishing the endpoints of the major and minor axes provides a basis for sketching an ellipse.

EXAMPLE 2 Graph $9x^2 + 4y^2 = 36$.

Solution Again let's find the x- and y-intercepts. Let $x = 0$; then

$$9(0)^2 + 4y^2 = 36$$
$$4y^2 = 36$$
$$y^2 = 9$$
$$y = \pm 3.$$

Thus, the points $(0, 3)$ and $(0, -3)$ are on the graph. Let $y = 0$; then

$$9x^2 + 4(0)^2 = 36$$
$$9x^2 = 36$$
$$x^2 = 4$$
$$x = \pm 2.$$

Thus, the points $(2, 0)$ and $(-2, 0)$ are also on the graph. The ellipse is sketched in Figure 7.56.

Figure 7.56

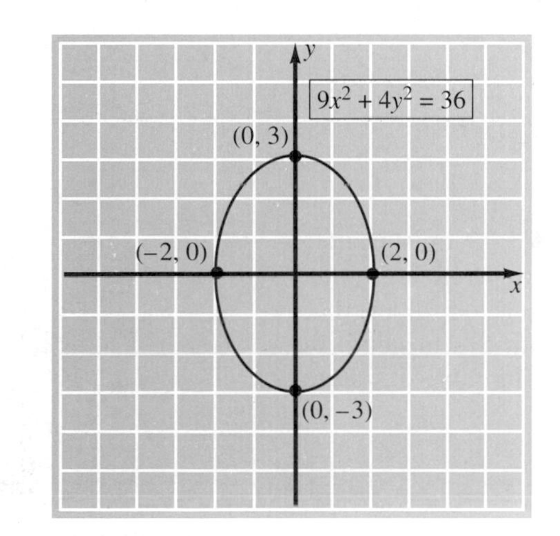

In Figure 7.56, the major axis has endpoints at $(0, -3)$ and $(0, 3)$, and the minor axis has endpoints at $(-2, 0)$ and $(2, 0)$. The ellipses in Figures 7.55 and 7.56 are symmetrical about the x-axis and about the y-axis. In other words, both the x-axis and the y-axis serve as **axes of symmetry**.

Hyperbolas

The graph of an equation of the form $Ax^2 + By^2 = C$, where A, B, and C are nonzero constants and A and B are of unlike signs, is a **hyperbola**. The next two examples illustrate the graphing of hyperbolas.

EXAMPLE 3 Graph $x^2 - y^2 = 9$.

Solution If we let $y = 0$, we obtain

$$x^2 - 0^2 = 9$$
$$x^2 = 9$$
$$x = \pm 3.$$

Thus, the points (3, 0) and (−3, 0) are on the graph. If we let $x = 0$, we obtain

$$0^2 - y^2 = 9$$
$$-y^2 = 9$$
$$y^2 = -9.$$

Since $y^2 = -9$ has no real number solutions, there are no points of the y-axis on this graph. That is to say, the graph does not intersect the y-axis. Now let's solve the given equation for y so that we have a more convenient form for finding other solutions.

$$x^2 - y^2 = 9$$
$$-y^2 = 9 - x^2$$
$$y^2 = x^2 - 9$$
$$y = \pm\sqrt{x^2 - 9}$$

Since the radicand, $x^2 - 9$, must be nonnegative, the values we chose for x must be greater than or equal to 3, or less than or equal to −3. With this in mind, we can form the following table of values.

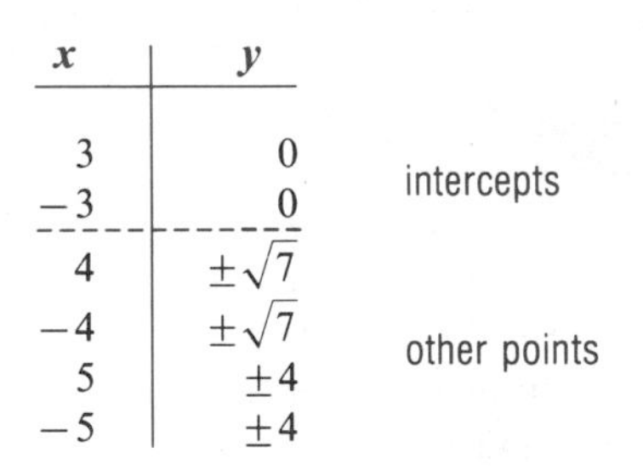

x	y	
3	0	intercepts
−3	0	
4	$\pm\sqrt{7}$	other points
−4	$\pm\sqrt{7}$	
5	± 4	
−5	± 4	

We plot these points and draw the hyperbola as in Figure 7.57 (This graph is also symmetrical about both axes.)

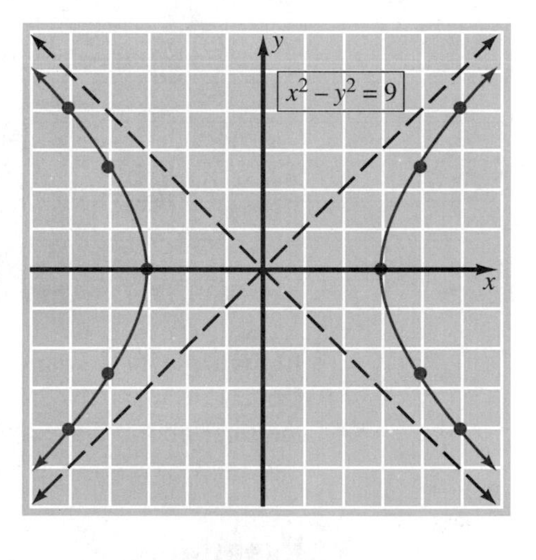

Figure 7.57 ■

Notice the dashed lines in Figure 7.57; they are called **asymptotes**. Each branch of the hyperbola approaches one of these lines, but does not intersect it. Therefore, the ability to sketch the asymptotes of a hyperbola is very helpful when graphing the hyperbola. Fortunately, the equations of the asymptotes are easy to determine. They can be found by replacing the constant term in the given equation of the hyperbola with 0 and solving for y. (The reason this works will become evident in a later course.) Thus, for the hyperbola in Example 3 we obtain

$$x^2 - y^2 = 0$$
$$y^2 = x^2$$
$$y = \pm x.$$

So, the two lines, $y = x$ and $y = -x$, are the asymptotes indicated by the dashed lines in Figure 7.57.

EXAMPLE 4 Graph $y^2 - 5x^2 = 4$.

Solution If we let $x = 0$, we obtain

$$y^2 - 5(0)^2 = 4$$
$$y^2 = 4$$
$$y = \pm 2.$$

The points $(0, 2)$ and $(0, -2)$ are on the graph. If we let $y = 0$, we obtain

$$0^2 - 5x^2 = 4$$
$$-5x^2 = 4$$
$$x^2 = -\frac{4}{5}.$$

Since $x^2 = -\frac{4}{5}$ has no real number solutions, we know that this hyperbola does not intersect the x-axis. Solving the given equation for y yields

$$y^2 - 5x^2 = 4$$
$$y^2 = 5x^2 + 4$$
$$y = \pm\sqrt{5x^2 + 4}.$$

The following table shows some additional solutions for the equation.

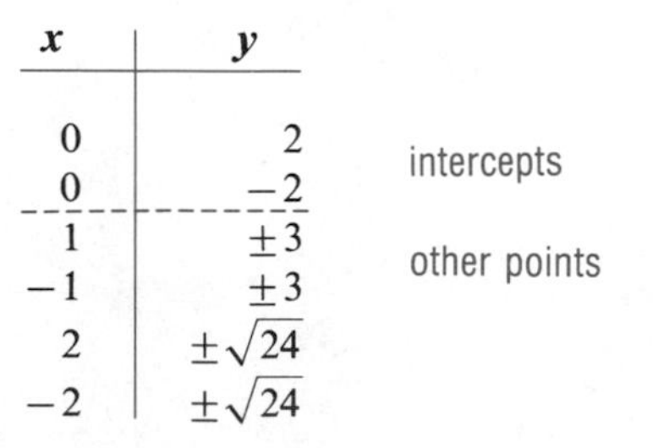

x	y	
0	2	intercepts
0	-2	
1	± 3	other points
-1	± 3	
2	$\pm\sqrt{24}$	
-2	$\pm\sqrt{24}$	

The equations of the asymptotes are determined as follows.

$$y^2 - 5x^2 = 0$$
$$y^2 = 5x^2$$
$$y = \pm\sqrt{5}x$$

Sketch the asymptotes and plot the points determined by the table of values to determine the hyperbola in Figure 7.58. (Notice that this hyperbola is also symmetrical about the x-axis and the y-axis.)

Figure 7.58

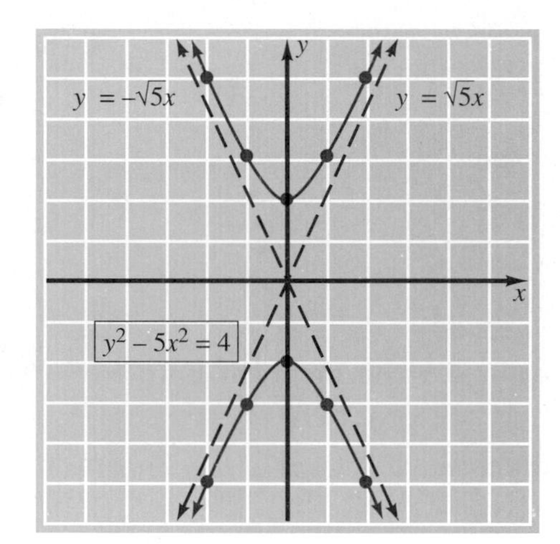

■

EXAMPLE 5 Graph $4x^2 - 9y^2 = 36$.

Solution If we let $x = 0$, we obtain

$$4(0)^2 - 9y^2 = 36$$
$$-9y^2 = 36$$
$$y^2 = -4.$$

Since $y^2 = -4$ has no real number solutions, we know that this hyperbola does not intersect the y-axis. If we let $y = 0$, we obtain

$$4x^2 - 9(0)^2 = 36$$
$$4x^2 = 36$$
$$x^2 = 9$$
$$x = \pm 3.$$

Thus, the points $(3, 0)$ and $(-3, 0)$ are on the graph.

Now let's solve the equation for y in terms of x and set up a table of values.

$$4x^2 - 9y^2 = 36$$
$$-9y^2 = 36 - 4x^2$$
$$9y^2 = 4x^2 - 36$$
$$y^2 = \frac{4x^2 - 36}{9}$$
$$y = \pm\frac{\sqrt{4x^2 - 36}}{3}$$

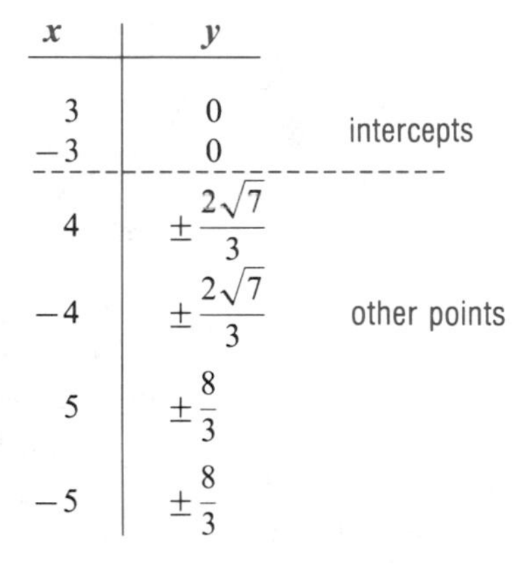

x	y	
3	0	intercepts
-3	0	
4	$\pm\frac{2\sqrt{7}}{3}$	
-4	$\pm\frac{2\sqrt{7}}{3}$	other points
5	$\pm\frac{8}{3}$	
-5	$\pm\frac{8}{3}$	

The equations of the asymptotes are found as follows.

$$4x^2 - 9y^2 = 0$$
$$-9y^2 = -4x^2$$
$$9y^2 = 4x^2$$
$$y^2 = \frac{4x^2}{9}$$
$$y = \pm\frac{2}{3}x$$

Sketch the asymptotes and plot the points determined by the table to determine the hyperbola as shown in Figure 7.59.

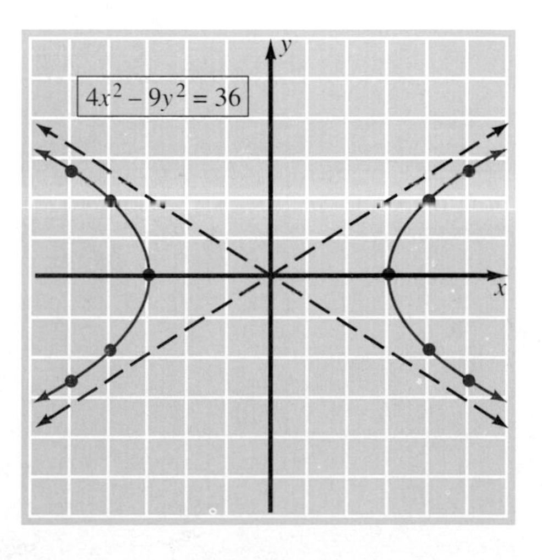

Figure 7.59 ■

As a way of summarizing, let's focus our attention on the continuity pattern used in these last two sections. In Section 7.7, we used the definition of a circle to generate a standard form for the equation of a circle. Then, in this section, ellipses and hyperbolas were discussed, not from a definition viewpoint, but by considering variations of the general equation of a circle with its center at the origin ($Ax^2 + By^2 = C$, where A, B, and C are of the same sign and $A = B$). In a subsequent mathematics course parabolas, ellipses, and hyperbolas will also be developed from a definition viewpoint. That is to say, each concept will be first defined and then the definition will be used to generate a standard form of its equation. In such a course you will also study ellipses and hyperbolas that have their centers at points other than the origin.

Figure 7.60

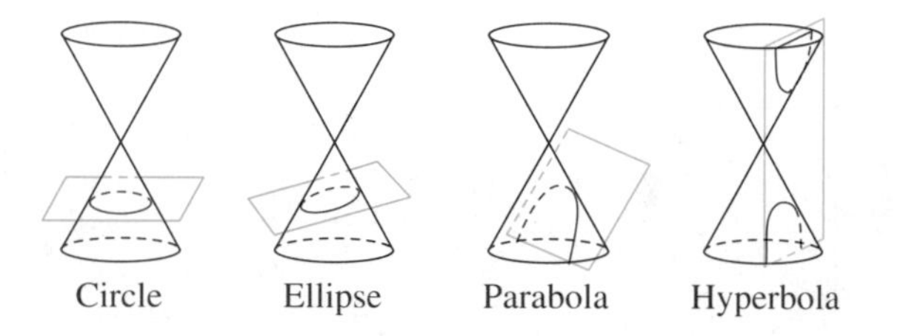

Parabolas, circles, ellipses, and hyperbolas can be formed by intersecting a plane with a conical surface as shown in Figure 7.60; we often refer to these curves as the **conic sections**. A flashlight produces a "cone of light" that can be cut by the plane of a wall to illustrate the conic sections. Try shining a flashlight against a wall at different angles to produce a circle, an ellipse, a parabola, and one branch of a hyperbola. (You may find it difficult to distinguish between a parabola and a branch of a hyperbola.)

Problem Set 7.8

Graph each of the following. If the equation represents a hyperbola, find its asymptotes and use them to help sketch the hyperbola.

1. $x^2 + 4y^2 = 36$ see page A54

2. $x^2 + 4y^2 = 16$ see page A54

3. $9x^2 + y^2 = 36$ see page A54

4. $16x^2 + 9y^2 = 144$ see page A55

5. $x^2 - y^2 = 1$ see page A55

6. $x^2 - y^2 = 4$ see page A55

7. $y^2 - 4x^2 = 9$ see page A55

8. $4y^2 - x^2 = 16$ see page A55

9. $4x^2 + 3y^2 = 12$ see page A55

10. $5x^2 + 4y^2 = 20$ see page A55

11. $5x^2 - 2y^2 = 20$ see page A55

12. $9x^2 - 4y^2 = 9$ see page A55

13. $25x^2 + 2y^2 = 50$ see page A55

14. $12x^2 + y^2 = 36$ see page A55

15. $y^2 - 16x^2 = 4$ see page A55

16. $y^2 - 9x^2 = 16$ see page A56

17. $-4x^2 + y^2 = -4$ see page A56

18. $-9x^2 + y^2 = -36$ see page A56

19. $25y^2 - 3x^2 = 75$ see page A56

20. $16y^2 - 5x^2 = 80$ see page A56

21. The graphs of equations of the form $xy = k$, where k is a nonzero constant, are also hyperbolas, sometimes referred to as rectangular hyperbolas. Graph each of the following.

(a) $xy = 3$ see page A56 **(b)** $xy = 5$ see page A56

(c) $xy = -2$ see page A56 **(d)** $xy = -4$ see page A56

22. What is the graph of $xy = 0$? Defend your answer. Since $xy = 0$ is equivalent to $x = 0$ or $y = 0$, the graph of $xy = 0$ is the two coordinate axes.

23. We have graphed various equations of the form $Ax^2 + By^2 = C$, where C is a nonzero constant. Now graph each of the following.

(a) $x^2 + y^2 = 0$ origin **(b)** $2x^2 + 3y^2 = 0$ origin

(c) $x^2 - y^2 = 0$ see page A56 **(d)** $4y^2 - x^2 = 0$ see page A56

Chapter 7 Summary

(7.1) The **Cartesian** (or **rectangular**) **coordinate system** is used to graph ordered pairs of real numbers. The first number, a, of the ordered pair (a, b) is called the **abscissa** and the second number, b, is called the **ordinate**; together they are referred to as the **coordinates** of a point.

Two basic kinds of problems exist in coordinate geometry:

1. Given an algebraic equation, find its geometric graph;
2. Given a set of conditions that pertains to a geometric figure, find its algebraic equation.

A **solution** of an equation in two variables is an ordered pair of real numbers that satisfies the equation.

The following suggestions are offered for **graphing an equation** in two variables.

1. Determine the type of symmetry that the equation exhibits.
2. Find the intercepts.
3. Solve the equation for y in terms of x or for x in terms of y if it is not already in such a form.
4. Set up a table of ordered pairs that satisfy the equation. The type of symmetry will affect your choice of values in the table.
5. Plot the points associated wth the ordered pairs from the table, and connect them with a smooth curve. Then, if appropriate, reflect this part of the curve according to the symmetry shown by the equation.

(7.2) Any equation of the form $Ax + By = C$, where A, B, and C are constants (A and B not both zero) and x and y are variables, is a **linear equation** and its graph is a **straight line**.

Any equation of the form $Ax + By = C$ where $C = 0$ is a straight line that contains the origin.

Any equation of the form $x = a$, where a is a constant, is a line parallel to the y-axis that has an x-intercept of a.

Any equation of the form $y = b$, where b is a constant, is a line parallel to the x-axis that has a y-intercept of b.

(7.3) Linear inequalities in two variables are of the form $Ax + By > C$ or $Ax + By < C$. To **graph a linear inequality**, we suggest the following steps.

1. First, graph the corresponding equality. Use a solid line if equality is included in the original statement and a dashed line if equality is not included.
2. Choose a test point not on the line and substitute its coordinates into the inequality.
3. The graph of the original inequality is:
   - **(a)** The half-plane that contains the test point if the inequality is satisfied by that point; or
   - **(b)** The half-plane that does not contain the test point if the inequality is not satisfied by the point.

(7.4) The distance between any two points (x_1, y_1) and (x_2, y_2) is given by the **distance formula,**

$$d = \sqrt{(x_2 - x_1)^2 + (y_2 - y_1)^2}.$$

The **slope** (denoted by m) of a line determined by the points (x_1, y_1) and (x_2, y_2) is given by the slope formula,

$$m = \frac{y_2 - y_1}{x_2 - x_1}, \qquad x_2 \neq x_1.$$

(7.5) The equation $y = mx + b$ is referred to as the **slope-intercept form** of the equation of a straight line. If the equation of a nonvertical line is written in this y-form, the coefficient of x is the slope of the line and the constant term is the y-intercept.

If two lines have slopes m_1 and m_2, respectively, then:

1. The two lines are parallel if and only if $m_1 = m_2$;
2. The two lines are perpendicular if and only if $(m_1)(m_2) = -1$.

(7.6) and (7.7) The graph of any quadratic equation of the form $y = ax^2 + bx + c$, where a, b, and c are real numbers and $a \neq 0$, is a **parabola**.

The following diagram summarizes the graphing of parabolas.

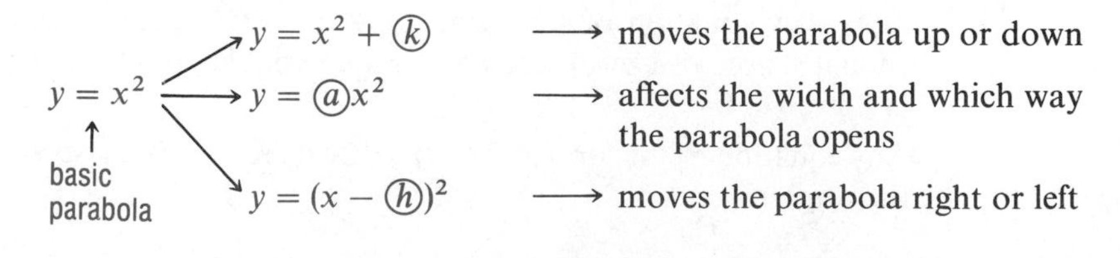

The **standard form of the equation of a circle** with its center at (h, k) and a radius of length r is

$$(x - h)^2 + (y - k)^2 = r^2.$$

The standard form of the equation of a circle with its center at the origin and a radius of length r is

$$x^2 + y^2 = r^2.$$

(7.8) The graph of an equation of the form

$$Ax^2 + By^2 = C,$$

where A, B, and C are nonzero constants of the same sign and $A \neq B$, is an **ellipse**.

The graph of an equation of the form

$$Ax^2 + By^2 = C,$$

where A, B, and C are nonzero constants and A and B are of unlike signs, is a **hyperbola**.

Circles, ellipses, parabolas, and hyperbolas are often referred to as **conic sections**.

Chapter 7 Review Problem Set

1. Find the slope of the line determined by each pair of points.
   (a) $(3, 4), (-2, -2)$ $\frac{6}{5}$ (b) $(-2, 3), (4, -1)$ $-\frac{2}{3}$
2. Find the slope of each of the following lines.
   (a) $4x + y = 7$ -4 (b) $2x - 7y = 3$ $\frac{2}{7}$
3. Find the lengths of the sides of a triangle whose vertices are at $(2, 3)$, $(5, -1)$, and $(-4, -5)$. 5, 10, and $\sqrt{97}$

For Problems 4–8, write the equation of each of the lines that satisfy the stated conditions. Express final equations in standard form.

4. containing the points $(-1, 2)$ and $(3, -5)$ $7x + 4y = 1$
5. having a slope of $-\frac{3}{7}$ and a y-intercept of 4 $3x + 7y = 28$
6. containing the point $(-1, -6)$ and having a slope of $\frac{2}{3}$ $2x - 3y = 16$
7. containing the point $(2, 5)$ and parallel to the line $x - 2y = 4$ $x - 2y = -8$
8. containing the point $(-2, -6)$ and perpendicular to the line $3x + 2y = 12$ $2x - 3y = 14$

For Problems 9–20, graph each of the equations.

9. $2x - y = 6$ see page A57
10. $y = -2x^2 - 1$ see page A57
11. $x^2 + y^2 = 1$ see page A57
12. $4x^2 + y^2 = 16$ see page A57
13. $y = -4x$ see page A57
14. $2x^2 - y^2 = 8$ see page A57

15. $y = x^2 + 4x - 1$ see page A57

16. $y = 4x^2 - 8x + 2$ see page A57

17. $xy^2 = -1$ see page A57

18. $3x^2 + 5y^2 = 30$ see page A57

19. $5y^2 - x^2 = 20$ see page A57

20. $y = -2x^2 + 12x - 16$ see page A57

For Problems 21 and 22, graph each of the inequalities.

21. $x + 2y \geq 4$ see page A58

22. $2x - 3y \leq 6$ see page A58

23. Find the center and length of a radius of the circle $x^2 + y^2 + 6x - 8y + 16 = 0$. $(-3, 4)$, $r = 3$

24. Find the coordinates of the vertex of the parabola $y = x^2 + 8x + 10$. $(-4, -6)$

25. Find the equations of the asymptotes of the hyperbola $9x^2 - 4y^2 = 72$, $y = \frac{3}{2}x$ and $y = -\frac{3}{2}x$

26. Find the length of the major axis of the ellipse $4x^2 + y^2 = 9$. 6

Thoughts into Words

27. Give a definition of analytic geometry that could be understood by a person who has never taken such a course.

28. Explain how you would go about graphing an equation such as $y = \dfrac{-1}{x^2 + 9}$.

29. Explain how you would go about graphing an equation such as $-2x - 5y = 10$.

30. Explain how you would go about graphing an inequality such as $-3 \leq x - 3y$.

31. What does it mean to say that two points determine a straight line? Do three points determine a straight line? Explain your answer.

32. Explain the concept of an asymptote.

Cumulative Review Problem Set

For Problems 1–6, express each radical in simplest radical form.

1. $\sqrt{32xy^3}$ $4y\sqrt{2xy}$

2. $\sqrt{48a^3b^4}$ $4ab^2\sqrt{3a}$

3. $\dfrac{3\sqrt{2}}{2\sqrt{3}}$ $\dfrac{\sqrt{6}}{2}$

4. $-\dfrac{2\sqrt{5}}{\sqrt{8}}$ $-\dfrac{\sqrt{10}}{2}$

5. $\sqrt[3]{48x^5}$ $2x\sqrt[3]{6x^2}$

6. $\dfrac{3}{4}\sqrt{108}$ $\dfrac{9\sqrt{3}}{2}$

For Problems 7 and 8, rationalize the denominator and simplify.

7. $\dfrac{3}{2\sqrt{5} - 1}$ $\dfrac{6\sqrt{5} + 3}{19}$

8. $\dfrac{2\sqrt{3}}{\sqrt{6} + 2\sqrt{2}}$ $-3\sqrt{2} + 2\sqrt{6}$

For Problems 9–18, perform the indicated operations and express answers in simplified form.

9. $(3\sqrt{6})(2\sqrt{8})$ $24\sqrt{3}$

10. $(\sqrt{x} + 2)(\sqrt{x} - 3)$ $x - \sqrt{x} - 6$

11. $(2\sqrt{3} + \sqrt{5})(\sqrt{3} - 2\sqrt{5})$ $-4 - 3\sqrt{15}$

12. $(x - 1)(2x^2 - x + 7)$ $2x^3 - 3x^2 + 8x - 7$

13. $\dfrac{9y^2}{x^2 + 12x + 36} \div \dfrac{12y}{x^2 + 6x}$ $\dfrac{3xy}{4(x + 6)}$

14. $\dfrac{x^2 - x}{4y} \cdot \dfrac{10xy^2}{2x - 2} \div \dfrac{3x^2 + 3x}{15x^2y^2}$ $\dfrac{25x^3y^3}{4(x + 1)}$

15. $\frac{3}{5x} - \frac{2}{3x} + \frac{5x}{6}$ $\frac{25x^2 - 2}{30x}$

16. $\frac{5}{x-9} + \frac{4}{x^2 - 3x - 54} - \frac{1}{x+6}$ $\frac{4x + 43}{(x-9)(x+6)}$

17. $(20x^2 - 39x + 18) \div (5x - 6)$ $4x - 3$

18. $\frac{4x^3 - 5x^2 + 2x - 6}{x^2 - 3x}$ $4x + 7 + \frac{23x - 6}{x^2 - 3x}$

For Problems 19–21, use scientific notation to help perform the indicated operations.

19. $\frac{(.00063)(960000)}{(3200)(.0000021)}$ 90,000

20. $(8000)^{\frac{2}{3}}$ 400

21. $\sqrt{.000009}$ .003

For Problems 22–27, find each of the indicated products or quotients and express answers in standard form.

22. $(3 - 2i)(5 + i)$ $17 - 7i$

23. $(-2 + 5i)(4 - 7i)$ $27 + 34i$

24. $(2 - 6i)^2$ $-32 - 24i$

25. $\frac{2}{5i}$ $0 - \frac{2}{5}i$

26. $\frac{3 + i}{2 - 4i}$ $\frac{1}{10} + \frac{7}{10}i$

27. $\frac{-1 - 4i}{-2 + 8i}$ $-\frac{15}{34} + \frac{4}{17}i$

For Problems 28–32, evaluate each of the numerical expressions.

28. $-8^{-\frac{1}{3}}$ $-\frac{1}{2}$

29. $(-8)^{\frac{1}{3}}$ -2

30. $\left(\frac{2}{5}\right)^{-2}$ $\frac{25}{4}$

31. $\sqrt[4]{16}$ 2

32. $2^0 + 2^{-1} + 2^{-2} + 2^{-3}$ $\frac{15}{8}$

For Problems 33–37, factor each of the algebraic expressions completely.

33. $9x^2 + 12xy + 4y^2$ $(3x + 2y)^2$

34. $27x^3 - 64y^3$ $(3x - 4y)(9x^2 + 12xy + 16y^2)$

35. $4x^4 - 13x^2 + 9$ $(2x + 3)(2x - 3)(x + 1)(x - 1)$

36. $3x^3 - 30x^2 - 72x$ $3x(x - 12)(x + 2)$

37. $x^2 - 4xy + 4y^2 - 4$ $(x - 2y - 2)(x - 2y + 2)$

For Problems 38–57, solve each of the equations. Some of these may have complex numbers as solutions.

38. $16n^2 - 40n + 25 = 0$ $\{\frac{5}{4}\}$

39. $x^3 = 8x$ $\{0, \pm 2\sqrt{2}\}$

40. $n^2 = -4n - 1$ $\{-2 \pm \sqrt{3}\}$

41. $(y + 2)^2 = -24$ $\{-2 \pm 2i\sqrt{6}\}$

42. $|3x - 2| = |-2x - 4|$ $\{-\frac{2}{5}, 6\}$

43. $x^3 = 8$ $\{2, -1 \pm i\sqrt{3}\}$

44. $(2x - 7)(x + 4) = 0$ $\{-4, \frac{7}{2}\}$

45. $(x - 5)(4x - 1) = 23$ $\{-\frac{3}{4}, 6\}$

46. $\frac{2n - 3}{3} + \frac{n + 1}{2} = 3$ $\{6\}$

47. $(x - 1)(x + 4) = (x + 1)(2x - 5)$ $\{3 \pm \sqrt{10}\}$

48. $.5(3x + .7) = 20.6$ $\{13.5\}$

49. $t^2 - 2t = -4$ $\{1 \pm i\sqrt{3}\}$

50. $4x^2 + 23x - 6 = 0$ $\{-6, \frac{1}{4}\}$

51. $x^2 - 4x - 192$ $\{-12, 16\}$

52. $\dfrac{3}{2x-8} - \dfrac{x-5}{x^2-2x-8} = \dfrac{7}{x+2}$ $\left\{\dfrac{72}{13}\right\}$

53. $\dfrac{3n}{n^2+n-6} + \dfrac{2}{n^2+4n+3} = \dfrac{n}{n^2-n-2}$ $\{-2, 1\}$

54. $\sqrt{x} + 6 = x$ $\{9\}$

55. $\sqrt{x+4} = \sqrt{x-1} + 1$ $\{5\}$

56. $x^4 + 5x^2 - 36 = 0$ $\{\pm 3i, \pm 2\}$

57. $\dfrac{2}{x-1} = \dfrac{x+4}{3}$ $\{-5, 2\}$

For Problems 58–61, solve each equation for the indicated variable.

58. $3x - 5y = 10$ for y $y = \dfrac{3x-10}{5}$

59. $\dfrac{3}{4} = \dfrac{y-1}{x-2}$ for y $y = \dfrac{3x-2}{4}$

60. $f = \dfrac{1}{\dfrac{1}{a} + \dfrac{1}{b}}$ for b $b = \dfrac{fa}{a-f}$

61. $V = C\left(1 - \dfrac{T}{N}\right)$ for T $T = \dfrac{NC - NV}{C}$

For Problems 62–70, solve each inequality and graph the solution set on a number line.

62. $|-2x - 1| > 3$ $(-\infty, -2) \cup (1, \infty)$

63. $|3x + 5| < 2$ $(-\frac{7}{3}, -1)$

64. $(x-2)(x+4) > 0$ $(-\infty, -4) \cup (2, \infty)$

65. $(x+1)(2x-3) < 0$ $(-1, \frac{3}{2})$

66. $\dfrac{x-1}{4} - \dfrac{x+2}{6} \le \dfrac{1}{8}$ $\left(-\infty, \dfrac{17}{2}\right]$

67. $\dfrac{x-3}{x-5} \ge 0$ $(-\infty, 3] \cup (5, \infty)$

68. $6x^2 + 13x - 5 > 0$ $(-\infty, -\frac{5}{2}) \cup (\frac{1}{3}, \infty)$

69. $\dfrac{2x}{x+3} > 4$ $(-6, -3)$

70. $\dfrac{3x+2}{x+4} \le 2$ $(-4, 6]$

For Problems 71–74, solve each problem by setting up and solving an appropriate equation.

71. Find two numbers whose sum is -2 and whose product is -35. -7 and 5

72. The sum of the lengths of the two legs of a right triangle is 9 centimeters. If the length of the hypotenuse is $3\sqrt{5}$ centimeters, find the length of each leg. 3 cm and 6 cm

73. A 3-by-5-inch picture is surrounded by a frame of uniform width. The area of the picture and frame together is 24 square inches. Find the width of the frame. $\frac{1}{2}$ inch

74. It takes Kent 2 hours longer to do a certain job than it takes Cindy. They worked together for 2 hours; then Cindy left to go shopping and Kent finished the job in 1 hour. How long would it take each of them to do the job alone? 4 hours for Cindy and 6 hours for Kent

Chapter 8

Functions

One of the fundamental concepts of mathematics is that of a function. Functions are used to unify mathematics and also to apply mathematics to many real-world problems. Functions provide a means of studying quantities that vary with one another, that is, when a change in one quantity causes a corresponding change in another.

This chapter will (1) introduce the basic ideas that pertain to the function concept, (2) review some concepts from Chapter 7 regarding functions, and (3) discuss some applications of functions.

8.1 Relations and Functions

Mathematically, a function is a special kind of **relation**, so we will begin our discussion with a simple definition of a relation.

DEFINITION 8.1

A **relation** is a set of ordered pairs.

Thus, a set of ordered pairs such as $\{(1, 2), (3, 7), (8, 14)\}$ is a relation. The set of all first components of the ordered pairs is the **domain** of the relation and the set of all second components is the **range** of the relation. The relation $\{(1, 2), (3, 7), (8, 14)\}$ has a domain of $\{1, 3, 8\}$ and a range of $\{2, 7, 14\}$.

The ordered pairs we refer to in Definition 8.1 may be generated by various means, such as a graph or a chart. However, one of the most common ways of generating ordered pairs is by use of equations. Since the solution set of an equation in two variables is a set of ordered pairs, such an equation describes a relation. Each of the following equations describes a relation between the variables x and y. We have listed *some* of the infinitely many ordered pairs (x, y) of each relation.

1. $x^2 + y^2 = 4$: $(1, \sqrt{3}), (1, -\sqrt{3}), (0, 2), (0, -2)$
2. $y^2 = x^3$: $(0, 0), (1, 1), (1, -1), (4, 8), (4, -8)$
3. $y = x + 2$: $(0, 2), (1, 3), (2, 4), (-1, 1), (5, 7)$
4. $y = \dfrac{1}{x - 1}$: $(0, -1), (2, 1), \left(3, \dfrac{1}{2}\right), \left(-1, -\dfrac{1}{2}\right), \left(-2, -\dfrac{1}{3}\right)$
5. $y = x^2$: $(0, 0), (1, 1), (2, 4), (-1, 1), (-2, 4)$

Now we direct your attention to the ordered pairs associated with equations 3, 4, and 5. Note that in each case no two ordered pairs have the same first component. Such a set of ordered pairs is called a **function**.

DEFINITION 8.2

A **function** is a relation in which no two ordered pairs have the same first component.

Stated another way, Definition 8.2 means that a function is a relation where each member of the domain is assigned *one and only one* member of the range. Thus, it is easy to determine that each of the following sets of ordered pairs is a function.

$$f = \{(x, y) \mid y = x + 2\},$$

$$g = \left\{(x, y) \,\middle|\, y = \frac{1}{x - 1}\right\},$$

$$h = \{(x, y) \mid y = x^2\}$$

In each case there is one and only one value of y (an element of the range) associated with each value of x (an element of the domain).

Notice that we named the previous functions f, g, and h. It is common to name functions by means of a single letter and the letters f, g, and h are often used. We would suggest more meaningful choices when functions are used to portray real-world situations. For example, if a problem involves a profit function, then naming the function p or even P would seem natural.

The symbol for a function can be used along with a variable that represents an element in the domain to represent the associated element in the range. For example, suppose that we have a function f specified in terms of the variable x. The symbol, $f(x)$, (read "f of x" or "the value of f at x") represents the element in the range associated with the element x from the domain. The function $f = \{(x, y) \mid y = x + 2\}$ can be written as $f = \{(x, f(x)) \mid f(x) = x + 2\}$ and this is usually shortened to read "f is the function determined by the equation $f(x) = x + 2$."

REMARK Be careful with the symbolism $f(x)$. As we stated above, it means the value of the function f at x. It does not mean f times x.

This **function notation** is very convenient when computing and expressing various values of the function. For example, the value of the function $f(x) = 3x - 5$ at $x = 1$ is

$$f(1) = 3(1) - 5 = -2.$$

Likewise, the functional values for $x = 2$, $x = -1$, and $x = 5$ are

$$f(2) = 3(2) - 5 = 1,$$

$$f(-1) = 3(-1) - 5 = -8, \quad \text{and}$$

$$f(5) = 3(5) - 5 = 10.$$

Thus, this function f contains the ordered pairs $(1, -2)$, $(2, 1)$, $(-1, -8)$, $(5, 10)$, and in general all ordered pairs of the form $(x, f(x))$, where $f(x) = 3x - 5$ and x is any real number.

It may be helpful for you to mentally picture the concept of a function in terms of a "function machine" as in Figure 8.1. Each time that a value of x is put into the

Figure 8.1

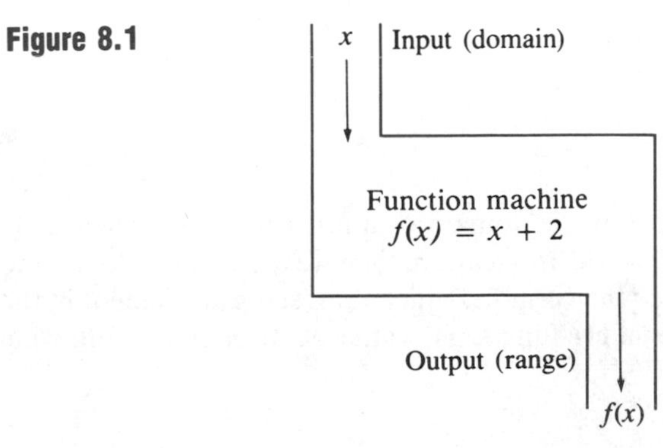

machine, the equation $f(x) = x + 2$ is used to generate one and only one value for $f(x)$ to be ejected from the machine. For example, if 3 is put into this machine, then $f(3) = 3 + 2 = 5$, and 5 is ejected. Thus, the ordered pair (3, 5) is one element of the function. Now let's look at some examples to help pull together some of the ideas about functions.

EXAMPLE 1 Determine whether the relation $\{(x, y) \mid y^2 = x\}$ is a function and specify its domain and range.

Solution Because $y^2 = x$ is equivalent to $y = \pm\sqrt{x}$, to each value of x there are assigned *two* values for y. Therefore, this relation is not a function.
The expression $\sqrt{x}$ requires that x be nonnegative; therefore, the domain (D) is

$$D = \{x \mid x \geq 0\}.$$

To each nonnegative real number, the relation assigns two real numbers, $\sqrt{x}$ and $-\sqrt{x}$. Thus, the range (R) is

$$R = \{y \mid y \text{ is a real number}\}.$$

■

EXAMPLE 2 For the function $f(x) = x^2$,

(a) specify its domain,

(b) determine its range, and

(c) evaluate $f(-2)$, $f(0)$, and $f(4)$.

Solution

(a) Any real number can be squared; therefore, the domain (D) is

$$D = \{x \mid x \text{ is a real number}\}.$$

(b) Squaring a real number always produces a nonnegative result. Thus, the range (R) is

$$R = \{f(x) \mid f(x) \geq 0\}.$$

(c) $f(-2) = (-2)^2 = 4$,
$f(0) = (0)^2 = 0$,
$f(4) = (4)^2 = 16$

■

For our purposes in this text, if the domain of a function is not specifically indicated or determined by a real-world application, then we assume the domain to be all **real number** replacements for the variable, which represents an element in the domain, that will produce **real number** functional values. Consider the following examples.

EXAMPLE 3 Specify the domain for each of the following.

(a) $f(x) = \dfrac{1}{x-1}$ (b) $f(t) = \dfrac{1}{t^2-4}$ (c) $f(s) = \sqrt{s-3}$

Solution

(a) We can replace x with any real number except 1, because 1 makes the denominator zero. Thus, the domain is given by

$$D = \{x \mid x \neq 1\}.$$

(b) We need to eliminate any value of t that will make the denominator zero. Thus, let's solve the equation $t^2 - 4 = 0$.

$$t^2 - 4 = 0$$
$$t^2 = 4$$
$$t = \pm 2$$

The domain is the set

$$D = \{t \mid t \neq -2 \text{ and } t \neq 2\}.$$

(c) The radicand, $s - 3$, must be nonnegative.

$$s - 3 \geq 0$$
$$s \geq 3$$

The domain is the set

$$D = \{s \mid s \geq 3\}.$$ ■

REMARK Certainly interval notation could be used to express the domains of functions such as in Example 3. However, we have chosen to use this section to give you a little more experience with set-builder notation.

EXAMPLE 4 If $f(x) = -2x + 7$ and $g(x) = x^2 - 5x + 6$, find $f(3)$, $f(-4)$, $g(2)$, and $g(-1)$.

Solution

$$f(x) = -2x + 7$$
$$f(3) = -2(3) + 7 = 1,$$
$$f(-4) = -2(-4) + 7 = 15$$

$$g(x) = x^2 - 5x + 6$$
$$g(2) = 2^2 - 5(2) + 6 = 0,$$
$$g(-1) = (-1)^2 - 5(-1) + 6 = 12$$ ■

In Example 4, notice that we are working with two different functions in the same problem. Thus, different names, f and g, are used.

The quotient $\dfrac{f(a+h) - f(a)}{h}$ is often called a **difference quotient** and we use it extensively with functions when studying the limit concept in calculus. The next two examples show how we found the difference quotient for two specific functions.

EXAMPLE 5 If $f(x) = 3x - 5$, find $\dfrac{f(a+h) - f(a)}{h}$.

Solution

$$\begin{aligned} f(a+h) &= 3(a+h) - 5 \\ &= 3a + 3h - 5 \end{aligned}$$

and

$$f(a) = 3a - 5$$

Therefore,

$$\begin{aligned} f(a+h) - f(a) &= (3a + 3h - 5) - (3a - 5) \\ &= 3a + 3h - 5 - 3a + 5 \\ &= 3h \end{aligned}$$

and

$$\frac{f(a+h) - f(a)}{h} = \frac{3h}{h} = 3.$$

■

EXAMPLE 6 If $f(x) = x^2 + 2x - 3$, find $\dfrac{f(a+h) - f(a)}{h}$.

Solution

$$\begin{aligned} f(a+h) &= (a+h)^2 + 2(a+h) - 3 \\ &= a^2 + 2ah + h^2 + 2a + 2h - 3, \end{aligned}$$

and

$$f(a) = a^2 + 2a - 3$$

Therefore,

$$\begin{aligned} f(a+h) - f(a) &= (a^2 + 2ah + h^2 + 2a + 2h - 3) - (a^2 + 2a - 3) \\ &= a^2 + 2ah + h^2 + 2a + 2h - 3 - a^2 - 2a + 3 \\ &= 2ah + h^2 + 2h, \end{aligned}$$

and

$$\begin{aligned} \frac{f(a+h) - f(a)}{h} &= \frac{2ah + h^2 + 2h}{h} \\ &= \frac{\not{h}(2a + h + 2)}{\not{h}} \\ &= 2a + h + 2. \end{aligned}$$

■

Functions and functional notation provide the basis for describing many real-world relationships. The next example illustrates this point.

EXAMPLE 7 Suppose a factory determines that the overhead for producing a quantity of a certain item is \$500 and the cost for each item is \$25. Express the total expenses as a function of the number of items produced and compute the expenses for producing 12, 25, 50, 75, and 100 items.

Solution Let n represent the number of items produced. Then $25n + 500$ represents the total expenses. Let's use E to represent the *expense function*, so that we have

$$E(n) = 25n + 500, \quad \text{where } n \text{ is a whole number,}$$

from which we obtain

$$E(12) = 25(12) + 500 = 800,$$
$$E(25) = 25(25) + 500 = 1125,$$
$$E(50) = 25(50) + 500 = 1750,$$
$$E(75) = 25(75) + 500 = 2375,$$
$$E(100) = 25(100) + 500 = 3000.$$

So the total expenses for producing 12, 25, 50, 75, and 100 items are \$800, \$1125, \$1750, \$2375, and \$3000, respectively. ■

Problem Set 8.1

For Problems 1–10, specify the domain and the range for each relation. Also state whether or not the relation is a function.

1. $\{(1,5), (2,8), (3,11), (4,14)\}$ $D = \{1,2,3,4\}$, $R = \{5,8,11,14\}$ It is a function.
2. $\{(0,0), (2,10), (4,20), (6,30), (8,40)\}$ $D = \{0,2,4,6,8\}$, $R = \{0,10,20,30,40\}$ It is a function.
3. $\{(0,5), (0,-5), (1,2\sqrt{6}), (1,-2\sqrt{6})\}$ $D = \{0,1\}$, $R = \{-2\sqrt{6},-5,5,2\sqrt{6}\}$ It is not a funotion.
4. $\{(1,1), (1,2), (1,-1), (1,-2), (1,3)\}$ $D = \{1\}$, $R = \{-1,-2,1,2,3\}$ It is not a function.
5. $\{(1,2), (2,5), (3,10), (4,17), (5,26)\}$ $D = \{1,2,3,4,5\}$, $R = \{2,5,10,17,26\}$ It is a function.
6. $\{(-1,5), (0,1), (1,-3), (2,-7)\}$ $D = \{-1,0,1,2\}$, $R = \{-7,-3,1,5\}$ It is a function.
7. $\{(x,y) \mid 5x - 2y = 6\}$ $D = \{\text{all reals}\}$, $R = \{\text{all reals}\}$ It is a function.
8. $\{(x,y) \mid y = -3x\}$ $D = \{\text{all reals}\}$, $R = \{\text{all reals}\}$ It is a function.
9. $\{(x,y) \mid x^2 = y^3\}$ $D = \{\text{all reals}\}$, $R = \{y \mid y \geq 0\}$ It is a function.
10. $\{(x,y) \mid x^2 - y^2 = 16\}$ $D = \{x \mid x \leq -4 \text{ or } x \geq 4\}$, $R = \{\text{all reals}\}$ It is not a function.

For Problems 11–36, specify the domain for each of the functions.

11. $f(x) = 7x - 2$ $\{\text{all reals}\}$
12. $f(x) = x^2 + 1$ $\{\text{all reals}\}$

13. $f(x) = \dfrac{1}{x-1}$ $\{x \mid x \neq 1\}$

14. $f(x) = \dfrac{-3}{x+4}$ $\{x \mid x \neq -4\}$

15. $g(x) = \dfrac{3x}{4x-3}$ $\left\{x \mid x \neq \dfrac{3}{4}\right\}$

16. $g(x) = \dfrac{5x}{2x+7}$ $\left\{x \mid x \neq -\dfrac{7}{2}\right\}$

17. $h(x) = \dfrac{2}{(x+1)(x-4)}$ $\{x \mid x \neq -1 \text{ and } x \neq 4\}$

18. $h(x) = \dfrac{-3}{(x-6)(2x+1)}$ $\left\{x \mid x \neq -\dfrac{1}{2} \text{ and } x \neq 6\right\}$

19. $f(x) = \dfrac{14}{x^2+3x-40}$ $\{x \mid x \neq -8 \text{ and } x \neq 5\}$

20. $f(x) = \dfrac{7}{x^2-8x-20}$ $\{x \mid x \neq -2 \text{ and } x \neq 10\}$

21. $f(x) = \dfrac{-4}{x^2+6x}$ $\{x \mid x \neq -6 \text{ and } x \neq 0\}$

22. $f(x) = \dfrac{9}{x^2-12x}$ $\{x \mid x \neq 0 \text{ and } x \neq 12\}$

23. $f(t) = \dfrac{4}{t^2+9}$ {all reals}

24. $f(t) = \dfrac{8}{t^2+1}$ {all reals}

25. $f(t) = \dfrac{3t}{t^2-4}$ $\{t \mid t \neq -2 \text{ and } t \neq 2\}$

26. $f(t) = \dfrac{-2t}{t^2-25}$ $\{t \mid t \neq -5 \text{ and } t \neq 5\}$

27. $h(x) = \sqrt{x+4}$ $\{x \mid x \geq -4\}$

28. $h(x) = \sqrt{5x-3}$ $\{x \mid \geq \frac{3}{5}\}$

29. $f(s) = \sqrt{4s-5}$ $\{s \mid s \geq \frac{5}{4}\}$

30. $f(s) = \sqrt{s-2} + 5$ $\{s \mid s \geq 2\}$

31. $f(x) = \sqrt{x^2-16}$ $\{x \mid x \leq -4 \text{ or } x \geq 4\}$

32. $f(x) = \sqrt{x^2-49}$ $\{x \mid x \leq -7 \text{ or } x \geq 7\}$

33. $f(x) = \sqrt{x^2-3x-18}$ $\{x \mid x \leq -3 \text{ or } x \geq 6\}$

34. $f(x) = \sqrt{x^2+4x-32}$ $\{x \mid x \leq -8 \text{ or } x \geq 4\}$

35. $f(x) = \sqrt{1-x^2}$ $\{x \mid -1 \leq x \leq 1\}$

36. $f(x) = \sqrt{9-x^2}$ $\{x \mid -3 \leq x \leq 3\}$

37. If $f(x) = 5x - 2$, find $f(0)$, $f(2)$, $f(-1)$, and $f(-4)$. $f(0) = -2, f(2) = 8, f(-1) = -7, f(-4) = -22$

38. If $f(x) = -3x - 4$, find $f(-2)$, $f(-1)$, $f(3)$, and $f(5)$. $f(-2) = 2, f(-1) = -1, f(3) = -13, f(5) = -19$

39. If $f(x) = \dfrac{1}{2}x - \dfrac{3}{4}$, find $f(-2)$, $f(0)$, $f\left(\dfrac{1}{2}\right)$, and $f\left(\dfrac{2}{3}\right)$. $f(-2) = -\dfrac{7}{4}$, $f(0) = -\dfrac{3}{4}$, $f\left(\dfrac{1}{2}\right) = -\dfrac{1}{2}$, $f\left(\dfrac{2}{3}\right) = -\dfrac{5}{12}$

40. If $g(x) = x^2 + 3x - 1$, find $g(1)$, $g(-1)$, $g(3)$, and $g(-4)$. $g(1) = 3, g(-1) = -3$ $g(3) = 17, g(-4) = 3$

41. If $g(x) = 2x^2 - 5x - 7$, find $g(-1)$, $g(2)$, $g(-3)$, and $g(4)$. $g(-1) = 0, g(2) = -9, g(-3) = 26, g(4) = 5$

42. If $h(x) = -x^2 - 3$, find $h(1)$, $h(-1)$, $h(-3)$, and $h(5)$. $h(1) = -4, h(-1) = -4, h(-3) = -12, h(5) = -28$

43. If $h(x) = -2x^2 - x + 4$, find $h(-2)$, $h(-3)$, $h(4)$, and $h(5)$. $h(-2) = -2, h(-3) = -11, h(4) = -32, h(5) = -51$

44. If $f(x) = \sqrt{x-1}$, find $f(1)$, $f(5)$, $f(13)$, and $f(26)$. $f(1) = 0, f(5) = 2, f(13) = 2\sqrt{3}, f(26) = 5$

45. If $f(x) = \sqrt{2x+1}$, find $f(3)$, $f(4)$, $f(10)$, and $f(12)$. $f(3) = \sqrt{7}, f(4) = 3, f(10) = \sqrt{21}, f(12) = 5$

46. If $f(x) = \dfrac{3}{x-2}$, find $f(3)$, $f(0)$, $f(-1)$, and $f(-5)$. $f(3) = 3, f(0) = -\dfrac{3}{2}, f(-1) = -1, f(-5) = -\dfrac{3}{7}$

47. If $f(x) = \dfrac{-4}{x+3}$, find $f(1)$, $f(-1)$, $f(3)$, and (-6). $f(1) = -1, f(-1) = -2, f(3) = -\dfrac{2}{3}, f(-6) = \dfrac{4}{3}$

48. If $f(x) = 2x^2 - 7$ and $g(x) = x^2 + x - 1$, find $f(-2)$, $f(3)$, $g(-4)$, and $g(5)$. $f(-2) = 1, f(3) = 11, g(-4) = 11, g(5) = 29$

49. If $f(x) = 5x^2 - 2x + 3$ and $g(x) = -x^2 + 4x - 5$, find $f(-2)$, $f(3)$, $g(-4)$, and $g(6)$. $f(-2) = 27, f(3) = 42, g(-4) = -37, g(6) = -17$

50. If $f(x) = |3x - 2|$ and $g(x) = |x| + 2$, find $f(1)$, $f(-1)$, $g(2)$, and $g(-3)$. $f(1) = 1, f(-1) = 5, g(2) = 4, g(-3) = 5$

51. If $f(x) = 3|x| - 1$ and $g(x) = -|x| + 1$, find $f(-2)$, $f(3)$, $g(-4)$, and $g(5)$. $f(-2) = 5, f(3) = 8, g(-4) = -3, g(5) = -4$

For Problems 52–59, find $\dfrac{f(a+h) - f(a)}{h}$ for each of the given functions.

52. $f(x) = 5x - 4$ 5

53. $f(x) = -3x + 6$ -3

54. $f(x) = x^2 + 5$ $2a + h$

55. $f(x) = -x^2 - 1$ $-2a - h$

56. $f(x) = x^2 - 3x + 7$ $2a - 3 + h$

57. $f(x) = 2x^2 - x + 8$ $4a - 1 + 2h$

58. $f(x) = -3x^2 + 4x - 1$ $-6a + 4 - 3h$

59. $f(x) = -4x^2 - 7x - 9$ $-8a - 7 - 4h$

60. Suppose that the cost function for producing a certain item is given by $C(n) = 3n + 5$, where n represents the number of items produced. Compute $C(150)$, $C(500)$, $C(750)$, and $C(1500)$. $C(150) = 455, C(500) = 1505, C(750) = 2255, C(1500) = 4505$

61 The height of a projectile fired vertically into the air (neglecting air resistance) at an initial velocity of 64 feet per second is a function of the time (t) and is given by the equation

$$h(t) = 64t - 16t^2.$$

Compute $h(1)$, $h(2)$, $h(3)$, and $h(4)$. $h(1) = 48, h(2) = 64, h(3) = 48, h(4) = 0$

62. The profit function for selling n items is given by $P(n) = -n^2 + 500n - 61{,}500$. Compute $P(200)$, $P(230)$, $P(250)$, and $P(260)$. $P(200) = -1500, P(230) = 600, P(250) = 1000, P(260) = 900$

63. A car rental agency charges \$50 per day plus \$0.32 a mile. Therefore, the daily charge for renting a car is a function of the number of miles traveled (m) and can be expressed as $C(m) = 50 + 0.32m$. Compute $C(75)$, $C(150)$, $C(225)$, and $C(650)$. $C(75) = \$74, C(150) = \$98, C(225) = \$122, C(650) = \$258$

64. The equation $A(r) = \pi r^2$ expresses the area of a circular region as a function of the length of a radius (r). Use 3.14 as an approximation for π and compute $A(2)$, $A(3)$, $A(12)$, and $A(17)$. $A(2) = 12.56$, $A(3) = 28.26$, $A(12) = 452.16$, $A(17) = 907.46$

65. The equation $I(r) = 500\,r$ expresses the amount of simple interest earned by an investment of \$500 for one year as a function of the rate of interest (r). Compute $I(0.11)$, $I(0.12)$, $I(0.135)$, and $I(0.15)$. $I(.11) = 55$, $I(.12) = 60$, $I(.135) = 67.5$, $I(.15) = 75$

8.2 Special Functions and Their Graphs

In Section 7.1 we used phrases such as "the graph of the solution set of the equation $y = x - 1$," or simply "the graph of the equation $y = x - 1$" is a line that contains the points $(0, -1)$ and $(1, 0)$. Because the equation $y = x - 1$ (which can be written as $f(x) = x - 1$) can be used to specify a function, the line previously referred to is also called the **graph of the function specified by the equation** or simply the **graph of the function**. Generally speaking, the graph of any equation that determines a function is also called the graph of the function. Thus, the graphing techniques discussed in Chapter 7 will continue to play an important role as we graph functions.

As we use the function concept in our study of mathematics, it is helpful to classify certain types of functions and become familiar with their equations, characteristics, and graphs. In this section we will discuss two special types of functions—**linear** and **quadratic functions**. These functions are merely an outgrowth of our earlier study of linear and quadratic equations.

Linear Functions

Any function that can be written in the form

$$f(x) = ax + b,$$

where a and b are real numbers, is called a **linear function**. The following are examples of linear functions.

$$f(x) = -3x + 6, \qquad f(x) = 2x + 4, \qquad f(x) = -\frac{1}{2}x - \frac{3}{4}$$

Graphing linear functions is quite easy because the graph of every linear function is a straight line. Therefore, all we need to do is determine two points of the graph and draw the line determined by those two points. You may want to continue using a third point as a check point.

EXAMPLE 1 Graph the function $f(x) = -3x + 6$.

Solution Because $f(0) = 6$, the point $(0, 6)$ is on the graph. Likewise, because $f(1) = 3$, the point $(1, 3)$ is on the graph. Plot these two points and draw the line determined by the two points to produce Figure 8.2.

Figure 8.2

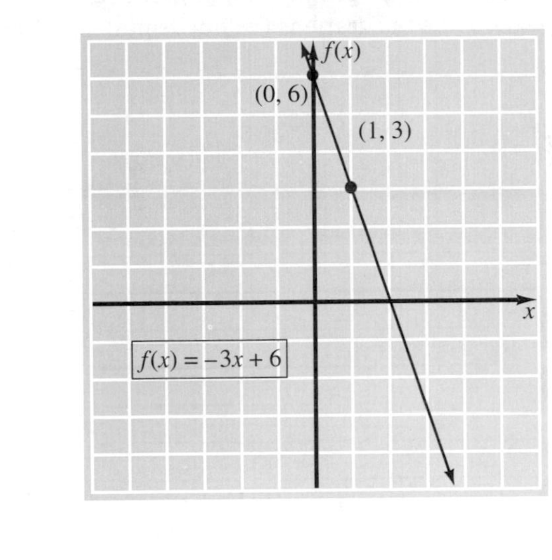

■

REMARK Note that in Figure 8.2 we labeled the vertical axis $f(x)$. We could also label it y, since $f(x) = -3x + 6$ and $y = -3x + 6$ mean the same thing. We will continue to use the $f(x)$ label in this chapter to help you adjust to the function notation.

EXAMPLE 2 Graph the function $f(x) = x$.

Solution The equation $f(x) = x$ can be written as $f(x) = 1x + 0$; thus, it is a linear function. Since $f(0) = 0$ and $f(2) = 2$, the points $(0, 0)$ and $(2, 2)$ determine the line in Figure 8.3. The function $f(x) = x$ is often called the **identity function**.

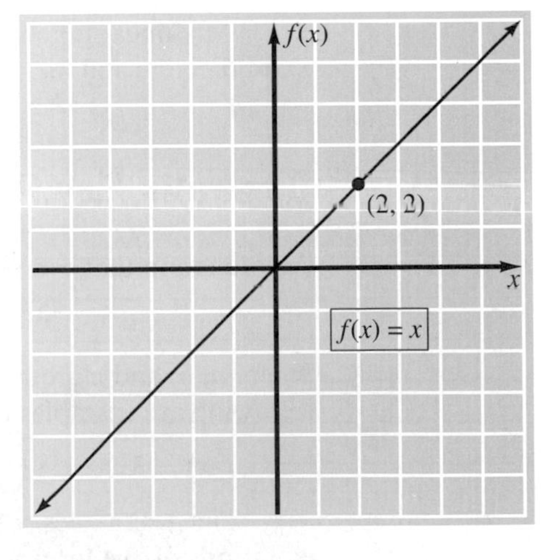

Figure 8.3

■

As you use function notation to graph functions, it is often helpful to think of the ordinate of every point on the graph as the value of the function at a specific value of x. Geometrically, this functional value is the directed distance of the point from the x-axis as illustrated in Figure 8.4, with the function $f(x) = 2x - 4$. For example, consider the graph of the function $f(x) = 2$. The function $f(x) = 2$ means that every functional value is 2, or geometrically, that every point on the graph is 2 units above the x-axis. Thus the graph is the horizontal line shown in Figure 8.5.

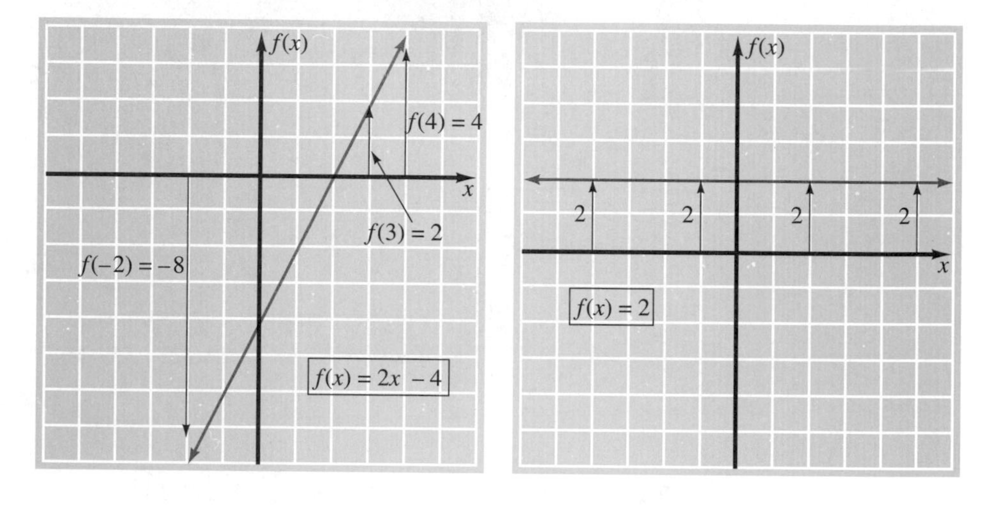

Figure 8.4

Figure 8.5

Any linear function of the form $f(x) = ax + b$, where $a = 0$, is called a **constant function** and its graph is a *horizontal line.*

Quadratic Functions

Any function that can be written in the form

$$f(x) = ax^2 + bx + c,$$

where a, b, and c are real numbers with $a \neq 0$, is called a **quadratic function**. The following are examples of quadratic functions.

$$f(x) = 3x^2, \qquad f(x) = -2x^2 + 5x, \qquad f(x) = 4x^2 - 7x + 1$$

The techniques we discussed in Chapter 7 relative to graphing quadratic equations of the form $y = ax^2 + bx + c$ provide the basis for graphing quadratic functions. Let's review some work from Chapter 7 with an example.

EXAMPLE 3 Graph the function $f(x) = 2x^2 - 4x + 5$.

Solution

$$\begin{aligned} f(x) &= 2x^2 - 4x + 5 \\ &= 2(x^2 - 2x + \underline{\quad}) + 5 \\ &= 2(x^2 - 2x + 1) + 5 - 2 \\ &= 2(x - 1)^2 + 3 \end{aligned}$$

Recall the process of completing the square!

From this form we can obtain the following information about the parabola.

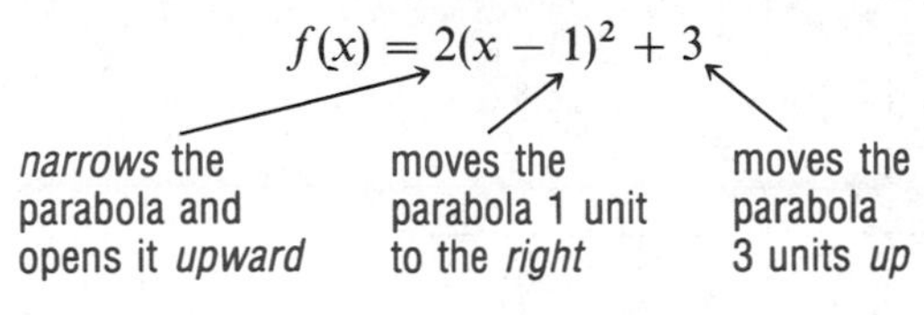

Thus, the parabola can be drawn as in Figure 8.6.

Figure 8.6

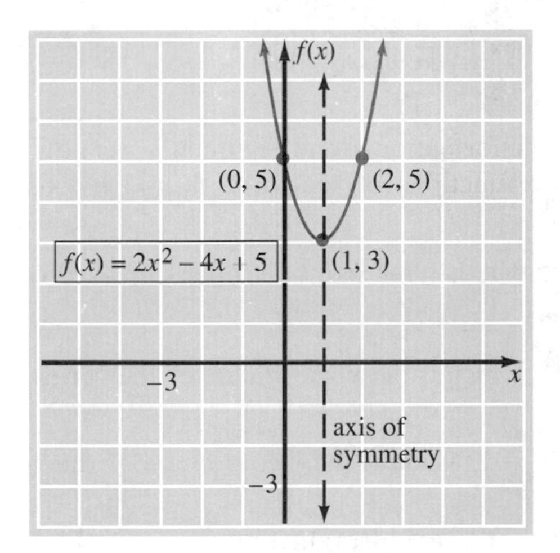

In general, if we complete the square on

$$f(x) = ax^2 + bx + c,$$

we obtain

$$\begin{aligned} f(x) &= a\left(x^2 + \frac{b}{a}x + \underline{\quad}\right) + c \\ &= a\left(x^2 + \frac{b}{a}x + \frac{b^2}{4a^2}\right) + c - \frac{b^2}{4a} \\ &= a\left(x + \frac{b}{2a}\right)^2 + \frac{4ac - b^2}{4a}. \end{aligned}$$

Therefore, the parabola associated with $f(x) = ax^2 + bx + c$ has its vertex at $\left(-\frac{b}{2a}, \frac{4ac - b^2}{4a}\right)$ and the equation of its axis of symmetry is $x = -\frac{b}{2a}$. These facts are illustrated in Figure 8.7.

Figure 8.7

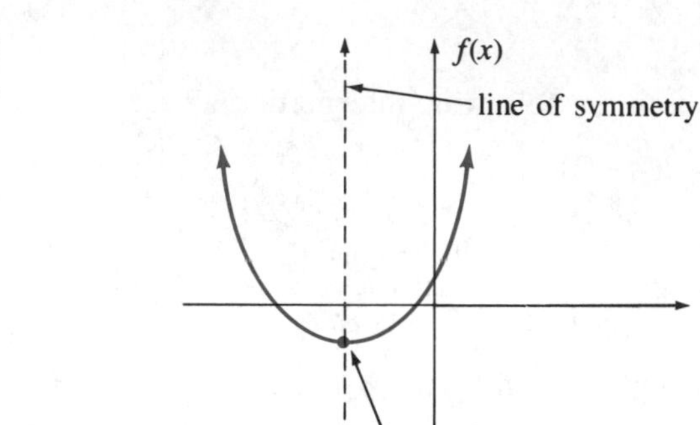

By using the information from Figure 8.7 we now have another way of graphing quadratic functions of the form $f(x) = ax^2 + bx + c$, shown by the following steps.

1. Determine whether the parabola opens upward (if $a > 0$) or downward (if $a < 0$).
2. Find $-\frac{b}{2a}$, which is the x-coordinate of the vertex.
3. Find $f\left(-\frac{b}{2a}\right)$, which is the y-coordinate of the vertex. $\left(\text{You could also find the } y\text{-coordinate by evaluating } \frac{4ac - b^2}{4a}.\right)$
4. Locate another point on the parabola and also locate its image across the line of symmetry, $x = -\frac{b}{2a}$.

The three points in Steps 2, 3, and 4 should determine the general shape of the parabola. Let's try two examples and use these steps.

EXAMPLE 4 Graph $f(x) = 3x^2 - 6x + 5$.

Solution

Step 1. Because $a = 3$, the parabola opens upward.

Step 2. $-\frac{b}{2a} = -\frac{-6}{6} = 1.$

Step 3. $f\left(-\dfrac{b}{2a}\right) = f(1) = 3 - 6 + 5 = 2$. Thus, the vertex is at $(1, 2)$.

Step 4. Letting $x = 2$, we obtain $f(2) = 12 - 12 + 5 = 5$. Thus $(2, 5)$ is on the graph and so is its reflection $(0, 5)$ across the line of symmetry $x = 1$.

The three points $(1, 2)$, $(2, 5)$ and $(0,5)$ are used to graph the parabola in Figure 8.8.

Figure 8.8

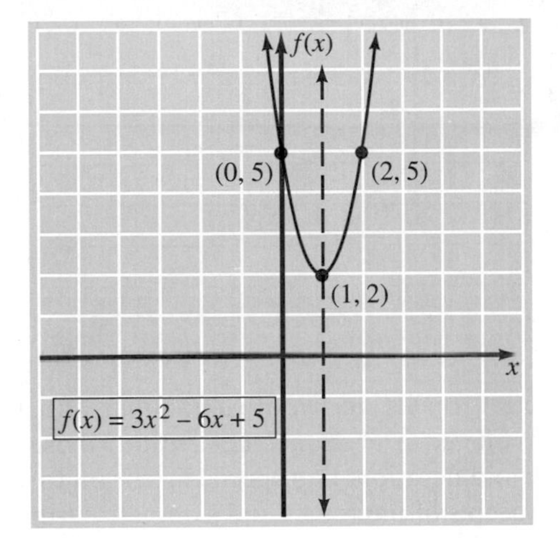

■

EXAMPLE 5 Graph $f(x) = -x^2 - 4x - 7$.

Solution

Step 1. Since $a = -1$, the parabola opens downward.

Step 2. $-\dfrac{b}{2a} = -\dfrac{-4}{-2} = -2$.

Step 3. $f\left(-\dfrac{b}{2a}\right) = f(-2) = -(-2)^2 - 4(-2) - 7 = -3$.

Step 4. Letting $x = 0$, we obtain $f(0) = -7$. Thus, $(0, -7)$ is on the graph and so is its reflection $(-4, -7)$ across the line of symmetry $x = -2$.

The three points $(-2, -3)$, $(0, -7)$ and $(-4, -7)$ are used to draw the parabola in Figure 8.9. See page 394. ■

In summary, to graph a quadratic function we basically have two methods.

1. We can express the function in the form $f(x) = a(x - h)^2 + k$ and use the values of a, h, and k to determine the parabola; *or*
2. We can express the function in the form $f(x) = ax^2 + bx + c$ and use the approach demonstrated in Examples 4 and 5.

Figure 8.9

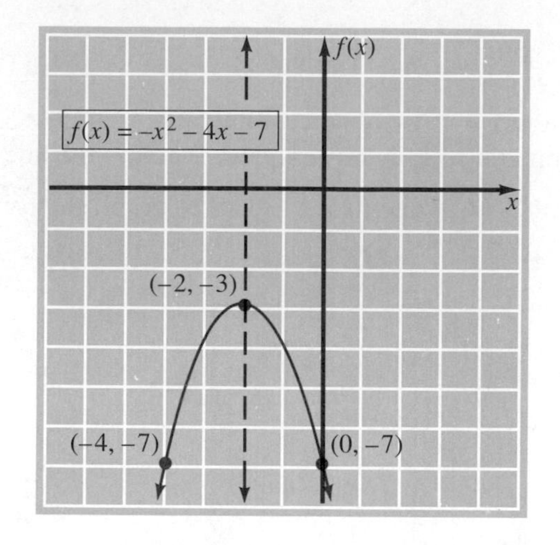

Other Functions

To graph a new function, that is, one we are unfamiliar with, we can use some of the graphing suggestions offered in Chapter 7. Let's restate those suggestions in terms of function vocabulary and notation. Pay special attention to steps 2 and 3 where we have restated the concepts of intercepts and symmetry using function notation.

1. Determine the domain of the function.
2. Determine any types of symmetry that the equation possesses. If $f(-x) = f(x)$, then the function exhibits y-axis symmetry. If $f(-x) = -f(x)$, then the function exhibits origin symmetry. (Note that the definition of a function rules out the possibility that the graph of a function has x-axis symmetry.)
3. Find the y-intercept (we are labeling the y-axis with $f(x)$) by evaluating $f(0)$. Find the x-intercept by finding the value(s) of x such that $f(x) = 0$.
4. Set up a table of ordered pairs that satisfy the equation. The type of symmetry and the domain will affect your choice of values of x in the table.
5. Plot the points associated with the ordered pairs and connect them with a smooth curve. Then, if appropriate, reflect this part of the curve according to any symmetries possessed by the graph.

Let's consider some examples in terms of these suggestions.

EXAMPLE 6 Graph $f(x) = \sqrt{x}$.

Solution Since the radicand must be nonnegative, the domain is the set of nonnegative real numbers. Since $x \geq 0$, $f(-x)$ is not a real number; so there is no symmetry for this graph. We see that $f(0) = 0$; so both intercepts are 0. That is to say, the origin $(0, 0)$ is a point of the graph. Now let's set up a table of values keeping in mind that $x \geq 0$. Plotting these points and connecting them with a smooth curve produces Figure 8.10.

x	$f(x)$
0	0
1	1
4	2
9	3

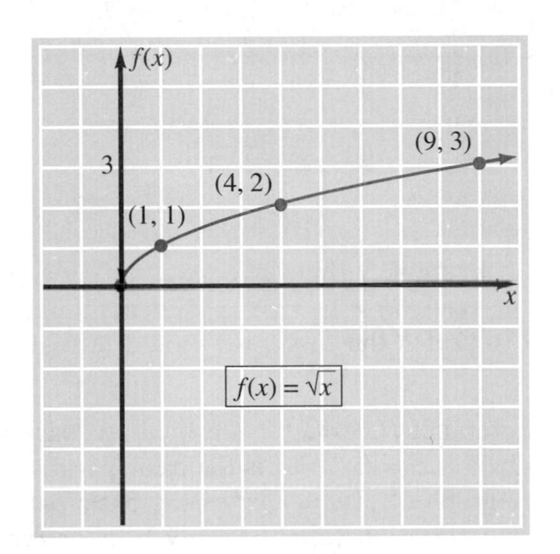

Sometimes a new function is defined in terms of old functions. In such cases, the definition plays an important role in the study of the new function. Consider the following example.

EXAMPLE 7 Graph the function $f(x) = |x|$.

Solution The concept of absolute value is defined for all real numbers as

$$|x| = x \quad \text{if } x \geq 0,$$

$$|x| = -x \quad \text{if } x < 0.$$

Therefore, we can express the absolute value function as

$$f(x) = |x| = \begin{Bmatrix} x & \text{if } x \geq 0 \\ -x & \text{if } x < 0 \end{Bmatrix}.$$

The graph of $f(x) = x$ for $x \geq 0$ is the ray in the first quadrant and the graph of $f(x) = -x$ for $x < 0$ is the half-line in the second quadrant as indicated in Figure 8.11.

Figure 8.11

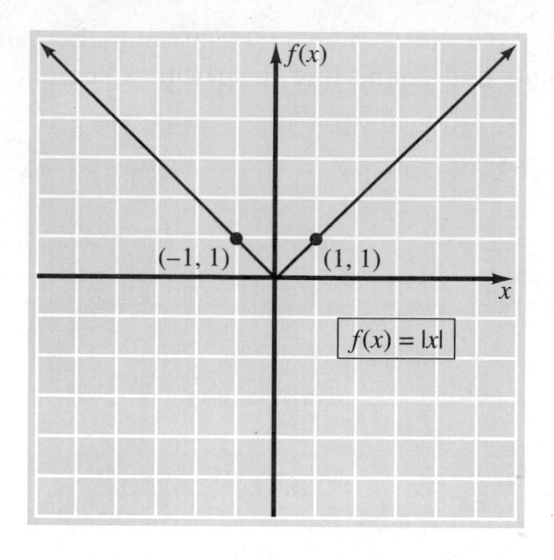

REMARK Note in Example 7 that the equation $f(x) = |x|$ does exhibit y-axis symmetry because $f(-x) = |-x| = |x|$. Even though we did not use the symmetry idea to sketch the curve, we should recognize that the symmetry does exist.

Recall that the graph of $f(x) = x^2 + k$ is the basic parabola moved vertically k units, and the graph of $f(x) = (x - h)^2$ is the basic parabola moved horizontally h units. Likewise, the graph of $f(x) = -x^2$ is the basic parabola reflected across the x-axis. These same variations apply to any basic curve. In Figure 8.12 the graph of $f(x) = \sqrt{x} + 2$ is shown as the graph of $f(x) = \sqrt{x}$ moved up two units. Likewise, in Figure 8.13 the graph of $f(x) = |x - 3|$ is shown as the graph of $f(x) = |x|$ moved three units to the right.

Figure 8.12

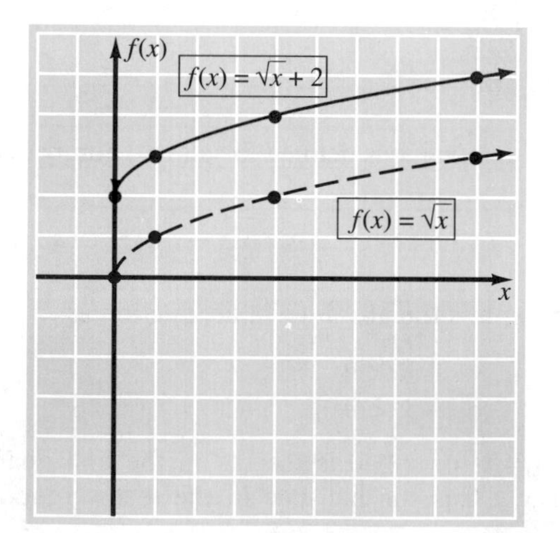

Figure 8.13

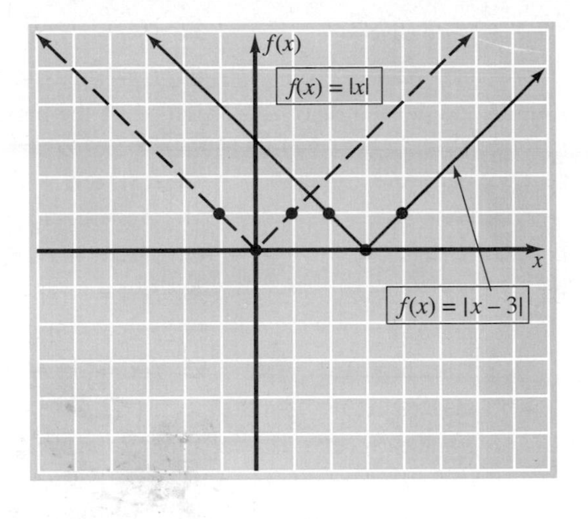

Problem Set 8.2

Graph each of the following linear and quadratic functions.

1. $f(x) = 2x - 4$ see page A59
2. $f(x) = 3x + 3$ see page A59
3. $f(x) = -2x^2$ see page A59
4. $f(x) = -4x^2$ see page A59
5. $f(x) = -3x$ see page A59
6. $f(x) = -4x$ see page A59
7. $f(x) = -(x + 1)^2 - 2$ see page A59
8. $f(x) = -(x - 2)^2 + 4$ see page A59
9. $f(x) = -x + 3$ see page A59
10. $f(x) = -2x - 4$ see page A59
11. $f(x) = x^2 + 2x - 2$ see page A59
12. $f(x) = x^2 - 4x - 1$ see page A59
13. $f(x) = -x^2 + 6x - 8$ see page A60
14. $f(x) = -x^2 - 8x - 15$ see page A60
15. $f(x) = -3$ see page A60
16. $f(x) = 1$ see page A60
17. $f(x) = 2x^2 - 20x + 52$ see page A60
18. $f(x) = 2x^2 + 12x + 14$ see page A60
19. $f(x) = -3x^2 + 6x$ see page A60
20. $f(x) = -4x^2 - 8x$ see page A60
21. $f(x) = x^2 - x + 2$ see page A60
22. $f(x) = x^2 + 3x + 2$ see page A60
23. $f(x) = 2x^2 + 10x + 11$ see page A60
24. $f(x) = 2x^2 - 10x + 15$ see page A60
25. $f(x) = -2x^2 - 1$ see page A61
26. $f(x) = -3x^2 + 2$ see page A61
27. $f(x) = -3x^2 + 12x - 7$ see page A61
28. $f(x) = -3x^2 - 18x - 23$ see page A61
29. $f(x) = -2x^2 + 14x - 25$ see page A61
30. $f(x) = -2x^2 - 10x - 14$ see page A61

For Problems 31–40, graph each curve as a variation of the basic square root curve determined by $f(x) = \sqrt{x}$. (See Figure 8.10 on page 395.)

31. $f(x) = \sqrt{x} - 1$ see page A61
32. $f(x) = \sqrt{x} + 3$ see page A61
33. $f(x) = \sqrt{x + 2}$ see page A61
34. $f(x) = \sqrt{x - 1}$ see page A61
35. $f(x) = -\sqrt{x - 2}$ see page A61
36. $f(x) = -\sqrt{x}$ see page A61

37. $f(x) = \sqrt{x + 3} + 1$ see page A62

38. $f(x) = -\sqrt{x - 2} - 2$ see page A62

39. $f(x) = 3\sqrt{x}$ see page A62

40. $f(x) = -2\sqrt{x}$ see page A62

For Problems 41–52, graph each curve as a variation of the basic absolute value curve determined by $f(x) = |x|$. (See Figure 8.11 on page 396.)

41. $f(x) = |x| - 1$ see page A62

42. $f(x) = |x| + 2$ see page A62

43. $f(x) = |x + 2|$ see page A62

44. $f(x) = |x - 1|$ see page A62

45. $f(x) = -|x - 2|$ see page A62

46. $f(x) = -|x + 3|$ see page A62

47. $f(x) = -|x| + 3$ see page A62

48. $f(x) = -|x| - 2$ see page A62

49. $f(x) = 2|x|$ see page A63

50. $f(x) = -3|x|$ see page A63

51. $f(x) = 3|x - 2| + 1$ see page A63

52. $f(x) = 2|x + 2| - 1$ see page A63

53. Graph $f(x) = x^3$. see page A63

For Problems 54–61, graph each curve as a variation of the basic cubic curve from Problem 53.

54. $f(x) = x^3 - 1$ see page A63

55. $f(x) = x^3 + 2$ see page A63

56. $f(x) = (x + 2)^3$ see page A63

57. $f(x) = (x - 3)^3$ see page A63

58. $f(x) = -x^3$ see page A63

59. $f(x) = -2x^3$ see page A63

60. $f(x) = (x - 1)^3 - 2$ see page A63

61. $f(x) = -(x + 3)^3 + 1$ see page A64

62. Graph $f(x) = \frac{1}{x}$. see page A64

For Problems 63–68, graph each curve as a variation of the basic curve from Problem 62.

63. $f(x) = \frac{1}{x} - 2$ see page A64

64. $f(x) = \frac{1}{x} + 3$ see page A64

65. $f(x) = \frac{1}{x + 1}$ see page A64

66. $f(x) = \frac{1}{x - 2}$ see page A64

67. $f(x) = -\frac{1}{x}$ see page A64

68. $f(x) = \frac{3}{x}$ see page A64

8.3 Problem Solving and the Composition of Functions

As we have seen, the vertex of the graph of a quadratic function is either the lowest or the highest point on the graph. Thus, the vocabulary *minimum value* or *maximum value* of a function is often used in applications of the parabola. The x-value of the vertex indicates where the minimum or maximum occurs and $f(x)$ yields the minimum or maximum value of the function. Let's consider some examples that use these ideas.

EXAMPLE 1 A farmer has 120 rods of fencing and wants to enclose a rectangular plot of land that requires fencing on only three sides, since it is bounded by a river on one side. Find the length and width of the plot that will maximize the area.

Solution Let x represent the width; then $120 - 2x$ represents the length, as indicated in Figure 8.14.

Figure 8.14

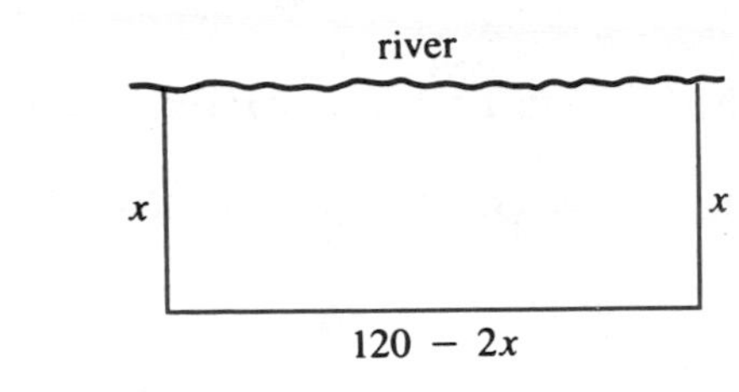

The function $A(x) = x(120 - 2x)$ represents the area of the plot in terms of the width x. Since

$$\begin{aligned} A(x) &= x(120 - 2x) \\ &= 120x - 2x^2 \\ &= -2x^2 + 120x, \end{aligned}$$

we have a quadratic function with $a = -2$, $b = 120$, and $c = 0$. Therefore, the x-value where the maximum value of the function is obtained is

$$-\frac{b}{2a} = -\frac{120}{2(-2)} = 30.$$

If $x = 30$, then $120 - 2x = 120 - 2(30) = 60$.
Thus, the farmer should make the plot 30 rods wide and 60 rods long to maximize the area at $(30)(60) = 1800$ square rods. ■

EXAMPLE 2 Find two numbers whose sum is 30, such that the sum of their squares is a minimum.

Solution Let x represent one of the numbers; then $30 - x$ represents the other number. By expressing the sum of the squares as a function of x we obtain

$$f(x) = x^2 + (30 - x)^2,$$

which can be simplified to

$$\begin{aligned} f(x) &= x^2 + 900 - 60x + x^2 \\ &= 2x^2 - 60x + 900. \end{aligned}$$

This is a quadratic function with $a = 2$, $b = -60$, and $c = 900$. Therefore, the x-value where the minimum occurs is

$$-\frac{b}{2a} = -\frac{-60}{4} = 15.$$

If $x = 15$, then $30 - x = 30 - (15) = 15$. Thus, the two numbers should both be 15. ■

EXAMPLE 3 A golf pro-shop operator finds that she can sell 30 sets of golf clubs at $500 per set in a year. Furthermore, she predicts that for each $25 decrease in price, three extra sets of golf clubs could be sold. At what price should she sell the clubs to maximize gross income?

Solution Sometimes in analyzing such a problem it helps to start setting up a table as follows.

	Number of sets	*Price per set*	*Income*
	30	$500	$15,000
3 additional sets can be sold for a $25 decrease in price →	33	$475	$15,675
	36	$450	$16,200

Let x represent the number of $25 decreases in price. Then we can express the income as a function of x as follows.

$$f(x) = \underbrace{(30 + 3x)}_{\text{number of sets}}\underbrace{(500 - 25x)}_{\text{price per set}}$$

Simplifying this, we obtain

$$\begin{aligned} f(x) &= 15{,}000 - 750x + 1500x - 75x^2 \\ &= -75x^2 + 750x + 15{,}000. \end{aligned}$$

Completing the square we obtain

$$\begin{aligned} f(x) &= -75x^2 + 750x + 15{,}000 \\ &= -75(x^2 - 10x + \underline{\quad}) + 15{,}000 \\ &= -75(x^2 - 10x + 25) + 15{,}000 + 1875 \\ &= -75(x - 5)^2 + 16{,}875. \end{aligned}$$

From this form we know that the vertex of the parabola is at (5,16875). So 5 decreases of $25 each, that is, a $125 reduction in price, will give a maximum income of $16,875. The golf clubs should be sold at $375 per set. ■

The Composition of Functions

The basic operations of addition, subtraction, multiplication, and division can be performed on functions. However, for our purposes in this text, there is an additional operation, called **composition**, that we will use in the next section. Here is the definition and an illustration of this operation.

DEFINITION 8.3

> The **composition** of functions f and g is defined by
>
> $$(f \circ g)(x) = f(g(x)),$$
>
> for all x in the domain of g such that $g(x)$ is in the domain of f.

The left side, $(f \circ g)(x)$, of the equation in Definition 8.3 can be read as "the composition of f and g" and the right side, $f(g(x))$, can be read as "f of g of x." It may also be helpful for you to mentally picture Definition 8.3 as two function machines *hooked together* to produce another function (often called a **composite function**) as illustrated in Figure 8.15. Notice that what comes out of the function g is substituted into the function f. Thus, composition is sometimes called the substitution of functions.

Figure 8.15

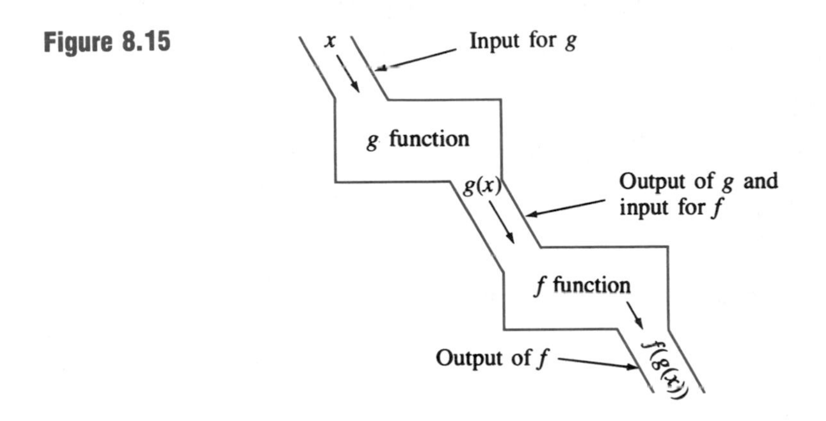

Figure 8.15 also vividly illustrates the fact that $f \circ g$ is defined *for all x in the domain of g such that $g(x)$ is in the domain of f*. In other words, what comes out of g must be capable of being fed into f. Let's consider some examples.

EXAMPLE 4 If $f(x) = x^2$ and $g(x) = x - 3$, find $(f \circ g)(x)$ and determine its domain.

Solution Applying Definition 8.3 we obtain

$$\begin{aligned}(f \circ g)(x) &= f(g(x)) \\ &= f(x - 3) \\ &= (x - 3)^2.\end{aligned}$$

Because g and f are both defined for all real numbers, so is $f \circ g$. ■

EXAMPLE 5 If $f(x) = \sqrt{x}$ and $g(x) = x - 4$, find $(f \circ g)(x)$ and determine its domain.

Solution Applying Definition 8.3 we obtain

$$\begin{aligned}(f \circ g)(x) &= f(g(x)) \\ &= f(x - 4) \\ &= \sqrt{x - 4}.\end{aligned}$$

The domain of g is all real numbers but the domain of f is only the nonnegative real numbers. Thus $g(x)$, which is $x - 4$, has to be nonnegative. So

$$x - 4 \geq 0$$

$$x \geq 4,$$

and the domain of $f \circ g$ is $D = \{x \mid x \geq 4\}$. ■

Definition 8.3, with f and g interchanged, defines the composition of g and f as $(g \circ f)(x) = g(f(x))$.

EXAMPLE 6 If $f(x) = x^2$ and $g(x) = x - 3$, find $(g \circ f)(x)$ and determine its domain.

Solution

$$\begin{aligned}(g \circ f)(x) &= g(f(x)) \\ &= g(x^2) \\ &= x^2 - 3\end{aligned}$$

Since f and g are both defined for all real numbers, the domain of $g \circ f$ is the set of all real numbers. ■

The results of Examples 4 and 6 demonstrate an important idea, namely, that the composition of functions is *not a commutative operation.* In other words, it is not true that $f \circ g = g \circ f$ for all functions f and g. However, as we will see in the next section, there is a special class of functions where $f \circ g = g \circ f$.

EXAMPLE 7 If $f(x) = 2x + 3$ and $g(x) = \sqrt{x - 1}$, determine each of the following.

(a) $(f \circ g)(x)$ **(b)** $(g \circ f)(x)$ **(c)** $(f \circ g)(5)$ **(d)** $(g \circ f)(7)$

Solution

(a) $$\begin{aligned}(f \circ g)(x) &= f(g(x)) \\ &= f(\sqrt{x - 1}) \\ &= 2\sqrt{x - 1} + 3 \qquad D = \{x \mid x \geq 1\}\end{aligned}$$

(b) $$\begin{aligned}(g \circ f)(x) &= g(f(x)) \\ &= g(2x + 3) \\ &= \sqrt{2x + 3 - 1} \\ &= \sqrt{2x + 2} \qquad D = \{x \mid x \geq -1\}\end{aligned}$$

(c) By using the composite function formed in part (a), we obtain

$$\begin{aligned}(f \circ g)(5) &= 2\sqrt{5-1} + 3 \\ &= 2\sqrt{4} + 3 \\ &= 2(2) + 3 = 7.\end{aligned}$$

(d) By using the composite function formed in part (b) we obtain

$$(g \circ f)(7) = \sqrt{2(7) + 2} = \sqrt{16} = 4.$$

■

Problem Set 8.3

1. Suppose that the cost function for a particular item is given by the equation $C(x) = 2x^2 - 320x + 12{,}920$, where x represents the number of items. How many items should be produced to minimize the cost? 80
2. Suppose that the equation $p(x) = -2x^2 + 280x - 1000$, where x represents the number of items sold, describes the profit function for a certain business. How many items should be sold to maximize the profit? 70
3. Find two numbers whose sum is 30, such that the sum of the square of one number plus ten times the other number is a minimum. 5 and 25
4. The height of a projectile fired vertically into the air (neglecting air resistance) at an initial velocity of 96 feet per second is a function of the time and is given by the equation $f(x) = 96x - 16x^2$, where x represents the time. Find the highest point reached by the projectile. 144
5. Two hundred and forty meters of fencing is available to enclose a rectangular playground. What should be the dimensions of the playground to maximize the area? 60 meters by 60 meters
6. Find two numbers whose sum is 50 and whose product is a maximum. 25 and 25
7. A Cable TV company has 1000 subscribers who each pay $15 per month. Based on a survey, they feel that for each decrease of $0.25 on the monthly rate, they could obtain 20 additional subscribers. At what rate will maximum revenue be obtained and how many subscribers will it take at that rate? 1100 subscribers at $13.75 per month
8. A motel advertises that they will provide dinner, a dance, and drinks at $50 per couple for a New Year's Eve party. They must have a guarantee of 30 couples. Furthermore, they will agree that for each couple in excess of 30, they will reduce the price per couple for all attending by $0.50. How many couples will it take to maximize the motel's revenue? 65 couples

For Problems 9–20, determine the indicated functional values.

9. If $f(x) = 9x - 2$ and $g(x) = -4x + 6$, find $(f \circ g)(-2)$ and $(g \circ f)(4)$. 124 and -130
10. If $f(x) = -2x - 6$ and $g(x) = 3x + 10$, find $(f \circ g)(5)$ and $(g \circ f)(-3)$. -56 and 10
11. If $f(x) = 4x^2 - 1$ and $g(x) = 4x + 5$, find $(f \circ g)(1)$ and $(g \circ f)(4)$. 323 and 257
12. If $f(x) = -5x + 2$ and $g(x) = -3x^2 + 4$, find $(f \circ g)(-2)$ and $(g \circ f)(-1)$. 42 and -143

13. If $f(x) = \frac{1}{x}$ and $g(x) = \frac{2}{x-1}$, find $(f \circ g)(2)$ and $(g \circ f)(-1)$. $\frac{1}{2}$ and -1

14. If $f(x) = \frac{2}{x-1}$ and $g(x) = -\frac{3}{x}$, find $(f \circ g)(1)$ and $(g \circ f)(-1)$. $-\frac{1}{2}$ and 3

15. If $f(x) = \frac{1}{x-2}$ and $g(x) = \frac{4}{x-1}$, find $(f \circ g)(3)$ and $(g \circ f)(2)$. undefined and undefined

16. If $f(x) = \sqrt{x+6}$ and $g(x) = 3x - 1$, find $(f \circ g)(-2)$ and $(g \circ f)(-2)$, undefined and 5

17. If $f(x) = \sqrt{3x-2}$ and $g(x) = -x + 4$, find $(f \circ g)(1)$ and $(g \circ f)(6)$. $\sqrt{7}$ and 0

18. If $f(x) = -5x + 1$ and $g(x) = \sqrt{4x+1}$, find $(f \circ g)(6)$ and $(g \circ f)(-1)$. -24 and 5

19. If $f(x) = |4x - 5|$ and $g(x) = x^3$, find $(f \circ g)(-2)$ and $(g \circ f)(2)$. 37 and 27

20. If $f(x) = -x^3$ and $g(x) = |2x + 4|$, find $(f \circ g)(-1)$ and $(g \circ f)(-3)$. -8 and 58

For Problems 21–38, determine $(f \circ g)(x)$ and $(g \circ f)(x)$ for each pair of functions. Also specify the domain of $(f \circ g)(x)$ and $(g \circ f)(x)$.

21. $f(x) = 3x$ and $g(x) = 5x - 1$ $(f \circ g)(x) = 15x - 3, D = \{\text{all reals})$
$(g \circ f)(x) = 15x - 1, D = \{\text{all reals}\}$

22. $f(x) = 4x - 3$ and $g(x) = -2x$ $(f \circ g((x) = -8x - 3, D = \{\text{all reals}\}$
$(g \circ f)(x) = -8x + 6, D = \{\text{all reals}\}$

23. $f(x) = -2x + 1$ and $g(x) = 7x + 4$ $(f \circ g)(x) = -14x - 7, D = \{\text{all reals}\}$
$(g \circ f)(x) = -14x + 11, D = \{\text{all reals}\}$

24. $f(x) = 6x - 5$ and $g(x) = -x + 6$ $(f \circ g)(x) = -6x + 31, D = \{\text{all reals}\}$
$(g \circ f)(x) = -6x + 11, D = \{\text{all reals}\}$

25. $f(x) = 3x + 2$ and $g(x) = x^2 + 3$ $(f \circ g)(x) = 3x^2 + 11, D = \{\text{all reals}\}$
$(g \circ f)(x) = 9x^2 + 12x + 7, D = \{\text{all reals}\}$

26. $f(x) = -2x + 4$ and $g(x) = 2x^2 - 1$ $(f \circ g)(x) = -4x^2 + 6, D = \{\text{all reals}\}$
$(g \circ f)(x) = 8x^2 - 32x + 31, D = \{\text{all reals}\}$

27. $f(x) = 2x^2 - x + 2$ and $g(x) = -x + 3$ $(f \circ g)(x) = 2x^2 - 11x + 17, D = \{\text{all reals}\}$
$(g \circ f)(x) = -2x^2 + x + 1, D = \{\text{all reals}\}$

28. $f(x) = 3x^2 - 2x - 4$ and $g(x) = -2x + 1$ $(f \circ g)(x) = 12x^2 - 8x - 3, D = \{\text{all reals}\}$
$(g \circ f)(x) = -6x^2 + 4x + 9, D = \{\text{all reals}\}$

29. $f(x) = \frac{3}{x}$ and $g(x) = 4x - 9$ $(f \circ g)(x) = \frac{3}{4x-9}, D = \left\{x \mid x \neq \frac{9}{4}\right\}$
$(g \circ f)(x) = \frac{12 - 9x}{x}, D = \{x \mid x \neq 0\}$

30. $f(x) = -\frac{2}{x}$ and $g(x) = -3x + 6$ $(f \circ g)(x) = \frac{2}{3x-6}, D = \{x \mid x \neq 2\}$
$(g \circ f)(x) = \frac{6 + 6x}{x}, D = \{x \mid x \neq 0\}$

31. $f(x) = \sqrt{x+1}$ and $g(x) = 5x + 3$ $(f \circ g)(x) = \sqrt{5x+4}, D = \{x \mid x \geq -\frac{4}{5}\}$
$(g \circ f)(x) = 5\sqrt{x+1} + 3, D = \{x \mid x \geq -1\}$

32. $f(x) = 7x - 2$ and $g(x) = \sqrt{2x-1}$ $(f \circ g)(x) = 7\sqrt{2x-1} - 2, D = \{x \mid x \geq \frac{1}{2}\}$
$(g \circ f)(x) = \sqrt{14x-5}, D = \{x \mid x \geq \frac{5}{14}\}$

33. $f(x) = \frac{1}{x}$ and $g(x) = \frac{1}{x-4}$ $(f \circ g)(x) = x - 4,\ D = \{x \mid x \neq 4\}$

$(g \circ f)(x) = \frac{x}{1-4x},\ D = \left\{x \mid x \neq 0 \text{ and } x \neq \frac{1}{4}\right\}$

34. $f(x) = \frac{2}{x+3}$ and $g(x) = -\frac{3}{x}$ $(f \circ g)(x) = \frac{2x}{3x-3},\ D = \{x \mid x \neq 0 \text{ and } x \neq 1\}$

$(g \circ f)(x) = \frac{-3x-9}{2},\ D = \{x \mid x \neq -3\}$

35. $f(x) = \sqrt{x}$ and $g(x) = \frac{4}{x}$ $(f \circ g)(x) = \frac{2\sqrt{x}}{x},\ D = \{x \mid x > 0\}$

$(g \circ f)(x) = \frac{4\sqrt{x}}{x},\ D = \{x \mid x > 0\}$

36. $f(x) = \frac{2}{x}$ and $g(x) = |x|$ $(f \circ g)(x) = \frac{2}{|x|},\ D = \{x \mid x \neq 0\}$

$(g \circ f)(x) = \left|\frac{2}{x}\right|,\ D = \{x \mid x \neq 0\}$

37. $f(x) = \frac{3}{2x}$ and $g(x) = \frac{1}{x+1}$ $(f \circ g)(x) = \frac{3x+3}{2},\ D = \{x \mid x \neq -1\}$

$(g \circ f)(x) = \frac{2x}{2x+3},\ D = \left\{x \mid x \neq 0 \text{ and } x \neq -\frac{3}{2}\right\}$

38. $f(x) = \frac{4}{x-2}$ and $g(x) = \frac{3}{4x}$ $(f \circ g)(x) = \frac{16x}{3-8x},\ D = \left\{x \mid x \neq 0 \text{ and } x \neq \frac{3}{8}\right\}$

$(g \circ f)(x) = \frac{3x-6}{16},\ D = \{x \mid x \neq 2\}$

For Problems 39–46, show that $(f \circ g)(x) = x$ and $(g \circ f)(x) = x$ for each pair of functions.

39. $f(x) = 3x$ and $g(x) = \frac{1}{3}x$

40. $f(x) = -2x$ and $g(x) = -\frac{1}{2}x$

41. $f(x) = 4x + 2$ and $g(x) = \frac{x-2}{4}$

42. $f(x) = 3x - 7$ and $g(x) = \frac{x+7}{3}$

43. $f(x) = \frac{1}{2}x + \frac{3}{4}$ and $g(x) = \frac{4x-3}{2}$

44. $f(x) = \frac{2}{3}x - \frac{1}{5}$ and $g(x) = \frac{3}{2}x + \frac{3}{10}$

45. $f(x) = -\frac{1}{4}x - \frac{1}{2}$ and $g(x) = -4x - 2$

46. $f(x) = -\frac{3}{4}x + \frac{1}{3}$ and $g(x) = -\frac{4}{3}x + \frac{4}{9}$

8.4 Inverse Functions

Graphically, the distinction between a relation and a function can be easily recognized. In Figure 8.16 we have sketched four graphs. Which of these are graphs of functions and which are graphs of relations that are not functions? Think in terms of "to each member of the domain there is assigned one and only one member of the range"; this is the basis for what is known as the **vertical line test for functions.** Because each value of x produces only one value of $f(x)$, any vertical line drawn through a **graph of a function must not intersect the graph in more than one point.** Therefore, parts (a) and (c) of Figure 8.16 are graphs of functions and parts (b) and (d) are graphs of relations that are not functions.

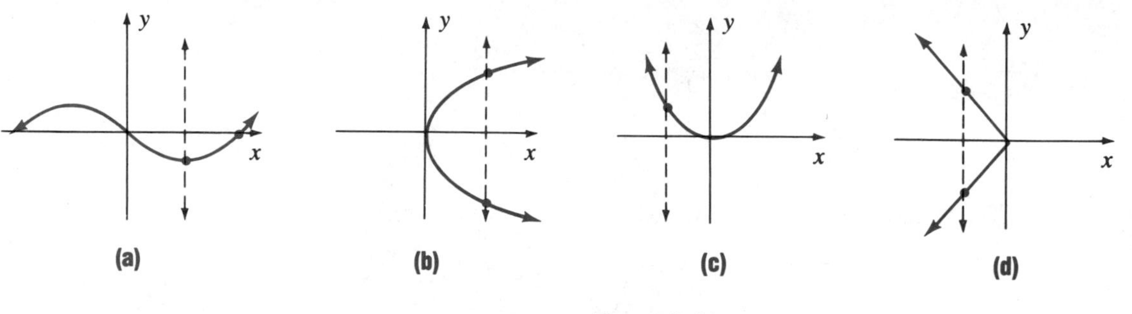

Figure 8.16

There is also a useful distinction made between two basic types of functions. Consider the graphs of the two functions $f(x) = 2x - 1$ and $f(x) = x^2$ in Figure 8.17. In part (a) any *horizontal line* will intersect the graph in no more than one point. Therefore, every value of $f(x)$ has only one value of x associated with it. Any function that has the additional property of having only one value of x associated with each value of $f(x)$ is called a **one-to-one function.** The function $f(x) = x^2$ is not a one-to-one function because the horizontal line in part (b) of Figure 8.17 intersects the parabola in two points.

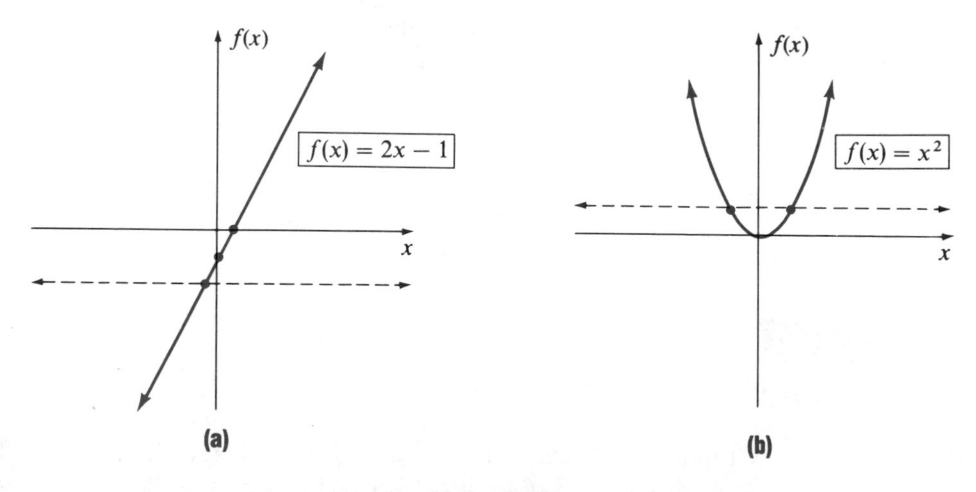

Figure 8.17

In terms of ordered pairs, a one-to-one function does not contain any ordered pairs that have the same second component. For example,

$$f = \{(1, 3), (2, 6), (4, 12)\}$$

is a one-to-one function, but

$$g = \{(1, 2), (2, 5), (-2, 5)\}$$

is not a one-to-one function.

If the components of each ordered pair of a given one-to-one function are interchanged, the resulting function and the given function are called **inverses** of each other. Thus,

$$\{(1, 3), (2, 6), (4, 12)\} \quad \text{and} \quad \{(3, 1), (6, 2), (12, 4)\}$$

are **inverse functions**. The inverse of a function f is denoted by f^{-1} (read "f inverse" or "the inverse of f"). If (a, b) is an ordered pair of f, then (b, a) is an ordered pair of f^{-1}. The domain and range of f^{-1} are the range and domain, respectively, of f.

REMARK Do not confuse the -1 in f^{-1} with a negative exponent. The symbol f^{-1} does not mean $\frac{1}{f^1}$, but refers to the inverse function of function f.

Graphically, two functions that are inverses of each other are mirror images with reference to the line $y = x$. This is due to the fact that ordered pairs (a, b) and (b, a) are mirror images with respect to the line $y = x$ as illustrated in Figure 8.18. Therefore, if we know the graph of a function f, as in Figure 8.19(a), then we can determine the graph of f^{-1} by reflecting f across the line $y = x$ (Figure 8.19(b)).

Another useful way of viewing inverse functions is in terms of composition. Basically, inverse functions *undo* each other and this can be more formally stated as follows. If f and g are inverses of each other, then

1. $(f \circ g)(x) = f(g(x)) = x$ for all x in domain of g; and
2. $(g \circ f)(x) = g(f(x)) = x$ for all x in domain of f.

Figure 8.18

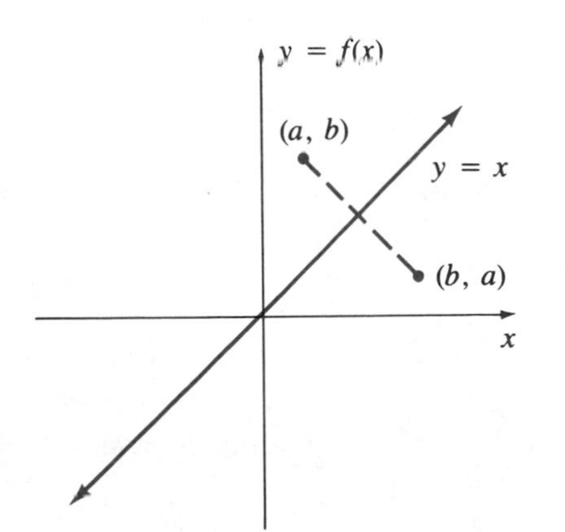

Figure 8.19

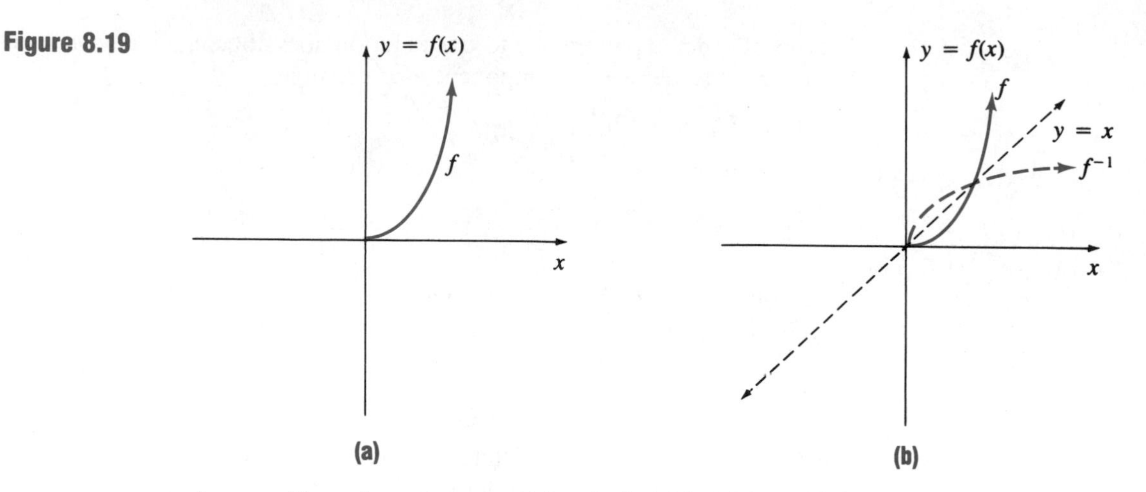

As we will see in a moment, this relationship of inverse functions can be used to verify whether two functions are indeed inverses of each other.

Finding Inverse Functions

The idea of inverse functions *undoing each other* provides the basis for a rather informal approach to finding the inverse of a function. Consider the function

$$f(x) = 2x + 1.$$

To each x this function assigns *twice x plus 1*. To *undo* this function, we could *subtract* 1 *and divide by* 2. So, the inverse should be

$$f^{-1}(x) = \frac{x-1}{2}.$$

Now let's verify that f and f^{-1} are inverses of each other.

$$(f \circ f^{-1})(x) = f(f^{-1}(x)) = f\left(\frac{x-1}{2}\right) = 2\left(\frac{x-1}{2}\right) + 1 = x$$

and

$$(f^{-1} \circ f)(x) = f^{-1}(f(x)) = f^{-1}(2x+1) = \frac{2x+1-1}{2} = x$$

Thus, the inverse of f is given by

$$f^{-1}(x) = \frac{x-1}{2}.$$

Let's consider another example of finding an inverse function by the *undoing* process.

EXAMPLE 1 Find the inverse of $f(x) = 3x - 5$.

Solution To each x, the function f assigns *three times x minus 5*. To *undo* this we can *add 5 and then divide by 3*. So, the inverse should be

$$f^{-1}(x) = \frac{x+5}{3}.$$

To verify that f and f^{-1} are inverses we can show that

$$(f \circ f^{-1})(x) = f(f^{-1}(x)) = f\left(\frac{x+5}{3}\right)$$
$$= 3\left(\frac{x+5}{3}\right) - 5 = x$$

and

$$(f^{-1} \circ f)(x) = f^{-1}(f(x)) = f^{-1}(3x-5)$$
$$= \frac{3x-5+5}{3} = x.$$

Thus, f and f^{-1} are inverses and we can write

$$f^{-1}(x) = \frac{x+5}{3}.$$

■

This informal approach may not work very well with more complex functions, but it does emphasize how inverse functions are related to each other. A more formal and systematic technique for finding the inverse of a function can be described as follows.

1. Replace the symbol $f(x)$ by y.
2. Interchange x and y.
3. Solve the equation for y in terms of x.
4. Replace y by the symbol $f^{-1}(x)$.

Now let's use two examples to illustrate this technique.

EXAMPLE 2 Find the inverse of $f(x) = -3x + 11$.

Solution Replacing $f(x)$ by y, the given equation becomes

$$y = -3x + 11.$$

Interchanging x and y produces

$$x = -3y + 11.$$

Now, solving for y we obtain

$$x = -3y + 11$$
$$3y = -x + 11$$
$$y = \frac{-x + 11}{3}.$$

Finally, replacing y by $f^{-1}(x)$ we can express the inverse function as

$$f^{-1}(x) = \frac{-x + 11}{3}.$$

■

EXAMPLE 3 Find the inverse of $f(x) = \frac{3}{2}x - \frac{1}{4}$.

Solution Replacing $f(x)$ by y, the given equation becomes

$$y = \frac{3}{2}x - \frac{1}{4}.$$

Interchanging x and y produces

$$x = \frac{3}{2}y - \frac{1}{4}.$$

Now, solving for y we obtain

$$x = \frac{3}{2}y - \frac{1}{4}$$
$$4x = 6y - 1$$
$$4x + 1 = 6y$$
$$\frac{4x + 1}{6} = y$$
$$\frac{2}{3}x + \frac{1}{6} = y.$$

Finally, replacing y by $f^{-1}(x)$ we can express the inverse function as

$$f^{-1}(x) = \frac{2}{3}x + \frac{1}{6}.$$

■

For both Examples 2 and 3 you should be able to show that $(f \circ f^{-1})(x) = x$ and $(f^{-1} \circ f)(x) = x$.

Problem Set 8.4

For Problems 1–8, identify each graph as (a) the graph of a function or (b) the graph of a relation that is not a function. Use the vertical line test.

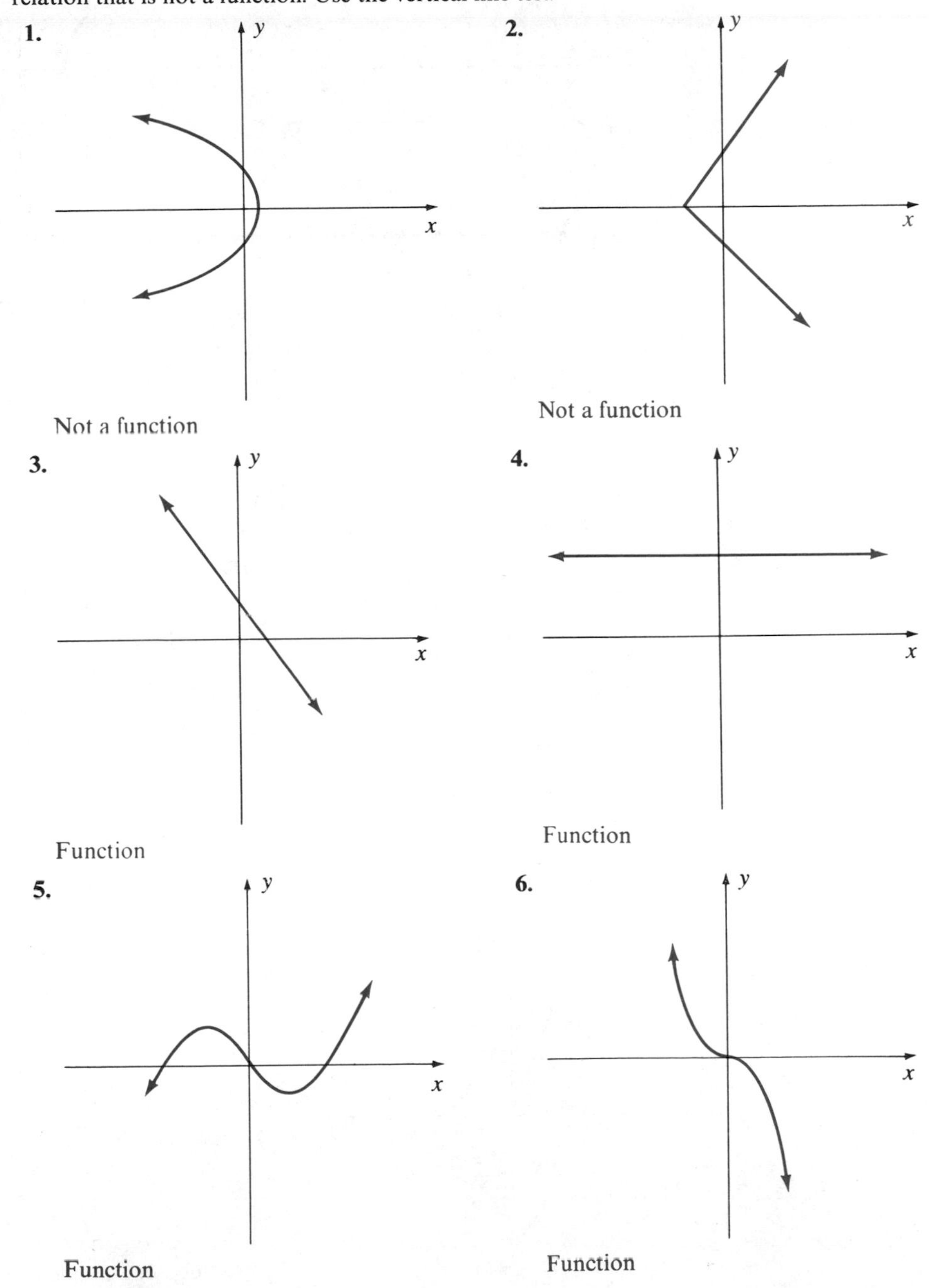

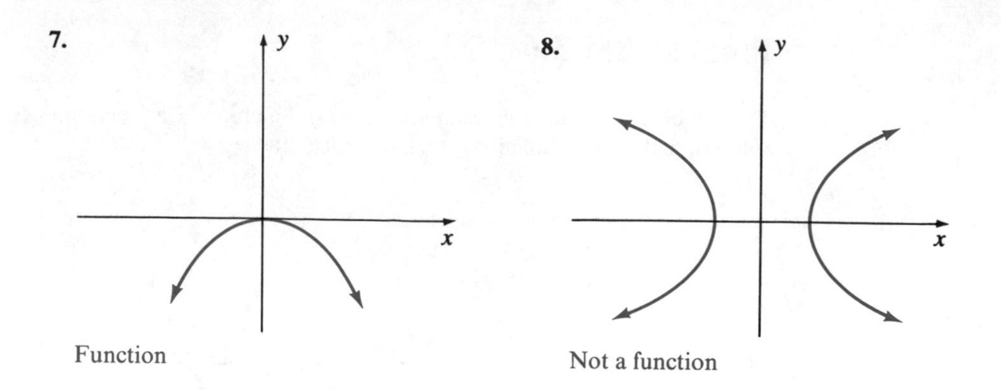

For Problems 9–16, identify each graph as (a) the graph of a one-to-one function or (b) the graph of a function that is not one-to-one. Use the horizontal line test.

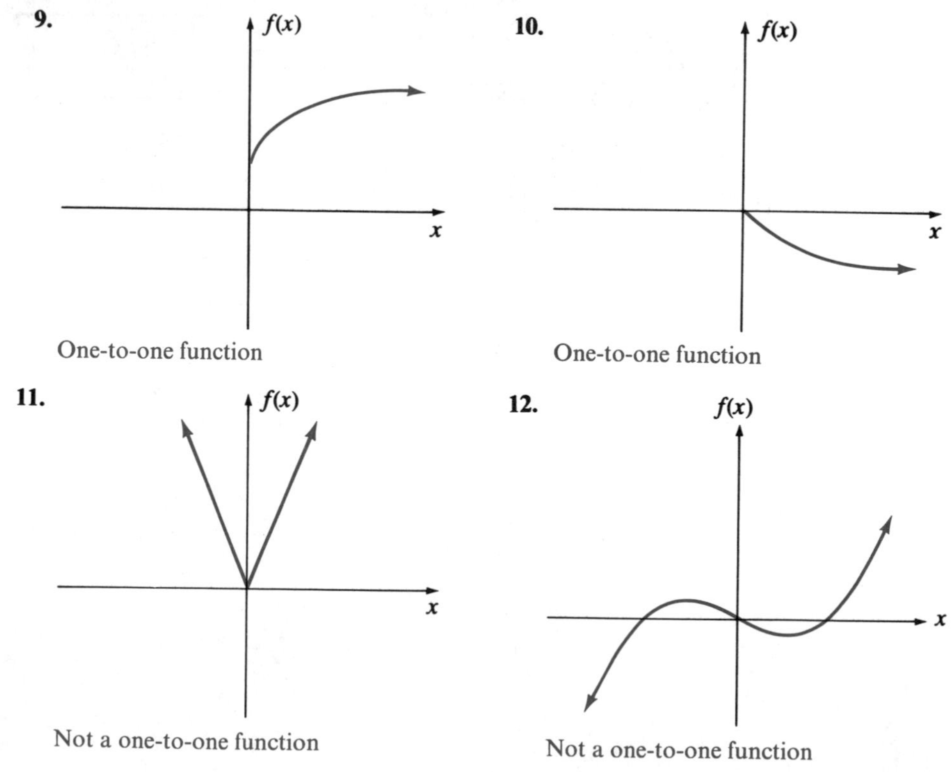

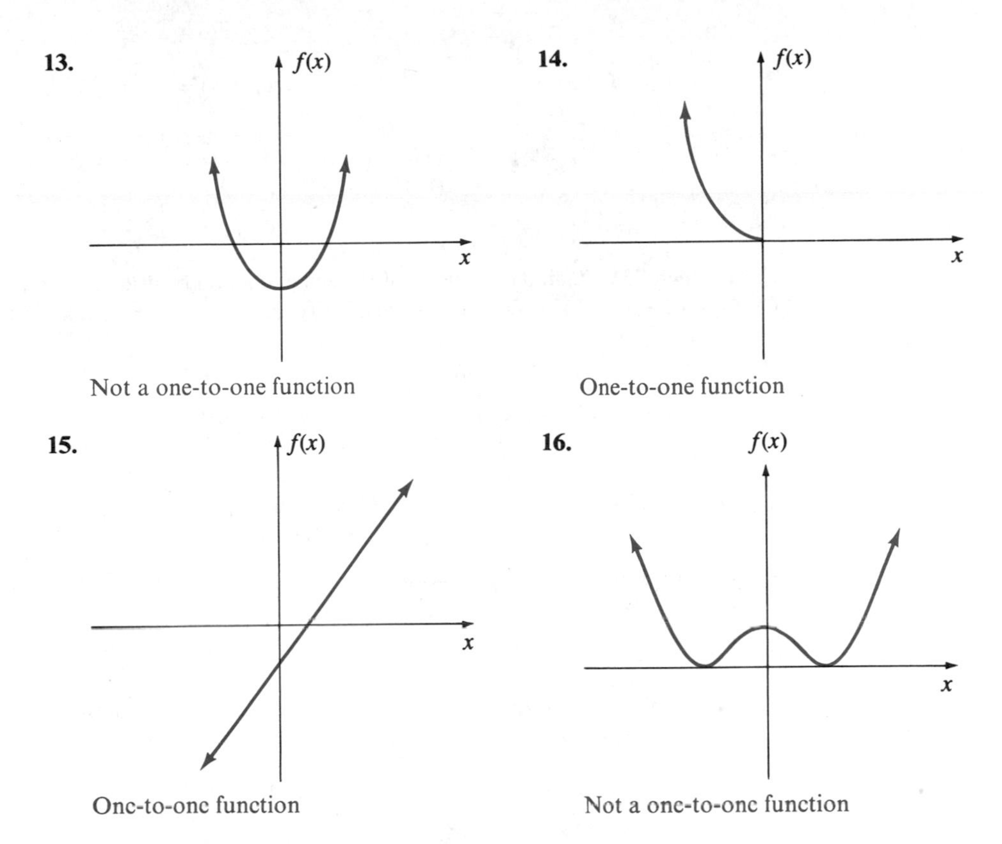

13. Not a one-to-one function

14. One-to-one function

15. Onc-to-onc function

16. Not a onc-to-onc function

For Problems 17–20, (a) list the domain and range of the given function, (b) form the inverse function, and (c) list the domain and range of the inverse function.

17. $f = \{(1, 3), (2, 6), (3, 11), (4, 18)\}$
Domain of f: $\{1, 2, 3, 4\}$
Range of f: $\{3, 6, 11, 18\}$
f^{-1}: $\{(3, 1), (6, 2), (11, 3), (18, 4)\}$
Domain of f^{-1}: $\{3, 6, 11, 18\}$
Range of f^{-1}: $\{1, 2, 3, 4\}$

18. $f = \{(0, -4), (1, -3), (4, -2)\}$
Domain of f: $\{0, 1, 4\}$
Range of f: $\{-4, -3, -2\}$
f^{-1}: $\{(-4, 0), (-3, 1), (-2, 4)\}$
Domain of f^{-1}: $\{-4, -3, -2\}$
Range of f^{-1}: $\{0, 1, 4\}$

19. $f = \{(-2, -1), (-1, 1), (0, 5), (5, 10)\}$
Domain of f: $\{-2, -1, 0, 5\}$
Range of f: $\{-1, 1, 5, 10\}$
f^{-1}: $\{(-1, -2), (1, -1), (5, 0), (10, 5)\}$
Domain of f^{-1}: $\{-1, 1, 5, 10\}$
Range of f^{-1}: $\{-2, -1, 0, 5\}$

20. $f = \{(-1, 1), (-2, 4), (1, 9), (2, 12)\}$
Domain of f: $\{-2, -1, 1, 2\}$
Range of f: $\{1, 4, 9, 12\}$
f^{-1}: $\{(1, -1), (4, -2), (9, 1), (12, 2)\}$
Domain of f^{-1}: $\{1, 4, 9, 12\}$
Range of f^{-1}: $\{-2, -1, 1, 2\}$

For Problems 21–30, find the inverse of the given function by using the "undoing process" and then verify that $(f \circ f^{-1})(x) = x$ and $(f^{-1} \circ f)(x) = x$.

21. $f(x) = 5x - 4$ $\quad f^{-1}(x) = \dfrac{x + 4}{5}$

22. $f(x) = 7x + 9$ $\quad f^{-1}(x) = \dfrac{x - 9}{7}$

23. $f(x) = -2x + 1$ $\quad f^{-1}(x) = \dfrac{1 - x}{2}$

24. $f(x) = -4x - 3$ $\quad f^{-1}(x) = \dfrac{-x - 3}{4}$

25. $f(x) = \frac{4}{5}x$ $f^{-1}(x) = \frac{5}{4}x$

26. $f(x) = -\frac{2}{3}x$ $f^{-1}(x) = -\frac{3}{2}x$

27. $f(x) = \frac{1}{2}x + 4$ $f^{-1}(x) = 2x - 8$

28. $f(x) = \frac{3}{4}x - 2$ $f^{-1}(x) = \frac{4x + 8}{3}$

29. $f(x) = \frac{1}{3}x - \frac{2}{5}$ $f^{-1}(x) = \frac{15x + 6}{5}$

30. $f(x) = \frac{2}{5}x + \frac{1}{3}$ $f^{-1}(x) = \frac{15x - 5}{6}$

For Problems 31–40, find the inverse of the given function by using the process illustrated in Examples 2 and 3 of this section and then verify that $(f \circ f^{-1})(x) = x$ and $(f^{-1} \circ f)(x) = x$.

31. $f(x) = 9x + 4$ $f^{-1}(x) = \frac{x - 4}{9}$

32. $f(x) = 8x - 5$ $f^{-1}(x) = \frac{x + 5}{8}$

33. $f(x) = -5x - 4$ $f^{-1}(x) = \frac{-x - 4}{5}$

34. $f(x) = -6x + 2$ $f^{-1}(x) = \frac{2 - x}{6}$

35. $f(x) = -\frac{2}{3}x + 7$ $f^{-1}(x) = \frac{-3x + 21}{2}$

36. $f(x) = -\frac{3}{5}x + 1$ $f^{-1}(x) = \frac{-5x + 5}{3}$

37. $f(x) = \frac{4}{3}x - \frac{1}{4}$ $f^{-1}(x) = \frac{3}{4}x + \frac{3}{16}$

38. $f(x) = \frac{5}{2}x + \frac{2}{7}$ $f^{-1}(x) = \frac{2}{5}x - \frac{4}{35}$

39. $f(x) = -\frac{3}{7}x - \frac{2}{3}$ $f^{-1}(x) = -\frac{7}{3}x - \frac{14}{9}$

40. $f(x) = -\frac{3}{5}x + \frac{3}{4}$ $f^{-1}(x) = -\frac{5}{3}x + \frac{5}{4}$

For Problems 41–50, (a) find the inverse of the given function, and (b) graph the given function and its inverse on the same set of axes.

41. $f(x) = 4x$ $f^{-1}(x) = \frac{1}{4}x$

42. $f(x) = \frac{2}{5}x$ $f^{-1}(x) = \frac{5}{2}x$

43. $f(x) = -\frac{1}{3}x$ $f^{-1}(x) = -3x$

44. $f(x) = -6x$ $f^{-1}(x) = -\frac{1}{6}x$

45. $f(x) = 3x - 3$ $f^{-1}(x) = \frac{x + 3}{3}$

46. $f(x) = 2x + 2$ $f^{-1}(x) = \frac{x - 2}{2}$

47. $f(x) = -2x - 4$ $f^{-1}(x) = \frac{-x - 4}{2}$

48. $f(x) = -3x + 9$ $f^{-1}(x) = \frac{9 - x}{3}$

49. $f(x) = x^2, \quad x \geq 0$ $f^{-1}(x) = \sqrt{x}, x \geq 0$

50. $f(x) = x^2 + 2, \quad x \geq 0$ $f^{-1}(x) = \sqrt{x - 2}, x \geq 2$

Miscellaneous Problems

51. Explain why every nonconstant linear function has an inverse. Every nonconstant linear function is a one-to-one function.

52. The composition idea can also be used to find the inverse of a function. For example, to find the inverse of $f(x) = 5x + 3$, we could proceed as follows.

$$f(f^{-1}(x)) = 5(f^{-1}(x)) + 3 \quad \text{and} \quad f(f^{-1}(x)) = x.$$

Therefore, equating the two expressions for $f(f^{-1}(x))$ we obtain

$$5(f^{-1}(x)) + 3 = x$$
$$5(f^{-1}(x)) = x - 3$$
$$f^{-1}(x) = \frac{x-3}{5}.$$

Use this approach to find the inverse of each of the following functions.

(a) $f(x) = 2x + 1$ $f^{-1}(x) = \frac{x-1}{2}$

(b) $f(x) = 3x - 2$ $f^{-1}(x) = \frac{x+2}{3}$

(c) $f(x) = -4x + 5$ $f^{-1}(x) = \frac{-x+5}{4}$

(d) $f(x) = -x + 1$ $f^{-1}(x) = -x + 1$

(e) $f(x) = 2x$ $f^{-1}(x) = \frac{1}{2}x$

(f) $f(x) = -5x$ $f^{-1}(x) = -\frac{1}{5}x$

8.5 Direct and Inverse Variations

"The distance a car travels at a fixed rate *varies directly* as the time." "At a constant temperature, the volume of an enclosed gas *varies inversely* as the pressure." Such statements illustrate two basic types of functional relationships, called **direct** and **inverse variation**, which are widely used, especially in the physical sciences. These relationships can be expressed by equations that specify functions. The purpose of this section is to investigate these special functions.

The statement *y varies directly as x* means

$$y = kx$$

where k is a nonzero constant, called the **constant of variation**. The phrase "y is directly proportional to x," is also used to indicate direct variation; k is then referred to as the **constant of proportionality**.

REMARK Notice that the equation $y = kx$ defines a function and could be written as $f(x) = kx$ by using function notation. However, in this section it is more convenient to avoid the function notation and use variables that are meaningful in terms of the physical entities involved in the problem.

Statements that indicate direct variation may also involve powers of x. For example, "y varies directly as the square of x" can be written as

$$y = kx^2.$$

In general, "y varies directly as the nth power of $x (n > 0)$" means

$$y = kx^n.$$

There are basically three types of problems that deal with direct variation, namely, (1) to translate an English statement into an equation that expresses the direct variation, (2) to find the constant of variation from given values of the variables, and (3) to find additional values of the variables once the constant of variation has been determined. Let's consider an example of each of these types of problems.

EXAMPLE 1 Translate the statement "the tension on a spring varies directly as the distance it is stretched" into an equation and use k as the constant of variation.

Solution Letting t represent the tension and d the distance, the equation becomes

$$t = kd.$$

■

EXAMPLE 2 If A varies directly as the square of s, and $A = 28$ when $s = 2$, find the constant of variation.

Solution Since A varies directly as the square of s, we have

$$A = ks^2.$$

Substitute $A = 28$ and $s = 2$, to obtain

$$28 = k(2)^2.$$

Solving this equation for k yields

$$28 = 4k$$

$$7 = k.$$

The constant of variation is 7.

■

EXAMPLE 3 If y is directly proportional to x and if $y = 6$ when $x = 9$, find the value of y when $x = 24$.

Solution The statement "y is directly proportional to x" translates into

$$y = kx.$$

Letting $y = 6$ and $x = 9$, the constant of variation becomes

$$6 = k(9)$$

$$6 = 9k$$

$$\frac{6}{9} = k$$

$$\frac{2}{3} = k.$$

So, the specific equation is $y = \frac{2}{3}x$. Now, letting $x = 24$, we obtain

$$y = \frac{2}{3}(24) = 16.$$

The required value of y is 16. ■

Inverse Variation

We define the second basic type of variation, called **inverse variation**, as follows. The statement *y varies inversely as x* means

$$y = \frac{k}{x}$$

where k is a nonzero constant; again we refer to it as the constant of variation. The phrase "y is inversely proportional to x" is also used to express inverse variation. As with direct variation, statements that indicate inverse variation may involve powers of x. For example, "y varies inversely as the square of x" can be written as

$$y = \frac{k}{x^2}.$$

In general, "y varies inversely as the nth power of $x (n > 0)$" means

$$y = \frac{k}{x^n}.$$

The following examples illustrate the three basic kinds of problems that we run across which involve inverse variation.

EXAMPLE 4 Translate the statement "the length of a rectangle of a fixed area varies inversely as the width" into an equation that uses k as the constant of variation.

Solution Let l represent the length and w the width and the equation is

$$l = \frac{k}{w}.$$

■

EXAMPLE 5 If y is inversely proportional to x and $y = 4$ when $x = 12$, find the constant of variation.

Solution Since y is inversely proportional to x, we have

$$y = \frac{k}{x}.$$

Substituting $y = 4$ and $x = 12$, we obtain

$$4 = \frac{k}{12}.$$

Solving this equation for k yields

$$k = 48.$$

The constant of variation is 48. ■

EXAMPLE 6 Suppose the number of days it takes to complete a construction job varies inversely as the number of people assigned to the job. If it takes 7 people 8 days to do the job, how long would it take 14 people to complete the job?

Solution Let d represent the number of days and p the number of people. The phrase "number of days . . . varies inversely as the number of people" translates into

$$d = \frac{k}{p}.$$

Let $d = 8$ when $p = 7$ and the constant of variation becomes

$$8 = \frac{k}{7}$$

$$k = 56.$$

So, the specific equation is

$$d = \frac{56}{p}.$$

Now, let $p = 14$, to obtain

$$d = \frac{56}{14}$$

$$d = 4.$$

It should take 14 people 4 days to complete the job. ■

The terms *direct* and *inverse*, as applied to variation, refer to the relative behavior of the variables involved in the equation. That is to say, in direct variation ($y = kx$) an assignment of *increasing absolute values* for x produces *increasing absolute values* for y. Whereas, in inverse variation $\left(y = \frac{k}{x}\right)$ an assignment of *increasing absolute values* for x produces *decreasing absolute values* for y.

Joint Variation

Variation may involve more than two variables. The following table illustrates some variation statements and their equivalent algebraic equations that use k as the constant of variation.

Variation statement	*Algebraic equation*
1. y varies jointly as x and z	$y = kxz$
2. y varies jointly as x, z, and w	$y = kxzw$
3. V varies jointly as h and the square of r	$V = khr^2$
4. h varies directly as V and inversely as w	$h = \frac{kV}{w}$
5. y is directly proportional to x and inversely proportional to the square of z	$y = \frac{kx}{z^2}$
6. y varies jointly as w and z, and inversely as x	$y = \frac{kwz}{x}$

Statements 1, 2, and 3 illustrate the concept of **joint variation**. Statements 4 and 5 show that both direct and inverse variation may occur in the same problem. Statement 6 combines joint variation with inverse variation.

The two final examples illustrate problems that involve some of these possible variation situations.

EXAMPLE 7 The length of a rectangular box with a fixed height varies directly as the volume and inversely as the width. If the length is 12 centimeters when the volume is 960 cubic centimeters and the width is 8 centimeters, find the length when the volume is 700 centimeters and the width is 5 centimeters.

Solution Use l for length, V for volume, and w for width and the phrase "length varies directly as the volume and inversely as the width" translates into

$$l = \frac{kV}{w}.$$

Substitute $l = 12$, $V = 960$, and $w = 8$ and the constant of variation becomes

$$12 = \frac{k(960)}{8}$$

$$12 = 120k$$

$$\frac{1}{10} = k.$$

So the specific equation is

$$l = \frac{\frac{1}{10}V}{w} = \frac{V}{10w}.$$

Now, let $V = 700$ and $w = 5$ to obtain

$$l = \frac{700}{10(5)} = \frac{700}{50} = 14.$$

The length is 14 centimeters. ■

EXAMPLE 8 Suppose that y varies jointly as x and z, and inversely as w. If $y = 154$ when $x = 6$, $z = 11$, and $w = 3$, find the constant of variation.

Solution The statement "y varies jointly as x and z, and inversely as w" translates into

$$y = \frac{kxz}{w}.$$

Substitute $y = 154$, $x = 6$, $z = 11$, and $w = 3$ to obtain

$$154 = \frac{k(6)(11)}{3}$$

$$154 = 22k$$

$$7 = k.$$

The constant of variation is 7. ■

Problem Set 8.5

For Problems 1–10, translate each statement of variation into an equation and use k as the constant of variation.

1. y varies inversely as the square of x. $y = \frac{k}{x^2}$
2. y varies directly as the cube of x. $y = kx^3$
3. C varies directly as g and inversely as the cube of t. $C = \frac{kg}{t^3}$
4. V varies jointly as l and w. $V = klw$
5. The volume (V) of a sphere is directly proportional to the cube of its radius (r). $V = kr^3$
6. At a constant temperature, the volume (V) of a gas varies inversely as the pressure (P). $V = \frac{k}{P}$
7. The surface area (S) of a cube varies directly as the square of the length of an edge (e). $S = ke^2$

8. The intensity of illumination (I) received from a source of light is inversely proportional to the square of the distance (d) from the source. $I = \frac{k}{d^2}$

9. The volume (V) of a cone varies jointly as its height and the square of its radius. $V = khr^2$

10. The volume (V) of a gas varies directly as the absolute temperature (T) and inversely as the pressure (P). $V = \frac{kT}{P}$

For Problems 11–24, find the constant of variation for each of the stated conditions.

11. y varies directly as x, and $y = 8$ when $x = 12$. $\frac{2}{3}$

12. y varies directly as x, and $y = 60$ when $x = 24$. $\frac{5}{2}$

13. y varies directly as the square of x, and $y = -144$ when $x = 6$. -4

14. y varies directly as the cube of x, and $y = 48$ when $x = -2$. -6

15. V varies jointly as B and h, and $V = 96$ when $B = 24$ and $h = 12$. $\frac{1}{3}$

16. A varies jointly as b and h, and $A = 72$ when $b = 16$ and $h = 9$. $\frac{1}{2}$

17. y varies inversely as x, and $y = -4$ when $x = \frac{1}{2}$. -2

18. y varies inversely as x, and $y = -6$ when $x = \frac{4}{3}$. -8

19. r varies inversely as the square of t, and $r = \frac{1}{8}$ when $t = 4$. 2

20. r varies inversely as the cube of t, and $r = \frac{1}{16}$ when $t = 4$. 4

21. y varies directly as x and inversely as z, and $y = 45$ when $x = 18$ and $z = 2$. 5

22. y varies directly as x and inversely as z, and $y = 24$ when $x = 36$ and $z = 18$. 12

23. y is directly proportional to x and inversely proportional to the square of z, and $y = 81$ when $x = 36$ and $z = 2$. 9

24. y is directly proportional to the square of x and inversely proportional to the cube of z, and $y = 4\frac{1}{2}$ when $x = 6$ and $z = 4$. 8

Solve each of the following problems.

25. If y is directly proportional to x, and $y = 36$ when $x = 48$, find the value of y when $x = 12$. 9

26. If y is directly proportional to x, and $y = 42$ when $x = 28$, find the value of y when $x = 38$. 57

27. If y is inversely proportional to x, and $y = \frac{1}{9}$ when $x = 12$, find the value of y when $x = 8$. $\frac{1}{6}$

28. If y is inversely proportional to x, and $y = \frac{1}{35}$ when $x = 14$, find the value of y when $x = 16$. $\frac{1}{40}$

29. If A varies jointly as b and h, and $A = 60$ when $b = 12$ and $h = 10$, find A when $b = 16$ and $h = 14$. 112

30. If V varies jointly as B and h, and $V = 51$ when $B = 17$ and $h = 9$, find V when $B = 19$ and $h = 12$. 76

31. The volume of a gas at a constant temperature varies inversely as the pressure. What is the volume of a gas under pressure of 25 pounds if the gas occupies 15 cubic centimeters under a pressure of 20 pounds? 12 cubic centimeters

32. The time required for a car to travel a certain distance varies inversely as the rate at which it travels. If it takes 4 hours at 50 miles per hour to travel the distance, how long will it take at 40 miles per hour? 5 hours

33. The volume (V) of a gas varies directly as the temperature (T) and inversely as the pressure (P). If $V = 48$ when $T = 320$ and $P = 20$, find V when $T = 280$ and $P = 30$. 28

34. The distance that a freely falling body falls varies directly as the square of the time it falls. If a body falls 144 feet in 3 seconds, how far will it fall in 5 seconds? 400 feet

35. The period (the time required for one complete oscillation) of a simple pendulum varies directly as the square root of its length. If a pendulum 12 feet long has a period of 4 seconds, find the period of a pendulum of length 3 feet. 2 seconds

36. The simple interest earned by a certain amount of money varies jointly as the rate of interest and the time (in years) that the money is invested. If \$120 is earned for the money invested at 12% for 2 years, how much is earned if the money is invested at 14% for 3 years? \$210

37. The electrical resistance of a wire varies directly as its length and inversely as the square of its diameter. If the resistance of 200 meters of wire that has a diameter of $\frac{1}{2}$ centimeter is 1.5 ohms, find the resistance of 400 meters of wire with a diameter of $\frac{1}{4}$ centimeter. 12 ohms

38. The volume of a cylinder varies jointly as its altitude and the square of the radius of its base. If the volume of a cylinder is 1386 cubic centimeters when the radius of the base is 7 centimeters and its altitude is 9 centimeters, find the volume of a cylinder that has a base of radius 14 centimeters; the altitude of the cylinder is 5 centimeters. 3080 cubic centimeters

Miscellaneous Problems

In the previous problems numbers were chosen to make computations reasonable without the use of a calculator. However, many times variation-type problems involve messy computations and the calculator becomes a very useful tool. Use your calculator to help solve the following problems.

39. The simple interest earned by a certain amount of money varies jointly as the rate of interest and the time (in years) that the money is invested.

    (a) If some money invested at 11% for 2 years earns \$385, how much would the same amount earn at 12% for 1 year? \$210

    (b) If some money invested at 12% for 3 years earns \$819, how much would the same amount earn at 14% for 2 years? \$637

    (c) If some money invested at 14% for 4 years earns \$1960, how much would the same amount earn at 15% for 2 years? \$1050

40. The period (the time required for one complete oscillation) of a simple pendulum varies directly as the square root of its length. If a pendulum 9 inches long has a period of 2.4 seconds, find the period of a pendulum of length 12 inches. Express your answer to the nearest one-tenth of a second. 2.8 seconds

41. The volume of a cylinder varies jointly as its altitude and the square of the radius of its base. If the volume of a cylinder is 549.5 cubic meters when the radius of the base is 5 meters and its altitude is 7 meters, find the volume of a cylinder that has a base of a radius 9 meters and an altitude of 14 meters. 3560.76 cubic meters

42. If y is directly proportional to x and inversely proportional to the square of z, and if $y = 0.336$ when $x = 6$ and $z = 5$, find the constant of variation. 1.4

43. If y is inversely proportional to the square root of x, and $y = 0.08$ when $x = 225$, find y when $x = 625$. .048

Chapter 8 Summary

(8.1) A **relation** is a set of ordered pairs; a **function** is a relation in which no two ordered pairs have the same first component. The **domain** of a relation (or function) is the set of all first components and the **range** is the set of all second components.

Single symbols such as f, g, and h are commonly used to name functions. The symbol $f(x)$ represents the element in the range associated with x from the domain. Thus, if $f(x) = 3x + 7$, then $f(1) = 3(1) + 7 = 10$.

(8.2) A summary of the three special types of functions follows.

Equation	*Type of function*	*Graph*
1. $f(x) = b$, where b is a real number	constant	horizontal line
2. $f(x) = ax + b$, where a and b are real numbers and $a \neq 0$	linear	straight line
3. $f(x) = ax^2 + bx + c$, where a, b, and c are real numbers and $a \neq 0$	quadratic	parabola

We list the following suggestions for graphing functions you are unfamiliar with.

1. Determine the domain of the function.
2. Determine the type of symmetry exhibited by the equation.
3. Find the intercepts.
4. Set up a table of values that satisfy the equation.
5. Plot the points associated with the ordered pairs and connect them with a smooth curve. Then, if appropriate, reflect this part of the curve according to any symmetry possessed by the graph.

A useful graphing tool is the ability to recognize variations of certain basic curves.

(8.3) We can solve some applications that involve maximum or minimum values with our knowledge of parabolas that are generated by quadratic functions.

The **composition** of two functions f and g is defined by

$$(f \circ g)(x) = f(g(x))$$

for all x in the domain of g such that $g(x)$ is in the domain of f.

(8.4) A **one-to-one function** is a function such that no two ordered pairs have the same second component.

If the components of each ordered pair of a given one-to-one function are interchanged, the resulting function and the given function are **inverses** of each other. The inverse of a function f is denoted by f^{-1}.

Graphically, two functions that are inverses of each other are mirror images with reference to the line $y = x$.

We can show that two functions f and f^{-1} are inverses of each other by verifying that

1. $(f^{-1} \circ f)(x) = x$ for all x in the domain of f,
2. $(f \circ f^{-1})(x) = x$ for all x in the domain of f^{-1}.

A technique for finding the inverse of a function follows.

1. Let $y = f(x)$.
2. Interchange x and y.
3. Solve the equation for y in terms of x.
4. $f^{-1}(x)$ is determined by the final equation.

(8.5) The equation $y = kx$ (k is a nonzero constant) defines a function called a **direct variation**. The equation $y = \frac{k}{x}$ defines a function called **inverse variation**. In both cases, k is called the **constant of variation**.

Chapter 8 Review Problem Set

For Problems 1–4, specify the domain of each function.

1. $f = \{(1,3), (2,5), (4,9)\}$ $D = \{1, 2, 4\}$
2. $f(x) = \dfrac{4}{x-5}$ $D = \{x \mid x \neq 5\}$
3. $f(x) = \dfrac{3}{x^2 + 4x}$ $D = \{x \mid x \neq 0 \text{ and } x \neq -4\}$
4. $f(x) = \sqrt{x^2 - 25}$ $D = \{x \mid x \geq 5 \text{ or } x \leq -5\}$
5. If $f(x) = x^2 - 2x - 1$, find $f(2)$, $f(-3)$, and $f(a)$. $f(2) = -1$; $f(-3) = 14$; $f(a) = a^2 - 2a - 1$
6. If $f(x) = 2x^2 + x - 7$, find $\dfrac{f(a+h) - f(a)}{h}$. $4a + 2h + 1$

For Problems 7–16, graph each of the functions.

7. $f(x) = 4$ see page A66

8. $f(x) = -3x + 2$ see page A66

9. $f(x) = x^2 + 2x + 2$ see page A66

10. $f(x) = |x| + 4$ see page A66

11. $f(x) = -|x - 2|$ see page A66

12. $f(x) = \sqrt{x - 2} - 3$ see page A66

13. $f(x) = \frac{1}{x^2}$ see page A66

14. $f(x) = -\frac{1}{2}x^2$ see page A66

15. $f(x) = -3x^2 + 6x - 2$ see page A66

16. $f(x) = -\sqrt{x + 1} - 2$ see page A66

17. Find the coordinates of the vertex and the equation of the line of symmetry for each of the following parabolas.

(a) $f(x) = x^2 + 10x - 3$ $(-5, -28)$; $x = -5$

(b) $f(x) = -2x^2 - 14x + 9$ $(-\frac{7}{2}, \frac{67}{2})$; $x = -\frac{7}{2}$

For Problems 18–20, determine $(f \circ g)(x)$ and $(g \circ f)(x)$ for each pair of functions.

18. $f(x) = 2x - 3$ and $g(x) = 3x - 4$ $(f \circ g)(x) = 6x - 11$ and $(g \circ f)(x) = 6x - 13$

19. $f(x) = x - 4$ and $g(x) = x^2 - 2x + 3$ $(f \circ g)(x) = x^2 - 2x - 1$ and $(g \circ f)(x) - x^2 - 10x + 27$

20. $f(x) = x^2 - 5$ and $g(x) = -2x + 5$ $(f \circ g)(x) = 4x^2 - 20x + 20$ and $(g \circ f)(x) = -2x^2 + 15$

For Problems 21–23, find the inverse (f^{-1}) of the given function.

21. $f(x) = 6x - 1$ $f^{-1}(x) = \frac{x+1}{6}$

22. $f(x) = \frac{2}{3}x + 7$ $f^{-1}(x) = \frac{3x - 21}{2}$

23. $f(x) = -\frac{3}{5}x - \frac{2}{7}$ $f^{-1}(x) = \frac{-35x - 10}{21}$

24. If y varies directly as x and inversely as z, and if $y = 21$ when $x = 14$ and $z = 6$, find the constant of variation. $k = 9$

25. If y varies jointly as x and the square root of z, and if $y = 60$ when $x = 2$ and $z = 9$, find y when $x = 3$ and $z = 16$. $y = 120$

26. The weight of a body above the surface of the earth varies inversely as the square of its distance from the center of the earth. Assume that the radius of the earth is 4000 miles. How much would a man weigh 1000 miles above the earth's surface if he weighs 200 pounds on the surface? 128 lbs

27. Find two numbers whose sum is 40 and whose product is a maximum. 20 and 20

28. Find two numbers whose sum is 50 such that the square of one number plus six times the other number is a minimum. 3 and 47

29. Suppose that 50 students are able to raise $250 for a party when each one contributes $5. Furthermore, they figure that for each additional student they can find to contribute, the cost per student will decrease by a nickel. How many additional students do they need to maximize the amount they will have for a party? 25 students

30. The surface area of a cube varies directly as the square of the length of an edge. If the surface area of a cube having edges 8 inches long is 384 square inches, find the surface area of a cube having edges 10 inches long. 600 sq. inches

Chapter 9

Systems of Equations

"Find two numbers whose sum is 125 and whose difference is 21." Such a problem translates naturally into the two equations: $x + y = 125$ and $x - y = 21$, where x represents the larger number and y the smaller number. The two equations considered together form a system of linear equations and the original problem can be solved by solving the system of equations. Throughout most of this chapter we will be considering systems of linear equations and their applications.

9.1

Systems of Two Linear Equations in Two Variables

In Chapter 7 we stated that any equation of the form $Ax + By = C$, where A, B, and C are real numbers (A and B not both zero) is a *linear equation* in the two variables x and y, and its graph is a straight line. Two linear equations in two variables considered together form a **system of two linear equations in two variables** as illustrated by the following.

$$\begin{pmatrix} x + y = 6 \\ x - y = 2 \end{pmatrix}, \quad \begin{pmatrix} 3x + 2y = 1 \\ 5x - 2y = 23 \end{pmatrix}, \quad \begin{pmatrix} 4x - 5y = 21 \\ 3x + 7y = -38 \end{pmatrix}$$

To **solve a system**, such as the ones above, means to find all of the ordered pairs that satisfy both equations in the system. For example, if we graph the two equations $x + y = 6$ and $x - y = 2$ on the same set of axes, as in Figure 9.1, then the ordered pair associated with the point of intersection of the two lines is the solution of the system. Thus, we say that $\{(4, 2)\}$ is the solution set of the system

$$\begin{pmatrix} x + y = 6 \\ x - y = 2 \end{pmatrix}.$$

Figure 9.1

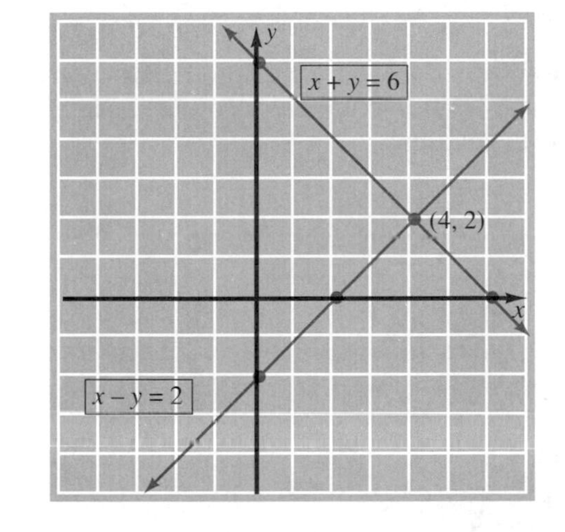

To check, we substitute 4 for x and 2 for y in the two equations and this yields:

$x + y$ becomes $4 + 2 = 6$ a true statement

$x - y$ becomes $4 - 2 = 2$ a true statement

Since the graph of a linear equation in two variables is a straight line, there are three possible situations that can occur when solving a system of two linear equations in two variables. We illustrate these cases in Figure 9.2.

Figure 9.2

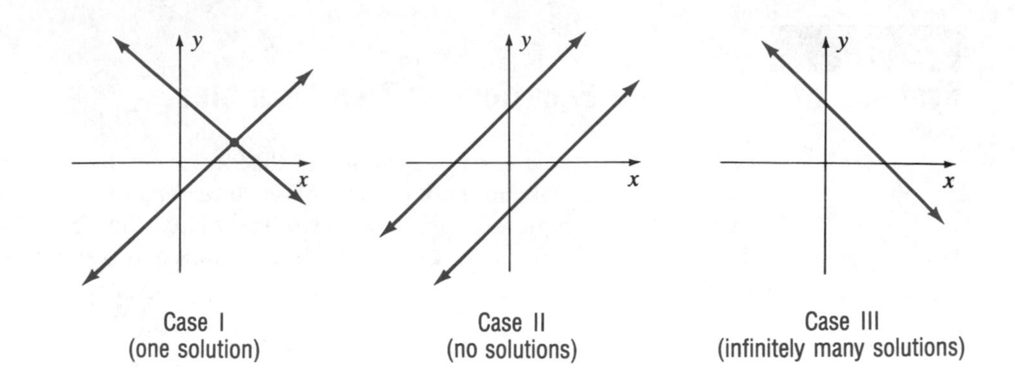

Case I
(one solution)

Case II
(no solutions)

Case III
(infinitely many solutions)

Case I. The graphs of the two equations are two lines intersecting in *one* point. There is *one solution* and the system is called a **consistent system**.

Case II. The graphs of the two equations are parallel lines. There is *no solution* and the system is called an **inconsistent system**.

Case III. The graphs of the two equations are the same line and there are *infinitely many solutions* to the system. Any pair of real numbers that satisfies one of the equations will also satisfy the other equation and we say that the equations are **dependent**.

Thus, as we solve a system of two linear equations in two variables we know what to expect. The system will have *no* solutions, *one* ordered pair as a solution, or *infinitely many* ordered pairs as solutions.

The Substitution Method

It should be evident that solving systems of equations by graphing requires accurate graphs. In fact, unless the solutions are integers, it is quite difficult to obtain exact solutions from a graph. Thus, we will consider some other methods for solving systems of equations.

The **substitution method**, which works quite well with systems of two linear equations in two unknowns, we describe as follows.

Step 1. Solve one of the equations for one variable in terms of the other variable if neither equation is in such a form. (If possible make a choice that will avoid fractions.)

Step 2. Substitute the expression obtained in step 1 into the other equation to produce an equation with one variable.

Step 3. Solve the equation obtained in step 2.

Step 4. Use the solution obtained in step 3, along with the expression obtained in step 1, to determine the solution of the system.

Now let's look at some examples that illustrate the substitution method.

EXAMPLE 1 Solve the system $\begin{pmatrix} x + y = 16 \\ y = x + 2 \end{pmatrix}$.

Solution Because the second equation states that y equals $x + 2$, we can substitute $x + 2$ for y in the first equation.

$$x + y = 16 \xrightarrow{\text{substitute } x+2 \text{ for } y} x + (x + 2) = 16$$

Now we have an equation with one variable that we can solve in the usual way.

$$\begin{aligned} x + (x + 2) &= 16 \\ 2x + 2 &= 16 \\ 2x &= 14 \\ x &= 7 \end{aligned}$$

Substituting 7 for x in one of the two original equations (let's use the second one) yields

$$y - 7 + 2 - 9.$$

To check, we can substitute 7 for x and 9 for y in both of the original equations.

$$7 + 9 = 16 \qquad \text{a true statement}$$

$$9 = 7 + 2 \qquad \text{a true statement}$$

The solution set is $\{(7, 9)\}$. ■

EXAMPLE 2 Solve the system $\begin{pmatrix} x = 3y - 25 \\ 4x + 5y = 19 \end{pmatrix}$.

Solution In this case the first equation states that x equals $3y - 25$. Therefore, we can substitute $3y - 25$ for x in the second equation.

$$4x + 5y = 19 \xrightarrow{\text{substitute } 3y-25 \text{ for } x} 4(3y - 25) + 5y = 19$$

Solving this equation yields

$$\begin{aligned} 4(3y - 25) + 5y &= 19 \\ 12y - 100 + 5y &= 19 \\ 17y &= 119 \\ y &= 7. \end{aligned}$$

Substituting 7 for y in the first equation produces

$$\begin{aligned} x &= 3(7) - 25 \\ &= 21 - 25 = -4. \end{aligned}$$

The solution set is $\{(-4, 7)\}$; check it. ■

EXAMPLE 3 Solve the system $\begin{pmatrix} 3x - 7y = 2 \\ x + 4y = 1 \end{pmatrix}$.

Solution Let's solve the second equation for x in terms of y.

$$x + 4y = 1$$
$$x = 1 - 4y$$

Now we can substitute $1 - 4y$ for x in the first equation.

$$3x - 7y = 2 \xrightarrow{\text{substitute } 1 - 4y \text{ for } x} 3(1 - 4y) - 7y = 2$$

Let's solve this equation for y.

$$3(1 - 4y) - 7y = 2$$
$$3 - 12y - 7y = 2$$
$$-19y = -1$$
$$y = \frac{1}{19}$$

Finally, we can substitute $\frac{1}{19}$ for y in the equation $x = 1 - 4y$.

$$x = 1 - 4\left(\frac{1}{19}\right)$$
$$= 1 - \frac{4}{19}$$
$$= \frac{15}{19}$$

The solution set is $\left\{\left(\frac{15}{19}, \frac{1}{19}\right)\right\}$. ■

EXAMPLE 4 Solve the system $\begin{pmatrix} 5x - 6y = -4 \\ 3x + 2y = -8 \end{pmatrix}$.

Solution Note that solving either equation for either variable will produce a fractional form. Let's solve the second equation for y in terms of x.

$$3x + 2y = -8$$
$$2y = -8 - 3x$$
$$y = \frac{-8 - 3x}{2}$$

Now we can substitute $\frac{-8-3x}{2}$ for y in the first equation.

$$5x - 6y = -4 \xrightarrow{\text{substitute } \frac{-8-3x}{2} \text{ for } y} 5x - 6\left(\frac{-8-3x}{2}\right) = -4$$

Solving this equation yields

$$\begin{aligned} 5x - 6\left(\frac{-8-3x}{2}\right) &= -4 \\ 5x - 3(-8-3x) &= -4 \\ 5x + 24 + 9x &= -4 \\ 14x &= -28 \\ x &= -2. \end{aligned}$$

Substituting -2 for x in $y = \frac{-8-3x}{2}$ yields

$$\begin{aligned} y &= \frac{-8-3(-2)}{2} \\ &= \frac{-8+6}{2} \\ &= \frac{-2}{2} \\ &= -1. \end{aligned}$$

The solution set is $\{(-2, -1)\}$. ■

Problem Solving

Many word problems that we have solved earlier in this text using one variable and one equation can also be solved using a system of two linear equations in two variables. In fact, in many of these problems you may find it much more natural to use two variables. Let's consider some examples.

PROBLEM 1 Anita invested some money at 8% and \$400 more than that amount at 9%. The yearly interest from the two investments was \$87. How much did Anita invest at each rate?

Solution Let x represent the amount invested at 8% and let y represent the amount invested at 9%. The problem translates into the following system.

The amount invested at 9% was \$400 more than at 8%.

The yearly interest from the two investments was \$87.

$$\begin{pmatrix} y = x + 400 \\ .08x + .09y = 87 \end{pmatrix}$$

From the first equation we can substitute $x + 400$ for y in the second equation and solve for x.

$$\begin{aligned} .08x + .09(x + 400) &= 87 \\ .08x + .09x + 36 &= 87 \\ .17x &= 51 \\ x &= 300 \end{aligned}$$

Therefore, \$300 is invested at 8% and \$300 + \$400 = \$700 is invested at 9%. ■

The two-variable expression $10t + u$ can be used to represent any two-digit number. The t represents the tens digit and the u represents the units digit. For example, if $t = 5$ and $u = 2$, then $10t + u$ becomes $10(5) + 2 = 52$. We use this general representation for a two-digit number in the next problem.

PROBLEM 2 The tens digit of a two-digit number is 2 more than twice the units digit. The number with the digits reversed is 45 less than the original number. Find the original number.

Solution Let u represent the units digit of the original number. Let t represent the tens digit of the original number. Then $10t + u$ represents the original number and $10u + t$ represents the number with the digits reversed. The problem translates into the following system.

$$\begin{pmatrix} t = 2u + 2 \\ 10u + t = 10t + u - 45 \end{pmatrix}$$

The tens digit is 2 more than twice the units digit.

The number with the digits reversed is 45 less than the original number.

Simplifying the second equation the system becomes

$$\begin{pmatrix} t = 2u + 2 \\ -9t + 9u = -45 \end{pmatrix}.$$

From the first equation we can *substitute* $2u + 2$ for t in the second equation and solve.

$$\begin{aligned} -9t + 9u &= -45 \\ -9(2u + 2) + 9u &= -45 \\ -18u - 18 + 9u &= -45 \\ -9u &= -27 \\ u &= 3 \end{aligned}$$

Substitute 3 for u in $t = 2u + 2$ to obtain

$$t = 2u + 2$$
$$= 2(3) + 2 = 8.$$

The tens digit is 8 and the units digit is 3, so the number is 83. (You should check to see if 83 satisfies the original conditions stated in the problem.) ■

Problem Set 9.1

For Problems 1–10, use the graphing approach to determine whether each of the systems is *consistent*, *inconsistent*, or *dependent*. If the system is consistent, find the solution set from the graph and check it.

1. $\begin{pmatrix} x - y = 1 \\ 2x + y = 8 \end{pmatrix}$ $\{(3, 2)\}$

2. $\begin{pmatrix} 3x + y = 0 \\ x - 2y = -7 \end{pmatrix}$ $\{(-1, 3)\}$

3. $\begin{pmatrix} 4x + 3y = -5 \\ 2x - 3y = -7 \end{pmatrix}$ $\{(-2, 1)\}$

4. $\begin{pmatrix} 2x - y = 9 \\ 4x - 2y = 11 \end{pmatrix}$ Inconsistent

5. $\begin{pmatrix} \frac{1}{2}x + \frac{1}{4}y = 9 \\ 4x + 2y = 72 \end{pmatrix}$ Dependent

6. $\begin{pmatrix} 5x + 2y = -9 \\ 4x - 3y = 2 \end{pmatrix}$ $\{(-1, -2)\}$

7. $\begin{pmatrix} \frac{1}{2}x - \frac{1}{3}y = 3 \\ x + 4y = -8 \end{pmatrix}$ $\{(4, -3)\}$

8. $\begin{pmatrix} 4x - 9y = -60 \\ \frac{1}{3}x - \frac{3}{4}y = -5 \end{pmatrix}$ Dependent

9. $\begin{pmatrix} x - \frac{1}{2}y = -4 \\ 8x - 4y = -1 \end{pmatrix}$ Inconsistent

10. $\begin{pmatrix} 3x - 2y = 7 \\ 6x + 5y = -4 \end{pmatrix}$ $\{(1, -2)\}$

For Problems 11–36, solve each system by using the substitution method.

11. $\begin{pmatrix} x + y = 20 \\ x = y - 4 \end{pmatrix}$ $\{(8, 12)\}$

12. $\begin{pmatrix} x + y = 23 \\ y = x - 5 \end{pmatrix}$ $\{(14, 9)\}$

13. $\begin{pmatrix} y = -3x - 18 \\ 5x - 2y = -8 \end{pmatrix}$ $\{(-4, -6)\}$

14. $\begin{pmatrix} 4x - 3y = 33 \\ x = -4y - 25 \end{pmatrix}$ $\{(3, -7)\}$

15. $\begin{pmatrix} x = -3y \\ 7x - 2y = -69 \end{pmatrix}$ $\{(-9, 3)\}$

16. $\begin{pmatrix} 9x - 2y = -38 \\ y = -5x \end{pmatrix}$ $\{(-2, 10)\}$

17. $\begin{pmatrix} 2x + 3y = 11 \\ 3x - 2y = -3 \end{pmatrix}$ $\{(1, 3)\}$

18. $\begin{pmatrix} 3x - 4y = -14 \\ 4x + 3y = 23 \end{pmatrix}$ $\{(2, 5)\}$

19. $\begin{pmatrix} 3x - 4y = 9 \\ x = 4y - 1 \end{pmatrix}$ $\left\{\left(5, \frac{3}{2}\right)\right\}$

20. $\begin{pmatrix} y = 3x - 5 \\ 2x + 3y = 6 \end{pmatrix}$ $\left\{\left(\frac{21}{11}, \frac{8}{11}\right)\right\}$

21. $\begin{pmatrix} y = \frac{2}{5}x - 1 \\ 3x + 5y = 4 \end{pmatrix}$ $\left\{\left(\frac{9}{5}, -\frac{7}{25}\right)\right\}$

22. $\begin{pmatrix} y = \frac{3}{4}x - 5 \\ 5x - 4y = 9 \end{pmatrix}$ $\left\{\left(\frac{11}{2}, -\frac{73}{8}\right)\right\}$

23. $\begin{pmatrix} 7x - 3y = -2 \\ x = \frac{3}{4}y + 1 \end{pmatrix}$ $\{(-2, -4)\}$

24. $\begin{pmatrix} 5x - y = 9 \\ x = \frac{1}{2}y - 3 \end{pmatrix}$ $\{(5, 16)\}$

25. $\begin{pmatrix} 2x + y = 12 \\ 3x - y = 13 \end{pmatrix}$ $\{(5, 2)\}$

26. $\begin{pmatrix} -x + 4y = -22 \\ x - 7y = 34 \end{pmatrix}$ $\{(6, -4)\}$

27. $\begin{pmatrix} 4x + 3y = -40 \\ 5x - y = -12 \end{pmatrix}$ $\{(-4, -8)\}$

28. $\begin{pmatrix} x - 5y = 33 \\ -4x + 7y = -41 \end{pmatrix}$ $\{(-2, -7)\}$

29. $\begin{pmatrix} 3x + y = 2 \\ 11x - 3y = 5 \end{pmatrix}$ $\left\{\left(\frac{11}{20}, \frac{7}{20}\right)\right\}$

30. $\begin{pmatrix} 2x - y = 9 \\ 7x + 4y = 1 \end{pmatrix}$ $\left\{\left(\frac{37}{15}, -\frac{61}{15}\right)\right\}$

31. $\begin{pmatrix} 3x + 5y = 22 \\ 4x - 7y = -39 \end{pmatrix}$ $\{(-1, 5)\}$

32. $\begin{pmatrix} 2x - 3y = -16 \\ 6x + 7y = 16 \end{pmatrix}$ $\{(-2, 4)\}$

33. $\begin{pmatrix} 4x - 5y = 3 \\ 8x + 15y = -24 \end{pmatrix}$ $\left\{\left(-\frac{3}{4}, -\frac{6}{5}\right)\right\}$

34. $\begin{pmatrix} 2x + 3y = 3 \\ 4x - 9y = -4 \end{pmatrix}$ $\left\{\left(\frac{1}{2}, \frac{2}{3}\right)\right\}$

35. $\begin{pmatrix} 6x - 3y = 4 \\ 5x + 2y = -1 \end{pmatrix}$ $\left\{\left(\frac{5}{27}, -\frac{26}{27}\right)\right\}$

36. $\begin{pmatrix} 7x - 2y = 1 \\ 4x + 5y = 2 \end{pmatrix}$ $\left\{\left(\frac{9}{43}, \frac{10}{43}\right)\right\}$

For Problems 37–50, solve each problem by setting up and solving an appropriate system of equations.

37. Doris invested some money at 7% and some money at 8%. She invested \$6000 more at 8% than she did at 7%. Her total yearly interest from the two investments was \$780. How much did Doris invest at each rate? \$2000 at 7% and \$8000 at 8%

38. Suppose that Gus invested a total of \$8000, part of it at 8% and the remainder at 9%. His yearly income from the two investments was \$690. How much did he invest at each rate? \$3000 at 8% and \$5000 at 9%

39. The sum of the digits of a two-digit number is 11. The tens digit is one more than four times the units digit. Find the number. 92

40. The units digit of a two-digit number is one less than three times the tens digit. If the sum of the digits is 11, find the number. 38

41. Find two numbers whose sum is 131 such that one number is 5 less than three times the other. 34 and 97

42. The difference of two numbers is 75. The larger number is three less than four times the smaller number. Find the numbers. 26 and 101

43. In a class of 50 students, the number of females is two more than five times the number of males. How many females are there in the class? 42 females

44. In a recent survey, one thousand registered voters were asked about their political preferences. The number of males in the survey was five less than one-half of the number of females. Find the number of males in the survey. 330 males

45. The perimeter of a rectangle is 94 inches. The length of the rectangle is 7 inches more than the width. Find the dimensions of the rectangle. 20 inches by 27 inches

46. Two angles are supplementary and the measure of one of them is 20° less than three times the measure of the other angle. Find the measure of each angle. 50° and 130°

47. A deposit slip listed \$700 in cash to be deposited. There were 100 bills, some of them five-

dollar bills and the remainder ten-dollar bills. How many bills of each denomination were deposited? 60 5-dollar bills and 40 10-dollar bills

48. Cindy has 30 coins, consisting of dimes and quarters, that total \$5.10. How many coins of each kind does she have? 16 dimes and 14 quarters

49. The income from a student production was \$10,000. The price of a student ticket was \$3 and nonstudent tickets were sold at \$5 each. Three thousand tickets were sold. How many tickets of each kind were sold? 2500 student tickets and 500 nonstudent tickets

50. Sue bought 3 packages of cookies and 2 sacks of potato chips for \$3.65. Later she bought 2 more packages of cookies and 5 additional sacks of potato chips for \$4.23. Find the price of a package of cookies. \$.89

9.2 The Elimination-by-Addition Method

We found in the previous section that the substitution method for solving a system of two equations and two unknowns works rather well. However, as the number of equations and unknowns increases, the substitution method becomes quite unwieldy. In this section we are going to introduce another method called the **elimination-by-addition** method. We shall introduce it here using systems of two linear equations in two unknowns and then extend its use to three linear equations in three unknowns in the next section.

The elimination-by-addition method involves the replacement of systems of equations with simpler equivalent systems until we obtain a system whereby we can easily extract the solutions. **Equivalent systems of equations are systems that have exactly the same solution set.** We can apply the following operations or transformations to a system of equations to produce an equivalent system.

1. Any two equations of the system can be interchanged.
2. Both sides of an equation of the system can be multiplied by any nonzero real number.
3. Any equation of the system can be replaced by the *sum* of that equation and a nonzero multiple of another equation.

Now let's see how to apply these operations to solve a system of two linear equations in two unknowns.

EXAMPLE 1 Solve the system $\begin{pmatrix} 3x + 2y = 1 \\ 5x - 2y = 23 \end{pmatrix}$. (1) (2)

Solution Let's replace equation (2) with an equation we form by multiplying equation (1) by 1 and then adding that result to equation (2).

$$\begin{pmatrix} 3x + 2y = 1 \\ 8x \quad\quad = 24 \end{pmatrix} \begin{matrix} (3) \\ (4) \end{matrix}$$

From equation (4) we can easily obtain the value of x.

$$8x = 24$$
$$x = 3$$

Then we can substitute 3 for x in equation (3).

$$3x + 2y = 1$$
$$3(3) + 2y = 1$$
$$2y = -8$$
$$y = -4$$

The solution set is $\{(3, -4)\}$. Check it! ■

EXAMPLE 2 Solve the system $\begin{pmatrix} x + 5y = -2 \\ 3x - 4y = -25 \end{pmatrix}$. $\begin{matrix} (1) \\ (2) \end{matrix}$

Solution Let's replace equation (2) with an equation we form by multiplying equation (1) by -3 and then adding that result to equation (2).

$$\begin{pmatrix} x + 5y = -2 \\ -19y = -19 \end{pmatrix} \begin{matrix} (3) \\ (4) \end{matrix}$$

From equation (4) we can obtain the value of y.

$$-19y = -19$$
$$y = 1$$

Now we can substitute 1 for y in equation (3).

$$x + 5y = -2$$
$$x + 5(1) = -2$$
$$x = -7$$

The solution set is $\{(-7, 1)\}$. ■

Notice that our objective has been to produce an equivalent system of equations whereby one of the variables can be *eliminated* from one equation. We accomplish this by multiplying one equation of the system by an appropriate number and then *adding* that result to the other equation. Thus, the method is called **elimination-by-addition**. Let's look at another example.

EXAMPLE 3 Solve the system $\begin{pmatrix} 2x + 5y = 4 \\ 5x - 7y = -29 \end{pmatrix}$. (1) (2)

Solution Let's form an equivalent system where the second equation has no x-term. First, we can multiply equation (2) by -2.

$$\begin{pmatrix} 2x + 5y = 4 \\ -10x + 14y = 58 \end{pmatrix} \quad \begin{matrix} (3) \\ (4) \end{matrix}$$

Now we can replace equation (4) with an equation that we form by multiplying equation (3) by 5 and then adding that result to equation (4).

$$\begin{pmatrix} 2x + 5y = 4 \\ 39y = 78 \end{pmatrix} \quad \begin{matrix} (5) \\ (6) \end{matrix}$$

From equation (6) we can find the value of y.

$$39y = 78$$
$$y = 2$$

Now we can substitute 2 for y in equation (5).

$$2x + 5y = 4$$
$$2x + 5(2) = 4$$
$$2x = -6$$
$$x = -3$$

The solution set is $\{(-3, 2)\}$. ■

EXAMPLE 4 Solve the system $\begin{pmatrix} 3x - 2y = 5 \\ 2x + 7y = 9 \end{pmatrix}$. (1) (2)

Solution We can start by multiplying equation (2) by -3.

$$\begin{pmatrix} 3x - 2y = 5 \\ -6x - 21y = -27 \end{pmatrix} \quad \begin{matrix} (3) \\ (4) \end{matrix}$$

Now we can replace equation (4) with an equation we form by multiplying equation (3) by 2 and then adding that result to equation (4).

$$\begin{pmatrix} 3x - 2y = 5 \\ -25y = -17 \end{pmatrix} \quad \begin{matrix} (5) \\ (6) \end{matrix}$$

From equation (6) we can find the value of y.

$$-25y = -17$$
$$y = \frac{17}{25}$$

Now we can substitute $\frac{17}{25}$ for y in equation (5).

$$3x - 2y = 5$$

$$3x - 2\left(\frac{17}{25}\right) = 5$$

$$3x - \frac{34}{25} = 5$$

$$3x = 5 + \frac{34}{25}$$

$$3x = \frac{125}{25} + \frac{34}{25}$$

$$3x = \frac{159}{25}$$

$$x = \left(\frac{159}{25}\right)\left(\frac{1}{3}\right) = \frac{53}{25}$$

The solution set is $\left\{\left(\frac{53}{25}, \frac{17}{25}\right)\right\}$. (Perhaps you should check this result!) ■

Which Method to Use

Both the elimination-by-addition and the substitution methods can be used to obtain exact solutions for any system of two linear equations in two unknowns. Sometimes the issue is that of deciding which method to use on a particular system. As we have seen with the examples thus far in this section and those of the previous section, many systems lend themselves to one or the other method by the original format of the equations. Let's emphasize that point with some more examples.

EXAMPLE 5 Solve the system $\begin{pmatrix} 4x - 3y = 4 \\ 10x + 9y = -1 \end{pmatrix}$. (1) (2)

Solution Because changing the form of either equation in preparation for the substitution method would produce a fractional form, we are probably better off using the elimination-by-addition method.

Let's replace equation (2) with an equation we form by multiplying equation (1) by 3 and then adding that result to equation (2).

$$\begin{pmatrix} 4x - 3y = 4 \\ 22x \quad\quad = 11 \end{pmatrix} \quad \begin{matrix} (3) \\ (4) \end{matrix}$$

From equation (4) we can determine the value of x.

$$22x = 11$$

$$x = \frac{11}{22} = \frac{1}{2}$$

Now we can substitute $\frac{1}{2}$ for x in equation (3).

$$\begin{aligned} 4x - 3y &= 4 \\ 4\left(\frac{1}{2}\right) - 3y &= 4 \\ 2 - 3y &= 4 \\ -3y &= 2 \\ y &= -\frac{2}{3} \end{aligned}$$

The solution set is $\left\{\left(\frac{1}{2}, -\frac{2}{3}\right)\right\}$. ■

EXAMPLE 6 Solve the system $\begin{pmatrix} 6x + 5y = -3 \\ y = -2x - 7 \end{pmatrix}$. (1) (2)

Solution Because the second equation is of the form y *equals*, let's use the substitution method. From the second equation we can substitute $-2x - 7$ for y in the first equation.

$$6x + 5y = -3 \xrightarrow{\text{substitute } -2x-7 \text{ for } y} 6x + 5(-2x - 7) = -3$$

Solving this equation yields

$$\begin{aligned} 6x + 5(-2x - 7) &= -3 \\ 6x - 10x - 35 &= -3 \\ -4x - 35 &= -3 \\ -4x &= 32 \\ x &= -8. \end{aligned}$$

Substitute -8 for x in the second equation to yield

$$y = -2(-8) - 7 = 16 - 7 = 9.$$

The solution set is $\{(-8, 9)\}$. ■

Sometimes we need to simplify the equations of a system before we can decide which method to use for solving the system. Let's consider an example of that type.

EXAMPLE 7 Solve the system

$$\begin{pmatrix} \frac{x-2}{4} + \frac{y+1}{3} = 2 \\ \frac{x+1}{7} + \frac{y-3}{2} = \frac{1}{2} \end{pmatrix}. \quad \begin{matrix} (1) \\ (2) \end{matrix}$$

Solution First, we need to simplify the two equations. Let's multiply both sides of equation (1) by 12 and simplify.

$$12\left(\frac{x-2}{4} + \frac{y+1}{3}\right) = 12(2)$$

$$3(x-2) + 4(y+1) = 24$$

$$3x - 6 + 4y + 4 = 24$$

$$3x + 4y - 2 = 24$$

$$3x + 4y = 26$$

Let's multiply both sides of equation (2) by 14.

$$14\left(\frac{x+1}{7} + \frac{y-3}{2}\right) = 14\left(\frac{1}{2}\right)$$

$$2(x+1) + 7(y-3) = 7$$

$$2x + 2 + 7y - 21 = 7$$

$$2x + 7y - 19 = 7$$

$$2x + 7y = 26$$

Now we have the following system to solve.

$$\begin{pmatrix} 3x + 4y = 26 \\ 2x + 7y = 26 \end{pmatrix} \quad \begin{matrix} (3) \\ (4) \end{matrix}$$

Probably the easiest approach is to use the elimination-by-addition method. We can start by multiplying equation (4) by -3.

$$\begin{pmatrix} 3x + 4y = 26 \\ -6x - 21y = -78 \end{pmatrix} \quad \begin{matrix} (5) \\ (6) \end{matrix}$$

Now we can replace equation (6) with an equation we form by multiplying equation (5) by 2 and then adding that result to equation (6).

$$\begin{pmatrix} 3x + 4y = 26 \\ -13y = -26 \end{pmatrix} \quad \begin{matrix} (7) \\ (8) \end{matrix}$$

From equation (8) we can find the value of y.

$$-13y = -26$$
$$y = 2$$

Now we can substitute 2 for y in equation (7).

$$3x + 4y = 26$$
$$3x + 4(2) = 26$$
$$3x = 18$$
$$x = 6$$

The solution set is $\{(6, 2)\}$. ■

REMARK Don't forget that to check a problem like Example 7 you must check the potential solutions back into the *original* equations.

In Section 9.1 we discussed the fact that you can tell whether a system of two linear equations in two unknowns has no solution, one solution, or infinitely many solutions by graphing the equations of the system. That is, the two lines may be parallel (no solution), or they may intersect in one point (one solution), or they may coincide (infinitely many solutions).

From a practical viewpoint, the systems that have one solution deserve most of our attention. However, we do need to be able to deal with the other situations since they do occur occasionally. Let's use two examples to illustrate the type of thing that happens when we hit a *no solution* or *infinitely many solutions* situation when using either the elimination-by-addition method or the substitution method.

EXAMPLE 8 Solve the system $\begin{pmatrix} y = 3x - 1 \\ -9x + 3y = 4 \end{pmatrix}$. (1) (2)

Solution Using the substitution method, we can proceed as follows.

$$-9x + 3y = 4 \xrightarrow{\text{substitute } 3x - 1 \text{ for } y} -9x + 3(3x - 1) = 4$$

Solving this equation yields

$$-9x + 3(3x - 1) = 4$$
$$-9x + 9x - 3 = 4$$
$$-3 = 4.$$

The *false numerical statement*, $-3 = 4$, implies that the system has *no solution*. (You may want to graph the two lines to verify this conclusion!) ■

EXAMPLE 9 Solve the system $\begin{pmatrix} 5x + y = 2 \\ 10x + 2y = 4 \end{pmatrix}$. (1) (2)

Solution Use the elimination-by-addition method and proceed as follows. Let's replace equation (2) with an equation we form by multiplying equation (1) by -2 and then adding that result to equation (2).

$$\begin{pmatrix} 5x + y = 2 \\ 0 + 0 = 0 \end{pmatrix} \quad \begin{matrix} (3) \\ (4) \end{matrix}$$

The *true numerical statement*, $0 + 0 = 0$, implies that the system has *infinitely many solutions*. Any ordered pair that satisfies one of the equations will also satisfy the other equation. Thus, the solution set can be expressed as

$$\{(x, y) \mid 5x + y = 2\}.$$

■

Problem Set 9.2

For Problems 1–16, use the elimination-by-addition method to solve each system.

1. $\begin{pmatrix} 2x + 3y = -1 \\ 5x - 3y = 29 \end{pmatrix}$ $\{(4, -3)\}$

2. $\begin{pmatrix} 3x - 4y = -30 \\ 7x + 4y = 10 \end{pmatrix}$ $\{(-2, 6)\}$

3. $\begin{pmatrix} 6x - 7y = 15 \\ 6x + 5y = -21 \end{pmatrix}$ $\{(-1, -3)\}$

4. $\begin{pmatrix} 5x + 2y = -4 \\ 5x - 3y = 6 \end{pmatrix}$ $\{(0, -2)\}$

5. $\begin{pmatrix} x - 2y = -12 \\ 2x + 9y = 2 \end{pmatrix}$ $\{(-8, 2)\}$

6. $\begin{pmatrix} x - 4y = 29 \\ 3x + 2y = -11 \end{pmatrix}$ $\{(1, -7)\}$

7. $\begin{pmatrix} 4x + 7y = -16 \\ 6x - y = -24 \end{pmatrix}$ $\{(-4, 0)\}$

8. $\begin{pmatrix} 6x + 7y = 17 \\ 3x + y = -4 \end{pmatrix}$ $\{(-3, 5)\}$

9. $\begin{pmatrix} 3x - 2y = 5 \\ 2x + 5y = -3 \end{pmatrix}$ $\{(1, -1)\}$

10. $\begin{pmatrix} 4x + 3y = -4 \\ 3x - 7y = 34 \end{pmatrix}$ $\{(2, -4)\}$

11. $\begin{pmatrix} 7x - 2y = 4 \\ 7x - 2y = 9 \end{pmatrix}$ Inconsistent

12. $\begin{pmatrix} 5x - y = 6 \\ 10x - 2y = 12 \end{pmatrix}$ Dependent

13. $\begin{pmatrix} 5x + 4y = 1 \\ 3x - 2y = -1 \end{pmatrix}$ $\left\{\left(-\frac{1}{11}, \frac{4}{11}\right)\right\}$

14. $\begin{pmatrix} 2x - 7y = -2 \\ 3x + y = 1 \end{pmatrix}$ $\left\{\left(\frac{5}{23}, \frac{8}{23}\right)\right\}$

15. $\begin{pmatrix} 8x - 3y = 13 \\ 4x + 9y = 3 \end{pmatrix}$ $\left\{\left(\frac{3}{2}, -\frac{1}{3}\right)\right\}$

16. $\begin{pmatrix} 10x - 8y = -11 \\ 8x + 4y = -1 \end{pmatrix}$ $\left\{\left(-\frac{1}{2}, \frac{3}{4}\right)\right\}$

For Problems 17–44, solve each system by using either the substitution or the elimination-by-addition method, whichever seems more appropriate.

17. $\begin{pmatrix} 5x + 3y = -7 \\ 7x - 3y = 55 \end{pmatrix}$ $\{(4, -9)\}$

18. $\begin{pmatrix} 4x - 7y = 21 \\ -4x + 3y = -9 \end{pmatrix}$ $\{(0, -3)\}$

19. $\begin{pmatrix} x = 5y + 7 \\ 4x + 9y = 28 \end{pmatrix}$ $\{(7, 0)\}$

20. $\begin{pmatrix} 11x - 3y = -60 \\ y = -38 - 6x \end{pmatrix}$ $\{(-6, -2)\}$

21. $\begin{pmatrix} x = -6y + 79 \\ x = 4y - 41 \end{pmatrix}$ $\{(7, 12)\}$

22. $\begin{pmatrix} y = 3x + 34 \\ y = -8x - 54 \end{pmatrix}$ $\{(-8, 10)\}$

23. $\begin{pmatrix} 4x - 3y = 2 \\ 5x - y = 3 \end{pmatrix}$ $\left\{\left(\frac{7}{11}, \frac{2}{11}\right)\right\}$

24. $\begin{pmatrix} 3x - y = 9 \\ 5x + 7y = 1 \end{pmatrix}$ $\left\{\left(\frac{32}{13}, -\frac{21}{13}\right)\right\}$

25. $\begin{pmatrix} 5x - 2y = 1 \\ 10x - 4y = 7 \end{pmatrix}$ Inconsistent

26. $\begin{pmatrix} 4x + 7y = 2 \\ 9x - 2y = 1 \end{pmatrix}$ $\left\{\left(\frac{11}{71}, \frac{14}{71}\right)\right\}$

27. $\begin{pmatrix} 3x - 2y = 7 \\ 5x + 7y = 1 \end{pmatrix}$ $\left\{\left(\frac{51}{31}, -\frac{32}{31}\right)\right\}$

28. $\begin{pmatrix} 2x - 3y = 4 \\ y = \frac{2}{3}x - \frac{4}{3} \end{pmatrix}$ Dependent

29. $\begin{pmatrix} -2x + 5y = -16 \\ x = \frac{3}{4}y + 1 \end{pmatrix}$ $\{(-2, -4)\}$

30. $\begin{pmatrix} y = \frac{2}{3}x - \frac{3}{4} \\ 2x + 3y = 11 \end{pmatrix}$ $\left\{\left(\frac{53}{16}, \frac{35}{24}\right)\right\}$

31. $\begin{pmatrix} y = \frac{2}{3}x - 4 \\ 5x - 3y = 9 \end{pmatrix}$ $\left\{\left(-1, -\frac{14}{3}\right)\right\}$

32. $\begin{pmatrix} 5x - 3y = 7 \\ x = \frac{3y}{4} - \frac{1}{3} \end{pmatrix}$ $\left\{\left(\frac{25}{3}, \frac{104}{9}\right)\right\}$

33. $\begin{pmatrix} \frac{x}{6} + \frac{y}{3} = 3 \\ \frac{5x}{2} - \frac{y}{6} = -17 \end{pmatrix}$ $\{(-6, 12)\}$

34. $\begin{pmatrix} \frac{3x}{4} - \frac{2y}{3} = 31 \\ \frac{7x}{5} + \frac{y}{4} = 22 \end{pmatrix}$ $\{(20, -24)\}$

35. $\begin{pmatrix} -(x - 6) + 6(y + 1) = 58 \\ 3(x + 1) - 4(y - 2) = -15 \end{pmatrix}$ $\{(2, 8)\}$

36. $\begin{pmatrix} -2(x + 2) + 4(y - 3) = -34 \\ 3(x + 4) - 5(y + 2) = 23 \end{pmatrix}$ $\{(-3, -6)\}$

37. $\begin{pmatrix} 5(x + 1) - (y + 3) = -6 \\ 2(x - 2) + 3(y - 1) = 0 \end{pmatrix}$ $\{(-1, 3)\}$

38. $\begin{pmatrix} 2(x - 1) - 3(y + 2) = 30 \\ 3(x + 2) + 2(y - 1) = -4 \end{pmatrix}$ $\{(4, -10)\}$

39. $\begin{pmatrix} \frac{1}{2}x - \frac{1}{3}y = 12 \\ \frac{3}{4}x + \frac{2}{3}y = 4 \end{pmatrix}$ $\{(16, -12)\}$

40. $\begin{pmatrix} \frac{2}{3}x + \frac{1}{5}y = 0 \\ \frac{3}{2}x - \frac{3}{10}y = 15 \end{pmatrix}$ $\{(-6, 20)\}$

41. $\begin{pmatrix} \frac{2x}{3} - \frac{y}{2} = -\frac{5}{4} \\ \frac{x}{4} + \frac{5y}{6} = \frac{17}{16} \end{pmatrix}$ $\left\{\left(-\frac{3}{4}, \frac{3}{2}\right)\right\}$

42. $\begin{pmatrix} \frac{x}{2} + \frac{y}{3} = \frac{5}{72} \\ \frac{x}{4} + \frac{5y}{2} = -\frac{17}{48} \end{pmatrix}$ $\left\{\left(\frac{1}{4}, -\frac{1}{6}\right)\right\}$

43. $\begin{pmatrix} \frac{3x + y}{2} + \frac{x - 2y}{5} = 8 \\ \frac{x - y}{3} - \frac{x + y}{6} = \frac{10}{3} \end{pmatrix}$ $\{(5, -5)\}$

44. $\begin{pmatrix} \frac{x - y}{4} - \frac{2x - y}{3} = -\frac{1}{4} \\ \frac{2x + y}{3} + \frac{x + y}{2} = \frac{17}{6} \end{pmatrix}$ $\{(1, 2)\}$

For Problems 45–57, solve each problem by setting up and solving an appropriate system of equations.

45. A 10% salt solution is to be mixed with a 20% salt solution to produce 20 gallons of a 17.5% salt solution. How many gallons of the 10% solution and how many gallons of the 20% solution will be needed? 5 gallons of 10% solution and 15 gallons of 20% solution

46. A small town library buys a total of 35 books that cost \$462. Some of the books cost \$12 each and the remainder cost \$14 per book. How many books of each price did they buy? 14 books at \$12 each and 21 books at \$14 each

47. Suppose that on a particular day the cost of 3 tennis balls and 2 golf balls is \$7. The cost of 6 tennis balls and 3 golf balls is \$12. Find the cost of 1 tennis ball and also the cost of 1 golf ball. \$1 for a tennis ball and \$2 for a golf ball

48. For moving purposes, the Hendersons bought 25 cardboard boxes for \$97.50. There were two kinds of boxes; the large ones cost \$7.50 per box and the small ones were \$3 per box. How many boxes of each kind did they buy? 5 large boxes and 20 small boxes

49. A motel in a suburb of Chicago rents double rooms for \$42 per day and single rooms for \$22 per day. If a total of 55 rooms were rented for \$2010, how many of each kind were rented? 40 double rooms and 15 single rooms

50. Suppose that one solution contains 50% alcohol and another solution contains 80% alcohol. How many liters of each solution should be mixed to make 10.5 liters of a 70% solution? 3.5 liters of the 50% solution and 7 liters of the 80% solution

51. Suppose that a fulcrum is placed so that weights of 40 pounds and 80 pounds are in balance. Furthermore, suppose that when 20 pounds is added to the 40-pound weight, the 80-pound weight must be moved $1\frac{1}{2}$ feet farther from the fulcrum to obtain a balance. Find the original distance between the two weights. 9 feet

52. If a certain two-digit number is divided by the sum of its digits, the quotient is 2. If the digits are reversed, the new number is 9 less than five times the original number. Find the original number. 18

53. If the numerator of a certain fraction is increased by 5 and the denominator is decreased by 1, the resulting fraction is $\frac{8}{3}$. However, if the numerator of the original fraction is doubled and the denominator is increased by 7, the resulting fraction is $\frac{6}{11}$. Find the original fraction. $\frac{3}{4}$

54. A man bought 2 pounds of coffee and 1 pound of butter for a total of \$7.75. A month later the prices had not changed (this makes it a fictitious problem) and he bought 3 pounds of coffee and 2 pounds of butter for \$12.50. Find the price per pound of both the coffee and the butter. \$3 per lb for coffee and \$1.75 per lb for butter

55. Suppose that we have a rectangular-shaped book cover. If the width is increased by 2 centimeters and the length is decreased by 1 centimeter, the area is increased by 28 square centimeters. However, if the width is decreased by 1 centimeter and the length is increased by 2 centimeters, then the area is increased by 10 square centimeters. Find the dimensions of the book cover. 18 centimeters by 24 centimeters

56. A blueprint indicates a master bedroom in the shape of a rectangle. If the width is increased by 2 feet and the length remains the same, then the area is increased by 36 square feet. However, if the width is increased by 1 foot and the length is increased by 2 feet, then the area is increased by 48 square feet. Find the dimensions of the room as indicated on the blueprint. 14 ft by 18 ft

57. A fulcrum is placed so that weights of 60 pounds and 100 pounds are in balance. If 20 pounds are subtracted from the 100-pound weight, then the 60-pound weight must be moved 1 foot closer to the fulcrum to preserve the balance. Find the original distance between the 60-pound and 100-pound weights. 8 ft

Miscellaneous Problems

58. There is another way of telling whether a system of two linear equations in two unknowns is consistent, inconsistent, or dependent without taking the time to graph each equation. It can be shown that any system of the form

$$a_1x + b_1y = c_1$$
$$a_2x + b_2y = c_2$$

has one and only one solution if

$$\frac{a_1}{a_2} \neq \frac{b_1}{b_2} \qquad \text{consistent}$$

has no solution if

$$\frac{a_1}{a_2} = \frac{b_1}{b_2} \neq \frac{c_1}{c_2} \qquad \text{inconsistent}$$

and has infinitely many solutions if

$$\frac{a_1}{a_2} = \frac{b_1}{b_2} = \frac{c_1}{c_2} \qquad \text{dependent}$$

Determine whether each of the following systems is consistent, inconsistent, or dependent.

(a) $\begin{pmatrix} 4x - 3y = 7 \\ 9x + 2y = 5 \end{pmatrix}$ Consistent

(b) $\begin{pmatrix} 5x - y = 6 \\ 10x - 2y = 19 \end{pmatrix}$ Inconsistent

(c) $\begin{pmatrix} 5x - 4y = 11 \\ 4x + 5y = 12 \end{pmatrix}$ Consistent

(d) $\begin{pmatrix} x + 2y = 5 \\ x - 2y = 9 \end{pmatrix}$ Consistent

(e) $\begin{pmatrix} x - 3y = 5 \\ 3x - 9y = 15 \end{pmatrix}$ Dependent

(f) $\begin{pmatrix} 4x + 3y = 7 \\ 2x - y = 10 \end{pmatrix}$ Consistent

(g) $\begin{pmatrix} 3x + 2y = 4 \\ y = -\frac{3}{2}x - 1 \end{pmatrix}$ Inconsistent

(h) $\begin{pmatrix} y = \frac{4}{3}x - 2 \\ 4x - 3y = 6 \end{pmatrix}$ Dependent

59. A system such as

$$\begin{pmatrix} \frac{3}{x} + \frac{2}{y} = 2 \\ \frac{2}{x} - \frac{3}{y} = \frac{1}{4} \end{pmatrix}$$

is not a system of linear equations but can be transformed into a linear system by changing variables. For example, substituting u for $\frac{1}{x}$ and v for $\frac{1}{y}$, the above system becomes

$$\begin{pmatrix} 3u + 2v = 2 \\ 2u - 3v = \frac{1}{4} \end{pmatrix}.$$

We can solve this "new" system by either elimination-by-addition or substitution (we will leave the details for you) to produce $u = \frac{1}{2}$ and $v = \frac{1}{4}$. Therefore, since $u = \frac{1}{x}$ and $v = \frac{1}{y}$, we have

$$\frac{1}{x} = \frac{1}{2} \quad \text{and} \quad \frac{1}{y} = \frac{1}{4}.$$

Solving these equations yields

$$x = 2 \quad \text{and} \quad y = 4.$$

The solution set of the original system is $\{(2, 4)\}$.

Solve each of the following systems.

(a) $\begin{pmatrix} \frac{1}{x} + \frac{2}{y} = \frac{7}{12} \\ \frac{3}{x} - \frac{2}{y} = \frac{5}{12} \end{pmatrix}$ $\{(4, 6)\}$

(b) $\begin{pmatrix} \frac{2}{x} + \frac{3}{y} = \frac{19}{15} \\ -\frac{2}{x} + \frac{1}{y} = -\frac{7}{15} \end{pmatrix}$ $\{(3, 5)\}$

(c) $\begin{pmatrix} \frac{3}{x} - \frac{2}{y} = \frac{13}{6} \\ \frac{2}{x} + \frac{3}{y} = 0 \end{pmatrix}$ $\{(2, -3)\}$

(d) $\begin{pmatrix} \frac{4}{x} + \frac{1}{y} = 11 \\ \frac{3}{x} - \frac{5}{y} = -9 \end{pmatrix}$ $\left\{\left(\frac{1}{2}, \frac{1}{3}\right)\right\}$

(e) $\begin{pmatrix} \frac{5}{x} - \frac{2}{y} = 23 \\ \frac{4}{x} + \frac{3}{y} = \frac{23}{2} \end{pmatrix}$ $\left\{\left(\frac{1}{4}, -\frac{2}{3}\right)\right\}$

(f) $\begin{pmatrix} \frac{2}{x} - \frac{7}{y} = \frac{9}{10} \\ \frac{5}{x} + \frac{4}{y} = -\frac{41}{20} \end{pmatrix}$ $\{(-4, -5)\}$

60. Solve the following system for x and y.

$$\begin{pmatrix} a_1x + b_1y = c_1 \\ a_2x + b_2y = c_2 \end{pmatrix} \quad x = \frac{b_1c_2 - b_2c_1}{a_2b_1 - a_1b_2} \quad \text{and} \quad y = \frac{a_2c_1 - a_1c_2}{a_2b_1 - a_1b_2}$$

9.3

Systems of Three Linear Equations in Three Variables

Consider a linear equation in three variables x, y, and z, such as $3x - 2y + z = 7$. Any **ordered triple** (x, y, z) that makes the equation a true numerical statement is said to be a solution of the equation. For example, the ordered triple $(2, 1, 3)$ is a solution because $3(2) - 2(1) + 3 = 7$. However, the ordered triple $(5, 2, 4)$ is not a solution because $3(5) - 2(2) + 4 \neq 7$. There are infinitely many solutions in the solution set.

REMARK The concept of a *linear* equation is generalized to include equations of more than two variables. Thus, an equation such as $5x - 2y + 9z = 8$ is called a linear equation in three variables; the equation $5x - 7y + 2z - 11w = 1$ is called a linear equation in four variables, and so on.

To *solve* a system of three linear equations in three variables, such as

$$\begin{pmatrix} 3x - y + 2z = 13 \\ 4x + 2y + 5z = 30 \\ 5x - 3y - z = 3 \end{pmatrix}$$

means to find all of the ordered triples that satisfy all three equations. In other words, the solution set of the system is the intersection of the solution sets of all three equations in the system.

The graph of a linear equation in three variables is a **plane**, not a line. In fact, graphing equations in three variables requires the use of a three-dimensional coordinate system. Thus, using a graphing approach to solve systems of three linear equations in three variables is not at all practical. However, a simple graphic analysis does provide us with some direction as to what we can expect as we begin solving such systems.

In general, since each linear equation in three variables produces a plane, a system of three such equations produces three planes. There are various ways that three planes can be related. For example, they may be mutually parallel, or two of the planes may be parallel and the third one intersect each of the two. (You may want to analyze all of the other possibilities for the three planes!) However, for our purposes at this time we need to realize that from a solution set viewpoint, a system of three linear equations in three variables produces one of the following possibilities.

1. There is *one ordered triple* that satisfies all three equations. The three planes have a common point of intersection as indicated in Figure 9.3.

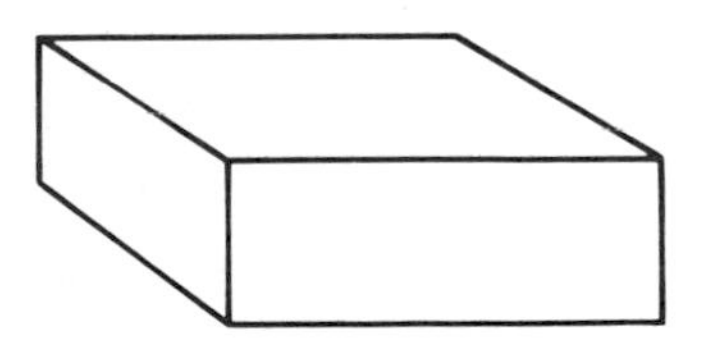

Figure 9.3

2. There are *infinitely many* ordered triples in the solution set, all of which are coordinates of points on a line common to the planes. This can happen if the three planes have a common line of intersection (Figure 9.4(a)), or if two of the planes coincide and the third plane intersects them (Figure 9.4(b)).

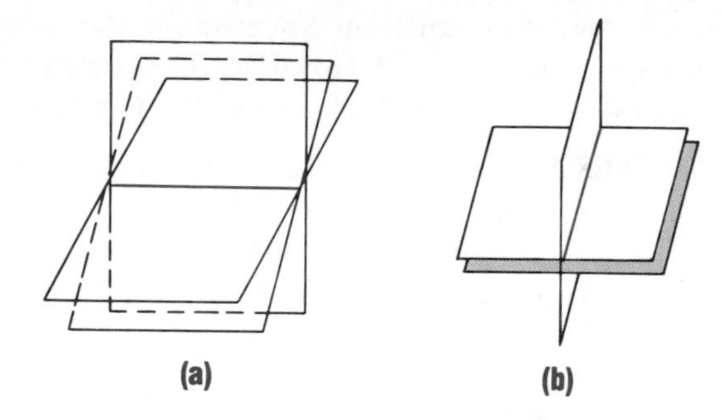

(a) (b)

Figure 9.4

3. There are *infinitely many* ordered triples in the solution set, all of which are coordinates of points on a plane. This can happen if the three planes coincide, as illustrated in Figure 9.5.

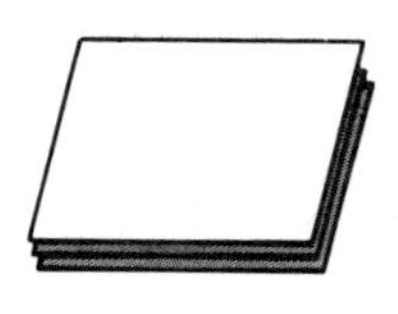

Figure 9.5

4. The solution set is *empty*; it is $\varnothing$. This can happen in various ways, as we see in Figure 9.6. Notice that in each situation there are no points common to all three planes.

Figure 9.6

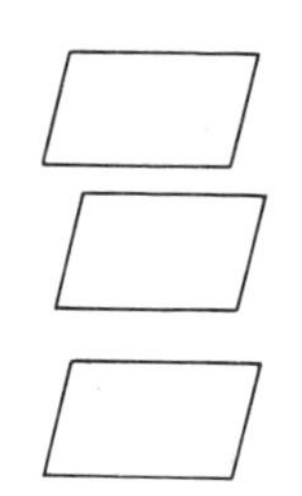

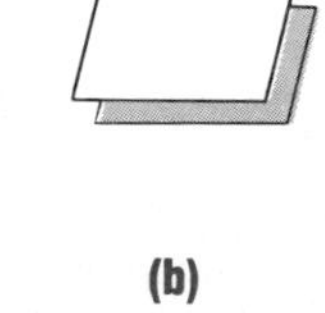

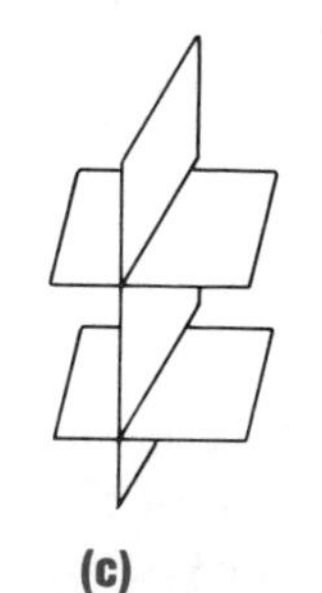

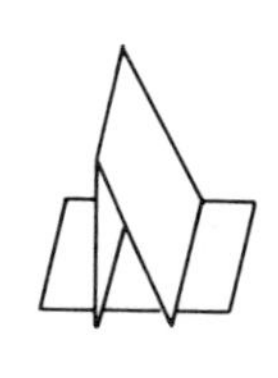

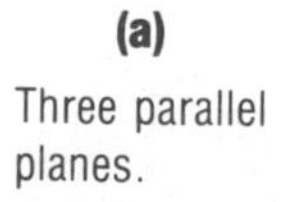

(a) Three parallel planes.

(b) Two planes coincide, and the third one is parallel to the coinciding planes.

(c) Two planes are parallel and the third one intersects them in parallel lines.

(d) No two planes are parallel, but two of them intersect in a line which is parallel to the third plane.

Now that we know what possibilities exist, let's consider finding the solution sets for some systems. Our approach will be the elimination-by-addition method, whereby systems are replaced with equivalent systems until a system is obtained where we can easily determine the solution set. Let's start with an example that allows us to determine the solution set without changing to another equivalent system.

EXAMPLE 1 Solve the system

$$\begin{pmatrix} 2x - 3y + 5z = -5 \\ 2y - 3z = 4 \\ 4z = -8 \end{pmatrix}. \quad \begin{matrix} (1) \\ (2) \\ (3) \end{matrix}$$

Solution From equation (3) we can find the value of z.

$$4z = -8$$
$$z = -2$$

Now we can substitute -2 for z in equation (2).

$$2y - 3z = 4$$
$$2y - 3(-2) = 4$$
$$2y + 6 = 4$$
$$2y = -2$$
$$y = -1$$

Finally, we can substitute -2 for z and -1 for y in equation (1).

$$2x - 3y + 5z = -5$$
$$2x - 3(-1) + 5(-2) = -5$$
$$2x + 3 - 10 = -5$$
$$2x - 7 = -5$$
$$2x = 2$$
$$x = 1$$

The solution set is $\{(1, -1, -2)\}$. ■

Notice the format of the equations in the system of Example 1. The first equation contains all three variables, the second equation has only two variables, and the third equation has only one variable. This allowed us to solve the third equation and then to use "back substitution" to find the values of the other variables. Let's consider another example where we have to make one replacement of an equivalent system.

EXAMPLE 2 Solve the system $\begin{pmatrix} 3x + 2y - 7z = -34 \\ y + 5z = 21 \\ 3y - 2z = -22 \end{pmatrix}$. (1) (2) (3)

Solution Let's replace equation (3) with an equation we form by multiplying equation (2) by -3 and then adding that result to equation (3).

$$\begin{pmatrix} 3x + 2y - 7z = -34 \\ y + 5z = 21 \\ -17z = -85 \end{pmatrix} \quad \begin{matrix} (4) \\ (5) \\ (6) \end{matrix}$$

From equation (6) we can find the value of z.

$$-17z = -85$$
$$z = 5$$

Now we can substitute 5 for z in equation (5).

$$y + 5z = 21$$
$$y + 5(5) = 21$$
$$y = -4$$

Finally, we can substitute 5 for z and -4 for y in equation (4).

$$3x + 2y - 7z = -34$$
$$3x + 2(-4) - 7(5) = -34$$
$$3x - 8 - 35 = -34$$
$$3x - 43 = -34$$
$$3x = 9$$
$$x = 3$$

The solution set is $\{(3, -4, 5)\}$. ■

Now let's consider some examples where we have to make more than one replacement of equivalent systems.

EXAMPLE 3 Solve the system $\begin{pmatrix} x - y + 4z = -29 \\ 3x - 2y - z = -6 \\ 2x - 5y + 6z = -55 \end{pmatrix}$. (1) (2) (3)

Solution Let's replace equation (2) with an equation we form by multiplying equation (1) by -3 and then adding that result to equation (2). Let's also replace equation (3) with an equation we form by multiplying equation (1) by -2 and then adding that result to equation (3).

$$\begin{pmatrix} x - y + 4z = -29 \\ y - 13z = 81 \\ -3y - 2z = 3 \end{pmatrix} \quad \begin{matrix} (4) \\ (5) \\ (6) \end{matrix}$$

Now let's replace equation (6) with an equation we form by multiplying equation (5) by 3 and then adding that result to equation (6).

$$\begin{pmatrix} x - y + 4z = -29 \\ y - 13z = 81 \\ -41z = 246 \end{pmatrix} \quad \begin{matrix} (7) \\ (8) \\ (9) \end{matrix}$$

From equation (9) we can determine the value of z.

$$-41z = 246$$
$$z = -6$$

Now we can substitute -6 for z in equation (8).

$$y - 13z = 81$$
$$y - 13(-6) = 81$$
$$y + 78 = 81$$
$$y = 3$$

Finally, we can substitute -6 for z and 3 for y in equation (7).

$$x - y + 4z = -29$$
$$x - 3 + 4(-6) = -29$$
$$x - 3 - 24 = -29$$
$$x - 27 = -29$$
$$x = -2$$

The solution set is $\{(-2, 3, -6)\}$. ■

EXAMPLE 4 Solve the system $\begin{pmatrix} 3x - 4y + z = 14 \\ 5x + 3y - 2z = 27 \\ 7x - 9y + 4z = 31 \end{pmatrix}$. $\begin{matrix} (1) \\ (2) \\ (3) \end{matrix}$

Solution A glance at the coefficients in the system indicates that eliminating the z-terms from equations (2) and (3) would be easy to do. Let's replace equation (2) with an equation we form by multiplying equation (1) by 2 and then adding that result to equation (2). Also let's replace equation (3) with an equation we form by multiplying equation (1) by -4 and then adding that result to equation (3).

$$\begin{pmatrix} 3x - 4y + z = 14 \\ 11x - 5y \quad = 55 \\ -5x + 7y \quad = -25 \end{pmatrix} \begin{matrix} (4) \\ (5) \\ (6) \end{matrix}$$

Now let's eliminate the y-terms from equations (5) and (6). First, let's multiply equation (6) by 5.

$$\begin{pmatrix} 3x - 4y + z = 14 \\ 11x - 5y \quad = 55 \\ -25x + 35y \quad = -125 \end{pmatrix} \begin{matrix} (7) \\ (8) \\ (9) \end{matrix}$$

Now we can replace equation (9) with an equation we form by multiplying equation (8) by 7 and then adding that result to equation (9).

$$\begin{pmatrix} 3x - 4y + z = 14 \\ 11x - 5y \quad = 55 \\ 52x \quad = 260 \end{pmatrix} \begin{matrix} (10) \\ (11) \\ (12) \end{matrix}$$

From equation (12) we can determine the value of x.

$$52x = 260$$
$$x = 5$$

Now we can substitute 5 for x in equation (11).

$$11x - 5y = 55$$
$$11(5) - 5y = 55$$
$$-5y = 0$$
$$y = 0$$

Finally, we can substitute 5 for x and 0 for y in equation (10).

$$3x - 4y + z = 14$$
$$3(5) - 4(0) + z = 14$$
$$15 - 0 + z = 14$$
$$z = -1$$

The solution set is $\{(5, 0, -1)\}$. ■

EXAMPLE 5 Solve the system $\begin{pmatrix} x - 2y + 3z = 1 \\ 3x - 5y - 2z = 4 \\ 2x - 4y + 6z = 7 \end{pmatrix}$. (1) (2) (3)

Solution A glance at the coefficients indicates that it should be easy to eliminate the x-terms from equations (2) and (3). We can replace equation (2) with an equation we form by multiplying equation (1) by -3 and then adding that result to equation (2). Likewise, we can replace equation (3) with an equation we form by multiplying equation (1) by -2 and then adding that result to equation (3).

$$\begin{pmatrix} x - 2y + 3z = 1 \\ y - 11z = 1 \\ 0 + 0 + 0 = 5 \end{pmatrix} \begin{matrix} (4) \\ (5) \\ (6) \end{matrix}$$

The false statement, $0 = 5$, indicates that the system is inconsistent and therefore the solution set is $\varnothing$. (If you were to graph this system, equations (1) and (3) would produce parallel planes. Thus, this is the situation depicted back in Figure 9.6(c).) ■

EXAMPLE 6 Solve the system

$$\begin{pmatrix} 2x - y + 4z = 1 \\ 3x + 2y - z = -5 \\ 5x - 6y + 17z = -1 \end{pmatrix}. \begin{matrix} (1) \\ (2) \\ (3) \end{matrix}$$

Solution A glance at the coefficients indicates that it is easy to eliminate the y-terms from equations (2) and (3). We can replace equation (2) with an equation we form by multiplying equation (1) by 2 and then adding that result to equation (2). Likewise, we can replace equation (3) with an equation we form by multiplying equation (1) by -6 and then adding that result to equation (3).

$$\begin{pmatrix} 2x - y + 4z = 1 \\ 7x + 7z = 7 \\ -7x - 7z = -7 \end{pmatrix} \begin{matrix} (4) \\ (5) \\ (6) \end{matrix}$$

Now let's replace equation (6) with an equation we form by multiplying equation (5) by 1 and then adding that result to equation (6).

$$\begin{pmatrix} 2x - y + 4z = 1 \\ 7x + 7z = 7 \\ 0 + 0 = 0 \end{pmatrix} \begin{matrix} (7) \\ (8) \\ (9) \end{matrix}$$

The true numerical statement, $0 + 0 = 0$, indicates that the system has *infinitely many solutions*. (The graph of this system is shown in Figure 9.4(a).) ■

REMARK It can be shown that the solutions for the system in Example 4 are of the form $(t, 3 - 2t, 1 - t)$, where t is any real number. For example, if we let $t = 2$,then we get the ordered triple $(2, -1, -1)$ and this triple will satisfy all three of the original equations. For our purposes in this text we shall simply indicate that such a system has *infinitely many solutions*.

Problem Set 9.3

Solve each of the following systems. If the solution set is $\varnothing$ or if it contains infinitely many solutions, then so indicate.

1. $\begin{pmatrix} x + 2y - 3z = 2 \\ 3y - z = 13 \\ 3y + 5z = 25 \end{pmatrix}$ $\{(-2, 5, 2)\}$

2. $\begin{pmatrix} 2x + 3y - 4z = -10 \\ 2y + 3z = 16 \\ 2y - 5z = -16 \end{pmatrix}$ $\{(0, 2, 4)\}$

3. $\begin{pmatrix} 3x + 2y - 2z = 14 \\ x - 6z = 16 \\ 2x + 5z = -2 \end{pmatrix}$ $\{(4, -1, -2)\}$

4. $\begin{pmatrix} 3x + 2y - z = -11 \\ 2x - 3y = -1 \\ 4x + 5y = -13 \end{pmatrix}$ $\{(-2, -1, 3)\}$

5. $\begin{pmatrix} 2x - y + z = 0 \\ 3x - 2y + 4z = 11 \\ 5x + y - 6z = -32 \end{pmatrix}$ $\{(-1, 3, 5)\}$

6. $\begin{pmatrix} x - 2y + 3z = 7 \\ 2x + y + 5z = 17 \\ 3x - 4y - 2z = 1 \end{pmatrix}$ $\{(3, 1, 2)\}$

7. $\begin{pmatrix} 4x - y + z = 5 \\ 3x + y + 2z = 4 \\ x - 2y - z = 1 \end{pmatrix}$ Infinitely many solutions

8. $\begin{pmatrix} 2x - y + 3z = -14 \\ 4x + 2y - z = 12 \\ 6x - 3y + 4z = -22 \end{pmatrix}$ $\{(\frac{1}{2}, 3, -4)\}$

9. $\begin{pmatrix} x - y + 2z = 4 \\ 2x - 2y + 4z = 7 \\ 3x - 3y + 6z = 1 \end{pmatrix}$ $\varnothing$

10. $\begin{pmatrix} x + y - z = 2 \\ 3x - 4y + 2z = 5 \\ 2x + 2y - 2z = 7 \end{pmatrix}$ $\varnothing$

11. $\begin{pmatrix} x - 2y + z = -4 \\ 2x + 4y - 3z = -1 \\ -3x - 6y + 7z = 4 \end{pmatrix}$ $\{(-2, \frac{3}{2}, 1)\}$

12. $\begin{pmatrix} 2x - y + 3z = 1 \\ 4x + 7y - z = 7 \\ x + 4y - 2z = 3 \end{pmatrix}$ Infinitely many solutions

13. $\begin{pmatrix} 3x - 2y + 4z = 6 \\ 9x + 4y - z = 0 \\ 6x - 8y - 3z = 3 \end{pmatrix}$ $\{(\frac{1}{3}, -\frac{1}{2}, 1)\}$

14. $\begin{pmatrix} 2x - y + 3z = 0 \\ 3x + 2y - 4z = 0 \\ 5x - 3y + 2z = 0 \end{pmatrix}$ $\{(0, 0, 0)\}$

15. $\begin{pmatrix} 3x - y + 4z = 9 \\ 3x + 2y - 8z = -12 \\ 9x + 5y - 12z = -23 \end{pmatrix}$ $\{(\frac{2}{3}, -4, \frac{3}{4})\}$

16. $\begin{pmatrix} 5x - 3y + z = 1 \\ 2x - 5y = -2 \\ 3x - 2y - 4z = -27 \end{pmatrix}$ $\{(-1, 0, 6)\}$

17. $\begin{pmatrix} 4x - y + 3z = -12 \\ 2x + 3y - z = 8 \\ 6x + y + 2z = -8 \end{pmatrix}$ $\{(-2, 4, 0)\}$

18. $\begin{pmatrix} x + 3y - 2z = 19 \\ 3x - y - z = 7 \\ -2x + 5y + z = 2 \end{pmatrix}$ $\{(1, 2, -6)\}$

19. $\begin{pmatrix} x + y + z = 1 \\ 2x - 3y + 6z = 1 \\ -x + y + z = 0 \end{pmatrix}$ $\{(\frac{1}{2}, \frac{1}{3}, \frac{1}{6})\}$

20. $\begin{pmatrix} 3x + 2y - 2z = -2 \\ x - 3y + 4z = -13 \\ -2x + 5y + 6z = 29 \end{pmatrix}$ $\{(-3, 4, \frac{1}{2})\}$

Solve each of the following problems by setting up and solving a system of three linear equations in three variables.

21. The sum of the digits of a three-digit number is 14. The number is 14 larger than twenty times the tens digit. The sum of the tens digit and the units digit is 12 larger than the hundreds digit. Find the number. 194

22. The sum of the digits of a three-digit number is 13. The sum of the hundreds digit and the tens digit is one less than the units digit. The sum of three times the hundreds digit and four times the units digit is 26 more than twice the tens digit. Find the number. 247

23. Two bottles of catsup, 2 jars of peanut butter, and 1 jar of pickles cost $4.20. Three bottles of catsup, 4 jars of peanut butter, and 2 jars of pickles cost $7.70. Four bottles of catsup, 3 jars of peanut butter, and 5 jars of pickles cost $9.80. Find the cost per bottle of catsup and per jar for peanut butter and pickles. $.70, $1, and $.80

24. Five pounds of potatoes, 1 pound of onions, and 2 pounds of apples cost $1.26. Two pounds of potatoes, 3 pounds of onions, and 4 pounds of apples cost $1.88. Three pounds of potatoes, 4 pounds of onions, and 1 pound of apples cost $1.24. Find the price per pound for each item. $.10, $.16, and $.30

25. The sum of three numbers is 20. The sum of the first and third numbers is 2 more than twice the second number. The third number minus the first yields three times the second number. Find the numbers. -2, 6, and 16

26. The sum of three numbers is 40. The third number is 10 less than the sum of the first two numbers. The second number is 1 larger than the first. Find the numbers. 12, 13, and 15

27. The sum of the measures of the angles of a triangle is 180°. The largest angle is twice the smallest angle. The sum of the smallest and the largest angle is twice the other angle. Find the measure of each angle. 40°, 60°, and 80°

28. A box contains $2 in nickels, dimes, and quarters. There are 19 coins in all with twice as many nickels as dimes. How many coins of each kind are there? 10 n, 5 d, and 4 q

29. Part of $3000 is invested at 12%, another part at 13%, and the remainder at 14%. The total yearly income from the three investments is $400. The sum of the amounts invested at 12% and 13% equals the amount invested at 14%. Determine how much is invested at each rate. $500 at 12%, $1000 at 13%, and $1500 at 14%

30. The perimeter of a triangle is 45 centimeters. The longest side is 4 centimeters less than twice the shortest side. The sum of the lengths of the shortest and longest sides is 7 centimeters less than three times the length of the remaining side. Find the lengths of all three sides of the triangle. 12 centimeters, 13 centimeters, and 20 centimeters

9.4 A Matrix Approach to Solving Systems

The primary objective of this chapter is to introduce a variety of techniques for solving systems of linear equations. The techniques we've discussed thus far lend themselves to "small" systems. As the number of equations and variables increases, the systems become more difficult to solve and require other techniques. In these next three sections we will continue to work with small systems for the sake of convenience but will learn some techniques that can be extended to larger systems. This section will introduce a matrix approach to solving systems.

A **matrix** is simply an array of numbers arranged in horizontal rows and vertical columns. For example, the matrix

$$\text{2 rows} \begin{matrix} \longrightarrow \\ \longrightarrow \end{matrix} \begin{bmatrix} 2 & 1 & -4 \\ 5 & -7 & 6 \end{bmatrix}$$

$$\uparrow \quad \uparrow \quad \uparrow$$

$$\text{3 columns}$$

has 2 rows and 3 columns, which we refer to as a 2×3 (read "two-by-three") matrix. Some additional examples of matrices (*matrices* is the plural of matrix) are as follows.

$$\underset{3 \times 2}{\begin{bmatrix} 3 & 2 \\ -1 & 4 \\ 5 & 7 \end{bmatrix}} \qquad \underset{2 \times 2}{\begin{bmatrix} 4 & 1 \\ 0 & -5 \end{bmatrix}} \qquad \underset{1 \times 4}{\begin{bmatrix} 1 & 2 & 6 & 8 \end{bmatrix}} \qquad \underset{5 \times 1}{\begin{bmatrix} 3 \\ 7 \\ 10 \\ 2 \\ -4 \end{bmatrix}}$$

In general, *a* matrix of m rows and n columns is called a matrix of dimension $m \times n$.

With every system of linear equations we can associate a matrix that consists of the coefficients and constant terms. For example, with the system

$$\begin{pmatrix} x - 3y = -17 \\ 2x + 7y = 31 \end{pmatrix}$$

we can associate the matrix

$$\left[\begin{array}{cc:c} 1 & -3 & -17 \\ 2 & 7 & 31 \end{array}\right],$$

which is called the **augmented matrix** of the system. The dashed line separates the coefficients from the constant terms; technically it is not necessary.

Since augmented matrices represent systems of equations, we can operate with them as we do with systems of equations. Our previous work with systems of equations was based on the following properties.

1. Any two equations of a system may be interchanged.

Example By interchanging the two equations, the system $\begin{pmatrix} 2x - 5y = 9 \\ x + 3y = 4 \end{pmatrix}$ is equivalent to the system $\begin{pmatrix} x + 3y = 4 \\ 2x - 5y = 9 \end{pmatrix}$.

2. Any equation of the system may be multiplied by a nonzero constant.

Example By multiplying the top equation by -2, the system

$$\begin{pmatrix} x + 3y = 4 \\ 2x - 5y = 9 \end{pmatrix}$$

is equivalent to the system $\begin{pmatrix} -2x - 6y = -8 \\ 2x - 5y = 9 \end{pmatrix}$.

3. Any equation of the system can be replaced by adding a nonzero multiple of another equation to that equation.

Example By adding -2 times the first equation to the second equation, the system $\begin{pmatrix} x + 3y = 4 \\ 2x - 5y = 9 \end{pmatrix}$ is equivalent to the system $\begin{pmatrix} x + 3y = 4 \\ -11y = 1 \end{pmatrix}$.

Each of the properties geared to solving a system of equations produces a corresponding property of the augmented matrix of the system. For example, exchanging two equations of a system corresponds to exchanging two rows of the augmented matrix that represents the system. We usually refer to these properties as *elementary row operations* and we can state them as follows.

Elementary Row Operations

1. Any two rows of an augmented matrix can be interchanged.
2. Any row can be multiplied by a nonzero constant.
3. Any row of the augmented matrix can be replaced by adding a nonzero multiple of another row to that row.

Using the elementary row operations on an augmented matrix provides a basis for solving systems of linear equations. Study the following examples very carefully; keep in mind that the general scheme, called **Gaussian elimination,** is one of using elementary row operations on a matrix to continue replacing a system of equations with an equivalent system until a system is obtained where the solutions are easily determined. We will use a format similar to the one we used in the previous section except that we will represent systems of equations by matrices.

EXAMPLE 1 Solve the system $\begin{pmatrix} x - 3y = -17 \\ 2x + 7y = 31 \end{pmatrix}$.

Solution The augmented matrix of the system is

$$\left[\begin{array}{cc|c} 1 & -3 & -17 \\ 2 & 7 & 31 \end{array}\right].$$

We can multiply row one by -2 and add this result to row two to produce a new row two.

$$\left[\begin{array}{cc|c} 1 & -3 & -17 \\ 0 & 13 & 65 \end{array}\right]$$

This matrix represents the system

$$\begin{pmatrix} x - 3y = -17 \\ 13y = 65 \end{pmatrix}.$$

From the last equation we can determine the value of y.

$$13y = 65$$
$$y = 5$$

Now we can substitute 5 for y in the equation $x - 3y = -17$.

$$x - 3(5) = -17$$
$$x - 15 = -17$$
$$x = -2$$

The solution set is $\{(-2, 5)\}$. ■

EXAMPLE 2 Solve the system $\begin{pmatrix} 3x + 2y = 3 \\ 30x - 6y = 17 \end{pmatrix}$.

Solution The augmented matrix of the system is

$$\left[\begin{array}{cc|c} 3 & 2 & 3 \\ 30 & -6 & 17 \end{array}\right].$$

We can multiply row one by -10 and add this result to row two to produce a new row two.

$$\left[\begin{array}{cc|c} 3 & 2 & 3 \\ 0 & -26 & -13 \end{array}\right]$$

This matrix represents the system

$$\begin{pmatrix} 3x + 2y = 3 \\ -26y = -13 \end{pmatrix}.$$

From the last equation we can determine the value of y.

$$-26y = -13$$
$$y = \frac{-13}{-26} = \frac{1}{2}$$

Now we can substitute $\frac{1}{2}$ for y in the equation $3x + 2y = 3$.

$$3x + 2\left(\frac{1}{2}\right) = 3$$
$$3x + 1 = 3$$
$$3x = 2$$
$$x = \frac{2}{3}$$

The solution set is $\left\{\left(\frac{2}{3}, \frac{1}{2}\right)\right\}$. ■

EXAMPLE 3 Solve the system $\begin{pmatrix} 2x - 3y - z = -2 \\ x - 2y + 3z = 9 \\ 3x + y - 5z = -8 \end{pmatrix}$.

Solution The augmented matrix of the system is

$$\left[\begin{array}{ccc|c} 2 & -3 & -1 & -2 \\ 1 & -2 & 3 & 9 \\ 3 & 1 & -5 & -8 \end{array}\right].$$

Let's begin by interchanging the top two rows.

$$\left[\begin{array}{ccc|c} 1 & -2 & 3 & 9 \\ 2 & -3 & -1 & -2 \\ 3 & 1 & -5 & -8 \end{array}\right]$$

Now we can multiply row one by -2 and add this result to row two to produce a new row two. Also we can multiply row one by -3 and add this result to row three to produce a new row three.

$$\left[\begin{array}{ccc|c} 1 & -2 & 3 & 9 \\ 0 & 1 & -7 & -20 \\ 0 & 7 & -14 & -35 \end{array}\right]$$

Now we can multiply row two by -7 and add this result to row three to produce a new row three.

$$\left[\begin{array}{ccc|c} 1 & -2 & 3 & 9 \\ 0 & 1 & -7 & -20 \\ 0 & 0 & 35 & 105 \end{array}\right]$$

This last matrix represents the system $\begin{pmatrix} x - 2y + 3z = 9 \\ y - 7z = -20 \\ 35z = 105 \end{pmatrix}$, which is said to be in **triangular form**. We can use the third equation to determine the value of z.

$$35z = 105$$
$$z = 3$$

Now we can substitute 3 for z in the second equation.

$$y - 7z = -20$$
$$y - 7(3) = -20$$
$$y - 21 = -20$$
$$y = 1$$

Finally, we can substitute 3 for z and 1 for y in the first equation.

$$x - 2y + 3z = 9$$
$$x - 2(1) + 3(3) = 9$$
$$x - 2 + 9 = 9$$
$$x + 7 = 9$$
$$x = 2$$

The solution set is $\{(2, 1, 3)\}$. ■

At this time it might be very helpful for you to look back at Example 3 of the previous section and then to take another look at Example 3 of this section. Notice that our approach to both problems is basically the same except that in this section we are using matrices to represent the systems of equations.

Problem Set 9.4

Solve each of the following systems and use matrices as we did in the examples of this section.

1. $\begin{pmatrix} x - 2y = 14 \\ 4x + 5y = 4 \end{pmatrix}$ $\{(6, -4)\}$

2. $\begin{pmatrix} x + 5y = -3 \\ 3x - 2y = -26 \end{pmatrix}$ $\{(-8, 1)\}$

3. $\begin{pmatrix} 3x + 7y = -40 \\ x + 4y = -20 \end{pmatrix}$ $\{(-4, -4)\}$

4. $\begin{pmatrix} 7x - 9y = 53 \\ x - 3y = 11 \end{pmatrix}$ $\{(5, -2)\}$

5. $\begin{pmatrix} x - 3y = 4 \\ 4x - 5y = 3 \end{pmatrix}$ $\left\{\left(-\frac{11}{7}, -\frac{13}{7}\right)\right\}$

6. $\begin{pmatrix} x + 3y = 7 \\ 2x - 4y = 9 \end{pmatrix}$ $\left\{\left(\frac{11}{2}, \frac{1}{2}\right)\right\}$

7. $\begin{pmatrix} 6x + 7y = -15 \\ 4x - 9y = 31 \end{pmatrix}$ $\{(1, -3)\}$

8. $\begin{pmatrix} 5x - 3y = -16 \\ 6x + 5y = -2 \end{pmatrix}$ $\{(-2, 2)\}$

9. $\begin{pmatrix} x + 3y - 4z = 5 \\ -2x - 5y + z = 9 \\ 7x - y - z = -2 \end{pmatrix}$ $\{(-1, -2, -3)\}$

10. $\begin{pmatrix} x - y + 5z = -2 \\ -3x + 2y + z = 17 \\ 4x - 5y - 3z = -36 \end{pmatrix}$ $\{(-2, 5, 1)\}$

11. $\begin{pmatrix} x - 2y - 3z = -11 \\ 2x - 3y + z = 7 \\ -3x - 5y + 7z = 14 \end{pmatrix}$ $\{(3, 1, 4)\}$

12. $\begin{pmatrix} x + y + 3z = -8 \\ 3x + 2y - 5z = 19 \\ 5x - y - 4z = 23 \end{pmatrix}$ $\{(2, -1, -3)\}$

13. $\begin{pmatrix} y + 3z = -3 \\ 2x - 5z = 18 \\ 3x - y + 2z = 5 \end{pmatrix}$ $\{(4, 3, -2)\}$

14. $\begin{pmatrix} x - z = -1 \\ -2x + y + 3z = 4 \\ 3x - 4y = 31 \end{pmatrix}$ $\{(5, -4, 6)\}$

15. $\begin{pmatrix} -x - 5y + 2z = -5 \\ 3x + 14y - z = 13 \\ 4x - 3y + 5z = -26 \end{pmatrix}$ $\{(-5, 2, 0)\}$

16. $\begin{pmatrix} -x - 3y + 4z = -3 \\ 3x + 8y - z = 27 \\ 5x - y + 2z = -5 \end{pmatrix}$ $\{(-1, 4, 2)\}$

17. $\begin{pmatrix} x + 2y - z = -5 \\ 3x + 4y + 2z = -8 \\ -2x - y + 5z = 10 \end{pmatrix}$ $\{(-2, -1, 1)\}$

18. $\begin{pmatrix} x - 3y + 2z = 0 \\ 2x - 4y - 3z = 19 \\ -3x - y + z = -11 \end{pmatrix}$ $\{(3, -1, -3)\}$

19. $\begin{pmatrix} -3x + 2y - z = 12 \\ 5x + 2y - 3z = 6 \\ x - y + 5z = -10 \end{pmatrix}$ $\{(-1, 4, -1)\}$

20. $\begin{pmatrix} -2x - 3y + 5z = 15 \\ 4x - y + 2z = -4 \\ x + y - 3z = -7 \end{pmatrix}$ $\{(-2, -2, 1)\}$

21. $\begin{pmatrix} -2x + 5y - z = -1 \\ 4x + y - 5z = 23 \\ x - 2y + 3z = -7 \end{pmatrix}$ $\{(2, 0, -3)\}$

22. $\begin{pmatrix} 2x + 5y + z = 1 \\ x + 2y - 3z = -13 \\ 3x - y - 2z = -4 \end{pmatrix}$ $\{(1, -1, 4)\}$

Miscellaneous Problems

23. Solve the system $\begin{pmatrix} x - 3y - 2z + w = -3 \\ -2x + 7y + z - 2w = -1 \\ 3x - 7y - 3z + 3w = -5 \\ 5x + y + 4z - 2w = 18 \end{pmatrix}$. $\{(1, -1, 2, -3)\}$

24. Solve the system $\begin{pmatrix} x - 2y + 2z - w = -2 \\ -3x + 5y - z - 3w = 2 \\ 2x + 3y + 3z + 5w = -9 \\ 4x - y - z - 2w = 8 \end{pmatrix}$. $\{(1, 0, -2, -1)\}$

25. Suppose that the augmented matrix of a system of three equations in three variables can be changed to the following matrix.

$$\begin{bmatrix} 1 & 1 & -2 & 4 \\ 0 & -5 & 11 & -13 \\ 0 & 0 & 0 & -9 \end{bmatrix}$$

What can be said about the solution set of the system? $\varnothing$

26. Suppose that the augmented matrix of a system of three linear equations in three variables can be changed to the following matrix.

$$\begin{bmatrix} 1 & 0 & 1 & 1 \\ 0 & 1 & -1 & 0 \\ 0 & 0 & 0 & 0 \end{bmatrix}$$

What can be said about the solution set of the system? Infinitely many solutions

9.5 Determinants

A **square matrix** is one that has the same number of rows as columns. Associated with each square matrix that has real number entries is a real number called the **determinant** of the matrix. For a 2×2 matrix

$$\begin{bmatrix} a_1 & b_1 \\ a_2 & b_2 \end{bmatrix},$$

the determinant is written as

$$\begin{vmatrix} a_1 & b_1 \\ a_2 & b_2 \end{vmatrix}$$

and defined by

$$\begin{vmatrix} a_1 & b_1 \\ a_2 & b_2 \end{vmatrix} = a_1 b_2 - a_2 b_1. \tag{1}$$

Notice that a determinant is simply a number and the determinant notation used on the left side of equation (1) is a way of expressing the number on the right side.

EXAMPLE 1 Find the determinant of the matrix $\begin{bmatrix} 3 & -2 \\ 5 & 8 \end{bmatrix}$.

Solution In this case, $a_1 = 3$, $b_1 = -2$, $a_2 = 5$, and $b_2 = 8$. Thus, we have

$$\begin{vmatrix} 3 & -2 \\ 5 & 8 \end{vmatrix} = 3(8) - 5(-2) = 24 + 10 = 34.$$ ■

Finding the determinant of a square matrix is commonly called *evaluating the determinant* and the matrix notation is sometimes omitted.

EXAMPLE 2 Evaluate $\begin{vmatrix} -3 & 5 \\ 1 & 2 \end{vmatrix}$.

Solution

$$\begin{vmatrix} -3 & 5 \\ 1 & 2 \end{vmatrix} = -3(2) - 1(5) = -11.$$ ■

Cramer's Rule

Determinants provide the basis for another method of solving linear systems. Consider the system

$$\begin{pmatrix} a_1x + b_1y = c_1 \\ a_2x + b_2y = c_2 \end{pmatrix}. \qquad \begin{matrix} (1) \\ (2) \end{matrix}$$

We shall solve this system by using the elimination method; observe that our solutions can be conveniently written in determinant form. To solve for x we can multiply equation (1) by b_2 and equation (2) by $-b_1$ and then add.

$$\begin{aligned} a_1b_2x + b_1b_2y &= c_1b_2 \\ -a_2b_1x - b_1b_2y &= -c_2b_1 \\ \hline a_1b_2x - a_2b_1x &= c_1b_2 - c_2b_1 \\ (a_1b_2 - a_2b_1)x &= c_1b_2 - c_2b_1 \\ x &= \frac{c_1b_2 - c_2b_1}{a_1b_2 - a_2b_1}. \qquad \text{if } a_1b_2 - a_2b_1 \neq 0 \end{aligned}$$

To solve for y we can multiply equation (1) by $-a_2$ and equation (2) by a_1 and add.

$$\begin{aligned} -a_1a_2x - a_2b_1y &= -a_2c_1 \\ a_1a_2x + a_1b_2y &= a_1c_2 \\ \hline a_1b_2y - a_2b_1y &= a_1c_2 - a_2c_1 \\ (a_1b_2 - a_2b_1)y &= a_1c_2 - a_2c_1 \\ y &= \frac{a_1c_2 - a_2c_1}{a_1b_2 - a_2b_1}. \qquad \text{if } a_1b_2 - a_2b_1 \neq 0 \end{aligned}$$

We can express the solutions for x and y in determinant form as follows.

$$x = \frac{c_1b_2 - c_2b_1}{a_1b_2 - a_2b_1} = \frac{\begin{vmatrix} c_1 & b_1 \\ c_2 & b_2 \end{vmatrix}}{\begin{vmatrix} a_1 & b_1 \\ a_2 & b_2 \end{vmatrix}}$$

$$y = \frac{a_1c_2 - a_2c_1}{a_1b_2 - a_2b_1} = \frac{\begin{vmatrix} a_1 & c_1 \\ a_2 & c_2 \end{vmatrix}}{\begin{vmatrix} a_1 & b_1 \\ a_2 & b_2 \end{vmatrix}}$$

For convenience, we shall denote the three determinants in the solution as

$$\begin{vmatrix} a_1 & b_1 \\ a_2 & b_2 \end{vmatrix} = D, \qquad \begin{vmatrix} c_1 & b_1 \\ c_2 & b_2 \end{vmatrix} = D_x, \qquad \begin{vmatrix} a_1 & c_1 \\ a_2 & c_2 \end{vmatrix} = D_y.$$

Notice that the elements of D are the coefficients of the variables in the given system. In D_x, we obtain the elements by replacing the coefficients of x with the respective constants. In D_y, we replace the coefficients of y with the respective constants. This method of using determinants to solve a system of two linear equations in two variables is called **Cramer's Rule** and can be stated as follows.

Cramer's Rule

Given the system

$$\begin{pmatrix} a_1x + b_1y = c_1 \\ a_2x + b_2y = c_2 \end{pmatrix} \quad \text{with } a_1b_2 - a_2b_1 \neq 0,$$

then

$$x = \frac{\begin{vmatrix} c_1 & b_1 \\ c_2 & b_2 \end{vmatrix}}{\begin{vmatrix} a_1 & b_1 \\ a_2 & b_2 \end{vmatrix}} = \frac{D_x}{D} \quad \text{and} \quad y = \frac{\begin{vmatrix} a_1 & c_1 \\ a_2 & c_2 \end{vmatrix}}{\begin{vmatrix} a_1 & b_1 \\ a_2 & b_2 \end{vmatrix}} = \frac{D_y}{D}.$$

Let's use Cramer's Rule to solve some systems.

EXAMPLE 3 Solve the system $\begin{pmatrix} x + 2y = 11 \\ 2x - y = 2 \end{pmatrix}$.

Solution Let's find D, D_x, and D_y.

$$D = \begin{vmatrix} 1 & 2 \\ 2 & -1 \end{vmatrix} = -1 - 4 = -5,$$

$$D_x = \begin{vmatrix} 11 & 2 \\ 2 & -1 \end{vmatrix} = -11 - 4 = -15,$$

$$D_y = \begin{vmatrix} 1 & 11 \\ 2 & 2 \end{vmatrix} = 2 - 22 = -20$$

Thus, we have

$$x = \frac{D_x}{D} = \frac{-15}{-5} = 3,$$

$$y = \frac{D_y}{D} = \frac{-20}{-5} = 4.$$

The solution set is $\{(3, 4)\}$ which can be verified, as always, by substituting back into the original equations. ■

REMARK Notice that Cramer's Rule has a restriction, namely, $a_1b_2 - a_2b_1 \neq 0$; that is, $D \neq 0$. Thus, it is a good idea to find D first. Then if $D = 0$, Cramer's Rule does not apply and you can revert to one of the other methods to determine whether the solution set is empty or has infinitely many solutions.

EXAMPLE 4 Solve the system $\begin{pmatrix} 2x - 3y = -8 \\ 3x + 5y = 7 \end{pmatrix}$.

Solution

$$D = \begin{vmatrix} 2 & -3 \\ 3 & 5 \end{vmatrix} = 10 - (-9) = 19,$$

$$D_x = \begin{vmatrix} -8 & -3 \\ 7 & 5 \end{vmatrix} = -40 - (-21) = -19,$$

$$D_y = \begin{vmatrix} 2 & -8 \\ 3 & 7 \end{vmatrix} = 14 - (-24) = 38$$

Thus, we obtain

$$x = \frac{D_x}{D} = \frac{-19}{19} = -1 \quad \text{and} \quad y = \frac{D_y}{D} = \frac{38}{19} = 2.$$

The solution set is $\{(-1, 2)\}$. ■

EXAMPLE 5 Solve the system $\begin{pmatrix} y = -2x - 2 \\ 4x - 5y = 17 \end{pmatrix}$.

Solution First, we must change the form of the first equation so that the system fits the form given in Cramer's Rule. The equation $y = -2x - 2$ can be written as $2x + y = -2$. The system now becomes

$$\begin{pmatrix} 2x + y = -2 \\ 4x - 5y = 17 \end{pmatrix}$$

and we can proceed as before.

$$D = \begin{vmatrix} 2 & 1 \\ 4 & -5 \end{vmatrix} = -10 - 4 = -14,$$

$$D_x = \begin{vmatrix} -2 & 1 \\ 17 & -5 \end{vmatrix} = 10 - 17 = -7,$$

$$D_y = \begin{vmatrix} 2 & -2 \\ 4 & 17 \end{vmatrix} = 34 - (-8) = 42$$

Thus, the solutions are

$$x = \frac{D_x}{D} = \frac{-7}{-14} = \frac{1}{2} \quad \text{and} \quad y = \frac{D_y}{D} = \frac{42}{-14} = -3.$$

The solution set is $\left\{\left(\frac{1}{2}, -3\right)\right\}$. ■

Problem Set 9.5

Evaluate each of the following determinants.

1. $\begin{vmatrix} 6 & 2 \\ 4 & 3 \end{vmatrix}$ 10

2. $\begin{vmatrix} 7 & 6 \\ 2 & 5 \end{vmatrix}$ 23

3. $\begin{vmatrix} 4 & 7 \\ 8 & 2 \end{vmatrix}$ -48

4. $\begin{vmatrix} 3 & 9 \\ 6 & 4 \end{vmatrix}$ -42

5. $\begin{vmatrix} -3 & 2 \\ 7 & 5 \end{vmatrix}$ -29

6. $\begin{vmatrix} 5 & 1 \\ 8 & -4 \end{vmatrix}$ -28

7. $\begin{vmatrix} 8 & -3 \\ 6 & 4 \end{vmatrix}$ 50

8. $\begin{vmatrix} 5 & 9 \\ -3 & 6 \end{vmatrix}$ 57

9. $\begin{vmatrix} -3 & 2 \\ 5 & -6 \end{vmatrix}$ 8

10. $\begin{vmatrix} -2 & 4 \\ 9 & -7 \end{vmatrix}$ -22

11. $\begin{vmatrix} 3 & -3 \\ -6 & 8 \end{vmatrix}$ 6

12. $\begin{vmatrix} 6 & -5 \\ -8 & 12 \end{vmatrix}$ 32

13. $\begin{vmatrix} -7 & 2 \\ -2 & 4 \end{vmatrix}$ -32

14. $\begin{vmatrix} 6 & -1 \\ -8 & -3 \end{vmatrix}$ -26

15. $\begin{vmatrix} -2 & -3 \\ -4 & -5 \end{vmatrix}$ -2

16. $\begin{vmatrix} -9 & -7 \\ -6 & -4 \end{vmatrix}$ -6

17. $\begin{vmatrix} \frac{1}{4} & -2 \\ \frac{3}{2} & 8 \end{vmatrix}$ 5

18. $\begin{vmatrix} -\frac{2}{3} & 10 \\ -\frac{1}{2} & 6 \end{vmatrix}$ 1

19. $\begin{vmatrix} \frac{3}{2} & -\frac{1}{2} \\ \frac{1}{2} & -\frac{2}{5} \end{vmatrix}$ $-\frac{7}{20}$

20. $\begin{vmatrix} -\frac{1}{4} & \frac{1}{3} \\ \frac{3}{4} & \frac{2}{3} \end{vmatrix}$ $-\frac{5}{12}$

Use Cramer's Rule to find the solution set for each of the following systems.

21. $\begin{pmatrix} 2x + y = 14 \\ 3x - y = 1 \end{pmatrix}$ $\{(3, 8)\}$

22. $\begin{pmatrix} 4x - y = 11 \\ 2x + 3y = 23 \end{pmatrix}$ $\{(4, 5)\}$

23. $\begin{pmatrix} -x + 3y = 17 \\ 4x - 5y = -33 \end{pmatrix}$ $\{(-2, 5)\}$

24. $\begin{pmatrix} 5x + 2y = -15 \\ 7x - 3y = 37 \end{pmatrix}$ $\{(1, -10)\}$

25. $\begin{pmatrix} 9x + 5y = -8 \\ 7x - 4y = -22 \end{pmatrix}$ $\{(-2, 2)\}$

26. $\begin{pmatrix} 8x - 11y = 3 \\ -x + 4y = -3 \end{pmatrix}$ $\{(-1, -1)\}$

27. $\begin{pmatrix} x + 5y = 4 \\ 3x + 15y = -1 \end{pmatrix}$ $\varnothing$

28. $\begin{pmatrix} 4x - 7y = 0 \\ 7x + 2y = 0 \end{pmatrix}$ $\{(0, 0)\}$

29. $\begin{pmatrix} 6x - y = 0 \\ 5x + 4y = 29 \end{pmatrix}$ $\{(1, 6)\}$

30. $\begin{pmatrix} 3x - 4y = 2 \\ 9x - 12y = 6 \end{pmatrix}$ Infinitely many solutions

31. $\begin{pmatrix} -4x + 3y = 3 \\ 4x - 6y = -5 \end{pmatrix}$ $\left\{\left(-\frac{1}{4}, \frac{2}{3}\right)\right\}$

32. $\begin{pmatrix} x - 2y = -1 \\ x = -6y + 5 \end{pmatrix}$ $\left\{\left(\frac{1}{2}, \frac{3}{4}\right)\right\}$

33. $\begin{pmatrix} 6x - 5y = 1 \\ 4x + 7y = 2 \end{pmatrix}$ $\left\{\left(\frac{17}{62}, \frac{4}{31}\right)\right\}$

34. $\begin{pmatrix} y = 3x + 5 \\ y = 6x + 6 \end{pmatrix}$ $\left\{\left(-\frac{1}{3}, 4\right)\right\}$

35. $\begin{pmatrix} 7x + 2y = -1 \\ y = -x + 2 \end{pmatrix}$ $\{(-1, 3)\}$

36. $\begin{pmatrix} 9x - y = -2 \\ y = 4 - 8x \end{pmatrix}$ $\left\{\left(\frac{2}{17}, \frac{52}{17}\right)\right\}$

37. $\begin{pmatrix} -\frac{2}{3}x + \frac{1}{2}y = -7 \\ \frac{1}{3}x - \frac{3}{2}y = 6 \end{pmatrix}$ $\{(9, -2)\}$

38. $\begin{pmatrix} \frac{1}{2}x + \frac{2}{3}y = -6 \\ \frac{1}{4}x - \frac{1}{3}y = -1 \end{pmatrix}$ $\{(-8, -3)\}$

39. $\begin{pmatrix} x + \frac{2}{3}y = -6 \\ -\frac{1}{4}x + 3y = -8 \end{pmatrix}$ $\{(-4, -3)\}$

40. $\begin{pmatrix} 3x - \frac{1}{2}y = 6 \\ -2x + \frac{1}{3}y = -4 \end{pmatrix}$ $\{(0, -12)\}$

Miscellaneous Problems

41. Verify each of the following. The variables represent real numbers.

(a) $\begin{vmatrix} a & b \\ a & b \end{vmatrix} = 0$

(b) $\begin{vmatrix} a & a \\ b & b \end{vmatrix} = 0$

(c) $\begin{vmatrix} a & b \\ c & d \end{vmatrix} = -\begin{vmatrix} b & a \\ d & c \end{vmatrix}$

(d) $\begin{vmatrix} a & b \\ c & d \end{vmatrix} = -\begin{vmatrix} c & d \\ a & b \end{vmatrix}$

(e) $k\begin{vmatrix} a & b \\ c & d \end{vmatrix} = \begin{vmatrix} ka & b \\ kc & d \end{vmatrix}$

(f) $k\begin{vmatrix} a & b \\ c & d \end{vmatrix} = \begin{vmatrix} ka & kb \\ c & d \end{vmatrix}$

9.6

3 × 3 Determinants and Systems of Three Linear Equations in Three Variables

This section will extend the concept of a determinant to include 3 × 3 determinants and then also extend the use of determinants to solve systems of three linear equations in three variables.

For a 3 × 3 matrix

$$\begin{bmatrix} a_1 & b_1 & c_1 \\ a_2 & b_2 & c_2 \\ a_3 & b_3 & c_3 \end{bmatrix},$$

the determinant is written as

$$\begin{vmatrix} a_1 & b_1 & c_1 \\ a_2 & b_2 & c_2 \\ a_3 & b_3 & c_3 \end{vmatrix}$$

and defined by

$$\begin{vmatrix} a_1 & b_1 & c_1 \\ a_2 & b_2 & c_2 \\ a_3 & b_3 & c_3 \end{vmatrix} = a_1b_2c_3 + b_1c_2a_3 + c_1a_2b_3 - a_3b_2c_1 - b_3c_2a_1 - c_3a_2b_1. \qquad (1)$$

It is evident that the definition given by equation (1) is a bit complicated to be very useful in practice. Fortunately, there is a method, called **expansion of a determinant by minors**, that we can use to calculate such a determinant.

The **minor** of an element in a determinant is the determinant that remains after deleting the row and column in which the element appears. For example, consider the determinant of equation (1).

The minor of a_1 is $\begin{vmatrix} b_2 & c_2 \\ b_3 & c_3 \end{vmatrix}$,

The minor of a_2 is $\begin{vmatrix} b_1 & c_1 \\ b_3 & c_3 \end{vmatrix}$;

The minor of a_3 is $\begin{vmatrix} b_1 & c_1 \\ b_2 & c_2 \end{vmatrix}$

Now let's consider the terms, in pairs, of the right side of equation (1) and show the tie-up with minors.

$$\begin{aligned} a_1b_2c_3 - b_3c_2a_1 &= a_1(b_2c_3 - b_3c_2). \\ &= a_1\begin{vmatrix} b_2 & c_2 \\ b_3 & c_3 \end{vmatrix}, \\ c_1a_2b_3 - c_3a_2b_1 &= -(c_3a_2b_1 - c_1a_2b_3) \\ &= -a_2(b_1c_3 - b_3c_1). \\ &= -a_2\begin{vmatrix} b_1 & c_1 \\ b_3 & c_3 \end{vmatrix}, \\ b_1c_2a_3 - a_3b_2c_1 &= a_3(b_1c_2 - b_2c_1). \\ &= a_3\begin{vmatrix} b_1 & c_1 \\ b_2 & c_2 \end{vmatrix} \end{aligned}$$

Therefore, we have

$$\begin{vmatrix} a_1 & b_1 & c_1 \\ a_2 & b_2 & c_2 \\ a_3 & b_3 & c_3 \end{vmatrix} = a_1\begin{vmatrix} b_2 & c_2 \\ b_3 & c_3 \end{vmatrix} - a_2\begin{vmatrix} b_1 & c_1 \\ b_3 & c_3 \end{vmatrix} + a_3\begin{vmatrix} b_1 & c_1 \\ b_2 & c_2 \end{vmatrix},$$

and this is called the **expansion of the determinant by minors about the first column.**

EXAMPLE 1 Evaluate $\begin{vmatrix} 1 & 2 & -1 \\ 3 & 1 & -2 \\ 2 & 4 & 3 \end{vmatrix}$ by expanding by minors about the first column.

Solution

$$\begin{aligned} \begin{vmatrix} 1 & 2 & -1 \\ 3 & 1 & -2 \\ 2 & 4 & 3 \end{vmatrix} &= 1\begin{vmatrix} 1 & -2 \\ 4 & 3 \end{vmatrix} - 3\begin{vmatrix} 2 & -1 \\ 4 & 3 \end{vmatrix} + 2\begin{vmatrix} 2 & -1 \\ 1 & -2 \end{vmatrix} \\ &= 1(3-(-8)) - 3(6-(-4)) + 2(-4-(-1)) \\ &= 1(11) - 3(10) + 2(-3) = -25 \end{aligned}$$

■

It is possible to expand a determinant by minors about *any row* or *any column*. To help determine the signs of the terms in the expansion, the following *sign array* is very useful.

$$\begin{matrix} + & - & + \\ - & + & - \\ + & - & + \end{matrix}$$

For example, let's expand the determinant in Example 1 by minors about the *second row*. The second row in the sign array is $- + -$. Therefore,

$$\begin{aligned}\begin{vmatrix} 1 & 2 & -1 \\ 3 & 1 & -2 \\ 2 & 4 & 3 \end{vmatrix} &= -3\begin{vmatrix} 2 & -1 \\ 4 & 3 \end{vmatrix} + 1\begin{vmatrix} 1 & -1 \\ 2 & 3 \end{vmatrix} - (-2)\begin{vmatrix} 1 & 2 \\ 2 & 4 \end{vmatrix} \\ &= -3(6 - (-4)) + 1(3 - (-2)) + 2(4 - 4) \\ &= -3(10) + 1(5) + 2(0) \\ &= -25.\end{aligned}$$

Your decision as to which row or column to use for expanding a particular determinant by minors may depend upon the numbers involved in the determinant. A row or column with one or more zeros is frequently a good choice, as the next example illustrates.

EXAMPLE 2 Evaluate $\begin{vmatrix} 3 & -1 & 4 \\ 5 & 2 & 0 \\ -2 & 6 & 0 \end{vmatrix}$.

Solution Since the third column has two zeros, we shall expand about it.

$$\begin{aligned}\begin{vmatrix} 3 & -1 & 4 \\ 5 & 2 & 0 \\ -2 & 6 & 0 \end{vmatrix} &= 4\begin{vmatrix} 5 & 2 \\ -2 & 6 \end{vmatrix} - 0\begin{vmatrix} 3 & -1 \\ -2 & 6 \end{vmatrix} + 0\begin{vmatrix} 3 & -1 \\ 5 & 2 \end{vmatrix} \\ &= 4(30 - (-4)) - 0 + 0 = 136\end{aligned}$$

(Notice that because of the zeros there is no need to evaluate the last two minors.) ■

REMARK 1 The expansion-by-minors method can be extended to determinants of size 4 × 4, 5 × 5, and so on. However, it should be obvious that it becomes increasingly more tedious with "bigger" determinants. Fortunately, the computer handles the calculation of such determinants with a different technique.

REMARK 2 There is another method for evaluating 3 × 3 determinants. This method is demonstrated in Problem 34 of the next problem set. If you choose to use that method, keep in mind *that it only works for 3 × 3 determinants.*

Without showing all of the details, we will simply state that *Cramer's Rule* also applies to solving systems of three linear equations in three variables. It can be stated as follows.

Cramer's Rule

Given the system

$$\begin{pmatrix} a_1x + b_1y + c_1z = d_1 \\ a_2x + b_2y + c_2z = d_2 \\ a_3x + b_3y + c_3z = d_3 \end{pmatrix}$$

$$\text{with } D = \begin{vmatrix} a_1 & b_1 & c_1 \\ a_2 & b_2 & c_2 \\ a_3 & b_3 & c_3 \end{vmatrix} \neq 0, \qquad D_x = \begin{vmatrix} d_1 & b_1 & c_1 \\ d_2 & b_2 & c_2 \\ d_3 & b_3 & c_3 \end{vmatrix},$$

$$D_y = \begin{vmatrix} a_1 & d_1 & c_1 \\ a_2 & d_2 & c_2 \\ a_3 & d_3 & c_3 \end{vmatrix}, \qquad D_z = \begin{vmatrix} a_1 & b_1 & d_1 \\ a_2 & b_2 & d_2 \\ a_3 & b_3 & d_3 \end{vmatrix},$$

then $x = \dfrac{D_x}{D}$, $y = \dfrac{D_y}{D}$, and $z = \dfrac{D_z}{D}$.

Notice that the elements of D are the coefficients of the variables in the given system. Then D_x, D_y, D_z are formed by replacing the elements in the x, y, or z column, respectively, by the constants of the system d_1, d_2, and d_3. Again, note the restriction $D \neq 0$. As before, if $D = 0$, then Cramer's Rule does not apply and you can use the elimination method to determine whether the system has *no solution* or *infinitely many* solutions.

EXAMPLE 3 Use Cramer's Rule to solve the system $\begin{pmatrix} x - 2y + z = -4 \\ 2x + y - z = 5 \\ 3x + 2y + 4z = 3 \end{pmatrix}$.

Solution To find D, let's expand about *row* 1.

$$D = \begin{vmatrix} 1 & -2 & 1 \\ 2 & 1 & -1 \\ 3 & 2 & 4 \end{vmatrix} = 1\begin{vmatrix} 1 & -1 \\ 2 & 4 \end{vmatrix} - (-2)\begin{vmatrix} 2 & -1 \\ 3 & 4 \end{vmatrix} + 1\begin{vmatrix} 2 & 1 \\ 3 & 2 \end{vmatrix}$$

$$= 1(4 - (-2)) + 2(8 - (-3)) + 1(4 - 3)$$

$$= 1(6) + 2(11) + 1(1) = 29$$

To find D_x, let's expand about *column* 3.

$$D_x = \begin{vmatrix} -4 & -2 & 1 \\ 5 & 1 & -1 \\ 3 & 2 & 4 \end{vmatrix} = 1\begin{vmatrix} 5 & 1 \\ 3 & 2 \end{vmatrix} - (-1)\begin{vmatrix} -4 & -2 \\ 3 & 2 \end{vmatrix} + 4\begin{vmatrix} -4 & -2 \\ 5 & 1 \end{vmatrix}$$

$$= 1(10 - 3) + 1(-8 - (-6)) + 4(-4 - (-10))$$
$$= 1(7) + 1(-2) + 4(6)$$
$$= 29$$

To find D_y, let's expand about *row* 1.

$$D_y = \begin{vmatrix} 1 & -4 & 1 \\ 2 & 5 & -1 \\ 3 & 3 & 4 \end{vmatrix} = 1\begin{vmatrix} 5 & -1 \\ 3 & 4 \end{vmatrix} - (-4)\begin{vmatrix} 2 & -1 \\ 3 & 4 \end{vmatrix} + 1\begin{vmatrix} 2 & 5 \\ 3 & 3 \end{vmatrix}$$

$$= 1(20 - (-3)) + 4(8 - (-3)) + 1(6 - 15)$$
$$= 1(23) + 4(11) + 1(-9)$$
$$= 58$$

To find D_z, let's expand about *column* 1.

$$D_z = \begin{vmatrix} 1 & -2 & -4 \\ 2 & 1 & 5 \\ 3 & 2 & 3 \end{vmatrix} = 1\begin{vmatrix} 1 & 5 \\ 2 & 3 \end{vmatrix} - 2\begin{vmatrix} -2 & -4 \\ 2 & 3 \end{vmatrix} + 3\begin{vmatrix} -2 & -4 \\ 1 & 5 \end{vmatrix}$$

$$= 1(3 - 10) - 2(-6 - (-8)) + 3(-10 - (-4))$$
$$= 1(-7) - 2(2) + 3(-6)$$
$$= -29$$

Thus,

$$x = \frac{D_x}{D} = \frac{29}{29} = 1,$$

$$y = \frac{D_y}{D} = \frac{58}{29} = 2,$$

$$z = \frac{D_z}{D} = \frac{-29}{29} = -1.$$

The solution set is $\{(1, 2, -1)\}$. (Be sure to check it!) ■

EXAMPLE 4 Use Cramer's Rule to solve the system $\begin{pmatrix} 2x - y + 3z = -17 \\ 3y + z = 5 \\ x - 2y - z = -3 \end{pmatrix}$.

Solution To find D, let's expand about column 1.

$$D = \begin{vmatrix} 2 & -1 & 3 \\ 0 & 3 & 1 \\ 1 & -2 & -1 \end{vmatrix} = 2\begin{vmatrix} 3 & 1 \\ -2 & -1 \end{vmatrix} - 0\begin{vmatrix} -1 & 3 \\ -2 & -1 \end{vmatrix} + 1\begin{vmatrix} -1 & 3 \\ 3 & 1 \end{vmatrix}$$
$$= 2(-3-(-2)) - 0 + 1(-1-9)$$
$$= 2(-1) - 0 - 10 = -12$$

To find D_x, let's expand about column 3.

$$D_x = \begin{vmatrix} -17 & -1 & 3 \\ 5 & 3 & 1 \\ -3 & -2 & -1 \end{vmatrix}$$
$$= 3\begin{vmatrix} 5 & 3 \\ -3 & -2 \end{vmatrix} - 1\begin{vmatrix} -17 & -1 \\ -3 & -2 \end{vmatrix} + (-1)\begin{vmatrix} -17 & -1 \\ 5 & 3 \end{vmatrix}$$
$$= 3(-10-(-9)) - 1(34-3) - 1(-51-(-5))$$
$$= 3(-1) - 1(31) - 1(-46) = 12$$

To find D_y, let's expand about column 1.

$$D_y = \begin{vmatrix} 2 & -17 & 3 \\ 0 & 5 & 1 \\ 1 & -3 & -1 \end{vmatrix} = 2\begin{vmatrix} 5 & 1 \\ -3 & -1 \end{vmatrix} - 0\begin{vmatrix} -17 & 3 \\ -3 & -1 \end{vmatrix} + 1\begin{vmatrix} -17 & 3 \\ 5 & 1 \end{vmatrix}$$
$$= 2(-5-(-3)) - 0 + 1(-17-15)$$
$$= 2(-2) - 0 + 1(-32) = -36$$

To find D_z, let's expand about column 1.

$$D_z = \begin{vmatrix} 2 & -1 & -17 \\ 0 & 3 & 5 \\ 1 & -2 & -3 \end{vmatrix} = 2\begin{vmatrix} 3 & 5 \\ -2 & -3 \end{vmatrix} - 0\begin{vmatrix} -1 & -17 \\ -2 & -3 \end{vmatrix} + 1\begin{vmatrix} -1 & -17 \\ 3 & 5 \end{vmatrix}$$
$$= 2(-9-(-10)) - 0 + 1(-5-(-51))$$
$$= 2(1) - 0 + 1(46) = 48$$

Thus,

$$x = \frac{D_x}{D} = \frac{12}{-12} = -1, \qquad y = \frac{D_y}{D} = \frac{-36}{-12} = 3,$$

$$z = \frac{D_z}{D} = \frac{48}{-12} = -4.$$

The solution set is $\{(-1, 3, -4)\}$. ■

Problem Set 9.6

Use expansion-by-minors to evaluate each of the following determinants.

1. $\begin{vmatrix} 2 & 7 & 5 \\ 1 & -1 & 1 \\ -4 & 3 & 2 \end{vmatrix}$ -57

2. $\begin{vmatrix} 2 & 4 & 1 \\ -1 & 5 & 1 \\ -3 & 6 & 2 \end{vmatrix}$ 13

3. $\begin{vmatrix} 3 & -2 & 1 \\ 2 & 1 & 4 \\ -1 & 3 & 5 \end{vmatrix}$ 14

4. $\begin{vmatrix} 1 & -1 & 2 \\ 2 & 1 & 3 \\ -1 & -2 & 1 \end{vmatrix}$ 6

5. $\begin{vmatrix} -3 & -2 & 1 \\ 5 & 0 & 6 \\ 2 & 1 & -4 \end{vmatrix}$ -41

6. $\begin{vmatrix} -5 & 1 & -1 \\ 3 & 4 & 2 \\ 0 & 2 & -3 \end{vmatrix}$ 83

7. $\begin{vmatrix} 3 & -4 & -2 \\ 5 & -2 & 1 \\ 1 & 0 & 0 \end{vmatrix}$ -8

8. $\begin{vmatrix} -6 & 5 & 3 \\ 2 & 0 & -1 \\ 4 & 0 & 7 \end{vmatrix}$ -90

9. $\begin{vmatrix} 4 & -2 & 7 \\ 1 & -1 & 6 \\ 3 & 5 & -2 \end{vmatrix}$ -96

10. $\begin{vmatrix} -5 & 2 & 6 \\ 1 & -1 & 3 \\ 4 & -2 & -4 \end{vmatrix}$ -6

Use Cramer's Rule to find the solution set for each of the following systems.

11. $\begin{pmatrix} 2x - y + 3z = -10 \\ x + 2y - 3z = 2 \\ 3x - 2y + 5z = -16 \end{pmatrix}$ $\{(-3, 1, -1)\}$

12. $\begin{pmatrix} -x + y - z = 1 \\ 2x + 3y - 4z = 10 \\ -3x - y + z = -5 \end{pmatrix}$ $\{(1, 0, -2)\}$

13. $\begin{pmatrix} x - y + 2z = -8 \\ 2x + 3y - 4z = 18 \\ -x + 2y - z = 7 \end{pmatrix}$ $\{(0, 2, -3)\}$

14. $\begin{pmatrix} x - 2y + z = 3 \\ 3x + 2y + z = -3 \\ 2x - 3y - 3z = -5 \end{pmatrix}$ $\{(-1, -1, 2)\}$

15. $\begin{pmatrix} 3x - 2y - 3z = -5 \\ x + 2y + 3z = -3 \\ -x + 4y - 6z = 8 \end{pmatrix}$ $\{(-2, \frac{1}{2}, -\frac{2}{3})\}$

16. $\begin{pmatrix} 2x - 3y + 3z = -3 \\ -2x + 5y - 3z = 5 \\ 3x - y + 6z = -1 \end{pmatrix}$ $\{(0, 1, 0)\}$

17. $\begin{pmatrix} -x + y + z = -1 \\ x - 2y + 5z = -4 \\ 3x + 4y - 6z = -1 \end{pmatrix}$ $\{(-1, -1, -1)\}$

18. $\begin{pmatrix} x - 2y + 3z = 1 \\ 2x + y + z = 4 \\ 4x - 3y + 7z = 6 \end{pmatrix}$ Infinitely many solutions

19. $\begin{pmatrix} x - y + 2z = 4 \\ 3x - 2y + 4z = 6 \\ 2x - 2y + 4z = -1 \end{pmatrix}$ $\varnothing$

20. $\begin{pmatrix} -x - 2y + z = 8 \\ 3x + y - z = 5 \\ 5x - y + 4z = 33 \end{pmatrix}$ $\{(4, -5, 2)\}$

21. $\begin{pmatrix} 2x - y + 3z = -5 \\ 3x + 4y - 2z = -25 \\ -x + z = 6 \end{pmatrix}$ $\{(-5, -2, 1)\}$

22. $\begin{pmatrix} 3x - 2y + z = 11 \\ 5x + 3y = 17 \\ x + y - 2z = 6 \end{pmatrix}$ $\{(\frac{15}{4}, -\frac{7}{12}, -\frac{17}{12})\}$

23. $\begin{pmatrix} 2y - z = 10 \\ 3x + 4y = 6 \\ x - y + z = -9 \end{pmatrix}$ $\{(-2, 3, -4)\}$

24. $\begin{pmatrix} 6x - 5y + 2z = 7 \\ 2x + 3y - 4z = -21 \\ 2y + 3z = 10 \end{pmatrix}$ $\{(-1, -1, 4)\}$

25. $\begin{pmatrix} -2x + 5y - 3z = -1 \\ 2x - 7y + 3z = 1 \\ 4x - y - 6z = -6 \end{pmatrix}$ $\{(-\frac{1}{2}, 0, \frac{2}{3})\}$

26. $\begin{pmatrix} 7x - 2y + 3z = -4 \\ 5x + 2y - 3z = 4 \\ -3x - 6y + 12z = -13 \end{pmatrix}$ $\{(0, \frac{3}{2}, -\frac{1}{3})\}$

27. $\begin{pmatrix} -x - y + 5z = 4 \\ x + y - 7z = -6 \\ 2x + 3y + 4z = 13 \end{pmatrix}$ $\{(-6, 7, 1)\}$

28. $\begin{pmatrix} x + 7y - z = -1 \\ -x - 9y + z = 3 \\ 3x + 4y - 6z = 5 \end{pmatrix}$ $\{(9, -1, 3)\}$

29. $\begin{pmatrix} 5x - y + 2z = 10 \\ 7x + 2y - 2z = -4 \\ -3x - y + 4z = 1 \end{pmatrix}$ $\{(1, -6, -\frac{1}{2})\}$

30. $\begin{pmatrix} 4x - y - 3z = -12 \\ 5x + y + 6z = 4 \\ 6x - y - 3z = -14 \end{pmatrix}$ $\{(-1, 7, \frac{1}{3})\}$

Miscellaneous Problems

31. Evaluate the following determinant by expanding about the *second column.*

$$\begin{vmatrix} a & e & a \\ b & f & b \\ c & g & c \end{vmatrix}$$

Make a conjecture about determinants that contain two identical columns. 0

32. Show that $\begin{vmatrix} 1 & -1 & 2 \\ 2 & 3 & -1 \\ -1 & 2 & 4 \end{vmatrix} = -\begin{vmatrix} -1 & 1 & 2 \\ 3 & 2 & -1 \\ 2 & -1 & 4 \end{vmatrix}$.

Make a conjecture about the result of interchanging two columns of a determinant.

33. **(a)** Show that $\begin{vmatrix} 2 & 1 & 2 \\ 4 & -1 & -2 \\ 6 & 3 & 1 \end{vmatrix} = 2\begin{vmatrix} 1 & 1 & 2 \\ 2 & -1 & -2 \\ 3 & 3 & 1 \end{vmatrix}$.

Make a conjecture about the result of factoring a common factor from each element of a column in a determinant.

(b) Use your conjecture from part (a) to help evaluate the following determinant.

$$\begin{vmatrix} 2 & 4 & -1 \\ -3 & -4 & -2 \\ 5 & 4 & 3 \end{vmatrix}. \quad -20$$

34. We can describe another technique for evaluating 3×3 determinants as follows. First, let's write the given determinant with its first two columns repeated on the right.

$$\begin{vmatrix} a_1 & b_1 & c_1 \\ a_2 & b_2 & c_2 \\ a_3 & b_3 & c_3 \end{vmatrix} \begin{matrix} a_1 & b_1 \\ a_2 & b_2 \\ a_3 & b_3 \end{matrix}$$

Then we can add the three products shown with +, and subtract the three products shown with −.

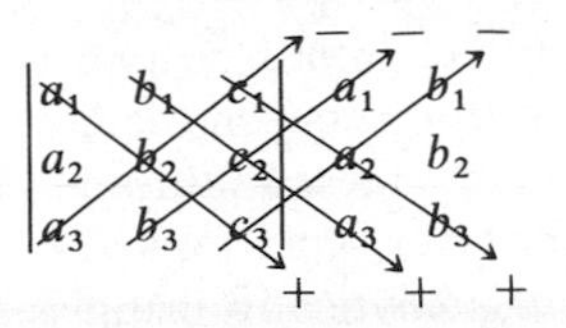

(a) Be sure that the previous description will produce equation (1) on page 467.

(b) Use this technique to do Problems 1–10.

9.7 Systems Involving Nonlinear Equations and Systems of Inequalities

Thus far in this chapter we have solved systems of *linear equations*. In this section, we shall consider some systems of **linear inequalities** and also some systems where at least one of the equations is *nonlinear*.

Let's begin by considering a system of one linear equation and one quadratic equation.

EXAMPLE 1 Solve the system $\begin{pmatrix} x^2 + y^2 = 17 \\ x + y = 5 \end{pmatrix}$.

Solution First, let's graph the system so that we can predict approximate solutions. From our previous graphing experiences in Chapters 7 and 8, we should recognize $x^2 + y^2 = 17$ as a circle and $x + y = 5$ as a straight line (Figure 9.7).

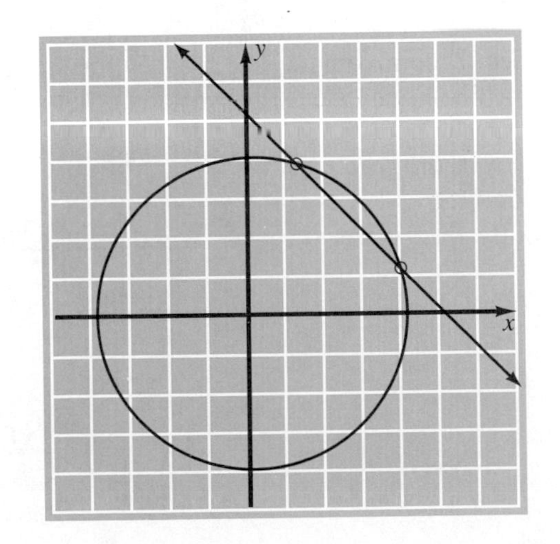

Figure 9.7

The graph indicates that there should be two ordered pairs with positive components (the points of intersection occur in the first quadrant) as solutions for this system. In fact, we could guess that these solutions are (1, 4) and (4, 1), and verify our guess by checking them in the given equations.

Let's also solve the system analytically using the substitution method as follows. Change the form of $x + y = 5$ to $y = 5 - x$ and substitute $5 - x$ for y in the first equation.

$$x^2 + y^2 = 17$$

$$x^2 + (5 - x)^2 = 17$$

$$x^2 + 25 - 10x + x^2 = 17$$

$$2x^2 - 10x + 8 = 0$$

$$x^2 - 5x + 4 = 0$$

$$(x - 4)(x - 1) = 0$$

$$x - 4 = 0 \quad \text{or} \quad x - 1 = 0$$

$$x = 4 \quad \text{or} \quad x = 1$$

Substitute 4 for x and then 1 for x in the second equation of the system to produce

$$x + y = 5 \qquad x + y = 5$$

$$4 + y = 5 \qquad 1 + y = 5$$

$$y = 1. \qquad y = 4.$$

Therefore, the solution set is $\{(1, 4), (4, 1)\}$. ■

EXAMPLE 2 Solve the system $\begin{pmatrix} y = -x^2 + 1 \\ y = x^2 - 2 \end{pmatrix}$.

Solution Again, let's get an idea of approximate solutions by graphing the system. Both equations produce parabolas as indicated in Figure 9.8. From the graph we can predict two nonintegral ordered pair solutions, one in the third quadrant and the other in the fourth quadrant.

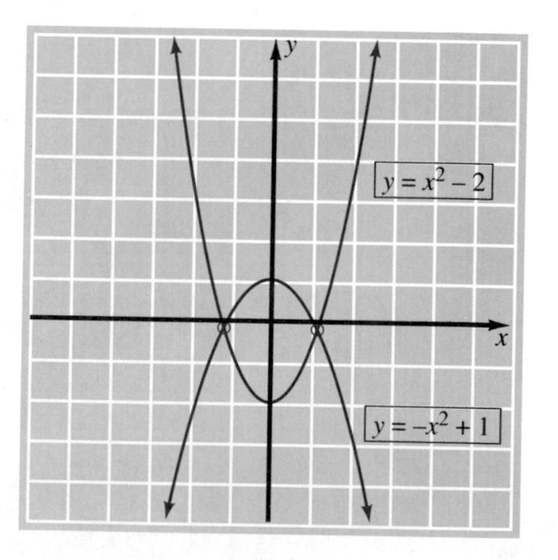

Figure 9.8

Substitute $-x^2 + 1$ for y in the second equation to obtain

$$\begin{aligned} y &= x^2 - 2 \\ -x^2 + 1 &= x^2 - 2 \\ 3 &= 2x^2 \\ \frac{3}{2} &= x^2 \\ \pm\sqrt{\frac{3}{2}} &= x \\ \pm\frac{\sqrt{6}}{2} &= x. \end{aligned}$$

Substitute $\frac{\sqrt{6}}{2}$ for x in the second equation to yield

$$\begin{aligned} y &= x^2 - 2 \\ y &= \left(\frac{\sqrt{6}}{2}\right)^2 - 2 \\ &= \frac{6}{4} - 2 \\ &= -\frac{1}{2}. \end{aligned}$$

Substitute $-\frac{\sqrt{6}}{2}$ for x in the second equation to yield

$$\begin{aligned} y &= x^2 - 2 \\ y &= \left(-\frac{\sqrt{6}}{2}\right)^2 - 2 \\ &= \frac{6}{4} - 2 = -\frac{1}{2}. \end{aligned}$$

The solution set is $\left\{\left(-\frac{\sqrt{6}}{2}, -\frac{1}{2}\right), \left(\frac{\sqrt{6}}{2}, -\frac{1}{2}\right)\right\}$. Check it! ■

EXAMPLE 3 Solve the system $\begin{pmatrix} x^2 + y^2 = 16 \\ -x^2 + y^2 = 4 \end{pmatrix}$.

Solution Graphing the system produces Figure 9.9, and indicates that there are four solutions to the system.

Figure 9.9

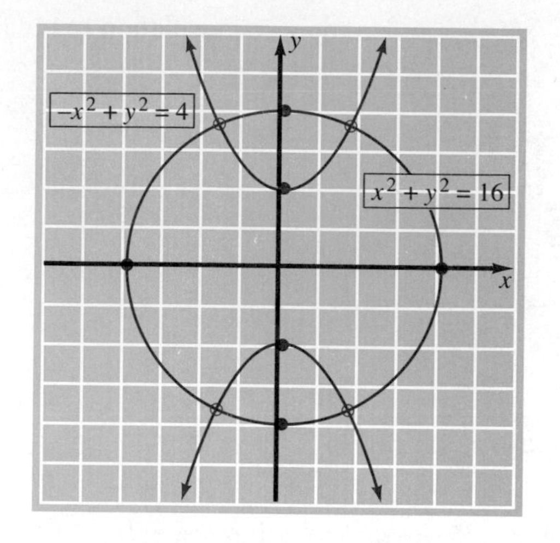

Solving the system by using the *elimination method* works nicely. We can simply add the two equations, which eliminates the x's.

$$\begin{aligned} x^2 + y^2 &= 16 \\ -x^2 + y^2 &= 4 \\ \hline 2y^2 &= 20 \\ y^2 &= 10 \\ y &= \pm\sqrt{10} \end{aligned}$$

Substitute $\sqrt{10}$ for y in the first equation to yield

$$\begin{aligned} x^2 + y^2 &= 16 \\ x^2 + (\sqrt{10})^2 &= 16 \\ x^2 + 10 &= 16 \\ x^2 &= 6 \\ x &= \pm\sqrt{6}. \end{aligned}$$

Thus, $(\sqrt{6}, \sqrt{10})$ and $(-\sqrt{6}, \sqrt{10})$ are solutions. Substitute $-\sqrt{10}$ for y in the first equation to yield

$$\begin{aligned} x^2 + y^2 &= 16 \\ x^2 + (-\sqrt{10})^2 &= 16 \\ x^2 + 10 &= 16 \\ x^2 &= 6 \\ x &= \pm\sqrt{6}. \end{aligned}$$

Thus, $(\sqrt{6}, -\sqrt{10})$ and $(-\sqrt{6}, -\sqrt{10})$ are solutions. The solution set is $\{(-\sqrt{6}, \sqrt{10}), (-\sqrt{6}, -\sqrt{10}), (\sqrt{6}, \sqrt{10}), (\sqrt{6}, -\sqrt{10})\}$. ■

REMARK The **graphing** of systems has been given additional emphasis in this section. Perhaps you will need to return to Chapters 7 and 8 to review the various graphing techniques discussed at that time. Be sure to include Section 7.3 in your review since the ideas from that section will be used in a moment.

Systems of Linear Inequalities

Finding solution sets for systems of **linear inequalities** relies heavily on the graphing approach. The solution set of a system of linear inequalities, such as

$$\begin{pmatrix} x + y > 2 \\ x - y < 2 \end{pmatrix},$$

is the intersection of the solution sets of the individual inequalities. In Figure 9.10(a) we have indicated the solution set for $x + y > 2$ and in Figure 9.10(b) we have indicated the solution set for $x - y < 2$. Then, in Figure 9.10(c) we have shaded the region that represents the intersection of the two solution sets from (a) and (b); thus, it is the graph of the system. Remember that dashed lines are used to indicate that the points on the lines are not included in the solution set.

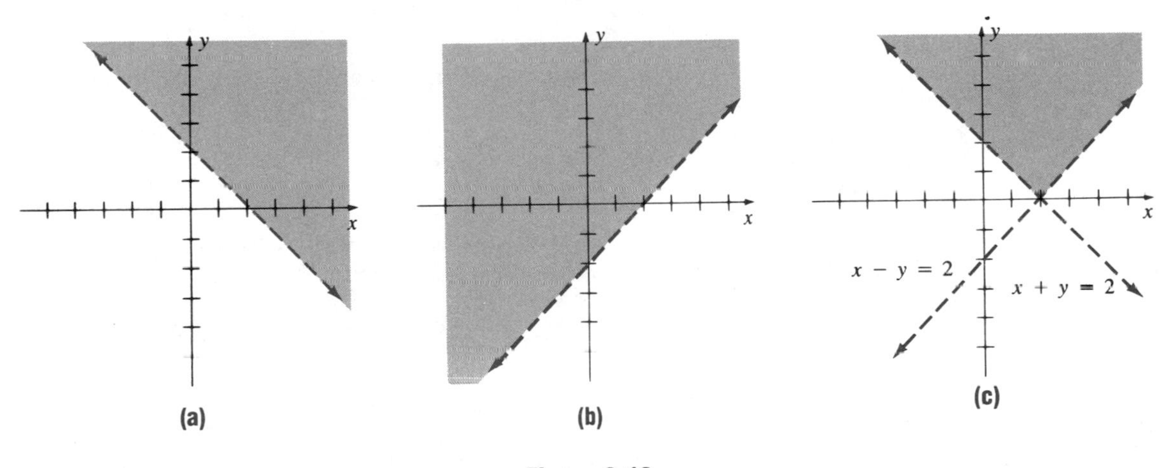

Figure 9.10

In the following examples, we have indicated only the final solution set for the system.

EXAMPLE 4 Solve the following system by graphing.

$$\begin{pmatrix} 2x - y \geq 4 \\ x + 2y < 2 \end{pmatrix}$$

Solution The graph of $2x - y \geq 4$ consists of all points *on or below* the line $2x - y = 4$. The graph of $x + 2y < 2$ consists of all points *below* the line $x + 2y = 2$. The graph of the system is indicated by the shaded region in Figure 9.11. Notice that all points in the shaded region are on or below the line $2x - y = 4$ *and* below the line $x + 2y = 2$.

Figure 9.11

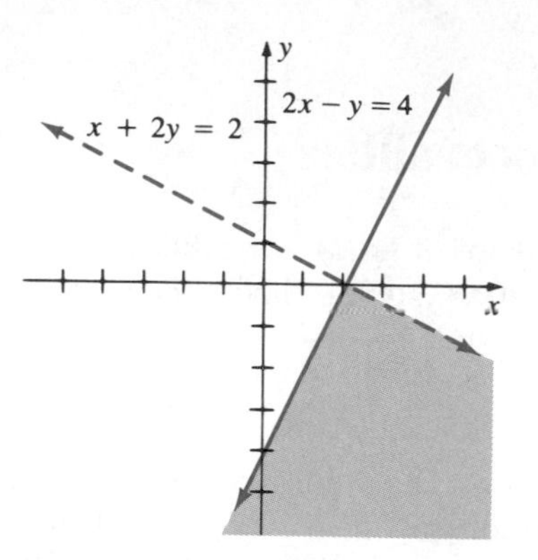

■

EXAMPLE 5 Solve the following system by graphing.

$$\begin{pmatrix} x \leq 2 \\ y \geq -1 \end{pmatrix}$$

Solution Remember that even though each inequality contains only one variable, we are working in a rectangular coordinate system that involves ordered pairs. That is to say, the system could be written as

$$\begin{pmatrix} x + 0(y) \leq 2 \\ 0(x) + y \geq -1 \end{pmatrix}.$$

The graph of the system is the shaded region in Figure 9.12. Notice that all points in the shaded region are *on or to the left of* the line $x = 2$ *and on or above* the line $y = -1$.

Figure 9.12

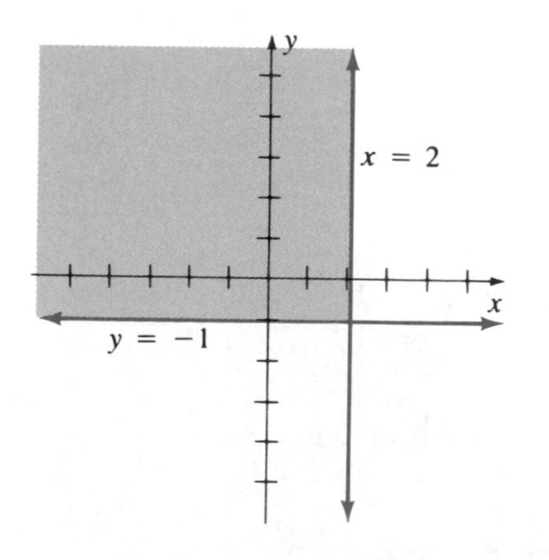

■

Problem Set 9.7

For each of the following systems, (a) graph the system so that approximate solutions can be predicted, and (b) solve the system by the substitution or elimination method.

1. $\begin{pmatrix} y = (x+2)^2 \\ y = -2x - 4 \end{pmatrix}$ $\{(-2, 0), (-4, 4)\}$

2. $\begin{pmatrix} y = x^2 \\ y = x + 2 \end{pmatrix}$ $\{(2, 4), (-1, 1)\}$

3. $\begin{pmatrix} x^2 + y^2 = 13 \\ 3x + 2y = 0 \end{pmatrix}$ $\{(2, -3), (-2, 3)\}$

4. $\begin{pmatrix} x^2 + y^2 = 26 \\ x + y = 6 \end{pmatrix}$ $\{(1, 5), (5, 1)\}$

5. $\begin{pmatrix} y = x^2 + 6x + 7 \\ 2x + y = -5 \end{pmatrix}$ $\{(-6, 7), (-2, -1)\}$

6. $\begin{pmatrix} y = x^2 - 4x + 5 \\ -x + y = 1 \end{pmatrix}$ $\{(1, 2), (4, 5)\}$

7. $\begin{pmatrix} y = x^2 \\ y = x^2 - 4x + 4 \end{pmatrix}$ $\{(1, 1)\}$

8. $\begin{pmatrix} y = -x^2 + 3 \\ y = x^2 + 1 \end{pmatrix}$ $\{(1, 2), (-1, 2)\}$

9. $\begin{pmatrix} x + y = -8 \\ x^2 - y^2 = 16 \end{pmatrix}$ $\{(-5, -3)\}$

10. $\begin{pmatrix} x - y = 2 \\ x^2 - y^2 = 16 \end{pmatrix}$ $\{(5, 3)\}$

11. $\begin{pmatrix} y = x^2 + 2x - 1 \\ y = x^2 + 4x + 5 \end{pmatrix}$ $\{(-3, 2)\}$

12. $\begin{pmatrix} 2x^2 + y^2 = 8 \\ x^2 + y^2 = 4 \end{pmatrix}$ $\{(2, 0), (-2, 0)\}$

13. $\begin{pmatrix} xy = 4 \\ y = x \end{pmatrix}$ $\{(2, 2), (-2, -2)\}$

14. $\begin{pmatrix} y = x^2 + 2 \\ y = 2x^2 + 1 \end{pmatrix}$ $\{(1, 3), (-1, 3)\}$

15. $\begin{pmatrix} x^2 + 2y^2 = 8 \\ x^2 - y^2 = 1 \end{pmatrix}$ $\left\{\left(\frac{\sqrt{30}}{3}, \frac{\sqrt{21}}{3}\right), \left(\frac{\sqrt{30}}{3}, -\frac{\sqrt{21}}{3}\right), \left(-\frac{\sqrt{30}}{3}, \frac{\sqrt{21}}{3}\right), \left(-\frac{\sqrt{30}}{3}, -\frac{\sqrt{21}}{3}\right)\right\}$

16. $\begin{pmatrix} x^2 - y^2 = 4 \\ x^2 + y^2 = 4 \end{pmatrix}$ $\{(2, 0), (-2, 0)\}$

For each of the following, indicate the solution set for each system of inequalities by shading the appropriate region.

17. $\begin{pmatrix} 3x - 4y \geq 0 \\ 2x + 3y \leq 0 \end{pmatrix}$ see page A68

18. $\begin{pmatrix} 3x + 2y \leq 6 \\ 2x - 3y \geq 6 \end{pmatrix}$ see page A68

19. $\begin{pmatrix} x - 3y < 6 \\ x + 2y \geq 4 \end{pmatrix}$ see page A68

20. $\begin{pmatrix} 2x - y \leq 4 \\ 2x + y > 4 \end{pmatrix}$ see page A68

21. $\begin{pmatrix} x + y < 4 \\ x - y > 2 \end{pmatrix}$ see page A68

22. $\begin{pmatrix} x + y > 1 \\ x - y < 1 \end{pmatrix}$ see page A68

23. $\begin{pmatrix} y < x + 1 \\ y \geq x \end{pmatrix}$ see page A69

24. $\begin{pmatrix} y > x - 3 \\ y < x \end{pmatrix}$ see page A69

25. $\begin{pmatrix} y > x \\ y > 2 \end{pmatrix}$ see page A69

26. $\begin{pmatrix} 2x + y > 6 \\ 2x + y < 2 \end{pmatrix}$ see page A69

27. $\begin{pmatrix} x \geq -1 \\ y < 4 \end{pmatrix}$ see page A69

28. $\begin{pmatrix} x < 3 \\ y > 2 \end{pmatrix}$ see page A69

29. $\begin{pmatrix} 2x - y > 4 \\ 2x - y > 0 \end{pmatrix}$ see page A69

30. $\begin{pmatrix} x + y > 4 \\ x + y > 6 \end{pmatrix}$ see page A69

31. $\begin{pmatrix} 3x - 2y < 6 \\ 2x - 3y < 6 \end{pmatrix}$ see page A69

32. $\begin{pmatrix} 2x + 5y > 10 \\ 5x + 2y > 10 \end{pmatrix}$ see page A69

Miscellaneous Problems

33. The following figure shows a graph of the system $\begin{pmatrix} y = x^2 \\ y = -1 \end{pmatrix}$.

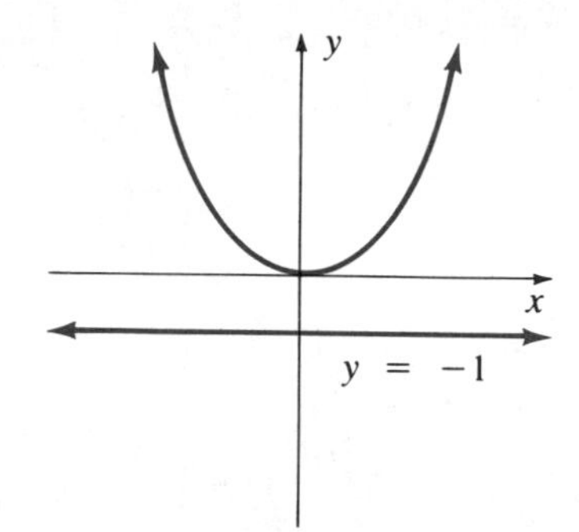

The graph indicates that the system has no solutions, *but* we must realize that we are graphing in the *real number plane.* Thus, the fact that the parabola and the line do not intersect simply indicates that the system has *no real number solutions.*

Using the substitution method, we can find some complex solutions as follows. Substitute x^2 for y in the second equation.

$y = -1$ becomes

$x^2 = -1$

$x = \pm\sqrt{-1} = \pm i.$

Substitute $\pm i$ for x in the first equation to yield

$$\begin{array}{ll} y = x^2 & y = x^2 \\ y = i^2 & y = (-i)^2 \\ y = -1. & y = i^2 \\ & y = -1. \end{array}$$

The solution set is $\{(i, -1), (-i, -1)\}$.

For each of the following systems, (a) graph the system to show that there are no real number solutions, and (b) solve the system by substitution or elimination to find the complex solutions.

(a) $\begin{pmatrix} y = x^2 + 1 \\ y = -3 \end{pmatrix}$ $\{(2i, -3), (-2i, -3)\}$

(b) $\begin{pmatrix} y = -x^2 + 1 \\ y = 3 \end{pmatrix}$ $\{(i\sqrt{2}, 3), (-i\sqrt{2}, 3)\}$

(c) $\begin{pmatrix} y = x^2 \\ x - y = 4 \end{pmatrix}$ $\left\{\left(\frac{1 + i\sqrt{15}}{2}, \frac{-7 + i\sqrt{15}}{2}\right), \left(\frac{1 - i\sqrt{15}}{2}, \frac{-7 - i\sqrt{15}}{2}\right)\right\}$

(d) $\begin{pmatrix} y = x^2 + 1 \\ y = -x^2 \end{pmatrix}$ $\left\{\left(\frac{i\sqrt{2}}{2}, \frac{1}{2}\right), \left(\frac{-i\sqrt{2}}{2}, \frac{1}{2}\right)\right\}$

(e) $\begin{pmatrix} x^2 + y^2 = 1 \\ x + y = 2 \end{pmatrix}$ $\left\{\left(\frac{2 + i\sqrt{2}}{2}, \frac{2 - i\sqrt{2}}{2}\right), \left(\frac{2 - i\sqrt{2}}{2}, \frac{2 + i\sqrt{2}}{2}\right)\right\}$

(f) $\begin{pmatrix} x^2 + y^2 = 2 \\ x^2 - y^2 = 6 \end{pmatrix}$ $\{(2, i\sqrt{2}), (2, -i\sqrt{2}), (-2, i\sqrt{2}), (-2, -i\sqrt{2})\}$

Chapter 9 Summary

(9.1) Graph a **system of two linear equations in two variables** to produce one of the following results.

1. The graphs of the two equations are two intersecting lines, which indicates that there is **one unique solution** of the system. Such a system is called a **consistent system.**
2. The graphs of the two equations are two parallel lines that indicate that there is **no solution** for the system and it is called an **inconsistent system.**
3. The graphs of the two equations are the same line, which indicates **infinitely many solutions** for the system. The equations are called **dependent** equations.

We can describe the **substitution method** of solving a system of equations as follows.

Step 1. Solve one of the equations for one variable in terms of the other variable if neither equation is in such a form. (If possible make a choice that will avoid fractions.)

Step 2. Substitute the expression obtained in step 1 into the other equation to produce an equation with one variable.

Step 3. Solve the equation obtained in step 2.

Step 4. Use the solution obtained in step 3, along with the expression obtained in step 1, to determine the solution of the system.

(9.2) The **elimination-by-addition method** involves the replacement of a system of equations with equivalent systems until a system is obtained whereby the solutions can be easily determined. The following operations or transformations can be performed on a system to produce an equivalent system.

1. Any two equations of the system can be interchanged.
2. Both sides of any equation of the system can be multiplied by any nonzero real number.
3. Any equation of the system can be replaced by the *sum* of that equation and a nonzero multiple of another equation.

(9.3) Solving **a system of three linear equations in three variables** produces one of the following results.

1. There is **one ordered triple** that satisfies all three equations.
2. There are **infinitely many ordered triples** in the solution set, all of which are coordinates of points on a line common to the planes.
3. There are **infinitely many ordered triples** in the solution set, all of which are coordinates on a plane.
4. The solution set is empty; it is $\varnothing$.

(9.4) A **matrix** is an array of numbers arranged in horizontal rows and vertical columns. A matrix of m rows and n columns is called an $m \times n$ ("m-by-n") matrix. The **augmented matrix** of the system

$$\begin{pmatrix} 5x - 2y - z = 4 \\ 3x - y - z = 7 \\ 2x + 3y + 7z = 9 \end{pmatrix}$$

is

$$\left[\begin{array}{rrr|r} 5 & -2 & -1 & 4 \\ 3 & -1 & -1 & 7 \\ 2 & 3 & 7 & 9 \end{array}\right].$$

The following **elementary row operations** provide the basis for transforming matrices.

1. Any two rows of an augmented matrix can be interchanged.
2. Any row can be multiplied by a nonzero constant.
3. Any row can be replaced by adding a nonzero multiple of another row to that row.

Transforming an augmented matrix to **triangular form** and then using back-substitution provides a systematic technique for solving systems of linear equations.

(9.5) A rectangular array of numbers is called a **matrix**. A **square matrix** has the same number of rows as columns. For a 2×2 matrix

$$\begin{bmatrix} a_1 & b_1 \\ a_2 & b_2 \end{bmatrix},$$

the **determinant** of the matrix is written as

$$\begin{vmatrix} a_1 & b_1 \\ a_2 & b_2 \end{vmatrix}$$

and defined by

$$\begin{vmatrix} a_1 & b_1 \\ a_2 & b_2 \end{vmatrix} = a_1b_2 - a_2b_1.$$

Cramer's Rule for solving a system of two linear equations in two variables is stated as follows.

Given the system $\begin{pmatrix} a_1x + b_1y = c_1 \\ a_2x + b_2y = c_2 \end{pmatrix}$

with

$$D = \begin{vmatrix} a_1 & b_1 \\ a_2 & b_2 \end{vmatrix} \neq 0, \qquad D_x = \begin{vmatrix} c_1 & b_1 \\ c_2 & b_2 \end{vmatrix}, \qquad D_y = \begin{vmatrix} a_1 & c_1 \\ a_2 & c_2 \end{vmatrix},$$

then

$$x = \frac{D_x}{D} \quad \text{and} \quad y = \frac{D_y}{D}.$$

(9.6) A 3×3 determinant is defined by

$$\begin{vmatrix} a_1 & b_1 & c_1 \\ a_2 & b_2 & c_2 \\ a_3 & b_3 & c_3 \end{vmatrix} = a_1b_2c_3 + b_1c_2a_3 + c_1a_2b_3 - a_3b_2c_1 - b_3c_2a_1 - c_3a_2b_1.$$

The **minor** of an element in a determinant is the determinant that remains after deleting the row and column in which the element appears. A determinant can be evaluated by **expansion-by-minors** of the elements of any row or any column.

Cramer's Rule for solving a system of three linear equations in three variables is stated as follows.

Given the system $\begin{pmatrix} a_1x + b_1y + c_1z = d_1 \\ a_2x + b_2y + c_2z = d_2 \\ a_3x + b_3y + c_3z = d_3 \end{pmatrix}$

with

$$D = \begin{vmatrix} a_1 & b_1 & c_1 \\ a_2 & b_2 & c_2 \\ a_3 & b_3 & c_3 \end{vmatrix} \neq 0, \qquad D_x = \begin{vmatrix} d_1 & b_1 & c_1 \\ d_2 & b_2 & c_2 \\ d_3 & b_3 & c_3 \end{vmatrix},$$

$$D_y = \begin{vmatrix} a_1 & d_1 & c_1 \\ a_2 & d_2 & c_2 \\ a_3 & d_3 & c_3 \end{vmatrix}, \qquad D_z = \begin{vmatrix} a_1 & b_1 & d_1 \\ a_2 & b_2 & d_2 \\ a_3 & b_3 & d_3 \end{vmatrix},$$

then $x = \frac{D_x}{D}$, $y = \frac{D_y}{D}$, and $z = \frac{D_z}{D}$.

(9.7) The substitution and elimination methods can also be used to solve systems involving **nonlinear equations.**

The solution set of a system of **linear inequalities** is the intersection of the solution sets of the individual inequalities.

Chapter 9 Review Problem Set

For Problems 1–4, solve each system of equations using **(a)** the substitution method, **(b)** the elimination method, **(c)** a matrix approach, and **(d)** Cramer's Rule.

1. $\begin{pmatrix} 3x - 2y = -6 \\ 2x + 5y = 34 \end{pmatrix}$ $\{(2, 6)\}$

2. $\begin{pmatrix} x + 4y = 25 \\ y = -3x - 2 \end{pmatrix}$ $\{(-3, 7)\}$

3. $\begin{pmatrix} x = 5y - 49 \\ 4x + 3y = -12 \end{pmatrix}$ $\{(-9, 8)\}$

4. $\begin{pmatrix} x - 6y = 7 \\ 3x + 5y = 9 \end{pmatrix}$ $\left\{\left(\frac{89}{23}, -\frac{12}{23}\right)\right\}$

For Problems 5–14, solve each system using the method that seems most appropriate to you.

5. $\begin{pmatrix} x - 3y = 25 \\ -3x + 2y = -26 \end{pmatrix}$ $\{(4, -7)\}$

6. $\begin{pmatrix} 5x - 7y = -66 \\ x + 4y = 30 \end{pmatrix}$ $\{(-2, 8)\}$

7. $\begin{pmatrix} 4x + 3y = -9 \\ 3x - 5y = 15 \end{pmatrix}$ $\{(0, -3)\}$

8. $\begin{pmatrix} 2x + 5y = 47 \\ 4x - 7y = -25 \end{pmatrix}$ $\{(6, 7)\}$

9. $\begin{pmatrix} 7x - 3y = 25 \\ y = 3x - 9 \end{pmatrix}$ $\{(1, -6)\}$

10. $\begin{pmatrix} x = -4 - 5y \\ y = 4x + 16 \end{pmatrix}$ $\{(-4, 0)\}$

11. $\begin{pmatrix} \frac{1}{2}x + \frac{2}{3}y = 6 \\ \frac{3}{4}x - \frac{5}{6}y = -24 \end{pmatrix}$ $\{(-12, 18)\}$

12. $\begin{pmatrix} \frac{3}{4}x - \frac{1}{2}y = 14 \\ \frac{5}{12}x + \frac{3}{4}y = 16 \end{pmatrix}$ $\{(24, 8)\}$

13. $\begin{pmatrix} 6x - 4y = 7 \\ 9x + 8y = 0 \end{pmatrix}$ $\left\{\left(\frac{2}{3}, -\frac{3}{4}\right)\right\}$

14. $\begin{pmatrix} 4x - 5y = -5 \\ 6x - 10y = -9 \end{pmatrix}$ $\left\{\left(-\frac{1}{2}, \frac{3}{5}\right)\right\}$

For Problems 15–18, evaluate each of the determinants.

15. $\begin{vmatrix} 3 & 5 \\ 6 & 4 \end{vmatrix}$ -18

16. $\begin{vmatrix} -1 & -5 \\ 4 & 9 \end{vmatrix}$ 11

17. $\begin{vmatrix} 4 & -1 & -3 \\ 2 & 1 & 4 \\ -3 & 2 & 2 \end{vmatrix}$ -29

18. $\begin{vmatrix} 5 & 3 & 4 \\ -2 & 0 & 1 \\ -1 & -2 & 6 \end{vmatrix}$ 59

For Problems 19 and 20, solve each system of equations using **(a)** the elimination method, **(b)** a matrix approach, and **(c)** Cramer's Rule.

19. $\begin{pmatrix} x - 2y + 4z = -14 \\ 3x - 5y + z = 20 \\ -2x + y - 5z = 22 \end{pmatrix}$ $\{(2, -4, -6)\}$

20. $\begin{pmatrix} x + 3y - 2z = 28 \\ 2x - 8y + 3z = -63 \\ 3x + 8y - 5z = 72 \end{pmatrix}$ $\{(-1, 5, -7)\}$

For Problems 21–24, solve each system using the method that seems most appropriate to you.

21. $\begin{pmatrix} x + y - z = -2 \\ 2x - 3y + 4z = 17 \\ -3x + 2y + 5z = -7 \end{pmatrix}$ $\{(2, -3, 1)\}$

22. $\begin{pmatrix} -x - y + z = -3 \\ 3x + 2y - 4z = 12 \\ 5x + y + 2z = 5 \end{pmatrix}$ $\{(2, -1, -2)\}$

23. $\begin{pmatrix} 3x + y - z = -6 \\ 3x + 2y + 3z = 9 \\ 6x - 2y + 2z = 8 \end{pmatrix}$ $\{(-\frac{1}{3}, -1, 4)\}$

24. $\begin{pmatrix} x - 3y + z = 2 \\ 2x - 5y - 3z = 22 \\ -4x + 3y + 5z = -26 \end{pmatrix}$ $\{(0, -2, -4)\}$

25. Graph the following system and then find the solution set by using either the substitution or the elimination method.

$\begin{pmatrix} y = 2x^2 - 1 \\ 2x + y = 3 \end{pmatrix}$ see page A70

26. Indicate the solution set of the following system of inequalities by graphing the system and shading the appropriate region.

$\begin{pmatrix} 3x + y > 6 \\ x - 2y \leq 4 \end{pmatrix}$ see page A70

For Problems 27–33, solve each problem by setting up and solving a system of two equations and two unknowns.

27. The sum of the squares of two numbers is 13. If one number is 1 larger than the other number, find the numbers. 2 and 3 or −3 and 2

28. The sum of the squares of two numbers is 34. The difference of the squares of the same two numbers is 16. Find the numbers. 5 and 3, or 5 and −3 or −5 and 3 or −5 and −3

29. A number is 1 larger than the square of another number. The sum of the two numbers is 7. Find the numbers. 2 and 5 or −3 and 10

30. The area of a rectangular region is 54 square meters and its perimeter is 30 meters. Find the length and width of the rectangle. 6 meters by 9 meters

31. At a local confectionery, 7 pounds of cashews and 5 pounds of Spanish peanuts cost \$88, and 3 pounds of cashews and 2 pounds of Spanish peanuts cost \$37. Find the price per pound for cashews and for Spanish peanuts. \$9 and \$5

32. We bought 2 cartons of pop and 4 pounds of candy for \$12. The next day we bought 3 cartons of pop and 2 pounds of candy for \$9. Find the price of a carton of pop and also the price of a pound of candy. \$1.50 and \$2.25

33. Suppose that a mail order company charges a fixed fee for shipping merchandise that weighs 1 pound or less, plus an additional fee for each pound over 1 pound. If the shipping charge for 5 pounds is \$2.40 and for 12 pounds is \$3.10, find the fixed fee and the additional fee. The fixed fee is \$2 and the additional fee is \$0.10 per lb

Thoughts into Words

34. Explain how you would solve the system $\begin{pmatrix} 2x + 3y = 5 \\ 5x - y = 9 \end{pmatrix}$ using the substitution method.

35. Explain how you would solve the system $\begin{pmatrix} 3x - 4y = -1 \\ 2x - 5y = 9 \end{pmatrix}$ using the elimination-by-addition method.

36. Why does the solution set for the system $\begin{pmatrix} x = 2y + 4 \\ 3x - 6y = 12 \end{pmatrix}$ contain infinitely many ordered pairs as solutions? Find at least five of these ordered pairs.

37. Explain how you would solve the system $\begin{pmatrix} 2x + 4y = 10 \\ 5x - 7y = -2 \end{pmatrix}$.

38. Describe how you would solve the system $\begin{pmatrix} x - 3z = 4 \\ 3x - 2y + 7z = -1 \\ 2x + z = 9 \end{pmatrix}$.

39. Explain the difference between a matrix and a determinant.

Cumulative Review Problem Set

For Problems 1–5, evaluate each algebraic expression for the given values of the variables.

1. $-5(x - 1) - 3(2x + 4) + 3(3x - 1)$ for $x = -2$ -6

2. $\dfrac{14a^3b^2}{7a^2b}$ for $a = -1$ and $b = 4$ -8

3. $\dfrac{2}{n} - \dfrac{3}{2n} + \dfrac{5}{3n}$ for $n = 4$ $\dfrac{13}{24}$

4. $-4\sqrt{2x - y} + 5\sqrt{3x + y}$ for $x = 16$ and $y = 16$ 24

5. $\dfrac{3}{x - 2} - \dfrac{5}{x + 3}$ for $x = 3$ $\dfrac{13}{6}$

For Problems 6–15, perform the indicated operations and express answers in simplified form.

6. $(-5\sqrt{6})(3\sqrt{12})$ $-90\sqrt{2}$

7. $(2\sqrt{x} - 3)(\sqrt{x} + 4)$ $2x + 5\sqrt{x} - 12$

8. $(3\sqrt{2} - \sqrt{6})(\sqrt{2} + 4\sqrt{6})$ $-18 + 22\sqrt{3}$

9. $(2x - 1)(x^2 + 6x - 4)$ $2x^3 + 11x^2 - 14x + 4$

10. $\dfrac{x^2 - x}{x + 5} \cdot \dfrac{x^2 + 5x + 4}{x^4 - x^2}$ $\dfrac{x + 4}{x(x + 5)}$

11. $\dfrac{16x^2y}{24xy^3} \div \dfrac{9xy}{8x^2y^2}$ $\dfrac{16x^2}{27y}$

12. $\dfrac{x + 3}{10} + \dfrac{2x + 1}{15} - \dfrac{x - 2}{18}$ $\dfrac{16x + 43}{90}$

13. $\dfrac{7}{12ab} - \dfrac{11}{15a^2}$ $\dfrac{35a - 44b}{60a^2b}$

14. $\dfrac{8}{x^2 - 4x} + \dfrac{2}{x}$ $\dfrac{2}{x - 4}$

15. $(8x^3 - 6x^2 - 15x + 4) \div (4x - 1)$ $2x^2 - x - 4$

For Problems 16–19, simplify each of the complex fractions.

16. $\dfrac{\frac{5}{x^2} - \frac{3}{x}}{\frac{1}{y} + \frac{2}{y^2}}$ $\dfrac{5y^2 - 3xy^2}{x^2y + 2x^2}$

17. $\dfrac{\frac{2}{x} - 3}{\frac{3}{y} + 4}$ $\dfrac{2y - 3xy}{3x + 4xy}$

18. $\dfrac{2 - \frac{1}{n+2}}{3 + \frac{4}{n+3}}$ $\dfrac{(2n-5)(n+3)}{(n-2)(3n+13)}$

19. $\dfrac{3a}{2 - \frac{1}{a}} - 1$ $\dfrac{3a^2 - 2a + 1}{2a - 1}$

For Problems 20–25, factor each of the algebraic expressions completely.

20. $20x^2 + 7x - 6$ $(5x-2)(4x+3)$

21. $16x^3 + 54$ $2(2x+3)(4x^2 - 6x + 9)$

22. $4x^4 - 25x^2 + 36$ $(2x+3)(2x-3)(x+2)(x-2)$

23. $12x^3 - 52x^2 - 40x$ $4x(3x+2)(x-5)$

24. $xy - 6x + 3y - 18$ $(y-6)(x+3)$

25. $10 + 9x - 9x^2$ $(5-3x)(2+3x)$

For Problems 26–33, evaluate each of the numerical expressions.

26. $\left(\dfrac{2}{3}\right)^{-4}$ $\dfrac{81}{16}$

27. $\dfrac{3}{\left(\frac{4}{3}\right)^{-1}}$ 4

28. $\sqrt[3]{-\dfrac{27}{64}}$ $-\dfrac{3}{4}$

29. $-\sqrt{.09}$ $-.3$

30. $(27)^{-\frac{4}{3}}$ $\frac{1}{81}$

31. $4^0 + 4^{-1} + 4^{-2}$ $\frac{21}{16}$

32. $\left(\dfrac{3^{-1}}{2^{-3}}\right)^{-2}$ $\dfrac{9}{64}$

33. $(2^{-3} - 3^{-2})^{-1}$ 72

For Problems 34–36, find the indicated products and quotients and express final answers with positive integral exponents only.

34. $(-3x^{-1}y^2)(4x^{-2}y^{-3})$ $-\frac{12}{x^3y}$

35. $\dfrac{48x^{-4}y^2}{6xy}$ $\dfrac{8y}{x^5}$

36. $\left(\dfrac{27a^{-4}b^{-3}}{-3a^{-1}b^{-4}}\right)^{-1}$ $-\dfrac{a^3}{9b}$

For Problems 37–44, express each radical expression in simplest radical form.

37. $\sqrt{80}$ $4\sqrt{5}$

38. $-2\sqrt{54}$ $-6\sqrt{6}$

39. $\sqrt{\dfrac{75}{81}}$ $\dfrac{5\sqrt{3}}{9}$

40. $\dfrac{4\sqrt{6}}{3\sqrt{8}}$ $\dfrac{2\sqrt{3}}{3}$

41. $\sqrt[3]{56}$ $2\sqrt[3]{7}$

42. $\dfrac{\sqrt[3]{3}}{\sqrt[3]{4}}$ $\dfrac{\sqrt[3]{6}}{2}$

43. $4\sqrt{52x^3y^2}$ $8xy\sqrt{13x}$

44. $\sqrt{\dfrac{2x}{3y}}$ $\dfrac{\sqrt{6xy}}{3y}$

For Problems 45–47, use the distributive property to help simplify each of the following.

45. $-3\sqrt{24} + 6\sqrt{54} - \sqrt{6}$ $11\sqrt{6}$

46. $\dfrac{\sqrt{8}}{3} - \dfrac{3\sqrt{18}}{4} - \dfrac{5\sqrt{50}}{2}$ $-\dfrac{169\sqrt{2}}{12}$

47. $8\sqrt[3]{3} - 6\sqrt[3]{24} - 4\sqrt[3]{81}$ $-16\sqrt[3]{3}$

For Problems 48 and 49, rationalize the denominator and simplify.

48. $\dfrac{\sqrt{3}}{\sqrt{6}-2\sqrt{2}}$ $\dfrac{-3\sqrt{2}-2\sqrt{6}}{2}$

49. $\dfrac{3\sqrt{5}-\sqrt{3}}{2\sqrt{3}+\sqrt{7}}$ $\dfrac{6\sqrt{15}-3\sqrt{35}-6+\sqrt{21}}{5}$

For Problems 50–52, use scientific notation to help perform the indicated operations.

50. $\dfrac{(.00016)(300)(.028)}{.064}$ .021

51. $\dfrac{.00072}{.0000024}$ 300

52. $\sqrt{.00000009}$ .0003

For Problems 53–56, find each of the indicated products or quotients and express answers in standard form.

53. $(5-2i)(4+6i)$ $32+22i$

54. $(-3-i)(5-2i)$ $-17+i$

55. $\dfrac{5}{4i}$ $0-\dfrac{5}{4}i$

56. $\dfrac{-1+6i}{7-2i}$ $-\dfrac{19}{53}+\dfrac{40}{53}i$

57. Find the slope of the line determined by the points $(2,-3)$ and $(-1,7)$. $-\frac{10}{3}$

58. Find the slope of the line determined by the equation $4x-7y=9$. $\frac{4}{7}$

59. Find the length of the line segment whose endpoints are $(4,5)$ and $(-2,1)$. $2\sqrt{13}$

60. Write the equation of the line that contains the points $(3,-1)$ and $(7,4)$. $5x-4y=19$

61. Write the equation of the line that is perpendicular to the line $3x-4y=6$ and contains the point $(-3,-2)$. $4x+3y=-18$

62. Find the center and the length of a radius of the circle $x^2+4x+y^2-12y+31=0$. $(-2,6)$ and $r=3$

63. Find the coordinates of the vertex of the parabola $y=x^2+10x+21$. $(-5,-4)$

64. Find the length of the major axis of the ellipse $x^2+4y^2=16$. 8 units

For Problems 65–70, graph each of the equations.

65. $-x+2y=-4$ see page A70

66. $x^2+y^2=9$ see page A70

67. $x^2-y^2=9$ see page A70

68. $x^2+2y^2=8$ see page A71

69. $y=-3x$ see page A71

70. $x^2y=4$ see page A71

For Problems 71–76, graph each of the functions.

71. $f(x)=-2x-4$ see page A71

72. $f(x)=-2x^2-2$ see page A71

73. $f(x)=x^2-2x-2$ see page A71

74. $f(x)=\sqrt{x+1}+2$ see page A71

75. $f(x)=2x^2+8x+9$ see page A71

76. $f(x)=-|x-2|+1$ see page A71

77. If $f(x)=x-3$ and $g(x)=2x^2-x-1$, find $(g\circ f)(x)$ and $(f\circ g)(x)$. $(g\circ f)(x)=2x^2-13x+20$ $(f\circ g)(x)=2x^2-x-4$

78. Find the inverse (f^{-1}) of $f(x)=3x-7$. $f^{-1}(x)=\frac{x+7}{3}$

79. Find the inverse of $f(x)=-\dfrac{1}{2}x+\dfrac{2}{3}$. $f^{-1}(x)=-2x+\dfrac{4}{3}$

80. Find the constant of variation if y varies directly as x, and $y=2$ when $x=-\dfrac{2}{3}$. $k=-3$

81. If y is inversely proportional to the square of x, and $y=4$ when $x=3$, find y when $x=6$. $y=1$

82. The volume of a gas at a constant temperature varies inversely as the pressure. What is the volume of a gas under a pressure of 25 pounds if the gas occupies 15 cubic centimeters under a pressure of 20 pounds? 12 cubic centimeters

For Problems 83 and 84, evaluate each of the determinants.

83. $\begin{vmatrix} -2 & 4 \\ 7 & 6 \end{vmatrix}$ -40

84. $\begin{vmatrix} 1 & -2 & -1 \\ 2 & 1 & 3 \\ -1 & -3 & 4 \end{vmatrix}$ 40

For Problems 85–105, solve each of the equations.

85. $3(2x-1)-2(5x+1)=4(3x+4)$ $\{-\frac{21}{16}\}$

86. $n+\dfrac{3n-1}{9}-4=\dfrac{3n+1}{3}$ $\left\{\dfrac{40}{3}\right\}$

87. $.92+.9(x-.3)=2x-5.95$ $\{6\}$

88. $|4x-1|=11$ $\{-\frac{5}{2},3\}$

89. $3x^2=7x$ $\{0,\frac{7}{3}\}$

90. $x^3-36x=0$ $\{-6,0,6\}$

91. $30x^2+13x-10=0$ $\{-\frac{5}{6},\frac{2}{5}\}$

92. $8x^3+12x^2-36x=0$ $\{-3,0,\frac{3}{2}\}$

93. $x^4+8x^2-9=0$ $\{\pm 1,\pm 3i\}$

94. $(n+4)(n-6)=11$ $\{-5,7\}$

95. $2-\dfrac{3x}{x-4}=\dfrac{14}{x+7}$ $\{-29,0\}$

96. $\dfrac{2n}{6n^2+7n-3}-\dfrac{n-3}{3n^2+11n-4}=\dfrac{5}{2n^2+11n+12}$ $\left\{\dfrac{7}{2}\right\}$

97. $\sqrt{3y}-y=-6$ $\{12\}$

98. $\sqrt{x+19}-\sqrt{x+28}=-1$ $\{-3\}$

99. $(3x-1)^2=45$ $\{\frac{1\pm 3\sqrt{5}}{3}\}$

100. $(2x+5)^2=-32$ $\{\frac{-5\pm 4i\sqrt{2}}{2}\}$

101. $2x^2-3x+4=0$ $\{\frac{3\pm i\sqrt{23}}{4}\}$

102. $3n^2-6n+2=0$ $\{\frac{3\pm\sqrt{3}}{3}\}$

103. $\dfrac{5}{n-3}-\dfrac{3}{n+3}=1$ $\{1\pm\sqrt{34}\}$

104. $12x^4-19x^2+5=0$ $\{\pm\frac{\sqrt{5}}{2},\pm\frac{\sqrt{3}}{3}\}$

105. $2x^2+5x+5=0$ $\{\frac{-5\pm i\sqrt{15}}{4}\}$

For Problems 106–115, solve each of the inequalities.

106. $-5(y-1)+3>3y-4-4y$ $(-\infty,3)$

107. $.06x+.08(250-x)\geq 19$ $(-\infty,50]$

108. $|5x-2|>13$ $(-\infty,-\frac{11}{5})\cup(3,\infty)$

109. $|6x+2|<8$ $(-\frac{5}{3},1)$

110. $\dfrac{x-2}{5}-\dfrac{3x-1}{4}\leq\dfrac{3}{10}$ $\left[-\dfrac{9}{11},\infty\right)$

111. $(x-2)(x+4)\leq 0$ $[-4,2]$

112. $(3x-1)(x-4)>0$ $(-\infty,\frac{1}{3})\cup(4,\infty)$

113. $x(x+5)<24$ $(-8,3)$

114. $\dfrac{x-3}{x-7}\geq 0$ $(-\infty,3]\cup(7,\infty)$

115. $\dfrac{2x}{x+3}>4$ $(-6,-3)$

For Problems 116–120, solve each of the systems of equations.

116. $\begin{pmatrix} 4x-3y=18 \\ 3x-2y=15 \end{pmatrix}$ $\{(9,6)\}$

117. $\begin{pmatrix} y=\dfrac{2}{5}x-1 \\ 3x+5y=4 \end{pmatrix}$ $\left\{\left(\dfrac{9}{5},-\dfrac{7}{25}\right)\right\}$

118. $\begin{pmatrix} \dfrac{x}{2}-\dfrac{y}{3}=1 \\ \dfrac{2x}{5}+\dfrac{y}{2}=2 \end{pmatrix}$ $\left\{\left(\dfrac{70}{23},\dfrac{36}{23}\right)\right\}$

119. $\begin{pmatrix} 4x-y+3z=-12 \\ 2x+3y-z=8 \\ 6x+y+2z=-8 \end{pmatrix}$ $\{(-2,4,0)\}$

120. $\begin{pmatrix} x-y+5z=-10 \\ 5x+2y-3z=6 \\ -3x+2y-z=12 \end{pmatrix}$ $\{(-1,4,-1)\}$

For Problems 121–135, set up an equation, an inequality, or a system of equations to help solve each problem.

121. Find three consecutive odd integers whose sum is 57. 17, 19, and 21

122. Suppose that Eric has a collection of 63 coins consisting of nickels, dimes, and quarters. The number of dimes is 6 more than the number of nickels and the number of quarters is one more than twice the number of nickels. How many coins of each kind are in the collection? 14 nickels, 20 dimes, 29 quarters

123. One of two supplementary angles is 4° more than one-third of the other angle. Find the measure of each of the angles. 48° and 132°

124. If a ring costs a jeweler $300, at what price should it be sold to make a profit of 50% on the selling price? $600

125. Last year Beth invested a certain amount of money at 8% and $300 more than that amount at 9%. Her total yearly interest was $316. How much did she invest at each rate? $1700 at 8% and $2000 at 9%

126. Two trains leave the same depot at the same time, one traveling east and the other traveling west. At the end of $4\frac{1}{2}$ hours they are 639 miles apart. If the rate of the train traveling east is 10 miles per hour faster than the other train, find their rates. 66 mph and 76 mph

127. Suppose that a 10-quart radiator contains a 50% solution of antifreeze. How much needs to be drained out and replaced with pure antifreeze to obtain a 70% antifreeze solution? 4 quarts

128. Sam shot rounds of 70, 73, and 76 on the first three days of a golf tournament. What must he shoot on the fourth day of the tournament to average 72 or less for the four days? 69 or less

129. The cube of a number equals nine times the same number. Find the number. -3, 0, or 3

130. A strip of uniform width is to be cut off of both sides and both ends of a sheet of paper that is 8 inches by 14 inches to reduce the size of the paper to an area of 72 square inches. Find the width of the strip. 1-inch strip

131. A sum of $2450 is to be divided between two people in the ratio of 3 to 4. How much does each person receive? $1050 and $1400

132. Working together Sue and Dean can complete a task in $1\frac{1}{5}$ hours. Dean can do the task by himself in 2 hours. How long would it take Sue to complete the task by herself? 3 hours

133. Dudley bought a number of shares of stock for $300. A month later he sold all but 10 shares at a profit of $5 per share and regained his original investment of $300. How many shares did he originally buy and at what price per share? 30 shares at $10 per share

134. The units digit of a two-digit number is one more than twice the tens digit. The sum of the digits is 10. Find the number. 37

135. The sum of the two smallest angles of a triangle is 40° less than the other angle. The sum of the smallest and largest angle is twice the other angle. Find the measures of the three angles of the triangle. 10°, 60°, and 110°

Chapter 10

Exponential and Logarithmic Functions

This chapter will extend the meaning of an exponent, and introduce the concept of a logarithm. We will (1) work with some exponential functions, (2) work with some logarithmic functions, and (3) use the concepts of exponent and logarithm to expand our capabilities for solving problems. Your calculator will be a valuable tool throughout this chapter.

10.1
Exponential Functions

In Chapter 1, the expression b^n was defined to mean n factors of b, where n is any positive integer and b is any real number. For example,

$$4^3 = 4 \cdot 4 \cdot 4 = 64,$$

$$\left(\frac{1}{2}\right)^4 = \left(\frac{1}{2}\right)\left(\frac{1}{2}\right)\left(\frac{1}{2}\right)\left(\frac{1}{2}\right) = \frac{1}{16},$$

$$(-0.3)^2 = (-0.3)(-0.3) = 0.09.$$

In Chapter 5, by defining $b^0 = 1$ and $b^{-n} = \frac{1}{b^n}$, where n is any positive integer and b is any nonzero real number, the concept of an exponent was extended to include all integers. For example,

$$2^{-3} = \frac{1}{2^3} = \frac{1}{2 \cdot 2 \cdot 2} = \frac{1}{8}, \qquad \left(\frac{1}{3}\right)^{-2} = \frac{1}{\left(\frac{1}{3}\right)^2} = \frac{1}{\frac{1}{9}} = 9,$$

$$(0.4)^{-1} = \frac{1}{(0.4)^1} = \frac{1}{0.4} = 2.5, \qquad (-0.98)^0 = 1.$$

Chapter 5 also provided for the use of all rational numbers as exponents by defining $b^{\frac{m}{n}} = \sqrt[n]{b^m}$, where n is a positive integer greater than one and b is a real number such that $\sqrt[n]{b}$ exists. For example,

$$8^{\frac{2}{3}} = \sqrt[3]{8^2} = \sqrt[3]{64} = 4, \qquad 16^{\frac{1}{4}} = \sqrt[4]{16^1} = 2, \qquad 32^{-\frac{1}{5}} = \frac{1}{32^{\frac{1}{5}}} = \frac{1}{\sqrt[5]{32}} = \frac{1}{2}.$$

To formally extend the concept of an exponent to include the use of irrational numbers requires some ideas from calculus and is therefore beyond the scope of this text. However, here's a glance at the general idea involved. Consider the number $2^{\sqrt{3}}$. By using the nonterminating and nonrepeating decimal representation 1.73205... for $\sqrt{3}$, form the sequence of numbers 2^1, $2^{1.7}$, $2^{1.73}$, $2^{1.732}$, $2^{1.7320}$, $2^{1.73205}$.... It would seem reasonable that each successive power gets closer to $2^{\sqrt{3}}$. This is precisely what happens if b^n, where n is irrational, is properly defined by using the concept of a limit.

So from now on we can use any real number as an exponent and the basic properties stated in Chapter 5 can be extended to include all real numbers as exponents. Let's restate those properties at this time with the restriction that the bases a and b are to be positive numbers (to avoid expressions such as $(-4)^{\frac{1}{2}}$, which do not represent real numbers).

PROPERTY 10.1

If a and b are positive real numbers and m and n are any real numbers, then

1. $b^n \cdot b^m = b^{n+m}$ product of two powers
2. $(b^n)^m = b^{mn}$ power of a power
3. $(ab)^n = a^n b^n$ power of a product
4. $\left(\frac{a}{b}\right)^n = \frac{a^n}{b^n}$ power of a quotient
5. $\frac{b^n}{b^m} = b^{n-m}$ quotient of two powers

Another property that can be used to solve certain types of equations involving exponents can be stated as follows.

PROPERTY 10.2

If $b > 0$, $b \neq 1$, and m and n are real numbers, then

$b^n = b^m$ if and only if $n = m$.

The following examples illustrate the use of Property 10.2.

EXAMPLE 1 Solve $2^x = 32$.

Solution

$$2^x = 32$$

$$2^x = 2^5 \qquad 32 = 2^5$$

$$x = 5 \qquad \text{Property 10.2}$$

The solution set is $\{5\}$. ■

EXAMPLE 2 Solve $3^{2x} = \frac{1}{9}$.

Solution

$$3^{2x} = \frac{1}{9} = \frac{1}{3^2}$$

$$3^{2x} = 3^{-2}$$

$$2x = -2 \qquad \text{Property 10.2}$$

$$x = -1$$

The solution set is $\{-1\}$. ■

EXAMPLE 3 Solve $\left(\frac{1}{5}\right)^{x-2} = \frac{1}{125}$.

Solution

$$\left(\frac{1}{5}\right)^{x-2} = \frac{1}{125} = \left(\frac{1}{5}\right)^3$$

$$x - 2 = 3 \qquad \text{Property 10.2}$$

$$x = 5$$

The solution set is $\{5\}$. ■

EXAMPLE 4 Solve $8^x = 32$.

Solution

$$8^x = 32$$

$$(2^3)^x = 2^5 \qquad 8 = 2^3$$

$$2^{3x} = 2^5$$

$$3x = 5 \qquad \text{Property 10.2}$$

$$x = \frac{5}{3}$$

The solution set is $\left\{\frac{5}{3}\right\}$. ■

EXAMPLE 5 Solve $(3^{x+1})(9^{x-2}) = 27$.

Solution

$$(3^{x+1})(9^{x-2}) = 27$$

$$(3^{x+1})(3^2)^{x-2} = 3^3$$

$$(3^{x+1})(3^{2x-4}) = 3^3$$

$$3^{3x-3} = 3^3$$

$$3x - 3 = 3 \qquad \text{Property 10.2}$$

$$3x = 6$$

$$x = 2$$

The solution set is $\{2\}$. ■

If b is any positive number, then the expression b^x designates exactly one real number for every real value of x. Thus, the equation $f(x) = b^x$ defines a function whose domain is the set of real numbers. Furthermore, if we place the additional restriction $b \neq 1$, then any equation of the form $f(x) = b^x$ describes a one-to-one function and is called an **exponential function**. This leads to the following definition.

DEFINITION 10.1

> If $b > 0$ and $b \neq 1$, then the function f defined by
>
> $$f(x) = b^x,$$
>
> where x is any real number, is called the **exponential function with base b.**

REMARK The function $f(x) = 1^x$ is a constant function whose graph is a horizontal line, and therefore it is not a one-to-one function. Remember from Chapter 8 that one-to-one functions have inverses; this becomes a key issue in a later section. Now let's consider graphing some exponential functions.

EXAMPLE 6 Graph the function $f(x) = 2^x$.

Solution First, let's set up a table of values.

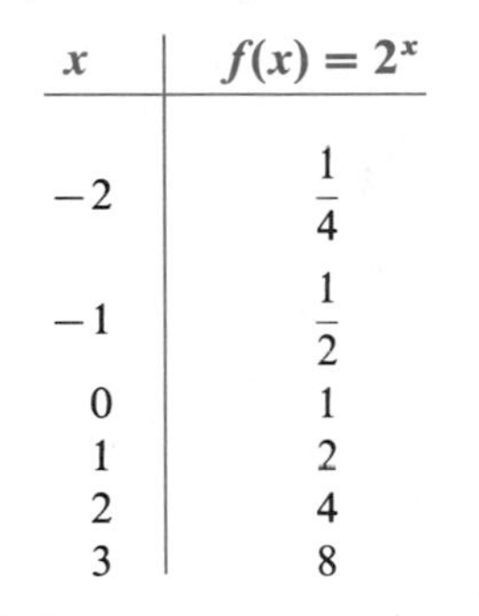

x	$f(x) = 2^x$
-2	$\frac{1}{4}$
-1	$\frac{1}{2}$
0	1
1	2
2	4
3	8

Plot these points and connect them with a smooth curve to produce Figure 10.1.

Figure 10.1

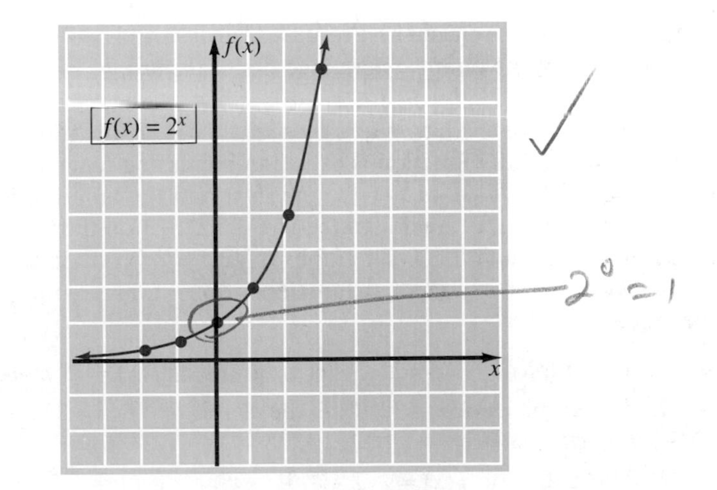

■

EXAMPLE 7 Graph $f(x) = \left(\frac{1}{2}\right)^x$

Solution Again, let's set up a table of values.

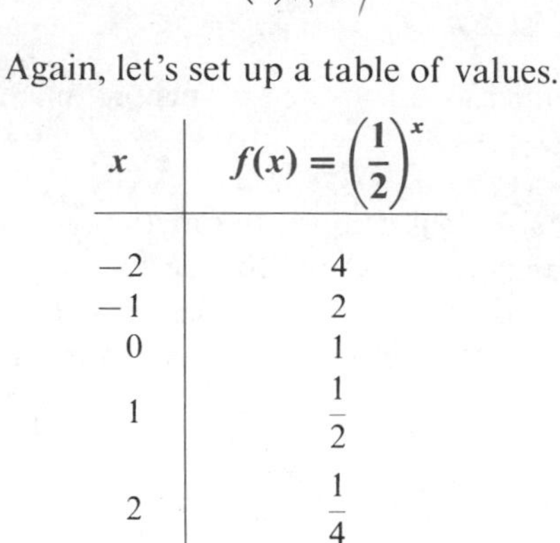

x	$f(x) = \left(\frac{1}{2}\right)^x$
-2	4
-1	2
0	1
1	$\frac{1}{2}$
2	$\frac{1}{4}$
3	$\frac{1}{8}$

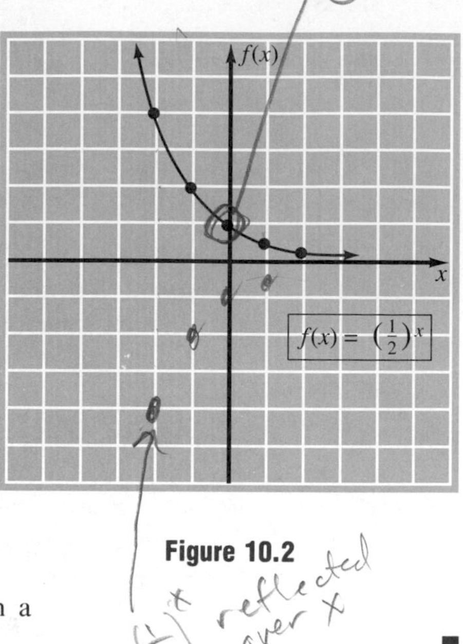

Figure 10.2

Plot these points and connect them with a smooth curve to produce Figure 10.2. ■

In the tables for Examples 6 and 7 we chose integral values for x to keep the computation simple. However, with the use of a calculator we could easily acquire functional values by using nonintegral exponents. Consider the following additional values for each of the tables.

$f(x) = 2^x$		$f(x) = \left(\frac{1}{2}\right)^x$	
$f(0.5) \approx 1.41$	$f(-0.5) \approx 0.71$	$f(0.7) \approx 0.62$	$f(-0.8) \approx 1.74$
$f(1.7) \approx 3.25$	$f(-2.6) \approx 0.16$	$f(2.3) \approx 0.20$	$f(-2.1) \approx 4.29$

Use your calculator to check these results. Also, it would be worthwhile for you to go back and see that the points determined do *fit the graphs* in Figures 10.1 and 10.2.

The graphs in Figures 10.1 and 10.2 illustrate a *general behavior* pattern of exponential functions. That is to say, if $b > 1$, then the graph of $f(x) = b^x$ *goes up to the right* and the function is called an **increasing function**. If $0 < b < 1$, then the graph of $f(x) = b^x$ *goes down to the right* and the function is called a **decreasing function**. These facts are illustrated in Figure 10.3 on the next page. Notice that because $b^0 = 1$ for any $b > 0$, all graphs of $f(x) = b^x$ contain the point (0, 1).

As you graph exponential functions, don't forget your previous graphing experiences.

1. The graph of $f(x) = 2^x - 4$ is the graph of $f(x) = 2^x$ *moved down four units.*
2. The graph of $f(x) = 2^{x+3}$ is the graph of $f(x) = 2^x$ *moved three units to the left.*
3. The graph of $f(x) = -2^x$ is the graph of $f(x) = 2^x$ *reflected across the x-axis.*

Figure 10.3

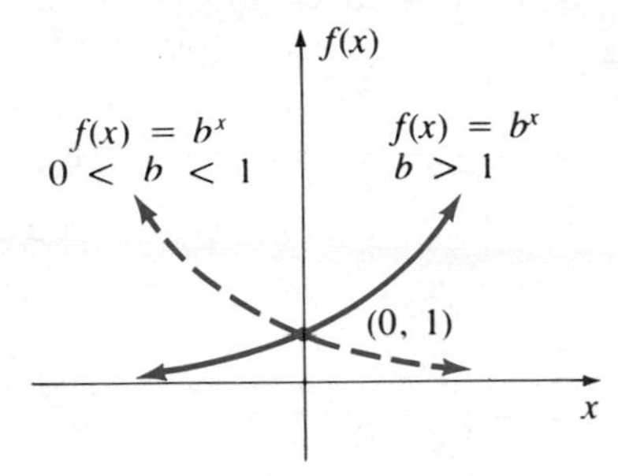

Problem Set 10.1

Solve each of the following equations.

1. $3^x = 27$ $\{3\}$
2. $2^x = 64$ $\{6\}$
3. $2^{2x} = 16$ $\{2\}$
4. $3^{2x} = 81$ $\{2\}$
5. $\left(\frac{1}{4}\right)^x = \frac{1}{256}$ $\{4\}$
6. $\left(\frac{1}{2}\right)^x = \frac{1}{128}$ $\{7\}$
7. $5^{x+2} = 125$ $\{1\}$
8. $4^{x-3} = 16$ $\{5\}$
9. $3^{-x} = \frac{1}{243}$ $\{5\}$
10. $5^{-x} = \frac{1}{25}$ $\{2\}$
11. $6^{3x-1} = 36$ $\{1\}$
12. $2^{2x+3} = 32$ $\{1\}$
13. $4^x = 8$ $\{\frac{3}{2}\}$
14. $16^x = 64$ $\{\frac{3}{2}\}$
15. $8^{2x} = 32$ $\{\frac{5}{6}\}$
16. $9^{3x} = 27$ $\{\frac{1}{2}\}$
17. $\left(\frac{1}{2}\right)^{2x} = 64$ $\{-3\}$
18. $\left(\frac{1}{3}\right)^{5x} = 243$ $\{-1\}$
19. $\left(\frac{3}{4}\right)^x = \frac{64}{27}$ $\{-3\}$
20. $\left(\frac{2}{3}\right)^x = \frac{9}{4}$ $\{-2\}$
21. $9^{4x-2} = \frac{1}{81}$ $\{0\}$
22. $8^{3x+2} = \frac{1}{16}$ $\left\{-\frac{10}{9}\right\}$
23. $6^{2x} + 3 = 39$ $\{1\}$
24. $5^{2x} - 2 = 123$ $\{\frac{3}{2}\}$
25. $10^x = .1$ $\{-1\}$
26. $10^x = .0001$ $\{-4\}$
27. $32^x = \frac{1}{4}$ $\left\{-\frac{2}{5}\right\}$
28. $9^x = \frac{1}{27}$ $\left\{-\frac{3}{2}\right\}$
29. $(2^{x+1})(2^x) = 64$ $\{\frac{5}{2}\}$
30. $(2^{2x-1})(2^{x+2}) = 32$ $\{\frac{4}{3}\}$
31. $(27)(3^x) = 9^x$ $\{3\}$
32. $(3^x)(3^{5x}) = 81$ $\{\frac{2}{3}\}$
33. $(4^x)(16^{3x-1}) = 8$ $\{\frac{1}{2}\}$
34. $(8^{2x})(4^{2x-1}) = 16$ $\{\frac{3}{5}\}$

Graph each of the following exponential functions.

35. $f(x) = 4^x$ see page A72
36. $f(x) = 3^x$ see page A72
37. $f(x) = 6^x$ see page A72
38. $f(x) = 5^x$ see page A72
39. $f(x) = \left(\frac{1}{3}\right)^x$ see page A72
40. $f(x) = \left(\frac{1}{4}\right)^x$ see page A72
41. $f(x) = \left(\frac{3}{4}\right)^x$ see page A72
42. $f(x) = \left(\frac{2}{3}\right)^x$ see page A72

43. $f(x) = 2^{x-2}$ see page A72

44. $f(x) = 2^{x+1}$ see page A72

45. $f(x) = 3^{-x}$ see page A72

46. $f(x) = 2^{-x}$ see page A72

47. $f(x) = 3^{2x}$ see page A73

48. $f(x) = 2^{2x}$ see page A73

49. $f(x) = 3^x - 2$ see page A73

50. $f(x) = 2^x + 1$ see page A73

51. $f(x) = 2^{-x-2}$ see page A73

52. $f(x) = 3^{-x+1}$ see page A73

10.2 Applications of Exponential Functions

Equations that describe exponential functions can represent many real-world situations that exhibit growth or decay. For example, suppose that an economist predicts an annual inflation rate of 5% for the next 10 years. This means an item that presently costs \$8 will cost 8(105%) = 8(1.05) = \$8.40 a year from now. The same item will cost (105%)[8(105%)] = $8(1.05)^2$ = \$8.82 in 2 years. In general, the equation

$$P = P_0(1.05)^t$$

yields the predicted price P of an item in t years at the annual inflation rate of 5%, where that item presently costs P_0. By using this equation, we can look at some future prices based on the prediction of a 5% inflation rate. For example,

1. A \$3.27 container of hot cocoa mix will cost $3.27(1.05)^3$ = \$3.79 in 3 years;
2. A \$4.07 jar of coffee will cost $4.07(1.05)^5$ = \$5.19 in 5 years;
3. A \$9500 car will cost $9500(1.05)^7$ = \$13,367 (rounded to the nearest dollar) in 7 years.

Compound interest provides another illustration of exponential growth. Suppose that \$500 (called the **principal**) is invested at an interest rate of 8% **compounded annually**. The interest earned the first year is \$500(0.08) = \$40 and this amount is added to the original \$500 to form a new principal of \$540 for the second year. The interest earned during the second year is \$540(0.08) = \$43.20 and this amount is added to \$540 to form a new principal of \$583.20 for the third year. Each year a new principal is formed by reinvesting the interest earned that year.

In general, suppose that a sum of money P (called the principal) is invested at an interest rate of r percent compounded annually. The interest earned the first year is Pr and the new principal for the second year is $P + Pr$ or $P(1 + r)$. Note that the new principal for the second year can be found by multiplying the original principal P by $(1 + r)$. In a like fashion, the new principal for the third year can be found by multiplying the previous principal $P(1 + r)$ by $(1 + r)$, thus obtaining $P(1 + r)^2$. If this process is continued, then after t years the *total amount of money accumulated, A,* is given by

$$A = P(1 + r)^t.$$

Consider the following examples of investments made at a certain rate of interest compounded annually.

1. \$750 invested for 5 years at 9% compounded annually produces

$$A = \$750(1.09)^5 = \$1153.97.$$

2. \$1000 invested for 10 years at 11% compounded annually produces

$$A = \$1000(1.11)^{10} = \$2839.42.$$

3. \$5000 invested for 20 years at 12% compounded annually produces

$$A = \$5000(1.12)^{20} = \$48{,}231.47.$$

If we invest money at a certain rate of interest to be *compounded* more than once a year, then we can adjust the basic formula, $A = P(1 + r)^t$, according to the number of compounding periods in the year. For example, for **compounding semiannually**, the formula becomes $A = P\left(1 + \frac{r}{2}\right)^{2t}$ and for **compounding quarterly**, the formula becomes $A = P\left(1 + \frac{r}{4}\right)^{4t}$. In general, we have the following formula for which n represents the number of compounding periods in a year.

$$A = P\left(1 + \frac{r}{n}\right)^{nt}$$

The following examples should clarify the use of this formula.

1. \$750 invested for 5 years at 9% compounded semiannually produces

$$A = \$750\left(1 + \frac{0.09}{2}\right)^{2(5)} = \$750(1.045)^{10} = \$1164.73.$$

2. \$1000 invested for 10 years at 11% compounded quarterly produces

$$A = \$1000\left(1 + \frac{0.11}{4}\right)^{4(10)} = \$1000(1.0275)^{40} = \$2959.87.$$

3. \$5000 invested for 20 years at 12% compounded monthly produces

$$A = \$5000\left(1 + \frac{0.12}{12}\right)^{12(20)} = \$5000(1.01)^{240} = \$54{,}462.77.$$

You may find it interesting to compare these results with those obtained earlier for compounding annually.

Exponential Decay

Suppose it is estimated that the value of a car depreciates 15% per year for the first five years. Thus a car that costs \$9500 will be worth $9500(100\% - 15\%) = 9500(85\%) = 9500(0.85) = \$8075$ in one year. In two years the value of the car will have depreciated to $9500(0.85)^2 = \$6864$(nearest dollar). The equation

$$V = V_0(0.85)^t$$

yields the value V of a car in t years at the annual depreciation rate of 15%, where the car intially cost V_0. By using this equation, we can estimate some car values to the nearest dollar, as follows.

1. A \$6900 car will be worth $6900(0.85)^3 = \$4237$ in 3 years.
2. A \$10,900 car will be worth $10{,}900(0.85)^4 = \$5690$ in 4 years.
3. A \$13,000 car will be worth $\$13{,}000(0.85)^5 = \$5768$ in 5 years.

Another example of exponential decay is associated with radioactive substances. We can describe the rate of decay exponentially, based on the half-life of a substance. The *half-life* of a radioactive substance is the amount of time that it takes for one-half of an initial amount of the substance to disappear as the result of decay. For example, suppose that we have 200 grams of a certain substance that has a half-life of 5 days. After 5 days, $200\left(\frac{1}{2}\right) = 100$ grams remain. After 10 days, $200\left(\frac{1}{2}\right)^2 = 50$ grams remain. After 15 days, $200\left(\frac{1}{2}\right)^3 = 25$ grams remain.

In general, after t days, $200\left(\frac{1}{2}\right)^{\frac{t}{5}}$ grams remain.

This discussion leads us to the following half-life formula. Suppose there is an initial amount, Q_0, of a radioactive substance with a half-life of h. The amount of substance remaining, Q, after a time period of t, is given by the formula

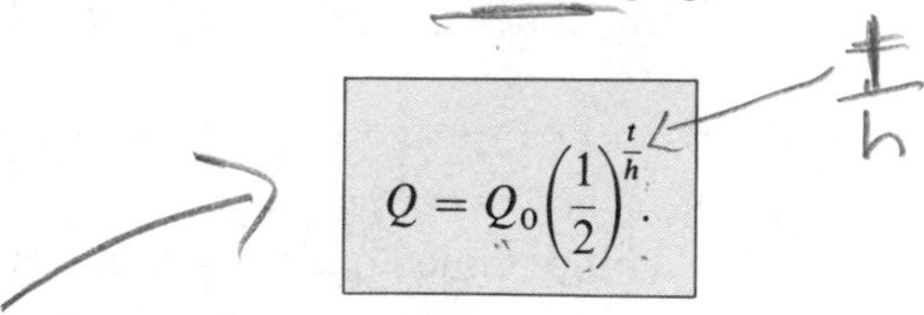

$$Q = Q_0\left(\frac{1}{2}\right)^{\frac{t}{h}}.$$

The units of measure for t and h must be the same.

EXAMPLE 1 Barium-140 has a half-life of 13 days. If there are 500 milligrams of barium initially, how many milligrams remain after 26 days? After 100 days?

Solution Using $Q_0 = 500$ and $h = 13$, the half-life formula becomes

$$Q = 500\left(\frac{1}{2}\right)^{\frac{t}{13}}.$$

If $t = 26$, then

$$
\begin{aligned}
Q &= 500\left(\frac{1}{2}\right)^{\frac{26}{13}} \\
&= 500\left(\frac{1}{2}\right)^{2} \\
&= 500\left(\frac{1}{4}\right) \\
&= 125.
\end{aligned}
$$

So, 125 milligrams remain after 26 days. If $t = 100$, then

$$
\begin{aligned}
Q &= 500\left(\frac{1}{2}\right)^{\frac{100}{13}} \\
&= 500(.5)^{\frac{100}{13}} \\
&= 2.4 \qquad \text{to the nearest tenth of a milligram.}
\end{aligned}
$$

So, approximately 2.4 milligrams remain after 100 days. ■

REMARK In the preface of this book we stated that the calculator is "useful at times, unnecessary at other times." Example 1 clearly demonstrates this point. We solved the first part of the problem very easily without the calculator but it certainly was helpful for the second part of the problem.

The Number *e*

An interesting situation occurs if we consider the compound interest formula for $P = \$1$, $r = 100\%$, and $t = 1$ year. The formula becomes $A = 1\left(1 + \frac{1}{n}\right)^n$. The following table shows some values, rounded to eight decimal places, of $\left(1 + \frac{1}{n}\right)^n$ for different values of n.

n	$\left(n + \frac{1}{n}\right)^n$
1	2.00000000
10	2.59374246
100	2.70481383
1000	2.71692393
10,000	2.71814593
100,000	2.71826824
1,000,000	2.71828047
10,000,000	2.71828169
100,000,000	2.71828181
1,000,000,000	2.71828183

The table suggests that as n increases, the value of $\left(1 + \frac{1}{n}\right)^n$ gets closer and closer to some fixed number. This does happen and the fixed number is called e. To five decimal places, $e = 2.71828$.

The function defined by the equation $f(x) = e^x$ is the **natural exponential function**. It has a great many real-world applications; some we will look at in a moment. First, however, let's get a picture of the natural exponential function. Since $2 < e < 3$, the graph of $f(x) = e^x$ must fall between the graphs of $f(x) = 2^x$ and $f(x) = 3^x$. To be more specific, let's use our calculator to determine a table of values. Use the $\boxed{e^x}$ key, and round the results to the nearest tenth to obtain the following table. Plot the points determined by this table and connect them with a smooth curve to produce Figure 10.4.

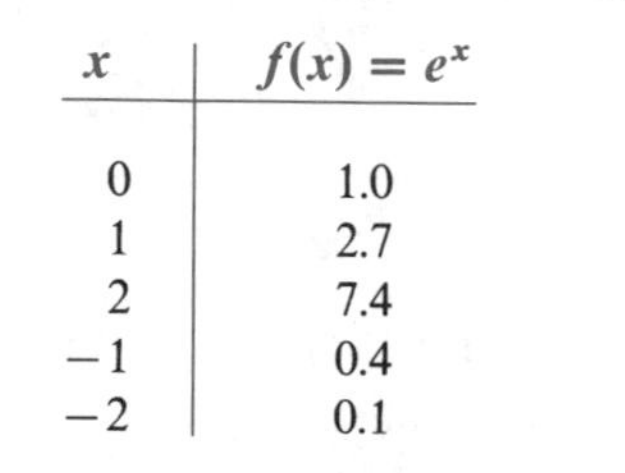

x	$f(x) = e^x$
0	1.0
1	2.7
2	7.4
−1	0.4
−2	0.1

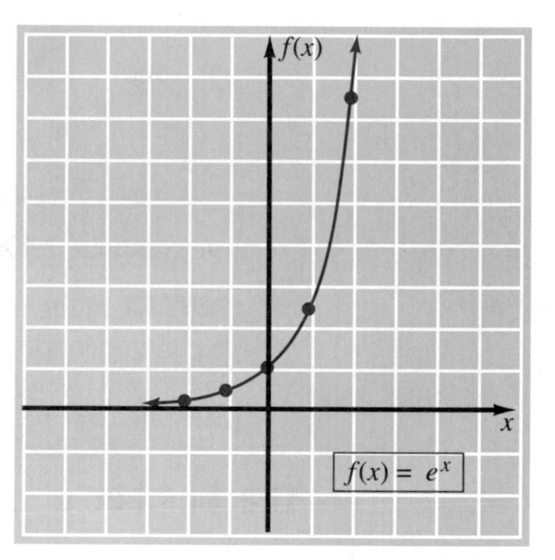

Figure 10.4

Let's return to the concept of compound interest. If the number of compounding periods in a year is increased indefinitely, we arrive at the concept of **compounding continuously**. Mathematically, this can be accomplished by applying the limit concept to the expression $P\left(1 + \frac{r}{n}\right)^{nt}$. We will not show the details here, but the following result is obtained. The formula

$$\boxed{A = Pe^{rt}}$$

yields the accumulated value, A, of a sum of money, P, that has been invested for t years at a rate of r percent compounded continuously. The following examples show the use of the formula.

1. \$750 invested for 5 years at 9% compounded continuously produces

$$A = 750e^{(.09)(5)} = 750e^{.45} = \$1176.23.$$

2. \$1000 invested for 10 years at 11% compounded continuously produces

$$A = 1000e^{(.11)(10)} = 1000e^{1.1} = \$3004.17.$$

3. \$5000 invested for 20 years at 12% compounded continuously produces

$$A = 5000e^{(.12)(20)} = 5000e^{2.4} = \$55{,}115.88.$$

Again you may find it interesting to compare these results with those you obtained earlier when using a different number of compounding periods.

The ideas behind compounding continously carry over to other growth situations. The law of exponential growth

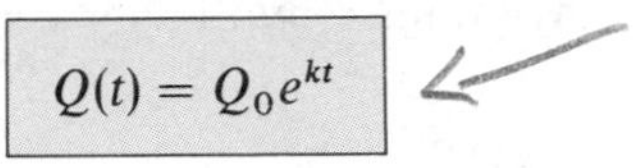

$$Q(t) = Q_0e^{kt}$$

is used as a mathematical model for numerous growth-and-decay applications. In this equation, $Q(t)$ represents the quantity of a given substance at any time t; Q_0 is the initial amount of the substance (when $t = 0$); and k is a constant that depends on the particular application. If $k < 0$, then $Q(t)$ decreases as t increases, and we refer to the model as the **law of decay**.

Let's consider some growth-and-decay applications.

EXAMPLE 2 Suppose that in a certain culture the equation $Q(t) = 15000e^{.3t}$ expresses the number of bacteria present as a function of the time t, where t is expressed in hours. Find (a) the initial number of bacteria, and (b) the number of bacteria after 3 hours.

Solution

(a) The initial number of bacteria is produced when $t = 0$.

$$\begin{aligned} Q(0) &= 15000e^{.3(0)} \\ &= 15000e^{0} \\ &= 15000 \qquad e^0 = 1 \end{aligned}$$

(b)

$$\begin{aligned} Q(3) &= 15000e^{.3(3)} \\ &= 15000e^{.9} \\ &= 36894 \qquad \text{nearest whole number} \end{aligned}$$

Therefore, there should be approximately 36,894 bacteria present after 3 hours. ■

EXAMPLE 3 Suppose the number of bacteria present in a certain culture after t minutes is given by the equation $Q(t) = Q_0e^{.05t}$, where Q_0 represents the intitial number of bacteria. If 5000 bacteria are present after 20 minutes, how many bacteria were present initially?

Solution If 5000 bacteria are present after 20 minutes, then $Q(20) = 5000$.

$$5000 = Q_0 e^{.05(20)}$$

$$5000 = Q_0 e^1$$

$$\frac{5000}{e} = Q_0$$

$$1839 = Q_0 \qquad \text{nearest whole number}$$

Therefore, there were approximately 1839 bacteria present initially. ■

EXAMPLE 4 The number of grams Q of a certain radioactive substance present after t seconds is given by $Q(t) = 200e^{-.3t}$. How many grams remain after 7 seconds?

Solution We need to evaluate $Q(7)$.

$$Q(7) = 200e^{-.3(7)}$$

$$= 200e^{-2.1}$$

$$= 24 \qquad \text{nearest whole number}$$

Thus, approximately 24 grams remain after 7 seconds. ■

Problem Set 10.2

1. Assuming that the rate of inflation is 4% per year, the equation $P = P_0(1.04)^t$ yields the predicted price P of an item in t years that presently costs P_0. Find the predicted price of each of the following items for the indicated years ahead.

(a) $0.55 can of soup in 3 years $.62

(b) $3.43 container of cocoa mix in 5 years $4.17

(c) $1.76 jar of coffee creamer in 4 years $2.06

(d) $0.44 can of beans and bacon in 10 years $.65

(e) $9000 car in 5 years (nearest dollar) $10,950

(f) $50,000 house in 8 years (nearest dollar) $68,428

(g) $500 TV set in 7 years (nearest dollar) $658

2. Suppose it is estimated that the value of a car depreciates 30% per year for the first 5 years. The equation $A = P_0(.7)^t$ yields the value (A) of a car after t years if the original price is P_0. Find the value (to the nearest dollar) of each of the following priced cars after the indicated time.

(a) $9000 car after 4 years $2161

(b) $5295 car after 2 years $2595

(c) $6395 car after 5 years $1075

(d) $15,595 car after 3 years $5349

For Problems 3–14, use the formula $A = P\left(1 + \frac{r}{n}\right)^{nt}$ to find the total amount of money accumulated at the end of the indicated time period for each of the following investments.

3. \$200 for 6 years at 6% compounded annually \$283.70
4. \$250 for 5 years at 7% compounded annually \$350.64
5. \$500 for 7 years at 8% compounded semiannually \$865.84
6. \$750 for 8 years at 8% compounded semiannually \$1404.74
7. \$800 for 9 years at 9% compounded quarterly \$1782.25
8. \$1200 for 10 years at 10% compounded quarterly \$3222.08
9. \$1500 for 5 years at 12% compounded monthly \$2725.05
10. \$2000 for 10 years at 9% compounded monthly \$4902.71
11. \$5000 for 15 years at 8.5% compounded annually \$16998.71
12. \$7500 for 20 years at 9.5% compounded semiannually \$47997.93
13. \$8000 for 10 years at 10.5% compounded quarterly \$22553.65
14. \$10,000 for 25 years at 9.25% compounded monthly \$100104.83

For Problems 15–23, use the formula $A = Pe^{rt}$ to find the total amount of money accumulated at the end of the indicated time period by compounding continuously.

15. \$400 for 5 years at 7% \$567.63
16. \$500 for 7 years at 6% \$760.98
17. \$750 for 8 years at 8% \$1422.36
18. \$1000 for 10 years at 9% \$2459.60
19. \$2000 for 15 years at 10% \$8963.38
20. \$5000 for 20 years at 11% \$45125.07
21. \$7500 for 10 years at 8.5% \$17547.35
22. \$10,000 for 25 years at 9.25% \$100996.42
23. \$15,000 for 10 years at 7.75% \$32558.88

24. Complete the following chart that illustrates what happens to \$1000 invested at various rates of interest for different lengths of time but always compounded continuously. Round your answers to the nearest dollar.

\$1000 compounded continuously

	8%	*10%*	*12%*	*14%*
5 years	\$1492	\$ 1649	\$ 1822	\$ 2014
10 years	2226	2718	3320	4055
15 years	3320	4482	6050	8166
20 years	4953	7389	11023	16445
25 years	7389	12182	20086	33115

25. Complete the following chart that illustrates what happens to \$1000 invested at 12% for different lengths of time and different numbers of compounding periods. Round all of your answers to the nearest dollar.

\$1000 at 12%

	1 year	*5 years*	*10 years*	*20 years*
Compounded annually	\$1120	\$1762	\$3106	\$ 9646
Compounded semiannually	1124	1791	3207	10286
Compounded quarterly	1126	1806	3262	10641
Compounded monthly	1127	1817	3300	10893
Compounded continuously	1127	1822	3320	11023

26. Complete the following chart that illustrates what happens to $1000 in 10 years based on different rates of interest and different numbers of compounding periods. Round your answers to the nearest dollar.

$1000 for 10 years

	8%	*10%*	*12%*	*14%*
Compounded annually	$2159	$2594	$3106	$3707
Compounded semiannually	2191	2653	3207	3870
Compounded quarterly	2208	2685	3262	3959
Compounded monthly	2220	2707	3300	4022
Compounded continuously	2226	2718	3320	4055

27. Suppose that Nora invested $500 at 8.25% compounded annually for 5 years and Patti invested $500 at 8% compounded quarterly for 5 years. At the end of 5 years, who will have the most money and by how much? Nora will have $.24 more

28. Two years ago Daniel invested some money at 8% interest compounded annually. Today it is worth $758.16. How much did he invest two years ago? $650

29. What rate of interest (nearest hundredth of a percent) is needed so that an investment of $2500 will yield $3000 in 2 years if the money is compounded annually? 9.54%

30. Suppose that a certain radioactive substance has a half-life of 20 years. If there are presently 2500 milligrams of the substance, how much, to the nearest milligram, will remain after 40 years? After 50 years? 625 milligrams; 442 milligrams

31. Strontium-90 has a half-life of 29 years. If there are 400 grams of strontium initially, how much, to the nearest gram, will remain after 87 years? After 100 years? 50 grams; 37 grams

32. The half-life of radium is approximately 1600 years. If the present amount of radium in a certain location is 500 grams, how much will remain after 800 years? Express your answer to the nearest gram. 354 grams

For Problems 33–38, graph each of the exponential functions.

33. $f(x) = e^x + 1$ see page A73

34. $f(x) = e^x - 2$ see page A73

35. $f(x) = 2e^x$ see page A73

36. $f(x) = -e^x$ see page A74

37. $f(x) = e^{2x}$ see page A74

38. $f(x) = e^{-x}$ see page A74

For Problems 39–44, express your answers to the nearest whole number.

39. Suppose that in a certain culture, the equation $Q(t) = 1000e^{.4t}$ expresses the number of bacteria present as a function of the time t, where t is expressed in hours. How many bacteria are present at the end of 2 hours? 3 hours? 5 hours? 2226; 3320; 7389

40. The number of bacteria present at a given time under certain conditions is given by the equation $Q = 5000e^{0.05t}$, where t is expressed in minutes. How many bacteria are present at the end of 10 minutes? 30 minutes? 1 hour? 8244; 22408, 100428

41. The number of bacteria present in a certain culture after t hours is given by the equation $Q = Q_0e^{0.3t}$, where Q_0 represents the initial number of bacteria. If 6640 bacteria are present after 4 hours, how many bacteria were present initially? 2000

42. The number of grams Q of a certain radioactive substance present after t seconds is given by the equation $Q = 1500e^{-0.4t}$. How many grams remain after 5 seconds? 10 seconds? 20 seconds? 203; 27; 1

43. Suppose that the present population of a city is 75,000. Using the equation $P(t) = 75000e^{.01t}$ to estimate future growth, estimate the population

(a) 10 years from now 82,888

(b) 15 years from now, and 87,138

(c) 25 years from now. 96, 302

44. Suppose that the present population of a city is 150,000. Use the equation $P(t) = 150000e^{.032t}$ to estimate future growth. Estimate the population

(a) 10 years from now, 206,569

(b) 20 years from now, and 284,472

(c) 30 years from now. 391,754

45. The atmospheric pressure, measured in pounds per square inch, is a function of the altitude above sea level. The equation $P(a) = 14.7e^{-.21a}$, where a is the altitude measured in miles, can be used to approximate atmospheric pressure. Find the atmospheric pressure at each of the following locations. Express each answer to the nearest tenth of a pound per square inch.

(a) Mount McKinley in Alaska—altitude of 3.85 miles; 6.5 lbs per sq. inch

(b) Denver, Colorado—the "mile-high" city; 11.9 lbs per sq. inch

(c) Asheville, North Carolina—altitude of 1985 feet (5280 feet = 1 mile); 13.6 lbs per sq. inch

(d) Phoenix, Arizona—altitude of 1090 feet. 14.1 lbs per sq. inch

10.3 Logarithms

In Sections 10.1 and 10.2, (1) we learned about exponential expressions of the form b^n, where b is any positive real number and n is any real number, (2) we used exponential expressions of the form b^n to define exponential functions, and (3) we used exponential functions to help solve problems. In the next three sections we will follow the same basic pattern with respect to a new concept, that of a **logarithm**. Let's begin with the following definition.

Definition 10.2

> If r is any positive real number, then the unique exponent t such that $b^t = r$ is called the **logarithm of r with base b** and is denoted by $\log_b r$.

According to Definition 10.2, the logarithm of 8 base 2 is the exponent t such that $2^t = 8$; thus, we can write $\log_2 8 = 3$. Likewise, we can write $\log_{10} 100 = 2$ because $10^2 = 100$. In general, we can remember Definition 10.2 in terms of the statement

> $\log_b r = t$ is equivalent to $b^t = r$.

Thus, we can easily switch back and forth between exponential and logarithmic forms of equations, as the next examples illustrate.

$$\log_3 81 = 4 \text{ is equivalent to } 3^4 = 81,$$

$$\log_{10} 100 = 2 \text{ is equivalent to } 10^2 = 100,$$

$$\log_{10} 0.001 = -3 \text{ is equivalent to } 10^{-3} = 0.001,$$

$$\log_2 128 = 7 \text{ is equivalent to } 2^7 = 128,$$

$$2^4 = 16 \text{ is equivalent to } \log_2 16 = 4,$$

$$5^2 = 25 \text{ is equivalent to } \log_5 25 = 2,$$

$$\left(\frac{1}{2}\right)^4 = \frac{1}{16} \text{ is equivalent to } \log_{\frac{1}{2}}\left(\frac{1}{16}\right) = 4,$$

$$10^{-2} = 0.01 \text{ is equivalent to } \log_{10} 0.01 = -2$$

We can conveniently calculate some logarithms by changing to exponential form as in the next examples.

EXAMPLE 1 Evaluate $\log_4 64$.

Solution Let $\log_4 64 = x$. Then by switching to exponential form we have $4^x = 64$, which we can solve as we did back in Section 10.1.

$$4^x = 64$$

$$4^x = 4^3$$

$$x = 3$$

Therefore, we can write $\log_4 64 = 3$. ■

EXAMPLE 2 Evaluate $\log_{10} 0.1$.

Solution Let $\log_{10} 0.1 = x$. Then by switching to exponential form we have $10^x = 0.1$, which we can solve as follows.

$$10^x = 0.1$$

$$10^x = \frac{1}{10}$$

$$10^x = 10^{-1}$$

$$x = -1$$

Thus, we obtain $\log_{10} 0.1 = -1$ ■

The link between logarithms and exponents also provides the basis for solving some equations that involve logarithms, as the next two examples illustrate.

EXAMPLE 3 Solve $\log_8 x = \frac{2}{3}$.

Solution

$$\log_8 x = \frac{2}{3}$$

$$8^{\frac{2}{3}} = x \quad \text{by switching to exponential form}$$

$$\sqrt[3]{8^2} = x$$

$$(\sqrt[3]{8})^2 = x$$

$$4 = x$$

The solution set is $\{4\}$. ■

EXAMPLE 4 Solve $\log_b 1000 = 3$.

Solution

$$\log_b 1000 = 3$$

$$b^3 = 1000$$

$$b = 10$$

The solution set is $\{10\}$. ■

Properties of Logarithms

There are some properties of logarithms that are a direct consequence of Definition 10.2 and our knowledge of exponents. For example, by writing the exponential equations $b^1 = b$ and $b^0 = 1$ in logarithmic form, the following property is obtained.

PROPERTY 10.3

For $b > 0$ and $b \neq 1$,

1. $\log_b b = 1$
2. $\log_b 1 = 0$.

Thus, we can write

$$\log_{10} 10 = 1,$$

$$\log_2 2 = 1,$$

$$\log_{10} 1 = 0,$$

$$\log_5 1 = 0.$$

By Definition 10.2, $\log_b r$ is the exponent t such that $b^t = r$. Therefore, raising b to the $\log_b r$ power must produce r. We state this fact in Property 10.4.

PROPERTY 10.4

> For $b > 0$, $b \neq 1$, and $r > 0$
>
> $$b^{\log_b r} = r.$$

The following examples illustrate Property 10.4.

$$10^{\log_{10} 19} = 19,$$

$$2^{\log_2 14} = 14,$$

$$e^{\log_e 5} = 5.$$

Because a logarithm is by definition an exponent, it would seem reasonable to predict that there are some properties of logarithms that correspond to the basic exponential properties. This is an accurate prediction; these properties provide a basis for computational work with logarithms. Let's state the first of these properties and show how it can be verified by using our knowledge of exponents.

PROPERTY 10.5

> For positive real numbers b, r, and s where $b \neq 1$,
>
> $$\log_b rs = \log_b r + \log_b s.$$

To verify Property 10.5 we can proceed as follows. Let $m = \log_b r$ and $n = \log_b s$. Change each of these equations to exponential form.

$m = \log_b r$ becomes $r = b^m$,

$n = \log_b s$ becomes $s = b^n$

Thus, the product rs becomes

$$rs = b^m \cdot b^n = b^{m+n}.$$

Now, by changing $rs = b^{m+n}$ back to logarithmic form, we obtain

$$\log_b rs = m + n.$$

Replacing m with $\log_b r$ and n with $\log_b s$ yields

$$\log_b rs = \log_b r + \log_b s.$$

The following three examples demonstrate a use of Property 10.5.

EXAMPLE 5 If $\log_2 5 = 2.3222$ and $\log_2 3 = 1.5850$, evaluate $\log_2 15$.

Solution Because $15 = 5 \cdot 3$, we can apply Property 10.5 as follows.

$$\begin{aligned} \log_2 15 &= \log_2(5 \cdot 3) \\ &= \log_2 5 + \log_2 3 \\ &= 2.3222 + 1.5850 = 3.9072 \end{aligned}$$

■

EXAMPLE 6 If $\log_{10} 178 = 2.2504$ and $\log_{10} 89 = 1.9494$, evaluate $\log_{10}(178 \cdot 89)$.

Solution

$$\begin{aligned} \log_{10}(178 \cdot 89) &= \log_{10} 178 + \log_{10} 89 \\ &= 2.2504 + 1.9494 = 4.1998 \end{aligned}$$

■

EXAMPLE 7 If $\log_3 8 = 1.8928$, then evaluate $\log_3 72$.

Solution

$$\begin{aligned} \log_3 72 &= \log_3(9 \cdot 8) \\ &= \log_3 9 + \log_3 8 \\ &= 2 + 1.8928 \qquad \log_3 9 = 2 \text{ because } 3^2 = 9 \\ &= 3.8928 \end{aligned}$$

■

Since $\frac{b^m}{b^n} = b^{m-n}$, we would expect a corresponding property pertaining to logarithms. There is such a property, Property 10.6.

PROPERTY 10.6

> For positive numbers b, r, and s, where $b \neq 1$,
>
> $$\log_b\left(\frac{r}{s}\right) = \log_b r - \log_b s.$$

This property can be verified by using an approach similar to the one we used to verify Property 10.5. We leave it for you to do in an exercise in the next problem set.

We can use Property 10.6 to change a division problem into a subtraction problem as in the next two examples.

EXAMPLE 8 If $\log_5 36 = 2.2265$ and $\log_5 4 = 0.8614$, evaluate $\log_5 9$.

Solution Since $9 = \frac{36}{4}$, we can use Property 10.6 as follows.

$$\log_5 9 = \log_5\left(\frac{36}{4}\right)$$
$$= \log_5 36 - \log_5 4$$
$$= 2.2265 - 0.8614 = 1.3651$$

■

EXAMPLE 9 Evaluate $\log_{10}\left(\frac{379}{86}\right)$ given that $\log_{10} 379 = 2.5786$ and $\log_{10} 86 = 1.9345$.

Solution

$$\log_{10}\left(\frac{379}{86}\right) = \log_{10} 379 - \log_{10} 86$$
$$= 2.5786 - 1.9345$$
$$= 0.6441$$

■

The next property of logarithms provides the basis for evaluating expressions such as $3^{\sqrt{2}}$, $(\sqrt{5})^{\frac{2}{3}}$, and $(0.076)^{\frac{3}{4}}$. Here follows the property, a basis for its justification, and illustrations of its use.

PROPERTY 10.7

> If r is a positive real number, b is a positive real number other than 1, and p is any real number, then
>
> $$\log_b r^p = p(\log_b r).$$

As you might expect, the exponential property $(b^n)^m = b^{mn}$ plays an important role in the verification of Property 10.7. This is an exercise for you in the next problem set. Let's look at some uses of Property 10.7.

EXAMPLE 10 Evaluate $\log_2 22^{\frac{1}{3}}$ given that $\log_2 22 = 4.4598$.

Solution

$$\log_2 22^{\frac{1}{3}} = \frac{1}{3}\log_2 22 \qquad \text{Property 10.7}$$
$$= \frac{1}{3}(4.4598)$$
$$= 1.4866$$

■

EXAMPLE 11 Evaluate $\log_{10}(8540)^{\frac{3}{5}}$ given that $\log_{10} 8540 = 3.9315$.

Solution

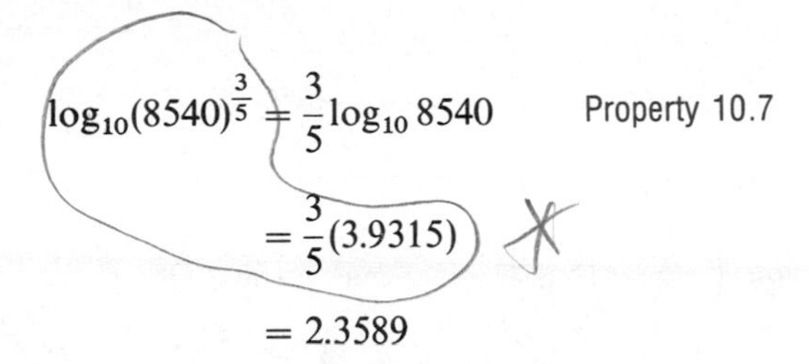

$$\log_{10}(8540)^{\frac{3}{5}} = \frac{3}{5}\log_{10} 8540 \qquad \text{Property 10.7}$$

$$= \frac{3}{5}(3.9315)$$

$$= 2.3589$$

■

Working together, the properties of logarithms allow us to change the forms of various logarithmic expressions. For example, an expression such as $\log_b \sqrt{\frac{xy}{z}}$ can be rewritten in terms of sums and differences of simpler logarithmic quantities as follows.

$$\log_b \sqrt{\frac{xy}{z}} = \log_b \left(\frac{xy}{z}\right)^{\frac{1}{2}}$$

$$= \frac{1}{2}\log_b\left(\frac{xy}{z}\right) \qquad \text{Property 10.7}$$

$$= \frac{1}{2}(\log_b xy - \log_b z) \qquad \text{Property 10.6}$$

$$= \frac{1}{2}(\log_b x + \log_b y - \log_b z) \qquad \text{Property 10.5}$$

Sometimes we need to change from an indicated sum or difference of logarithmic quantities to an indicated product or quotient. This is especially helpful when solving certain kinds of equations that involve logarithms. Note in these next two examples how we can use the properties, along with the process of changing from logarithmic form to exponential form, to solve some equations.

EXAMPLE 12 Solve $\log_{10} x + \log_{10}(x + 9) = 1$.

Solution

$$\log_{10} x + \log_{10}(x + 9) = 1$$

$$\log_{10}[x(x + 9)] = 1 \qquad \text{Property 10.5}$$

$$10^1 = x(x + 9) \qquad \text{change to exponential form}$$

$$10 = x^2 + 9x$$

$$0 = x^2 + 9x - 10$$

$$0 = (x + 10)(x - 1)$$

$$x + 10 = 0 \quad \text{or} \quad x - 1 = 0$$

$$x = -10 \quad \text{or} \quad x = 1$$

Since logarithms are defined only for positive numbers, x and $x + 9$ have to be positive. Therefore, the solution of -10 must be discarded. The solution set is $\{1\}$.

■

EXAMPLE 13 Solve $\log_5(x + 4) - \log_5 x = 2$.

Solution

$$\log_5(x + 4) - \log_5 x = 2$$

$$\log_5\left(\frac{x + 4}{x}\right) = 2 \qquad \text{Property 10.6}$$

$$5^2 = \frac{x + 4}{x} \qquad \text{change to exponential form}$$

$$25 = \frac{x + 4}{x}$$

$$25x = x + 4$$

$$24x = 4$$

$$x = \frac{4}{24} = \frac{1}{6}$$

The solution set is $\left\{\frac{1}{6}\right\}$. ■

Problem Set 10.3

Write each of the following in exponential form. For example, $\log_2 8 = 3$ becomes $2^3 = 8$ in logarithmic form.

1. $2^7 = 128$ $\log_2 128 = 7$
2. $3^3 = 27$ $\log_3 27 = 3$
3. $5^3 = 125$ $\log_5 125 = 3$
4. $2^6 = 64$ $\log_2 64 = 6$
5. $10^3 = 1000$ $\log_{10} 1000 = 3$
6. $10^1 = 10$ $\log_{10} 10 = 1$
7. $2^{-2} = \frac{1}{4}$ $\log_2\left(\frac{1}{4}\right) = -2$
8. $3^{-4} = \frac{1}{81}$ $\log_3\left(\frac{1}{81}\right) = -4$
9. $10^{-1} = .1$ $\log_{10} .1 = -1$
10. $10^{-2} = .01$ $\log_{10} .01 = -2$

Write each of the following in exponential form. For example, $\log_2 8 = 3$ becomes $2^3 = 8$ in exponential form.

11. $\log_3 81 = 4$ $3^4 = 81$
12. $\log_2 256 = 8$ $2^8 = 256$
13. $\log_4 64 = 3$ $4^3 = 64$
14. $\log_5 25 = 2$ $5^2 = 25$
15. $\log_{10} 10000 = 4$ $10^4 = 10000$
16. $\log_{10} 100000 = 5$ $10^5 = 100000$
17. $\log_2\left(\frac{1}{16}\right) = -4$ $2^{-4} = \frac{1}{16}$
18. $\log_5\left(\frac{1}{125}\right) = -3$ $5^{-3} = \frac{1}{125}$
19. $\log_{10} .001 = -3$ $10^{-3} = .001$
20. $\log_{10} .000001 = -6$ $10^{-6} = .000001$

Evaluate each of the following.

21. $\log_2 16$ 4
22. $\log_3 9$ 2
23. $\log_3 81$ 4
24. $\log_2 512$ 9
25. $\log_6 216$ 3
26. $\log_4 256$ 4
27. $\log_7 \sqrt{7}$ $\frac{1}{2}$
28. $\log_2 \sqrt[3]{2}$ $\frac{1}{3}$

29. $\log_{10} 1$ 0

30. $\log_{10} 10$ 1

31. $\log_{10} .1$ -1

32. $\log_{10} .0001$ -4

33. $10^{\log_{10} 5}$ 5

34. $10^{\log_{10} 14}$ 14

35. $\log_2\left(\frac{1}{32}\right)$ -5

36. $\log_5\left(\frac{1}{25}\right)$ -2

37. $\log_5(\log_2 32)$ 1

38. $\log_2(\log_4 16)$ 1

39. $\log_{10}(\log_7 7)$ 0

40. $\log_2(\log_5 5)$ 0

Solve each of the following equations.

41. $\log_7 x = 2$ $\{49\}$

42. $\log_2 x = 5$ $\{32\}$

43. $\log_8 x = \frac{4}{3}$ $\{16\}$

44. $\log_{16} x = \frac{3}{2}$ $\{64\}$

45. $\log_9 x = \frac{3}{2}$ $\{27\}$

46. $\log_8 x = -\frac{2}{3}$ $\left\{\frac{1}{4}\right\}$

47. $\log_4 x = -\frac{3}{2}$ $\left\{\frac{1}{8}\right\}$

48. $\log_9 x = -\frac{5}{2}$ $\left\{\frac{1}{243}\right\}$

49. $\log_x 2 = \frac{1}{2}$ $\{4\}$

50. $\log_x 3 = \frac{1}{2}$ $\{9\}$

Given that $\log_2 5 = 2.3219$ and $\log_2 7 = 2.8074$, evaluate each of the following by using Properties 10.5–10.7.

51. $\log_2 35$ 5.1293

52. $\log_2\left(\frac{7}{5}\right)$ .4855

53. $\log_2 125$ 6.9657

54. $\log_2 49$ 5.6148

55. $\log_2 \sqrt{7}$ 1.4037

56. $\log_2 \sqrt[3]{5}$ .7740

57. $\log_2 175$ 7.4512

58. $\log_2 56$ 5.8074

59. $\log_2 80$ 6.3219

Given that $\log_8 5 = .7740$ and $\log_8 11 = 1.1531$, evaluate each of the following using Properties 10.5–10.7.

60. $\log_8 55$ 1.9271

61. $\log_8\left(\frac{5}{11}\right)$ $-.3791$

62. $\log_8 25$ 1.5480

63. $\log_8 \sqrt{11}$ .5766

64. $\log_8 (5)^{\frac{2}{3}}$ .5160

65. $\log_8 88$ 2.1531

66. $\log_8 320$ 2.7740

67. $\log_8\left(\frac{25}{11}\right)$ .3949

68. $\log_8\left(\frac{121}{25}\right)$ .7582

Express each of the following as the sum or difference of simpler logarithmic quantities. Assume that all variables represent positive real numbers. For example,

$$\log_b \frac{x^3}{y^2} = \log_b x^3 - \log_b y^2$$

$$= 3\log_b x - 2\log_b y$$

69. $\log_b xyz$ $\log_b x + \log_b y + \log_b z$

70. $\log_b 5x$ $\log_b 5 + \log_b x$

71. $\log_b\left(\frac{y}{z}\right)$ $\log_b y - \log_b z$

72. $\log_b\left(\frac{x^2}{y}\right)$ $2\log_b x - \log_b y$

73. $\log_b y^3 z^4$ $3\log_b y + 4\log_b z$

74. $\log_b x^2 y^3$ $2\log_b x + 3\log_b y$

75. $\log_b\left(\frac{x^{\frac{1}{2}} y^{\frac{1}{3}}}{z^4}\right)$ $\frac{1}{2}\log_b x + \frac{1}{3}\log_b y - 4\log_b z$

76. $\log_b x^{\frac{2}{3}} y^{\frac{3}{4}}$ $\frac{2}{3}\log_b x + \frac{3}{4}\log_b y$

77. $\log_b \sqrt[3]{x^2 z}$ $\frac{2}{3}\log_b x + \frac{1}{3}\log_b z$

78. $\log_b \sqrt{xy}$ $\frac{1}{2}\log_b x + \frac{1}{2}\log_b y$

79. $\log_b\left(x\sqrt{\frac{x}{y}}\right)$ $\frac{3}{2}\log_b x - \frac{1}{2}\log_b y$

80. $\log_b \sqrt{\frac{x}{y}}$ $\frac{1}{2}\log_b x - \frac{1}{2}\log_b y$

For Problems 81–92, solve each of the equations.

81. $\log_3 x + \log_3 4 = 2$ $\{\frac{9}{4}\}$

82. $\log_7 5 + \log_7 x = 1$ $\{\frac{7}{5}\}$

83. $\log_{10} x + \log_{10}(x - 21) = 2$ $\{25\}$

84. $\log_{10} x + \log_{10}(x - 3) = 1$ $\{5\}$

85. $\log_2 x + \log_2 (x - 3) = 2$ $\{4\}$

86. $\log_3 x + \log_3 (x - 2) = 1$ $\{3\}$

87. $\log_{10} (2x - 1) - \log_{10} (x - 2) = 1$ $\{\frac{19}{8}\}$

88. $\log_{10} (9x - 2) = 1 + \log_{10} (x - 4)$ $\{38\}$

89. $\log_5 (3x - 2) = 1 + \log_5 (x - 4)$ $\{9\}$

90. $\log_6 x + \log_6 (x + 5) = 2$ $\{4\}$

91. $\log_8 (x + 7) + \log_8 x = 1$ $\{1\}$

92. $\log_6 (x + 1) + \log_6 (x - 4) = 2$ $\{8\}$

93. Verify Property 10.6.

94. Verify Property 10.7.

10.4 Logarithmic Functions

We can now use the concept of a logarithm to define a new function as follows.

DEFINITION 10.3

If $b > 0$ and $b \neq 1$, then the function f defined by

$$f(x) = \log_b x,$$

where x is any positive real number, is called the **logarithmic function with base *b*.**

We can obtain the graph of a specific logarithmic function in various ways. For example, we can change the equation $y = \log_2 x$ to the exponential equation $2^y = x$, where we can determine a table of values. The next set of exercises asks you to graph some logarithmic functions with this approach.

We can obtain the graph of a logarithmic function by setting up a table of values directly from the logarithmic equation. Example 1 illustrates this approach.

EXAMPLE 1 Graph $f(x) = \log_2 x$.

Solution Let's choose some values for x where the corresponding values for $\log_2 x$ are easily determined. (Remember that logarithms are only defined for the positive real numbers.)

x	$f(x)$	
$\frac{1}{8}$	-3	$\log_2 \frac{1}{8} = -3$ because $2^{-3} = \frac{1}{2^3} = \frac{1}{8}$
$\frac{1}{4}$	-2	
$\frac{1}{2}$	-1	
1	0	$\log_2 1 = 0$ because $2^0 = 1$
2	1	
4	2	
8	3	

Plot these points and connect them with a smooth curve to produce Figure 10.5.

Figure 10.5

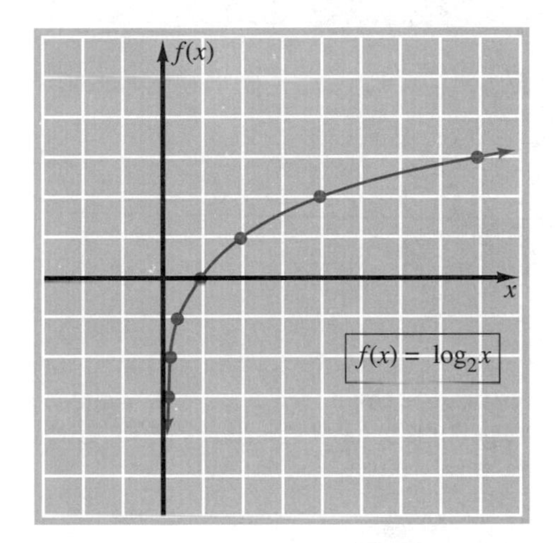

■

Suppose that we consider two functions f and g as follows.

$f(x) = b^x$	*Domain*: all real numbers *Range*: positive real numbers.
$g(x) = \log_b x$	*Domain*: positive real numbers *Range*: all real numbers.

Furthermore, suppose that we consider the composition of f and g, and the composition of g and f.

$$(f \circ g)(x) = f(g(x)) = f(\log_b x) = b^{\log_b x} = x,$$
$$(g \circ f)(x) = g(f(x)) = g(b^x) = \log_b b^x = x \log_b b = x(1) = x$$

Therefore, because the domain of f is the range of g, the range of f is the domain of g, $f(g(x)) = x$, and $g(f(x)) = x$, *the two functions f and g are inverses of each other.*

Remember also from Chapter 8 that the graphs of a function and its inverse are reflections of each other through the line $y = x$. Thus, the graph of a logarithmic function can also be determined by reflecting the graph of its inverse exponential function through the line $y = x$. We see this in Figure 10.6, where the graph of $y = 2^x$ has been reflected across the line $y = x$ to produce the graph of $y = \log_2 x$.

Figure 10.6

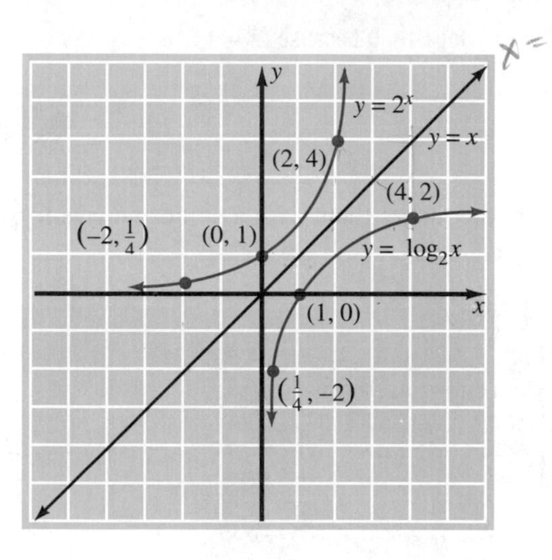

The *general behavior* patterns of exponential functions were illustrated by two graphs back in Figure 10.3. We can now reflect each of those graphs through the line $y = x$ and observe the general behavior patterns of logarithmic functions as shown in Figure 10.7.

Figure 10.7

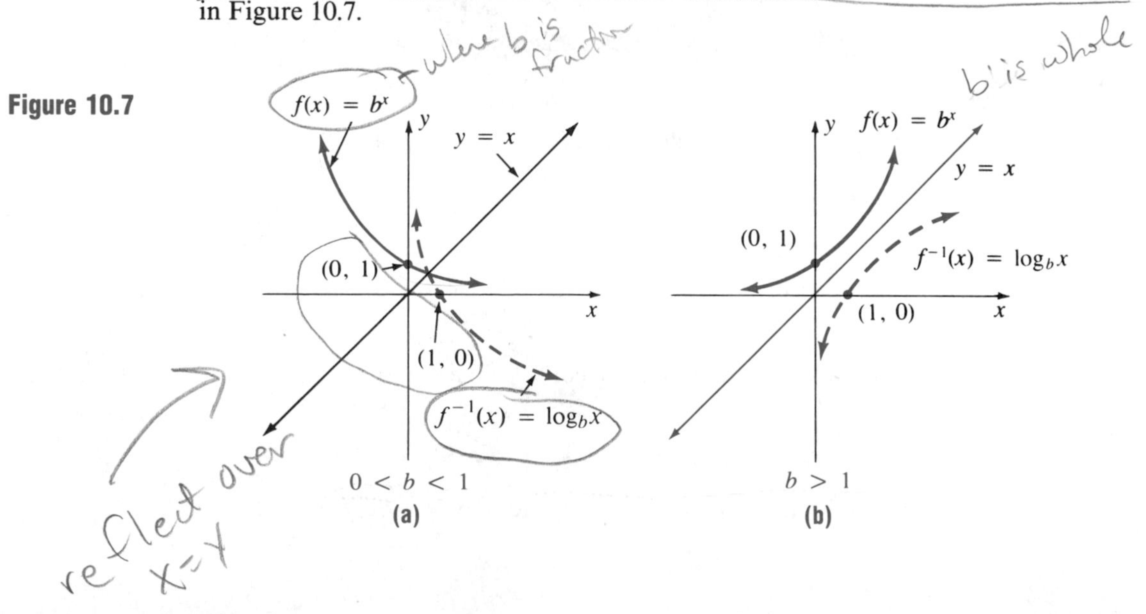

Finally, when graphing logarithmic functions, don't forget about variations of the basic curves.

1. The graph of $f(x) = 3 + \log_2 x$ is the graph of $f(x) = \log_2 x$ *moved up three units.* (Since $\log_2 x + 3$ is apt to be confused with $\log_2(x + 3)$, we commonly write $3 + \log_2 x$.)
2. The graph of $f(x) = \log_2(x - 4)$ is the graph of $f(x) = \log_2 x$ *moved four units to the right.*
3. The graph of $f(x) = -\log_2 x$ is the graph of $f(x) = \log_2 x$ *reflected across the x-axis.*

Common Logarithms—Base 10

The properties of logarithms we discussed in Section 10.3 are true for any valid base. For example, since the Hindu-Arabic numeration system that we use is a base-10 system, logarithms to base 10 have historically been used for computational purposes. Base-10 logarithms are called **common logarithms**.

Originally, common logarithms were developed to assist in complicated numerical calculations that involved products, quotients, and powers of real numbers. Today they are seldom used for that purpose because the calculator and computer can much more effectively handle the messy computational problems. However, common logarithms do still occur in applications; they are deserving of our attention.

REMARK In Appendix B we have included a short discussion relative to the computational aspects of common logarithms. You may find it interesting to browse through this material. It should enhance your appreciation of the calculator.

As we know from earlier work, the definition of a logarithm provides the basis for evaluating $\log_{10} x$ for values of x that are integral powers of 10. Consider the following examples.

$$\log_{10} 1000 = 3 \quad \text{because } 10^3 = 1000,$$

$$\log_{10} 100 = 2 \quad \text{because } 10^2 = 100,$$

$$\log_{10} 10 = 1 \quad \text{because } 10^1 = 10,$$

$$\log_{10} 1 = 0 \quad \text{because } 10^0 = 1,$$

$$\log_{10} 0.1 = -1 \quad \text{because } 10^{-1} = \frac{1}{10} = 0.1,$$

$$\log_{10} 0.01 = -2 \quad \text{because } 10^{-2} = \frac{1}{10^2} = 0.01,$$

$$\log_{10} 0.001 = -3 \quad \text{because } 10^{-3} = \frac{1}{10^3} = 0.001$$

When working exclusively with base 10 logarithms, it is customary to omit writing the numeral 10 to designate the base. Thus, the expression $\log_{10} x$ is written as $\log x$ and a statement such as $\log_{10} 1000 = 3$ becomes $\log 1000 = 3$. We will follow this practice from now on in this chapter, but don't forget that the base is understood to be 10.

$$\log_{10} x = \log x$$

To find the common logarithm of a positive number that is not an integral power of 10, we can use an appropriately equipped calculator or a table such as the one that appears inside the back cover of this text. Using a calculator equipped with a common logarithm function (ordinarily a key labeled log is used), we obtained the following results rounded to four decimal places.

$$\log 1.75 = 0.2430,$$
$$\log 23.8 = 1.3766,$$
$$\log 134 = 2.1271,$$
$$\log 0.192 = -0.7167,$$
$$\log 0.0246 = -1.6091$$

(Be sure that you can use a calculator and obtain these results.)

In order to use logarithms to solve problems we sometimes need to be able to determine a number when the logarithm of the number is known. That is to say, we may need to determine x if $\log x$ is known. Let's consider an example.

EXAMPLE 2 Find x if $\log x = .2430$.

Solution If $\log x = .2430$, then by changing to exponential form we have $10^{.2430} = x$. Therefore, using the 10^x key we can find x.

$$x = 10^{.2430} \approx 1.749846689$$

Therefore, $x = 1.7498$, rounded to five significant digits. ■

Be sure that you can use your calculator to obtain the following results. We have rounded the values for x to 5 significant digits.

If $\log x = .7629$, then $x = 10^{.7629} = 5.7930$.

If $\log x = 1.4825$, then $x = 10^{1.4825} = 30.374$.

If $\log x = 4.0214$, then $x = 10^{4.0214} = 10505$.

If $\log x = -1.5162$, then $x = 10^{-1.5162} = .030465$.

If $\log x = -3.8921$, then $x = 10^{-3.8921} = .00012820$.

The **common logarithmic function** is defined by the equation $f(x) = \log x$. It should now be a simple matter to set up a table of values and sketch the function.

We will have you do this in the next set of exercises. Remember that $f(x) = 10^x$ and $g(x) = \log x$ are inverses of each other. Therefore, we could also get the graph of $g(x) = \log x$ by reflecting the exponential curve $f(x) = 10^x$ across the line $y = x$.

Natural Logarithms—Base *e*

In many practical applications of logarithms, the number e (remember $e \approx 2.71828$) is used as a base. Logarithms with a base of e are called **natural logarithms** and the symbol $\ln x$ is commonly used instead of $\log_e x$.

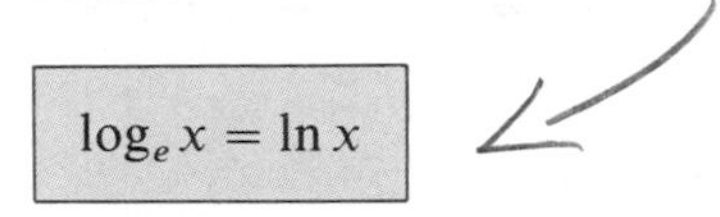

$$\log_e x = \ln x$$

Natural logarithms can be found with an appropriately equipped calculator or with a table of natural logarithms. (A table of natural logarithms is provided in Appendix C.) Using a calculator with a natural logarithm function (ordinarily a key labeled [ln x]), we can obtain the following results rounded to four decimal places.

$$\ln 3.21 = 1.1663,$$
$$\ln 47.28 = 3.8561,$$
$$\ln 842 = 6.7358,$$
$$\ln 0.21 = -1.5606,$$
$$\ln 0.0046 = -5.3817,$$
$$\ln 10 = 2.3026$$

Be sure that you can use your calculator to obtain these results. Keep in mind the significance of a statement such as $\ln 3.21 = 1.1663$. By changing to exponential form we are claiming that e raised to the 1.1663 power is approximately 3.21. Using a calculator we obtain $e^{1.1663} = 3.210093293$.

Let's do a few more problems and find x when given $\ln x$. Be sure that you agree with these results.

If $\ln x = 2.4156$, then $x = e^{2.4156} = 11.196$.
If $\ln x = .9847$, then $x = e^{.9847} = 2.6770$.
If $\ln x = 4.1482$, then $x = e^{4.1482} = 63.320$.
If $\ln x = 1.7654$, then $x = e^{-1.7654} = .17112$.

The **natural logarithmic function** is defined by the equation $f(x) = \ln x$. It is the inverse of the natural exponential function $f(x) = e^x$. Thus, one way to graph $f(x) = \ln x$ is to reflect the graph of $f(x) = e^x$ across the line $y = x$. We will have you do this in the next set of problems.

Problem Set 10.4

For Problems 1–10, use a calculator to find each **common logarithm**. Express answers to four decimal places.

1. $\log 7.24$ 0.8597
2. $\log 2.05$ 0.3118
3. $\log 52.23$ 1.7179
4. $\log 825.8$ 2.9169
5. $\log 3214.1$ 3.5071
6. $\log 14{,}189$ 4.1520
7. $\log 0.729$ -0.1373
8. $\log 0.04376$ -1.3589
9. $\log 0.00034$ -3.4685
10. $\log 0.000069$ -4.1612

For Problems 11–20, use your calculator to find x when given $\log x$. Express answers to five significant digits.

11. $\log x = 2.6143$ 411.43
12. $\log x = 1.5263$ 33.597
13. $\log x = 4.9547$ 90095
14. $\log x = 3.9335$ 8580.3
15. $\log x = 1.9006$ 79.543
16. $\log x = 0.5517$ 3.5620
17. $\log x = -1.3148$ 0.048440
18. $\log x = -0.1452$ 0.71581
19. $\log x = -2.1928$ 0.0064150
20. $\log x = -2.6542$ 0.0022172

For Problems 21–30, use your calculator to find each **natural logarithm**. Express answers to four decimal places.

21. $\ln 5$ 1.6094
22. $\ln 18$ 2.8904
23. $\ln 32.6$ 3.4843
24. $\ln 79.5$ 4.3758
25. $\ln 430$ 6.0638
26. $\ln 371.8$ 5.9184
27. $\ln 0.46$ -0.7765
28. $\ln 0.524$ -0.6463
29. $\ln 0.0314$ -3.4609
30. $\ln 0.008142$ -4.8107

For Problems 31–40, use your calculator to find x when given $\ln x$. Express answers to five significant digits.

31. $\ln x = 0.4721$ 1.6034
32. $\ln x = 0.9413$ 2.5633
33. $\ln x = 1.1425$ 3.1346
34. $\ln x = 2.7619$ 15.830
35. $\ln x = 4.6873$ 108.56
36. $\ln x = 3.0259$ 20.613
37. $\ln x = -0.7284$ 0.48268
38. $\ln x = -1.6246$ 0.19699
39. $\ln x = -3.3244$ 0.035994
40. $\ln x = -2.3745$ 0.093061

41. **(a)** Complete the following table and then graph $f(x) = \log x$. (Express the values for $\log x$ to the nearest tenth.)

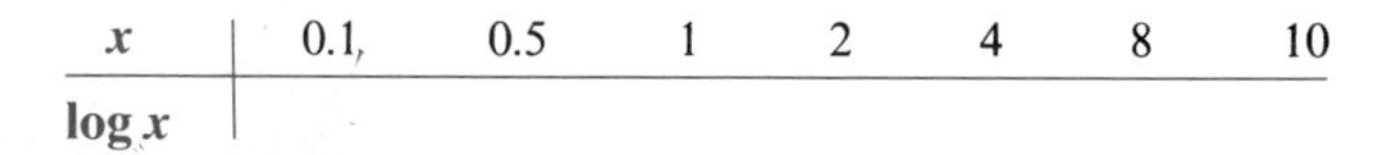

x	0.1	0.5	1	2	4	8	10
$\log x$							

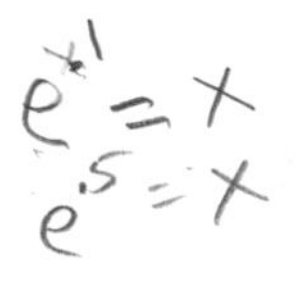

(b) Complete the following table and express values for 10^x to the nearest tenth.

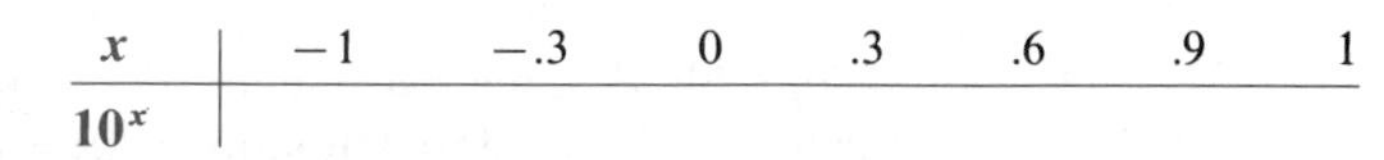

x	-1	$-.3$	0	.3	.6	.9	1
10^x							

Then graph $f(x) = 10^x$ and reflect it across the line $y = x$ to produce the graph for $f(x) = \log x$. see page A74

42. **(a)** Complete the following table and then graph $f(x) = \ln x$. (Express the values for $\ln x$ to the nearest tenth.)

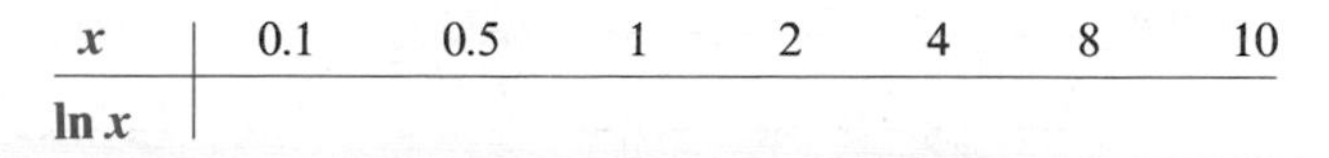

x	0.1	0.5	1	2	4	8	10
$\ln x$							

(b) Complete the following table and express values for e^x to the nearest tenth.

x	-2.3	$-.7$	0	.7	1.4	2.1	2.3
e^x							

Then graph $f(x) = e^x$ and reflect it across the line $y = x$ to produce the graph for $f(x) = \ln x$. see page A74

43. Graph $y = \log_{\frac{1}{2}} x$ by graphing $\left(\frac{1}{2}\right)^y = x$. see page A75

44. Graph $y = \log_2 x$ by graphing $2^y = x$. This graph is the same as Figure 10.5 in the text.

45. Graph $f(x) = \log_3 x$ by reflecting the graph of $g(x) = 3^x$ across the line $y = x$. see page A75

46. Graph $f(x) = \log_4 x$ by reflecting the graph of $g(x) = 4^x$ across the line $y = x$. see page A75

For Problems 47–53, graph each of the functions. Remember that the graph of $f(x) = \log_2 x$ is given in Figure 10.5.

47. $f(x) = 3 + \log_2 x$ see page A75

48. $f(x) = -2 + \log_2 x$ see page A75

49. $f(x) = \log_2(x + 3)$ see page A75

50. $f(x) = \log_2(x - 2)$ see page A75

51. $f(x) = \log_2 2x$ see page A75

52. $f(x) = -\log_2 x$ see page A76

53. $f(x) = 2\log_2 x$ see page A76

For Problems 54–61, perform the following calculations and express answers to the nearest hundredth. (These calculations are in preparation for our work in the next section.)

54. $\dfrac{\log 7}{\log 3}$ 1.77

55. $\dfrac{\ln 2}{\ln 7}$ 0.36

56. $\dfrac{2\ln 3}{\ln 8}$ 1.06

57. $\dfrac{\ln 5}{2\ln 3}$ 0.73

58. $\dfrac{\ln 3}{0.04}$ 27.47

59. $\dfrac{\ln 2}{0.03}$ 23.10

60. $\dfrac{\log 2}{5\log 1.02}$ 7.00

61. $\dfrac{\log 5}{3\log 1.07}$ 7.93

10.5

Exponential Equations, Logarithmic Equations, and Problem Solving

In Section 10.1 we solved exponential equations such as $3^x = 81$ when we expressed both sides of the equation as a power of 3 and then applied the property "if $b^n = b^m$, then $n = m$." However, if we try to use this same approach with an equation such as

$3^x = 5$, we face the difficulty of expressing 5 as a power of 3. We can solve this type of problem by using the properties of logarithms and the following property of equality.

PROPERTY 10.8

> If $x > 0$, $y > 0$, and $b \neq 1$, then
>
> $x = y$ if and only if $\log_b x = \log_b y$.

Property 10.8 is stated in terms of any valid base b; however, for most applications either common logarithms (base 10) or natural logarithms (base e) are used. Let's consider some examples.

EXAMPLE 1 Solve $3^x = 5$ to the nearest hundredth.

Solution By using common logarithms we can proceed as follows.

$$3^x = 5$$

$$\log 3^x = \log 5 \qquad \text{Property 10.8}$$

$$x \log 3 = \log 5 \qquad \log r^p = p \log r$$

$$x = \frac{\log 5}{\log 3}$$

$$x = 1.46 \qquad \text{nearest hundredth}$$

CHECK Since $3^{1.46} \approx 4.972754647$, we say that, to the nearest hundredth, the solution set for $3^x = 5$ is $\{1.46\}$. ■

EXAMPLE 2 Solve $e^{x+1} = 5$ to the nearest hundredth.

Solution Since the base of e is used in the exponential expression, let's use natural logarithms to help solve this equation.

$$e^{x+1} = 5$$

$$\ln e^{x+1} = \ln 5 \qquad \text{Property 10.8}$$

$$(x+1)\ln e = \ln 5 \qquad \ln r^2 = p \ln r$$

$$(x+1)(1) = \ln 5 \qquad \ln e = 1$$

$$x = \ln 5 - 1$$

$$x \approx 0.609437912$$

$$x = 0.61 \qquad \text{nearest hundredth}$$

The solution set is $\{0.61\}$. Check it! ■

Logarithmic Equations

In Example 12 of Section 10.3 we solved the logarithmic equation

$$\log_{10} x + \log_{10}(x + 9) = 1$$

by simplifying the left side of the equation to $\log_{10}[x(x + 9)]$ and then changing the equation to exponential form to complete the solution. Now, using Property 10.8, we can solve such a logarithmic equation another way and also expand our equation solving capabilities. Let's consider some examples.

EXAMPLE 3 Solve $\log x + \log(x - 15) = 2$.

Solution Since $\log 100 = 2$, the given equation becomes

$$\log x + \log(x - 15) = \log 100.$$

Now, simplify the left side, apply Property 10.8 and proceed as follows.

$$\log(x)(x - 15) = \log 100$$

$$x(x - 15) = 100$$

$$x^2 - 15x - 100 = 0$$

$$(x - 20)(x + 5) = 0$$

$$x - 20 = 0 \quad \text{or} \quad x + 5 = 0$$

$$x = 20 \quad \text{or} \quad x = -5$$

The domain of a logarithmic function must contain only positive numbers, so x and $x - 15$ must be positive in this problem. Therefore, we discard the solution of -5; the solution set is $\{20\}$. ■

EXAMPLE 4 Solve $\ln(x + 2) = \ln(x - 4) + \ln 3$.

Solution

$$\ln(x + 2) = \ln(x - 4) + \ln 3$$

$$\ln(x + 2) = \ln[3(x - 4)]$$

$$x + 2 = 3(x - 4)$$

$$x + 2 = 3x - 12$$

$$14 = 2x$$

$$7 = x$$

The solution set is $\{7\}$. ■

Problem Solving

In Section 10.2 we used the compound interest formula

$$A = P\left(1 + \frac{r}{n}\right)^{nt}$$

to determine the amount of money (A) accumulated at the end of t years if P dollars is invested at r rate of interest compounded n times per year. Now let's use this formula to solve other types of problems that deal with compound interest.

EXAMPLE 5 How long will \$500 take to double itself if invested at 12% interest compounded quarterly?

Solution To *double itself* means that the \$500 will grow into \$1000. Thus,

$$\begin{aligned} 1000 &= 500\left(1 + \frac{0.12}{4}\right)^{4t} \\ &= 500(1 + 0.03)^{4t} \\ &= 500(1.03)^{4t}. \end{aligned}$$

Multiply both sides of $1000 = 500(1.03)^{4t}$ by $\frac{1}{500}$ to yield

$$2 = (1.03)^{4t}.$$

Therefore,

$$\begin{aligned} \log 2 &= \log(1.03)^{4t} && \text{Property 10.8} \\ &= 4t \log 1.03. && \log r^p = p \log r \end{aligned}$$

Solve for t to obtain

$$\log 2 = 4t \log 1.03$$

$$\frac{\log 2}{\log 1.03} = 4t$$

$$\frac{\log 2}{4 \log 1.03} = t \qquad \text{Multiply both sides by } \frac{1}{4}.$$

$$t \approx 5.862443063$$

$$t = 5.9. \qquad \text{nearest tenth}$$

Therefore, we are claiming that \$500 invested at 12% interest compounded quarterly will double itself in approximately 5.9 years.

CHECK \$500 invested at 12% compounded quarterly for 5.9 years will produce

$$A = \$500\left(1 + \frac{0.12}{4}\right)^{4(5.9)}$$
$$= \$500(1.03)^{23.6}$$
$$= \$1004.45.$$

■

In Section 10.2, we also used the formula $A = Pe^{rt}$ when money was to be compounded continuously. At this time, with the help of natural logarithms, we can extend our use of this formula.

EXAMPLE 6 How long will it take \$100 to triple itself if it is invested at 8% interest compounded continuously?

Solution To *triple itself* means that the \$100 will grow into \$300. Thus, using the formula for interest that is compounded continuously, we can proceed as follows.

$$A = Pe^{rt}$$
$$\$300 = \$100e^{(0.08)t}$$
$$3 = e^{0.08t}$$
$$\ln 3 = \ln e^{0.08t} \qquad \text{Property 10.8}$$
$$\ln 3 = 0.08t \ln e \qquad \ln r^p = p \ln r$$
$$\ln 3 = 0.08t \qquad \ln e = 1$$
$$\frac{\ln 3}{0.08} = t$$
$$t \approx 13.73265361$$
$$t = 13.7 \qquad \text{nearest tenth}$$

Therefore, in approximately 13.7 years, \$100 will triple itself at 8% interest compounded continuously.

CHECK \$100 invested at 8% compounded continuously for 13.7 years produces

$$A = Pe^{rt}$$
$$= \$100e^{0.08(13.7)}$$
$$= \$100e^{1.096}$$
$$\approx \$299.22.$$

■

Seismologists use the Richter scale to measure and report the magnitude of earthquakes. The equation

$$R = \log \frac{I}{I_0}$$

R is called a Richter number.

compares the intensity I of an earthquake to a minimal or reference intensity I_0. The reference intensity is the smallest earth movement that can be recorded on a seismograph. Suppose that the intensity of an earthquake was determined to be 50,000 times the reference intensity. In this case, $I = 50{,}000\ I_0$ and the Richter number would be calculated as follows.

$$R = \log \frac{50{,}000\ I_0}{I_0}$$

$$R = \log 50{,}000$$

$$R \approx 4.698970004$$

Thus, a Richter number of 4.7 would be reported. Let's consider two more examples that involve Richter numbers.

EXAMPLE 7 An earthquake in San Francisco in 1989 was reported to have a Richter number of 6.9. How did its intensity compare to the reference intensity?

Solution

$$6.9 = \log \frac{I}{I_0}$$

$$10^{6.9} = \frac{I}{I_0}$$

$$I = (10^{6.9})(I_0)$$

$$I \approx 7943282\ I_0$$

So its intensity was a little less than 8 million times the reference intensity. ■

EXAMPLE 8 An earthquake in Iran in 1990 had a Richter number of 7.7. Compare the intensity level of that earthquake to the one in San Francisco (Example 7).

Solution From Example 7 we have $I = (10^{6.9})(I_0)$ for the earthquake in San Francisco. Then using a Richter number of 7.7 we obtain $I = (10^{7.7})(I_0)$ for the earthquake in Iran. Therefore, by comparison,

$$\frac{(10^{7.7})(I_0)}{(10^{6.9})(I_0)} = 10^{7.7-6.9} = 10^{0.8} \approx 6.3.$$

The earthquake in Iran was about 6 times as intense as the one in San Francisco. ■

Logarithms with Bases Other Than 10 or *e*

Now let's use either common or natural logarithms to evaluate logarithms that have bases other than 10 or e. Consider the following example.

EXAMPLE 9 Evaluate $\log_3 41$.

Solution Let $x = \log_3 41$. Change to exponential form to obtain

$$3^x = 41.$$

Now we can apply Property 10.8 and proceed as follows.

$$\log 3^x = \log 41$$

$$x \log 3 = \log 41$$

$$x = \frac{\log 41}{\log 3}$$

$$x = \frac{1.6128}{0.4771}$$

$$x = 3.3804 \quad \text{rounded to four decimal places}$$

■

Using the method of Example 9 to evaluate $\log_a r$ produces the following formula, which we often refer to as the **change-of-base** formula for logarithms.

PROPERTY 10.9

If a, b, and r are positive numbers with $a \neq 1$ and $b \neq 1$, then

$$\log_a r = \frac{\log_b r}{\log_b a}.$$

By using Property 10.9 we can easily determine a relationship between logarithms of a different base. For example, suppose that in Property 10.9 we let $a = 10$ and $b = e$. Then

$$\log_a r = \frac{\log_b r}{\log_b a}$$

becomes

$$\log_{10} r = \frac{\log_e r}{\log_e 10},$$

which can be written as

$$\log_e r = (\log_e 10)(\log_{10} r).$$

Since $\log_e 10 = 2.3026$, rounded to four decimal places, we have

$$\log_e r = (2.3026)(\log_{10} r).$$

Thus, the natural logarithm of any positive number is approximately equal to the common logarithm of the number times 2.3026.

Problem Set 10.5

For Problems 1–14, solve each exponential equation and express solutions to the nearest hundredth.

1. $3^x = 32$ {3.15}
2. $2^x = 40$ {5.32}
3. $4^x = 21$ {2.20}
4. $5^x = 73$ {2.67}
5. $3^{x-2} = 11$ {4.18}
6. $2^{x+1} = 7$ {1.81}
7. $5^{3x+1} = 9$ {0.12}
8. $7^{2x-1} = 35$ {1.41}
9. $e^x = 5.4$ {1.69}
10. $e^x = 45$ {3.81}
11. $e^{x-2} = 13.1$ {4.57}
12. $e^{x-1} = 8.2$ {3.10}
13. $3e^x = 35.1$ {2.46}
14. $4e^x - 2 = 26$ {1.95}

For Problems 15–22, solve each logarithmic equation.

15. $\log x + \log(x + 21) = 2$ {4}
16. $\log x + \log(x + 3) = 1$ {2}
17. $\log(3x - 1) = 1 + \log(5x - 2)$ $\{\frac{19}{47}\}$
18. $\log(2x - 1) - \log(x - 3) = 1$ $\{\frac{29}{8}\}$
19. $\log(x + 2) - \log(2x + 1) = \log x$ {1}
20. $\log(x + 1) - \log(x + 2) = \log\frac{1}{x}$ $\{\sqrt{2}\}$
21. $\ln(2t + 5) = \ln 3 + \ln(t - 1)$ {8}
22. $\ln(3t - 4) - \ln(t + 1) = \ln 2$ {6}

For Problems 23–32, approximate each of the following logarithms to three decimal places.

23. $\log_2 23$ 4.524
24. $\log_3 32$ 3.155
25. $\log_6 .214$ −.860
26. $\log_5 1.4$ .209
27. $\log_7 421$ 3.105
28. $\log_8 514$ 3.002
29. $\log_9 .0017$ −2.902
30. $\log_4 .00013$ −6.455
31. $\log_3 720$ 5.989
32. $\log_2 896$ 9.807

For Problems 33–41, solve each problem and express answers to the nearest tenth.

33. How long will it take \$750 to be worth \$1000 if it is invested at 12% interest compounded quarterly? 2.4 years

34. How long will it take \$1000 to double itself if it is invested at 9% interest compounded semiannually? 7.9 years

35. How long will it take \$2000 to double itself if it is invested at 13% interest compounded continuously? 5.3 years

36. How long will it take \$500 to triple itself if it is invested at 9% interest compounded continuously? 12.2 years

37. For a certain strain of bacteria, the number present after t hours is given by the equation $Q = Q_0 e^{0.34t}$, where Q_0 represents the initial number of bacteria. How long will it take 400 bacteria to increase to 4000 bacteria? 6.8 hours

38. A piece of machinery valued at \$30,000 depreciates at a rate of 10% yearly. How long will it take until it has a value of \$15,000? 6.6 years

39. The number of grams of a certain radioactive substance present after t hours is given by the equation $Q = Q_0e^{-0.45t}$, where Q_0 represents the initial number of grams. How long would it take 2500 grams to be reduced to 1250 grams? 1.5 hours

40. For a certain culture the equation $Q(t) = Q_0e^{0.4t}$, where Q_0 is an initial number of bacteria and t is time measured in hours, yields the number of bacteria as a function of time. How long will it take 500 bacteria to increase to 2000? 3.5 hours

41. Suppose that the equation $P(t) = P_0e^{0.02t}$, where P_0 represents an initial population and t is the time in years, is used to predict population growth. How long would it take a city of 50,000 to double its population? 34.7 years

Solve each of the Problems 42–46.

42. The equation $P(a) = 14.7e^{-0.21a}$, where a is the altitude above sea level measured in miles, yields the atmospheric pressure in pounds per square inch. If the atmospheric pressure at Cheyenne, Wyoming is approximately 11.53 pounds per square inch, find that city's altitude above sea level. Express your answer to the nearest hundred feet. 6100 feet

43. An earthquake in Los Angeles in 1971 had an intensity of approximately five million times the reference intensity. What was the Richter number associated with that earthquake? 6.7

44. An earthquake in San Francisco in 1906 was reported to have a Richter number of 8.3. How did its intensity compare to the reference intensity? approximately 200 million times the reference intensity

45. Calculate how many times more intense an earthquake with a Richter number of 7.3 is than an earthquake with a Richter number of 6.4. approximately 8 times

46. Calculate how many times more intense an earthquake with a Richter number of 8.9 is than an earthquake with a Richter number of 6.2. approximately 500 times

Chapter 10 Summary

(10.1) If a and b are positive real numbers, and m and n are any real numbers, then

1. $b^n \cdot b^m = b^{n+m}$ — product of two powers
2. $(b^n)^m = b^{mn}$ — power of a power
3. $(ab)^n = a^nb^n$ — power of a product
4. $\left(\frac{a}{b}\right)^n = \frac{a^n}{b^n}$ — power of a quotient
5. $\frac{b^n}{b^m} = b^{n-m}$ — quotient of two powers

If $b > 0$, $b \neq 1$, and m and n are real numbers, then

$$b^n = b^m \quad \text{if and only if } n = m.$$

A function defined by an equation of the form

$$f(x) = b^x, \qquad b > 0 \text{ and } b \neq 1$$

is called an **exponential function**.

(10.2) A general formula for any principal, P, being compounded n times per year for any number of years (t) at a rate of r percent is

$$A = P\left(1 + \frac{r}{n}\right)^{nt}$$

where A represents the total amount of money accumulated at the end of the t years. The value of $\left(1 + \frac{1}{n}\right)^n$, as n gets infinitely large, approaches the number e, where e equals 2.71828 to five decimal places.

The formula

$$A = Pe^{rt}$$

yields the accumulated value, A, of a sum of money, P, that has been invested for t years at a rate of r percent **compounded continuously**.

The equation

$$Q(t) = Q_0e^{kt}$$

is used as a mathematical model for many growth-and-decay applications.

(10.3) If r is any positive real number, then the unique exponent t such that $b^t = r$ is called the **logarithm of *r* with base *b*** and is denoted by $\log_b r$.

For $b > 0$ and $b \neq 1$, and $r > 0$,

1. $\log_b b = 1$,

2. $\log_b 1 = 0$,

3. $r = b^{\log_b r}$.

The following properties of logarithms are derived from the definition of a logarithm and the properties of exponents.

For positive real numbers b, r, and s where $b \neq 1$,

1. $\log_b rs = \log_b r + \log_b s$.

2. $\log_b\left(\frac{r}{s}\right) = \log_b r - \log_b s$.

3. $\log_b r^p = p\log_b r$. $\quad$ p is any real number.

(10.4) A function defined by an equation of the form

$$f(x) = \log_b x, \qquad b > 0 \text{ and } b \neq 1$$

is called a **logarithmic function**. The equation $y = \log_b x$ is equivalent to $x = b^y$. The two functions $f(x) = b^x$ and $g(x) = \log_b x$ are inverses of each other.

Logarithms with a base of 10 are called **common logarithms**. The expression $\log_{10} x$ is commonly written as $\log x$.

Natural logarithms are logarithms that have a base of e, where e is an irrational number whose decimal approximation to eight digits is 2.7182818. Natural logarithms are denoted by $\log_e x$ or $\ln x$.

(10.5) The properties of equality along with the properties of exponents and logarithms merge to help us solve a variety of exponential and logarithmic equations. Due to these properties we can now solve problems that deal with various applications including compound interest and growth problems.

The formula

$$\log_a r = \frac{\log_b r}{\log_b a}$$

is often called the **change-of-base** formula.

Chapter 10 Review Problem Set

For Problems 1–8, evaluate each expression without using a calculator.

1. $\log_2 128$ 7

2. $\log_4 64$ 3

3. $\log 10{,}000$ 4

4. $\log .001$ -3

5. $\ln e^2$ 2

6. $5^{\log_5 13}$ 13

7. $\log(\log_3 3)$ 0

8. $\log_2\left(\frac{1}{4}\right)$ -2

For Problems 9–16, solve each equation without using your calculator.

9. $2^x = \frac{1}{16}$ $\{-4\}$

10. $3^x - 4 = 23$ $\{3\}$

11. $16^x = \frac{1}{8}$ $\left\{-\frac{3}{4}\right\}$

12. $\log_8 x = \frac{2}{3}$ $\{4\}$

13. $\log_x 3\ \frac{1}{2}$ $\{9\}$

14. $\log_5 2 + \log_5(3x + 1) = 1$ $\{\frac{1}{2}\}$

15. $\log_2(x + 5) - \log_2 x = 1$ $\{5\}$

16. $\log_2 x + \log_2(x - 4) = 5$ $\{8\}$

For Problems 17–20, use your calculator to find each logarithm. Express answers to four decimal places.

17. $\log 73.14$ 1.8642

18. $\ln 114.2$ 4.7380

19. $\ln 0.014$ -4.2687

20. $\log 0.00235$ -2.6289

For Problems 21–24, use your calculator to find x when given $\log x$ or $\ln x$. Express answers to five significant digits.

21. $\ln x = 0.1724$ 1.1882

22. $\log x = 3.4215$ 2639.4

23. $\log x = -1.8765$ .013289

24. $\ln x = -2.5614$ .077197

For Problems 25–28, use your calculator to help solve each equation. Express solutions to the nearest hundredth.

25. $3^x = 42$ $\{3.40\}$

26. $2e^x = 14$ $\{1.95\}$

27. $2^{x+1} = 79$ $\{5.30\}$

28. $e^{x-2} = 37$ $\{5.61\}$

For Problems 29–32, graph each of the functions.

29. $f(x) = 2^x - 3$ see page A76

30. $f(x) = -3^x$ see page A76

31. $f(x) = -1 + \log_2 x$ see page A77

32. $f(x) = \log_2(x + 1)$ see page A77

33. Approximate a value for $\log_5 97$ to three decimal places. 2.842

34. Suppose that \$800 is invested at 9% interest compounded annually. How much money has accumulated at the end of 15 years? \$2913.99

35. If \$2500 is invested at 10% interest compounded quarterly, how much money has accumulated at the end of 12 years? \$8178.72

36. If \$3500 is invested at 8% interest compounded continuously, how much money will accumulate in 6 years? \$5656.26

37. Suppose that a certain radioactive substance has a half-life of 40 days. If there are presently 750 grams of the substance, how much, to the nearest gram, will remain after 100 days? 133 grams

38. How long will it take \$100 to double itself if it is invested at 14% interest compounded annually? approximately 5.3 years

39. How long will it take \$1000 to be worth \$3500 if it is invested at 10.5% interest compounded quarterly? approximately 12.1 years

40. Suppose that the present population of a city is 50,000. Use the equation $P(t) = P_0e^{0.02t}$, where P_0 represents an initial population, to estimate future populations. Estimate the population of that city in 10 years, 15 years, and 20 years. 61,070; 67,493; 74,591

41. The number of bacteria present in a certain culture after t hours is given by the equation $Q = Q_0e^{0.29t}$, where Q_0 represents the initial number of bacteria. How long will it take 500 bacteria to increase to 2000 bacteria? approximately 4.8 hours

42. An earthquake in Mexico City in 1985 had an intensity level about 125,000,000 times the reference intensity. Find the Richter number for that earthquake. 8.1

Thoughts into Words

43. Why is the base of an exponential function restricted to positive numbers not including one?

44. Explain how you would solve the equation $(2^{x+1})(8^{2x-3}) = 64$.

45. Explain the difference between simple interest and compound interest.

46. Explain without using Property 10.4 why $4^{\log_4 9}$ equals 9.

47. Explain the concept of an antilogarithm.

48. Explain how to determine $\log_3 746$ without using Property 10.9.

Chapter 11

Sequences and Series

Suppose that Math University had an enrollment of 7500 students in 1977 and each year thereafter through 1981 the enrollment increased by 500 students. The numbers

$$7500, 8000, 8500, 9000, 9500$$

represent the enrollment figures for the years 1977 through 1981. This list of numbers, where there is a constant difference of 500 between any two successive numbers in the list, is called an arithmetic sequence.

Suppose a woman's present yearly salary is \$20,000 and she expects a 10% raise for each of the next four years. The numbers

$$20{,}000, 22{,}000, 24{,}200, 26{,}620, 29{,}282$$

represent her present salary and her salary for each of the next four years. This list of numbers, where each number after the first one is 1.1 times the previous number in the list, is called a geometric sequence.

Arithmetic and geometric sequences are the focus of our attention in this chapter.

11.1
Arithmetic Sequences

An **infinite sequence** is a function whose domain is the set of positive integers. For example, consider the function defined by the equation

$$f(n) = 3n + 2$$

where the domain is the set of positive integers. Furthermore, let's substitute the numbers of the domain, in order, starting with 1. The resulting ordered pairs can be listed as

$$(1, 5), \quad (2, 8), \quad (3, 11), \quad (4, 14), \quad (5, 17),$$

and so on. Since we have agreed to use the domain of positive integers, in order starting with 1, there is no need to use ordered pairs. We can simply express the infinite sequence as

$$5, 8, 11, 14, 17, \ldots.$$

Frequently, the letter a is used to represent *sequential functions* and the functional value at n is written as a_n (read "a sub n") instead of $a(n)$. The sequence is then expressed as

$$a_1, a_2, a_3, a_4, \ldots, a_n, \ldots,$$

where a_1 is the *first term*, a_2 the *second term*, a_3 the *third term*, and so on. The expression a_n which defines the sequence, is called the **general term** of the sequence. Knowing the general term of a sequence allows us to find as many terms of the sequence as needed and also to find any specific terms. Consider the following examples.

EXAMPLE 1 Find the first five terms of each of the following sequences.

(a) $a_n = n^2 + 1$. **(b)** $a_n = 2^n$.

Solution **(a)** The first five terms are found by replacing n with 1, 2, 3, 4, and 5.

$$a_1 = 1^2 + 1 = 2,$$

$$a_2 = 2^2 + 1 = 5,$$

$$a_3 = 3^2 + 1 = 10,$$

$$a_4 = 4^2 + 1 = 17,$$

$$a_5 = 5^2 + 1 = 26$$

Thus, the first five terms are 2, 5, 10, 17, and 26.

(b)

$$a_1 = 2^1 = 2,$$
$$a_2 = 2^2 = 4,$$
$$a_3 = 2^3 = 8,$$
$$a_4 = 2^4 = 16,$$
$$a_5 = 2^5 = 32$$

The first five terms are 2, 4, 8, 16, and 32. ■

EXAMPLE 2 Find the 12th and 25th terms of the sequence $a_n = 5n - 1$.

Solution

$$a_{12} = 5(12) - 1 = 59,$$
$$a_{25} = 5(25) - 1 = 124$$

The 12th term is 59 and the 25th term is 124. ■

An **arithmetic sequence** (also called an *arithmetic progression*) is a sequence where there is a **common difference** between successive terms. The following are examples of arithmetic sequences.

1. 1, 4, 7, 10, 13,...
2. 6, 11, 16, 21, 26,...
3. 14, 25, 36, 47, 58,...
4. 4, 2, 0, -2, -4,...
5. -1, -7, -13, -19, -25,...

The common difference in number 1 of the previous list is 3. That is to say, $4 - 1 = 3$, $7 - 4 = 3$, $10 - 7 = 3$, $13 - 10 = 3$, and so on. The common differences for numbers 2, 3, 4, and 5 are 5, 11, -2, and -6, respectively. It is sometimes stated that arithmetic sequences exhibit *constant growth*. This is an accurate description if we keep in mind that the *growth* may be in a negative direction as illustrated by numbers 4 and 5 above.

In a more general setting we say that the sequence

$$a_1, a_2, a_3, a_4, \ldots, a_n, \ldots$$

is an arithmetic sequence if and only if there is a real number d such that

$$a_{k+1} - a_k = d \tag{1}$$

for every positive integer k. The number d is called the **common difference**.

From equation (1) we see that $a_{k+1} = a_k + d$. In other words, we can generate an arithmetic sequence that has a common difference of d by starting with a first term of a_1 and then simply adding d to each successive term as follows.

first term	a_1	
second term	$a_1 + d$	
third term	$a_1 + 2d$	$(a_1 + d) + d = a_1 + 2d$
fourth term	$a_1 + 3d$	$(a_1 + 2d) + d = a_1 + 3d$
⋮		
nth term	$a_1 + (n - 1)d$	

Thus, the *general term of an arithmetic sequence* is given by

$$a_n = a_1 + (n - 1)d$$

where a_1 is the first term and d the common difference. This general term formula provides the basis for doing a variety of problems that involve arithmetic sequences.

EXAMPLE 3 Find the general term for each of the following arithmetic sequences.

(a) 1, 5, 9, 13, ...

(b) 5, 2, −1, −4, ...

Solution

(a) The common difference is 4 and the first term is 1. Substituting these values into $a_n = a_1 + (n - 1)d$ and simplifying, we obtain

$$\begin{aligned} a_n &= a_1 + (n - 1)d \\ &= 1 + (n - 1)4 \\ &= 1 + 4n - 4 \\ &= 4n - 3. \end{aligned}$$

(Perhaps you should verify that the general term $a_n = 4n - 3$ does produce the sequence 1, 5, 9, 13, ...)

(b) Since the first term is 5 and the common difference is -3, we obtain

$$\begin{aligned} a_n &= a_1 + (n - 1)d \\ &= 5 + (n - 1)(-3) \\ &= 5 - 3n + 3 = -3n + 8. \end{aligned}$$

■

EXAMPLE 4 Find the 50th term of the arithmetic sequence 2, 6, 10, 14, ...

Solution

Certainly, we could simply continue to write the terms of the given sequence until the 50th term is reached. However, let's use the general term formula, $a_n = a_1 + (n - 1)d$, to find the 50th term without determining all of the other terms.

$$\begin{aligned} a_{50} &= 2 + (50 - 1)4 \\ &= 2 + 49(4) = 2 + 196 = 198. \end{aligned}$$

■

EXAMPLE 5 Find the first term of the arithmetic sequence where the 3rd term is 13 and the 10th term is 62.

Solution Using $a_n = a_1 + (n - 1)d$ with $a_3 = 13$ (the third term is 13) and $a_{10} = 62$ (the 10th term is 62) we have

$$13 = a_1 + (3 - 1)d = a_1 + 2d,$$
$$62 = a_1 + (10 - 1)d = a_1 + 9d.$$

Solve the system of equations

$$\begin{pmatrix} a_1 + 2d = 13 \\ a_1 + 9d = 62 \end{pmatrix}$$

to yield $a_1 = -1$. Thus, the first term is -1. ■

REMARK Perhaps you can think of another way to solve the problem in Example 5 without using a system of equations. [*Hint*: How many "differences" are there between the 3rd and 10th terms of an arithmetic sequence?]

Phrases such as "the set of odd whole numbers," "the set of even whole numbers," and "the set of positive multiples of five" are commonly used in mathematical literature to refer to various subsets of the whole numbers. Though no specific ordering of the numbers is implied by these phrases, most of us would probably react with a natural ordering. For example, if we were asked to list the set of *odd whole numbers*, our answer probably would be 1, 3, 5, 7,.... Using such an ordering, we can think of the set of odd whole numbers as an arithmetic sequence. Therefore, we can formulate a general representation for the set of odd whole numbers by using the general term formula. Thus,

$$\begin{aligned} a_n &= a_1 + (n - 1)d \\ &= 1 + (n - 1)2 \\ &= 1 + 2n - 2 \\ &= 2n - 1. \end{aligned}$$

The final example of this section illustrates the use of an arithmetic sequence to solve a problem that deals with the *constant growth* of a man's yearly salary.

EXAMPLE 6 A man started to work in 1970 at an annual salary of \$9500. He received a \$700 raise each year. How much was his annual salary in 1991?

Solution The following arithmetic sequence represents the annual salary beginning in 1970.

9500, 10,200, 10,900, 11,600,....

The general term of this sequence is

$$\begin{aligned} a_n &= a_1 + (n-1)d \\ &= 9500 + (n-1)700 \\ &= 9500 + 700n - 700 \\ &= 700n + 8800. \end{aligned}$$

The man's 1991 salary is the 22nd term of this sequence. Thus,

$$a_{22} = 700(22) + 8800 = 24{,}200.$$

His salary in 1991 was $24,200. ■

Problem Set 11.1

For Problems 1–14, write the first five terms of each sequence that has the indicated general term.

1. $a_n = 3n - 4$ $-1, 2, 5, 8, 11$

2. $a_n = 2n + 3$ $5, 7, 9, 11, 13$

3. $a_n = -2n + 5$ $3, 1, -1, -3, -5$

4. $a_n = -3n - 2$ $-5, -8, -11, -14, -17$

5. $a_n = n^2 - 2$ $-1, 2, 7, 14, 23$

6. $a_n = n^2 + 3$ $4, 7, 12, 19, 28$

7. $a_n = -n^2 + 1$ $0, -3, -8, -15, -24$

8. $a_n = -n^2 - 2$ $-3, -6, -11, -18, -27$

9. $a_n = 2n^2 - 3$ $-1, 5, 15, 29, 47$

10. $a_n = 3n^2 + 2$ $5, 14, 29, 50, 77$

11. $a_n = 2^{n-2}$ $\frac{1}{2}, 1, 2, 4, 8$

12. $a_n = 3^{n+1}$ $9, 27, 81, 243, 729$

13. $a_n = -2(3)^{n-2}$ $-\frac{2}{3}, -2, -6, -18, -54$

14. $a_n = -3(2)^{n-3}$ $-\frac{3}{4}, -\frac{3}{2}, -3, -6, -12$

15. Find the 8th and 12th terms of the sequence where $a_n = n^2 - n - 2$. $a_8 = 54$ and $a_{12} = 130$

16. Find the 10th and 15th terms of the sequence where $a_n = -n^2 - 2n + 3$. $a_{10} = -117$, and $a_{15} = -252$

17. Find the 7th and 8th terms of the sequence where $a_n = (-2)^{n-2}$. $a_7 = -32$ and $a_8 = 64$

18. Find the 6th and 7th terms of the sequence where $a_n = -(3)^{n-3}$. $a_6 = -27$ and $a_7 = -81$

For Problems 19–28, find the general term (nth term) for each of the arithmetic sequences.

19. $1, 3, 5, 7, 9, \ldots$ $a_n = 2n - 1$

20. $2, 4, 6, 8, 10, \ldots$ $a_n = 2n$

21. $-2, 2, 6, 10, 14, \ldots$ $a_n = 4n - 6$

22. $-3, 2, 7, 12, 17, \ldots$ $a_n = 5n - 8$

23. $5, 3, 1, -1, -3, \ldots$ $a_n = -2n + 7$

24. $2, -1, -4, -7, -10, \ldots$ $a_n = -3n + 5$

25. $-7, -10, -13, -16, -19, \ldots$ $a_n = -3n - 4$

26. $-3, -5, -7, -9, -11, \ldots$ $a_n = -2n - 1$

27. $1, \frac{3}{2}, 2, \frac{5}{2}, 3, \ldots$ $a_n = \frac{1}{2}(n + 1)$

28. $\frac{3}{2}, 3, \frac{9}{2}, 6, \frac{15}{2}, \ldots$ $a_n = \frac{3}{2}n$

For Problems 29–34, find the indicated term of the arithmetic sequence.

29. The 10th term of $7, 10, 13, 16, \ldots$ 34

30. The 12th term of $9, 11, 13, 15, \ldots$ 31

31. The 20th term of 2, 6, 10, 14,... 78

32. The 50th term of $-1, 4, 9, 14, \ldots$ 244

33. The 75th term of $-7, -9, -11, -13, \ldots$ -155

34. The 100th term of $-7, -10, -13, -16, \ldots$ -304

For Problems 35–40, find the number of terms of each of the finite arithmetic sequences.

35. 1, 3, 5, 7,..., 211 106

36. 2, 4, 6, 8,..., 312 156

37. 10, 13, 16, 19,..., 157 50

38. 9, 13, 17, 21,..., 849 211

39. $-7, -9, -11, -13, \ldots, -345$ 170

40. $-4, -7, -10, -13, \ldots, -331$ 110

41. If the 6th term of an arithmetic sequence is 24 and the 10th term is 44, find the first term. -1

42. If the 5th term of an arithmetic sequence is 26 and the 12th term is 75, find the first term. -2

43. If the 4th term of an arithmetic sequence is -9 and the 9th term is -29, find the 5th term. -13

44. If the 6th term of an arithmetic sequence is -4 and the 14th term is -20, find the 10th term. -12

45. In the arithmetic sequence .97, 1.00, 1.03, 1.06,..., which term is 5.02? 136

46. In the arithmetic sequence 1, 1.2, 1.4, 1.6,..., which term is 35.4? 173

For Problems 47–50, set up an arithmetic sequence and use $a_n = a_1 + (n - 1)d$ to solve each problem.

47. A woman started to work in 1975 at an annual salary of $12,500. She received a $900 raise each year. How much was her annual salary in 1992? $27,800

48. Math University had an enrollment of 8500 students in 1976. Each year the enrollment has increased by 350 students. What was the enrollment in 1990? $13,400

49. Suppose you are offered a job starting at $900 a month with a guaranteed increase of $30 a month every 6 months for the next 5 years. What will your monthly salary be for the last six months of the 5th year of your employment? $1170

50. Between 1976 and 1990 a person invested $500 at 14% simple interest at the beginning of each year. By the end of 1990, how much interest had been earned by the $500 that was invested at the beginning of 1982? $630

11.2 Arithmetic Series

Let's solve a problem to begin this section. Study the solution very carefully.

PROBLEM 1 Find the sum of the first one hundred positive integers.

Solution We are asked to find the sum of $1 + 2 + 3 + 4 + \cdots + 100$. Rather than use a calculator, let's find the sum in the following way.

$$\begin{array}{r}
1 + \quad 2 + \quad 3 + \quad 4 + \cdots + 100 \\
100 + \ 99 + \ 98 + \ 97 + \cdots + \quad 1 \\
\hline
101 + 101 + 101 + 101 + \cdots + 101
\end{array}$$

$$\frac{\overset{50}{\cancel{100}}(101)}{\cancel{2}} = 5050.$$

Note that we simply wrote the indicated sum *forward and backward* and then added the results. In so doing, 100 sums of 101 are produced, but half of them are "repeats." For example, (100 + 1) and (1 + 100) are both counted in this process. Thus, we divided the product (100)(101) by 2, which yielded the final result of 5050. ■

Now let's introduce some terminology that applies to problems like the one above. The indicated sum of a sequence is called a **series**. Associated with the finite sequence

$$a_1, a_2, a_3, \ldots, a_n$$

is a **finite series**

$$a_1 + a_2 + a_3 + \cdots + a_n.$$

Likewise, from the infinite sequence

$$a_1, a_2, a_3, \ldots$$

we can form the **infinite series**

$$a_1 + a_2 + a_3 + \cdots.$$

In this section we will direct our attention to working with **arithmetic series**; that is, the indicated sums of arithmetic sequences.

Problem 1 above could have been stated as "find the sum of the first one hundred terms of the arithmetic series that has a general term (*n*th term) of $a_n = n$." In fact, the *forward-backward* approach used to solve that problem can be applied to the general arithmetic series

$$a_1 + a_2 + a_3 + \cdots + a_n$$

to produce a formula for finding the sum of the first n terms of any arithmetic series. Use S_n to represent the sum of the first n terms and proceed as follows.

$$S_n = a_1 + (a_1 + d) + (a_1 + 2d) + \cdots + (a_n - 2d) + (a_n - d) + a_n.$$

Now write this sum in reverse as

$$S_n = a_n + (a_n - d) + (a_n - 2d) + \cdots + (a_1 + 2d) + (a_1 + d) + a_1.$$

Add the two equations to produce

$$\begin{aligned} 2S_n = {} & (a_1 + a_n) + (a_1 + a_n) + (a_1 + a_n) + \cdots + (a_1 + a_n) \\ & + (a_1 + a_n) + (a_1 + a_n). \end{aligned}$$

That is, we have n sums of $(a_1 + a_n)$, so

$$2S_n = n(a_1 + a_n)$$

from which we obtain

$$S_n = \frac{n(a_1 + a_n)}{2}.$$

Using the nth term formula $a_n = a_1 + (n - 1)d$ and the sum formula $\frac{n(a_1 + a_n)}{2}$, we can solve a variety of problems that involve arithmetic series.

EXAMPLE 1 Find the sum of the first 50 terms of the series

Solution

$$2 + 5 + 8 + 11 + \cdots.$$

Using $a_n = a_1 + (n - 1)d$, we find the 50th term to be

$$a_{50} = 2 + 49(3) = 149.$$

Then using $S_n = \frac{n(a_1 + a_n)}{2}$, we obtain

$$S_{50} = \frac{50(2 + 149)}{2} = 3775.$$

■

EXAMPLE 2 Find the sum of all odd numbers between 7 and 433, inclusive.

Solution We need to find the sum $7 + 9 + 11 + \cdots + 433$.

To use $S_n = \frac{n(a_1 + a_n)}{2}$, the number of terms, n, is needed. Perhaps you could figure this out without a formula (try it), but suppose we use the nth term formula.

$$a_n = a_1 + (n - 1)d$$

$$433 = 7 + (n - 1)2$$

$$433 = 7 + 2n - 2$$

$$433 = 2n + 5$$

$$428 = 2n$$

$$214 = n.$$

Then use $n = 214$ in the sum formula to yield

$$S_{214} = \frac{214(7 + 433)}{2} = 47{,}080.$$

■

EXAMPLE 3 Find the sum of the first 75 terms of the series that has a general term of $a_n = -5n + 9$.

Solution Using $a_n = -5n + 9$ we can generate the series as follows.

$$a_1 = -5(1) + 9 = 4,$$
$$a_2 = -5(2) + 9 = -1,$$
$$a_3 = -5(3) + 9 = -6,$$
$$\vdots$$
$$a_{75} = -5(75) + 9 = -366$$

Thus, we have the series

$$4 + (-1) + (-6) + \cdots + (-366).$$

Using the sum formula we obtain

$$S_{75} = \frac{75(4 + (-366))}{2} = -13{,}575.$$

■

EXAMPLE 4 Sue is saving quarters. She saves 1 quarter the first day, 2 quarters the second day, 3 quarters the third day, and so on. How much money will she have saved in 30 days?

Solution The total number of quarters will be the sum of the series

$$1 + 2 + 3 \cdots + 30.$$

Using the sum formula yields

$$S_{30} = \frac{30(1 + 30)}{2} = 465.$$

So Sue will have saved (465)(0.25) = $116.25 at the end of 30 days. ■

REMARK The sum formula, $S_n = \dfrac{n(a_1 + a_n)}{2}$, was developed by using the **forward-backward** technique that we previously used on a specific problem. Now that we have the sum formula, we have two choices as we meet problems where the formula applies. We can either memorize the formula and use it as it applies, or disregard the formula and use the forward-backward approach. However, even if you choose to use the formula and some day your memory fails you and you forget the formula, use the forward-backward technique. In other words, once you understand the development of a formula you can do some problems even though the formula itself is forgotten.

Problem Set 11.2

For Problems 1–10, find the sum of the indicated number of terms of the given series.

1. First 50 terms of $2 + 4 + 6 + 8 + \cdots$ 2550

2. First 45 terms of $1 + 3 + 5 + 7 + \cdots$ 2025

3. First 60 terms of $3 + 8 + 13 + 18 + \cdots$ 9030

4. First 80 terms of $2 + 6 + 10 + 14 + \cdots$ 12800

5. First 65 terms of $(-1) + (-3) + (-5) + (-7) + \cdots$ -4225

6. First 100 terms of $(-1) + (-4) + (-7) + (-10) + \cdots$ -14950

7. First 40 terms of $\frac{1}{2} + 1 + \frac{3}{2} + 2 + \cdots$ 410

8. First 50 terms of $1 + \frac{5}{2} + 4 + \frac{11}{2} + \cdots$ 1887.5

9. First 75 terms of $7 + 10 + 13 + 16 + \cdots$ 8850

10. First 90 terms of $(-8) + (-1) + 6 + 13 + \cdots$ 27315

For Problems 11–16, find the sum of each finite arithmetic series.

11. $4 + 8 + 12 + 16 + \cdots + 212$ 5724

12. $7 + 9 + 11 + 13 + \cdots + 179$ 8091

13. $(-4) + (-1) + 2 + 5 + \cdots + 173$ 5070

14. $5 + 10 + 15 + 20 + \cdots + 495$ 24750

15. $2.5 + 3.0 + 3.5 + 4.0 + \cdots + 18.5$ 346.5

16. $1 + (-6) + (-13) + (-20) + \cdots + (-202)$ -3015

For Problems 17–22, find the sum of the indicated number of terms of the series with the given nth term.

17. First 50 terms of series with $a_n = 3n - 1$ 3775

18. First 150 terms of series with $a_n = 2n - 7$ 21600

19. First 125 terms of series with $a_n = 5n + 1$ 39500

20. First 75 terms of series with $a_n = 4n + 3$ 11625

21. First 65 terms of series with $a_n = -4n - 1$ -8645

22. First 90 terms of series with $a_n = -3n - 2$ -12465

For Problems 23–34, use arithmetic sequences and series to help solve the problem.

23. Find the sum of the first 350 positive even whole numbers. 122,850

24. Find the sum of the first 200 odd whole numbers. 40,000

25. Find the sum of all odd whole numbers between 15 and 397, inclusive. 39,552

26. Find the sum of all even whole numbers between 14 and 286, inclusive. 20,552

27. An auditorium has 20 seats in the front row, 24 seats in the second row, 28 seats in the third row, and so on, for 15 rows. How many seats are there in the last row? How many seats are there in the auditorium? 76 seats; 720 seats

28. A pile of wood has 15 logs in the bottom row, 14 logs in the next to the bottom row, and so on, with one less log in each row until the top row, which consists of 1 log. How many logs are there in the pile? 120 logs

29. A raffle is organized so that the amount paid for each ticket is determined by a number on the ticket. The tickets are numbered with the consecutive odd whole numbers, 1, 3, 5, 7,.... Each participant pays as many cents as the number on the ticket drawn. How much money will the raffle take in if 1000 tickets are sold? $10,000

30. A woman invests $700 at 13% simple interest at the beginning of each year for a period of 15 years. Find the total accumulated value of all of the investments at the end of the 15-year period. $21,420

31. A man started to work in 1970 at an annual salary of $18,500. He received a $1500 raise each year through 1982. What were his total earnings for the 13-year period? $357,500

32. A well-driller charges $9.00 per foot for the first 10 feet, $9.10 per foot for the next 10 feet, $9.20 per foot for the next 10 feet, and so on; he continues to increase the price by $0.10 per foot for succeeding intervals of 10 feet. How much would it cost to drill a well with a depth of 150 feet? $1455

33. A display in a grocery store has cans stacked with 25 cans in the bottom row, 23 cans in the second row from the bottom, 21 cans in the third row from the bottom, and so on until there is only 1 can in the top row. How many cans are there in the display? 169

34. Suppose that a person starts on the first day of August and saves a dime the first day, $0.20 the second day, $0.30 the third day, and continues to save $0.10 more per day than the previous day. How much could be saved in the 31 days of August? $49.60

Miscellaneous Problems

35. We can express a series where the general term is known in a convenient and compact form using the symbol $\sum$ along with the general term expression. For example, consider the finite arithmetic series

$$1 + 3 + 5 + 7 + 9 + 11$$

where the general term is $a_n = 2n - 1$. This series can be expressed in *summation notation* as

$$\sum_{i=1}^{6} (2i - 1)$$

where the letter i is used as the **index of summation**. The individual terms of the series can be generated by successively replacing i in the expression $(2i - 1)$ with the numbers 1, 2, 3, 4, 5, and 6. Thus, the first term is $2(1) - 1 = 1$, the second term $2(2) - 1 = 3$, the third term is $2(3) - 1 = 5$, and so on. Write out the terms and find the sum of each of the following series.

(a) $\sum_{i=1}^{3} (5i + 2)$ $7 + 12 + 17; 36$

(b) $\sum_{i=1}^{4} (6i - 7)$ $-1 + 5 + 11 + 17; 32$

(c) $\sum_{i=1}^{6} (-2i - 1)$ $(-3) + (-5) + (-7) + (-9) + (-11) + (-13); -48$

(d) $\sum_{i=1}^{5} (-3i + 4)$ $1 + (-2) + (-5) + (-8) + (-11); -25$

(e) $\sum_{i=1}^{5} 3i$ $3 + 6 + 9 + 12 + 15; 45$

(f) $\sum_{i=1}^{6} -4i$ $-4 + (-8) + (-12) + (-16) + (-20) + (-24); -84$

36. Write each of the following in summation notation. For example, since $3 + 8 + 13 + 18 + 23 + 28$ is an arithmetic series, the general term formula $a_n = a_1 + (n - 1)d$ yields

$$\begin{aligned} a_n &= 3 + (n - 1)5 \\ &= 3 + 5n - 5 \\ &= 5n - 2. \end{aligned}$$

Now using i as an index of summation, we can write

$$\sum_{i=1}^{6} (5i - 2).$$

(a) $2 + 5 + 8 + 11 + 14$ $\sum_{i=1}^{5}(3i - 1)$

(b) $8 + 15 + 22 + 29 + 36 + 43$ $\sum_{i=1}^{6}(7i + 1)$

(c) $1 + (-1) + (-3) + (-5) + (-7)$ $\sum_{i=1}^{5}(-2i + 3)$

(d) $(-5) + (-9) + (-13) + (-17) + (-21) + (-25) + (-29)$ $\sum_{i=1}^{7}(-4i - 1)$

11.3 Geometric Sequences and Series

A **geometric sequence** or **geometric progression** is a sequence in which each term after the first is obtained by multiplying the preceding term by a common multiplier. The common multiplier is called the **common ratio** of the sequence. The following geometric sequences have common ratios of 2, 3, $\frac{1}{2}$, and -4, respectively.

$$1, 2, 4, 8, 16, \ldots, \qquad 3, 9, 27, 81, 243, \ldots,$$

$$8, 4, 2, 1, \frac{1}{2}, \ldots, \qquad 1, -4, 16, -64, 256, \ldots$$

We find the common ratio of a geometric sequence by dividing a term (other than the first term) by the preceding term.

In a more general setting we say that the sequence

$$a_1, a_2, a_3, a_4, \ldots, a_n, \ldots$$

is a geometric sequence if and only if there is a nonzero real number r, such that

$$a_{k+1} = ra_k \qquad (1)$$

for every positive integer k. The nonzero real number r is called the common ratio.

Equation (1) can be used to generate a general geometric sequence that has a_1 as a first term and r as a common ratio. We can proceed as follows.

first term	a_1	
second term	$a_1 r$	
third term	$a_1 r^2$	$(a_1 r)(r) = a_1 r^2$
fourth term	$a_1 r^3$	$(a_1 r^2)(r) = a_1 r^3$
$\vdots$		
nth term	$a_1 r^{n-1}$	

Thus, the *general term of a geometric sequence* is given by

$$a_n = a_1 r^{n-1}$$

where a_1 is the first term and r is the common ratio.

EXAMPLE 1 Find the general term for the geometric sequence 2, 4, 8, 16, ...

Solution Using $a_n = a_1 r^{n-1}$, we obtain

$$a_n = 2(2)^{n-1} \qquad r = \frac{4}{2} = \frac{8}{4} = \frac{16}{8} = 2$$

$$= 2^n. \qquad 2^1(2)^{n-1} = 2^{1+n-1} = 2^n$$

■

EXAMPLE 2 Find the 10th term of the geometric sequence 9, 3, 1, ...

Solution Using $a_n = a_1 r^{n-1}$, we can find the 10th term as follows.

$$a_{10} = 9\left(\frac{1}{3}\right)^{10-1}$$

$$= 9\left(\frac{1}{3}\right)^{9}$$

$$= 9\left(\frac{1}{19{,}683}\right)$$

$$= \frac{1}{2187}$$

■

A **geometric series** is the indicated sum of a geometric sequence. The following are examples of geometric series.

$$1 + 2 + 4 + 8 + \cdots,$$

$$3 + 9 + 27 + 81 + \cdots,$$

$$8 + 4 + 2 + 1 + \cdots,$$

$$1 + (-4) + 16 + (-64) + \cdots$$

Before we develop a general formula for finding the sum of a geometric series, let's consider a specific example.

EXAMPLE 3 Find the sum of $1 + 2 + 4 + 8 + \cdots + 512$.

Solution Let S represent the sum and we can proceed as follows.

$$S = 1 + 2 + 4 + 8 + \cdots + 512, \qquad (2)$$

$$2S = \quad 2 + 4 + 8 + \cdots + 512 + 1024 \qquad (3)$$

Equation (3) is the result of multiplying both sides of equation (2) by 2. Subtract equation (2) from equation (3) to yield

$$S = 1024 - 1 = 1023.$$

■

Now let's consider the general geometric series

$$a_1 + a_1r + a_1r^2 + \cdots + a_1r^{n-1}.$$

By applying a procedure similar to the one used in Example 3 we can develop a formula for finding the sum of the first n terms of any geometric series.

Let S_n represent the sum of the first n terms. Thus,

$$S_n = a_1 + a_1r + a_1r^2 + \cdots + a_1r^{n-1}. \qquad (4)$$

Multiply both sides of equation (4) by the common ratio r to produce

$$rS_n = a_1r + a_1r^2 + a_1r^3 + \cdots + a_1r^{n-1} + a_1r^n. \qquad (5)$$

Subtract equation (4) from equation (5) to yield

$$rS_n - S_n = a_1r^n - a_1.$$

Apply the distributive property on the left side and then solve for S_n to obtain

$$S_n(r - 1) = a_1r^n - a_1$$

$$S_n = \frac{a_1r^n - a_1}{r - 1}, \qquad r \neq 1.$$

Therefore, the sum of the first n terms of a geometric series that has a first term of a_1 and a common ratio of r is given by

$$S_n = \frac{a_1r^n - a_1}{r - 1}, \qquad r \neq 1.$$

EXAMPLE 4 Find the sum of the first 7 terms of the geometric series $2 + 6 + 18 + \cdots$.

Solution Use the sum formula to obtain

$$\begin{aligned} S_7 &= \frac{2(3)^7 - 2}{3 - 1} \\ &= \frac{2(3^7 - 1)}{2} \\ &= 3^7 - 1 \\ &= 2187 - 1 \\ &= 2186. \end{aligned}$$

■

If the common ratio of a geometric series is less than 1, it may be more convenient to change the form of the sum formula. That is, we can change the fraction $\frac{a_1 r^n - a_1}{r - 1}$ to $\frac{a_1 - a_1 r^n}{1 - r}$ by multiplying both the numerator and denominator by -1. Thus, using $S_n = \frac{a_1 - a_1 r^n}{1 - r}$ when $r < 1$ can sometimes avoid unnecessary work with negative numbers, as the next example demonstrates. ■

EXAMPLE 5 Find the sum of the geometric series $1 + \frac{1}{2} + \frac{1}{4} + \cdots + \frac{1}{256}$.

Solution A To use the sum formula we need to know the number of terms, which can be found by simply counting them or applying the nth term formula as follows.

$$\begin{aligned} a_n &= a_1 r^{n-1} \\ \frac{1}{256} &= 1\left(\frac{1}{2}\right)^{n-1} \\ \left(\frac{1}{2}\right)^8 &= \left(\frac{1}{2}\right)^{n-1} \\ 8 &= n - 1 \\ 9 &= n \end{aligned}$$

Remember that "if $b^n = b^m$, then $n = m$."

Using $n = 9$, $a_1 = 1$, and $r = \frac{1}{2}$ in the form of the sum formula

$$S_n = \frac{a_1 - a_1 r^n}{1 - r},$$

we obtain

$$S_9 = \frac{1 - 1\left(\frac{1}{2}\right)^9}{1 - \frac{1}{2}}$$

$$= \frac{1 - \frac{1}{512}}{\frac{1}{2}} = \frac{\frac{511}{512}}{\frac{1}{2}}$$

$$= \left(\frac{511}{512}\right)\left(\frac{2}{1}\right) = \frac{511}{256} \quad \text{or} \quad 1\frac{255}{256}.$$

You should realize that a problem such as Example 5 can be done *without* using the sum formula; you can apply the general technique used to develop the formula. Solution B illustrates this approach.

Solution B Let S represent the desired sum. Thus,

$$S = 1 + \frac{1}{2} + \frac{1}{4} + \cdots + \frac{1}{256}.$$

Multiply both sides by $\frac{1}{2}$ (the common ratio).

$$\frac{1}{2}S = \frac{1}{2} + \frac{1}{4} + \cdots + \frac{1}{256} + \frac{1}{512}$$

Subtract the second equation from the first equation to produce

$$\frac{1}{2}S = 1 - \frac{1}{512}$$

$$\frac{1}{2}S = \frac{511}{512}$$

$$S = \frac{511}{256} = 1\frac{255}{256}.$$

■

EXAMPLE 6 Suppose your employer agrees to pay you a penny for your first day's wages and then agrees to double your pay on each succeeding day. How much will you earn on the 15th day? What will be your total earnings for the first 15 days?

Solution The terms of the geometric series $1 + 2 + 4 + 8 + \cdots$ depict your daily wages and the sum of the first 15 terms is your total earnings for the 15 days. The formula $a_n = a_1r^{n-1}$ can be used to find the 15th day's wages.

$$a_{15} = (1)(2)^{14} = 16{,}384.$$

Since the terms of the series are expressed in cents, your wages for the 15th day would be \$163.84. Now using the sum formula we can find your total earnings as follows.

$$S_n = \frac{a_1 r^n - a_1}{r - 1}$$

$$S_{15} = \frac{1(2)^{15} - 1}{1} = 32{,}768 - 1 = 32{,}767.$$

Thus, for the 15 days you would earn a total of \$327.67. ■

Problem Set 11.3

For Problems 1–12, find the general term (nth term) of each geometric sequence.

1. $1, 3, 9, 27, \ldots$ $a_n = 3^{n-1}$
2. $1, 2, 4, 8, \ldots$ $a_n = 2^{n-1}$
3. $2, 8, 32, 128, \ldots$ $a_n = 2(4)^{n-1}$ or $a_n = 2^{2n-1}$
4. $3, 9, 27, 81, \ldots$ $a_n = 3(3)^{n-1}$ or $a_n = 3^n$
5. $1, \frac{1}{3}, \frac{1}{9}, \frac{1}{27}, \ldots$ $a_n = \left(\frac{1}{3}\right)^{n-1}$ or $a_n = \frac{1}{3^{n-1}}$
6. $\frac{1}{2}, \frac{1}{4}, \frac{1}{8}, \frac{1}{16}, \ldots$ $a_n = \left(\frac{1}{2}\right)^n$ or $a_n = \frac{1}{2^n}$
7. $.2, .04, .008, .0016, \ldots$ $a_n = .2(.2)^{n-1}$ or $a_n = (.2)^n$
8. $1, .3, .09, .027, \ldots$ $a_n = (.3)^{n-1}$
9. $9, 6, 4, \frac{8}{3}, \ldots$ $a_n = 9\left(\frac{2}{3}\right)^{n-1}$ or $a_n = (3^{-n+3})(2)^{n-1}$
10. $6, 2, \frac{2}{3}, \frac{2}{9}, \ldots$ $a_n = 6\left(\frac{1}{3}\right)^{n-1}$ or $a_n = (2)(3)^{-n+2}$
11. $1, -4, 16, -64, \ldots$ $a_n = (-4)^{n-1}$
12. $1, -2, 4, -8, \ldots$ $a_n = (-2)^{n-1}$

For Problems 13–18, find the indicated term of the geometric sequence.

13. 12th term of $\frac{1}{9}, \frac{1}{3}, 1, 3, \ldots$ 19,683
14. 9th term of $2, 4, 8, 16, \ldots$ 512
15. 10th term of $1, -2, 4, -8, \ldots$ -512
16. 8th term of $\frac{1}{2}, \frac{1}{8}, \frac{1}{32}, \frac{1}{128}, \ldots$ $\frac{1}{32{,}768}$

17. 9th term of $-1, -\frac{3}{2}, -\frac{9}{4}, -\frac{27}{8}, \ldots$ $-\frac{6561}{256}$

18. 11th term of $1, \frac{2}{3}, \frac{4}{9}, \frac{8}{27}, \ldots$ $\frac{1024}{59{,}049}$

For Problems 19–24, find the sum of the indicated number of terms of each geometric series.

19. First 10 terms of $\frac{1}{2} + \frac{3}{2} + \frac{9}{2} + \frac{27}{2} + \cdots$ 14,762

20. First 9 terms of $1 + 2 + 4 + 8 + \cdots$ 511

21. First 9 terms of $-2 + 6 + (-18) + 54 + \cdots$ -9842

22. First 10 terms of $-4 + 8 + (-16) + 32 + \cdots$ 1364

23. First 7 terms of $1 + 3 + 9 + 27 + \cdots$ 1093

24. First 8 terms of $4 + 2 + 1 + \frac{1}{2} + \cdots$ $\frac{255}{32}$

For Problems 25–30, find the sum of the indicated number of terms of the geometric series with the given nth term.

25. First 9 terms of series where $a_n = 2^{n-1}$ 511

26. First 8 terms of series where $a_n = 3^n$ 9840

27. First 8 terms of series where $a_n = 2(3)^n$ 19,680

28. First 10 terms of series where $a_n = \frac{1}{2^{n-4}}$ $\frac{1023}{64}$

29. First 12 terms of series where $a_n = (-2)^n$ 2730

30. First 9 terms of series where $a_n = (-3)^{n-1}$ 4221

For Problems 31–36, find the sum of each finite geometric series.

31. $1 + 3 + 9 + \cdots + 729$ 1093

32. $2 + 8 + 32 + \cdots + 2048$ 2730

33. $1 + \frac{1}{2} + \frac{1}{4} + \cdots + \frac{1}{1024}$ $1\frac{1023}{1024}$

34. $1 + (-2) + 4 + \cdots + (-128)$ -85

35. $8 + 4 + 2 + \cdots + \frac{1}{32}$ $15\frac{31}{32}$

36. $2 + 6 + 18 + \cdots + 4374$ 6560

For Problems 37–48, use geometric sequences and series to help solve the problem.

37. Find the common ratio of a geometric sequence if the 2nd term is $\frac{1}{6}$ and the 5th term is $\frac{1}{48}$. $\frac{1}{2}$

38. Find the first term of a geometric sequence if the 5th term is $\frac{32}{3}$ and the common ratio is 2. $\frac{2}{3}$

39. Find the sum of the first 16 terms of the geometric series where $a_n = (-1)^n$. Also find the sum of the first 19 terms. 0; -1

40. A fungus culture growing under controlled conditions doubles in size each day. How many units will the culture contain after 7 days if it originally contained 5 units? 640 units

41. A tank contains 16,000 liters of water. Each day one-half of the water in the tank is removed and not replaced. How much water remains in the tank at the end of the 7th day? 125 liters

42. Suppose that you save 25 cents the first day of a week, 50 cents the second day, \$1 the third day, and continue to double your savings each day. How much will you save on the 7th day? What will be your total savings for the week? \$16; \$31.75

43. Suppose you save a nickel the first day of a month, a dime the second day, 20 cents the third day, and continue to double your savings each day. How much will you save on the 12th day of the month? What will be your total savings for the first 12 days? \$102.40; \$204.75

44. Suppose an element has a half-life of 3 hours. This means that if n grams of it exist at a specific time, then only $\frac{1}{2}n$ grams remain 3 hours later. If at a particular moment we have 40 grams of the element, how much of it remains 24 hours later? $\frac{5}{32}$ gram

45. A rubber ball is dropped from a height of 486 meters and each time it rebounds one-third of the height from which it last fell. How far has the ball traveled by the time it strikes the ground for the 7th time? $971.\bar{3}$ meters

46. A pump is attached to a container for the purpose of creating a vacuum. For each stroke of the pump, one-fourth of the air remaining in the container is removed. $\frac{729}{4096}$

(a) Form a geometric sequence where each term represents the fractional part of the air *that still remains* in the container after each stroke. Then use this sequence to find out how much of the air remains after 6 strokes.

(b) Form a geometric sequence where each term represents the fractional part of the air *being removed* from the container on each stroke of the pump. Then use this sequence (or the associated *series*) to find out how much of the air remains after 6 strokes.

47. If you pay \$9500 for a car and its value depreciates 10% per year, how much will it be worth in 5 years? \$5609.66

48. Suppose that you could get a job that pays only a penny for the first day of employment, but doubles your wages each succeeding day. How much would you be earning on the 31st day of your employment? \$10,737,418

Miscellaneous Problems

For Problems 49–54, use your calculator to help find the indicated term of each geometric sequence.

49. 20th term of 2, 4, 8, 16,... 1,048,576

50. 15th term of 3, 9, 27, 81,... 14,348,907

51. 12th term of $\frac{2}{3}, \frac{4}{9}, \frac{8}{27}, \frac{16}{81}, \ldots$ $\frac{4096}{531{,}441}$

52. 10th term of $-\frac{3}{4}, \frac{9}{16}, -\frac{27}{64}, \frac{81}{256}, \ldots$ $\frac{59{,}049}{1{,}048{,}576}$

53. 11th term of $-\frac{3}{2}, \frac{9}{4}, -\frac{27}{8}, \frac{81}{16}, \ldots$ $-\frac{177{,}147}{2048}$

54. 6th term of the sequence where $a_n = (0.1)^n$ .000001

55. In Problem 35 of Problem Set 11.2 we introduced the summation notation, which can also be used with geometric series. For example, the series

$$2 + 4 + 8 + 16 + 32$$

can be expressed as

$$\sum_{i=1}^{5} 2^i.$$

Write out the terms and find the sum of each of the following.

(a) $\sum_{i=1}^{6} 2^i$ 126 **(b)** $\sum_{i=1}^{5} 3^i$ 363 **(c)** $\sum_{i=1}^{5} 2^{i-1}$ 31

(d) $\sum_{i=1}^{6} \left(\frac{1}{2}\right)^{i+1}$ $\frac{63}{128}$ **(e)** $\sum_{i=1}^{4} \left(\frac{2}{3}\right)^i$ $1\frac{49}{81}$ **(f)** $\sum_{i=1}^{5} \left(-\frac{3}{4}\right)^i$ $-\frac{543}{1024}$

11.4
Infinite Geometric Series

In Section 11.3 we used the formula

$$S_n = \frac{a_1 - a_1 r^n}{1 - r}, \qquad r \neq 1 \tag{1}$$

to find the sum of the first n terms of a geometric series. By using the property, $\frac{a - b}{c} = \frac{a}{c} - \frac{b}{c}$, we can express the right side of equation (1) in terms of two fractions as follows.

$$S_n = \frac{a_1 - a_1 r^n}{1 - r} = \frac{a_1}{1 - r} - \frac{a_1 r^n}{1 - r}, \qquad r \neq 1 \tag{2}$$

Now let's examine the behavior of r^n for $|r| < 1$, that is, for $-1 < r < 1$. For example, suppose that $r = \frac{1}{3}$; then

$$r^2 = \left(\frac{1}{3}\right)^2 = \frac{1}{9}, \qquad r^3 = \left(\frac{1}{3}\right)^3 = \frac{1}{27},$$

$$r^4 = \left(\frac{1}{3}\right)^4 = \frac{1}{81}, \qquad r^5 = \left(\frac{1}{3}\right)^5 = \frac{1}{243},$$

and so on. We can make $\left(\frac{1}{3}\right)^n$ as close to 0 as we please by taking sufficiently large values for n. In general, for values of r such that $|r| < 1$, the expression r^n will approach 0 as n increases. Therefore, in equation (2) the fraction $\frac{a_1 r^n}{1-r}$ will approach 0 as n increases and we say that the *sum of an infinite geometric series* is given by

$$S_\infty = \frac{a_1}{1-r}, \qquad |r| < 1.$$

EXAMPLE 1 Find the sum of the infinite geometric series

$$1 + \frac{2}{3} + \frac{4}{9} + \frac{8}{27} + \cdots.$$

Solution Since $a_1 = 1$ and $r = \frac{2}{3}$, we obtain

$$S_\infty = \frac{1}{1 - \frac{2}{3}}$$

$$= \frac{1}{\frac{1}{3}} = 3.$$

■

In Example 1, by stating $S_\infty = 3$ we mean that as we add more and more terms, the sum approaches 3 as follows.

first term: 1

sum of first 2 terms: $1 + \frac{2}{3} = 1\frac{2}{3}$

sum of first 3 terms: $1 + \frac{2}{3} + \frac{4}{9} = 2\frac{1}{9}$

sum of first 4 terms: $1 + \frac{2}{3} + \frac{4}{9} + \frac{8}{27} = 2\frac{11}{27}$

sum of first 5 terms: $1 + \frac{2}{3} + \frac{4}{9} + \frac{8}{27} + \frac{16}{81} = 2\frac{49}{81}$, etc.

EXAMPLE 2 Find the sum of the infinite geometric series

$$\frac{1}{2} - \frac{1}{4} + \frac{1}{8} - \frac{1}{16} + \cdots.$$

Solution Since $a_1 = \frac{1}{2}$ and $r = -\frac{1}{2}$, we obtain

$$S_\infty = \frac{\frac{1}{2}}{1 - \left(-\frac{1}{2}\right)} = \frac{\frac{1}{2}}{\frac{3}{2}} = \frac{1}{3}.$$

■

If $|r| > 1$, the absolute value of r^n increases without bound as n increases. Consider the following two examples and notice the unbounded growth of the absolute value of r^n.

Let r = 2	*Let r = −3*	
$r^2 = 2^2 = 4,$	$r^2 = (-3)^2 = 9,$	
$r^3 = 2^3 = 8,$	$r^3 = (-3)^3 = -27,$	$\lvert -27 \rvert = 27,$
$r^4 = 2^4 = 16,$	$r^4 = (-3)^4 = 81,$	
$r^5 = 2^5 = 32,$	$r^5 = (-3)^5 = -243,$	$\lvert -243 \rvert = 243,$
etc.	etc.	

If $r = 1$, then $S_n = na_1$, and as n increases without bound, $|S_n|$ also increases without bound. If $r = -1$, then S_n will either be a_1 or 0. Therefore, we say that the sum of any infinite geometric sequence where $|r| \geq 1$ *does not exist.*

Repeating Decimals as Infinite Geometric Series

In Section 1.1 we learned that a rational number is a number that has either a terminating or repeating decimal representation. For example,

$$0.23, \quad 0.147, \quad 0.\overline{3}, \quad 0.1\overline{4}, \quad \text{and} \quad 0.\overline{581}$$

are examples of rational numbers. (Remember that $0.\overline{3}$ means 0.333. . . .) Our knowledge of place value provides the basis for changing terminating decimals such as 0.23 and 0.147 to $\frac{a}{b}$ form, where a and b are integers, $b \neq 0$.

$$0.23 = \frac{23}{100},$$

$$0.147 = \frac{147}{1000}$$

However, changing repeating decimals to $\frac{a}{b}$ form requires a different technique and our work with infinite geometric series provides the basis for one such approach. Consider the following examples.

EXAMPLE 3 Change $0.\overline{3}$ to $\frac{a}{b}$ form, where a and b are integers, $b \neq 0$.

Solution We can write the repeating decimal $0.\overline{3}$ as the infinite geometric series

$$0.3 + 0.03 + 0.003 + 0.0003 + \cdots$$

with $a_1 = 0.3$ and $r = 0.1$. Therefore, we can use the sum formula and obtain

$$S_\infty = \frac{a_1}{1 - r} = \frac{0.3}{1 - 0.1} = \frac{0.3}{0.9} = \frac{3}{9} = \frac{1}{3}.$$

So, $0.\overline{3} = \frac{1}{3}$. ■

EXAMPLE 4 Change $0.\overline{14}$ to $\frac{a}{b}$ form, where a and b are integers, $b \neq 0$.

Solution We can write the repeating decimal $0.\overline{14}$ as the infinite geometric series

$$0.14 + 0.0014 + 0.000014 + \cdots$$

with $a_1 = 0.14$ and $r = 0.01$. The sum formula produces

$$S_\infty = \frac{0.14}{1 - 0.01} = \frac{0.14}{0.99} = \frac{14}{99}.$$

Thus, $0.\overline{14} = \frac{14}{99}$. ■

If the repeating block of digits does not begin immediately after the decimal point we can make a slight adjustment, as the final example illustrates.

EXAMPLE 5 Change $0.5\overline{81}$ to $\frac{a}{b}$ form, where a and b are integers, $b \neq 0$.

Solution We can write the repeating decimal $0.5\overline{81}$ as

$$[0.5] + [0.081 + 0.00081 + 0.0000081 + \cdots]$$

where

$$0.081 + 0.00081 + 0.0000081 + \cdots$$

is an infinite geometric series with $a_1 = 0.081$ and $r = 0.01$. Thus,

$$S_\infty = \frac{0.081}{1 - 0.01} = \frac{0.081}{0.99} = \frac{81}{990} = \frac{9}{110}.$$

Therefore,

$$\begin{aligned} 0.5\overline{81} &= 0.5 + \frac{9}{110} \\ &= \frac{5}{10} + \frac{9}{110} \\ &= \frac{55}{110} + \frac{9}{110} \\ &= \frac{64}{110} \\ &= \frac{32}{55}. \end{aligned}$$

■

Problem Set 11.4

For Problems 1–20, find the sum of the infinite geometric series. If the series has no sum, so state.

1. $1 + \frac{3}{4} + \frac{9}{16} + \frac{27}{64} + \cdots$ 4

2. $\frac{2}{3} + \frac{2}{9} + \frac{2}{27} + \frac{2}{81} + \cdots$ 1

3. $\frac{1}{2} + \frac{1}{4} + \frac{1}{8} + \frac{1}{16} + \cdots$ 1

4. $1 + \frac{1}{2} + \frac{1}{4} + \frac{1}{8} + \cdots$ 2

5. $\frac{2}{3} + \frac{4}{9} + \frac{8}{27} + \frac{16}{81} + \cdots$ 2

6. $\frac{1}{3} + \frac{1}{9} + \frac{1}{27} + \frac{1}{81} + \cdots$ $\frac{1}{2}$

7. $1 - \frac{1}{2} + \frac{1}{4} - \frac{1}{8} + \cdots$ $\frac{2}{3}$

8. $1 + 2 + 4 + 8 + \cdots$ No sum

9. $6 + 2 + \frac{2}{3} + \frac{2}{9} + \cdots$ 9

10. $4 + (-2) + 1 + \left(-\frac{1}{2}\right) + \cdots$ $\frac{8}{3}$

11. $2 + (-6) + 18 + (-54) + \cdots$ No sum

12. $4 + 2 + 1 + \frac{1}{2} + \cdots$ 8

13. $1 + \left(-\frac{3}{4}\right) + \frac{9}{16} + \left(-\frac{27}{64}\right) + \cdots$ $\frac{4}{7}$

14. $9 - 3 + 1 - \frac{1}{3} + \cdots$ $\frac{27}{4}$

15. $8 - 4 + 2 - 1 + \cdots$ $\frac{16}{3}$

16. $5 + 3 + \frac{9}{5} + \frac{27}{25} + \cdots$ $\frac{25}{2}$

17. $1 + \frac{3}{2} + \frac{9}{4} + \frac{27}{8} + \cdots$ No sum

18. $1 - \frac{4}{3} + \frac{16}{9} - \frac{64}{27} + \cdots$ No sum

19. $27 + 9 + 3 + 1 + \cdots$ $\frac{81}{2}$

20. $9 + 3 + 1 + \frac{1}{3} + \cdots$ $\frac{27}{2}$

For Problems 21–34, change each repeating decimal to $\frac{a}{b}$ form, where a and b are integers, $b \neq 0$. Express $\frac{a}{b}$ in reduced form.

21. $.\overline{4}$ $\frac{4}{9}$ **22.** $.\overline{7}$ $\frac{7}{9}$ **23.** $.\overline{47}$ $\frac{47}{99}$ **24.** $.\overline{23}$ $\frac{23}{99}$ **25.** $.\overline{45}$ $\frac{5}{11}$

26. $.72$ $\frac{8}{11}$ **27.** $.\overline{427}$ $\frac{427}{999}$ **28.** $.\overline{129}$ $\frac{43}{333}$ **29.** $.4\overline{6}$ $\frac{7}{15}$ **30.** $.8\overline{6}$ $\frac{13}{15}$

31. $2.\overline{18}$ $\frac{72}{33}$ **32.** $2.\overline{96}$ $\frac{98}{33}$ **33.** $.4\overline{27}$ $\frac{47}{110}$ **34.** $.2\overline{36}$ $\frac{13}{55}$

11.5

Binomial Expansions

In Chapter 3, we used the pattern $(x + y)^2 = x^2 + 2xy + y^2$ to square binomials and we used the pattern $(x + y)^3 = x^3 + 3x^2y + 3xy^2 + y^3$ to cube binomials. At this time, we can extend those ideas to arrive at a pattern that will allow us to write the expansion of $(x + y)^n$, where n is *any* positive integer. Let's begin by looking at some specific expansions that we can verify by direct multiplication.

$$(x + y)^1 = x + y,$$
$$(x + y)^2 = x^2 + 2xy + y^2,$$
$$(x + y)^3 = x^3 + 3x^2y + 3xy^2 + y^3,$$
$$(x + y)^4 = x^4 + 4x^3y + 6x^2y^2 + 4xy^3 + y^4,$$
$$(x + y)^5 = x^5 + 5x^4y + 10x^3y^2 + 10x^2y^3 + 5xy^4 + y^5$$

First, note the patterns of the exponents for x and y on a term-by-term basis. The exponents of x begin with the exponent of the binomial and term-by-term decrease by 1 until the last term has x^0, which is 1. The exponents of y begin with $0(y^0 = 1)$ and term-by-term increase by 1 until the last term contains y to the power of the original binomial. In other words, the variables in the expansion of $(x + y)^n$ have the following pattern.

$$x^n, \quad x^{n-1}y, \quad x^{n-2}y^2, \quad x^{n-3}y^3, \quad \ldots, \quad xy^{n-1}, \quad y^n.$$

Notice that the sum of the exponents of x and y for each term is n.

Next, let's arrange the **coefficients** in the following triangular formation that yields an easy-to-remember pattern.

```
        1   1
      1   2   1
    1   3   3   1
  1   4   6   4   1
1   5  10  10   5   1
```

The number of the row of the formation contains the coefficients of the expansion of $(x + y)$ to that power. For example, the 5th row contains 1 5 10 10 5 1, which are the coefficients of the terms of the expansion of $(x + y)^5$. Furthermore, each row can be formed from the previous row as follows.

1. Start and end each row with 1.
2. All other entries result from adding the two numbers in the row immediately above, one number to the left and one number to the right.

Thus, from row 5 we can form row 6 as follows.

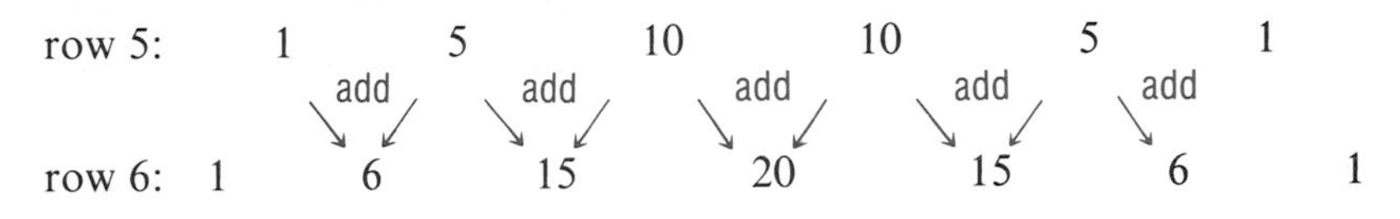

We can use the row 6 coefficients and our previous discussion relative to the exponents and write out the expansion for $(x + y)^6$.

$$(x + y)^6 = x^6 + 6x^5y + 15x^4y^2 + 20x^3y^3 + 15x^2y^4 + 6xy^5 + y^6$$

REMARK We often refer to the triangular formation of numbers that we have been discussing as Pascal's triangle. This is in honor of Blaise Pascal, a 17th century mathematician, to whom the discovery of this pattern is attributed.

Although Pascal's triangle will work for any positive integral power of a binomial, it does become somewhat impractical for large powers. So we need another technique for determining the coefficients. Let's look at the following notational agreements. $n!$ (read "n factorial") means $n(n - 1)(n - 2)\cdots 1$, where n is any positive integer. For example,

$3!$ means $3 \cdot 2 \cdot 1 = 6$,

$5!$ means $5 \cdot 4 \cdot 3 \cdot 2 \cdot 1 = 120$.

We also agree that $0! = 1$. (Note that both $0!$ and $1!$ equal 1.)

Let us now use the factorial notation and state the expansion of the general case $(x + y)^n$, where n is any positive integer.

$$(x + y)^n = x^n + nx^{n-1}y + \frac{n(n-1)}{2!}x^{n-2}y^2 + \frac{n(n-1)(n-2)}{3!}x^{n-3}y^3 + \cdots + y^n$$

The binomial expansion for the general case may look a little confusing, but actually it is quite easy to apply once you try it a few times on some specific examples. Remember the decreasing pattern for the exponents of x and the increasing pattern for the exponents of y. Furthermore, note the following pattern of the coefficients.

$$1, \quad n, \quad \frac{n(n-1)}{2!}, \quad \frac{n(n-1)(n-2)}{3!}, \quad \text{etc.}$$

Keep these ideas in mind as you study the following examples.

EXAMPLE 1 Expand $(x + y)^7$.

Solution

$$(x + y)^7 = x^7 + 7x^6y + \frac{7 \cdot 6}{2!}x^5y^2 + \frac{7 \cdot 6 \cdot 5}{3!}x^4y^3 + \frac{7 \cdot 6 \cdot 5 \cdot 4}{4!}x^3y^4 + \frac{7 \cdot 6 \cdot 5 \cdot 4 \cdot 3}{5!}x^2y^5 + \frac{7 \cdot 6 \cdot 5 \cdot 4 \cdot 3 \cdot 2}{6!}xy^6 + y^7$$

$$= x^7 + 7x^6y + 21x^5y^2 + 35x^4y^3 + 35x^3y^4 + 21x^2y^5 + 7xy^6 + y^7$$

■

EXAMPLE 2 Expand $(x - y)^5$.

Solution We shall treat $(x - y)^5$ as $(x + (-y))^5$.

$$(x + (-y))^5 = x^5 + 5x^4(-y) + \frac{5 \cdot 4}{2!}x^3(-y)^2 + \frac{5 \cdot 4 \cdot 3}{3!}x^2(-y)^3 + \frac{5 \cdot 4 \cdot 3 \cdot 2}{4!}x(-y)^4 + (-y)^5$$

$$= x^5 - 5x^4y + 10x^3y^2 - 10x^2y^3 + 5xy^4 - y^5$$

■

EXAMPLE 3 Expand and simplify $(2a + 3b)^4$.

Solution Let $x = 2a$ and $y = 3b$.

$$(2a + 3b)^4 = (2a)^4 + 4(2a)^3(3b) + \frac{4 \cdot 3}{2!}(2a)^2(3b)^2 + \frac{4 \cdot 3 \cdot 2}{3!}(2a)(3b)^3 + (3b)^4$$

$$= 16a^4 + 96a^3b + 216a^2b^2 + 216ab^3 + 81b^4$$

■

Finding Specific Terms

Sometimes it is convenient to find a specific term of a binomial expansion without writing out the entire expansion. For example, suppose that we need the 6th term of the expansion $(x + y)^{12}$. We could proceed as follows.

The 6th term will contain y^5. (Note in the general expansion that the **exponent of y is always one less than the number of the term**.) Since the sum of the exponents for x and y must be 12 (the exponent of the binomial), the 6th term will also contain x^7. Again looking back at the general binomial expansion, note that the **denominators of the coefficients** are of the form $r!$ where the value of r agrees with the exponent of y for each term. Thus, if we have y^5, the denominator of the coefficient is 5! In the general expansion each **numerator of a coefficient** contains r factors where the first factor is the exponent of the binomial and each succeeding factor is one less than the preceding one. Thus, the 6th term of $(x + y)^{12}$ is $\frac{12 \cdot 11 \cdot 10 \cdot 9 \cdot 8}{5!} x^7 y^5$, which simplifies to $792x^7y^5$.

EXAMPLE 4 Find the 4th term of $(3a + 2b)^7$.

Solution The 4th term will contain $(2b)^3$ and therefore $(3a)^4$. The coefficient is $\frac{7 \cdot 6 \cdot 5}{3!}$. Therefore, the 4th term is

$$\frac{7 \cdot 6 \cdot 5}{3!}(3a)^4(2b)^3,$$

which simplifies to $22{,}680a^4b^3$. ■

Problem Set 11.5

For Problems 1–6, use Pascal's triangle to help expand each of the following.

1. $(x + y)^8$ $x^8 + 8x^7y + 28x^6y^2 + 56x^5y^3 + 70x^4y^4 + 56x^3y^5 + 28x^2y^6 + 8xy^7 + y^8$
2. $(x + y)^7$ $x^7 + 7x^6y + 21x^5y^2 + 35x^4y^3 + 35x^3y^4 + 21x^2y^5 + 7xy^6 + y^7$
3. $(3x + y)^4$ $81x^4 + 108x^3y + 54x^2y^2 + 12xy^3 + y^4$
4. $(x + 2y)^4$ $x^4 + 8x^3y + 24x^2y^2 + 32xy^3 + 16y^4$
5. $(x - y)^5$ $x^5 - 5x^4y + 10x^3y^2 - 10x^2y^3 + 5xy^4 - y^5$
6. $(x - y)^4$ $x^4 - 4x^3y + 6x^2y^2 - 4xy^3 + y^4$

For Problems 7–20, expand and simplify.

7. $(x + y)^{10}$ $x^{10} + 10x^9y + 45x^8y^2 + 120x^7y^3 + 210x^6y^4 + 252x^5y^5 + 210x^4y^6 + 120x^3y^7 + 45x^2y^8 + 10xy^9 + y^{10}$
8. $(x + y)^9$ $x^9 + 9x^8y + 36x^7y^2 + 84x^6y^3 + 126x^5y^4 + 126x^4y^5 + 84x^3y^6 + 36x^2y^7 + 9xy^8 + y^9$
9. $(2x + y)^6$ $64x^6 + 192x^5y + 240x^4y^2 + 160x^3y^3 + 60x^2y^4 + 12xy^5 + y^6$
10. $(x + 3y)^5$ $x^5 + 15x^4y + 90x^3y^2 + 270x^2y^3 + 405xy^4 + 243y^5$
11. $(x - 3y)^5$ $x^5 - 15x^4y + 90x^3y^2 - 270x^2y^3 + 405xy^4 - 243y^5$
12. $(2x - y)^6$ $64x^6 - 192x^5y + 240x^4y^2 - 160x^3y^3 + 60x^2y^4 - 12xy^5 + y^6$
13. $(3a - 2b)^5$ $243a^5 - 810a^4b + 1080a^3b^2 - 720a^2b^3 + 240ab^4 - 32b^5$
14. $(2a - 3b)^4$ $16a^4 - 96a^3b + 216a^2b^2 - 216ab^3 + 81b^4$

15. $(x + y^3)^6$ $x^6 + 6x^5y^3 + 15x^4y^6 + 20x^3y^9 + 15x^2y^{12} + 6xy^{15} + y^{18}$

16. $(x^2 + y)^5$ $x^{10} + 5x^8y + 10x^6y^2 + 10x^4y^3 + 5x^2y^4 + y^5$

17. $(x + 2)^7$ $x^7 + 14x^6 + 84x^5 + 280x^4 + 560x^3 + 672x^2 + 448x + 128$

18. $(x + 3)^6$ $x^6 + 18x^5 + 135x^4 + 540x^3 + 1215x^2 + 1458x + 729$

19. $(x - 3)^4$ $x^4 - 12x^3 + 54x^2 - 108x + 81$

20. $(x - 1)^9$ $x^9 - 9x^8 + 36x^7 - 84x^6 + 126x^5 - 126x^4 + 84x^3 - 36x^2 + 9x - 1$

For Problems 21–24, write the first four terms of the expansion.

21. $(x + y)^{15}$ $x^{15} + 15x^{14}y + 105x^{13}y^2 + 455x^{12}y^3$

22. $(x + y)^{12}$ $x^{12} + 12x^{11}y + 66x^{10}y^2 + 220x^9y^3$

23. $(a - 2b)^{13}$ $a^{13} - 26a^{12}b + 312a^{11}b^2 - 2288a^{10}b^3$

24. $(x - y)^{20}$ $x^{20} - 20x^{19}y + 190x^{18}y^2 - 1140x^{17}y^3$

For Problems 25–30, find the indicated term of the expansion.

25. 7th term of $(x + y)^{11}$ $462x^5y^6$

26. 4th term of $(x + y)^8$ $56x^5y^3$

27. 4th term of $(x - 2y)^6$ $-160x^3y^3$

28. 5th term of $(x - y)^9$ $126x^5y^4$

29. 3rd term of $(2x - 5y)^5$ $2000x^3y^2$

30. 6th term of $(3a + b)^7$ $189a^2b^5$

Chapter 11 Summary

(11.1) An **infinite sequence** is a function whose domain is the set of positive integers. We frequently express a general infinite sequence as

$$a_1, a_2, a_3, \ldots, a_n, \ldots$$

where a_1 is the first term, a_2 the second term, and so on, and a_n represents the general or nth term.

An **arithmetic sequence** is a sequence where there is a **common difference** between successive terms.

The **general term of an arithmetic sequence** is given by

$$a_n = a_1 + (n - 1)d$$

where a_1 is the first term and d is the common difference.

(11.2) The indicated sum of a sequence is called a **series**. The sum of the first n terms of an arithmetic series is given by

$$S_n = \frac{n(a_1 + a_n)}{2}.$$

(11.3) A **geometric sequence** is a sequence in which each term after the first is obtained by multiplying the preceding term by a common multiplier. The common multiplier is called the **common ratio** of the sequence.

The **general term of a geometric sequence** is given by

$$a_n = a_1 r^{n-1}$$

where a_1 is the first term and r is the common ratio.
The sum of the first n terms of a geometric series is given by

$$S_n = \frac{a_1 r^n - a_1}{r - 1}, \qquad r \neq 1.$$

(11.4) The sum of an infinite geometric series is given by

$$S_\infty = \frac{a_1}{1 - r}, \qquad |r| < 1.$$

Any infinite geometric series where $|r| \geq 1$ has no sum.

This sum formula can be used to change repeating decimals to $\frac{a}{b}$ form.

(11.5) The expansion of $(x + y)^n$, where n is a positive integer is given by

$$(x + y)^n = x^n + nx^{n-1}y + \frac{n(n-1)}{2!}x^{n-2}y^2 + \frac{n(n-1)(n-2)}{3!}x^{n-3}y^3 + \cdots + y^n.$$

To find a specific term of a binomial expansion review Example 4.

Chapter 11 Review Problem Set

1. Find the first five terms of the sequence where the general term is given by $a_n = 2^n - 3$.
$-1, 1, 5, 13, 29$

2. Find the general term of each of the following sequences.

(a) $2, 8, 14, 20, \ldots$ $a_n = 6n - 4$

(b) $3, -2, -7, -12, \ldots$ $a_n = -5n + 8$

(c) $4, 12, 36, 108, \ldots$ $a_n = 4(3)^{n-1}$

(d) $5, \frac{5}{2}, \frac{5}{4}, \frac{5}{8}, \ldots$ $a_n = 5\left(\frac{1}{2}\right)^{n-1}$

3. Find the indicated term of each of the following sequences.

(a) The 8th term of $\frac{1}{2}, \frac{1}{6}, \frac{1}{18}, \frac{1}{54}, \ldots$ $\frac{1}{4374}$

(b) The 50th term of $8, 11, 14, 17, \ldots$ 155

4. Find the sum of each of the following series.

(a) $2 + 5 + 8 + \cdots + 149$ 3775

(b) $4 + 8 + 16 + \cdots + 2048$ 4092

(c) $-5 + (-9) + (-13) + \cdots + (-101)$ -1325

5. Find the sum of the first 300 odd whole numbers. 90,000

6. Find the sum of the first 6 terms of the series where $a_n = \left(\frac{1}{4}\right)^n$. $\frac{1365}{4096}$

7. Find the sum of the first 25 terms of the series $7 + 11 + 15 + 19 + \cdots$. 1375

8. Find the sum of all even numbers between 18 and 286 , inclusive. 20,520

9. A tank contains 60,750 liters of water. Each day one-third of the water in the tank is removed and not replaced. How much water is in the tank at the end of the 6th day? $5333.\overline{3}$ liters

10. An object falling from rest in a vacuum falls 16 feet the first second, 48 feet the second second, 80 feet the third second, 112 feet the fourth second, and so on. How far will the object fall in 15 seconds? 3600 feet

11. Suppose you save a penny the first day of a month, 2 cents the second day, 3 cents the third day, and continue to increase your savings per day by a penny. What will be your total savings for the first 20 days? $2.10

12. Suppose you save a nickel the first day of a month, a dime the second day, 20 cents the third day, and continue to double your daily savings each day. What will be your total savings at the end of 13 days? $409.55

13. Find the sum of the infinite geometric series $16 + 8 + 4 + 2 \cdots$. 32

14. Express $0.\overline{29}$ in $\frac{a}{b}$ form, where a and b are integers, $b \neq 0$. $\frac{29}{99}$

15. Expand and simplify $(2x + y)^7$. $128x^7 + 448x^6y + 672x^5y^2 + 560x^4y^3 + 280x^3y^4 + 84x^2y^5 + 14xy^6 + y^7$

16. Find the 6th term of $(a + b)^{10}$. $252a^5b^5$

Thoughts into Words

17. Explain the difference between an arithmetic sequence and a geometric sequence.

18. What does it mean to say that the sum of the infinite geometric series $1 + \frac{1}{2} + \frac{1}{4} + \frac{1}{8} + \cdots$ is 2?

19. What do we mean when we state that the infinite geometric series $1 + 2 + 4 + 8 + \cdots$ has no sum?

Appendixes

A

Synthetic Division and the Factor Theorem

In Section 4.5 we discussed the process of dividing polynomials. If the divisor is of the form $x - k$, then the typical long division algorithm can be conveniently simplified into a process called synthetic division.

First, let's consider an example and use the usual division process. Then, in a step-by-step fashion, we can observe some shortcuts that will lead us into the synthetic division procedure. Consider the division problem $(2x^4 + x^3 - 17x^2 + 13x + 2) \div (x - 2)$.

$$
\begin{array}{r|l}
 & 2x^3 + 5x^2 - 7x - 1 \\
\cline{2-2}
x - 2 & 2x^4 + x^3 - 17x^2 + 13x + 2 \\
 & 2x^4 - 4x^3 \\
\cline{2-2}
 & \quad 5x^3 - 17x^2 \\
 & \quad 5x^3 - 10x^2 \\
\cline{2-2}
 & \qquad -7x^2 + 13x \\
 & \qquad -7x^2 + 14x \\
\cline{2-2}
 & \qquad\qquad -x + 2 \\
 & \qquad\qquad -x + 2
\end{array}
$$

Notice that since the dividend $(2x^4 + x^3 - 17x^2 + 13x + 2)$ is written in descending powers of x, the quotient $(2x^3 + 5x^2 - 7x - 1)$ is produced, also in descending powers of x. In other words, the numerical coefficients are the *key issues.* So let's rewrite the above problem in terms of its coefficients.

$$
\begin{array}{r|rrrrr}
 & 2 & +5 & -7 & -1 & \\
\cline{2-6}
1 - 2 & 2 & +1 & -17 & +13 & +2 \\
 & \textcircled{2} & -4 & & & \\
\cline{2-3}
 & & 5 & \textcircled{-17} & & \\
 & & \textcircled{5} & -10 & & \\
\cline{3-4}
 & & & -7 & +\textcircled{13} & \\
 & & & \textcircled{-7} & +14 & \\
\cline{4-5}
 & & & & -1 & +\textcircled{2} \\
 & & & & \textcircled{-1} & +2 \\
\cline{5-6}
\end{array}
$$

Now observe that the numbers circled are simply repetitions of the numbers directly above them in the format. Therefore, we can write the process in a more compact form as

$$
\begin{array}{r|rrrrr}
 & 2 & 5 & -7 & -1 & \\
\cline{2-6}
-2 & 2 & 1 & -17 & -13 & 2 \\
 & & -4 & -10 & 14 & 2 \\
\cline{3-6}
 & & 5 & -7 & -1 & 0
\end{array}
\qquad
\begin{array}{r}
(1) \\ (2) \\ (3) \\ (4)
\end{array}
$$

where the repetitions are omitted and where 1, the coefficient of x in the divisor, is omitted.

Notice that line (4) reveals all of the coefficients of the quotient (line (1)), except for the first coefficient of 2. Thus, we can begin line (4) with the first coefficient and then use the following form.

$$\begin{array}{r|rrrrrl} -2 & 2 & 1 & -17 & 13 & 2 & \quad (5)\\ & & -4 & -10 & 14 & 2 & \quad (6)\\ \hline & 2 & 5 & -7 & -1 & 0 & \quad (7)\end{array}$$

Line (7) contains the coefficients of the quotient where the 0 indicates the remainder.

Finally, by changing the constant in the divisor to 2 (instead of -2) we can add the corresponding entries in lines (5) and (6) rather than subtract. Thus, the final synthetic division form for this problem is as follows.

$$\begin{array}{r|rrrrr} 2 & 2 & 1 & -17 & 13 & 2\\ & & 4 & 10 & -14 & -2\\ \hline & 2 & 5 & -7 & -1 & 0\end{array}$$

Now let's consider another problem that indicates a step-by-step procedure for carrying out the synthetic division process. Suppose that we want to divide $3x^3 - 2x^2 + 6x - 5$ by $x + 4$.

Step 1. Write the coefficients of the dividend as follows.

$$\begin{array}{r|rrrr} & 3 & -2 & 6 & -5\end{array}$$

Step 2. In the divisor, $(x + 4)$, use -4 instead of 4 so that later we can add rather than subtract.

$$\begin{array}{r|rrrr} -4 & 3 & -2 & 6 & -5\end{array}$$

Step 3. Bring down the first coefficient of the dividend (3).

$$\begin{array}{r|rrrr} -4 & 3 & -2 & 6 & -5\\ & & & & \\ \hline & 3 & & & \end{array}$$

Step 4. Multiply $(3)(-4)$, which yields -12; this result is to be added to the second coefficient of the dividend (-2).

$$\begin{array}{r|rrrr} -4 & 3 & -2 & 6 & -5\\ & & -12 & & \\ \hline & 3 & -14 & & \end{array}$$

Step 5. Multiply $(-14)(-4)$, which yields 56; this result is to be added to the third coefficient of the dividend (6).

$$\begin{array}{r|rrrr} -4 & 3 & -2 & 6 & -5\\ & & -12 & 56 & \\ \hline & 3 & -14 & 62 & \end{array}$$

Step 6. Multiply (62)(−4), which yields −248; this result is added to the last term of the dividend (−5).

$$\begin{array}{r|rrrr} -4 & 3 & -2 & 6 & -5 \\ & & -12 & 56 & -248 \\ \hline & 3 & -14 & 62 & -253 \end{array}$$

The last row indicates a quotient of $3x^2 - 14x + 62$ and a remainder of −253. Thus, we have

$$\frac{3x^3 - 2x^2 + 6x - 5}{x + 4} = 3x^2 - 14x + 62 - \frac{253}{x + 4}.$$

Let's consider one more example showing only the final compact form for synthetic division.

EXAMPLE 1 Find the quotient and remainder for $(4x^4 - 2x^3 + 6x - 1) \div (x - 1)$.

Solution

$$\begin{array}{r|rrrrr} 1 & 4 & -2 & 0 & 6 & -1 \\ & & 4 & 2 & 2 & 8 \\ \hline & 4 & 2 & 2 & 8 & 7 \end{array}$$

Notice that a zero has been inserted as the coefficient of the missing x^2 term.

Therefore,

$$\frac{4x^4 - 2x^3 + 6x - 1}{x - 1} = 4x^3 + 2x^2 + 2x + 8 + \frac{7}{x - 1}.$$ ■

Let us very briefly look at how the concept of synthetic division is used in subsequent mathematics courses. In general, if a polynomial, $P(x)$, is divided by $x - k$, then we can write

$$P(x) = (x - k)Q(x) + R$$

where $Q(x)$ is the quotient polynomial and R is the real number remainder. If k is substituted for x in the above relationship, we obtain

$$\begin{aligned} P(k) &= (k - k)Q(k) + R \\ &= 0 + R \\ &= R. \end{aligned}$$

In other words, the functional value $P(k)$ is equal to the remainder when $P(x)$ is divided by $x - k$. This conclusion is commonly referred to as the **Remainder Theorem**.

Remainder Theorem

If $P(x)$ is divided by $(x - k)$, the remainder is $P(k)$.

In Example 1, we obtained a remainder of 7 when dividing $4x^4 - 2x^3 + 6x - 1$ by $x - 1$. Therefore, according to the Remainder Theorem we know that the functional value of $4x^4 - 2x^3 + 6x - 1$ at $x = 1$ is 7. (Perhaps you should check this result by actually substituting 1 for x in $4x^4 - 2x^3 + 6x - 1$.) A direct consequence of the Remainder Theorem is the Factor Theorem, which can be stated as follows.

Factor Theorem

If $P(k) = 0$, then $(x - k)$ is a factor of $P(x)$.

Let's consider one final example where we use synthetic division, the Factor Theorem, and some previous equation solving ideas to find a solution set.

EXAMPLE 2 Show that $(x - 1)$, $(x + 2)$, and $(x - 3)$ are factors of $x^3 - 2x^2 - 5x + 6$, and therefore 1, -2, and 3 are solutions of $x^3 - 2x^2 - 5x + 6 = 0$.

Solution Using synthetic division we can show that $(x - 1)$, $(x + 2)$, and $(x - 3)$ are factors of $x^3 - 2x^2 - 5x + 6$ as follows.

$$\begin{array}{r|rrrr}
1 & 1 & -2 & -5 & 6 \\
 & & 1 & -1 & -6 \\
\hline
 & 1 & -1 & -6 & 0 \longleftarrow \text{remainder}
\end{array}$$

$$\begin{array}{r|rrrr}
-2 & 1 & -2 & -5 & 6 \\
 & & -2 & 8 & -6 \\
\hline
 & 1 & -4 & 3 & 0 \longleftarrow \text{remainder}
\end{array}$$

$$\begin{array}{r|rrrr}
3 & 1 & -2 & -5 & 6 \\
 & & 3 & 3 & -6 \\
\hline
 & 1 & 1 & -2 & 0 \longleftarrow \text{remainder}
\end{array}$$

Once we know that $(x - 1)$, $(x + 2)$, and $(x - 3)$ are factors of $x^3 - 2x^2 - 5x + 6$, then the solution set for $x^3 - 2x^2 - 5x + 6 = 0$ is easily obtained as follows.

$$x^3 - 2x^2 - 5x + 6 = 0$$

$$(x - 1)(x + 2)(x - 3) = 0$$

$$x - 1 = 0 \quad \text{or} \quad x + 2 = 0 \quad \text{or} \quad x - 3 = 0$$

$$x = 1 \quad \text{or} \quad x = -2 \quad \text{or} \quad x = 3$$

The solution set is $\{1, -2, 3\}$. ■

Practice Exercises

Use synthetic division to determine the quotient and remainder for each of the following.

1. $(x^2 - 8x + 12) \div (x - 2)$

2. $(x^2 + 9x + 18) \div (x + 3)$

3. $(x^2 + 2x - 10) \div (x - 4)$

4. $(x^2 - 10x + 15) \div (x - 8)$

5. $(x^3 - 2x^2 - x + 2) \div (x - 2)$

6. $(x^3 - 5x^2 + 2x + 8) \div (x + 1)$

7. $(x^3 - 7x - 6) \div (x + 2)$

8. $(x^3 + 6x^2 - 5x - 1) \div (x - 1)$

9. $(2x^3 - 5x^2 - 4x + 6) \div (x - 2)$

10. $(3x^4 - x^3 + 2x^2 - 7x - 1) \div (x + 1)$

11. $(x^4 + 4x^3 - 7x - 1) \div (x - 3)$

12. $(2x^4 + 3x^2 + 3) \div (x + 2)$

13. Use synthetic division and the Factor Theorem to show that for each of the following the given binomials are factors of the polynomial.

(**a**) $x + 1$, $x + 2$, and $x + 3$ are factors of $x^3 + 6x^2 + 11x + 6$

(**b**) $x - 2$, $x + 3$, and $x - 5$ are factors of $x^3 - 4x^2 - 11x + 30$

B

Common Logarithms

Using a table to find a common logarithm is relatively easy but it does require a little more effort than pushing a button as you would with a calculator. Let's consider a small part of the table that appears in the back of the book. Each number in the column headed n represents the first two significant digits of a number between 1 and 10 and each of the column headings 0 through 9 represents the third significant digit.

Table of Common Logarithms

n	0	1	2	3	4	5	6	7	8	9
1.0	0.0000	0.0043	0.0086	0.0128	0.0170	0.0212	0.0253	0.0294	0.0334	0.0374
1.1	0.0414	0.0453	0.0492	0.0531	0.0569	0.0607	0.0645	0.0682	0.0719	0.0755
1.2	0.0792	0.0828	0.0864	0.0899	0.0934	0.0969	0.1004	0.1038	0.1072	0.1106
1.3	0.1139	0.1173	0.1206	0.1239	0.1271	0.1303	0.1335	0.1367	0.1399	0.1430
1.4	0.1461	0.1492	0.1523	0.1553	0.1584	0.1614	0.1644	0.1673	0.1703	0.1732
1.5	0.1761	0.1790	0.1818	0.1847	0.1875	0.1903	0.1931	0.1959	0.1987	0.2014
1.6	0.2041	0.2068	0.2095	0.2122	0.2148	0.2175	0.2201	0.2227	0.2253	0.2279
1.7	0.2304	0.2330	0.2355	0.2380	0.2405	0.2430	0.2455	0.2480	0.2504	0.2529
1.8	0.2553	0.2577	0.2601	0.2625	0.2648	0.2672	0.2695	0.2718	0.2742	0.2765
1.9	0.2788	0.2810	0.2833	0.2856	0.2878	0.2900	0.2923	0.2945	0.2967	0.2989
2.0	0.3010	0.3032	0.3054	0.3075	0.3096	0.3118	0.3139	0.3160	0.3181	0.3201
2.1	0.3222	0.3243	0.3263	0.3284	0.3304	0.3324	0.3345	0.3365	0.3385	0.3404
2.2	0.3424	0.3444	0.3464	0.3483	0.3502	0.3522	0.3541	0.3560	0.3579	0.3598
2.3	0.3617	0.3636	0.3655	0.3674	0.3692	0.3711	0.3929	0.3747	0.3766	0.3784
2.4	0.3802	0.3820	0.3838	0.3856	0.3874	0.3892	0.3909	0.3927	0.3945	0.3962

To find the logarithm of a number such as 1.75, we look at the intersection of the row that contains 1.7 and the column headed 5. Thus, we obtain

$$\log 1.75 = 0.2430.$$

Similarly, we can find that

$$\log 2.09 = 0.3201 \qquad \text{and} \qquad \log 2.40 = 0.3802;$$

keep in mind that these values are also rounded to four decimal places.

Now suppose that we want to use the table to find the logarithm of a positive number greater than 10 or less than 1. To accomplish this we represent the number in scientific notation and then apply the property $\log rs = \log r + \log s$. For example, to find log 134 we can proceed as follows.

$$\begin{aligned}\log 134 &= \log(1.34 \cdot 10^2)\\ &= \log 1.34 + \log 10^2\\ &= 0.1271 + 2 = 2.1271\end{aligned}$$

By inspection we know that the common logarithm of 10^2 is 2 (the exponent), and the common logarithm of 1.34 can be found in the table.

The decimal part (0.1271) of the logarithm 2.1271 is called the **mantissa**, and the integral part, (2), is called the **characteristic**. Thus, we can find the characteristic of a common logarithm by inspection (since it is the exponent of 10 when the number is written in scientific notation) and the mantissa we can get from a table. Let's consider two more examples.

$$\begin{aligned}\log 23.8 &= \log(2.38 \cdot 10^1)\\ &= \log 2.38 + \log 10^1\\ &= 0.3766 + 1\end{aligned}$$

(0.3766: from the table; 1: exponent of 10)

$$= 1.3766$$

$$\begin{aligned}\log 0.192 &= \log(1.92 \cdot 10^{-1})\\ &= \log 1.92 + \log 10^{-1}\\ &= 0.2833 + (-1)\end{aligned}$$

(0.2833: from the table; -1: exponent of 10)

$$= 0.2833 + (-1)$$

Notice that in the last example we expressed the logarithm of 0.192 as $0.2833 + (-1)$; we did not add 0.2833 and -1. This is normal procedure when using a table of common logarithms because the mantissas given in the table are positive numbers. However, you should recognize that adding 0.2833 and -1 produces -0.7167, which agrees with the result obtained earlier with a calculator.

We can also use the table to find a number when given the common logarithm of the number. That is to say, given $\log x$ we can determine x from the table. Traditionally, x is referred to as the **antilogarithm** (abbreviated **antilog**) of $\log x$. Let's consider some examples.

EXAMPLE 1 Determine antilog 1.3365.

Solution Finding an antilogarithm simply reverses the process used before for finding a logarithm. Thus, antilog 1.3365 means that 1 is the characteristic and 0.3365 the mantissa. We look for 0.3365 in the body of the common logarithm table and we find that it is located at the intersection of the 2.1-row and the 7-column. Therefore, the antilogarithm is

$$2.17 \cdot 10^1 = 21.7.$$

■

EXAMPLE 2 Determine antilog $(0.1523 + (-2))$.

Solution The mantissa, 0.1523, is located at the intersection of the 1.4-row and 2-column. The characteristic is -2 and therefore the antilogarithm is

$$1.42 \cdot 10^{-2} = 0.0142.$$

■

EXAMPLE 3 Determine antilog -2.6038.

Solution The mantissas given in a table are *positive* numbers. Thus, we need to express -2.6038 in terms of a positive mantissa and this can be done by adding and subtracting 3 as follows.

$$(-2.6038 + 3) - 3 = 0.3962 + (-3)$$

Now we can look for 0.3962 and find it at the intersection of the 2.4-row and 9-column. Therefore, the antilogarithm is

$$2.49 \cdot 10^{-3} = 0.00249.$$

■

Linear Interpolation

Now suppose that we want to determine log 2.774 from the table in the inside back cover. Because the table contains only logarithms of numbers with, at most, three significant digits, we have a problem. However, by a process called **linear interpolation** we can extend the capabilities of the table to include numbers with four significant digits.

First, let's consider a geometric basis of linear interpolation and then we will use a systematic procedure for carrying out the necessary calculations. A portion of the graph of $y = \log x$, with the curvature exaggerated to help illustrate the principle involved, is shown in Figure B.1. The line segment that joins points P and Q is used to approximate the curve from P to Q. The actual value of log 2.744 is the ordinate of the point C, that is, the length of $\overline{AC}$. This cannot be determined from the table. Instead we will use the ordinate of point B (the length of $\overline{AB}$) as an approximation for log 2.744.

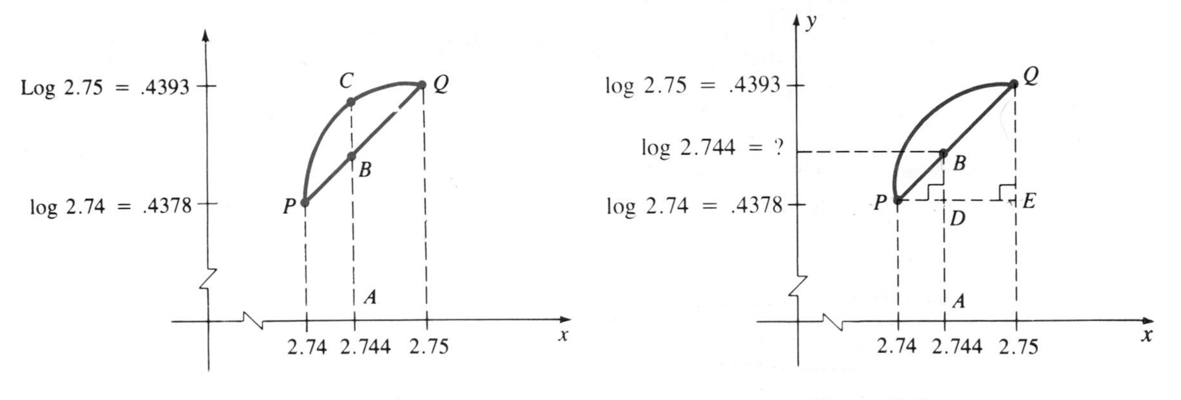

Figure B.1 **Figure B.2**

Consider Figure B.2 where line segments $\overline{DB}$ and $\overline{EQ}$ are drawn perpendicular to $\overline{PE}$. The right triangles formed $\triangle PDB$ and $\triangle PEQ$, are similar and therefore the lengths of their corresponding sides are proportional. Thus, we can write

$$\frac{PD}{PE} = \frac{DB}{EQ}. \tag{1}$$

From Figure B.2 we see that

$$PD = 2.744 - 2.74 = 0.004$$

$$PE = 2.75 - 2.74 = 0.01$$

$$EQ = 0.4393 - 0.4378 = 0.0015.$$

Therefore, the proportion (1) becomes

$$\frac{0.004}{0.01} = \frac{DB}{0.0015}.$$

Solving this proportion for DB yields

$$DB = 0.0006.$$

Since $AB = AD + DB$, we have

$$AB = 0.4378 + 0.0006 = 0.4384.$$

Thus, we obtain $\log 2.744 = 0.4384$.

Now let's suggest an abbreviated format for carrying out the calculations necessary to find $\log 2.744$.

$$\begin{array}{c|c}
x & \log x \\
\hline
\left.\begin{array}{r} 4\left\{\begin{array}{l} 2.740 \\ 2.744 \end{array}\right. \\ \\ 2.750 \end{array}\right\} 10 & \left.\begin{array}{r} k\left\{\begin{array}{l} 0.4378 \\ \quad ? \end{array}\right. \\ \\ 0.4393 \end{array}\right\} 0.0015
\end{array}$$

Notice that we have used 4 and 10 for the differences for values of x instead of 0.004 and 0.01 because the ratio $\frac{0.004}{0.01}$ equals $\frac{4}{10}$. Setting up a proportion and solving for k yields

$$\frac{4}{10} = \frac{k}{0.0015}$$

$$10k = 4(0.0015) = 0.0060$$

$$k = 0.0006.$$

Thus, $\log 2.744 = 0.4378 + 0.0006 = 0.4384$.

Let's do another example to make sure of the process.

EXAMPLE 4 Find $\log 617.6$.

Solution

$$\begin{aligned} \log 617.6 &= \log(6.176 \cdot 10^2) \\ &= \log 6.176 + \log 10^2 \end{aligned}$$

Thus, the characteristic is 2 and we can approximate the mantissa by using interpolation from the table as follows.

$$\begin{array}{c|c}
x & \log x \\
\hline
\left.\begin{array}{r} 6\left\{\begin{array}{l} 6.170 \\ \\ \\ 6.176 \end{array}\right. \\ 6.180 \end{array}\right\} 10 & \left.\begin{array}{r} k\left\{\begin{array}{l} 0.7903 \\ \\ \\ \quad ? \end{array}\right. \\ 0.7910 \end{array}\right\} 0.0007
\end{array}$$

$$\frac{6}{10} = \frac{k}{0.0007}$$

$$10k = 6(0.0007) = 0.0042$$

$$k = 0.00042 \approx 0.0004$$

Therefore, $\log 6.176 = 0.7903 + 0.0004 = 0.7907$ and we can complete the solution for $\log 617.6$ as follows.

$$\begin{aligned}\log 617.6 &= \log(6.176 \cdot 10^2)\\ &= \log 6.176 + \log 10^2\\ &= 0.7907 + 2\\ &= 2.7907\end{aligned}$$

■

The process of linear interpolation can also be used to approximate an antilogarithm when the mantissa is in between two values in the table. The following example illustrates this procedure.

EXAMPLE 5 Find antilog 1.6157.

Solution From the table we see that the mantissa, 0.6157, is between 0.6149 and 0.6160. We can carry out the interpolation as follows.

$$h\left\{\begin{array}{c}4.120\\ \\ ?\end{array}\right.\left.\begin{array}{c}\\ \\ \\ 4.130\end{array}\right\}0.010 \qquad 8\left\{\begin{array}{c}0.6149\\ \\ 0.6157\end{array}\right.\left.\begin{array}{c}\\ \\ \\ 0.6160\end{array}\right\}11 \qquad \frac{0.0008}{0.0011} = \frac{8}{11}$$

$$\frac{h}{0.010} = \frac{8}{11}$$

$$h = 8(0.010) = 0.080$$

$$h = \frac{1}{11}(0.080) = 0.007 \qquad \text{nearest thousandth}$$

Thus, $\text{antilog}\, 0.6157 = 4.120 + 0.007 = 4.127$.
Therefore,

$$\begin{aligned}\text{antilog}\, 1.6157 &= \text{antilog}(0.6157 + 1)\\ &= 4.127 \cdot 10^1\\ &= 41.27.\end{aligned}$$

■

Computation with Common Logarithms

Let's first restate the basic properties of logarithms in terms of **common logarithms**. (Remember that we are writing $\log x$ instead of $\log_{10} x$.)

If x and y are positive real numbers, then

1. $\log xy = \log x + \log y$;
2. $\log \frac{x}{y} = \log x - \log y$;
3. $\log x^p = p \log x$. p is any real number

The following two properties of equality that pertain to logarithms will also be used.

4. If $x = y$ (x and y are positive), then $\log x = \log y$.
5. If $\log x = \log y$, then $x = y$.

EXAMPLE 6 Find the product (49.1)(876).

Solution Let $N = (49.1)(876)$. By Property 4.

$$\log N = \log(49.1)(876)$$

By Property 1.

$$\log N = \log 49.1 + \log 876$$

From the table at the back of the book we find that $\log 49.1 = 1.6911$ and that $\log 876 = 2.9425$. Thus,

$$\begin{aligned} \log N &= 1.6911 + 2.9425 \\ &= 4.6336. \end{aligned}$$

Therefore,

$$N = \text{antilog}\, 4.6336.$$

By using linear interpolation, we can determine antilog 0.6336 to four significant digits. Thus, we obtain

$$\begin{aligned} N &= \text{antilog}(0.6336 + 4) \\ &= 4.301 \cdot 10^4 \\ &= 43{,}010. \end{aligned}$$

CHECK By using a calculator we obtain

$$N = (49.1)(876) = 43011.6.$$

■

EXAMPLE 7 Find the quotient $\frac{942}{64.8}$.

Solution Let $N = \frac{94.2}{64.8}$. Therefore,

$$\log N = \log \frac{942}{64.8}$$

$$= \log 942 - \log 64.8 \qquad \log \frac{x}{y} = \log x - \log y$$

$$= 2.9741 - 1.8116 \qquad \text{from the table}$$

$$= 1.1625.$$

Therefore,

$$\begin{aligned} N &= \text{antilog}\, 1.1625 \\ &= \text{antilog}(0.01625 + 1) \\ &= 1.454 \cdot 10^1 \\ &= 14.54. \end{aligned}$$

CHECK By using a calculator we obtain

$$N = \frac{942}{64.8} = 14.537037.$$

■

EXAMPLE 8 Evaluate $\frac{(571.4)(8.236)}{71.68}$.

Solution Let $N = \frac{(571.4)(8.236)}{71.68}$. Therefore,

$$\log N = \log \frac{(571.4)(8.236)}{71.68}$$

$$\begin{aligned} &= \log 571.4 + \log 8.236 - \log 71.68 \\ &= 2.7569 + 0.9157 - 1.8554 \\ &= 1.8172. \end{aligned}$$

Therefore,

$$\begin{aligned} N &= \text{antilog}\, 1.8172 \\ &= \text{antilog}(0.8172 + 1) \\ &= 6.564 \cdot 10^1 \\ &= 65.64. \end{aligned}$$

CHECK By using a calculator we obtain

$$N = \frac{(571.4)(8.236)}{71.68} = 65.653605.$$

■

EXAMPLE 9 Evaluate $\sqrt[3]{3770}$.

Solution Let $N = \sqrt[3]{3770} = (3770)^{\frac{1}{3}}$. Therefore,

$$\begin{aligned} \log N &= \log(3770)^{\frac{1}{3}} \\ &= \frac{1}{3}\log 3770 \qquad \log x^p = p \log x \\ &= \frac{1}{3}(3.5763) \\ &= 1.1921. \end{aligned}$$

Therefore,

$$\begin{aligned} N &= \text{antilog}\, 1.1921 \\ &= \text{antilog}(0.1921 + 1) \\ &= 1.556 \cdot 10^1 \\ &= 15.56. \end{aligned}$$

CHECK By using a calculator we obtain

$$N = \sqrt[3]{3370} = 15.563733.$$

■

When using tables of logarithms, it is sometimes necessary to change the form of writing a logarithm so that the decimal part (mantissa) is positive. The next example illustrates this idea.

EXAMPLE 10 Find the quotient $\frac{1.73}{5.08}$.

Solution Let $N = \frac{1.73}{5.08}$. Therefore,

$$\begin{aligned} \log N &= \log \frac{1.73}{5.08} \\ &= \log 1.73 - \log 5.08 \\ &= 0.2380 - 0.7059 = -0.4679. \end{aligned}$$

Now by adding 1 and subtracting 1, which changes the form but not the value, we obtain

$$\begin{aligned} \log N &= -0.4679 + 1 - 1 \\ &= 0.5321 - 1 \\ &= 0.5321 + (-1). \end{aligned}$$

Therefore,

$$N = \text{antilog}(0.5321 + (-1))$$
$$= 3.405 \cdot 10^{-1} = 0.3405.$$

CHECK By using a calculator we obtain

$$N = \frac{1.73}{5.08} = 0.34055118.$$ ■

Sometimes it is also necessary to change the form of a logarithm so that a subsequent calculation will produce an **integer for the characteristic part of the logarithm**. Let's consider an example to illustrate this idea.

EXAMPLE 11 Evaluate $\sqrt[4]{0.0767}$.

Solution Let $N = \sqrt[4]{0.0767} = (0.0767)^{\frac{1}{4}}$. Therefore,

$$\log N = \log(0.0767)^{\frac{1}{4}}$$
$$= \frac{1}{4}\log 0.0767$$
$$= \frac{1}{4}(0.8848 + (-2))$$
$$= \frac{1}{4}(-2 + 0.8848).$$

At this stage we recognize that applying the distributive property will produce a nonintegral characteristic, namely, $-\frac{1}{2}$. Therefore, let's add 4 and subtract 4 inside the parentheses, which will change the form as follows.

$$\log N = \frac{1}{4}(-2 + 0.8848 + 4 - 4)$$
$$= \frac{1}{4}(4 - 2 + 0.8848 - 4)$$
$$= \frac{1}{4}(2.8848 - 4)$$

Now applying the distributive property we obtain

$$\log N = \frac{1}{4}(2.8848) - \frac{1}{4}(4)$$
$$= 0.7212 - 1 = 0.7212 + (-1).$$

Therefore,

$$N = \text{antilog}(0.7212 + (-1))$$
$$= 5.262 \cdot 10^{-1} = 0.5262.$$

CHECK By using a calculator we obtain

$$N = \sqrt[4]{0.0767} = 0.5262816.$$

■

Practice Exercises

Use the table at the back of the book and linear interpolation to find each of the following common logarithms.

1. $\log 4.327$ **2.** $\log 27.43$ **3.** $\log 128.9$

4. $\log 3526$ **5.** $\log 0.8761$ **6.** $\log 0.07692$

7. $\log 0.005186$ **8.** $\log 0.0002558$

Use the table at the back of the book and linear interpolation to find each of the following antilogarithms to four significant digits.

9. antilog 0.4690 **10.** antilog 1.7971

11. antilog 2.1925 **12.** antilog 3.7225

13. antilog$(0.5026 + (-1))$ **14.** antilog$(0.9397 + (-2))$

Use common logarithms and linear interpolation to help evaluate each of the following. Express your answers with four significant digits. Check your answers by using a calculator.

15. $(294)(71.2)$ **16.** $(192.6)(4.017)$ **17.** $\dfrac{23.4}{4.07}$

18. $\dfrac{718.5}{8.248}$ **19.** $(17.3)^5$ **20.** $(48.02)^3$

21. $\dfrac{(108)(76.2)}{13.4}$ **22.** $\dfrac{(126.3)(24.32)}{8.019}$ **23.** $\sqrt[5]{0.821}$

24. $\sqrt[4]{645.3}$ **25.** $(79.3)^{\frac{3}{5}}$ **26.** $(176.8)^{\frac{3}{4}}$

27. $\sqrt{\dfrac{(7.05)(18.7)}{0.521}}$ **28.** $\sqrt[3]{\dfrac{(41.3)(0.271)}{8.05}}$

Natural Logarithms

The following table contains the natural logarithms for numbers between 0.1 and 10, inclusive, at intervals of 0.1. Be sure that you agree with the following values taken directly from the table.

$\ln 1.6 = 0.4700$

$\ln 0.5 = -0.6931$

$\ln 4.8 = 1.5686$

$\ln 9.2 = 2.2192$

Table of Natural Logarithms

n	$\ln n$	n	$\ln n$	n	$\ln n$	n	$\ln n$
0.1	−2.3026	2.6	0.9555	5.1	1.6292	7.6	2.0281
0.2	−1.6094	2.7	0.9933	5.2	1.6487	7.7	2.0412
0.3	−1.2040	2.8	1.0296	5.3	1.6677	7.8	2.0541
0.4	−0.9163	2.9	1.0647	5.4	1.6864	7.9	2.0669
0.5	−0.6931	3.0	1.0986	5.5	1.7047	8.0	2.0794
0.6	−0.5108	3.1	1.1314	5.6	1.7228	8.1	2.0919
0.7	−0.3567	3.2	1.1632	5.7	1.7405	8.2	2.1041
0.8	−0.2231	3.3	1.1939	5.8	1.7579	8.3	2.1163
0.9	−0.1054	3.4	1.2238	5.9	1.7750	8.4	2.1282
1.0	0.0000	3.5	1.2528	6.0	1.7918	8.5	2.1401
1.1	0.0953	3.6	1.2809	6.1	1.8083	8.6	2.1518
1.2	0.1823	3.7	1.3083	6.2	1.8245	8.7	2.1633
1.3	0.2624	3.8	1.3350	6.3	1.8405	8.8	2.1748
1.4	0.3365	3.9	1.3610	6.4	1.8563	8.9	2.1861
1.5	0.4055	4.0	1.3863	6.5	1.8718	9.0	2.1972
1.6	0.4700	4.1	1.4110	6.6	1.8871	9.1	2.2083
1.7	0.5306	4.2	1.4351	6.7	1.9021	9.2	2.2192
1.8	0.5878	4.3	1.4586	6.8	1.9169	9.3	2.2300
1.9	0.6419	4.4	1.4816	6.9	1.9315	9.4	2.2407
2.0	0.6931	4.5	1.5041	7.0	1.9459	9.5	2.2513
2.1	0.7419	4.6	1.5261	7.1	1.9601	9.6	2.2618
2.2	0.7885	4.7	1.5476	7.2	1.9741	9.7	2.2721
2.3	0.8329	4.8	1.5686	7.3	1.9879	9.8	2.2824
2.4	0.8755	4.9	1.5892	7.4	2.0015	9.9	2.2925
2.5	0.9163	5.0	1.6094	7.5	2.0149	10	2.3026

When using a table, the natural logarithm of a positive number less than 0.1 or greater than 10 can be approximated by using the property $\ln rs = \ln r + \ln s$ as follows.

$$\begin{aligned}\ln 190 &= \ln(1.9 \cdot 10^2)\\ &= \ln 1.9 + \ln 10^2\\ &= \ln 1.9 + 2\ln 10\\ &= \underset{\substack{\uparrow\\ \text{from the table}}}{0.6419} + 2(\underset{\substack{\uparrow\\ \text{from the table}}}{2.3026})\\ &= 5.2471.\end{aligned}$$

$$\begin{aligned}
\ln 0.0084 &= \ln(8.4 \cdot 10^{-3}) \\
&= \ln 8.4 + \ln 10^{-3} \\
&= \ln 8.4 + (-3)(\ln 10) \\
&= \underset{\substack{\uparrow \\ \text{from the} \\ \text{table}}}{2.1282} - 3(\underset{\substack{\uparrow \\ \text{from the} \\ \text{table}}}{2.3026}) \\
&= 2.1282 - 6.9078 = -4.7796.
\end{aligned}$$

Answers to Odd-Numbered Problems and All Chapter Review Problems

CHAPTER 1

Problem Set 1.1 (page 9)

1. True **3.** False **5.** True **7.** False **9.** True **11.** 0 and 14
13. $0, 14, \frac{2}{3}, -\frac{11}{14}, 2.34, 3.2\overline{1}, 6\frac{7}{8}, -19$, and -2.6 **15.** 0 and 14 **17.** All of them **19.** $\nsubseteq$
21. $\subseteq$ **23.** $\nsubseteq$ **25.** $\subseteq$ **27.** $\nsubseteq$ **29.** $\subseteq$ **31.** $\nsubseteq$ **33.** $\{1, 2\}$
35. $\{0, 1, 2, 3, 4, 5\}$ **37.** $\{\ldots, -1, 0, 1, 2\}$ **39.** $\varnothing$ **41.** $\{0, 1, 2, 3, 4\}$ **43.** -6 **45.** 2
47. $3x + 1$ **49.** $5x$ **51.** 26 **53.** 84 **55.** 23 **57.** 65 **59.** 60 **61.** 33
63. 1320 **65.** 20 **67.** 119 **69.** 18 **71.** 4 **73.** 31

Problem Set 1.2 (page 18)

1. -7 **3.** -19 **5.** -22 **7.** -7 **9.** 108 **11.** -70 **13.** 14 **15.** -7
17. 28 **19.** 36 **21.** -204 **23.** -6 **25.** 0 **27.** undefined **29.** -60
31. -17 **33.** 72 **35.** -360 **37.** -126 **39.** -8 **41.** -12 **43.** -24

45. 15 **47.** 15 **49.** -17 **51.** 21 **53.** 5 **55.** 14 **57.** 26 **59.** 6 **61.** 25
63. 78 **65.** -10 **67.** 5 **69.** -5 **71.** -620 **73.** -4 **75.** -12.6 **77.** 6.6
79. -4.8 **81.** 7.35 **83.** -38.88 **85.** .2 **87.** -14 **89.** -6.5 **91.** -5.87
93. -6.78 **97.** (a) 2 (c) -3 (e) -6 (g) -8 (i) 9

Problem Set 1.3 (page 27)

1. Associative property of addition **3.** Commutative property of addition
5. Additive inverse property **7.** Multiplication property of negative one
9. Commutative property of multiplication **11.** Distributive property
13. Associative property of multiplication **15.** 18 **17.** 2 **19.** -1300 **21.** 1700
23. -47 **25.** 3200 **27.** -19 **29.** -41 **31.** -17 **33.** -39 **35.** 24
37. 20 **39.** 55 **41.** 16 **43.** 64 **45.** -216 **47.** -14 **49.** -8 **51.** -16
53. 6 **57.** 2187 **59.** -2048 **61.** $-15{,}625$ **63.** 3.9525416

Problem Set 1.4 (page 35)

1. $4x$ **3.** $-a^2$ **5.** $-6n$ **7.** $-5x + 2y$ **9.** $6a^2 + 5b^2$ **11.** $21x - 13$
13. $-2a^2b - ab^2$ **15.** $8x + 21$ **17.** $-5a + 2$ **19.** $-5n^2 + 11$ **21.** $-7x^2 + 32$
23. $22x - 3$ **25.** $-14x - 7$ **27.** $-10n^2 + 4$ **29.** $4x - 30y$ **31.** $-13x - 31$
33. $-21x - 9$ **35.** -17 **37.** 12 **39.** 4 **41.** 3 **43.** -38 **45.** -14
47. -58 **49.** 104 **51.** 5 **53.** 4 **55.** -54 **57.** 25 **59.** 221.6 **61.** 1092.4
63. 1420.5 **65.** $n + 12$ **67.** $n - 5$ **69.** $50n$ **71.** $\frac{1}{2}n - 4$ **73.** $\frac{n}{8}$ **75.** $2n - 9$
77. $10(n - 6)$ **79.** $n + 20$ **81.** $2t - 3$ **83.** $n + 47$ **85.** $8y$ **87.** $25m$ **89.** $\frac{c}{25}$
91. $n + 2$ **93.** $\frac{c}{5}$ **95.** $12d$ **97.** $3y + f$ **99.** $5280m$

Chapter 1 Review Problem Set (page 39)

1. (a) 67 (b) 0, -8, and 67 (c) 0 and 67 (d) $0, \frac{3}{4}, -\frac{5}{6}, 8\frac{1}{3}, -8, 0.34, 0.2\overline{3}$, 67, and $\frac{9}{7}$
(e) $\sqrt{2}$ and $-\sqrt{3}$ **2.** Associative property for addition **3.** Substitution property for equality
4. Multiplication property of negative one **5.** Distributive property
6. Associative property for multiplication **7.** Commutative property for addition
8. Distributive property **9.** Multiplicative inverse property **10.** Symmetric property of equality
11. -6 **12.** -6 **13.** -10 **14.** -15 **15.** 20 **16.** 49 **17.** -56 **18.** -24
19. 6 **20.** 4 **21.** -1000 **22.** 8 **23.** $-4a^2 - 5b^2$ **24.** $3x - 2$ **25.** $5ab^2 + 3a^2b$
26. $6x + 31$ **27.** $10n^2 - 17$ **28.** $-13a + 4$ **29.** $-2n + 2$ **30.** $-7x - 29y$
31. $-7a - 9$ **32.** $-9x^2 + 7$ **33.** -31 **34.** -5 **35.** -55 **36.** 144 **37.** -16
38. -44 **39.** 53 **40.** 53 **41.** 6 **42.** -221 **43.** $4 + 2n$ **44.** $3n - 50$
45. $\frac{2}{3}n - 6$ **46.** $10(n - 14)$ **47.** $5n + 8$ **48.** $\frac{n}{n - 3}$ **49.** $5(n + 2) - 3$ **50.** $\frac{3}{4}(n + 12)$
51. $37 - n$ **52.** $\frac{w}{60}$ **53.** $2y - 7$ **54.** $n + 3$ **55.** $p + 5n + 25q$ **56.** $\frac{i}{48}$
57. $24f + 72y$ **58.** $10d$ **59.** $12f + i$ **60.** $25 - c$

CHAPTER 2

Problem Set 2.1 (page 48)

1. $\{4\}$ **3.** $\{-3\}$ **5.** $\{-14\}$ **7.** $\{6\}$ **9.** $\left\{\frac{19}{3}\right\}$ **11.** $\{1\}$ **13.** $\left\{-\frac{10}{3}\right\}$ **15.** $\{4\}$
17. $\left\{-\frac{13}{3}\right\}$ **19.** $\{3\}$ **21.** $\{8\}$ **23.** $\{-9\}$ **25.** $\{-3\}$ **27.** $\{0\}$ **29.** $\left\{-\frac{7}{2}\right\}$
31. $\{-2\}$ **33.** $\left\{-\frac{5}{3}\right\}$ **35.** $\left\{\frac{33}{2}\right\}$ **37.** $\{-35\}$ **39.** $\left\{\frac{1}{2}\right\}$ **41.** $\left\{\frac{1}{6}\right\}$ **43.** $\{5\}$
45. $\{-1\}$ **47.** $\left\{-\frac{21}{16}\right\}$ **49.** $\left\{\frac{12}{7}\right\}$ **51.** 14 **53.** 13, 14, and 15 **55.** 9, 11, and 13
57. 14 and 81 **59.** \$11 per hour **61.** 30 pennies, 50 nickels, and 70 dimes **63.** \$300
65. 20 three-bedroom, 70 two-bedroom, and 140 one-bedroom **67.** **(a)** $\varnothing$ **(c)** $\{0\}$ **(e)** $\varnothing$

Problem Set 2.2 (page 56)

1. $\{12\}$ **3.** $\left\{-\frac{3}{5}\right\}$ **5.** $\{3\}$ **7.** $\{-2\}$ **9.** $\{-36\}$ **11.** $\left\{\frac{20}{9}\right\}$ **13.** $\{3\}$ **15.** $\{3\}$
17. $\{-2\}$ **19.** $\left\{\frac{8}{5}\right\}$ **21.** $\{-3\}$ **23.** $\left\{\frac{48}{17}\right\}$ **25.** $\left\{\frac{103}{6}\right\}$ **27.** $\{3\}$ **29.** $\left\{\frac{40}{3}\right\}$
31. $\left\{-\frac{20}{7}\right\}$ **33.** $\left\{\frac{24}{5}\right\}$ **35.** $\{-10\}$ **37.** $\left\{-\frac{25}{4}\right\}$ **39.** $\{0\}$ **41.** 18
43. 16 inches long and 5 inches wide **45.** 14, 15, and 16 **47.** 8 feet
49. Angie is 22 and her mother is 42. **51.** 80, 90, and 94 **53.** 48° and 132° **55.** 78°

Problem Set 2.3 (page 63)

1. $\{20\}$ **3.** $\{50\}$ **5.** $\{40\}$ **7.** $\{12\}$ **9.** $\{6\}$ **11.** $\{400\}$ **13.** $\{400\}$ **15.** $\{38\}$
17. $\{6\}$ **19.** $\{3000\}$ **21.** $\{3000\}$ **23.** $\{400\}$ **25.** $\{14\}$ **27.** $\{15\}$ **29.** \$90
31. \$54.40 **33.** \$48 **35.** \$400 **37.** 65% **39.** \$32,500
41. \$3000 at 10% and \$4500 at 11% **43.** \$1000 **45.** 8 pennies, 15 nickels, and 18 dimes
47. 15 dimes, 45 quarters, and 10 half-dollars **49.** $\{1.5\}$ **51.** $\{225\}$ **53.** $\{3.6\}$ **55.** $\{14.4\}$
57. $\{12.9\}$

Problem Set 2.4 (page 72)

1. \$120 **3.** 3 years **5.** 6% **7.** \$800 **9.** \$1600 **11.** 8% **13.** \$200
15. 6 feet; 14 feet; 10 feet; 20 feet; 7 feet; 2 feet **17.** $h = \frac{V}{B}$ **19.** $h = \frac{V}{\pi r^2}$ **21.** $r = \frac{C}{2\pi}$
23. $C = \frac{100M}{I}$ **25.** $C = \frac{5}{9}(F - 32)$ or $C = \frac{5F - 160}{9}$ **27.** $x = \frac{y - b}{m}$ **29.** $x = \frac{y - y_1 + mx_1}{m}$
31. $x = \frac{ab + bc}{b - a}$ **33.** $x = a + bc$ **35.** $x = \frac{3b - 6a}{2}$ **37.** $x = \frac{5y + 7}{2}$ **39.** $y = -7x - 4$
41. $x = \frac{6y + 4}{3}$ **43.** $x = \frac{cy - ac - b^2}{b}$ **45.** $y = \frac{x - a + 1}{a - 3}$

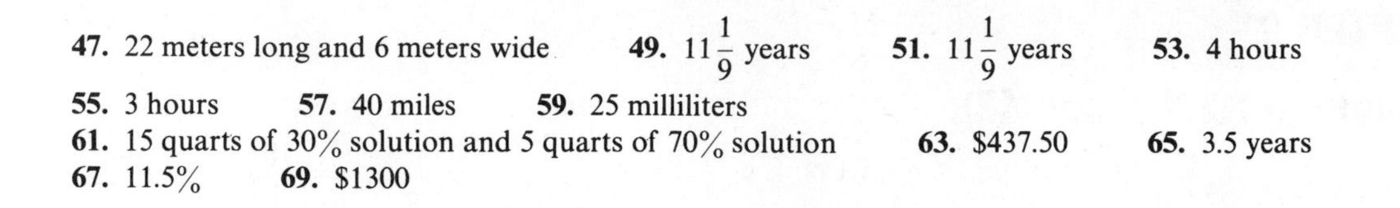

47. 22 meters long and 6 meters wide **49.** $11\frac{1}{9}$ years **51.** $11\frac{1}{9}$ years **53.** 4 hours
55. 3 hours **57.** 40 miles **59.** 25 milliliters
61. 15 quarts of 30% solution and 5 quarts of 70% solution **63.** \$437.50 **65.** 3.5 years
67. 11.5% **69.** \$1300

Problem Set 2.5 (page 81)

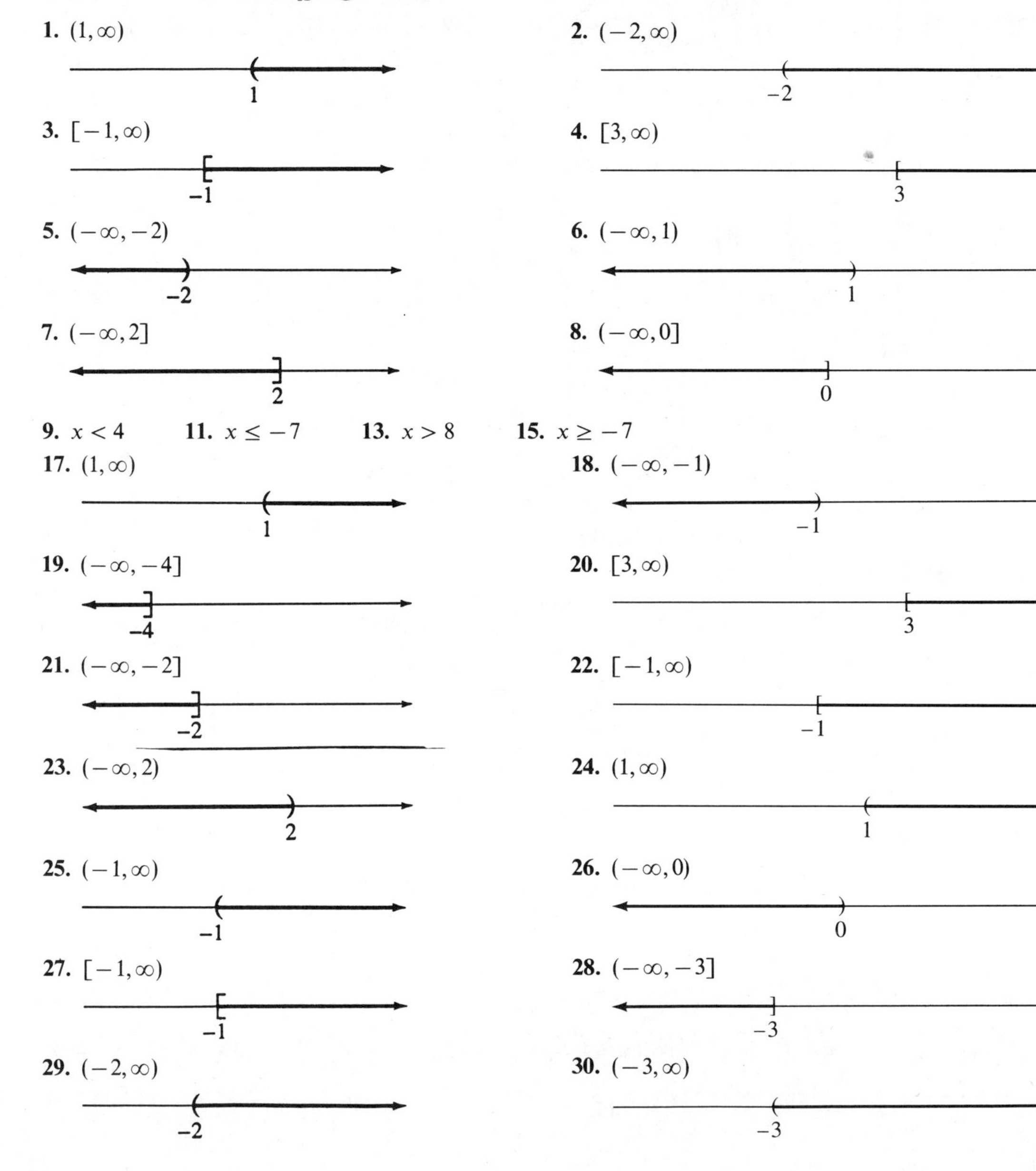

1. $(1, \infty)$

2. $(-2, \infty)$

3. $[-1, \infty)$

4. $[3, \infty)$

5. $(-\infty, -2)$

6. $(-\infty, 1)$

7. $(-\infty, 2]$

8. $(-\infty, 0]$

9. $x < 4$ **11.** $x \leq -7$ **13.** $x > 8$ **15.** $x \geq -7$

17. $(1, \infty)$

18. $(-\infty, -1)$

19. $(-\infty, -4]$

20. $[3, \infty)$

21. $(-\infty, -2]$

22. $[-1, \infty)$

23. $(-\infty, 2)$

24. $(1, \infty)$

25. $(-1, \infty)$

26. $(-\infty, 0)$

27. $[-1, \infty)$

28. $(-\infty, -3]$

29. $(-2, \infty)$

30. $(-3, \infty)$

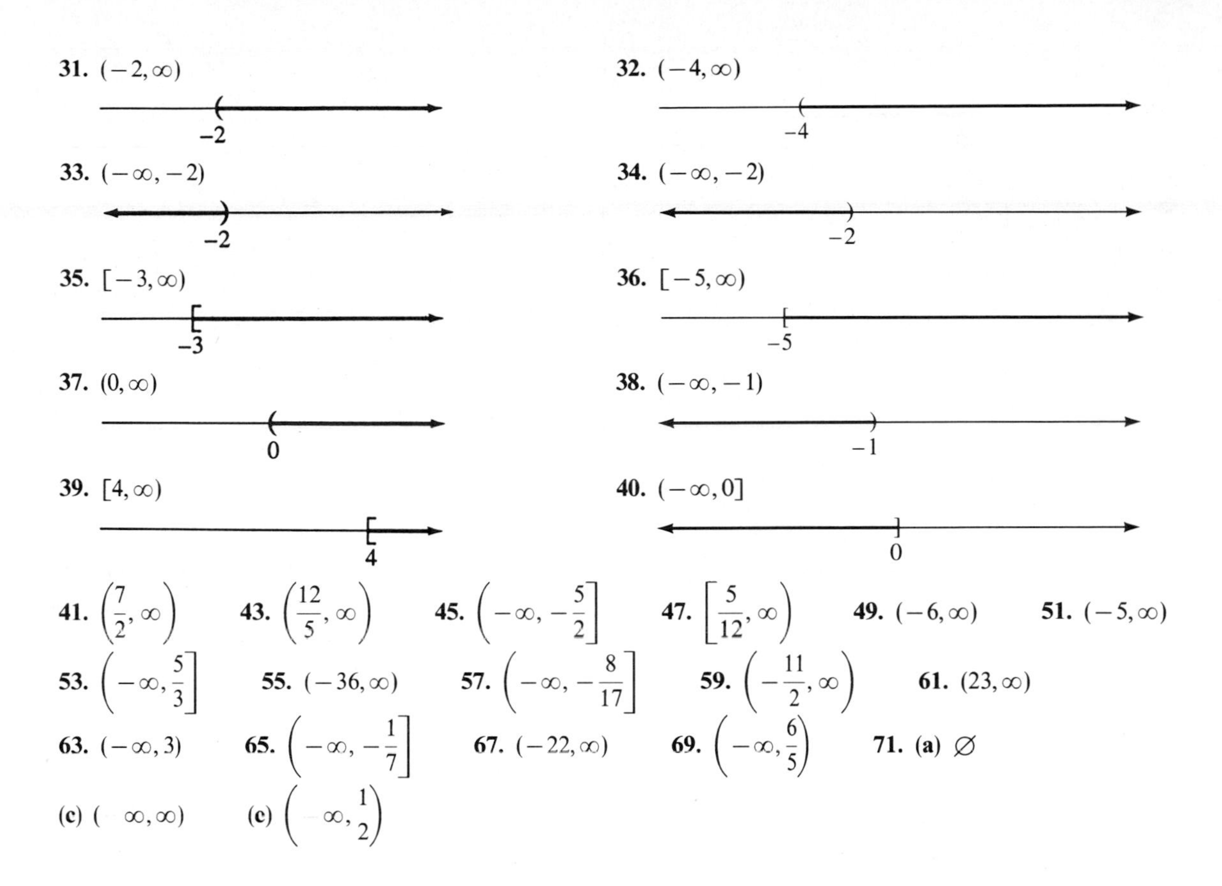

31. $(-2, \infty)$

32. $(-4, \infty)$

33. $(-\infty, -2)$

34. $(-\infty, -2)$

35. $[-3, \infty)$

36. $[-5, \infty)$

37. $(0, \infty)$

38. $(-\infty, -1)$

39. $[4, \infty)$

40. $(-\infty, 0]$

41. $\left(\frac{7}{2}, \infty\right)$ **43.** $\left(\frac{12}{5}, \infty\right)$ **45.** $\left(-\infty, -\frac{5}{2}\right]$ **47.** $\left[\frac{5}{12}, \infty\right)$ **49.** $(-6, \infty)$ **51.** $(-5, \infty)$

53. $\left(-\infty, \frac{5}{3}\right]$ **55.** $(-36, \infty)$ **57.** $\left(-\infty, -\frac{8}{17}\right]$ **59.** $\left(-\frac{11}{2}, \infty\right)$ **61.** $(23, \infty)$

63. $(-\infty, 3)$ **65.** $\left(-\infty, -\frac{1}{7}\right]$ **67.** $(-22, \infty)$ **69.** $\left(-\infty, \frac{6}{5}\right)$ **71. (a)** $\varnothing$

(c) $(-\infty, \infty)$ **(e)** $\left(-\infty, \frac{1}{2}\right)$

Problem Set 2.6 (page 90)

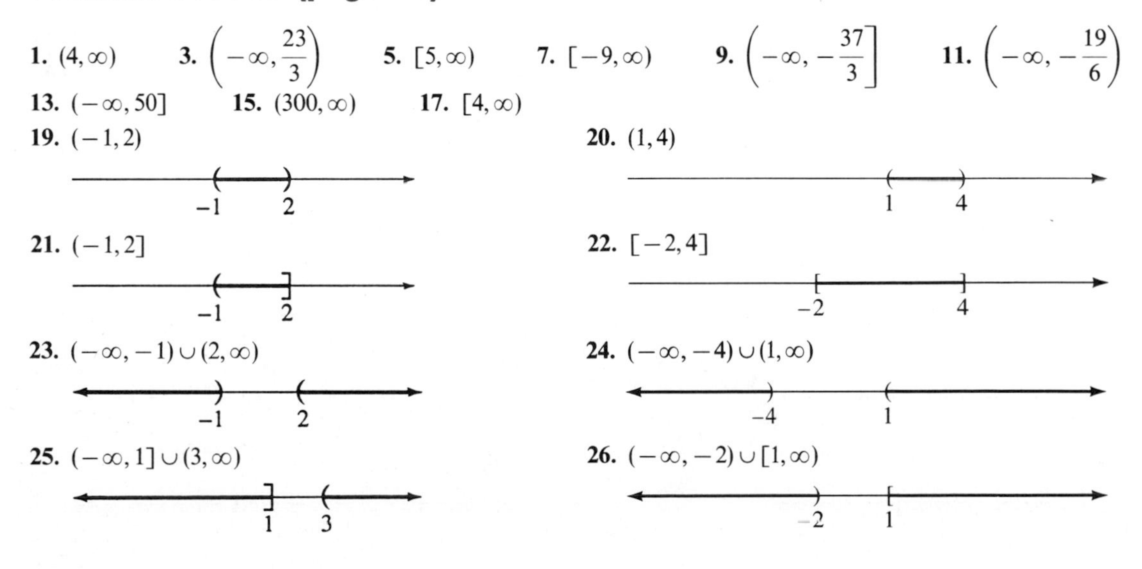

1. $(4, \infty)$ **3.** $\left(-\infty, \frac{23}{3}\right)$ **5.** $[5, \infty)$ **7.** $[-9, \infty)$ **9.** $\left(-\infty, -\frac{37}{3}\right]$ **11.** $\left(-\infty, -\frac{19}{6}\right)$

13. $(-\infty, 50]$ **15.** $(300, \infty)$ **17.** $[4, \infty)$

19. $(-1, 2)$

20. $(1, 4)$

21. $(-1, 2]$

22. $[-2, 4]$

23. $(-\infty, -1) \cup (2, \infty)$

24. $(-\infty, -4) \cup (1, \infty)$

25. $(-\infty, 1] \cup (3, \infty)$

26. $(-\infty, -2) \cup [1, \infty)$

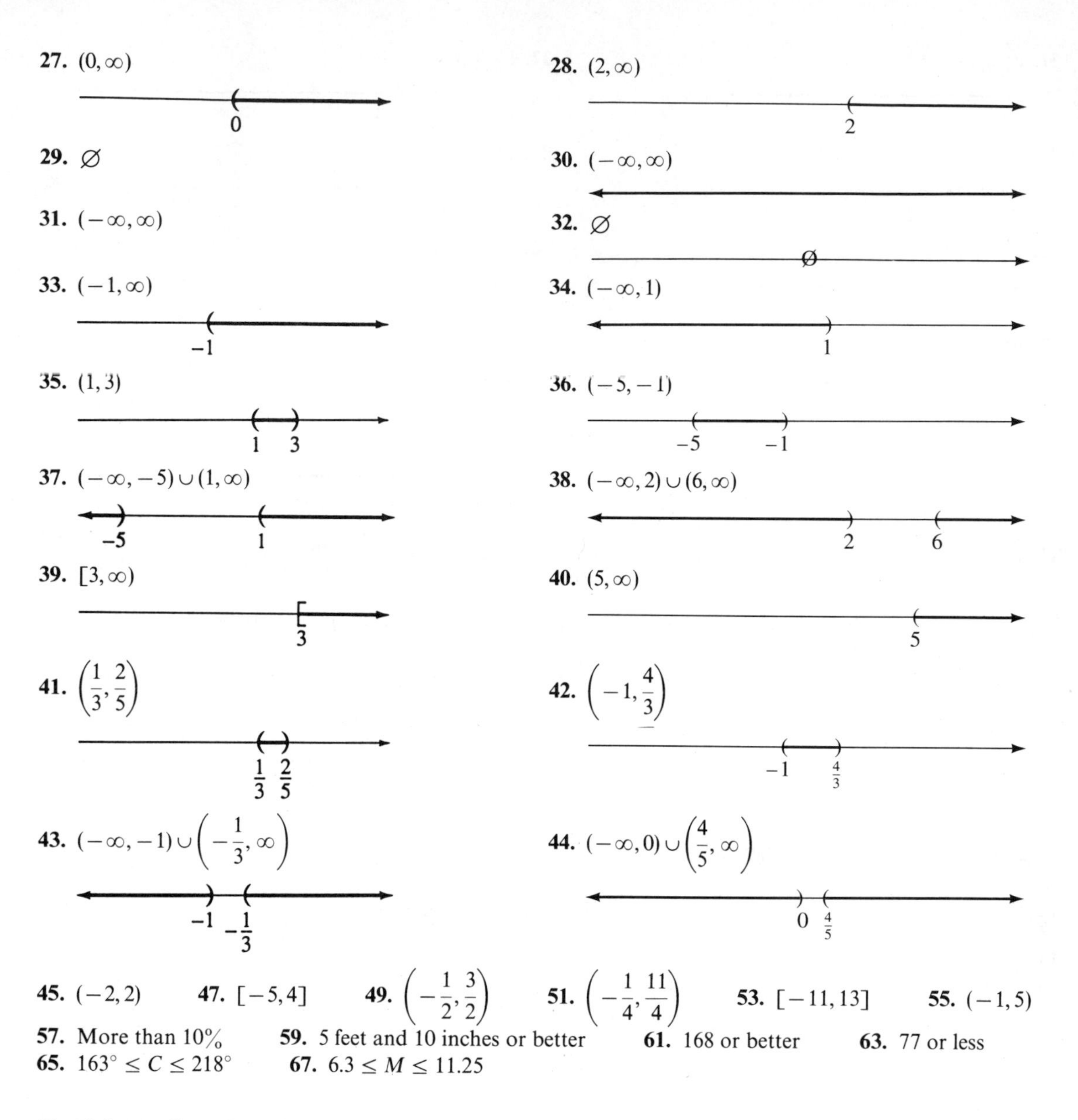

27. $(0, \infty)$ 28. $(2, \infty)$
29. $\varnothing$ 30. $(-\infty, \infty)$
31. $(-\infty, \infty)$ 32. $\varnothing$
33. $(-1, \infty)$ 34. $(-\infty, 1)$
35. $(1, 3)$ 36. $(-5, -1)$
37. $(-\infty, -5) \cup (1, \infty)$ 38. $(-\infty, 2) \cup (6, \infty)$
39. $[3, \infty)$ 40. $(5, \infty)$
41. $\left(\frac{1}{3}, \frac{2}{5}\right)$ 42. $\left(-1, \frac{4}{3}\right)$
43. $(-\infty, -1) \cup \left(-\frac{1}{3}, \infty\right)$ 44. $(-\infty, 0) \cup \left(\frac{4}{5}, \infty\right)$

45. $(-2, 2)$ 47. $[-5, 4]$ 49. $\left(-\frac{1}{2}, \frac{3}{2}\right)$ 51. $\left(-\frac{1}{4}, \frac{11}{4}\right)$ 53. $[-11, 13]$ 55. $(-1, 5)$
57. More than 10% 59. 5 feet and 10 inches or better 61. 168 or better 63. 77 or less
65. $163^\circ \leq C \leq 218^\circ$ 67. $6.3 \leq M \leq 11.25$

Problem Set 2.7 (page 96)

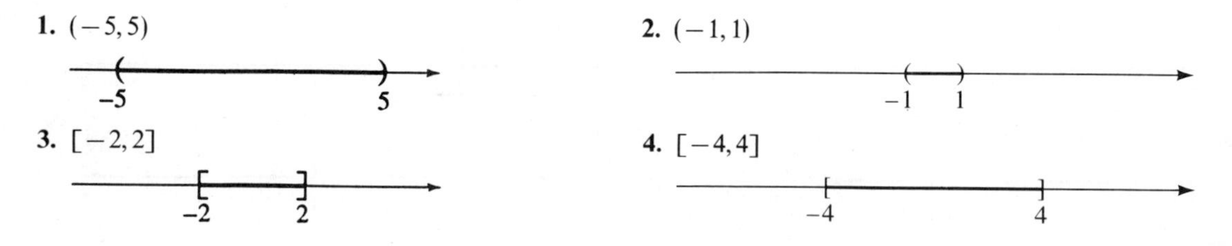

1. $(-5, 5)$ 2. $(-1, 1)$
3. $[-2, 2]$ 4. $[-4, 4]$

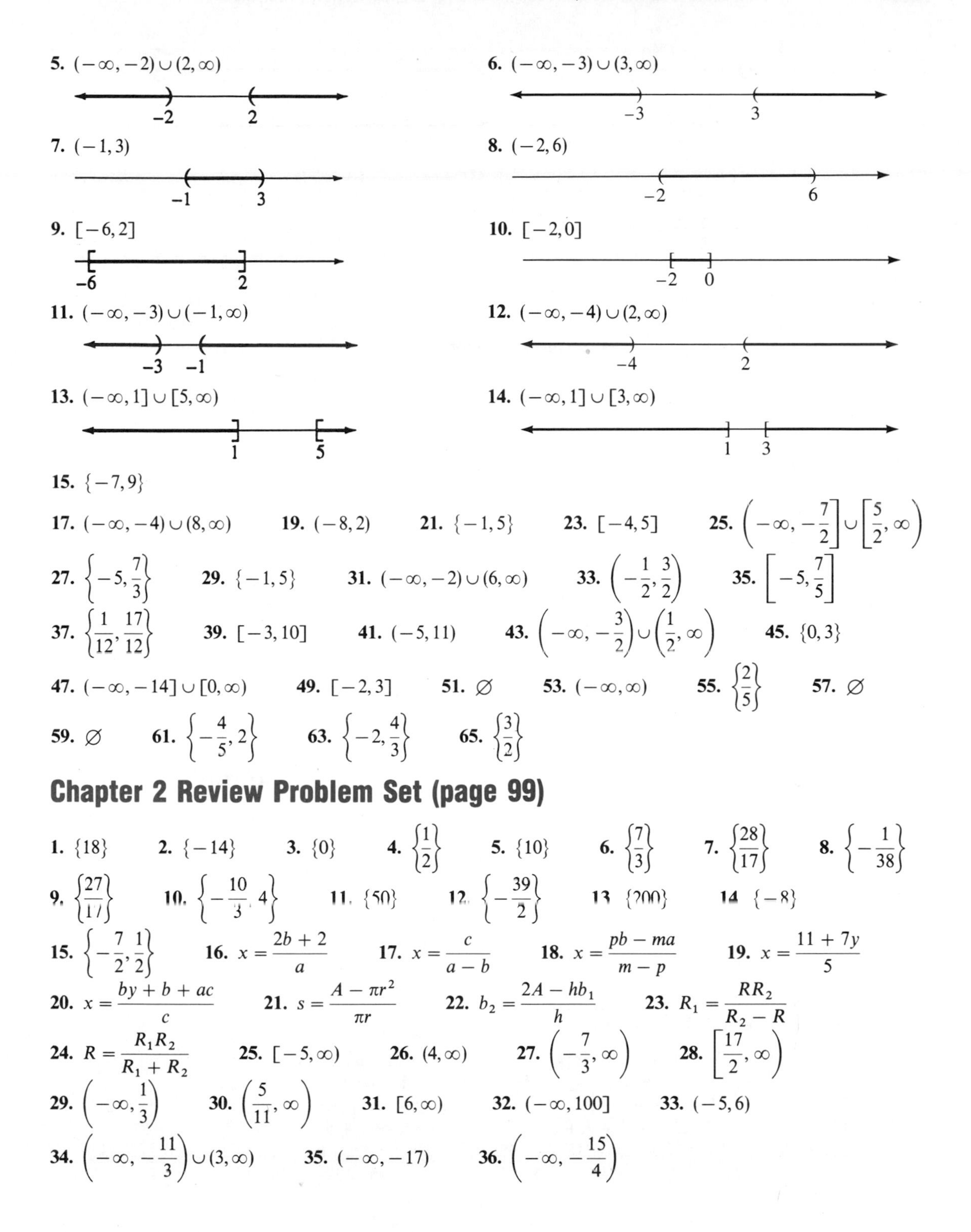

5. $(-\infty, -2) \cup (2, \infty)$

6. $(-\infty, -3) \cup (3, \infty)$

7. $(-1, 3)$

8. $(-2, 6)$

9. $[-6, 2]$

10. $[-2, 0]$

11. $(-\infty, -3) \cup (-1, \infty)$

12. $(-\infty, -4) \cup (2, \infty)$

13. $(-\infty, 1] \cup [5, \infty)$

14. $(-\infty, 1] \cup [3, \infty)$

15. $\{-7, 9\}$

17. $(-\infty, -4) \cup (8, \infty)$ **19.** $(-8, 2)$ **21.** $\{-1, 5\}$ **23.** $[-4, 5]$ **25.** $\left(-\infty, -\frac{7}{2}\right] \cup \left[\frac{5}{2}, \infty\right)$

27. $\left\{-5, \frac{7}{3}\right\}$ **29.** $\{-1, 5\}$ **31.** $(-\infty, -2) \cup (6, \infty)$ **33.** $\left(-\frac{1}{2}, \frac{3}{2}\right)$ **35.** $\left[-5, \frac{7}{5}\right]$

37. $\left\{\frac{1}{12}, \frac{17}{12}\right\}$ **39.** $[-3, 10]$ **41.** $(-5, 11)$ **43.** $\left(-\infty, -\frac{3}{2}\right) \cup \left(\frac{1}{2}, \infty\right)$ **45.** $\{0, 3\}$

47. $(-\infty, -14] \cup [0, \infty)$ **49.** $[-2, 3]$ **51.** $\varnothing$ **53.** $(-\infty, \infty)$ **55.** $\left\{\frac{2}{5}\right\}$ **57.** $\varnothing$

59. $\varnothing$ **61.** $\left\{-\frac{4}{5}, 2\right\}$ **63.** $\left\{-2, \frac{4}{3}\right\}$ **65.** $\left\{\frac{3}{2}\right\}$

Chapter 2 Review Problem Set (page 99)

1. $\{18\}$ **2.** $\{-14\}$ **3.** $\{0\}$ **4.** $\left\{\frac{1}{2}\right\}$ **5.** $\{10\}$ **6.** $\left\{\frac{7}{3}\right\}$ **7.** $\left\{\frac{28}{17}\right\}$ **8.** $\left\{-\frac{1}{38}\right\}$

9. $\left\{\frac{27}{17}\right\}$ **10.** $\left\{-\frac{10}{3}, 4\right\}$ **11.** $\{50\}$ **12.** $\left\{-\frac{39}{2}\right\}$ **13.** $\{200\}$ **14.** $\{-8\}$

15. $\left\{-\frac{7}{2}, \frac{1}{2}\right\}$ **16.** $x = \frac{2b+2}{a}$ **17.** $x = \frac{c}{a-b}$ **18.** $x = \frac{pb-ma}{m-p}$ **19.** $x = \frac{11+7y}{5}$

20. $x = \frac{by+b+ac}{c}$ **21.** $s = \frac{A-\pi r^2}{\pi r}$ **22.** $b_2 = \frac{2A-hb_1}{h}$ **23.** $R_1 = \frac{RR_2}{R_2-R}$

24. $R = \frac{R_1R_2}{R_1+R_2}$ **25.** $[-5, \infty)$ **26.** $(4, \infty)$ **27.** $\left(-\frac{7}{3}, \infty\right)$ **28.** $\left[\frac{17}{2}, \infty\right)$

29. $\left(-\infty, \frac{1}{3}\right)$ **30.** $\left(\frac{5}{11}, \infty\right)$ **31.** $[6, \infty)$ **32.** $(-\infty, 100]$ **33.** $(-5, 6)$

34. $\left(-\infty, -\frac{11}{3}\right) \cup (3, \infty)$ **35.** $(-\infty, -17)$ **36.** $\left(-\infty, -\frac{15}{4}\right)$

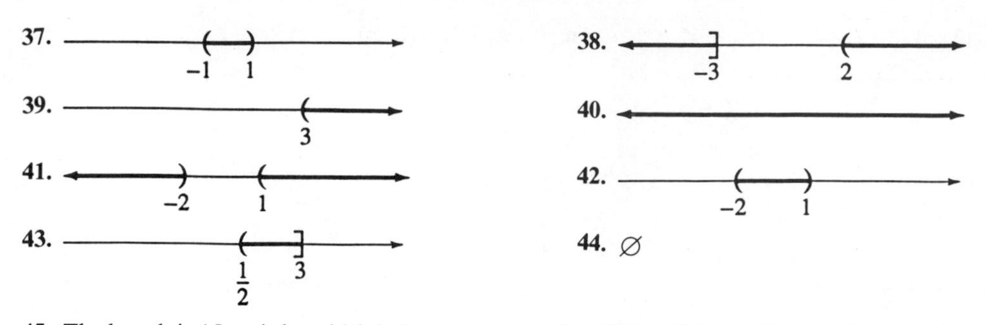

45. The length is 15 and the width is 7 meters. **46.** \$200 at 7% and \$300 at 8% **47.** 88 or better
48. 4, 5, and 6 **49.** \$10.50 per hour **50.** 20 nickels, 50 dimes, and 75 quarters **51.** 80°
52. \$45.60 **53.** 30 or more **54.** 55 miles per hour
55. Sonya for $3\frac{1}{4}$ hours and Rita for $4\frac{1}{2}$ hours **56.** $6\frac{1}{4}$ cups

CHAPTER 3

Problem Set 3.1 (page 106)

1. 2 **3.** 3 **5.** 2 **7.** 6 **9.** 0 **11.** $10x - 3$ **13.** $-11t + 5$ **15.** $-x^2 + 2x - 2$
17. $17a^2b^2 - 5ab$ **19.** $-9x + 7$ **21.** $-2x + 6$ **23.** $10a + 7$ **25.** $4x^2 + 10x + 6$
27. $-6a^2 + 12a + 14$ **29.** $3x^3 + x^2 + 13x - 11$ **31.** $7x + 8$ **33.** $-3x - 16$
35. $2x^2 - 2x - 8$ **37.** $-3x^3 + 5x^2 - 2x + 9$ **39.** $5x^2 - 4x + 11$ **41.** $-6x^2 + 9x + 7$
43. $-2x^2 + 9x + 4$ **45.** $-10n^2 + n + 9$ **47.** $8x - 2$ **49.** $8x - 14$ **51.** $-9x^2 - 12x + 4$
53. $10x^2 + 13x - 18$ **55.** $-n^2 - 4n - 4$ **57.** $-x + 6$ **59.** $6x^2 - 4$ **61.** $-7n^2 + n + 6$
63. $t^2 - 4t + 8$ **65.** $4n^2 - n - 12$ **67.** $-4x - 2y$ **69.** $-x^3 - x^2 + 3x$

Problem Set 3.2 (page 112)

1. $36x^4$ **3.** $-12x^5$ **5.** $4a^3b^4$ **7.** $-3x^3y^2z^6$ **9.** $-30xy^4$ **11.** $27a^4b^5$
13. $-m^3n^3$ **15.** $\frac{3}{10}x^3y^6$ **17.** $-\frac{3}{20}a^3b^4$ **19.** $-\frac{1}{6}x^3y^4$ **21.** $30x^6$ **23.** $-18x^9$
25. $-3x^6y^6$ **27.** $-24y^9$ **29.** $-56a^4b^2$ **31.** $-18a^3b^3$ **33.** $-10x^7y^7$ **35.** $50x^5y^2$
37. $27x^3y^6$ **39.** $-32x^{10}y^5$ **41.** $x^{16}y^{20}$ **43.** $a^6b^{12}c^{18}$ **45.** $64a^{12}b^{18}$ **47.** $81x^2y^8$
49. $81a^4b^{12}$ **51.** $-16a^4b^4$ **53.** $-x^6y^{12}z^{18}$ **55.** $-125a^6b^6c^3$ **57.** $-x^7y^{28}z^{14}$
59. $3x^3y^3$ **61.** $-5x^3y^2$ **63.** $9bc^2$ **65.** $-18xyz^4$ **67.** $-a^2b^3c^2$ **69.** 9 **71.** $-b^2$
73. $-18x^3$ **75.** $6x^{3n}$ **77.** a^{5n+3} **79.** x^{4n} **81.** a^{5n+1} **83.** $-10x^{2n}$ **85.** $12a^{n+4}$
87. $6x^{3n+2}$ **89.** $12x^{n+2}$

Problem Set 3.3 (page 117)

1. $10x^2y^3 + 6x^3y^4$ **3.** $-12a^3b^3 + 15a^5b$ **5.** $24a^4b^5 - 16a^4b^6 + 32a^5b^6$
7. $-6x^3y^3 - 3x^4y^4 + x^5y^2$ **9.** $ax + ay + 2bx + 2by$ **11.** $ac + 4ad - 3bc - 12bd$
13. $x^2 + 16x + 60$ **15.** $y^2 + 6y - 55$ **17.** $n^2 - 5n - 14$ **19.** $x^2 - 36$ **21.** $x^2 - 12x + 36$
23. $x^2 - 14x + 48$ **25.** $x^3 - 4x^2 + x + 6$ **27.** $x^3 - x^2 - 9x + 9$ **29.** $t^2 + 18t + 81$
31. $y^2 - 14y + 49$ **33.** $4x^2 + 33x + 35$ **35.** $9y^2 - 1$ **37.** $14x^2 + 3x - 2$
39. $5 + 3t - 2t^2$ **41.** $9t^2 + 42t + 49$ **43.** $4 - 25x^2$ **45.** $49x^2 - 56x + 16$

47. $18x^2 - 39x - 70$ **49.** $2x^2 + xy - 15y^2$ **51.** $25x^2 - 4a^2$ **53.** $t^3 - 14t - 15$
55. $x^3 + x^2 - 24x + 16$ **57.** $2x^3 + 9x^2 + 2x - 30$ **59.** $12x^3 - 7x^2 + 25x - 6$
61. $x^4 + 5x^3 + 11x^2 + 11x + 4$ **63.** $2x^4 - x^3 - 12x^2 + 5x + 4$ **65.** $x^3 + 6x^2 + 12x + 8$
67. $x^3 - 12x^2 + 48x - 64$ **69.** $8x^3 + 36x^2 + 54x + 27$ **71.** $64x^3 - 48x^2 + 12x - 1$
73. $125x^3 + 150x^2 + 60x + 8$ **75.** $x^{2n} - 16$ **77.** $x^{2a} + 4x^a - 12$ **79.** $6x^{2n} + x^n - 35$
81. $x^{4a} - 10x^{2a} + 21$ **83.** $4x^{2n} + 20x^n + 25$
85. **(a)** $a^6 + 6a^5b + 15a^4b^2 + 20a^3b^3 + 15a^2b^4 + 6ab^5 + b^6$
(c) $a^8 + 8a^7b + 28a^6b^2 + 56a^5b^3 + 70a^4b^4 + 56a^3b^5 + 28a^2b^6 + 8ab^7 + b^8$

Problem Set 3.4 (page 126)

1. Composite **3.** Prime **5.** Composite **7.** Composite **9.** Prime **11.** $2 \cdot 2 \cdot 7$
13. $2 \cdot 2 \cdot 11$ **15.** $2 \cdot 2 \cdot 2 \cdot 7$ **17.** $2 \cdot 2 \cdot 2 \cdot 3 \cdot 3$ **19.** $3 \cdot 29$ **21.** $3(2x + y)$
23. $2x(3x + 7)$ **25.** $4y(7y - 1)$ **27.** $5x(4y - 3)$ **29.** $x^2(7x + 10)$ **31.** $9ab(2a + 3b)$
33. $3x^3y^3(4y - 13x)$ **35.** $4x^2(2x^2 + 3x - 6)$ **37.** $x(5 + 7x + 9x^3)$ **39.** $5xy^2(3xy + 4 + 7x^2y^2)$
41. $(y + 2)(x + 3)$ **43.** $(2a + b)(3x - 2y)$ **45.** $(x + 2)(x + 5)$ **47.** $(a + 4)(x + y)$
49. $(a - 2b)(x + y)$ **51.** $(a - b)(3x - y)$ **53.** $(a + 1)(2x + y)$ **55.** $(a - 1)(x^2 + 2)$
57. $(a + b)(2c + 3d)$ **59.** $(a + b)(x - y)$ **61.** $(x + 9)(x + 6)$ **63.** $(x + 4)(2x + 1)$
65. $\{-7, 0\}$ **67.** $\{0, 1\}$ **69.** $\{0, 5\}$ **71.** $\left\{-\frac{1}{2}, 0\right\}$ **73.** $\left\{-\frac{7}{3}, 0\right\}$ **75.** $\left\{0, \frac{5}{4}\right\}$
77. $\left\{0, \frac{1}{4}\right\}$ **79.** $\{-12, 0\}$ **81.** $\left\{0, \frac{3a}{5b}\right\}$ **83.** $\left\{-\frac{3a}{2b}, 0\right\}$ **85.** $\{a, -2b\}$ **87.** 0 or 7
89. 6 units **91.** $\frac{4}{\pi}$ units **93.** The square is 100 feet by 100 feet and the rectangle is 50 feet by 100 feet.
95. 6 units **97.** $x^a(2x^a - 3)$ **99.** $y^{2m}(y^m + 5)$ **101.** $x^{4a}(2x^{2a} - 3x^a + 7)$

Problem Set 3.5 (page 133)

1. $(x + 1)(x - 1)$ **3.** $(4x + 5)(4x - 5)$ **5.** $(3x + 5y)(3x - 5y)$ **7.** $(5xy + 6)(5xy - 6)$
9. $(2x + y^2)(2x - y^2)$ **11.** $(1 + 12n)(1 - 12n)$ **13.** $(x + 2 + y)(x + 2 - y)$
15. $(2x + y + 1)(2x - y - 1)$ **17.** $(3a + 2b + 3)(3a - 2b - 3)$ **19.** $-5(2x + 9)$
21. $9(x + 2)(x - 2)$ **23.** $5(x^2 + 1)$ **25.** $8(y + 2)(y - 2)$ **27.** $ab(a + 3)(a - 3)$
29. Not factorable **31.** $(n + 3)(n - 3)(n^2 + 9)$ **33.** $3x(x^2 + 9)$ **35.** $4xy(x + 4y)(x - 4y)$
37. $6x(1 + x)(1 - x)$ **39.** $(1 + xy)(1 - xy)(1 + x^2y^2)$ **41.** $4(x + 4y)(x - 4y)$
43. $3(x + 2)(x - 2)(x^2 + 4)$ **45.** $(a - 4)(a^2 + 4a + 16)$ **47.** $(x + 1)(x^2 - x + 1)$
49. $(3x + 4y)(9x^2 - 12xy + 16y^2)$ **51.** $(1 - 3a)(1 + 3a + 9a^2)$ **53.** $(xy - 1)(x^2y^2 + xy + 1)$
55. $(x + y)(x - y)(x^4 + x^2y^2 + y^4)$ **57.** $\{-5, 5\}$ **59.** $\left\{-\frac{7}{3}, \frac{7}{3}\right\}$ **61.** $\{-2, 2\}$
63. $\{-1, 0, 1\}$ **65.** $\{-2, 2\}$ **67.** $\{-3, 3\}$ **69.** $\{0\}$ **71.** -3, 0, or 3 **73.** 4 and 8
75. 10 meters long and 5 meters wide **77.** 6 inches **79.** 8 yards

Problem Set 3.6 (page 141)

1. $(x + 5)(x + 4)$ **3.** $(x - 4)(x - 7)$ **5.** $(a + 9)(a - 4)$ **7.** $(y + 6)(y + 14)$
9. $(x - 7)(x + 2)$ **11.** Not factorable **13.** $(6 - x)(1 + x)$ **15.** $(x + 3y)(x + 12y)$
17. $(a - 8b)(a + 7b)$ **19.** $(3x + 1)(5x + 6)$ **21.** $(4x - 3)(3x + 2)$ **23.** $(2a + 3)(4a - 9)$
25. $(n - 4)(12n + 5)$ **27.** Not factorable **29.** $(2n - 7)(10n + 3)$ **31.** $(8x - 5)(2x + 9)$
33. $(1 - 6x)(6 + 7x)$ **35.** $(5y + 9)(4y - 1)$ **37.** $(12n + 5)(2n - 1)$ **39.** $(5n + 3)(n + 6)$
41. $(x + 10)(x + 15)$ **43.** $(n - 16)(n - 20)$ **45.** $(t + 15)(t - 12)$ **47.** $(t^2 - 3)(t^2 - 2)$

49. $(2x^2-1)(5x^2+4)$ **51.** $(x+1)(x-1)(x^2-8)$ **53.** $(3n+1)(3n-1)(2n^2+3)$
55. $(x+1)(x-1)(x+4)(x-4)$ **57.** $2(t+2)(t-2)$ **59.** $(4x+5y)(3x-2y)$
61. $3n(2n+5)(3n-1)$ **63.** $(n-12)(n-5)$ **65.** $(6a-1)^2$ **67.** $6(x^2+9)$
69. Not factorable **71.** $(x+y-7)(x-y+7)$ **73.** $(1+4x^2)(1+2x)(1-2x)$
75. $(4n+9)(3n+8)$ **77.** $n(n+7)(n-7)$ **79.** $(x-8)(x+1)$ **81.** $3x(x-3)(x^2+3x+9)$
83. $(x^2+3)^2$ **85.** $(x+3)(x-3)(x^2+4)$ **87.** $(6w-7)(3w+5)$ **89.** Not factorable
91. $2n(n^2+7n-10)$ **93.** $(2x+1)(y+3)$ **95.** $(x^a-4)(x^a+6)$ **97.** $(2x^a-1)(3x^a-2)$
99. $(3x^n+4)(4x^n-3)$ **101.** $(x+1)(x+3)$ **103.** $2(x+4)(2x-3)$ **105.** $(2x-5)(3x-13)$

Problem Set 3.7 (page 148)

1. $\{-3,-1\}$ **3.** $\{-12,-6\}$ **5.** $\{4,9\}$ **7.** $\{-6,2\}$ **9.** $\{-1,5\}$ **11.** $\{-13,-12\}$
13. $\left\{-5,\frac{1}{3}\right\}$ **15.** $\left\{-\frac{7}{2},-\frac{2}{3}\right\}$ **17.** $\{0,4\}$ **19.** $\left\{\frac{1}{6},2\right\}$ **21.** $\{-6,0,6\}$ **23.** $\{-4,6\}$
25. $\{-2,2\}$ **27.** $\{-11,4\}$ **29.** $\{-5,5\}$ **31.** $\left\{-\frac{5}{3},-\frac{3}{5}\right\}$ **33.** $\left\{-\frac{1}{8},6\right\}$ **35.** $\left\{\frac{3}{7},\frac{5}{4}\right\}$
37. $\left\{-\frac{2}{7},\frac{4}{5}\right\}$ **39.** $\left\{-7,\frac{2}{3}\right\}$ **41.** $\{-20,18\}$ **43.** $\left\{-2,-\frac{1}{3},\frac{1}{3},2\right\}$ **45.** $\{-4,4\}$
47. $\{-1,1\}$ **49.** $\left\{-\frac{5}{2},-\frac{4}{3},0\right\}$ **51.** $\left\{-1,\frac{5}{3}\right\}$ **53.** $\left\{-\frac{3}{2},\frac{1}{2}\right\}$ **55.** 8 and 9 or -9 and -8
57. 7 and 15 **59.** 10 inches by 6 inches **61.** -7 and -6 or 6 and 7
63. 4 centimeters by 4 centimeters and 6 centimeters by 8 centimeters **65.** 3, 4, and 5 units
67. 9 inches and 12 inches **69.** An altitude of 4 inches and a side of 14 inches long.

Chapter 3 Review Problem Set (page 152)

1. $5x-3$ **2.** $3x^2+12x-2$ **3.** $12x^2-x+5$ **4.** $-20x^5y^7$ **5.** $-6a^5b^5$
6. $15a^4-10a^3-5a^2$ **7.** $24x^2+2xy-15y^2$ **8.** $3x^3+7x^2-21x-4$ **9.** $256x^8y^{12}$
10. $9x^2-12xy+4y^2$ **11.** $-8x^6y^9z^3$ **12.** $-13x^2y$ **13.** $2x+y-2$
14. $x^4+x^3-18x^2-x+35$ **15.** $21+26x-15x^2$ **16.** $-12a^5b^7$ **17.** $-8a^7b^3$
18. $7x^2+19x-36$ **19.** $6x^3-11x^2-7x+2$ **20.** $6x^{4n}$ **21.** $4x^2+20xy+25y^2$
22. $x^3-6x^2+12x-8$ **23.** $8x^3+60x^2+150x+125$ **24.** $(x+7)(x-4)$
25. $2(t+3)(t-3)$ **26.** Not factorable **27.** $(4n-1)(3n-1)$ **28.** $x^2(x^2+1)(x+1)(x-1)$
29. $x(x-12)(x+6)$ **30.** $2a^2b(3a+2b-c)$ **31.** $(x-y+1)(x+y-1)$ **32.** $4(2x^2+3)$
33. $(4x+7)(3x-5)$ **34.** $(4n-5)^2$ **35.** $4n(n-2)$ **36.** $3w(w^2+6w-8)$
37. $(5x+2y)(4x-y)$ **38.** $16a(a-4)$ **39.** $3x(x+1)(x-6)$ **40.** $(n+8)(n-16)$
41. $(t+5)(t-5)(t^2+3)$ **42.** $(5x-3)(7x+2)$ **43.** $(3-x)(5-3x)$
44. $(4n-3)(16n^2+12n+9)$ **45.** $2(2x+5)(4x^2-10x+25)$ **46.** $\{-3,3\}$ **47.** $\{-6,1\}$
48. $\left\{\frac{2}{7}\right\}$ **49.** $\left\{-\frac{2}{5},\frac{1}{3}\right\}$ **50.** $\left\{-\frac{1}{3},3\right\}$ **51.** $\{-3,0,3\}$ **52.** $\{-1,0,1\}$ **53.** $\{-7,9\}$
54. $\left\{-\frac{4}{7},\frac{2}{7}\right\}$ **55.** $\left\{-\frac{4}{5},\frac{5}{6}\right\}$ **56.** $\{-2,2\}$ **57.** $\left\{\frac{5}{3}\right\}$ **58.** $\{-8,6\}$ **59.** $\left\{-5,\frac{2}{7}\right\}$
60. $\{-8,5\}$ **61.** $\{-12,1\}$ **62.** $\varnothing$ **63.** $\left\{-5,\frac{6}{5}\right\}$ **64.** $\{0,1,8\}$ **65.** $\left\{-10,\frac{1}{4}\right\}$
66. 8, 9, and 10 or -1, 0, and 1 **67.** -6 and 8 **68.** 13 and 15 **69.** 12 miles and 16 miles
70. 4 meters by 12 meters **71.** 9 rows and 16 chairs per row

72. The side is 13 feet long and the altitude is 6 feet. **73.** 3 feet
74. 5 centimeters by 5 centimeters and 8 centimeters by 8 centimeters **75.** 6 inches

CHAPTER 4

Problem Set 4.1 (page 162)

1. $\frac{3}{4}$ **3.** $\frac{5}{6}$ **5.** $-\frac{2}{5}$ **7.** $\frac{2}{7}$ **9.** $\frac{2x}{7}$ **11.** $\frac{2a}{5b}$ **13.** $-\frac{y}{4x}$ **15.** $-\frac{9c}{13d}$ **17.** $\frac{5x^2}{3y^3}$
19. $\frac{x-2}{x}$ **21.** $\frac{3x+2}{2x-1}$ **23.** $\frac{a+5}{a-9}$ **25.** $\frac{n-3}{5n-1}$ **27.** $\frac{5x^2+7}{10x}$ **29.** $\frac{3x+5}{4x+1}$
31. $\frac{3x}{x^2+4x+16}$ **33.** $\frac{x+6}{3x-1}$ **35.** $\frac{x(2x+7)}{y(x+9)}$ **37.** $\frac{y+4}{5y-2}$ **39.** $\frac{3x(x-1)}{x^2+1}$
41. $\frac{2(x+3y)}{3x(3x+y)}$ **43.** $\frac{3n-4}{7n+2}$ **45.** $\frac{4-x}{5+3x}$ **47.** $\frac{9x^2+3x+1}{2(x+2)}$ **49.** $\frac{-2(x-1)}{x+1}$
51. $\frac{y+b}{y+c}$ **53.** $\frac{x+2y}{2x+y}$ **55.** $\frac{x+1}{x-6}$ **57.** $\frac{2s+5}{3s+1}$ **59.** -1 **61.** $-n-7$
63. $-\frac{2}{x+1}$ **65.** -2 **67.** $-\frac{n+3}{n+5}$

Problem Set 4.2 (page 168)

1. $\frac{1}{10}$ **3.** $-\frac{4}{15}$ **5.** $\frac{3}{16}$ **7.** $-\frac{5}{6}$ **9.** $-\frac{2}{3}$ **11.** $\frac{10}{11}$ **13.** $-\frac{5x^3}{12y^2}$ **15.** $\frac{2a^3}{3b}$
17. $\frac{3x^3}{4}$ **19.** $\frac{25x^3}{108y^2}$ **21.** $\frac{ac^2}{2b^2}$ **23.** $\frac{3x}{4y}$ **25.** $\frac{3(x^2+4)}{5y(x+8)}$ **27.** $\frac{5(a+3)}{a(a-2)}$ **29.** $\frac{3}{2}$
31. $\frac{3xy}{4(x+6)}$ **33.** $\frac{5(x-2y)}{7y}$ **35.** $\frac{5+n}{3-n}$ **37.** $\frac{x^2+1}{x^2-10}$ **39.** $\frac{6x+5}{3x+4}$ **41.** $\frac{2t^2+5}{2(t^2+1)(t+1)}$
43. $\frac{t(t+6)}{4t+5}$ **45.** $\frac{n+3}{n(n-2)}$ **47.** $\frac{25x^3y^3}{4(x+1)}$ **49.** $\frac{2(a-2b)}{a(3a-2b)}$

Problem Set 4.3 (page 175)

1. $\frac{13}{12}$ **3.** $\frac{11}{40}$ **5.** $\frac{19}{20}$ **7.** $\frac{49}{75}$ **9.** $\frac{17}{30}$ **11.** $-\frac{11}{84}$ **13.** $\frac{2x+4}{x-1}$ **15.** 4
17. $\frac{7y-10}{7y}$ **19.** $\frac{5x+3}{6}$ **21.** $\frac{12a+1}{12}$ **23.** $\frac{n+14}{18}$ **25.** $-\frac{11}{15}$ **27.** $\frac{3x-25}{30}$
29. $\frac{43}{40x}$ **31.** $\frac{20y-77x}{28xy}$ **33.** $\frac{16y+15x-12xy}{12xy}$ **35.** $\frac{21+22x}{30x^2}$ **37.** $\frac{10n-21}{7n^2}$
39. $\frac{45-6n+20n^2}{15n^2}$ **41.** $\frac{11x-10}{6x^2}$ **43.** $\frac{42t+43}{35t^3}$ **45.** $\frac{20b^2-33a^3}{96a^2b}$
47. $\frac{14-24y^3+45xy}{18xy^3}$ **49.** $\frac{2x^2+3x-3}{x(x-1)}$ **51.** $\frac{a^2-a-8}{a(a+4)}$ **53.** $\frac{-41n-55}{(4n+5)(3n+5)}$

55. $\frac{-3x+17}{(x+4)(7x-1)}$ **57.** $\frac{-x+74}{(3x-5)(2x+7)}$ **59.** $\frac{38x+13}{(3x-2)(4x+5)}$ **61.** $\frac{5x+5}{2x+5}$ **63.** $\frac{x+15}{x-5}$

65. $\frac{-2x-4}{2x+1}$ **67. (a)** -1 **(c)** 0

Problem Set 4.4 (page 184)

1. $\frac{7x+20}{x(x+4)}$ **3.** $\frac{-x-3}{x(x+7)}$ **5.** $\frac{6x-5}{(x+1)(x-1)}$ **7.** $\frac{1}{a+1}$ **9.** $\frac{5n+15}{4(n+5)(n-5)}$

11. $\frac{x^2+60}{x(x+6)}$ **13.** $\frac{11x+13}{(x+2)(x+7)(2x+1)}$ **15.** $\frac{-3a+1}{(a-5)(a+2)(a+9)}$

17. $\frac{9a^2+17a+1}{(5a+1)(4a-3)(3a+4)}$ **19.** $\frac{3x^2+20x-111}{(x^2+3)(x+7)(x-3)}$ **21.** $\frac{-7y-14}{(y+8)(y-2)}$

23. $\frac{-2x^2-4x+3}{(x+2)(x-2)}$ **25.** $\frac{2x^2+14x-19}{(x+10)(x-2)}$ **27.** $\frac{2n+1}{n-6}$ **29.** $\frac{2x^2-32x+16}{(x+1)(2x-1)(3x-2)}$

31. $\frac{1}{(n^2+1)(n+1)}$ **33.** $\frac{-16x}{(5x-2)(x-1)}$ **35.** $\frac{t+1}{t-2}$ **37.** $\frac{2}{11}$ **39.** $-\frac{7}{27}$ **41.** $\frac{x}{4}$

43. $\frac{3y-2x}{4x-7}$ **45.** $\frac{6ab^2-5a^2}{12b^2+2a^2b}$ **47.** $\frac{2y-3xy}{3x+4xy}$ **49.** $\frac{3n+14}{5n+19}$ **51.** $\frac{5n-17}{4n-13}$

53. $\frac{-x+5y-10}{3y-10}$ **55.** $\frac{-x+15}{-2x-1}$ **57.** $\frac{3a^2-2a+1}{2a-1}$ **59.** $\frac{-x^2+6x-4}{3x-2}$

Problem Set 4.5 (page 190)

1. $3x^3+6x^2$ **3.** $-6x^4+9x^6$ **5.** $3a^2-5a-8$ **7.** $-13x^2+17x-28$

9. $-3xy+4x^2y-8xy^2$ **11.** $x-13$ **13.** $x+20$ **15.** $2x+1-\frac{3}{x-1}$ **17.** $5x-1$

19. $3x^2-2x-7$ **21.** x^2+5x-6 **23.** $4x^2+7x+12+\frac{30}{x-2}$ **25.** x^3-4x^2-5x+3

27. $x^2+5x+25$ **29.** $x^2-x+1+\frac{63}{x+1}$ **31.** $2x^2-4x+7-\frac{20}{x+2}$ **33.** $4a-4b$

35. $4x+7+\frac{23x-6}{x^2-3x}$ **37.** $8y-9+\frac{8y+5}{y^2+y}$ **39.** $2x-1$ **41.** $x-3$

43. $5a-8+\frac{42a-41}{a^2+3a-4}$ **45.** $2n^2+3n-4$ **47.** $x^4+x^3+x^2+x+1$ **49.** x^3-x^2+x-1

51. $3x^2+x+1+\frac{7}{x^2-1}$

Problem Set 4.6 (page 197)

1. $\{2\}$ **3.** $\{-3\}$ **5.** $\{6\}$ **7.** $\left\{-\frac{85}{18}\right\}$ **9.** $\left\{\frac{7}{10}\right\}$ **11.** $\{5\}$ **13.** $\{58\}$

15. $\left\{\frac{1}{4}, 4\right\}$ **17.** $\left\{-\frac{2}{5}, 5\right\}$ **19.** $\{-16\}$ **21.** $\left\{-\frac{13}{3}\right\}$ **23.** $\{-3, 1\}$ **25.** $\left\{-\frac{5}{2}\right\}$

27. $\{-51\}$ **29.** $\left\{-\frac{5}{3}, 4\right\}$ **31.** $\varnothing$ **33.** $\left\{-\frac{11}{8}, 2\right\}$ **35.** $\{-29, 0\}$ **37.** $\{-9, 3\}$

39. $\left\{-2, \frac{23}{8}\right\}$ **41.** $\left\{\frac{11}{23}\right\}$ **43.** $\left\{3, \frac{7}{2}\right\}$ **45.** \$750 and \$1000 **47.** 48° and 72°

49. $\frac{2}{7}$ or $\frac{7}{2}$ **51.** \$1080 **53.** \$69 for Tammy and \$51.75 for Laura **55.** 8 and 82

57. 14 feet and 6 feet **59.** 690 females and 460 males

Problem Set 4.7 (page 206)

1. $\{-21\}$ **3.** $\{-1, 2\}$ **5.** $\{2\}$ **7.** $\left\{\frac{37}{15}\right\}$ **9.** $\{-1\}$ **11.** $\{-1\}$ **13.** $\left\{0, \frac{13}{2}\right\}$

15. $\left\{-2, \frac{19}{2}\right\}$ **17.** $\{-2\}$ **19.** $\left\{-\frac{1}{5}\right\}$ **21.** $\varnothing$ **23.** $\left\{\frac{7}{2}\right\}$ **25.** $\{-3\}$ **27.** $\left\{-\frac{7}{9}\right\}$

29. $\left\{-\frac{7}{6}\right\}$ **31.** $x = \frac{18y - 4}{15}$ **33.** $y = \frac{-5x + 22}{2}$ **35.** $M = \frac{IC}{100}$ **37.** $R = \frac{ST}{S + T}$

39. $y = \frac{bx - x - 3b + a}{a - 3}$ **41.** $y = \frac{ab - bx}{a}$ **43.** $y = \frac{-2x - 9}{3}$

45. 50 miles per hour for Dave and 54 miles per hour for Kent **47.** 60 minutes

49. 60 words per minute for Connie and 40 words per minute for Katie

51. Plane B could travel at 400 miles per hour for 5 hours and plane A at 350 miles per hour for 4 hours or plane B could travel at 250 miles per hour for 8 hours and plane A at 200 miles per hour for 7 hours.

53. 60 minutes for Nancy and 120 minutes for Amy **55.** 3 hours

57. 16 miles per hour on the way out and 12 miles per hour on the way back or 12 miles per hour out and 8 miles per hour back

Chapter 4 Review Problem Set (page 210)

1. $\frac{2y}{3x^2}$ **2.** $\frac{a - 3}{a}$ **3.** $\frac{n - 5}{n - 1}$ **4.** $\frac{x^2 + 1}{x}$ **5.** $\frac{2x + 1}{3}$ **6.** $\frac{x^2 - 10}{2x^2 + 1}$ **7.** $\frac{3}{22}$

8. $\frac{18y + 20x}{48y - 9x}$ **9.** $\frac{3x + 2}{3x - 2}$ **10.** $\frac{x - 1}{2x - 1}$ **11.** $\frac{2x}{7y^2}$ **12.** $3b$ **13.** $\frac{n(n + 5)}{n - 1}$

14. $\frac{x(x - 3y)}{x^2 + 9y^2}$ **15.** $\frac{23x - 6}{20}$ **16.** $\frac{57 - 2n}{18n}$ **17.** $\frac{3x^2 - 2x - 14}{x(x + 7)}$ **18.** $\frac{2}{x - 5}$

19. $\frac{5n - 21}{(n - 9)(n + 4)(n - 1)}$ **20.** $\frac{6y - 23}{(2y + 3)(y - 6)}$ **21.** $6x - 1$ **22.** $3x^2 - 7x + 22 - \frac{90}{x + 4}$

23. $\left\{\frac{4}{13}\right\}$ **24.** $\left\{\frac{3}{16}\right\}$ **25.** $\varnothing$ **26.** $\{-17\}$ **27.** $\left\{\frac{2}{7}, \frac{7}{2}\right\}$ **28.** $\{22\}$ **29.** $\left\{-\frac{6}{7}, 3\right\}$

30. $\left\{\frac{3}{4}, \frac{5}{2}\right\}$ **31.** $\left\{\frac{9}{7}\right\}$ **32.** $\left\{-\frac{5}{4}\right\}$ **33.** $y = \frac{3x + 27}{4}$ **34.** $y = \frac{bx - ab}{a}$

35. \$525 and \$875 **36.** 20 minutes for Don and 30 minutes for Dan

37. 50 miles per hour and 55 miles per hour **38.** 9 hours **39.** 80 hours **40.** 13 miles per hour

CHAPTER 5

Problem Set 5.1 (page 219)

1. $\frac{1}{27}$ **3.** $-\frac{1}{100}$ **5.** 81 **7.** -27 **9.** -8 **11.** 1 **13.** $\frac{9}{49}$ **15.** 16
17. $\frac{1}{1000}$ **19.** $\frac{1}{1000}$ **21.** 27 **23.** $\frac{1}{125}$ **25.** $\frac{9}{8}$ **27.** $\frac{256}{25}$ **29.** $\frac{2}{25}$ **31.** $\frac{81}{4}$
33. 81 **35.** $\frac{1}{10000}$ **37.** $\frac{13}{36}$ **39.** $\frac{1}{2}$ **41.** $\frac{72}{17}$ **43.** $\frac{1}{x^6}$ **45.** $\frac{1}{a^3}$ **47.** $\frac{1}{a^8}$
49. $\frac{y^6}{x^2}$ **51.** $\frac{c^8}{a^4b^{12}}$ **53.** $\frac{y^{12}}{8x^9}$ **55.** $\frac{x^3}{y^{12}}$ **57.** $\frac{4a^4}{9b^2}$ **59.** $\frac{1}{x^2}$ **61.** a^5b^2 **63.** $\frac{6y^3}{x}$
65. $7b^2$ **67.** $\frac{7x}{y^2}$ **69.** $-\frac{12b^3}{a}$ **71.** $\frac{x^5y^5}{5}$ **73.** $\frac{b^{20}}{81}$ **75.** $\frac{x+1}{x^3}$ **77.** $\frac{y-x^3}{x^3y}$
79. $\frac{3b+4a^2}{a^2b}$ **81.** $\frac{1-x^2y}{xy^2}$ **83.** $\frac{2x-3}{x^2}$

Problem Set 5.2 (page 228)

1. 8 **3.** -10 **5.** 3 **7.** -4 **9.** 3 **11.** $\frac{4}{5}$ **13.** $-\frac{6}{7}$ **15.** $\frac{1}{2}$ **17.** $\frac{3}{4}$
19. 8 **21.** $3\sqrt{3}$ **23.** $4\sqrt{2}$ **25.** $4\sqrt{5}$ **27.** $4\sqrt{10}$ **29.** $12\sqrt{2}$ **31.** $-12\sqrt{5}$
33. $2\sqrt{3}$ **35.** $3\sqrt{6}$ **37.** $-\frac{5}{3}\sqrt{7}$ **39.** $\frac{\sqrt{19}}{2}$ **41.** $\frac{3\sqrt{3}}{4}$ **43.** $\frac{5\sqrt{3}}{9}$ **45.** $\frac{\sqrt{14}}{7}$
47. $\frac{\sqrt{6}}{3}$ **49.** $\frac{\sqrt{15}}{6}$ **51.** $\frac{\sqrt{66}}{12}$ **53.** $\frac{\sqrt{6}}{3}$ **55.** $\sqrt{5}$ **57.** $\frac{2\sqrt{21}}{7}$ **59.** $-\frac{8\sqrt{15}}{5}$
61. $\frac{\sqrt{6}}{4}$ **63.** $-\frac{12}{25}$ **65.** $2\sqrt[3]{2}$ **67.** $6\sqrt[3]{3}$ **69.** $\frac{2\sqrt[3]{3}}{3}$ **71.** $\frac{3\sqrt[3]{2}}{2}$ **73.** $\frac{\sqrt[3]{12}}{2}$
75. **(a)** 1.414 **(c)** 12.490 **(e)** 57.000 **(g)** .374 **(i)** .930

Problem Set 5.3 (page 233)

1. $13\sqrt{2}$ **3.** $54\sqrt{3}$ **5.** $-30\sqrt{2}$ **7.** $-\sqrt{5}$ **9.** $-21\sqrt{6}$ **11.** $-\frac{7\sqrt{7}}{12}$ **13.** $\frac{37\sqrt{10}}{10}$
15. $\frac{41\sqrt{2}}{20}$ **17.** $-9\sqrt[3]{3}$ **19.** $10\sqrt[3]{2}$ **21.** $4\sqrt{2x}$ **23.** $5x\sqrt{3}$ **25.** $2x\sqrt{5y}$
27. $8xy^3\sqrt{xy}$ **29.** $3a^2b\sqrt{6b}$ **31.** $3x^3y^4\sqrt{7}$ **33.** $4a\sqrt{10a}$ **35.** $\frac{8y}{3}\sqrt{6xy}$ **37.** $\frac{\sqrt{10xy}}{5y}$
39. $\frac{\sqrt{15}}{6x^2}$ **41.** $\frac{5\sqrt{2y}}{6y}$ **43.** $\frac{\sqrt{14xy}}{4y^3}$ **45.** $\frac{3y\sqrt{2xy}}{4x}$ **47.** $\frac{2\sqrt{42ab}}{7b^2}$ **49.** $2\sqrt[3]{3y}$
51. $2x\sqrt[3]{2x}$ **53.** $2x^2y^2\sqrt[3]{7y^2}$ **55.** $\frac{\sqrt[3]{21x}}{3x}$ **57.** $\frac{\sqrt[3]{12x^2y}}{4x^2}$ **59.** $\frac{\sqrt[3]{4x^2y^2}}{xy^2}$ **61.** $2\sqrt{2x+3y}$
63. $4\sqrt{x+3y}$ **65.** $33\sqrt{x}$ **67.** $-30\sqrt{2x}$ **69.** $7\sqrt{3n}$ **71.** $-40\sqrt{ab}$ **73.** $-7x\sqrt{2x}$
75. **(a)** .707 **(c)** .707

Problem Set 5.4 (page 239)

1. $6\sqrt{2}$ **3.** $18\sqrt{2}$ **5.** $-24\sqrt{10}$ **7.** $24\sqrt{6}$ **9.** 120 **11.** 24 **13.** $56\sqrt[3]{3}$
15. $\sqrt{6}+\sqrt{10}$ **17.** $6\sqrt{10}-3\sqrt{35}$ **19.** $24\sqrt{3}-60\sqrt{2}$ **21.** $-40-32\sqrt{15}$
23. $15\sqrt{2x}+3\sqrt{xy}$ **25.** $5xy-6x\sqrt{y}$ **27.** $2\sqrt{10xy}+2y\sqrt{15y}$ **29.** $-25\sqrt{6}$
31. $-25-3\sqrt{3}$ **33.** $23-9\sqrt{5}$ **35.** $6\sqrt{35}+3\sqrt{10}-4\sqrt{21}-2\sqrt{6}$
37. $8\sqrt{3}-36\sqrt{2}+6\sqrt{10}-18\sqrt{15}$ **39.** $11+13\sqrt{30}$ **41.** $141-51\sqrt{6}$ **43.** -10
45. -8 **47.** $2x-3y$ **49.** $10\sqrt[3]{12}+2\sqrt[3]{18}$ **51.** $12-36\sqrt[3]{2}$ **53.** $\frac{\sqrt{7}-1}{3}$
55. $\frac{-3\sqrt{2}-15}{23}$ **57.** $\frac{\sqrt{7}-\sqrt{2}}{5}$ **59.** $\frac{2\sqrt{5}+\sqrt{6}}{7}$ **61.** $\frac{\sqrt{15}-2\sqrt{3}}{2}$ **63.** $\frac{6\sqrt{7}+4\sqrt{6}}{13}$
65. $\sqrt{3}-\sqrt{2}$ **67.** $\frac{2\sqrt{x}-8}{x-16}$ **69.** $\frac{x+5\sqrt{x}}{x-25}$ **71.** $\frac{x-8\sqrt{x}+12}{x-36}$ **73.** $\frac{x-2\sqrt{xy}}{x-4y}$
75. $\frac{6\sqrt{xy}+9y}{4x-9y}$

Problem Set 5.5 (page 244)

1. $\{20\}$ **3.** $\varnothing$ **5.** $\left\{\frac{25}{4}\right\}$ **7.** $\left\{\frac{4}{9}\right\}$ **9.** $\{5\}$ **11.** $\left\{\frac{39}{4}\right\}$ **13.** $\varnothing$ **15.** $\{1\}$
17. $\left\{\frac{3}{2}\right\}$ **19.** $\{3\}$ **21.** $\left\{\frac{61}{25}\right\}$ **23.** $\{-3, 3\}$ **25.** $\{-9, -4\}$ **27.** $\{0\}$ **29.** $\{3\}$
31. $\{4\}$ **33.** $\{-4, -3\}$ **35.** $\{12\}$ **37.** $\{25\}$ **39.** $\{29\}$ **41.** $\{-15\}$ **43.** $\left\{-\frac{1}{3}\right\}$
45. $\{-3\}$ **47.** $\{0\}$ **49.** $\{5\}$ **51.** $\{2, 6\}$

Problem Set 5.6 (page 249)

1. 9 **3.** 3 **5.** -2 **7.** -5 **9.** $\frac{1}{6}$ **11.** 3 **13.** 8 **15.** 81 **17.** -1
19. -32 **21.** $\frac{81}{16}$ **23.** 4 **25.** $\frac{1}{128}$ **27.** -125 **29.** 625 **31.** $\sqrt[3]{x^4}$ **33.** $3\sqrt{x}$
35. $\sqrt[3]{2y}$ **37.** $\sqrt{2x-3y}$ **39.** $\sqrt[3]{(2a-3b)^2}$ **41.** $\sqrt[3]{x^2y}$ **43.** $3\sqrt[5]{xy^2}$ **45.** $5^{\frac{1}{2}}y^{\frac{1}{2}}$
47. $3y^{\frac{1}{2}}$ **49.** $x^{\frac{1}{3}}y^{\frac{2}{3}}$ **51.** $a^{\frac{1}{2}}b^{\frac{3}{4}}$ **53.** $(2x-y)^{\frac{3}{5}}$ **55.** $5xy^{\frac{1}{2}}$ **57.** $-(x+y)^{\frac{1}{3}}$
59. $12x^{\frac{13}{20}}$ **61.** $y^{\frac{5}{12}}$ **63.** $\frac{4}{x^{\frac{1}{10}}}$ **65.** $16xy^2$ **67.** $2x^2y$ **69.** $4x^{\frac{4}{15}}$ **71.** $\frac{4}{b^{\frac{5}{12}}}$
73. $\frac{36x^{\frac{4}{5}}}{49y^{\frac{4}{3}}}$ **75.** $\frac{y^{\frac{3}{2}}}{x}$ **77.** $4x^{\frac{1}{6}}$ **79.** $\frac{16}{a^{\frac{11}{10}}}$ **81.** $\sqrt[6]{243}$ **83.** $\sqrt[4]{216}$ **85.** $\sqrt[12]{3}$ **87.** $\sqrt{2}$
89. $\sqrt[4]{3}$ **91.** (a) 12 (c) 7 (e) 11 **93.** (a) 1024 (c) 512 (e) 49

Problem Set 5.7 (page 255)

1. $(8.9)(10)^1$ **3.** $(4.29)(10)^3$ **5.** $(6.12)(10)^6$ **7.** $(4)(10)^7$ **9.** $(3.764)(10)^2$
11. $(3.47)(10)^{-1}$ **13.** $(2.14)(10)^{-2}$ **15.** $(5)(10)^{-5}$ **17.** $(1.94)(10)^{-9}$ **19.** 23 **21.** 4190

23. 500,000,000 **25.** 31,400,000,000 **27.** .43 **29.** .000914 **31.** .00000005123
33. .000000074 **35.** .77 **37.** 300,000,000,000 **39.** .000000004 **41.** 1000 **43.** 1000
45. 3000 **47.** 20 **49.** 27,000,000 **51.** (a) 7000 (c) 120 (e) 30
53. (a) $(4.385)(10)^{14}$ (c) $(2.322)(10)^{17}$ (e) $(3.052)(10)^{12}$

Chapter 5 Review Problem Set (page 258)

1. $\frac{1}{64}$ **2.** $\frac{9}{4}$ **3.** 3 **4.** -2 **5.** $\frac{2}{3}$ **6.** 32 **7.** 1 **8.** $\frac{4}{9}$ **9.** -64 **10.** 32
11. 1 **12.** 27 **13.** $3\sqrt{6}$ **14.** $4x\sqrt{3xy}$ **15.** $2\sqrt{2}$ **16.** $\frac{\sqrt{15x}}{6x^2}$ **17.** $2\sqrt[3]{7}$
18. $\frac{\sqrt[3]{6}}{3}$ **19.** $\frac{3\sqrt{5}}{5}$ **20.** $\frac{x\sqrt{21x}}{7}$ **21.** $3xy^2\sqrt[3]{4xy^2}$ **22.** $\frac{15\sqrt{6}}{4}$ **23.** $2y\sqrt{5xy}$
24. $2\sqrt{x}$ **25.** $24\sqrt{10}$ **26.** 60 **27.** $24\sqrt{3} - 6\sqrt{14}$ **28.** $x - 2\sqrt{x} - 15$ **29.** 17
30. $12 - 8\sqrt{3}$ **31.** $6a - 5\sqrt{ab} - 4b$ **32.** 70 **33.** $\frac{2(\sqrt{7} + 1)}{3}$ **34.** $\frac{2\sqrt{6} - \sqrt{15}}{3}$
35. $\frac{3\sqrt{5} - 2\sqrt{3}}{11}$ **36.** $\frac{6\sqrt{3} + 3\sqrt{5}}{7}$ **37.** $\frac{x^6}{y^8}$ **38.** $\frac{27a^3b^{12}}{8}$ **39.** $20x^{\frac{7}{10}}$ **40.** $7a^{\frac{5}{12}}$
41. $\frac{y^{\frac{4}{3}}}{x}$ **42.** $\frac{x^{12}}{9}$ **43.** $\sqrt{5}$ **44.** $5\sqrt[3]{3}$ **45.** $\frac{29\sqrt{6}}{5}$ **46.** $-15\sqrt{3x}$ **47.** $\frac{y + x^2}{x^2y}$
48. $\frac{b - 2a}{a^2b}$ **49.** $\left\{\frac{19}{7}\right\}$ **50.** $\{4\}$ **51.** $\{8\}$ **52.** $\varnothing$ **53.** $\{14\}$ **54.** $\{-10, 1\}$
55. $\{2\}$ **56.** $\{8\}$ **57.** .000000006 **58.** 36,000,000,000 **59.** 6 **60.** .15
61. .000028 **62.** .002 **63.** .002 **64.** 8,000,000,000

CHAPTER 6

Problem Set 6.1 (page 268)

1. False **3.** True **5.** True **7.** True **9.** $10 + 8i$ **11.** $-6 + 10i$ **13.** $-2 - 5i$
15. $-12 + 5i$ **17.** $-1 - 23i$ **19.** $-4 - 5i$ **21.** $1 + 3i$ **23.** $\frac{5}{3} - \frac{5}{12}i$ **25.** $-\frac{17}{9} + \frac{23}{30}i$
27. $9i$ **29.** $i\sqrt{14}$ **31.** $\frac{4}{5}i$ **33.** $3i\sqrt{2}$ **35.** $5i\sqrt{3}$ **37.** $6i\sqrt{7}$ **39.** $-8i\sqrt{5}$
41. $36i\sqrt{10}$ **43.** -8 **45.** $-\sqrt{15}$ **47.** $-3\sqrt{6}$ **49.** $-5\sqrt{3}$ **51.** $-3\sqrt{6}$
53. $4i\sqrt{3}$ **55.** $\frac{5}{2}$ **57.** $2\sqrt{2}$ **59.** $2i$ **61.** $-20 + 0i$ **63.** $42 + 0i$ **65.** $15 + 6i$
67. $-42 + 12i$ **69.** $7 + 22i$ **71.** $40 - 20i$ **73.** $-3 - 28i$ **75.** $-3 - 15i$
77. $-9 + 40i$ **79.** $-12 + 16i$ **81.** $85 + 0i$ **83.** $5 + 0i$ **85.** $\frac{3}{5} + \frac{3}{10}i$ **87.** $\frac{5}{17} - \frac{3}{17}i$
89. $2 + \frac{2}{3}i$ **91.** $0 - \frac{2}{7}i$ **93.** $\frac{22}{25} - \frac{4}{25}i$ **95.** $-\frac{18}{41} + \frac{39}{41}i$ **97.** $\frac{9}{2} - \frac{5}{2}i$ **99.** $\frac{4}{13} - \frac{1}{26}i$

Problem Set 6.2 (page 276)

1. $\{0, 9\}$ **3.** $\{-3, 0\}$ **5.** $\{-4, 0\}$ **7.** $\left\{0, \frac{9}{5}\right\}$ **9.** $\{-6, 5\}$ **11.** $\{7, 12\}$
13. $\left\{-8, -\frac{3}{2}\right\}$ **15.** $\left\{-\frac{7}{3}, \frac{2}{5}\right\}$ **17.** $\left\{\frac{3}{5}\right\}$ **19.** $\left\{-\frac{3}{2}, \frac{7}{3}\right\}$ **21.** $\{1, 4\}$ **23.** $\{8\}$
25. $\{12\}$ **27.** $\{0, 5k\}$ **29.** $\{0, 16k^2\}$ **31.** $\{5k, 7k\}$ **33.** $\left\{\frac{k}{2}, -3k\right\}$ **35.** $\{\pm 1\}$
37. $\{\pm 6i\}$ **39.** $\{\pm\sqrt{14}\}$ **41.** $\{\pm 2\sqrt{7}\}$ **43.** $\{\pm 3\sqrt{2}\}$ **45.** $\left\{\pm\frac{\sqrt{14}}{2}\right\}$ **47.** $\left\{\pm\frac{2\sqrt{3}}{3}\right\}$
49. $\left\{\pm\frac{2i\sqrt{30}}{5}\right\}$ **51.** $\left\{\pm\frac{\sqrt{6}}{2}\right\}$ **53.** $\{-1, 5\}$ **55.** $\{-8, 2\}$ **57.** $\{-6 \pm 2i\}$ **59.** $\{1, 2\}$
61. $\{4 \pm \sqrt{5}\}$ **63.** $\{-5 \pm 2\sqrt{3}\}$ **65.** $\left\{\frac{2 \pm 3i\sqrt{3}}{3}\right\}$ **67.** $\{-12, -2\}$ **69.** $\left\{\frac{2 \pm \sqrt{10}}{5}\right\}$
71. $2\sqrt{13}$ centimeters **73.** $4\sqrt{5}$ inches **75.** 8 yards **77.** $6\sqrt{2}$ inches
79. $a = b = 4\sqrt{2}$ meters **81.** $b = 3\sqrt{3}$ inches and $c = 6$ inches
83. $a = 7$ centimeters and $b = 7\sqrt{3}$ centimeters **85.** $a = \frac{10\sqrt{3}}{3}$ feet and $c = \frac{20\sqrt{3}}{3}$ feet

Problem Set 6.3 (page 281)

1. $\{-6, 10\}$ **3.** $\{4, 10\}$ **5.** $\{-5, 10\}$ **7.** $\{-8, 1\}$ **9.** $\left\{-\frac{5}{2}, 3\right\}$ **11.** $\left\{-3, \frac{2}{3}\right\}$
13. $\{-16, 10\}$ **15.** $\{-2 \pm \sqrt{6}\}$ **17.** $\{-3 \pm 2\sqrt{3}\}$ **19.** $\{5 \pm \sqrt{26}\}$ **21.** $\{4 \pm i\}$
23. $\{-6 \pm 3\sqrt{3}\}$ **25.** $\{-1 \pm i\sqrt{5}\}$ **27.** $\left\{\frac{-3 \pm \sqrt{17}}{2}\right\}$ **29.** $\left\{\frac{-5 \pm \sqrt{21}}{2}\right\}$
31. $\left\{\frac{7 \pm \sqrt{37}}{2}\right\}$ **33.** $\left\{\frac{-2 \pm \sqrt{10}}{2}\right\}$ **35.** $\left\{\frac{3 \pm i\sqrt{6}}{3}\right\}$ **37.** $\left\{\frac{-5 \pm \sqrt{37}}{6}\right\}$ **39.** $\{-12, 4\}$
41. $\left\{\frac{4 \pm \sqrt{10}}{2}\right\}$ **43.** $\left\{-\frac{9}{2}, \frac{1}{3}\right\}$ **45.** $\{-3, 8\}$ **47.** $\{3 \pm 2\sqrt{3}\}$ **49.** $\left\{\frac{3 \pm i\sqrt{3}}{3}\right\}$
51. $\{-20, 12\}$ **53.** $\left\{-6, -\frac{11}{3}\right\}$ **55.** $\left\{-\frac{7}{3}, -\frac{3}{2}\right\}$ **57.** $\{-6 \pm 2\sqrt{10}\}$ **59.** $\left\{\frac{1}{4}, \frac{1}{3}\right\}$
61. $\left\{\frac{-b \pm \sqrt{b^2 - 4ac}}{2a}\right\}$ **63.** $x = \frac{a\sqrt{b^2 - y^2}}{b}$ **65.** $r = \frac{\sqrt{A\pi}}{\pi}$ **67.** $\{2a, 3a\}$ **69.** $\left\{\frac{a}{2}, -\frac{2a}{3}\right\}$
71. $\left\{\frac{2b}{3}\right\}$

Problem Set 6.4 (page 289)

1. Two real solutions; $\{-7, 3\}$ **3.** One real solution; $\left\{\frac{1}{3}\right\}$ **5.** Two complex solutions; $\left\{\frac{7 \pm i\sqrt{3}}{2}\right\}$
7. Two real solutions; $\left\{-\frac{4}{3}, \frac{1}{5}\right\}$ **9.** Two real solutions; $\left\{\frac{-2 \pm \sqrt{10}}{3}\right\}$ **11.** $\{-1 \pm \sqrt{2}\}$

13. $\left\{\frac{-5 \pm \sqrt{37}}{2}\right\}$ **15.** $\{4 \pm 2\sqrt{5}\}$ **17.** $\left\{\frac{-5 \pm i\sqrt{7}}{2}\right\}$ **19.** $\{8, 10\}$ **21.** $\left\{\frac{9 \pm \sqrt{61}}{2}\right\}$
23. $\left\{\frac{-1 \pm \sqrt{33}}{4}\right\}$ **25.** $\left\{\frac{-1 \pm i\sqrt{3}}{4}\right\}$ **27.** $\left\{\frac{4 \pm \sqrt{10}}{3}\right\}$ **29.** $\left\{-1, \frac{5}{2}\right\}$ **31.** $\left\{-5, -\frac{4}{3}\right\}$
33. $\left\{\frac{5}{6}\right\}$ **35.** $\left\{\frac{1 \pm \sqrt{13}}{4}\right\}$ **37.** $\left\{0, \frac{13}{5}\right\}$ **39.** $\left\{\pm\frac{\sqrt{15}}{3}\right\}$ **41.** $\left\{\frac{-1 \pm \sqrt{73}}{12}\right\}$
43. $\{-18, -14\}$ **45.** $\left\{\frac{11}{4}, \frac{10}{3}\right\}$ **47.** $\left\{\frac{2 \pm i\sqrt{2}}{2}\right\}$ **49.** $\left\{\frac{1 \pm \sqrt{7}}{6}\right\}$ **51.** $\{-1.359, 7.359\}$
53. $\{-10.280, 4.280\}$ **55.** $\{-.259, -7.742\}$ **57.** $\{.191, 1.309\}$ **59.** $\{-.422, 5.922\}$
61. $k > 1$ **63.** Any real number value for k

Problem Set 6.5 (page 298)

1. $\{2 \pm \sqrt{10}\}$ **3.** $\left\{-9, \frac{4}{3}\right\}$ **5.** $\{9 \pm 3\sqrt{10}\}$ **7.** $\left\{\frac{3 \pm i\sqrt{23}}{4}\right\}$ **9.** $\{-15, -9\}$
11. $\{-8, 1\}$ **13.** $\left\{\frac{2 \pm i\sqrt{10}}{2}\right\}$ **15.** $\{9 \pm \sqrt{66}\}$ **17.** $\left\{-\frac{5}{4}, \frac{2}{5}\right\}$ **19.** $\left\{\frac{-1 \pm \sqrt{2}}{2}\right\}$
21. $\left\{\frac{3}{4}, 4\right\}$ **23.** $\left\{\frac{11 \pm \sqrt{109}}{2}\right\}$ **25.** $\left\{\frac{3}{7}, 4\right\}$ **27.** $\left\{\frac{7 \pm \sqrt{129}}{10}\right\}$ **29.** $\left\{-\frac{10}{7}, 3\right\}$
31. $\{1 \pm \sqrt{34}\}$ **33.** $\{\pm\sqrt{6}, \pm 2\sqrt{3}\}$ **35.** $\left\{\pm 3, \pm\frac{2\sqrt{6}}{3}\right\}$ **37.** $\left\{\pm\frac{i\sqrt{15}}{3}, \pm 2i\right\}$
39. $\left\{\pm\frac{\sqrt{14}}{2}, \pm\frac{2\sqrt{3}}{3}\right\}$ **41.** 8 and 9 **43.** 9 and 12 **45.** $5 + \sqrt{3}$ and $5 - \sqrt{3}$ **47.** 3 and 6
49. 9 inches and 12 inches **51.** 1 meter **53.** 8 inches by 14 inches
55. 20 miles per hour for Lorraine and 25 miles per hour for Charlotte or 45 miles per hour for Lorraine and 50 miles per hour for Charlotte
57. 55 miles per hour **59.** 6 hours for Tom and 8 hours for Terry **61.** 30 hours **63.** 8 people
65. 40 shares at \$20 per share **67.** 50 numbers **69.** 9% **71.** $\{9, 36\}$ **73.** $\{1\}$
75. $\left\{-\frac{8}{27}, \frac{27}{8}\right\}$ **77.** $\left\{-4, \frac{3}{5}\right\}$

Problem Set 6.6 (page 305)

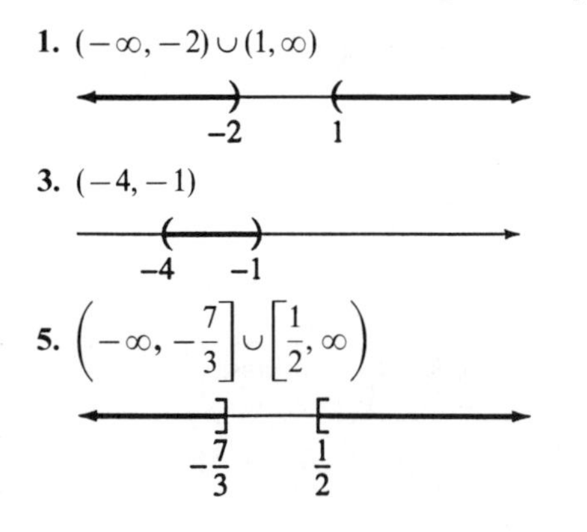

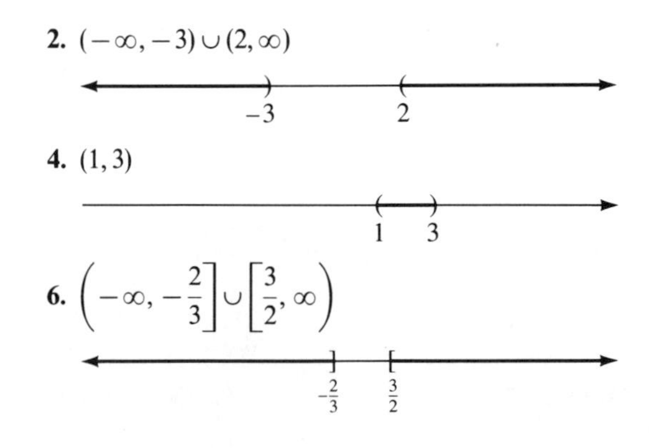

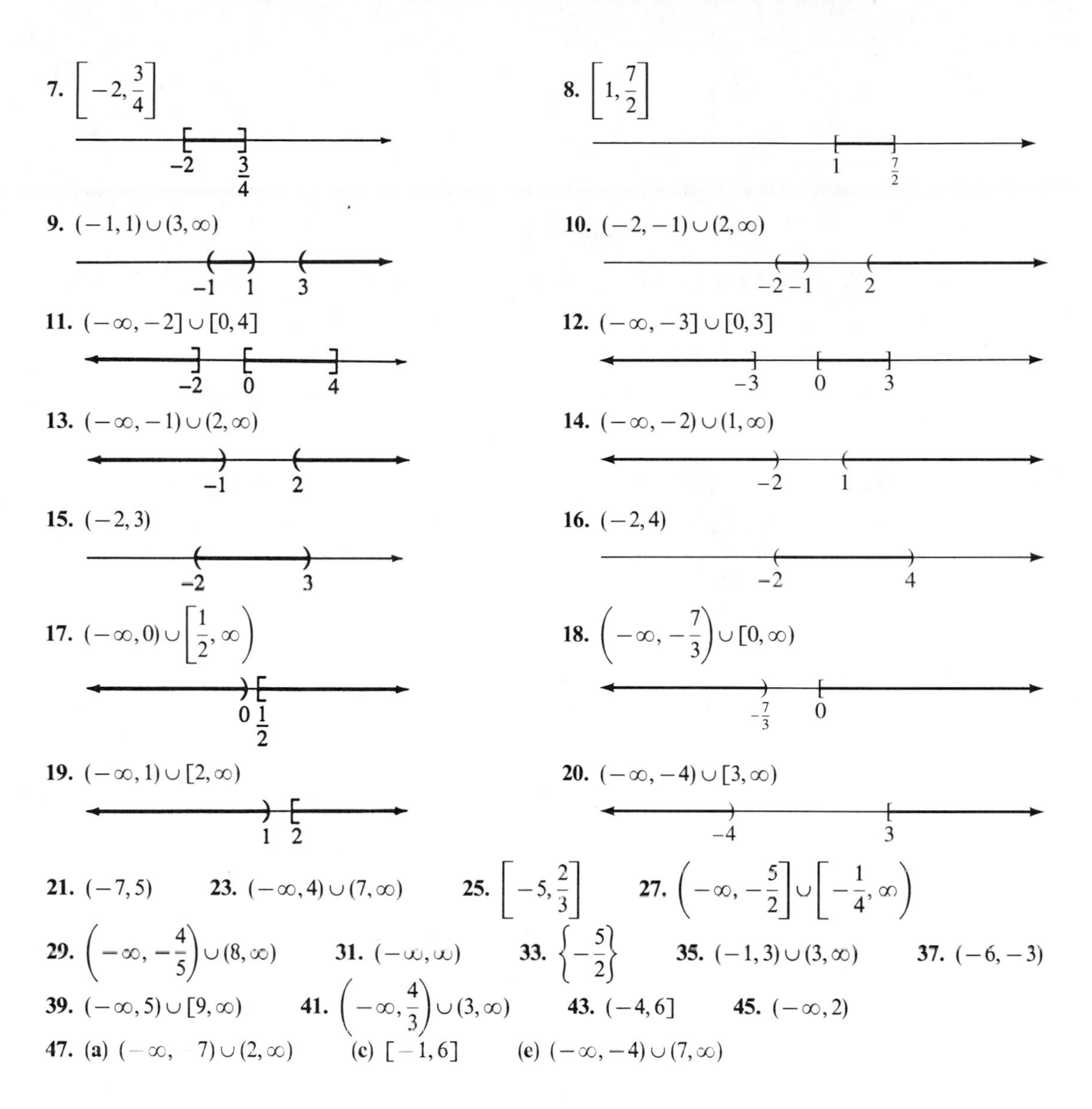

7. $\left[-2, \frac{3}{4}\right]$

8. $\left[1, \frac{7}{2}\right]$

9. $(-1, 1) \cup (3, \infty)$

10. $(-2, -1) \cup (2, \infty)$

11. $(-\infty, -2] \cup [0, 4]$

12. $(-\infty, -3] \cup [0, 3]$

13. $(-\infty, -1) \cup (2, \infty)$

14. $(-\infty, -2) \cup (1, \infty)$

15. $(-2, 3)$

16. $(-2, 4)$

17. $(-\infty, 0) \cup \left[\frac{1}{2}, \infty\right)$

18. $\left(-\infty, -\frac{7}{3}\right) \cup [0, \infty)$

19. $(-\infty, 1) \cup [2, \infty)$

20. $(-\infty, -4) \cup [3, \infty)$

21. $(-7, 5)$ **23.** $(-\infty, 4) \cup (7, \infty)$ **25.** $\left[-5, \frac{2}{3}\right]$ **27.** $\left(-\infty, -\frac{5}{2}\right] \cup \left[-\frac{1}{4}, \infty\right)$

29. $\left(-\infty, -\frac{4}{5}\right) \cup (8, \infty)$ **31.** $(-\infty, \infty)$ **33.** $\left\{-\frac{5}{2}\right\}$ **35.** $(-1, 3) \cup (3, \infty)$ **37.** $(-6, -3)$

39. $(-\infty, 5) \cup [9, \infty)$ **41.** $\left(-\infty, \frac{4}{3}\right) \cup (3, \infty)$ **43.** $(-4, 6]$ **45.** $(-\infty, 2)$

47. (a) $(-\infty, -7) \cup (2, \infty)$ (c) $[-1, 6]$ (e) $(-\infty, -4) \cup (7, \infty)$

Chapter 6 Review Problem Set (page 308)

1. $2 - 2i$ **2.** $-3 - i$ **3.** $30 + 15i$ **4.** $86 - 2i$ **5.** $-32 + 4i$ **6.** 25 **7.** $\frac{9}{20} + \frac{13}{20}i$

8. $-\frac{3}{29} + \frac{7}{29}i$ **9.** Two equal real solutions **10.** Two nonreal complex solutions

11. Two unequal real solutions **12.** Two unequal real solutions **13.** $\{0, 17\}$ **14.** $\{-4, 8\}$

15. $\left\{\frac{1 \pm 8i}{2}\right\}$ **16.** $\{-3, 7\}$ **17.** $\{-1 \pm \sqrt{10}\}$ **18.** $\{3 \pm 5i\}$ **19.** $\{25\}$ **20.** $\left\{-4, \frac{2}{3}\right\}$

21. $\{-10, 20\}$ **22.** $\left\{\frac{-1 \pm \sqrt{61}}{6}\right\}$ **23.** $\left\{\frac{1 \pm i\sqrt{11}}{2}\right\}$ **24.** $\left\{\frac{5 \pm i\sqrt{23}}{4}\right\}$ **25.** $\left\{\frac{-2 \pm \sqrt{14}}{2}\right\}$

26. $\{-9, 4\}$ **27.** $\{-2 \pm i\sqrt{5}\}$ **28.** $\{-6, 12\}$ **29.** $\{1 \pm \sqrt{10}\}$ **30.** $\left\{\pm\frac{\sqrt{14}}{2}, \pm 2\sqrt{2}\right\}$

31. $\left\{\frac{-3 \pm \sqrt{97}}{2}\right\}$ **32.** $(-\infty, -5) \cup (2, \infty)$ **33.** $\left[-\frac{7}{2}, 3\right]$ **34.** $(-\infty, -6) \cup [4, \infty)$

35. $\left(-\frac{5}{2}, -1\right)$ **36.** $3 + \sqrt{7}$ and $3 - \sqrt{7}$ **37.** 20 shares at $15 per share

38. 45 miles per hour and 52 miles per hour **39.** 8 units **40.** 8 and 10 **41.** 7 inches by 12 inches

42. 4 hours for Janet and 6 hours for Billy **43.** 10 meters

CHAPTER 7

Problem Set 7.1 (page 321)

1. (a) III (c) II **3.** (a) I and III (c) II and III **5.** $(-3, -1)$; $(3, 1)$; $(3, -1)$

7. $(7, 2)$; $(-7, -2)$; $(-7, 2)$ **9.** $(5, 0)$; $(-5, 0)$; $(-5, 0)$ **11.** x-axis **13.** y-axis

15. x-axis, y-axis, and origin **17.** x-axis **19.** none **21.** origin **23.** y-axis

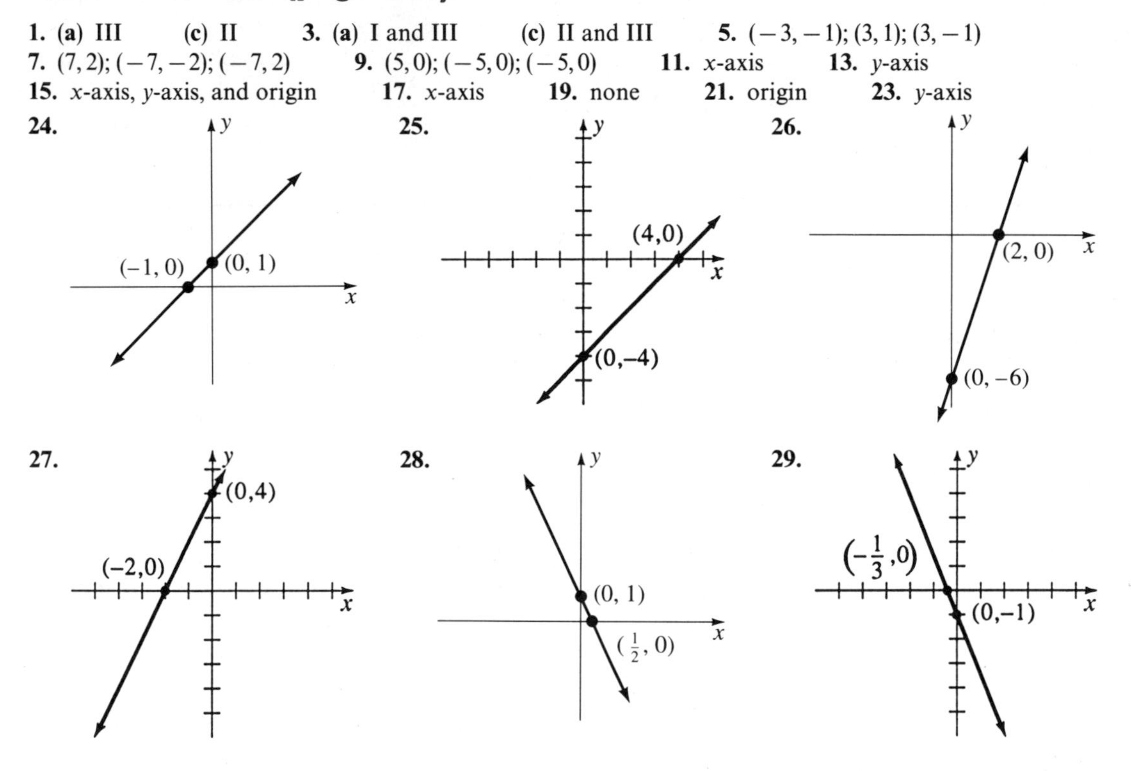

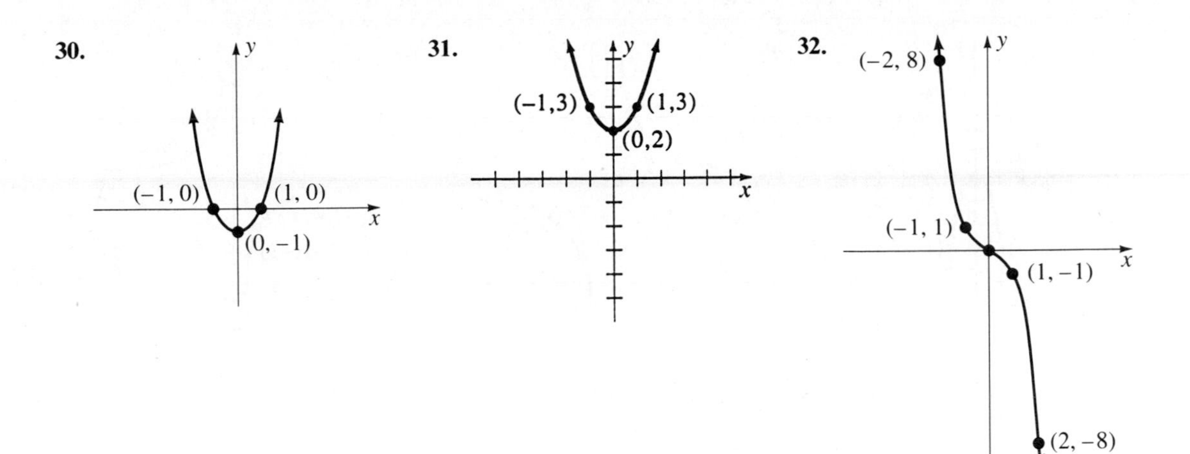
30.
y
(−1, 0)
(1, 0)
x
(0, −1)
31.
y
(−1,3)
(1,3)
(0,2)
x
32.
y
(−2, 8)
(−1, 1)
(1, −1)
x
(2, −8)

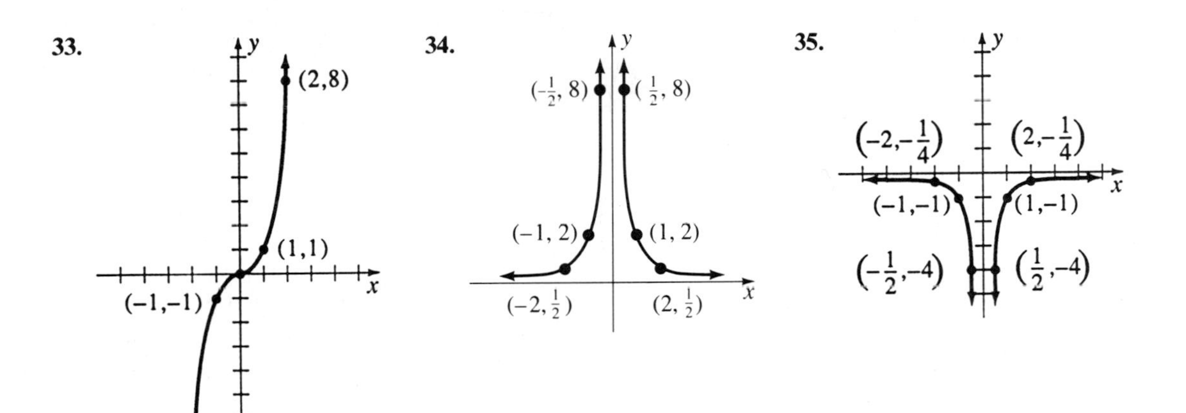
33.
y
(2,8)
(1,1)
x
(−1,−1)
(−2,−8)
34.
y
(−1/2, 8)
(1/2, 8)
(−1, 2)
(1, 2)
(−2, 1/2)
(2, 1/2)
x
35.
y
(−2,−1/4)
(2,−1/4)
x
(−1,−1)
(1,−1)
(−1/2,−4)
(1/2,−4)

36.

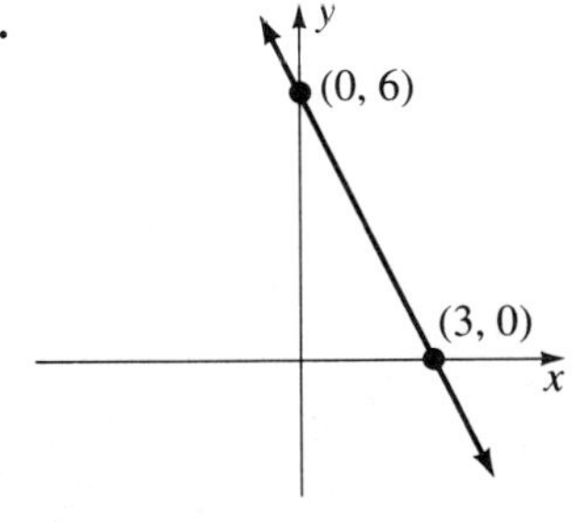
y
(0, 6)
(3, 0)
x

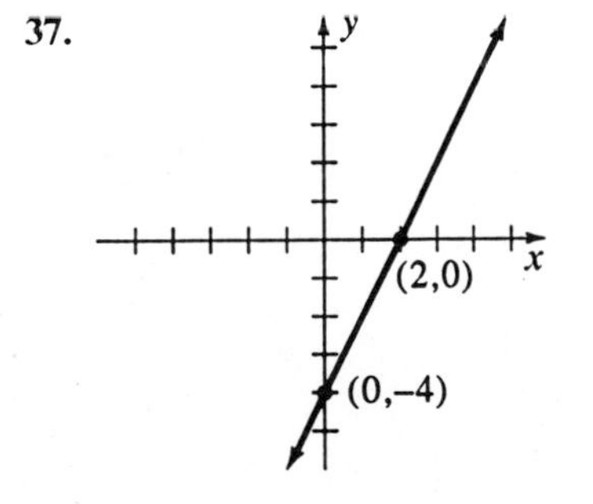
37.
y
(2,0)
x
(0,−4)

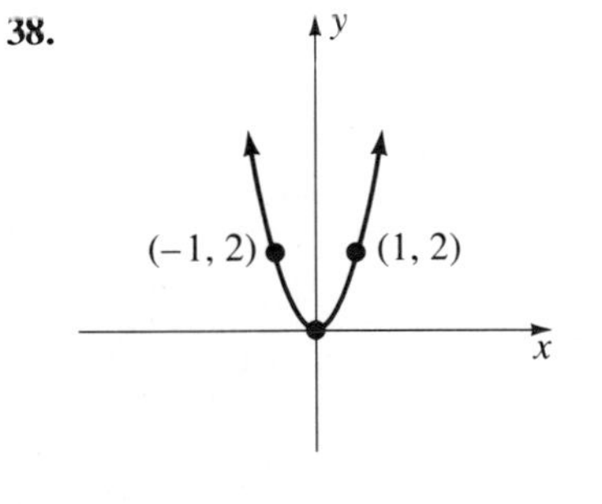
38.
y
(−1, 2)
(1, 2)
x

39. **40.** **41.**

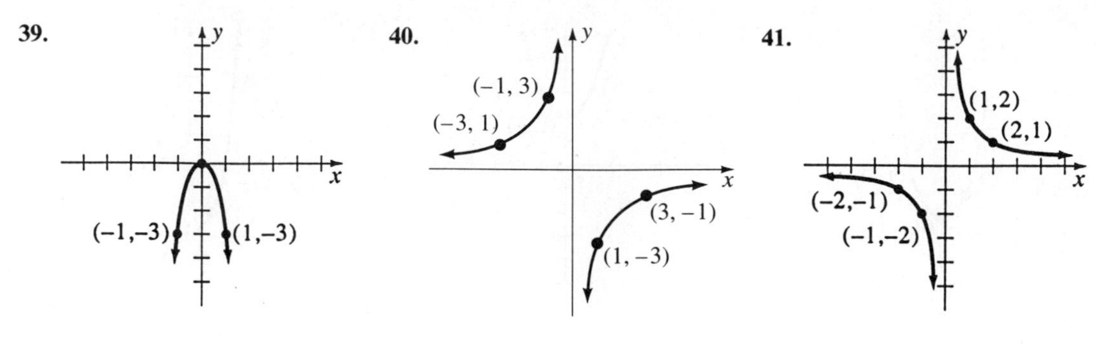

42.

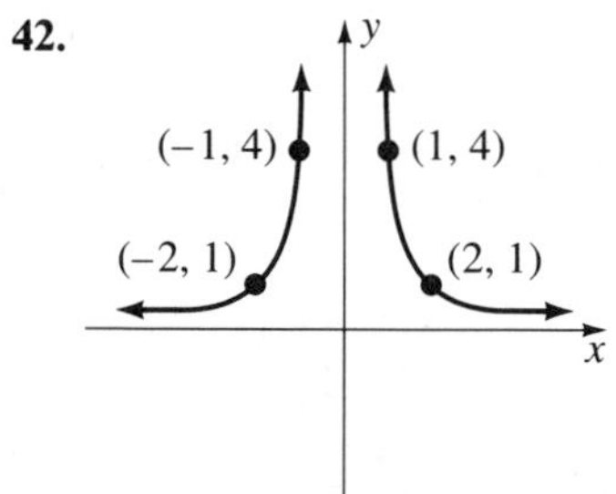

43.

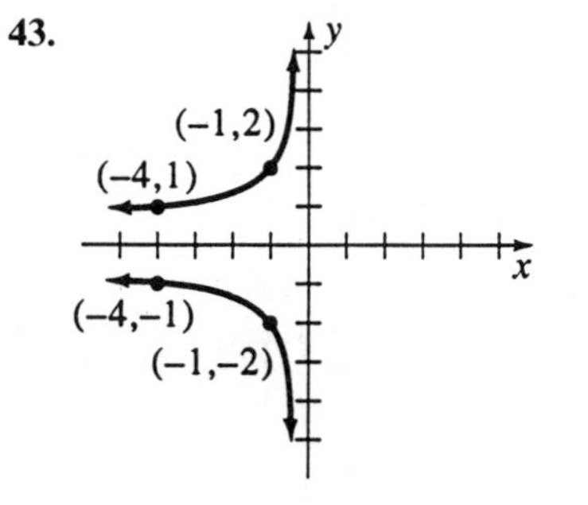

44.

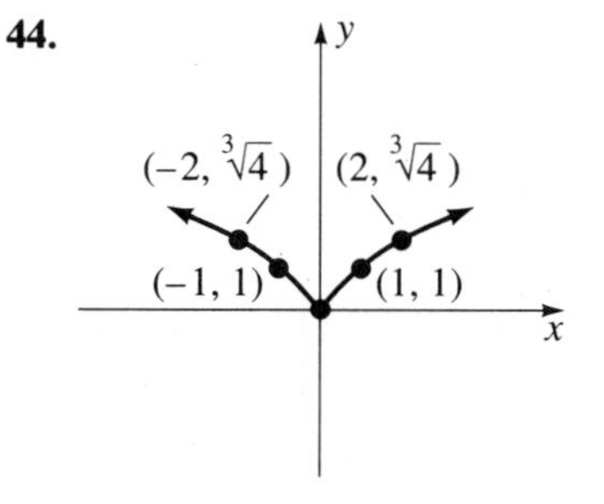

45.

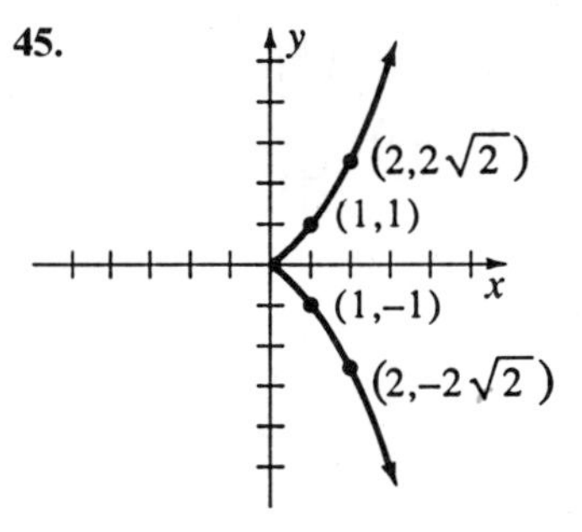

46.

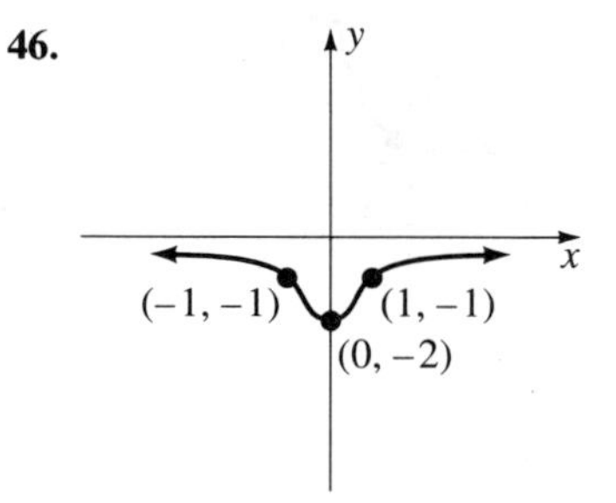

47.

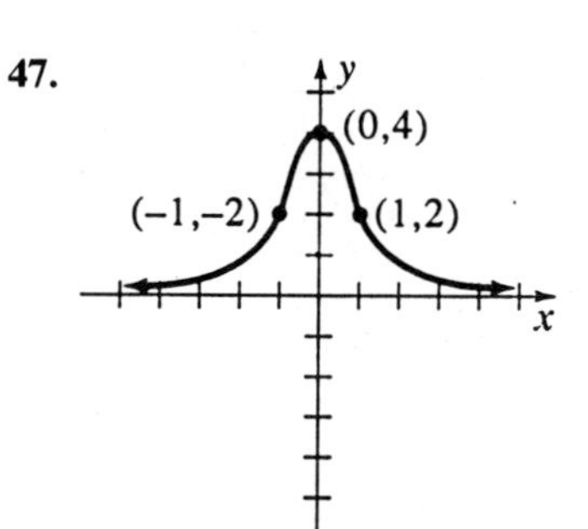

48.

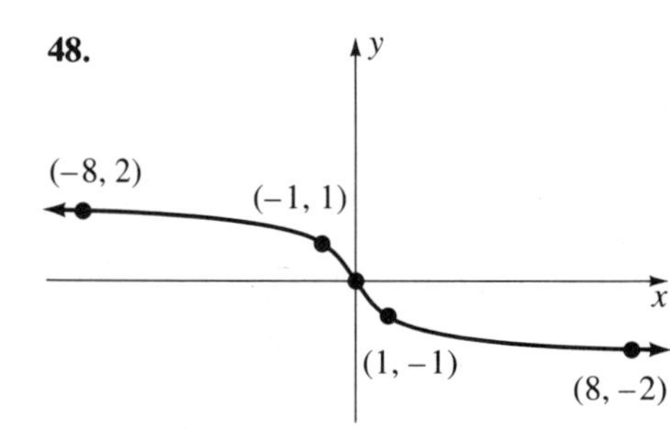

49.

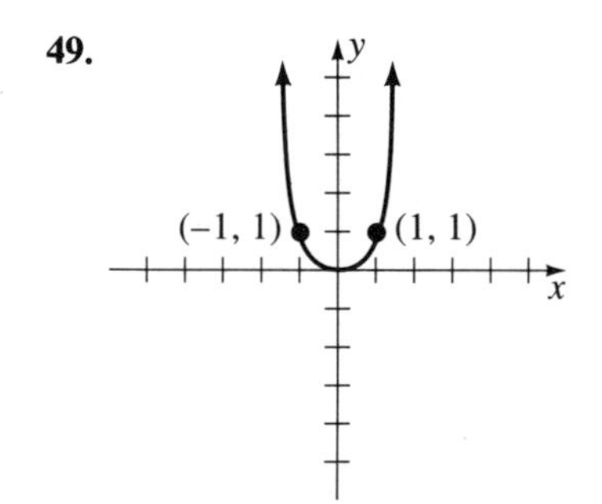

50.

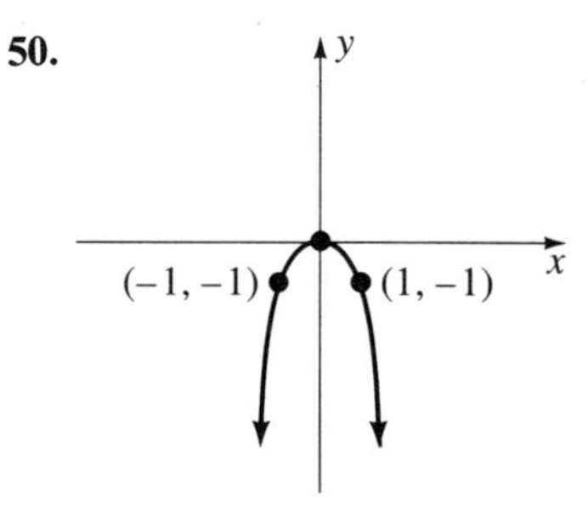

51.

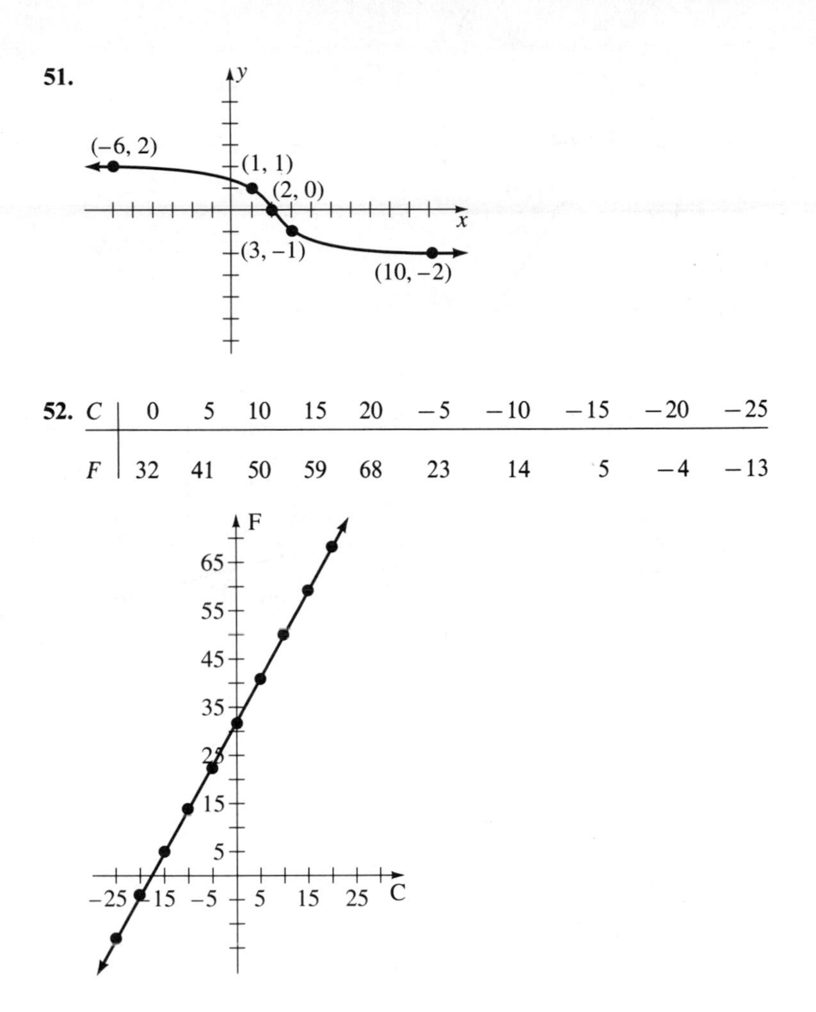

52.

C	0	5	10	15	20	−5	−10	−15	−20	−25
F	32	41	50	59	68	23	14	5	−4	−13

53. \$41.76; \$43.20; \$44.40; \$46.50; \$46.92

Problem Set 7.2 (page 327)

1. **2.** **3.**

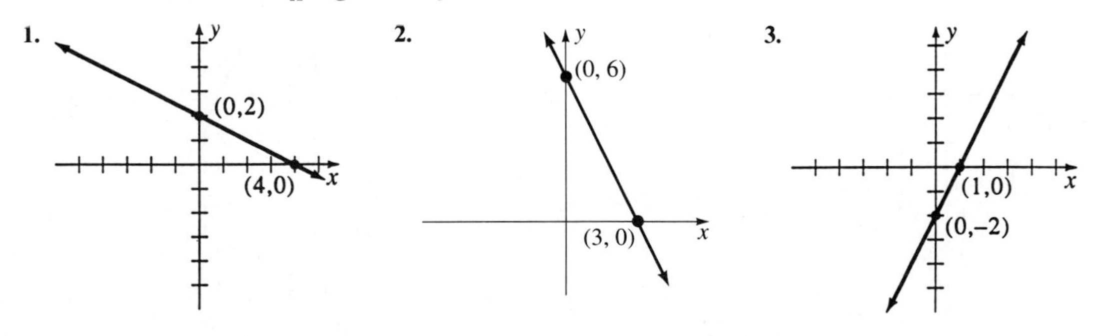

4. **5.** **6.**

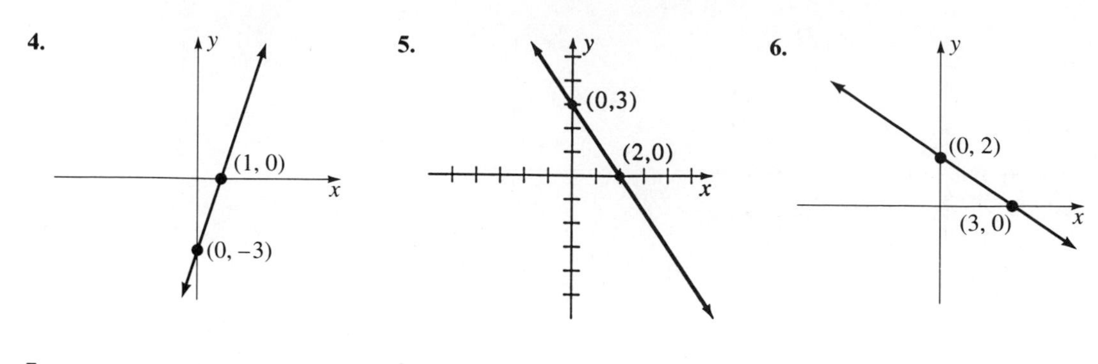

7.

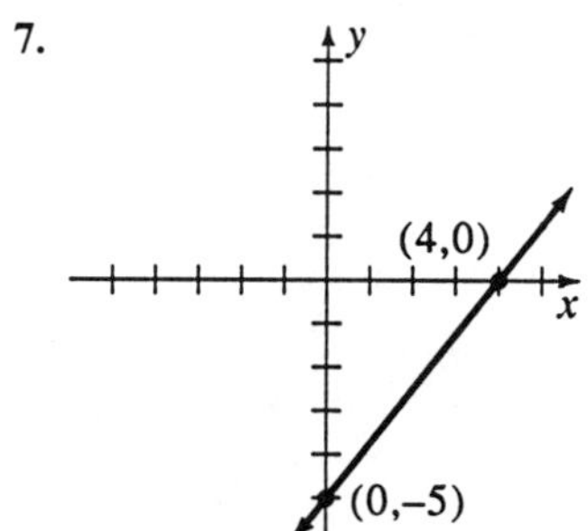

8.

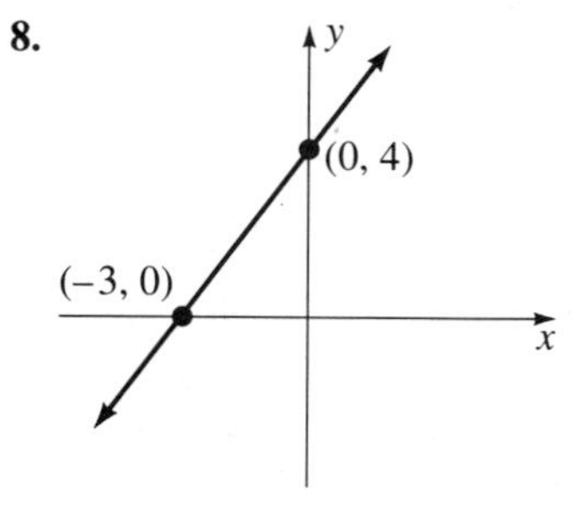

9.

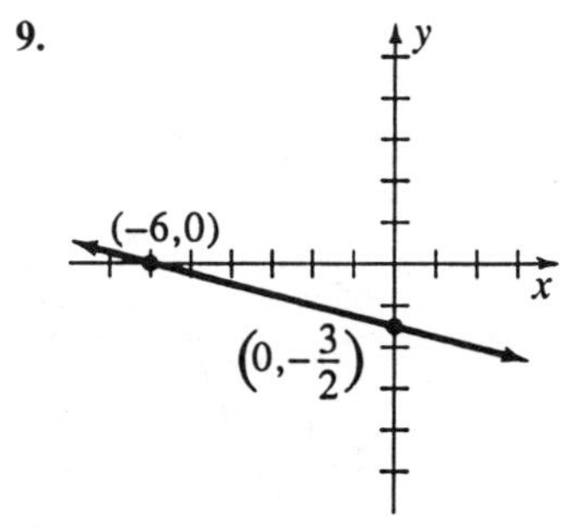

10.

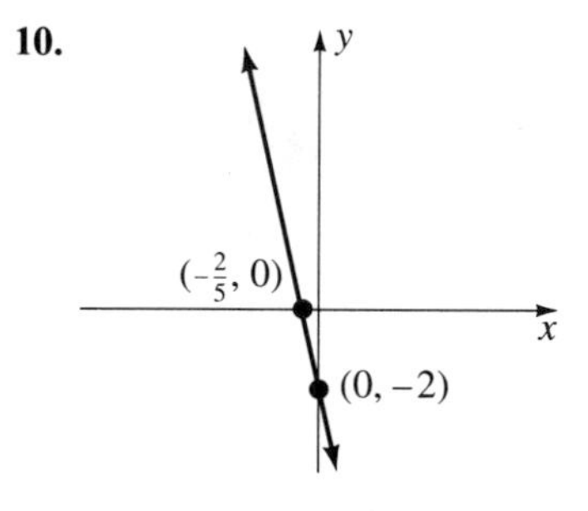

11.

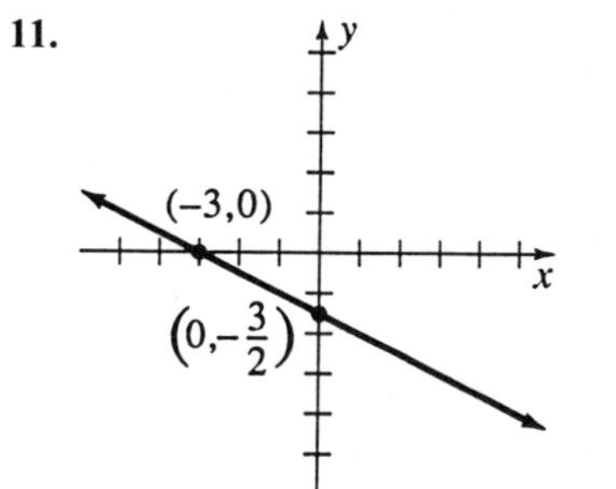

12.

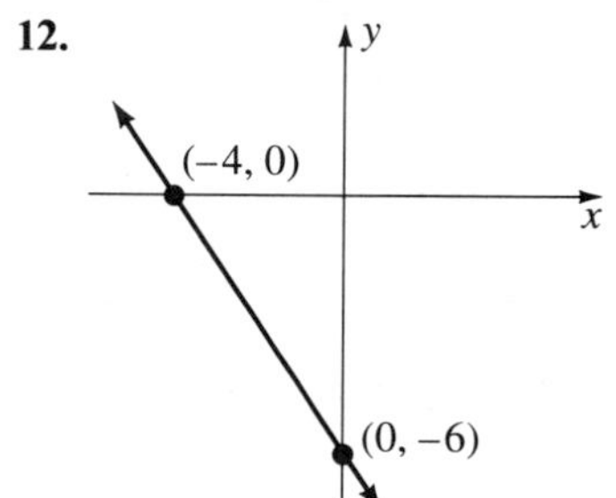

13.

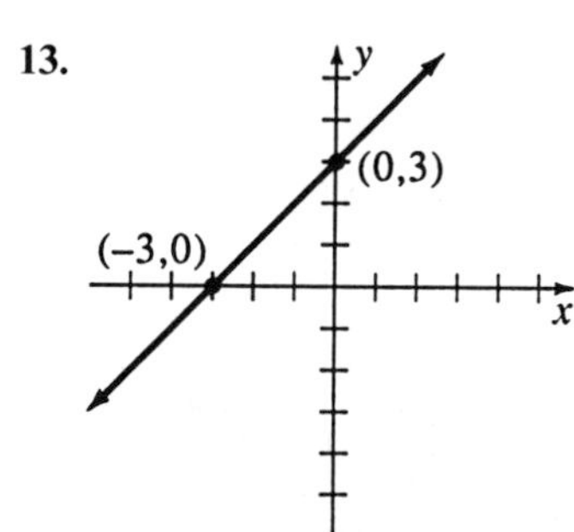

14.

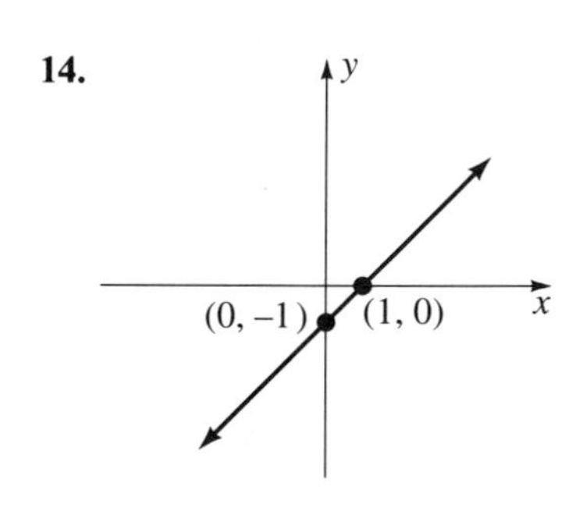

15.

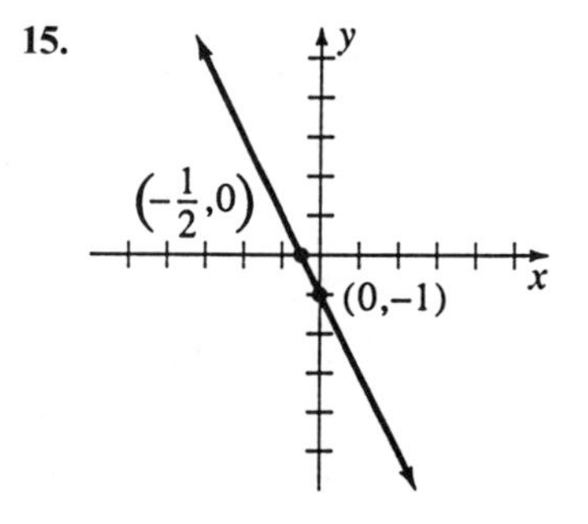

16. **17.** **18.**

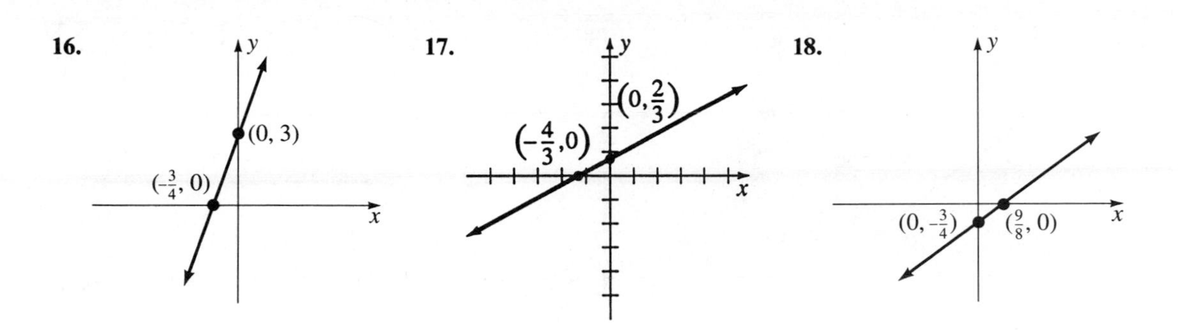

19.

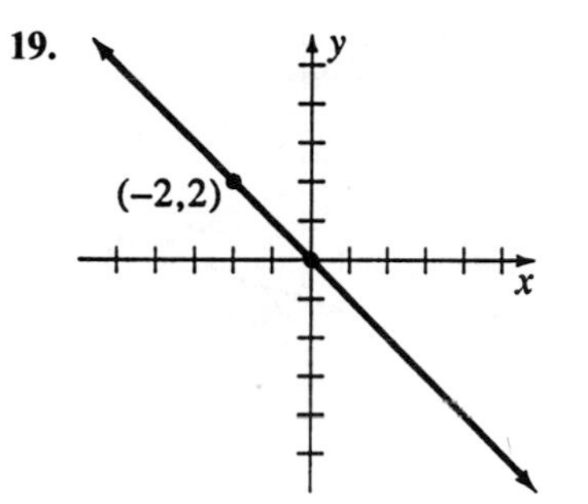

20.

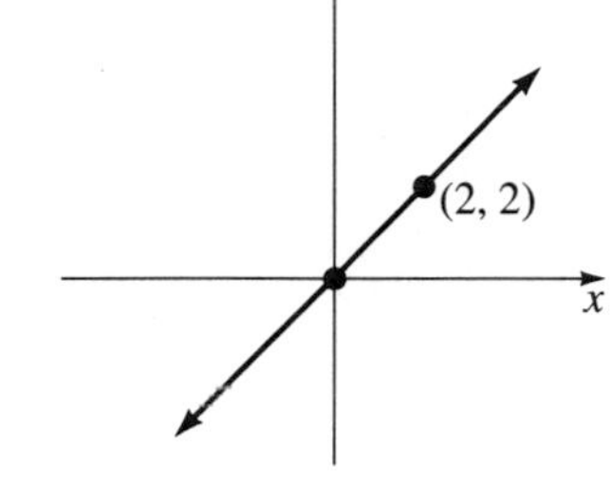

21.

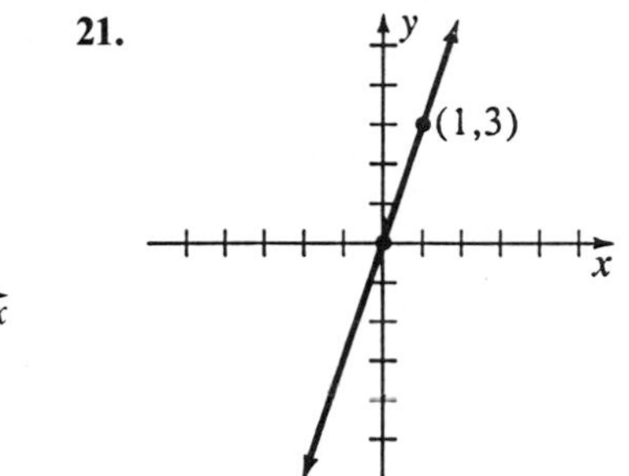

22.

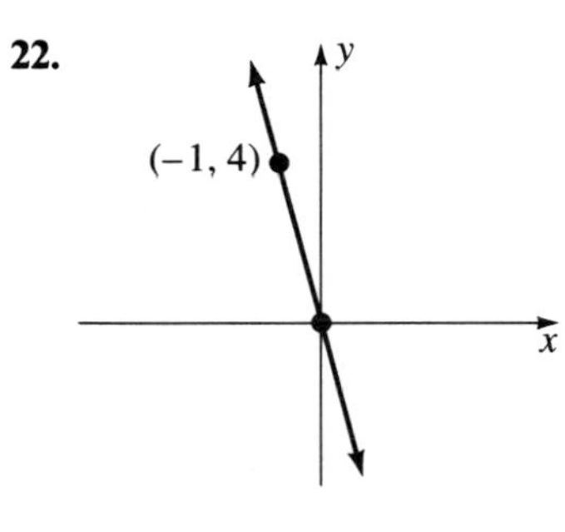

23.

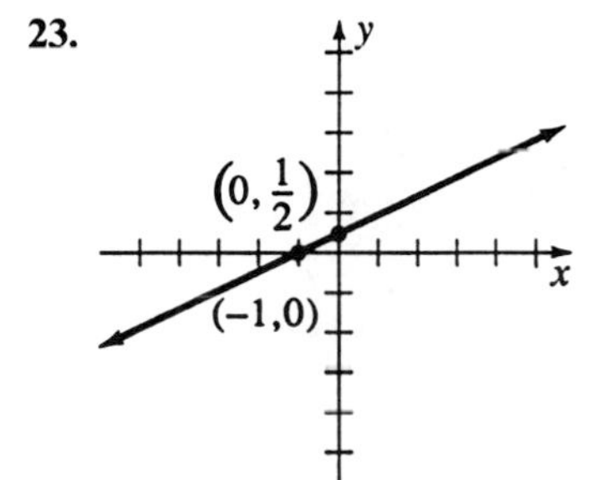

24.

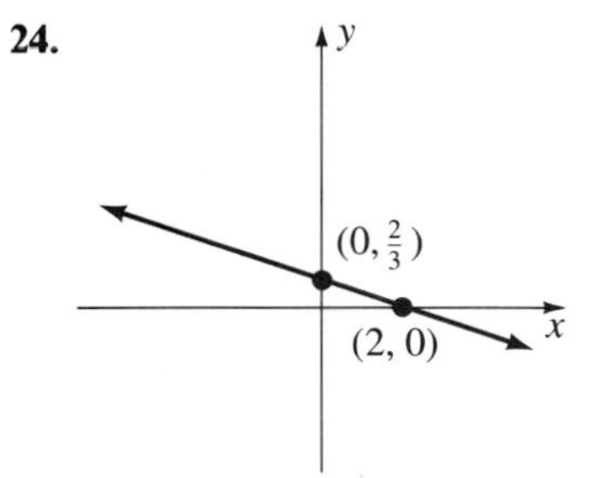

25.

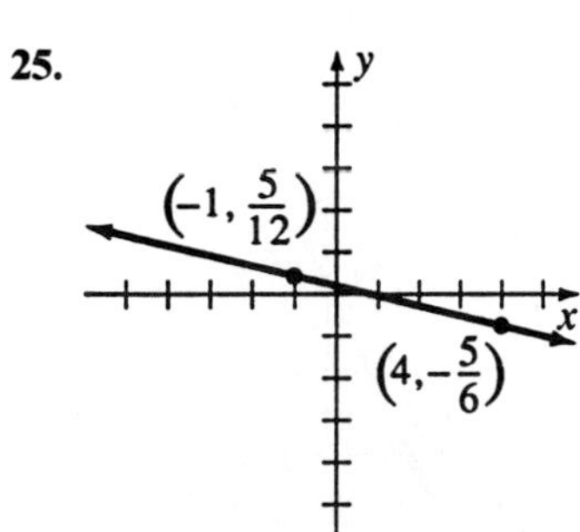

26.

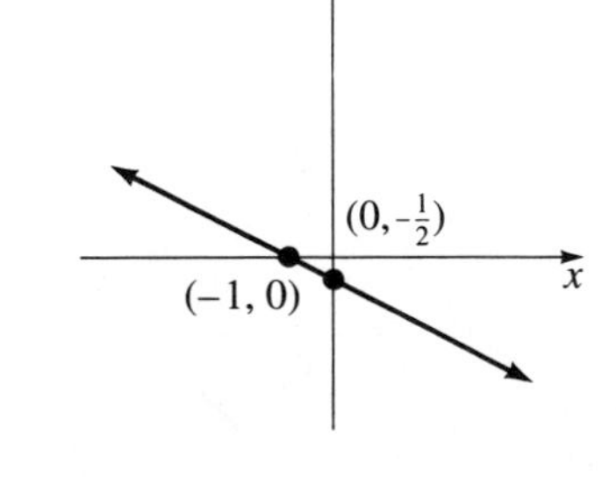

27.

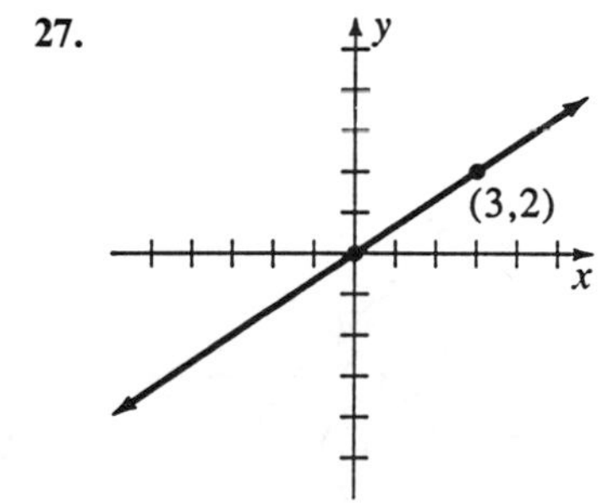

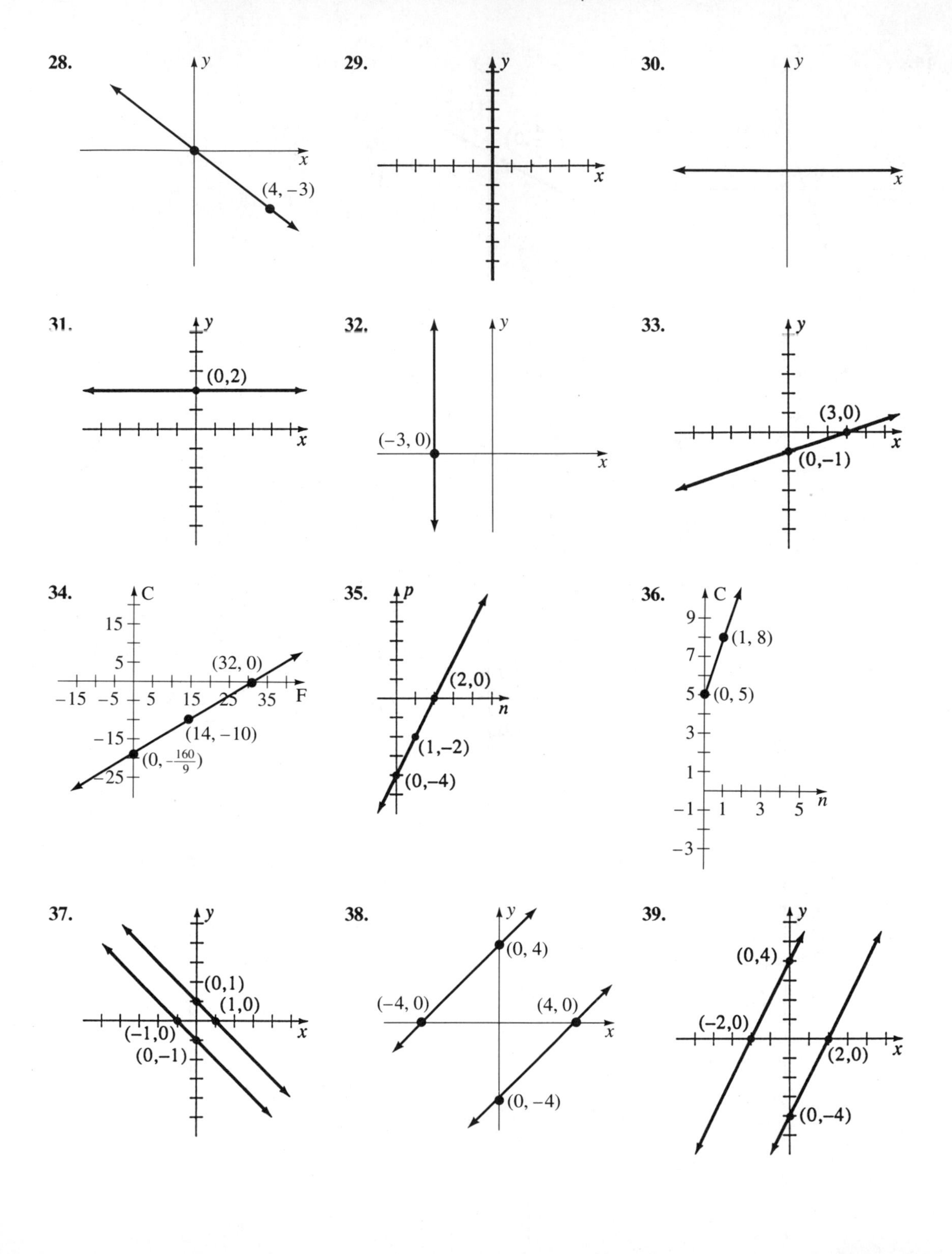

28.
y
x
(4, –3)
29.
y
x
30.
y
x
31.
y
(0,2)
x
32.
y
(–3, 0)
x
33.
y
(3,0)
(0,–1)
x
34.
C
15
5
(32, 0)
–15
–5
5
15
25
35
F
(14, –10)
–15
(0, –160/9)
–25
35.
p
(2,0)
n
(1,–2)
(0,–4)
36.
C
9
(1, 8)
7
5
(0, 5)
3
1
–1
1
3
5
n
–3
37.
y
(0,1)
(1,0)
(–1,0)
(0,–1)
x
38.
y
(0, 4)
(–4, 0)
(4, 0)
x
(0, –4)
39.
y
(0,4)
(–2,0)
(2,0)
x
(0,–4)

40.

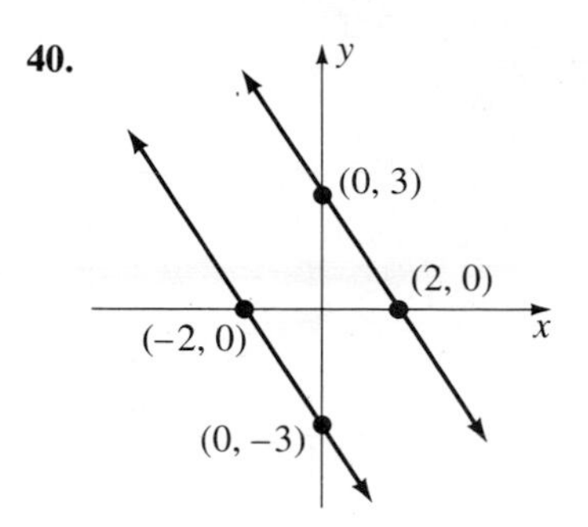

Problem Set 7.3 (page 332)

1.

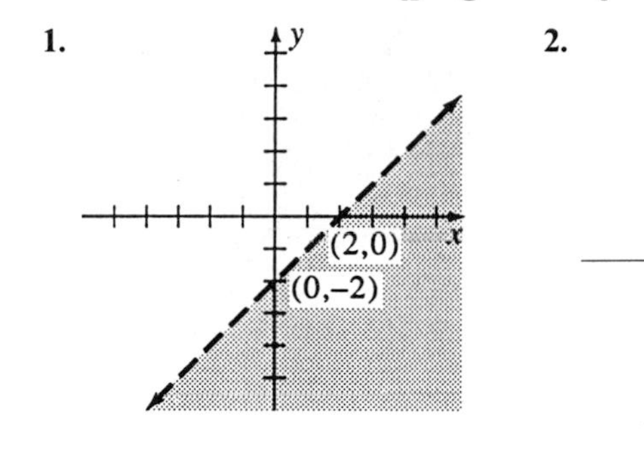

2.

(0, 4)
(4, 0)

3.

(0,1)
(3,0)

4.

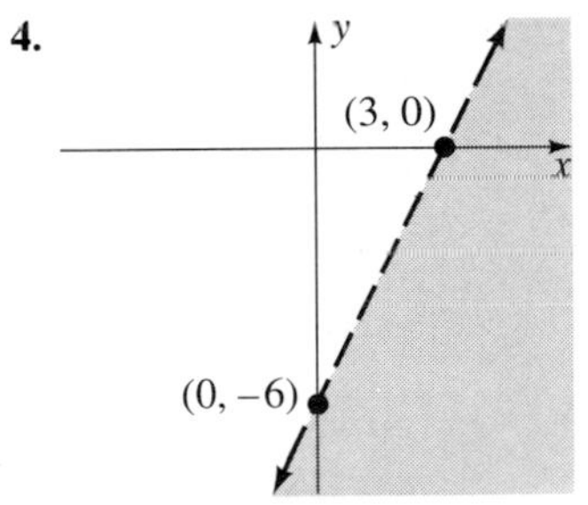

5.

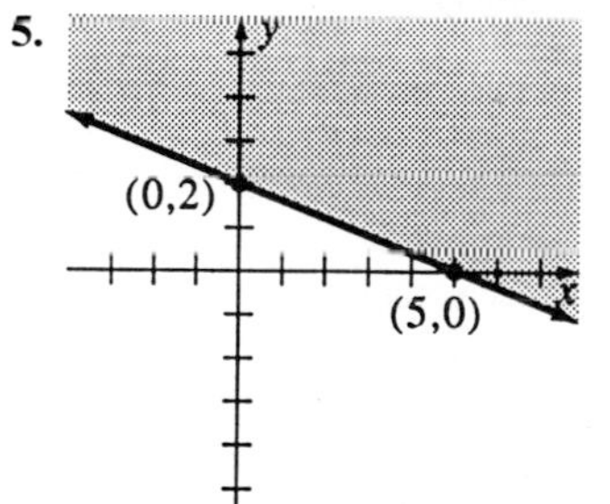

6.

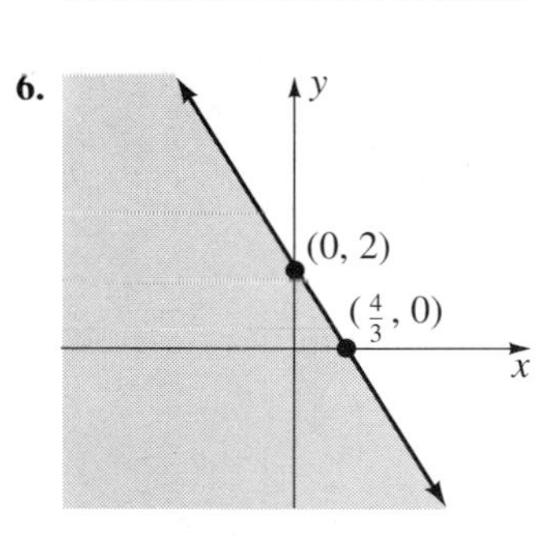

7.

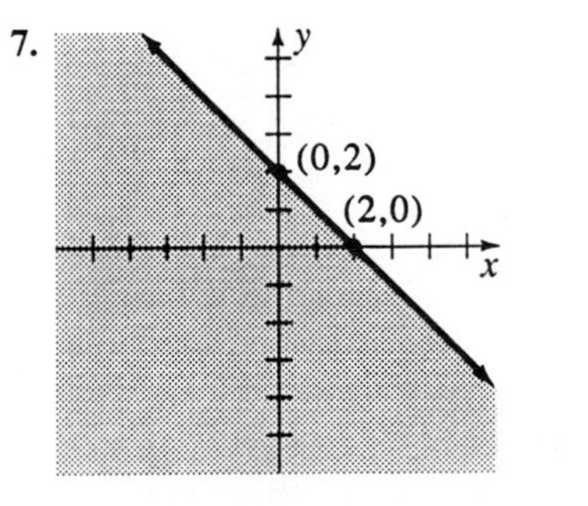

8.

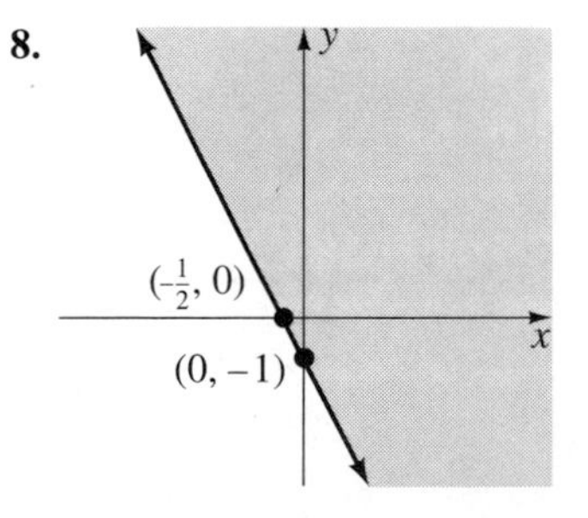

9.

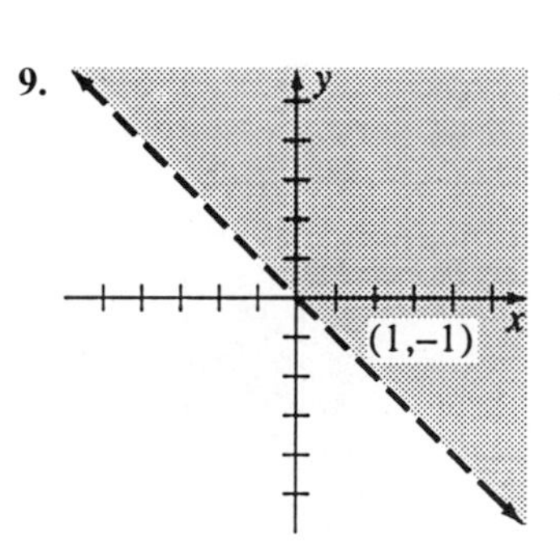

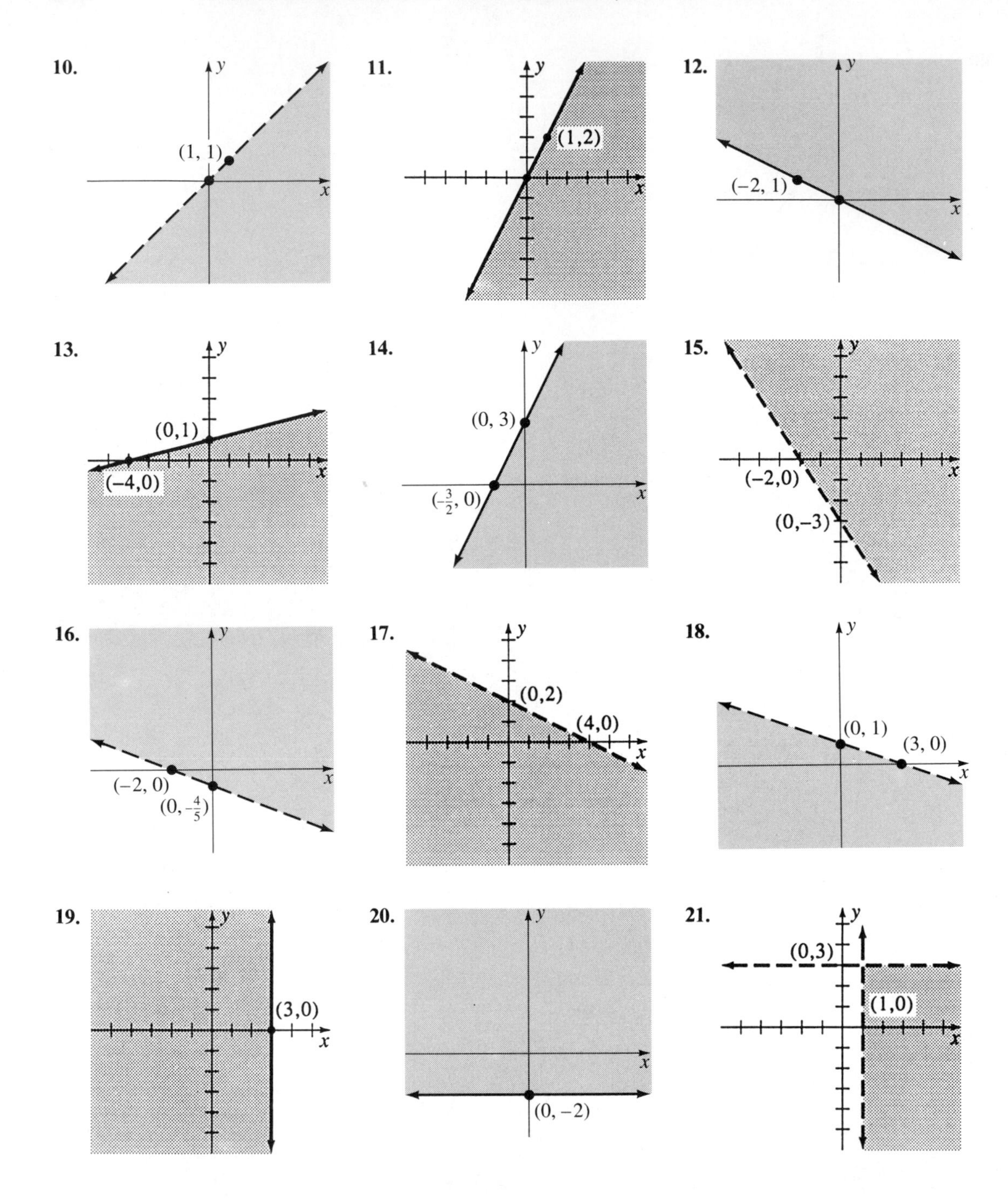
10.
(1, 1)
11.
(1,2)
12.
(−2, 1)
13.
(0,1)
(−4,0)
14.
(0, 3)
$(-\frac{3}{2}, 0)$
15.
(−2,0)
(0,−3)
16.
(−2, 0)
$(0, -\frac{4}{5})$
17.
(0,2)
(4,0)
18.
(0, 1)
(3, 0)
19.
(3,0)
20.
(0, −2)
21.
(0,3)
(1,0)

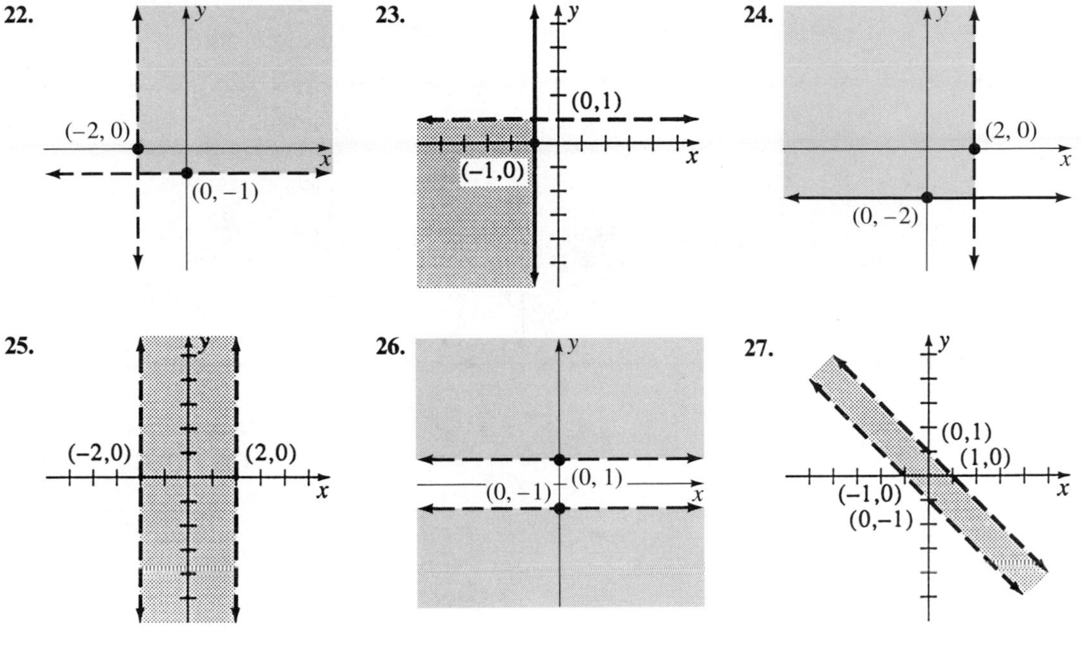

28.

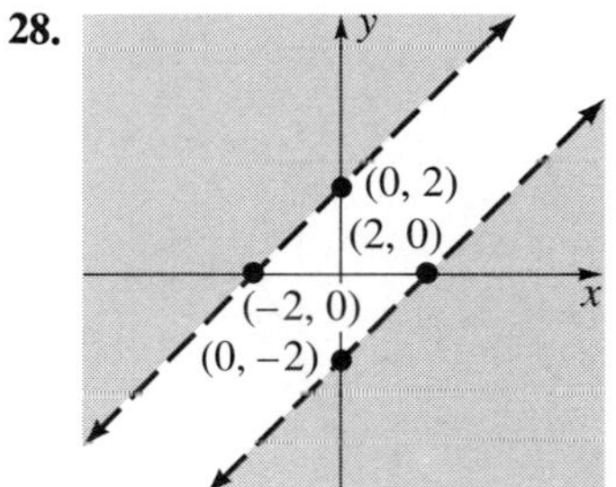

Problem Set 7.4 (page 338)

1. 15 **3.** $\sqrt{13}$ **5.** $3\sqrt{2}$ **7.** $3\sqrt{5}$ **9.** 6 **11.** $3\sqrt{10}$

13. The lengths of the sides are 10, $5\sqrt{5}$, and 5. Since $10^2 + 5^2 = (5\sqrt{5})^2$, it is a right triangle.

15. The distances between (3, 6) and (7, 12), between (7, 12) and (11, 18), and between (11, 18) and (15, 24) are all $2\sqrt{13}$ units.

17. $\frac{4}{3}$ **19.** $-\frac{7}{3}$ **21.** -2 **23.** $\frac{3}{5}$ **25.** 0 **27.** $\frac{1}{2}$ **29.** 7 **31.** -2

33–39. Answers will vary. **41.** $-\frac{2}{3}$ **43.** $\frac{1}{2}$ **45.** $\frac{4}{7}$ **47.** 0 **49.** -5

51. **(a)** 105.6 feet **53.** 1.0 feet

Problem Set 7.5 (page 347)

1. $x - 2y = -7$ **3.** $3x - y = -10$ **5.** $3x + 4y = -15$ **7.** $5x - 4y = 28$ **9.** $x - y = 1$

11. $5x - 2y = -4$ **13.** $x + 7y = 11$ **15.** $x + 2y = -9$ **17.** $7x - 5y = 0$ **19.** $y = \frac{3}{7}x + 4$

21. $y = 2x - 3$ **23.** $y = -\frac{2}{5}x + 1$ **25.** $y = 0(x) - 4$ **27.** $2x - y = 4$

29. $5x + 8y = -15$ **31.** $x + 0(y) = 2$ **33.** $0(x) + y = 6$ **35.** $x + 5y = 16$

37. $4x - 7y = 0$ **39.** $x + 2y = 5$ **41.** $3x + 2y = 0$

43.

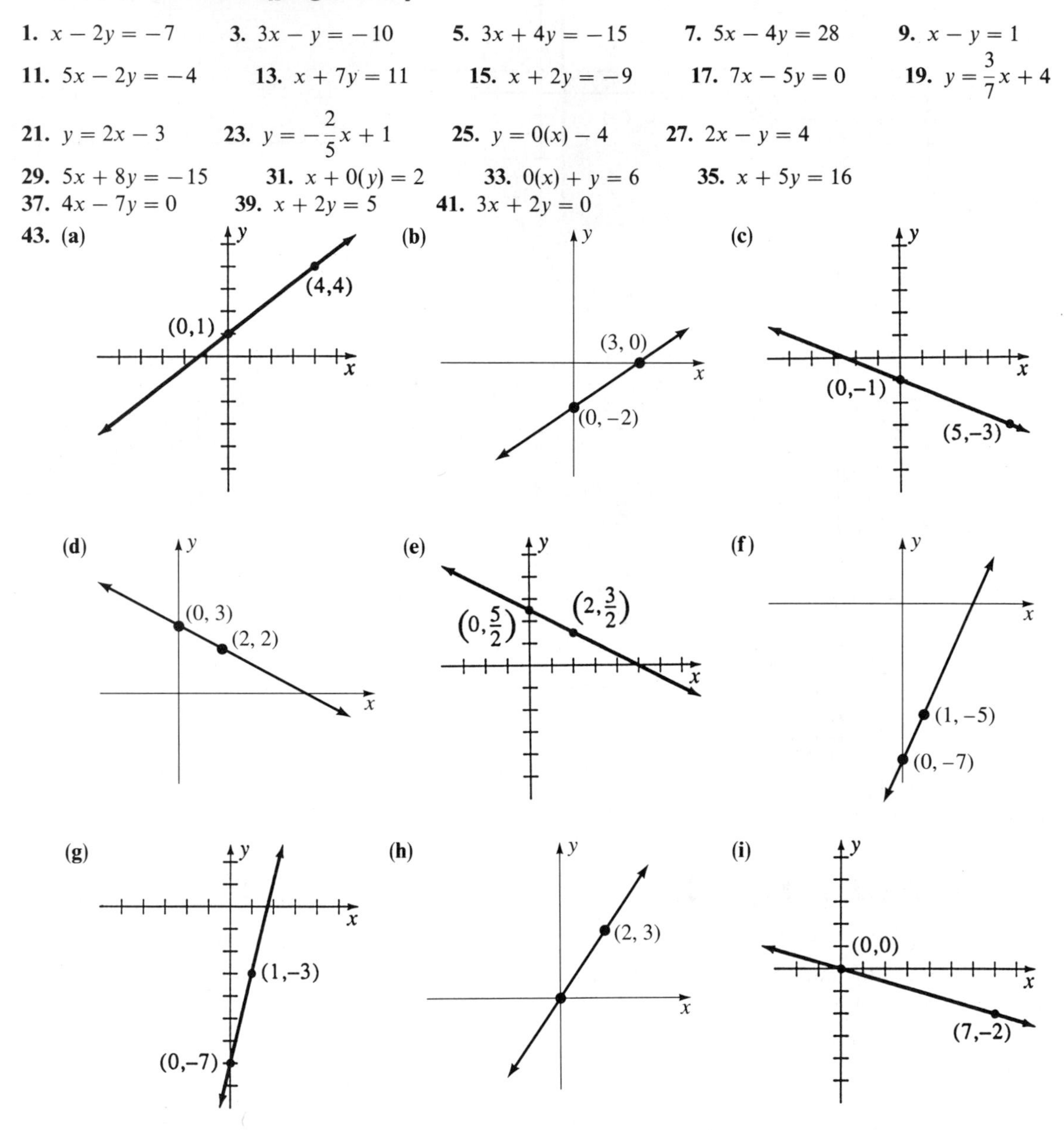

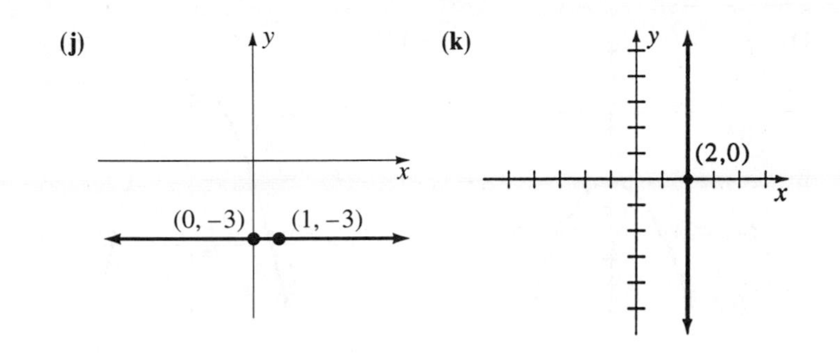

45. (a) $x - 4y = -3$ (c) $2x + 3y = 9$ **47.** (a) $2x + 3y = 26$ (c) $5x + 3y = -11$

Problem Set 7.6 (page 358)

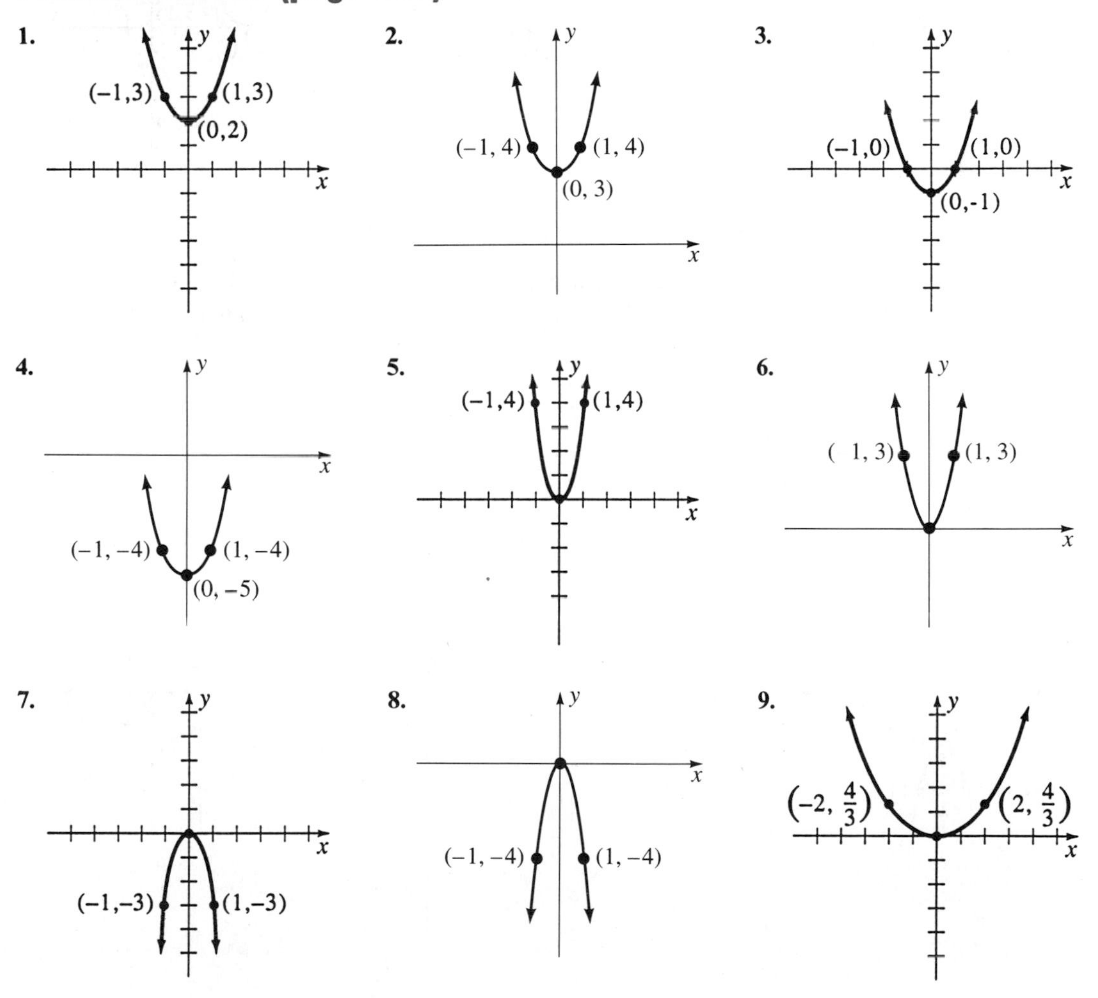

10. 11. 12.

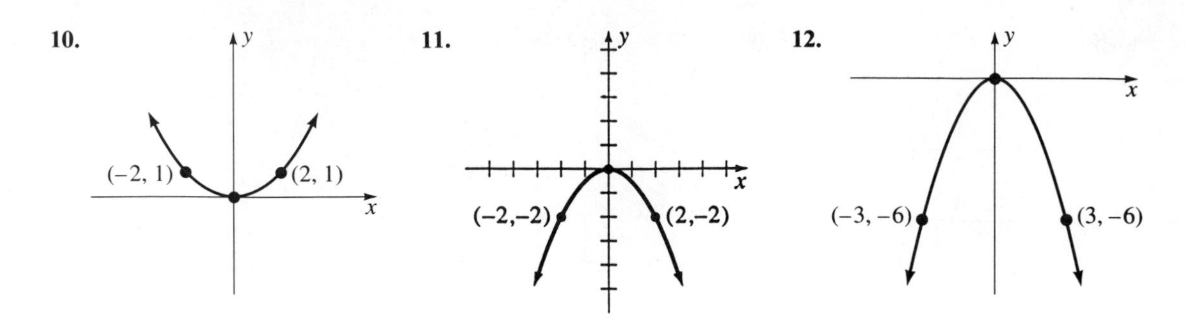

13.

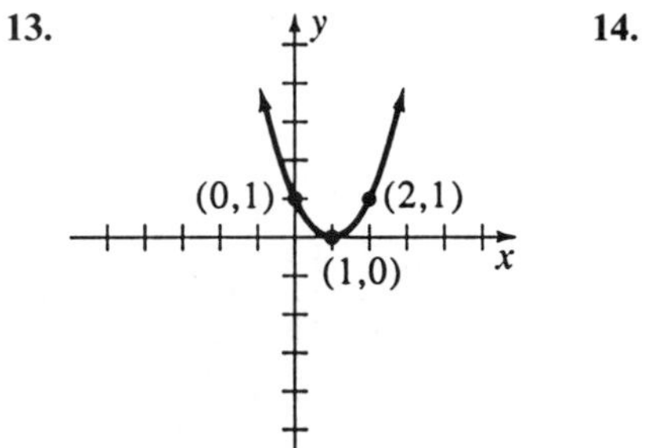

14.

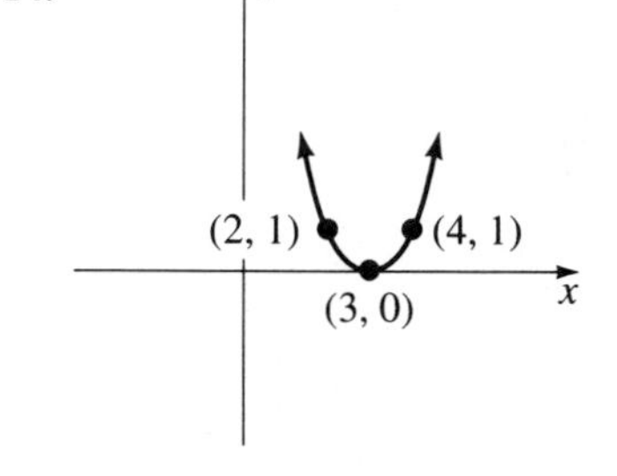

15.

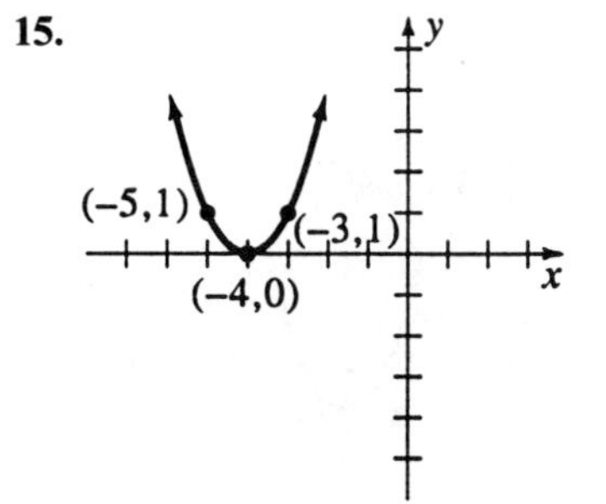

16.

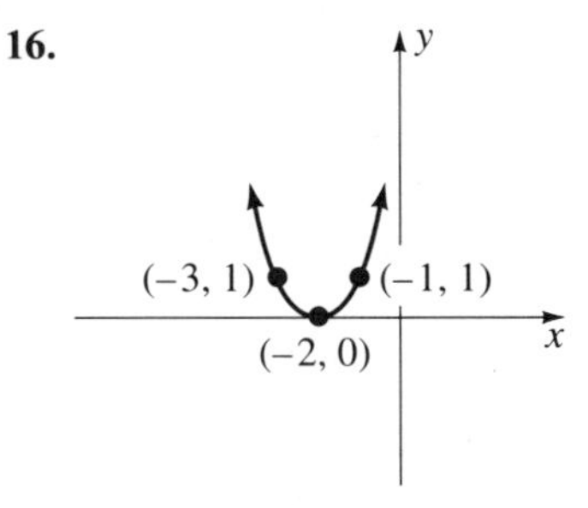

17.

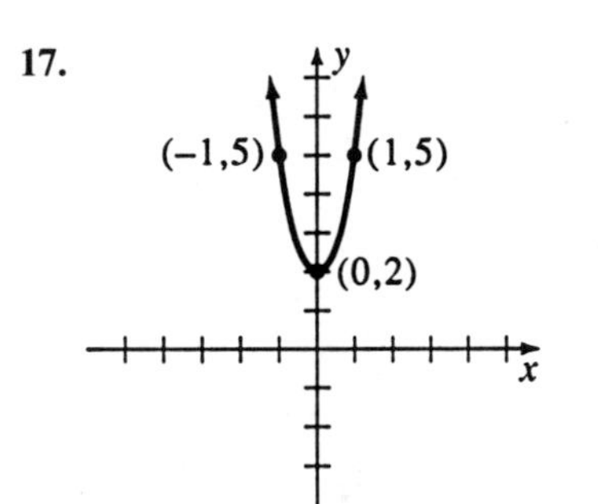

18.

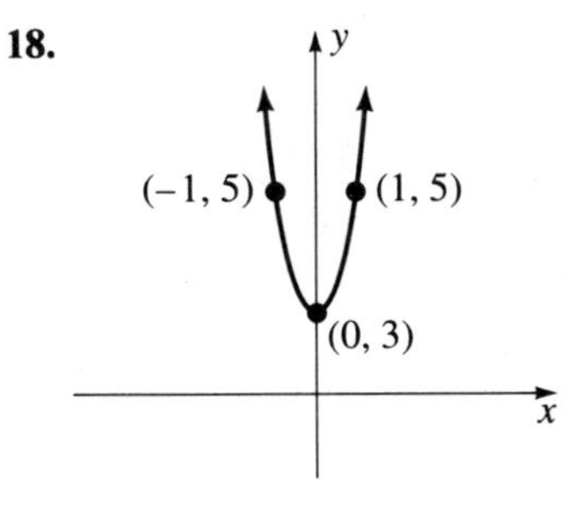

19.

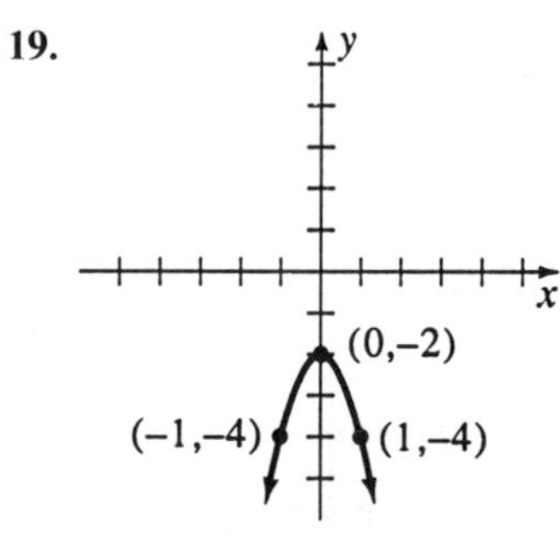

20.

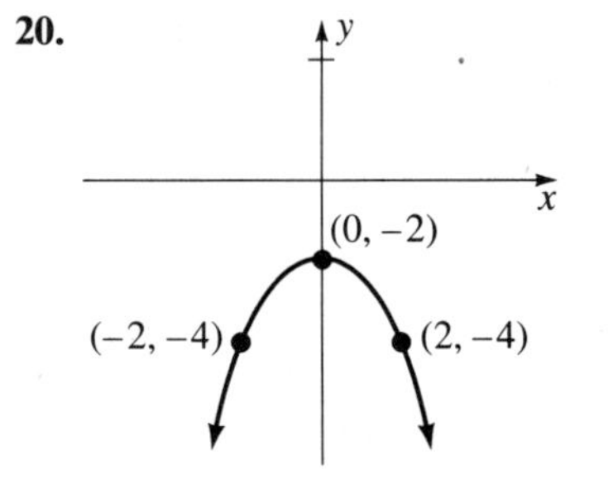

21.

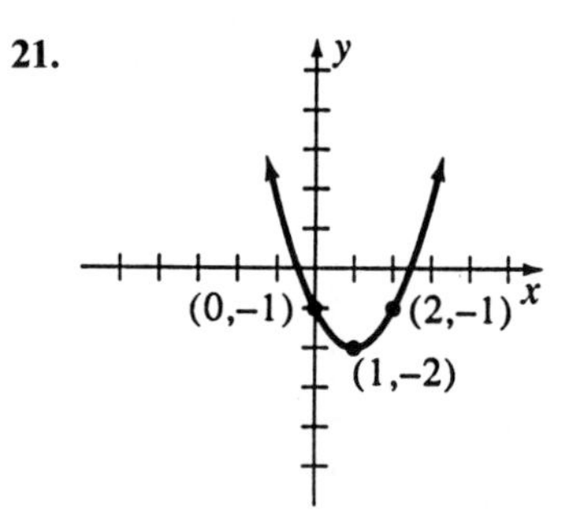

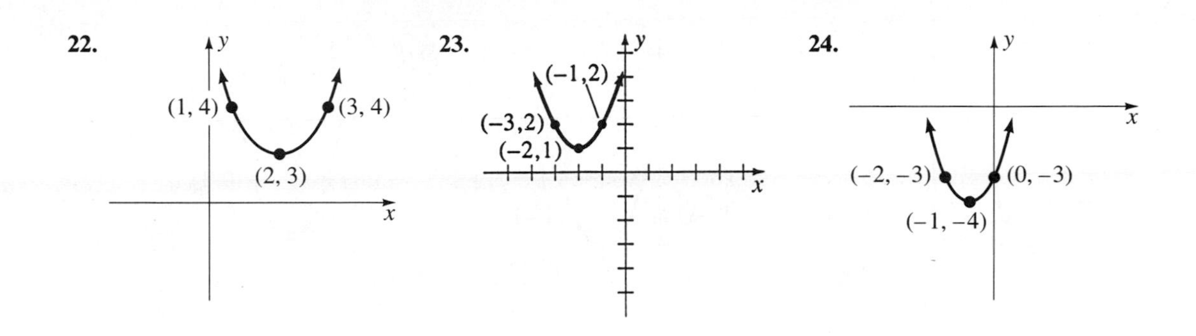

25.

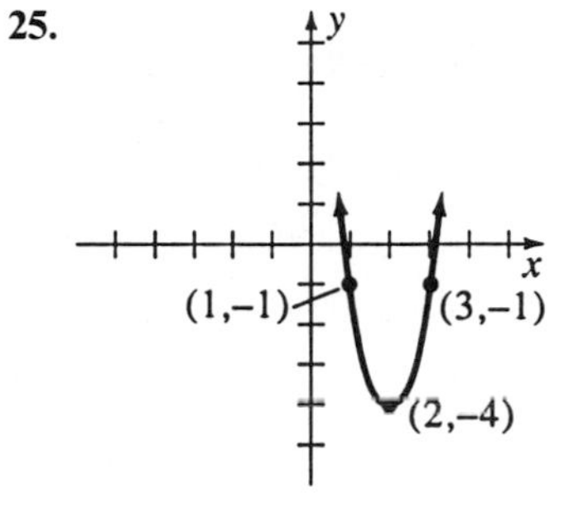

26.

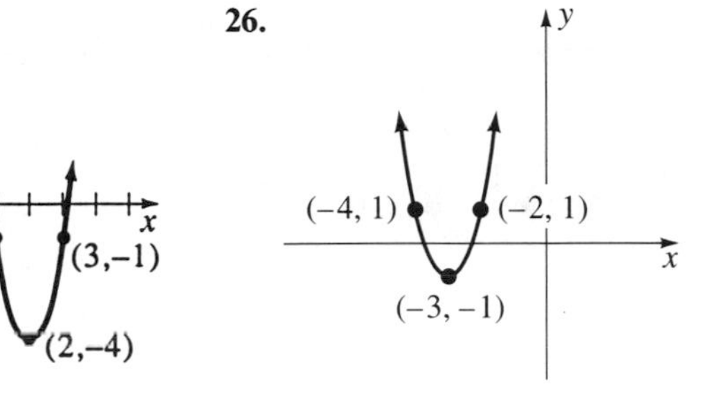

27.

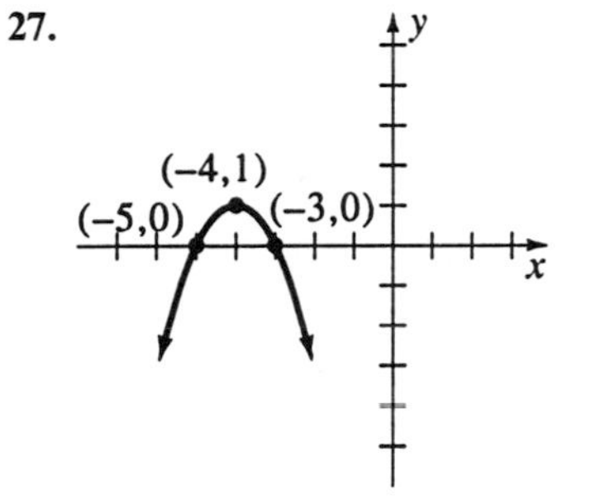

28.

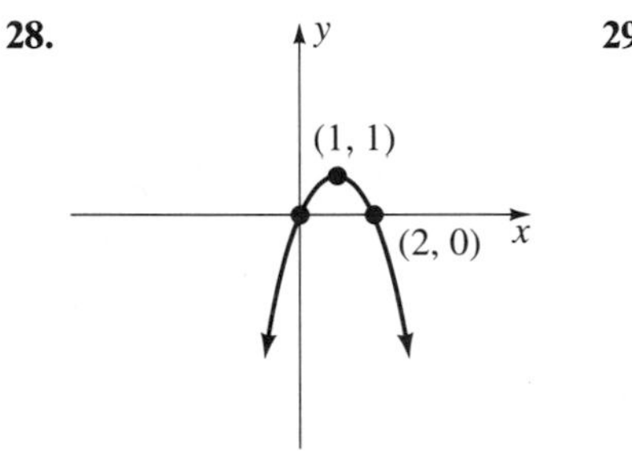

29.

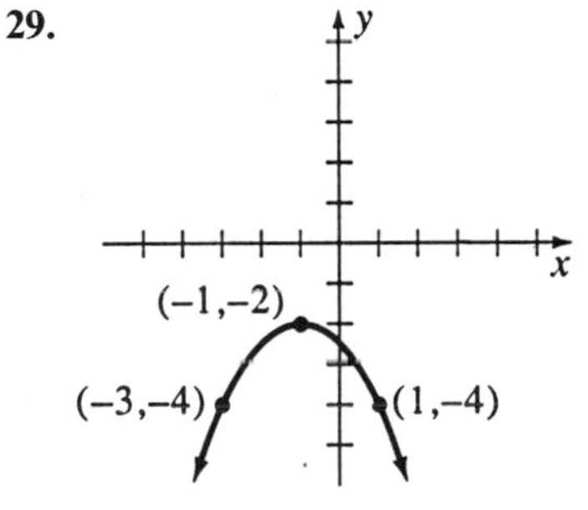

30.

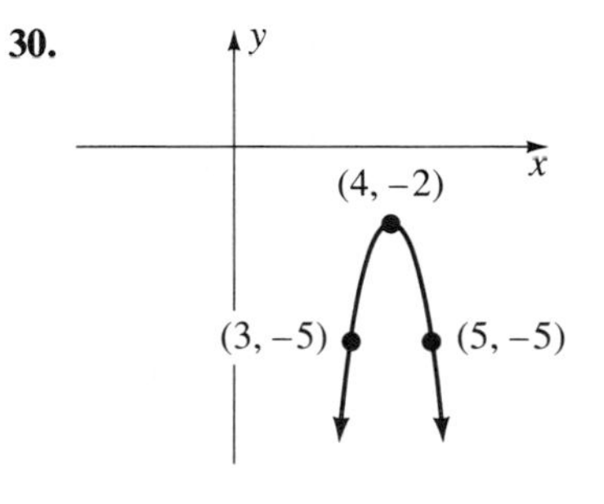

Problem Set 7.7 (page 363)

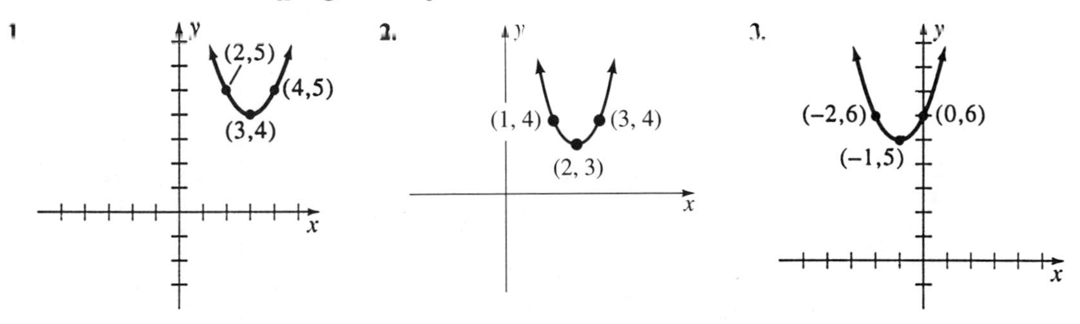

4.

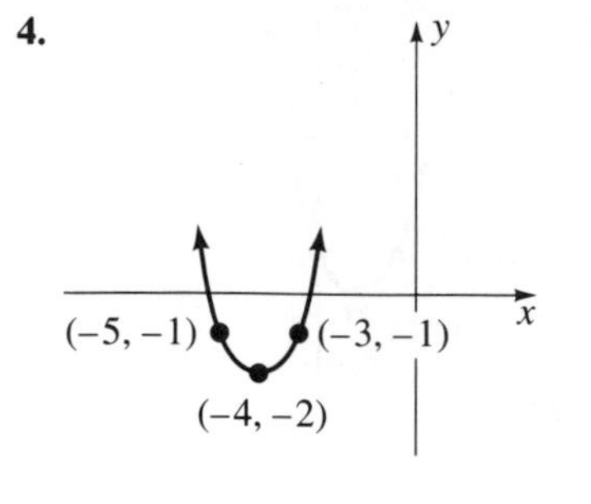

5.

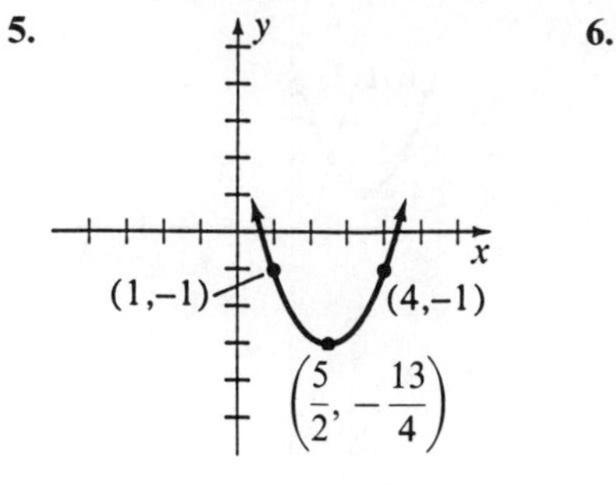

6.

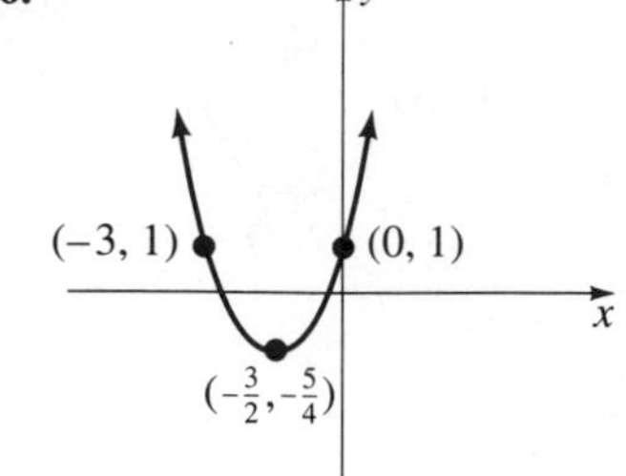

7.

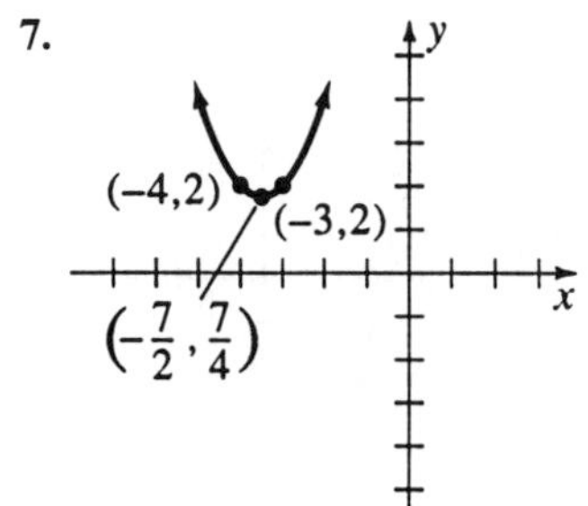

8.

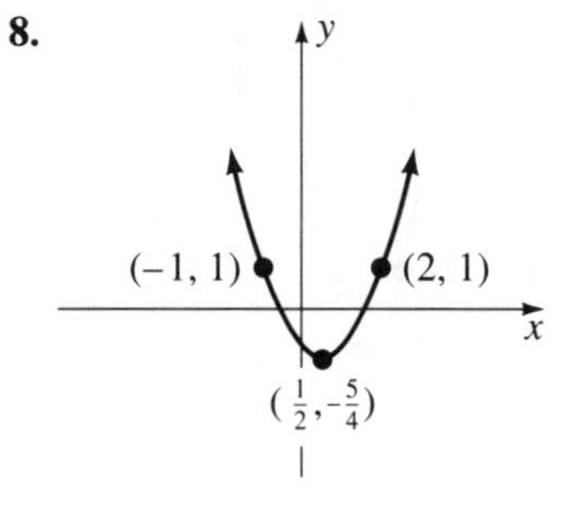

9.

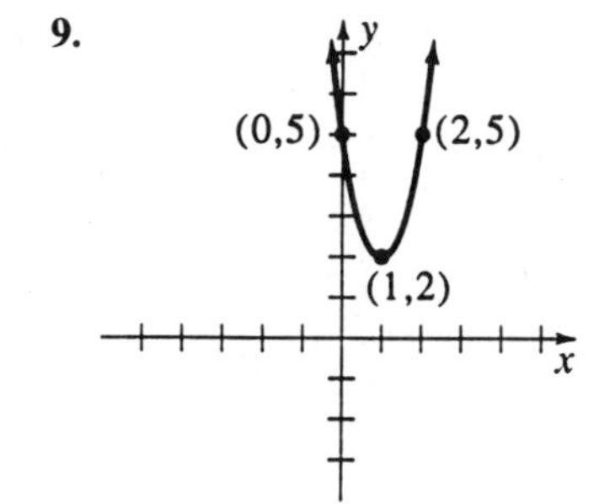

10.

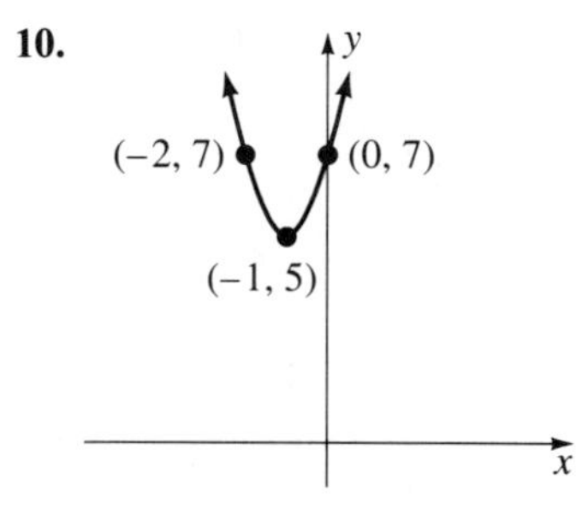

11.

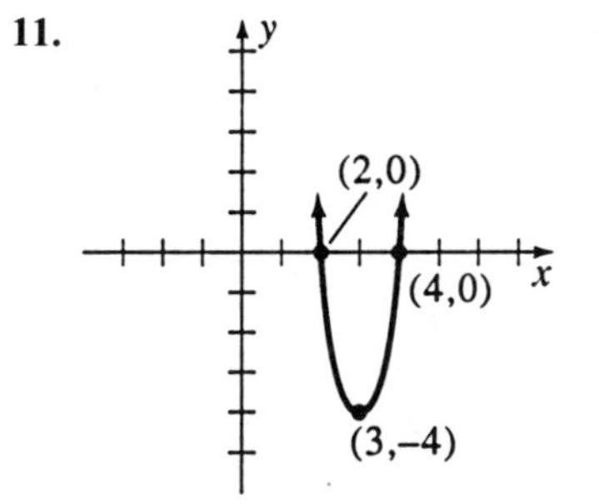

12.

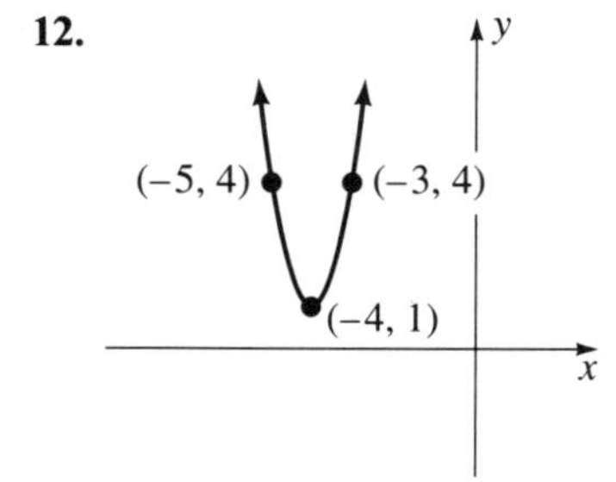

13.

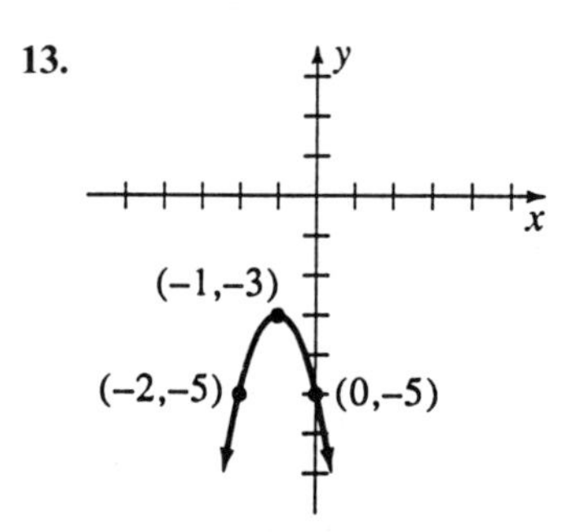

14. **15.**

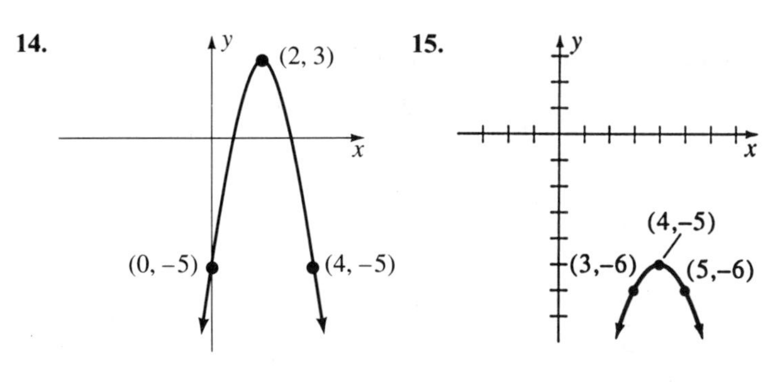

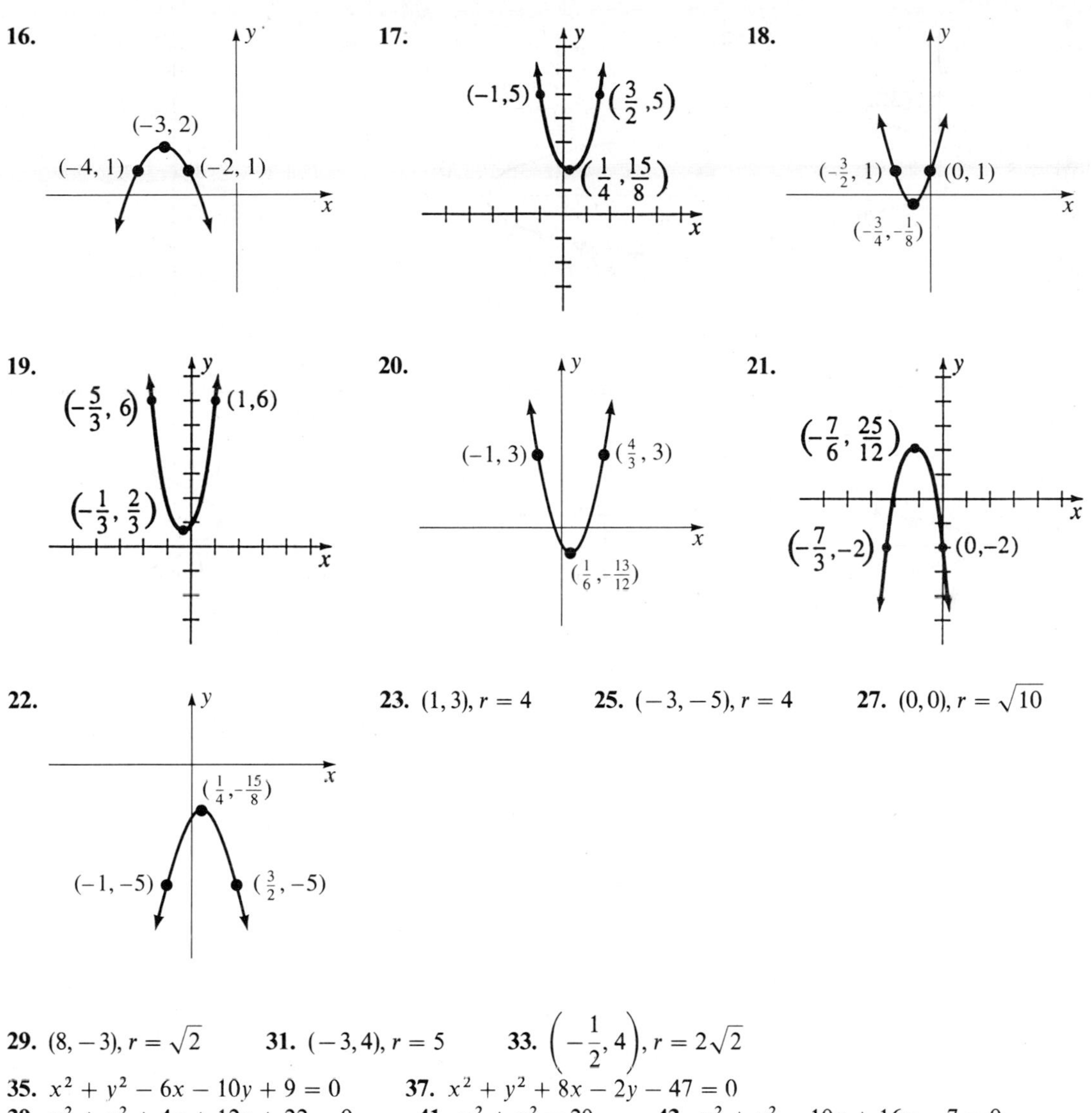

23. $(1, 3), r = 4$ **25.** $(-3, -5), r = 4$ **27.** $(0, 0), r = \sqrt{10}$

29. $(8, -3), r = \sqrt{2}$ **31.** $(-3, 4), r = 5$ **33.** $\left(-\frac{1}{2}, 4\right), r = 2\sqrt{2}$

35. $x^2 + y^2 - 6x - 10y + 9 = 0$ **37.** $x^2 + y^2 + 8x - 2y - 47 = 0$

39. $x^2 + y^2 + 4x + 12y + 22 = 0$ **41.** $x^2 + y^2 = 20$ **43.** $x^2 + y^2 - 10x + 16y - 7 = 0$

45. $x^2 + y^2 - 8y = 0$ **47.** $x^2 + y^2 + 8x - 6y = 0$

49. **(a)** **(b)**

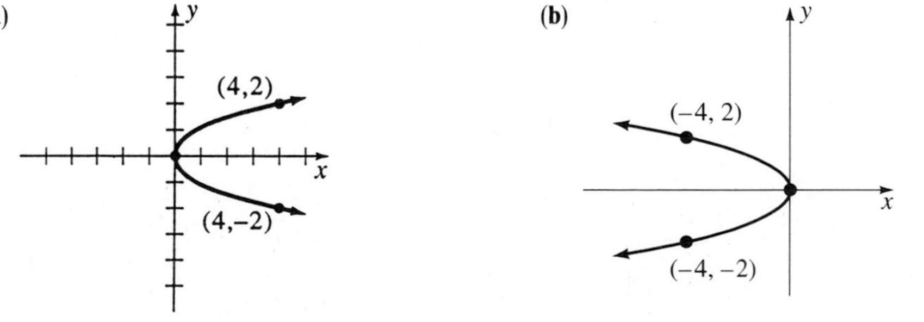

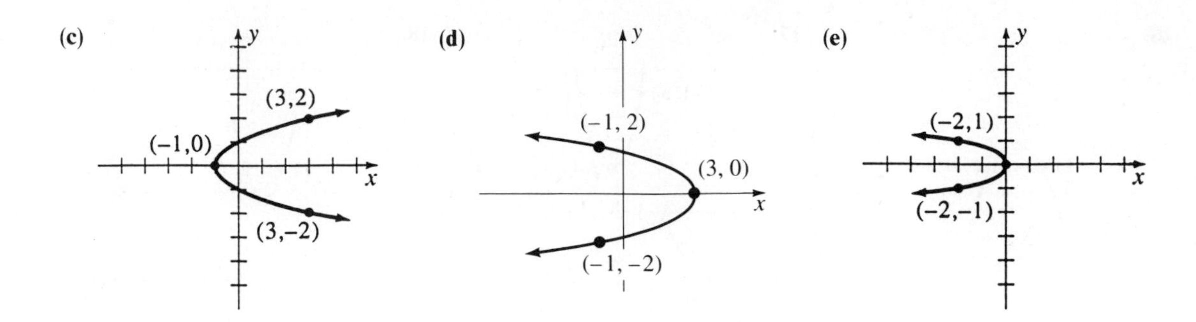

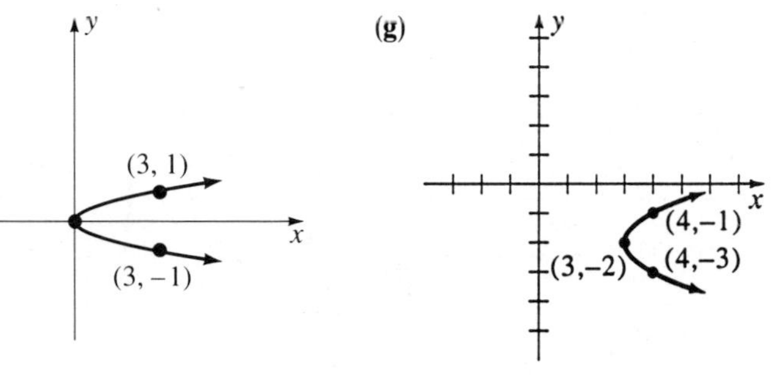

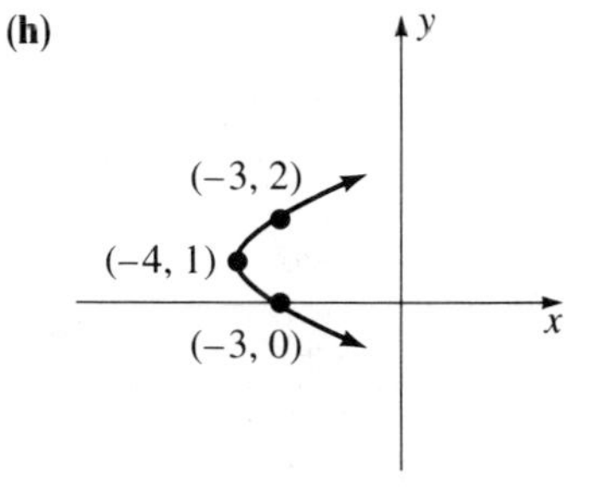

50. **(a)** $(1,4)$, $r = 3$ **(b)** $(-2,7)$, $r = 2$  **(c)** $(-6,-4)$, $r = 8$ **(d)** $(8,-10)$, $r = 7$
(e) $(0,6)$, $r = 9$ **(f)** $(-7,0)$, $r = 7$

Problem Set 7.8 (page 372)

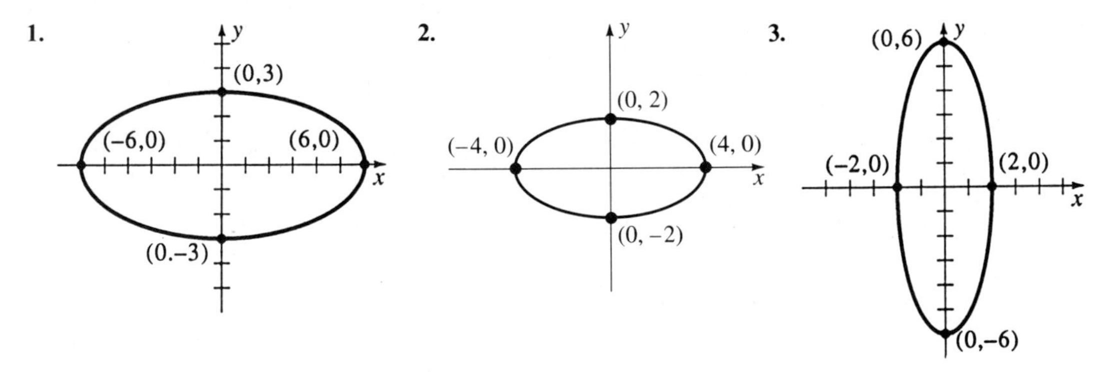

4.

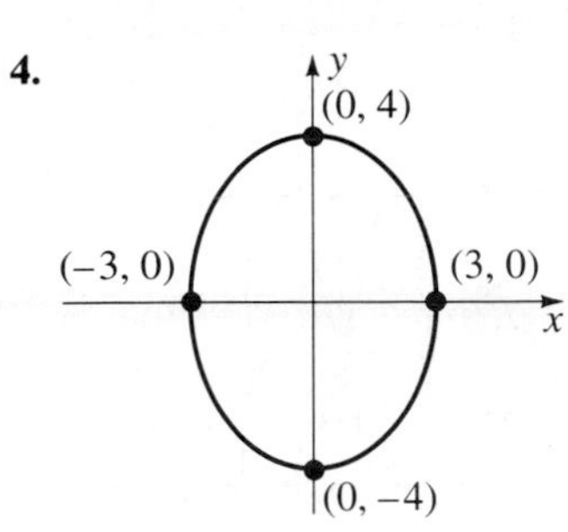

5.

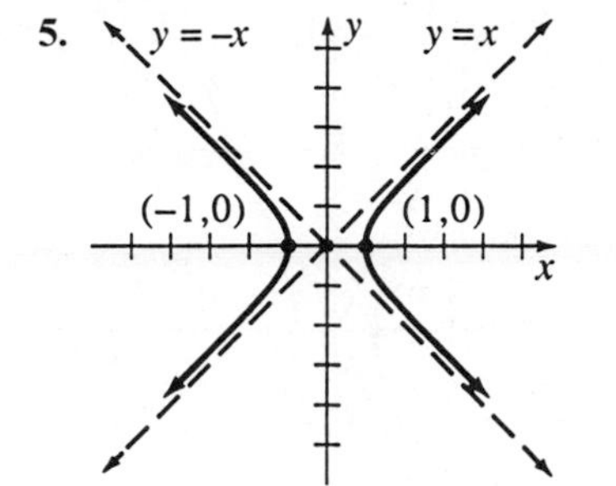

6.

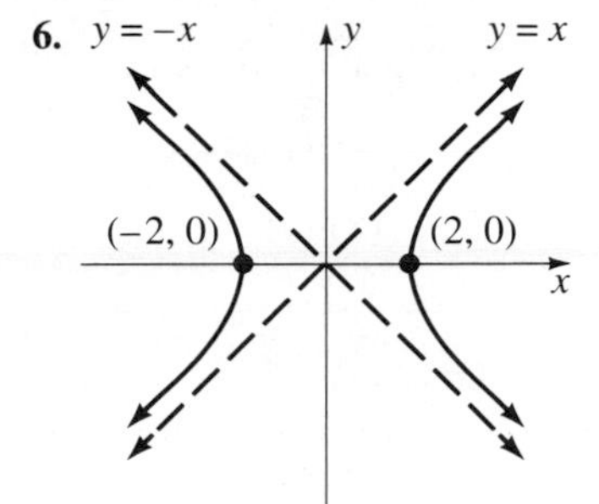

7.

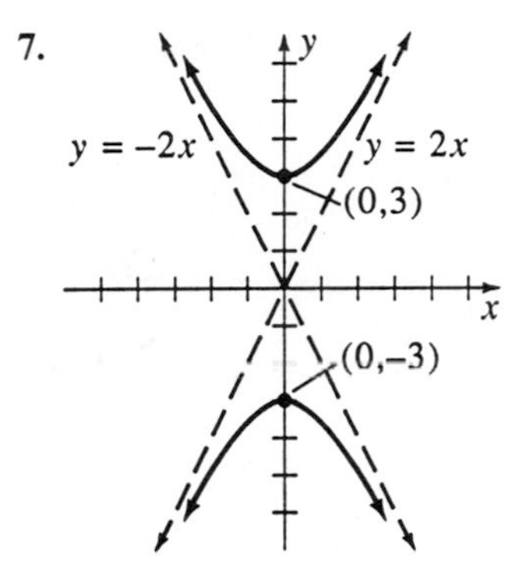

8.

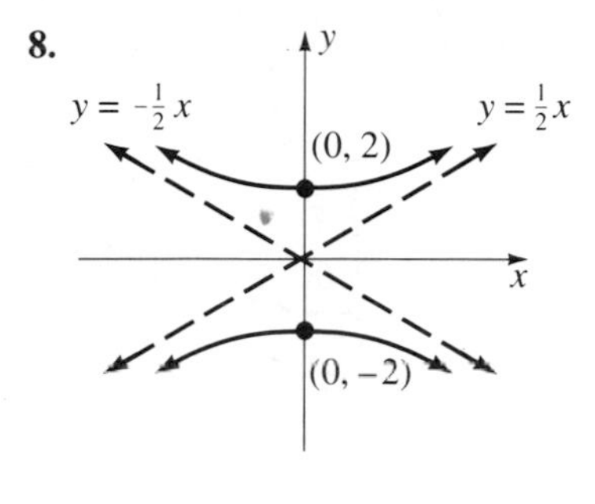

9.

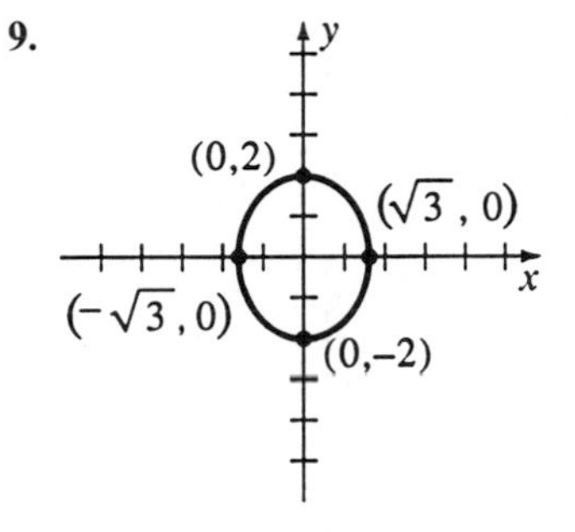

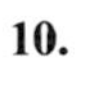

10.

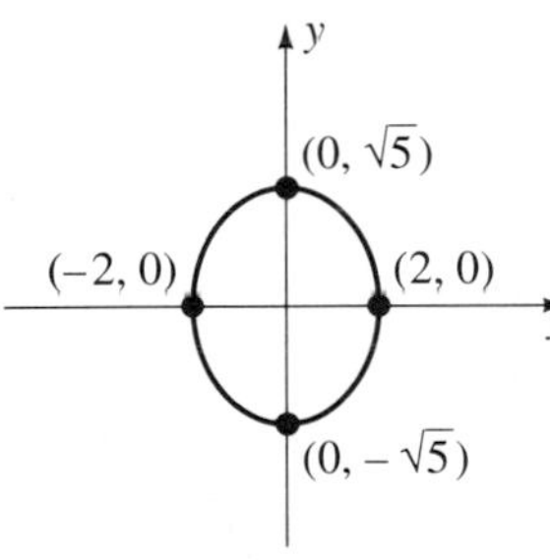

11.

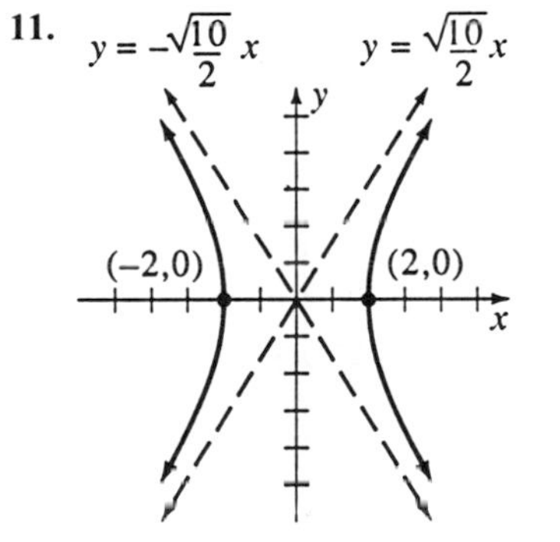

12.

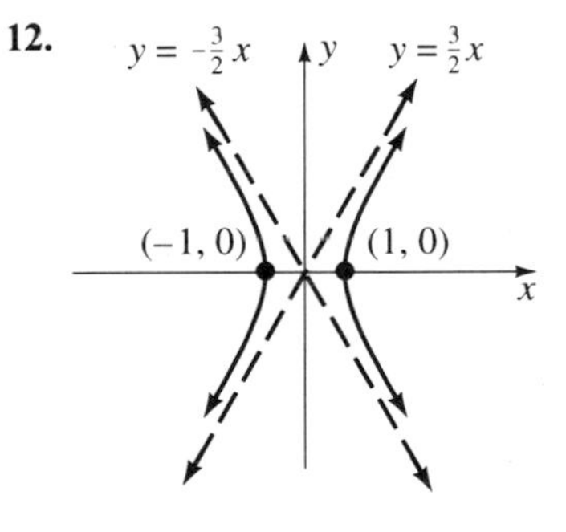

13.

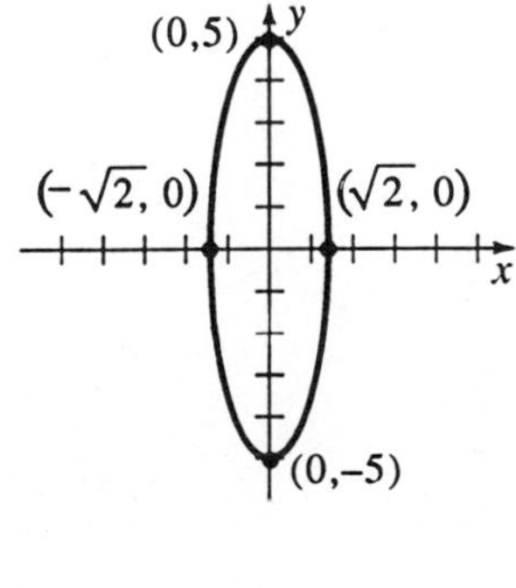

14.

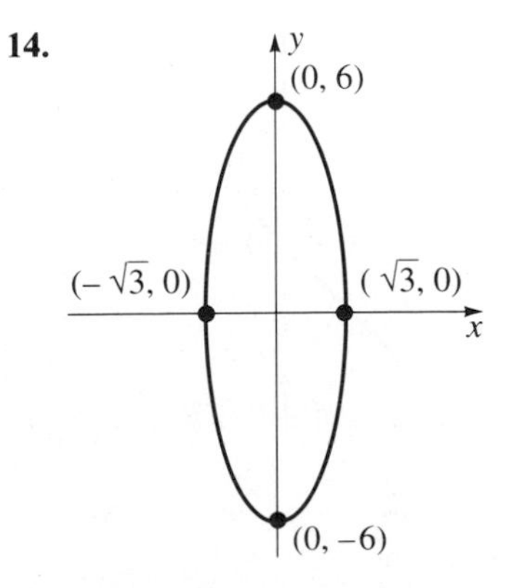

15.

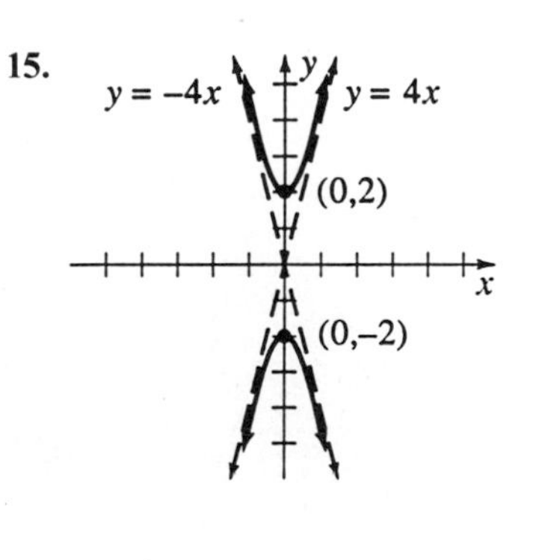

16. **17.** **18.**

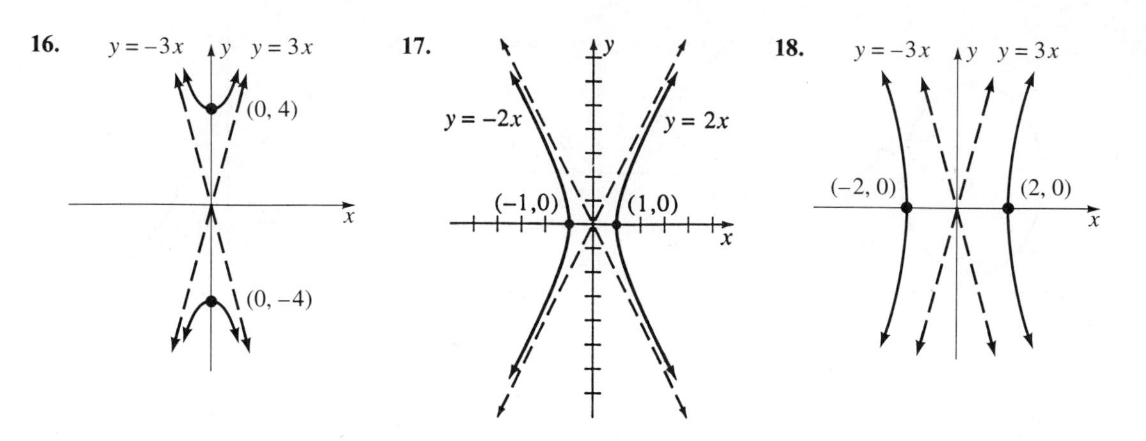

19.

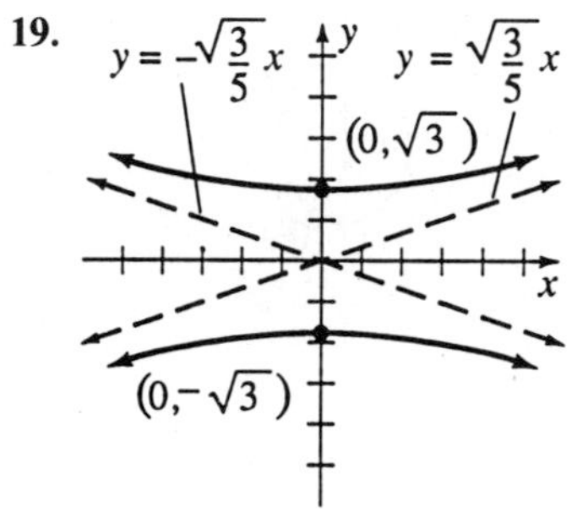

20.

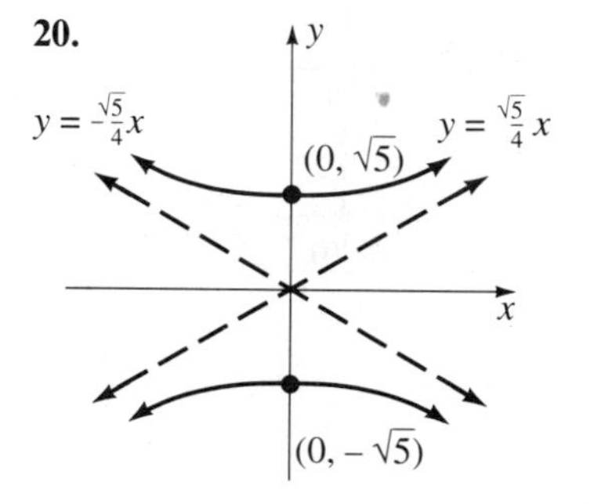

21. **(a)**

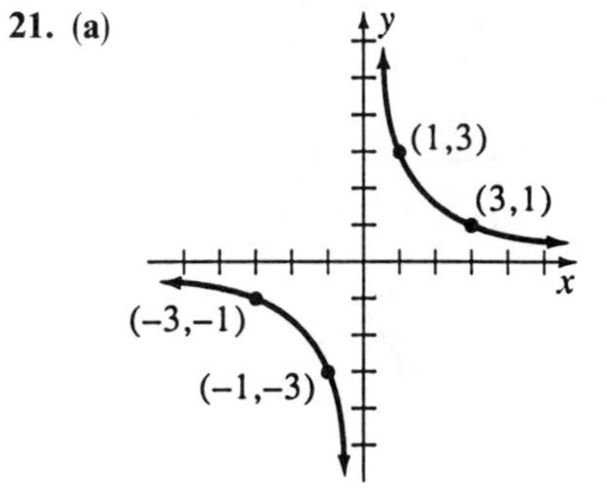

(b)

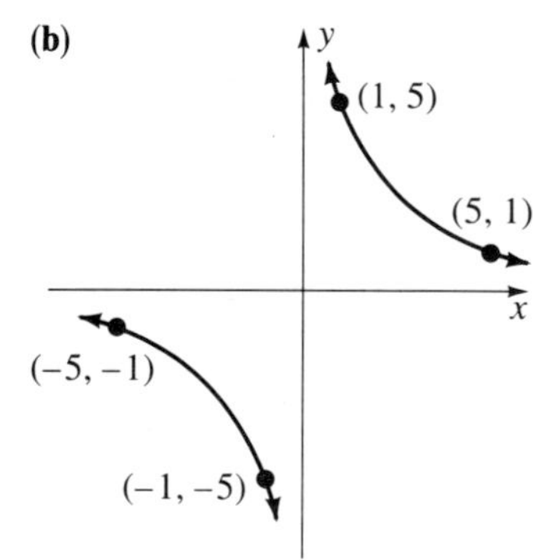

(c)

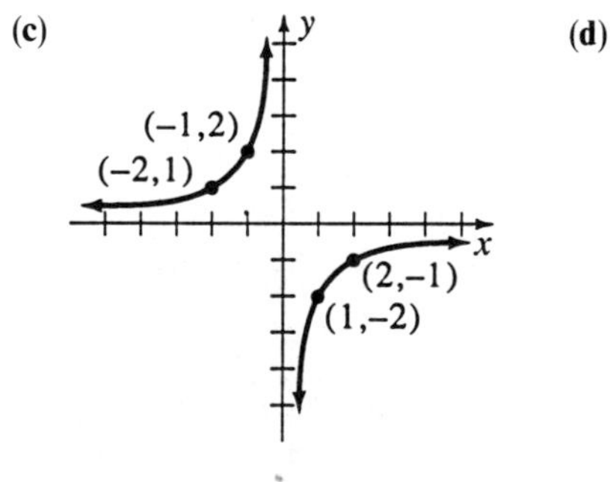

(d)

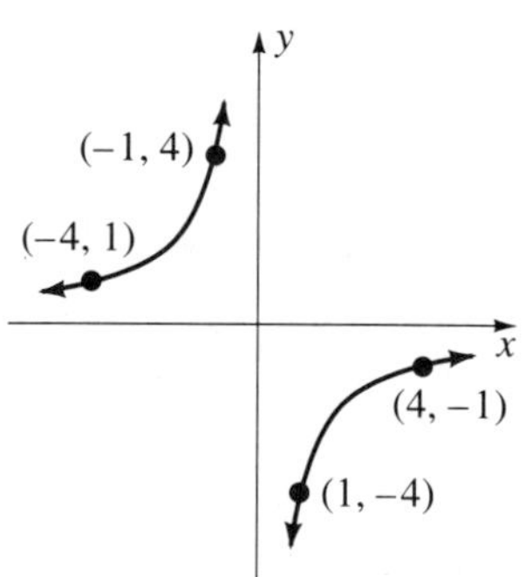

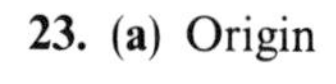

23. **(a)** Origin

(c)

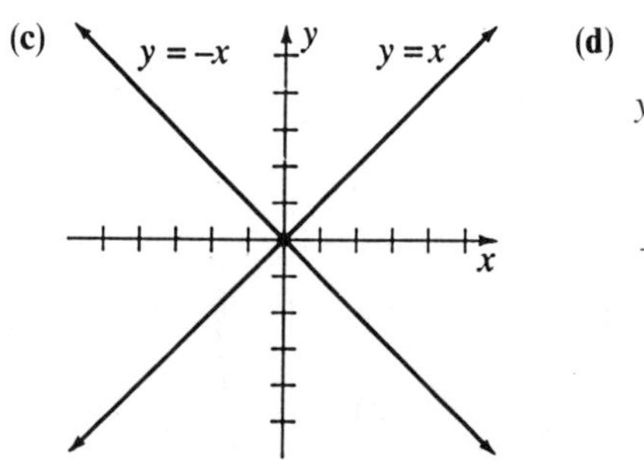

(d)

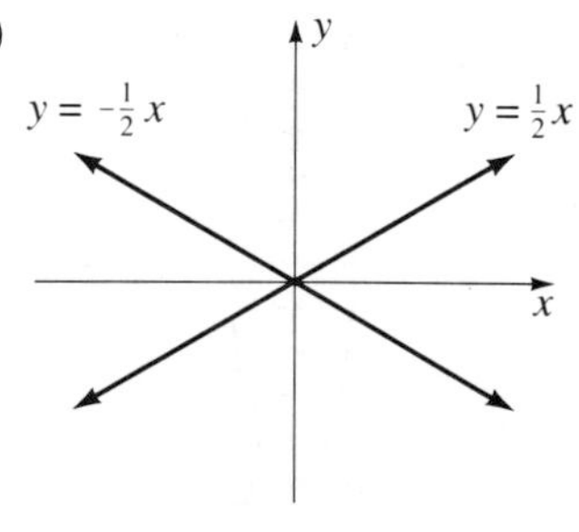

Chapter 7 Review Problem Set (page 375)

1. (a) $\frac{6}{5}$ (b) $-\frac{2}{3}$ **2.** (a) -4 (b) $\frac{2}{7}$ **3.** 5, 10, and $\sqrt{97}$ **4.** $7x + 4y = 1$

5. $3x + 7y = 28$ **6.** $2x - 3y = 16$ **7.** $x - 2y = -8$ **8.** $2x - 3y = 14$

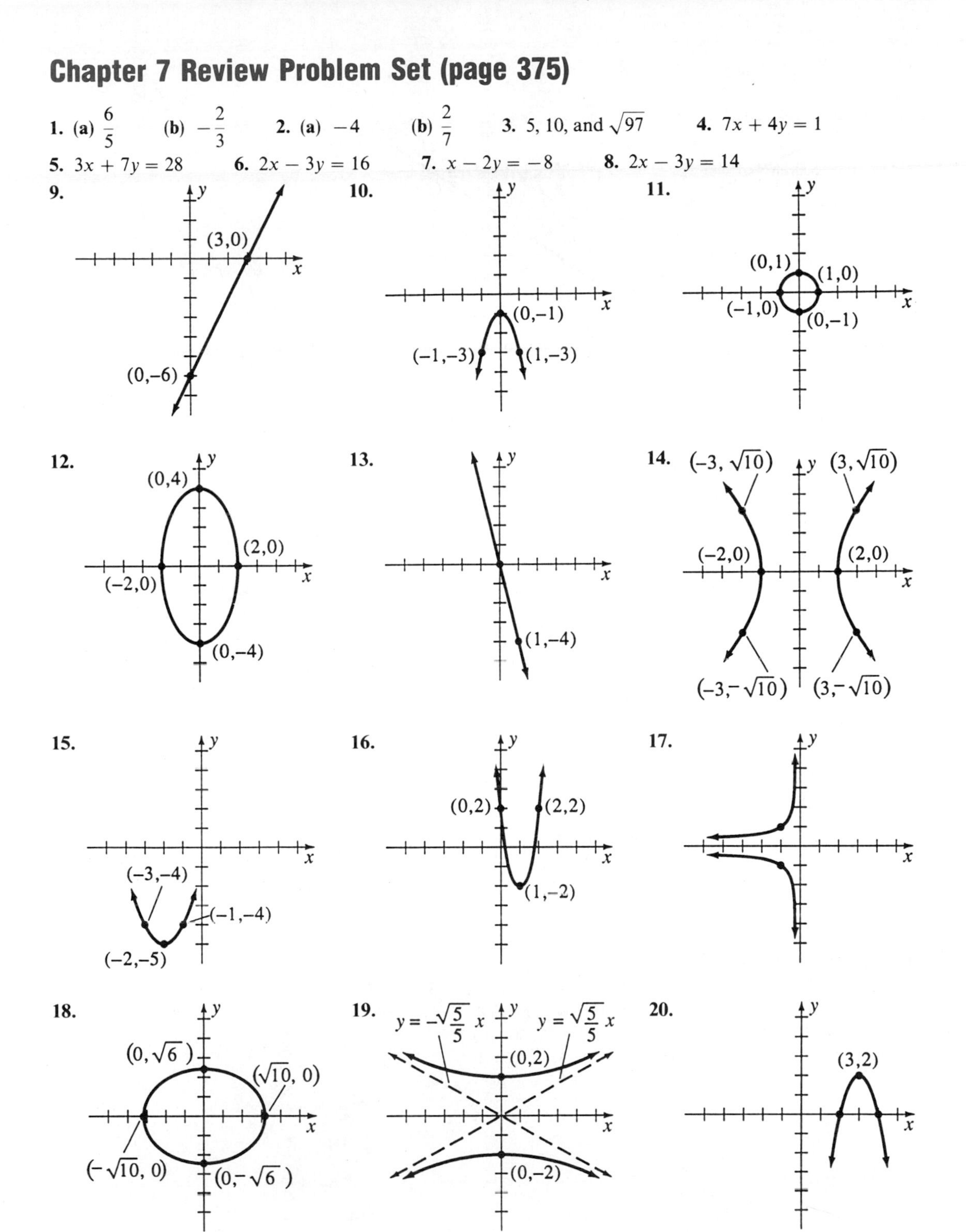

21. 22.

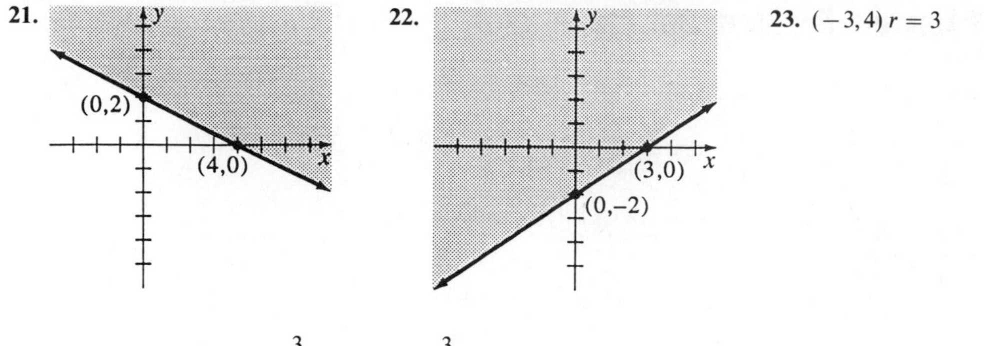

23. $(-3, 4)\ r = 3$

24. $(-4, -6)$ 25. $y = \frac{3}{2}x$ and $y = -\frac{3}{2}x$ 26. 6

CHAPTER 8

Problem Set 8.1 (page 385)

1. $D = \{1, 2, 3, 4\}$, $R = \{5, 8, 11, 14\}$ It is a function.
3. $D = \{0, 1\}$, $R = \{-2\sqrt{6}, -5, 5, 2\sqrt{6}\}$ It is not a function.
5. $D = \{1, 2, 3, 4, 5\}$, $R = \{2, 5, 10, 17, 26\}$ It is a function.
7. $D = \{\text{all reals}\}$, $R = \{\text{all reals}\}$ It is a function. 9. $D = \{\text{all reals}\}$, $R = \{y \mid y \geq 0\}$ It is a function.
11. {all reals} 13. $\{x \mid x \neq 1\}$ 15. $\left\{x \mid x \neq \frac{3}{4}\right\}$ 17. $\{x \mid x \neq -1 \text{ and } x \neq 4\}$
19. $\{x \mid x \neq -8 \text{ and } x \neq 5\}$ 21. $\{x \mid x \neq -6 \text{ and } x \neq 0\}$ 23. {all reals}
25. $\{t \mid t \neq -2 \text{ and } t \neq 2\}$ 27. $\{x \mid x \geq -4\}$ 29. $\left\{s \mid s \geq \frac{5}{4}\right\}$ 31. $\{x \mid x \leq -4 \text{ or } x \geq 4\}$
33. $\{x \mid x \leq -3 \text{ or } x \geq 6\}$ 35. $\{x \mid -1 \leq x \leq 1\}$
37. $f(0) = -2$, $f(2) = 8$, $f(-1) = -7$, $f(-4) = -22$
39. $f(-2) = -\frac{7}{4}$, $f(0) = -\frac{3}{4}$, $f\left(\frac{1}{2}\right) = -\frac{1}{2}$, $f\left(\frac{2}{3}\right) = -\frac{5}{12}$
41. $g(-1) = 0$; $g(2) = -9$; $g(-3) = 26$; $g(4) = 5$
43. $h(-2) = -2$; $h(-3) = -11$; $h(4) = -32$; $h(5) = -51$
45. $f(3) = \sqrt{7}$; $f(4) = 3$; $f(10) = \sqrt{21}$; $f(12) = 5$
47. $f(1) = -1$; $f(-1) = -2$; $f(3) = -\frac{2}{3}$; $f(-6) = \frac{4}{3}$
49. $f(-2) = 27$; $f(3) = 42$; $g(-4) = -37$; $g(6) = -17$
51. $f(-2) = 5$; $f(3) = 8$; $g(-4) = -3$; $g(5) = -4$ 53. -3 55. $-2a - h$ 57. $4a - 1 + 2h$
59. $-8a - 7 - 4h$ 61. $h(1) = 48$; $h(2) = 64$; $h(3) = 48$; $h(4) = 0$
63. $C(75) = \$74$; $C(150) = \$98$; $C(225) = \$122$; $C(650) = \$258$
65. $I(.11) = 55$; $I(.12) = 60$; $I(.135) = 67.5$; $I(.15) = 75$

Problem Set 8.2 (page 397)

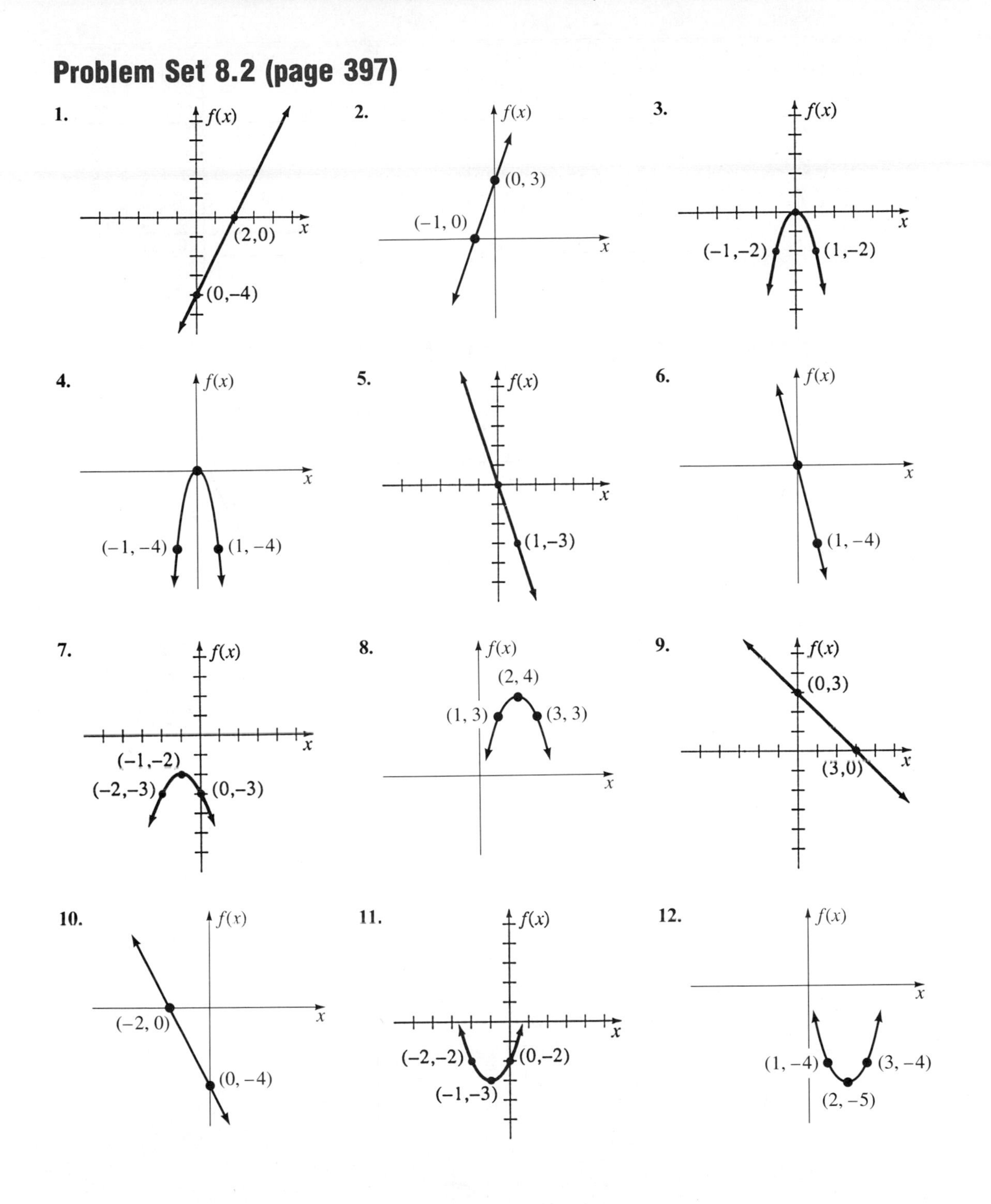

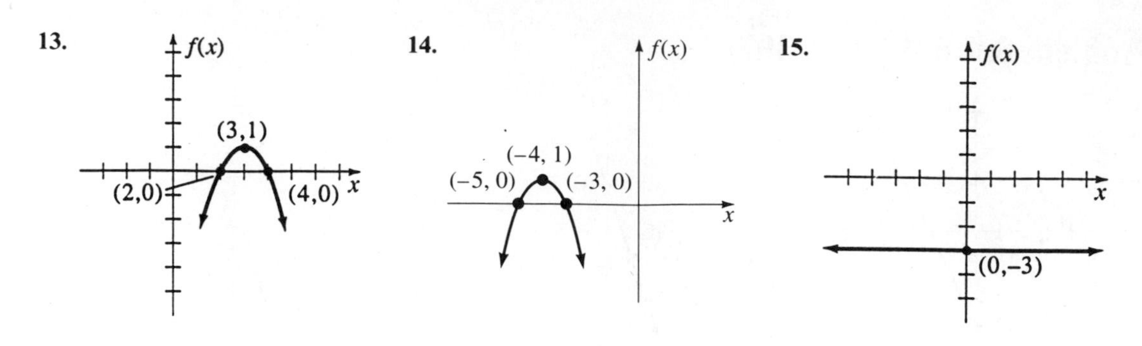
13.
f(x)
(3,1)
(2,0)
(4,0)
x
14.
f(x)
(−4, 1)
(−5, 0)
(−3, 0)
x
15.
f(x)
x
(0,−3)

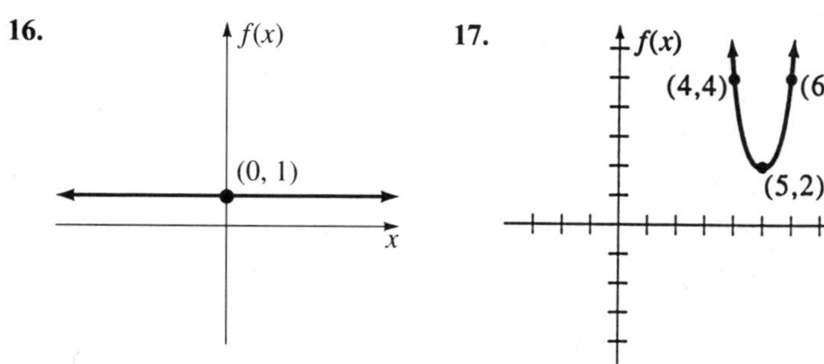
16.
f(x)
(0, 1)
x
17.
f(x)
(4,4)
(6,4)
(5,2)
x

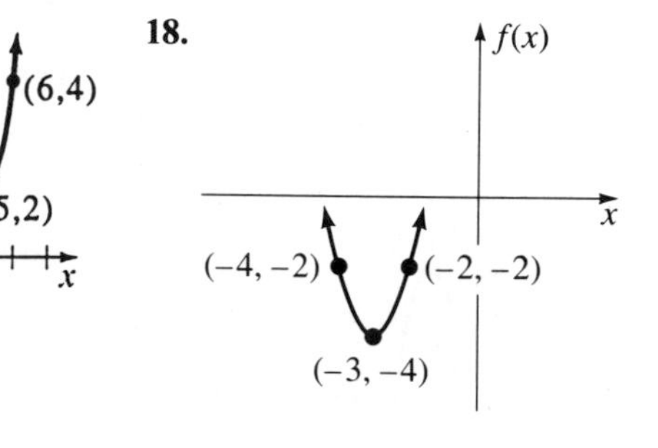
18.
f(x)
x
(−4, −2)
(−2, −2)
(−3, −4)

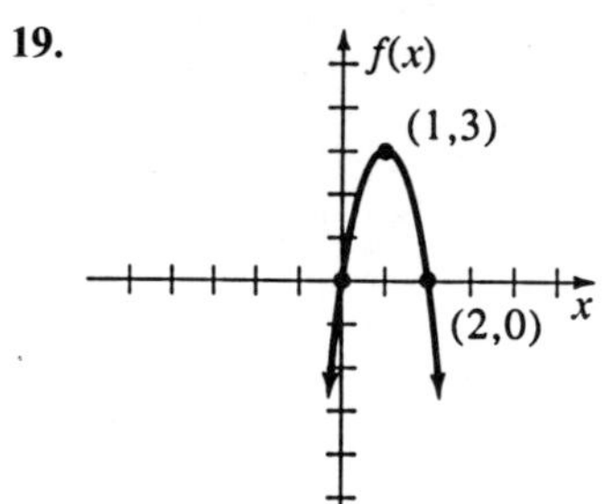
19.
f(x)
(1,3)
(2,0)
x

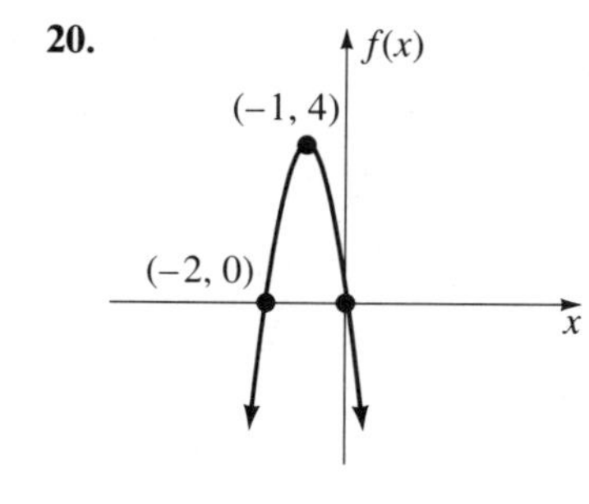
20.
f(x)
(−1, 4)
(−2, 0)
x

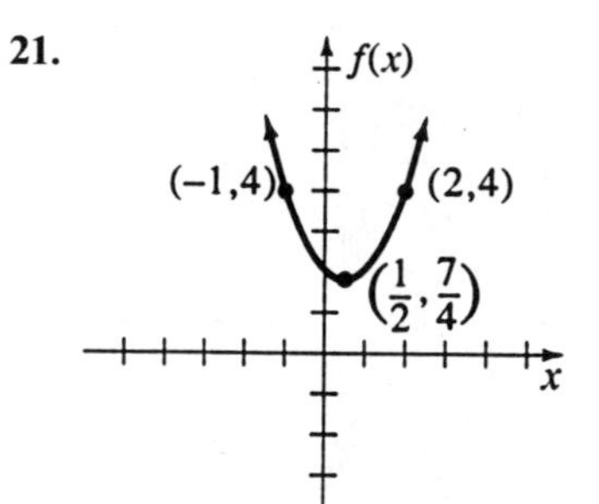
21.
f(x)
(−1,4)
(2,4)
$\left(\frac{1}{2}, \frac{7}{4}\right)$
x

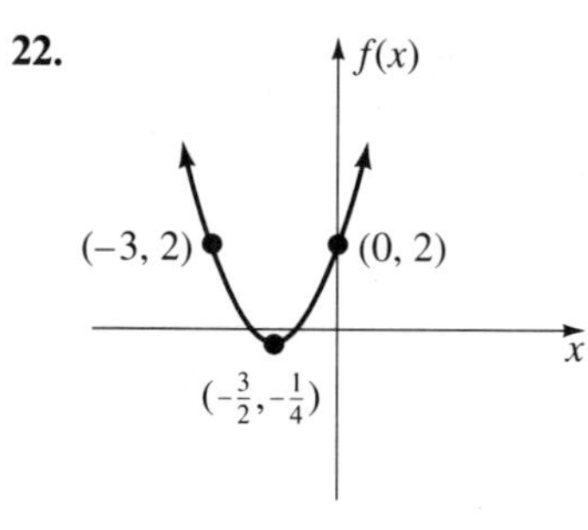
22.
f(x)
(−3, 2)
(0, 2)
$\left(-\frac{3}{2}, -\frac{1}{4}\right)$
x

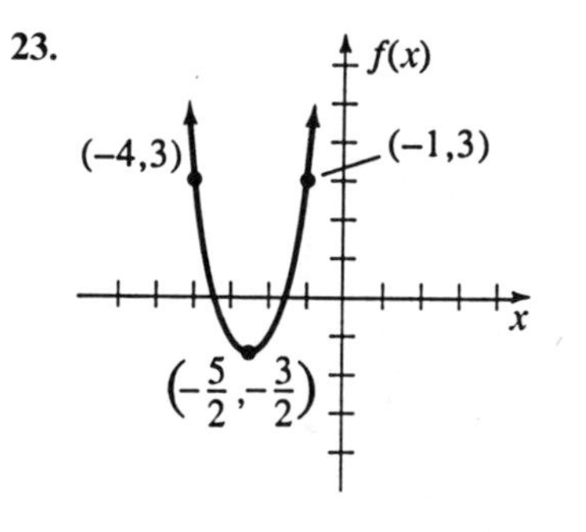
23.
f(x)
(−4,3)
(−1,3)
$\left(-\frac{5}{2}, -\frac{3}{2}\right)$
x

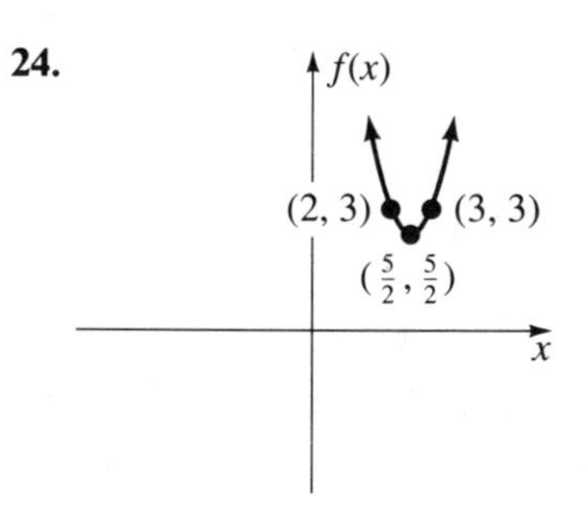
24.
f(x)
(2, 3)
(3, 3)
$\left(\frac{5}{2}, \frac{5}{2}\right)$
x

25. **26.** **27.**

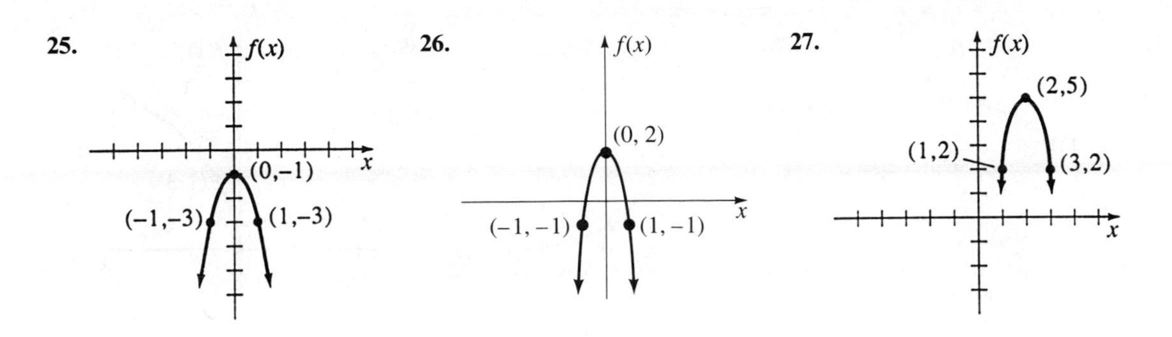

28.

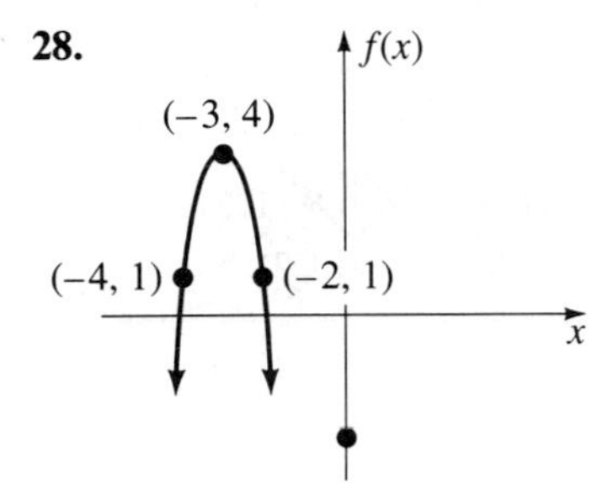

29.

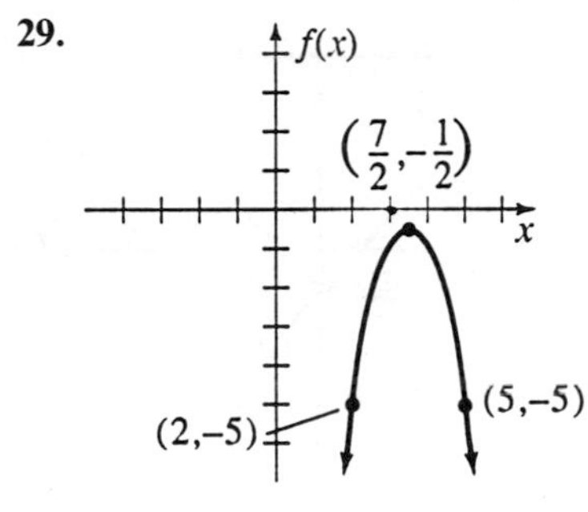

30.

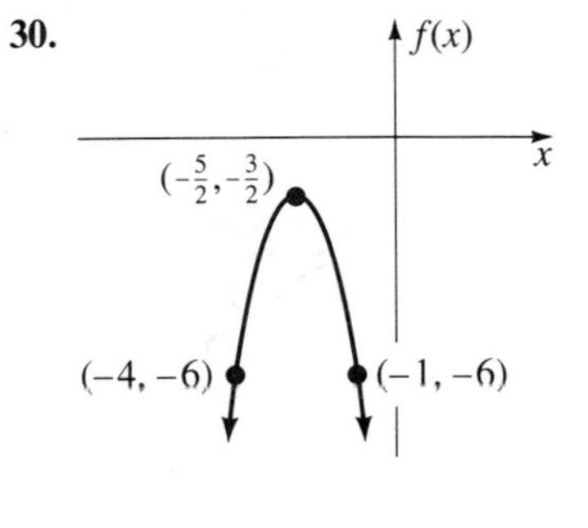

31.

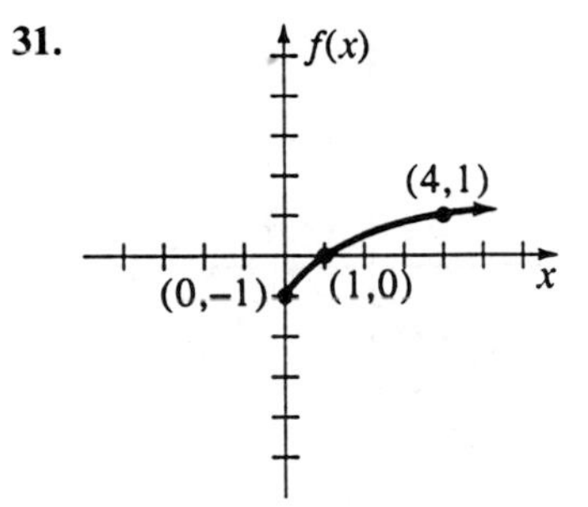

32.

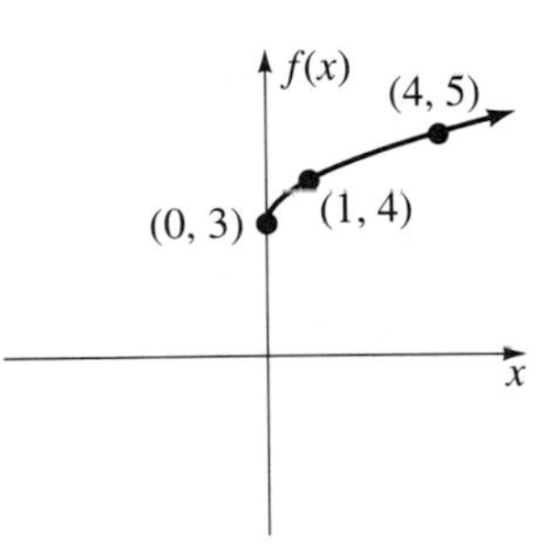

33.

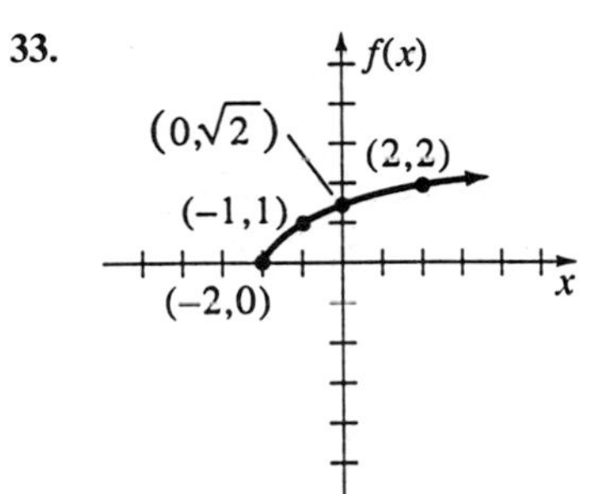

34.

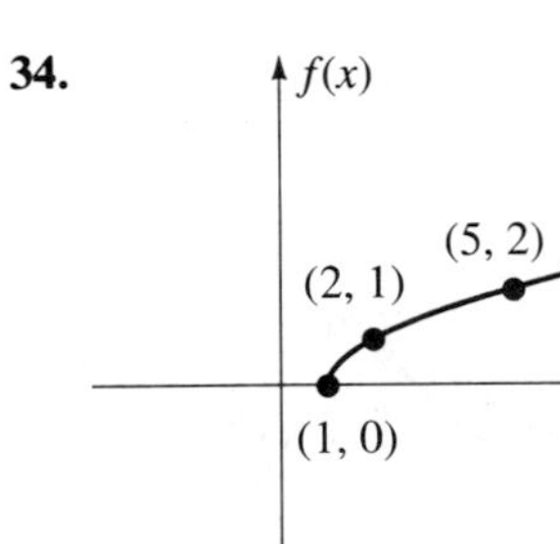

35.

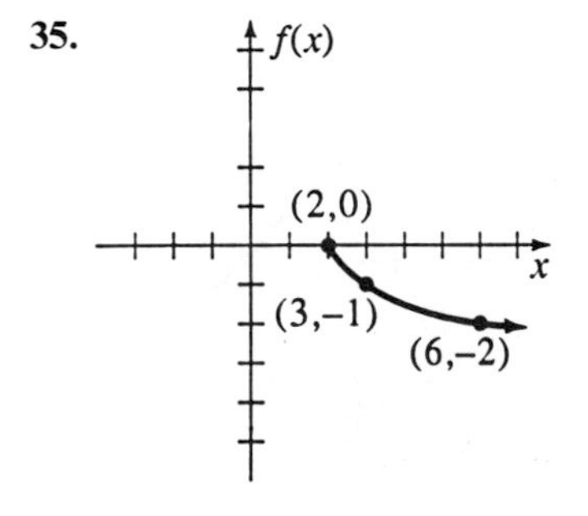

36.

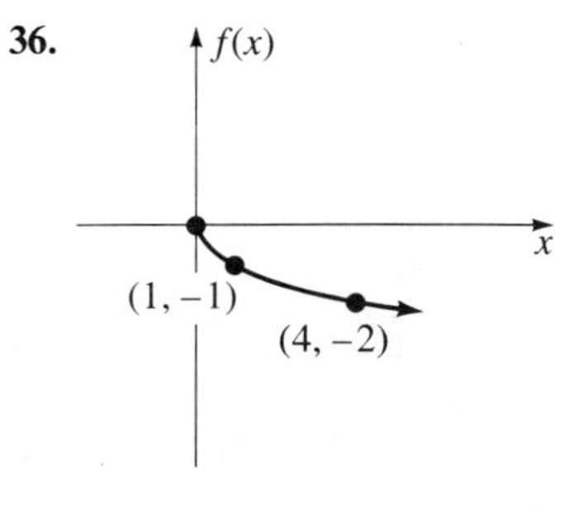

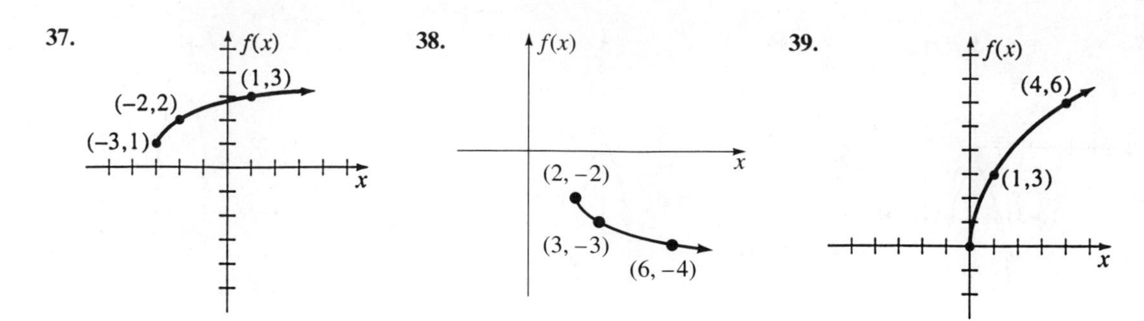

40.

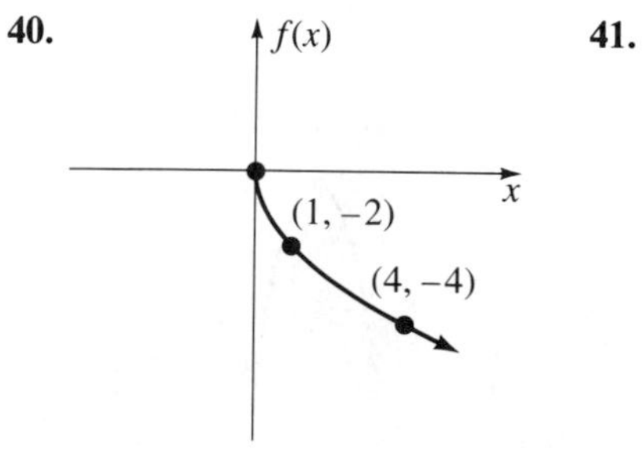

41.

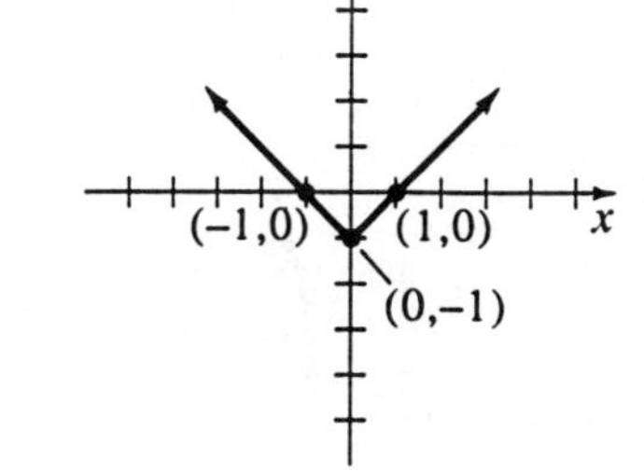

42.

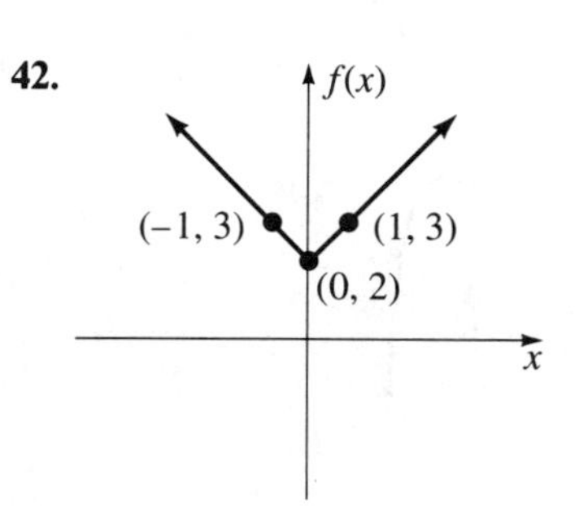

43.

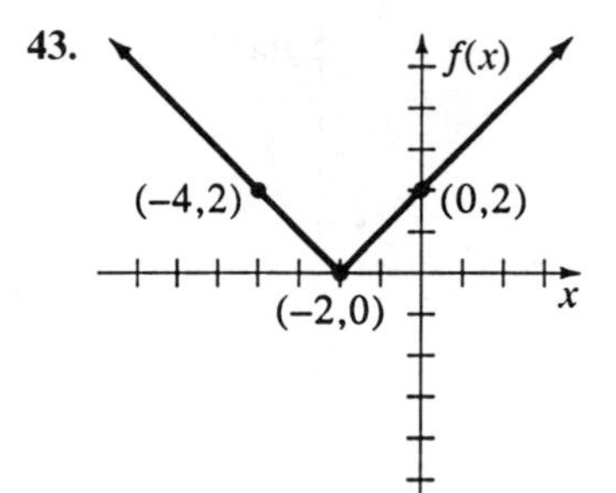

44.

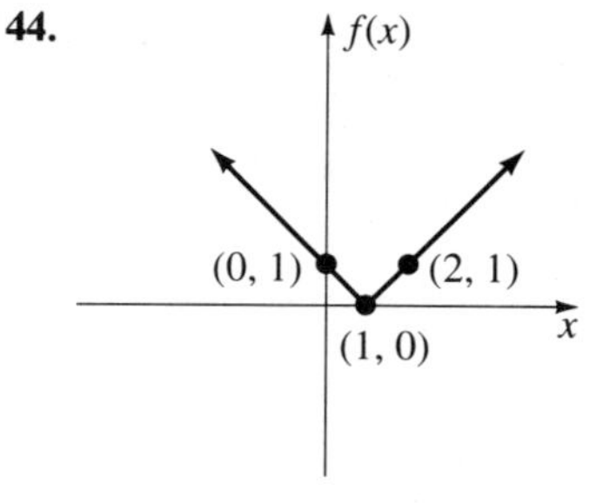

45.

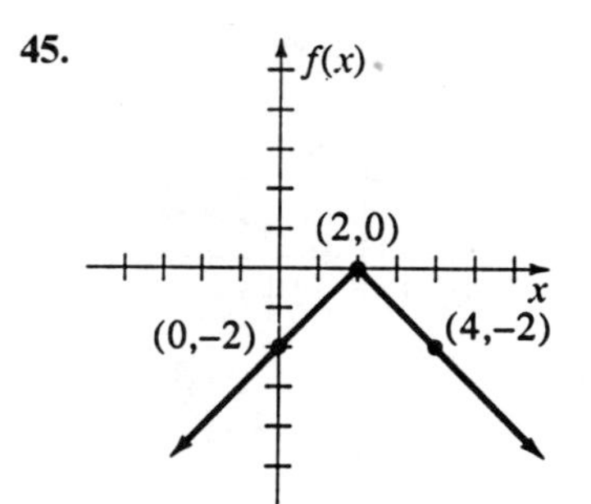

46.

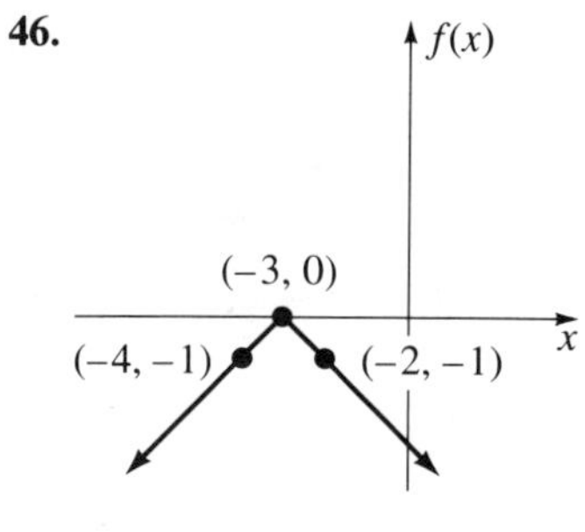

47.

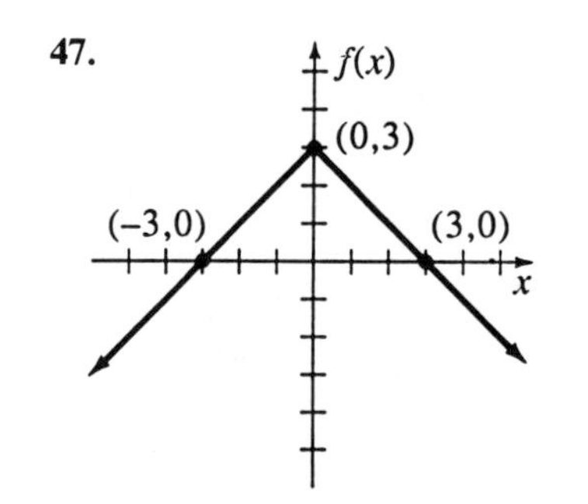

48.

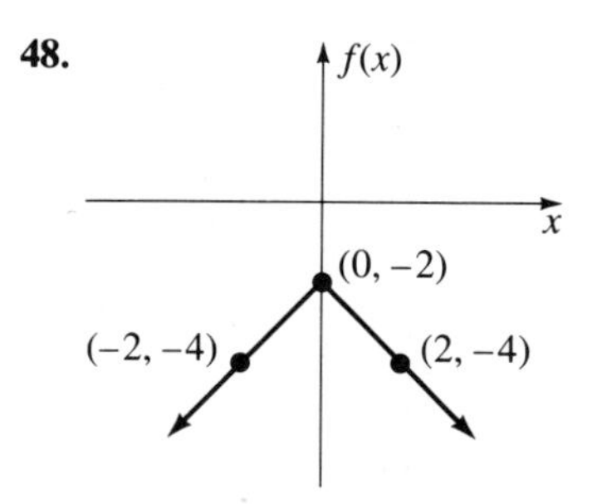

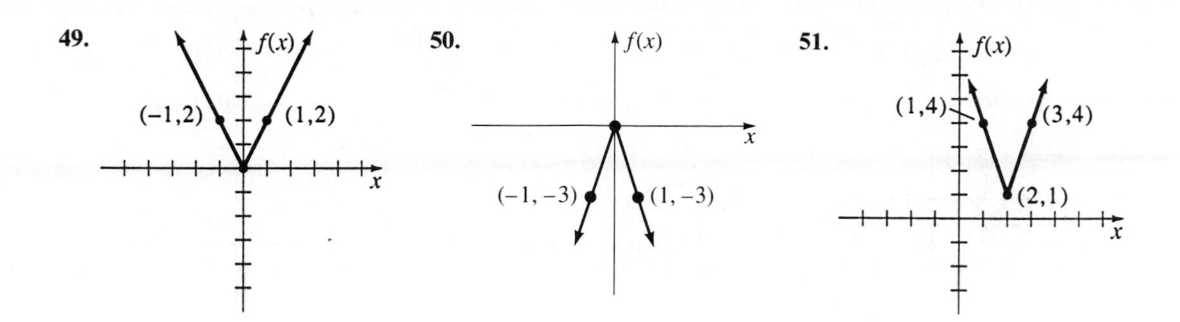
49.
f(x)
(−1,2)
(1,2)
x
50.
f(x)
x
(−1, −3)
(1, −3)
51.
f(x)
(1,4)
(3,4)
(2,1)
x

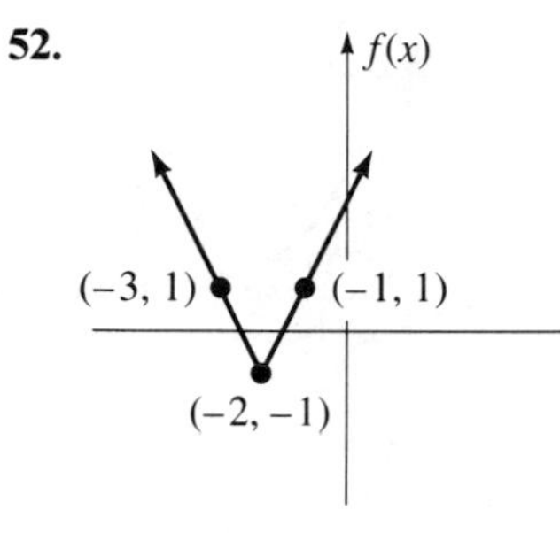
52.
f(x)
(−3, 1)
(−1, 1)
x
(−2, −1)

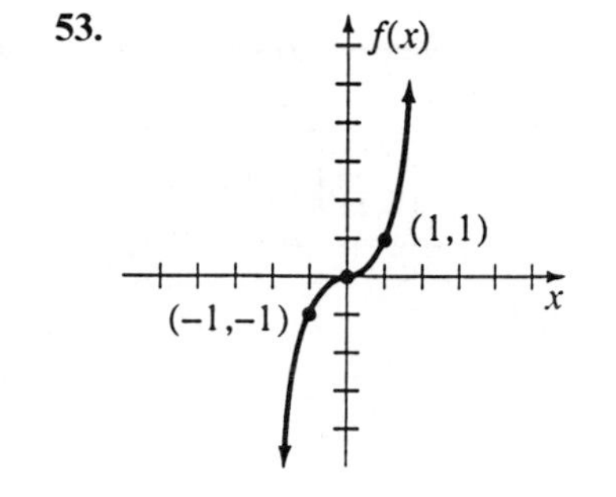
53.
f(x)
(1,1)
x
(−1,−1)

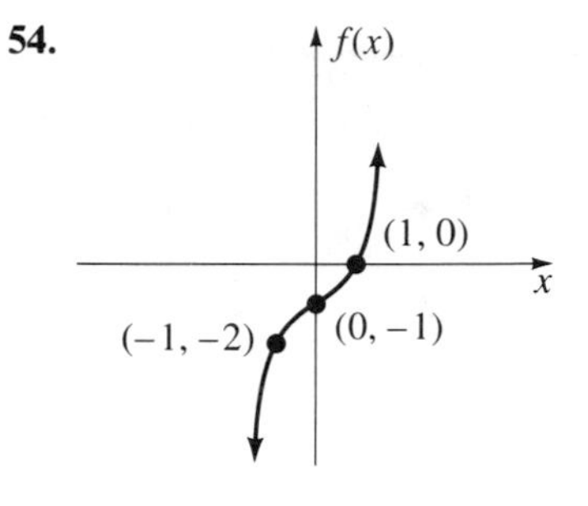
54.
f(x)
(1, 0)
x
(−1, −2)
(0, −1)

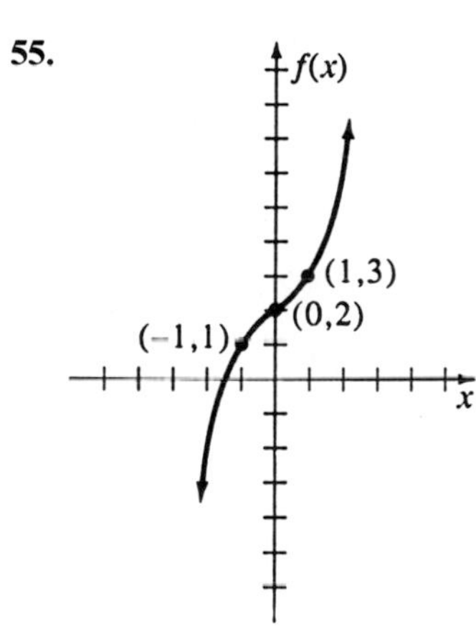
55.
f(x)
(1,3)
(0,2)
(−1,1)
x

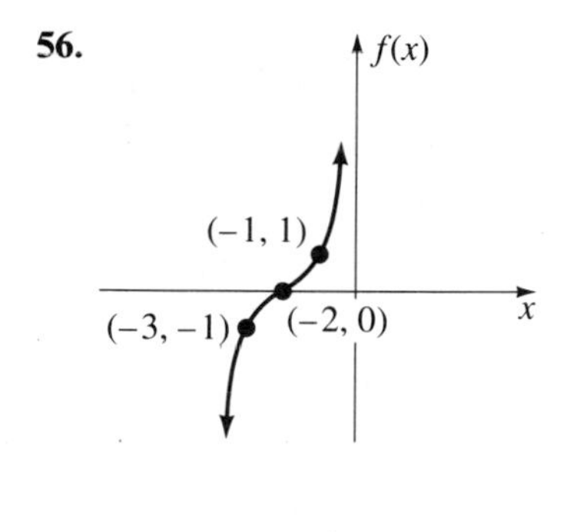
56.
f(x)
(−1, 1)
(−3, −1)
(−2, 0)
x

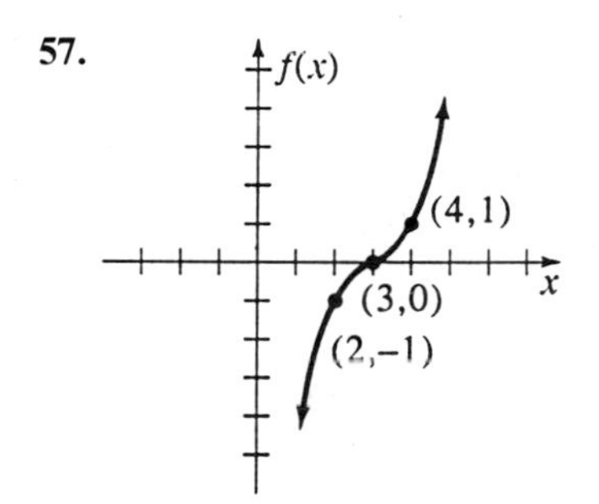
57.
f(x)
(4,1)
x
(3,0)
(2,−1)

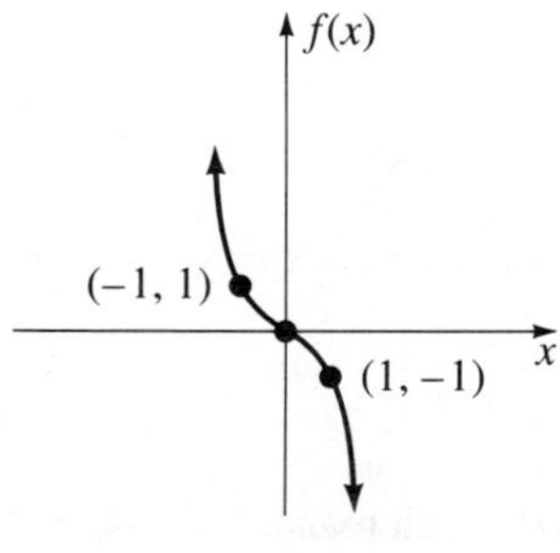
58.
f(x)
(−1, 1)
(1, −1)
x

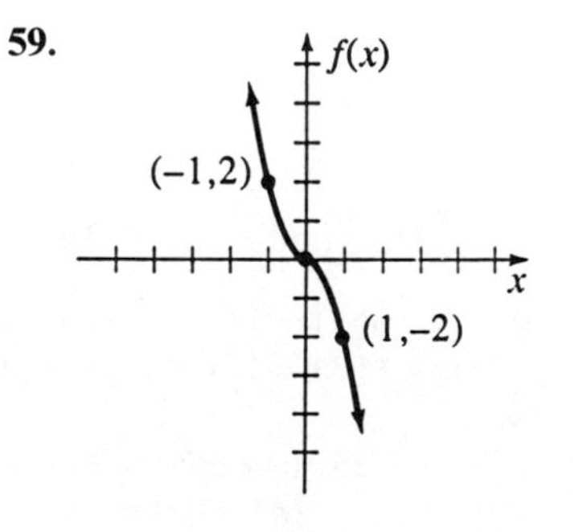
59.
f(x)
(−1,2)
x
(1,−2)

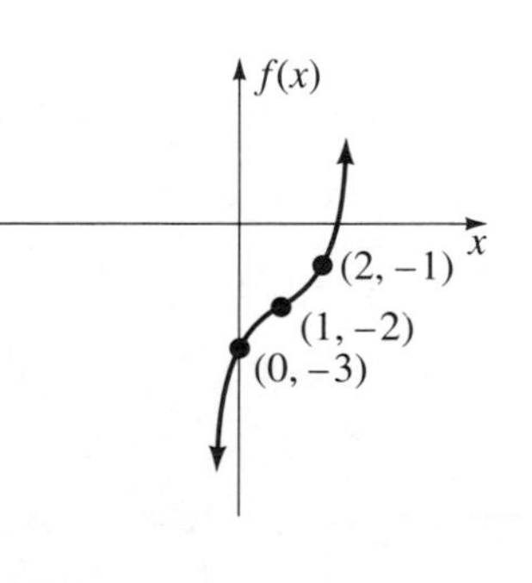
60.
f(x)
x
(2, −1)
(1, −2)
(0, −3)

61. **62.** **63.**

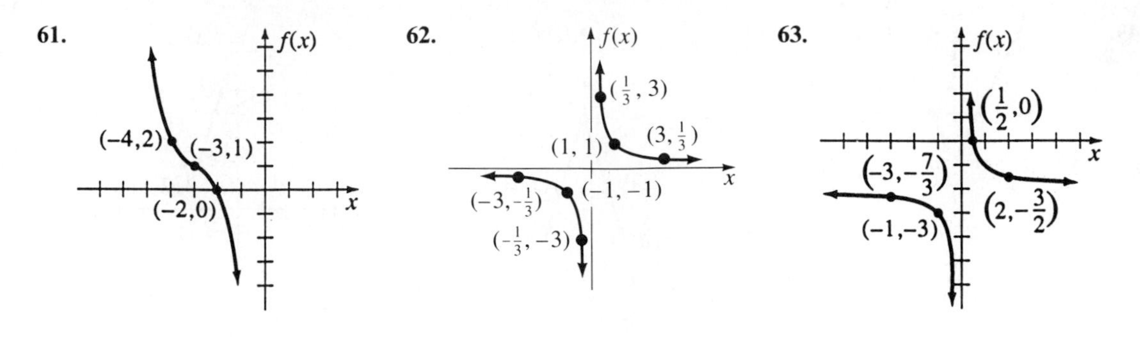

64.

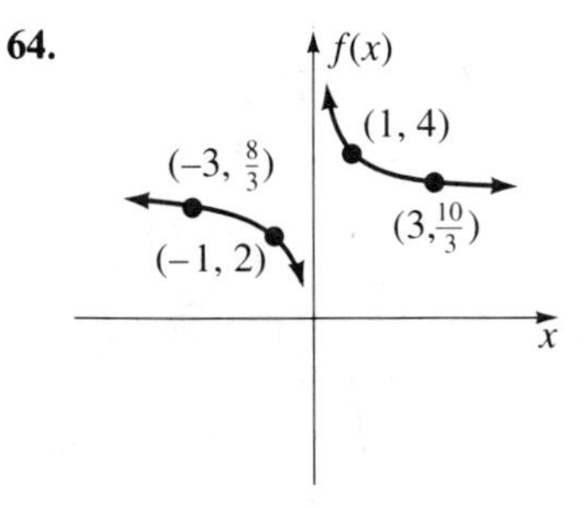

65.

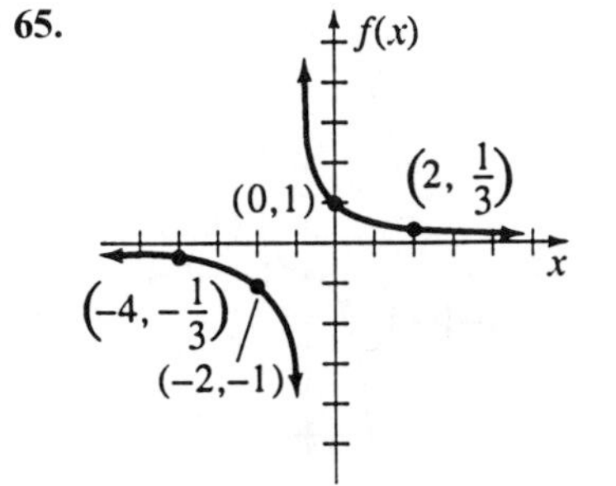

66.

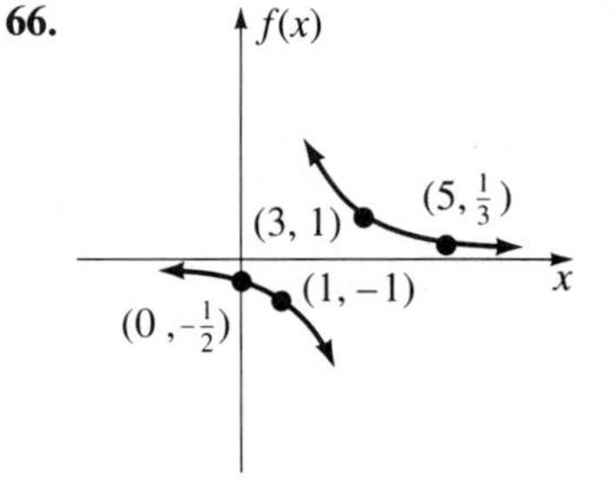

67. **68.**

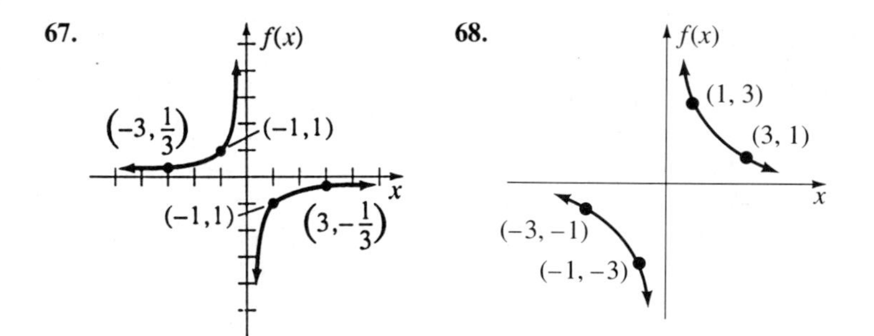

Problem Set 8.3 (page 403)

1. 80 **3.** 5 and 25 **5.** 60 meters by 60 meters **7.** 1100 subscribers at \$13.75 per month

9. 124 and -130 **11.** 323 and 257 **13.** $\frac{1}{2}$ and -1 **15.** undefined and undefined

17. $\sqrt{7}$ and 0 **19.** 37 and 27 **21.** $(f \circ g)(x) = 15x - 3$, $D = \{\text{all reals}\}$
$(g \circ f)(x) = 15x - 1$, $D = \{\text{all reals}\}$

23. $(f \circ g)(x) = -14x - 7$, $D = \{\text{all reals}\}$
$(g \circ f)(x) = -14x + 11$, $D = \{\text{all reals}\}$

25. $(f \circ g)(x) = 3x^2 + 11$, $D = \{\text{all reals}\}$
$(g \circ f)(x) = 9x^2 + 12x + 7$, $D = \{\text{all reals}\}$

27. $(f \circ g)(x) = 2x^2 - 11x + 17$, $D = \{\text{all reals}\}$
$(g \circ f)(x) = -2x^2 + x + 1$, $D = \{\text{all reals}\}$

29. $(f \circ g)(x) = \dfrac{3}{4x - 9}$, $D = \left\{x \mid x \neq \dfrac{9}{4}\right\}$
$(g \circ f)(x) = \dfrac{12 - 9x}{x}$, $D = \{x \mid x \neq 0\}$

31. $(f \circ g)(x) = \sqrt{5x + 4}$, $D = \left\{x \mid x \geq -\dfrac{4}{5}\right\}$
$(g \circ f)(x) = 5\sqrt{x + 1} + 3$, $D = \{x \mid x \geq -1\}$

33. $(f \circ g)(x) = x - 4$, $D = \{x \mid x \neq 4\}$
$(g \circ f)(x) = \dfrac{x}{1 - 4x}$, $D = \left\{x \mid x \neq 0 \text{ and } x \neq \dfrac{1}{4}\right\}$

35. $(f \circ g)(x) = \dfrac{2\sqrt{x}}{x}$, $D = \{x \mid x > 0\}$
$(g \circ f)(x) = \dfrac{4\sqrt{x}}{x}$, $D = \{x \mid x > 0\}$

37. $(f \circ g)(x) = \dfrac{3x + 3}{2}$, $D = \{x \mid x \neq -1\}$
$(g \circ f)(x) = \dfrac{2x}{2x + 3}$, $D = \left\{x \mid x \neq 0 \;\; x \neq -\dfrac{3}{2}\right\}$

Problem Set 8.4 (page 411)

1. Not a function **3.** Function **5.** Function **7.** Function **9.** One-to-one function
11. Not a one-to-one function **13.** Not a one-to-one function **15.** One-to-one function

17. Domain of f: $\{1, 2, 3, 4\}$
Range of f: $\{3, 6, 11, 18\}$
f^{-1}: $\{(3, 1), (6, 2), (11, 3), (18, 4)\}$
Domain of f^{-1}: $\{3, 6, 11, 18\}$
Range of f^{-1}: $\{1, 2, 3, 4\}$

19. Domain of f: $\{-2, -1, 0, 5\}$
Range of f: $\{-1, 1, 5, 10\}$
f^{-1}: $\{(-1, -2), (1, -1), (5, 0), (10, 5)\}$
Domain of f^{-1}: $\{-1, 1, 5, 10\}$
Range of f^{-1}: $\{-2, -1, 0, 5\}$

21. $f^{-1}(x) = \dfrac{x + 4}{5}$ **23.** $f^{-1}(x) = \dfrac{1 - x}{2}$

25. $f^{-1}(x) = \dfrac{5}{4}x$ **27.** $f^{-1}(x) = 2x - 8$ **29.** $f^{-1}(x) = \dfrac{15x + 6}{5}$ **31.** $f^{-1}(x) = \dfrac{x - 4}{9}$

33. $f^{-1}(x) = \dfrac{-x - 4}{5}$ **35.** $f^{-1}(x) = \dfrac{-3x + 21}{2}$ **37.** $f^{-1}(x) = \dfrac{3}{4}x + \dfrac{3}{16}$

39. $f^{-1}(x) = -\dfrac{7}{3}x - \dfrac{14}{9}$ **41.** $f^{-1}(x) = \dfrac{1}{4}x$ **43.** $f^{-1}(x) = -3x$ **45.** $f^{-1}(x) = \dfrac{x + 3}{3}$

47. $f^{-1}(x) = \dfrac{-x - 4}{2}$ **49.** $f^{-1}(x) = \sqrt{x}$, $x \geq 0$

51. Every nonconstant linear function is a one-to-one function.

Problem Set 8.5 (page 420)

1. $y = \dfrac{k}{x^2}$ **3.** $C = \dfrac{kg}{t^3}$ **5.** $V = kr^3$ **7.** $S = ke^2$ **9.** $V = khr^2$ **11.** $\dfrac{2}{3}$ **13.** -4

15. $\dfrac{1}{3}$ **17.** -2 **19.** 2 **21.** 5 **23.** 9 **25.** 9 **27.** $\dfrac{1}{6}$ **29.** 112

31. 12 cubic centimeters **33.** 28 **35.** 2 seconds **37.** 12 ohms **39.** **(a)** \$210
(c) \$1050 **41.** 3560.76 cubic meters **43.** .048

Chapter 8 Review Problem Set (page 424)

1. $D = \{1, 2, 4\}$ **2.** $D = \{x \mid x \neq 5\}$ **3.** $D = \{x \mid x \neq 0 \text{ and } x \neq -4\}$
4. $D = \{x \mid x \geq 5 \text{ or } x \leq -5\}$ **5.** $f(2) = -1$, $f(-3) = 14$; $f(a) = a^2 - 2a - 1$ **6.** $4a + 2h + 1$

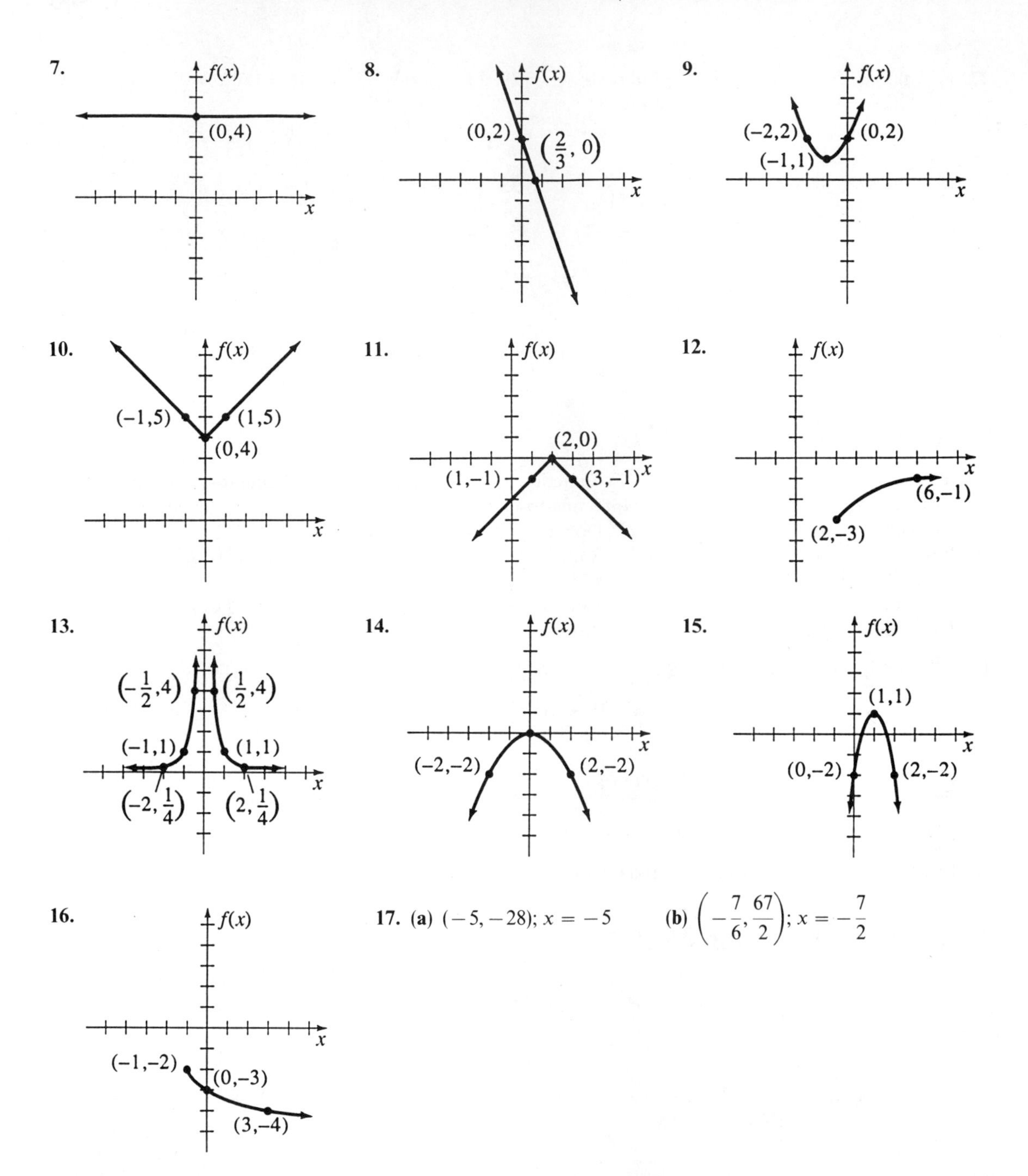

17. **(a)** $(-5, -28)$; $x = -5$ **(b)** $\left(-\frac{7}{6}, \frac{67}{2}\right)$; $x = -\frac{7}{2}$

18. $(f \circ g)(x) = 6x - 11$ and $(g \circ f)(x) = 6x - 13$

19. $(f \circ g)(x) = x^2 - 2x - 1$ and $(g \circ f)(x) = x^2 - 10x + 27$

20. $(f \circ g)(x) = 4x^2 - 20x + 20$ and $(g \circ f)(x) = -2x^2 + 15$

21. $f^{-1}(x) = \dfrac{x+1}{6}$

22. $f^{-1}(x) = \frac{3x - 21}{2}$ **23.** $f^{-1}(x) = \frac{-35x - 10}{21}$ **24.** $k = 9$ **25.** $y = 120$ **26.** 128 pounds
27. 20 and 20 **28.** 3 and 47 **29.** 25 students **30.** 600 square inches

CHAPTER 9

Problem Set 9.1 (page 433)

1. $\{(3, 2)\}$ **3.** $\{(-2, 1)\}$ **5.** Dependent **7.** $\{(4, -3)\}$ **9.** Inconsistent **11.** $\{(8, 12)\}$
13. $\{(-4, -6)\}$ **15.** $\{(-9, 3)\}$ **17.** $\{(1, 3)\}$ **19.** $\left\{\left(5, \frac{3}{2}\right)\right\}$ **21.** $\left\{\left(\frac{9}{5}, -\frac{7}{25}\right)\right\}$
23. $\{(-2, -4)\}$ **25.** $\{(5, 2)\}$ **27.** $\{(-4, -8)\}$ **29.** $\left\{\left(\frac{11}{20}, \frac{7}{20}\right)\right\}$ **31.** $\{(-1, 5)\}$
33. $\left\{\left(-\frac{3}{4}, -\frac{6}{5}\right)\right\}$ **35.** $\left\{\left(\frac{5}{27}, -\frac{26}{27}\right)\right\}$ **37.** \$2000 at 7% and \$8000 at 8% **39.** 92
41. 34 and 97 **43.** 42 females **45.** 20 inches by 27 inches
47. 60 five-dollar bills and 40 ten-dollar bills **49.** 2500 student tickets and 500 nonstudent tickets

Problem Set 9.2 (page 442)

1. $\{(4, -3)\}$ **3.** $\{(-1, -3)\}$ **5.** $\{(-8, 2)\}$ **7.** $\{(-4, 0)\}$ **9.** $\{(1, -1)\}$ **11.** Inconsistent
13. $\left\{\left(-\frac{1}{11}, \frac{4}{11}\right)\right\}$ **15.** $\left\{\left(\frac{3}{2}, -\frac{1}{3}\right)\right\}$ **17.** $\{(4, -9)\}$ **19.** $\{(7, 0)\}$ **21.** $\{(7, 12)\}$
23. $\left\{\left(\frac{7}{11}, \frac{2}{11}\right)\right\}$ **25.** Inconsistent **27.** $\left\{\left(\frac{51}{31}, -\frac{32}{31}\right)\right\}$ **29.** $\{(-2, -4)\}$ **31.** $\left\{\left(-1, -\frac{14}{3}\right)\right\}$
33. $\{(-6, 12)\}$ **35.** $\{(2, 8)\}$ **37.** $\{(-1, 3)\}$ **39.** $\{(16, -12)\}$ **41.** $\left\{\left(-\frac{3}{4}, \frac{3}{2}\right)\right\}$
43. $\{(5, -5)\}$ **45.** 5 gallons of 10% solution and 15 gallons of 20% solution
47. \$1 for a tennis ball and \$2 for a golf ball **49.** 40 double rooms and 15 single rooms **51.** 9 feet
53. $\frac{3}{4}$ **55.** 18 centimeters by 24 centimeters **57.** 8 feet **59.** **(a)** $\{(4, 6)\}$ **(c)** $\{(2, -3)\}$
(e) $\left\{\left(\frac{1}{4}, -\frac{2}{3}\right)\right\}$

Problem Set 9.3 (page 454)

1. $\{(-2, 5, 2)\}$ **3.** $\{(4, -1, -2)\}$ **5.** $\{(-1, 3, 5)\}$ **7.** Infinitely many solutions **9.** $\varnothing$
11. $\left\{\left(-2, \frac{3}{2}, 1\right)\right\}$ **13.** $\left\{\left(\frac{1}{3}, -\frac{1}{2}, 1\right)\right\}$ **15.** $\left\{\left(\frac{2}{3}, -4, \frac{3}{4}\right)\right\}$ **17.** $\{(-2, 4, 0)\}$ **19.** $\left\{\left(\frac{1}{2}, \frac{1}{3}, \frac{1}{6}\right)\right\}$
21. 194 **23.** \$.70 per bottle for catsup, \$1 per jar of peanut butter, and \$.80 per jar of pickles
25. -2, 6, and 16 **27.** 40°, 60°, and 80° **29.** \$500 at 12%, \$1000 at 13%, and \$1500 at 14%

Problem Set 9.4 (page 460)

1. $\{(6, -4)\}$ **3.** $\{(-4, -4)\}$ **5.** $\left\{\left(-\frac{11}{7}, -\frac{13}{7}\right)\right\}$ **7.** $\{(1, -3)\}$ **9.** $\{(-1, -2, -3)\}$
11. $\{(3, 1, 4)\}$ **13.** $\{(4, 3, -2)\}$ **15.** $\{(-5, 2, 0)\}$ **17.** $\{(-2, -1, 1)\}$ **19.** $\{(-1, 4, -1)\}$
21. $\{(2, 0, -3)\}$ **23.** $\{(1, -1, 2, -3)\}$ **25.** $\varnothing$

Problem Set 9.5 (page 466)

1. 10 **3.** -48 **5.** -29 **7.** 50 **9.** 8 **11.** 6 **13.** -32 **15.** -2 **17.** 5

19. $-\frac{7}{20}$ **21.** $\{(3,8)\}$ **23.** $\{(-2,5)\}$ **25.** $\{(-2,2)\}$ **27.** $\varnothing$ **29.** $\{(1,6)\}$

31. $\left\{\left(-\frac{1}{4},\frac{2}{3}\right)\right\}$ **33.** $\left\{\left(\frac{17}{62},\frac{4}{31}\right)\right\}$ **35.** $\{(-1,3)\}$ **37.** $\{(9,-2)\}$ **39.** $\{(-4,-3)\}$

Problem Set 9.6 (page 473)

1. -57 **3.** 14 **5.** -41 **7.** -8 **9.** -96 **11.** $\{(-3,1,-1)\}$ **13.** $\{(0,2,-3)\}$

15. $\left\{\left(-2,\frac{1}{2},-\frac{2}{3}\right)\right\}$ **17.** $\{(-1,-1,-1)\}$ **19.** $\varnothing$ **21.** $\{(-5,-2,1)\}$ **23.** $\{(-2,3,-4)\}$

25. $\left\{\left(-\frac{1}{2},0,\frac{2}{3}\right)\right\}$ **27.** $\{(-6,7,1)\}$ **29.** $\left\{\left(1,-6,-\frac{1}{2}\right)\right\}$ **31.** 0 **33.** **(b)** -20

Problem Set 9.7 (page 481)

1. $\{(-2,0),(-4,4)\}$ **3.** $\{\equiv 2,-3),(-2,3)\}$ **5.** $\{(-6,7),(-2,-1)\}$ **7.** $\{(1,1)\}$ **9.** $\{(-5,-3)\}$

11. $\{(-3,2)\}$ **13.** $\{(2,2),(-2,-2)\}$

15. $\left\{\left(\frac{\sqrt{30}}{3},\frac{\sqrt{21}}{3}\right),\left(\frac{\sqrt{30}}{3},-\frac{\sqrt{21}}{3}\right),\left(-\frac{\sqrt{30}}{3},\frac{\sqrt{21}}{3}\right),\left(-\frac{\sqrt{30}}{3},-\frac{\sqrt{21}}{3}\right)\right\}$

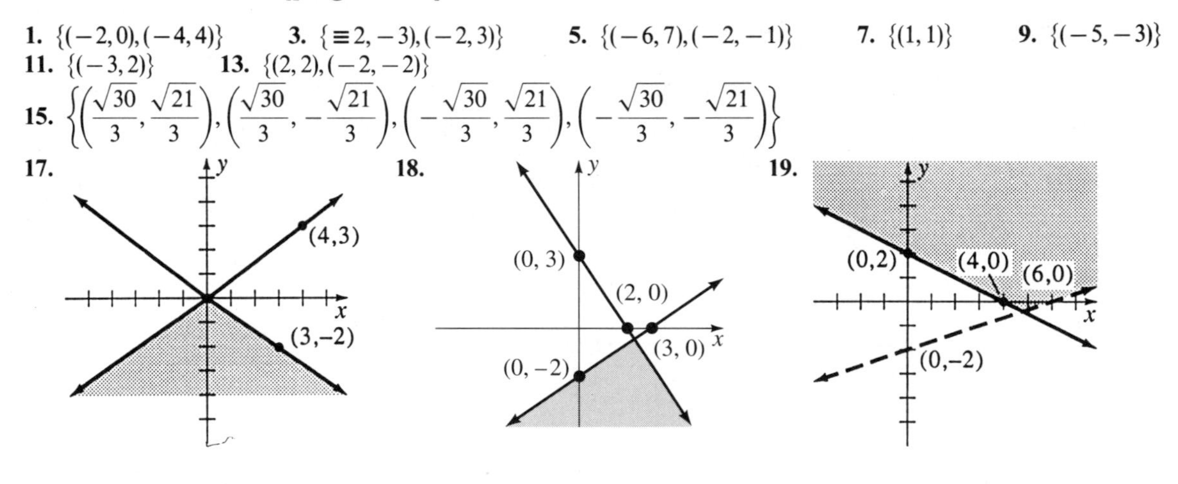

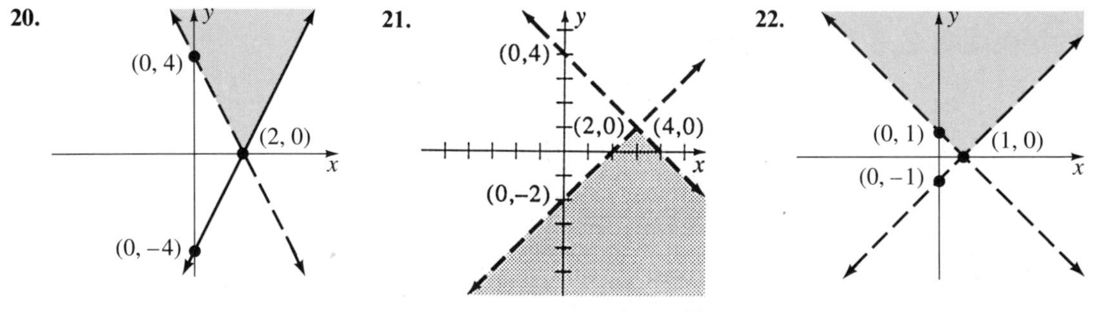

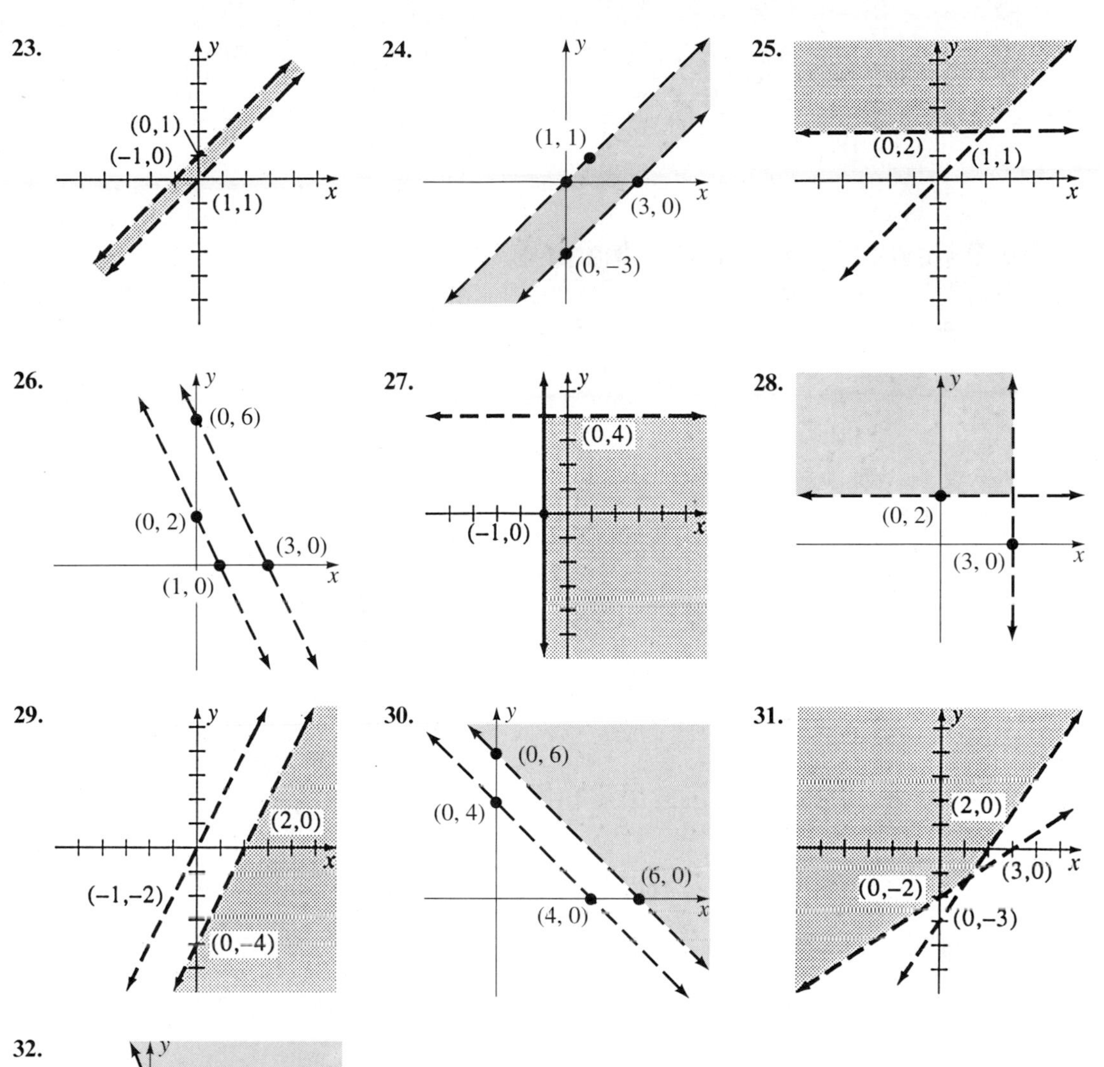
23.
(0,1)
(−1,0)
(1,1)
24.
(1, 1)
(3, 0)
(0, −3)
25.
(0,2)
(1,1)
26.
(0, 6)
(0, 2)
(3, 0)
(1, 0)
27.
(0,4)
(−1,0)
28.
(0, 2)
(3, 0)
29.
(2,0)
(−1,−2)
(0,−4)
30.
(0, 6)
(0, 4)
(6, 0)
(4, 0)
31.
(2,0)
(3,0)
(0,−2)
(0,−3)

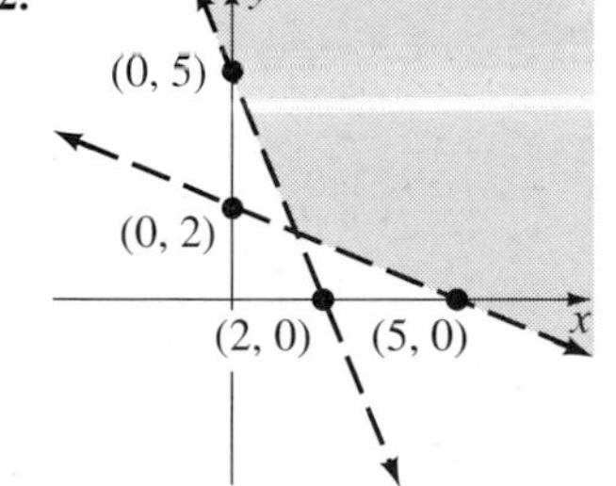
32.
(0, 5)
(0, 2)
(2, 0)
(5, 0)

33. **(a)** $\{(2i, -3), (-2i, -3)\}$ **(c)** $\left\{\left(\frac{1+i\sqrt{15}}{2}, \frac{-7+i\sqrt{15}}{2}\right), \left(\frac{1-i\sqrt{15}}{2}, \frac{-7-i\sqrt{15}}{2}\right)\right\}$

(e) $\left\{\left(\frac{2+i\sqrt{2}}{2}, \frac{2-i\sqrt{2}}{2}\right)\right\}, \left(\frac{2-i\sqrt{2}}{2}, \frac{2+i\sqrt{2}}{2}\right)\right\}$

Chapter 9 Review Problem Set (page 486)

1. $\{(2, 6)\}$ **2.** $\{(-3, 7)\}$ **3.** $\{(-9, 8)\}$ **4.** $\left\{\left(\frac{89}{23}, -\frac{12}{23}\right)\right\}$ **5.** $\{(4, -7)\}$ **6.** $\{(-2, 8)\}$

7. $\{(0, 3)\}$ **8.** $\{(6, 7)\}$ **9.** $\{(1, -6)\}$ **10.** $\{(-4, 0)\}$ **11.** $\{(-12, 18)\}$ **12.** $\{(24, 8)\}$

13. $\left\{\left(\frac{2}{3}, -\frac{3}{4}\right)\right\}$ **14.** $\left\{\left(-\frac{1}{2}, \frac{3}{5}\right)\right\}$ **15.** -18 **16.** 11 **17.** -29 **18.** 59

19. $\{(2, -4, -6)\}$ **20.** $\{(-1, 5, -7)\}$ **21.** $\{(2, -3, 1)\}$ **22.** $\{(2, -1, -2)\}$

23. $\left\{\left(-\frac{1}{3}, -1, 4\right)\right\}$ **24.** $\{(0, -2, -4)\}$

25. $\{(1, 1), (-2, 7)\}$ **26.**

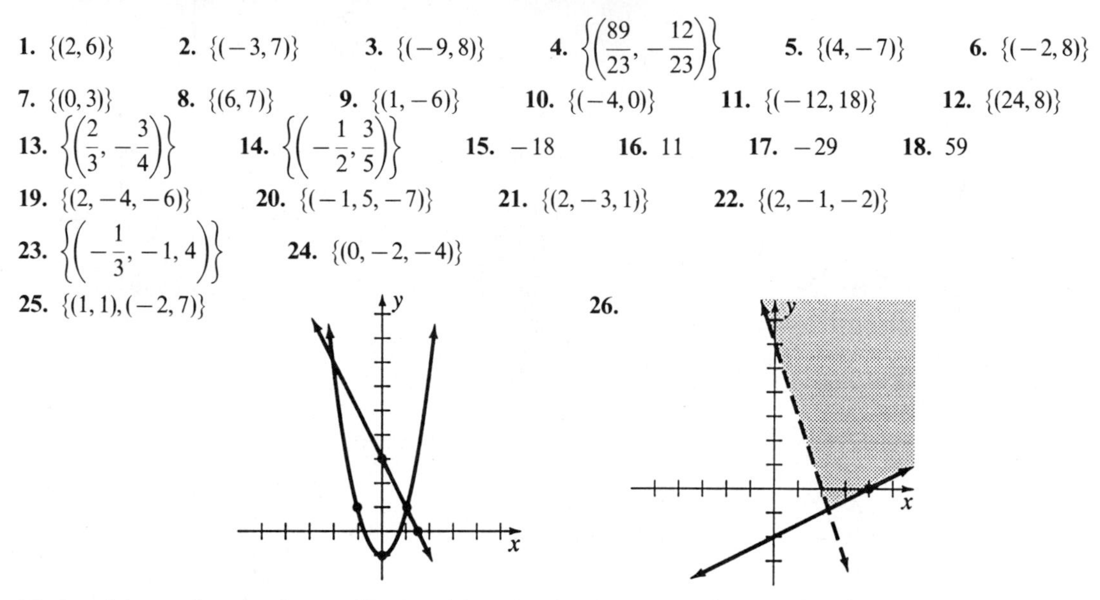

27. 2 and 3 or -3 and -2 **28.** 5 and 3 or 5 and -3 or -5 and 3 or -5 and -3

29. 2 and 5 or -3 and 10 **30.** 6 meters by 9 meters

31. \$9 per pound for cashews and \$5 per pound for Spanish peanuts

32. \$1.50 for a carton of pop and \$2.25 for a pound of candy

33. The fixed fee is \$2 and the additional fee is \$.10 per pound.

Cumulative Review Problem Set

65. **66.** **67.**

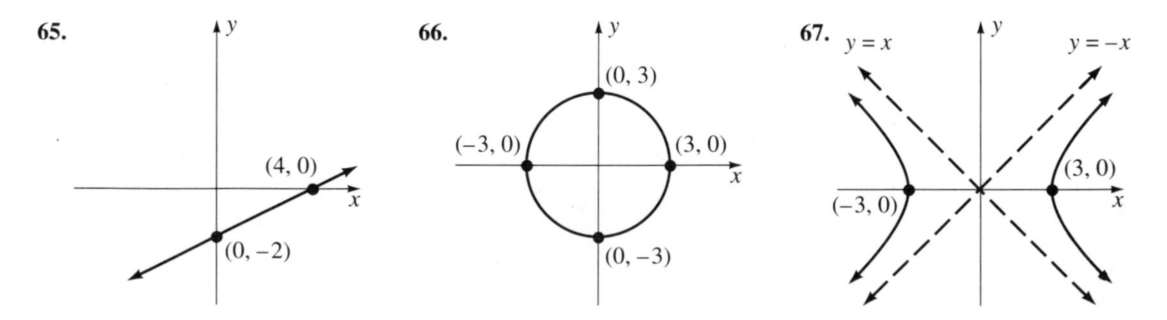

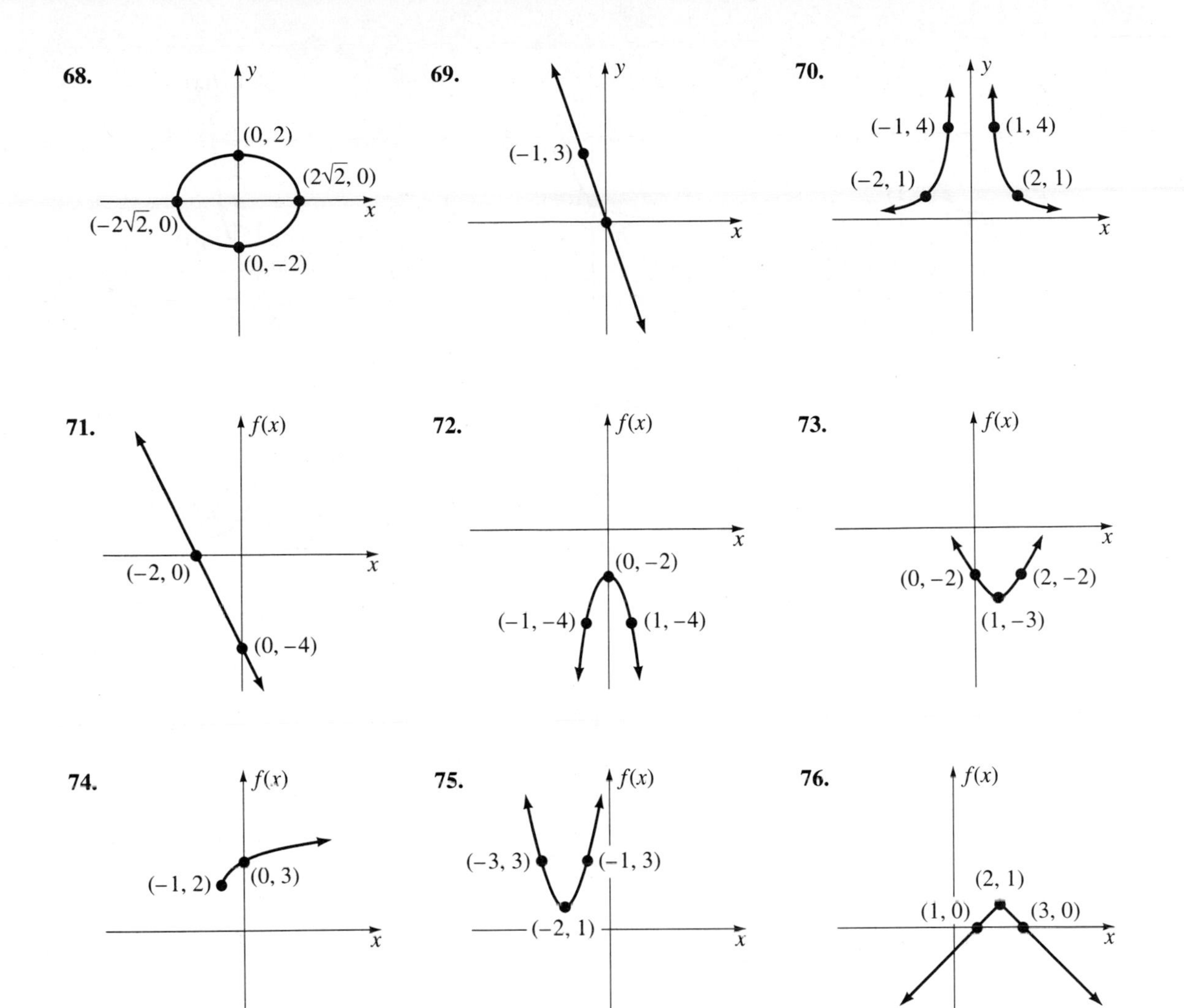

CHAPTER 10

Problem Set 10.1 (page 499)

1. $\{3\}$ **3.** $\{2\}$ **5.** $\{4\}$ **7.** $\{1\}$ **9.** $\{5\}$ **11.** $\{1\}$ **13.** $\left\{\frac{3}{2}\right\}$ **15.** $\left\{\frac{5}{6}\right\}$

17. $\{-3\}$ **19.** $\{-3\}$ **21.** $\{0\}$ **23.** $\{1\}$ **25.** $\{-1\}$ **27.** $\left\{-\frac{2}{5}\right\}$ **29.** $\left\{\frac{5}{2}\right\}$

31. $\{3\}$ **33.** $\left\{\frac{1}{2}\right\}$

35. **36.** **37.**

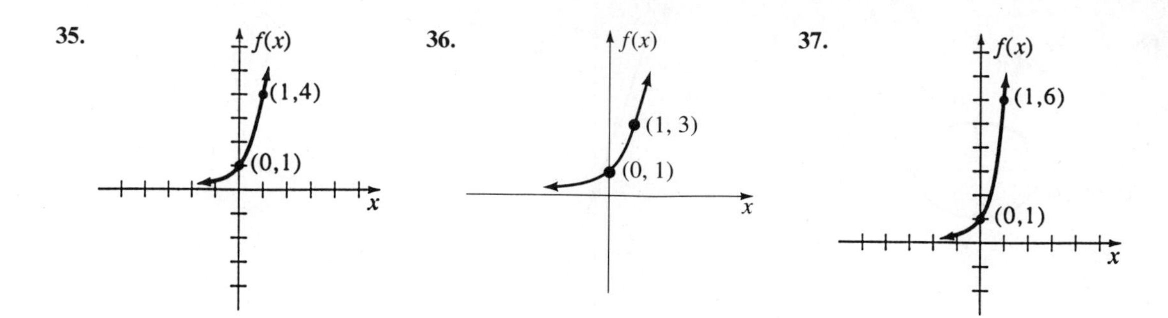

38.

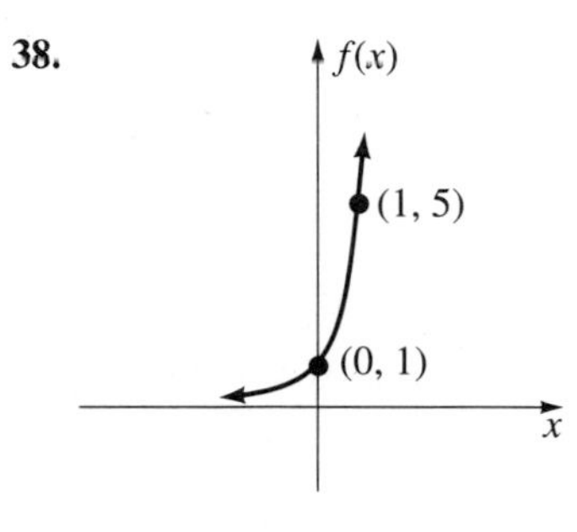

39.

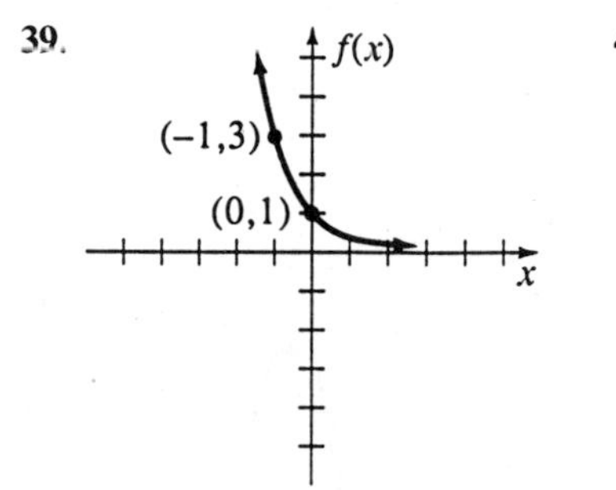

40.

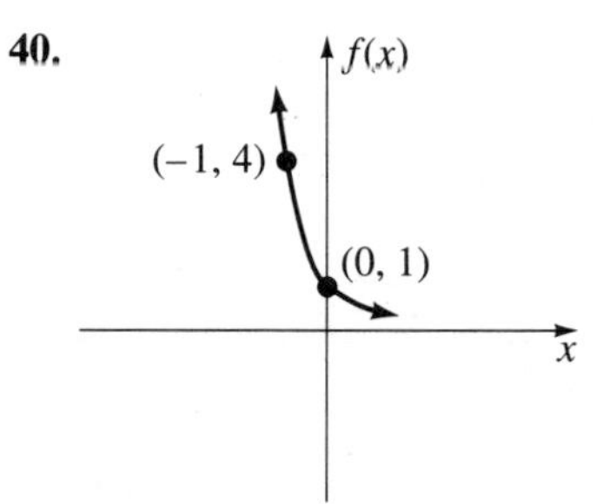

41.

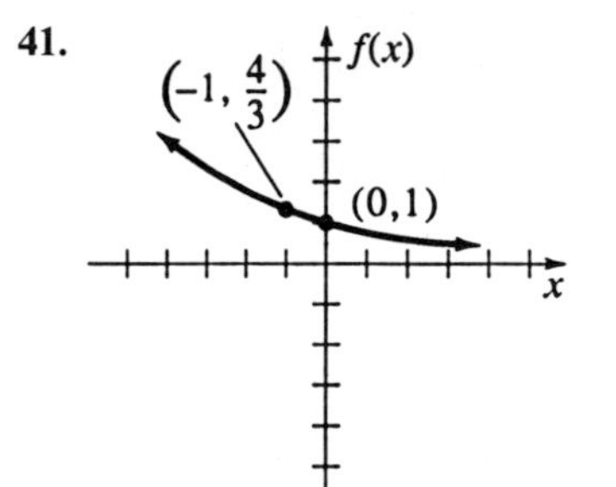

42.

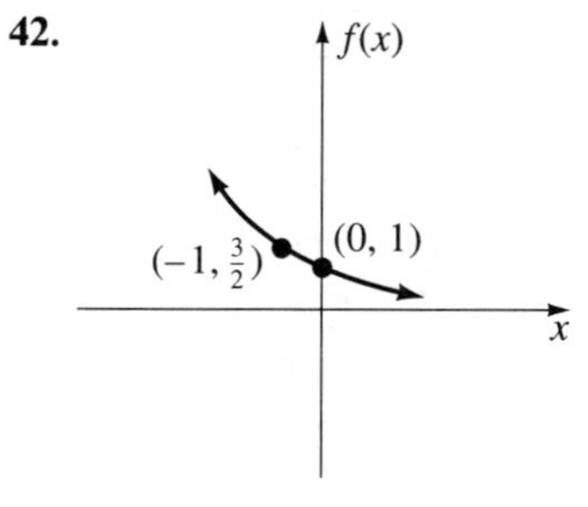

43.

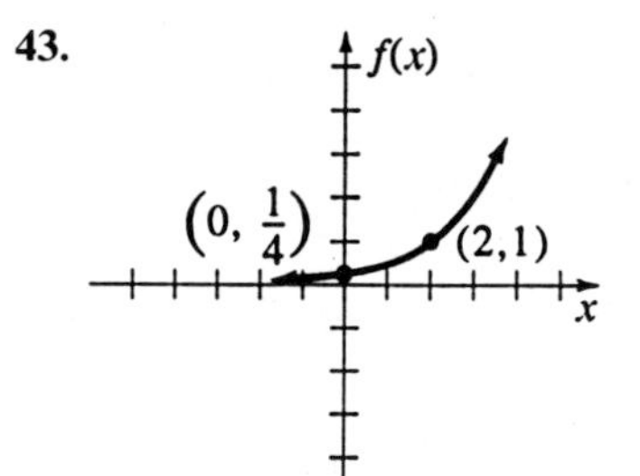

44.

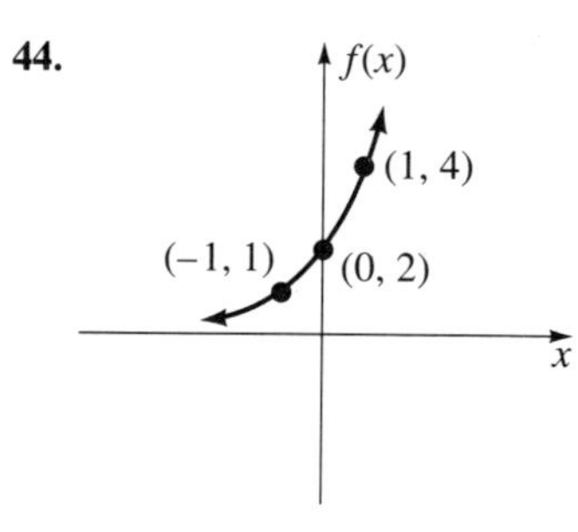

45.

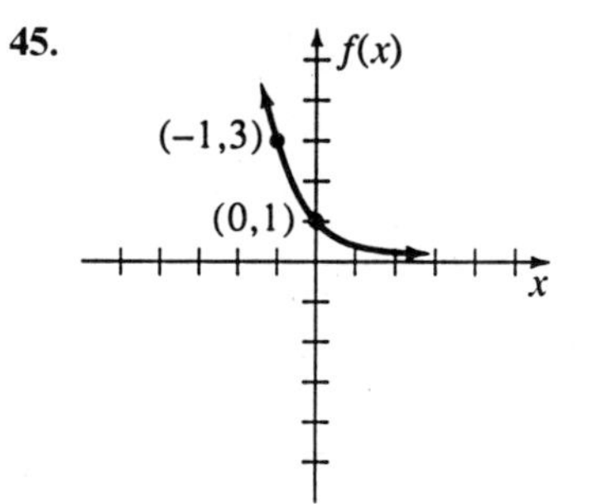

46.

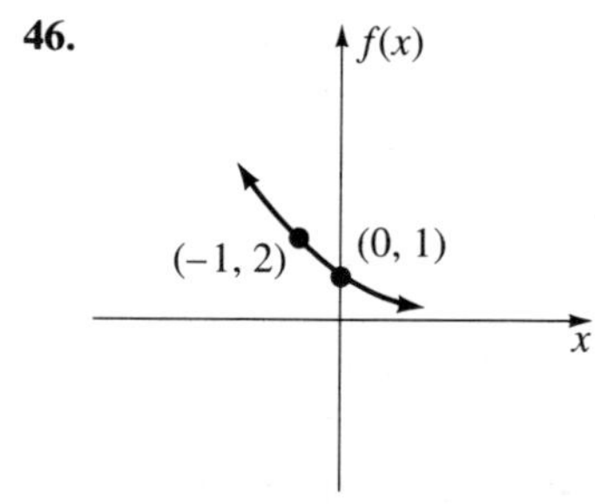

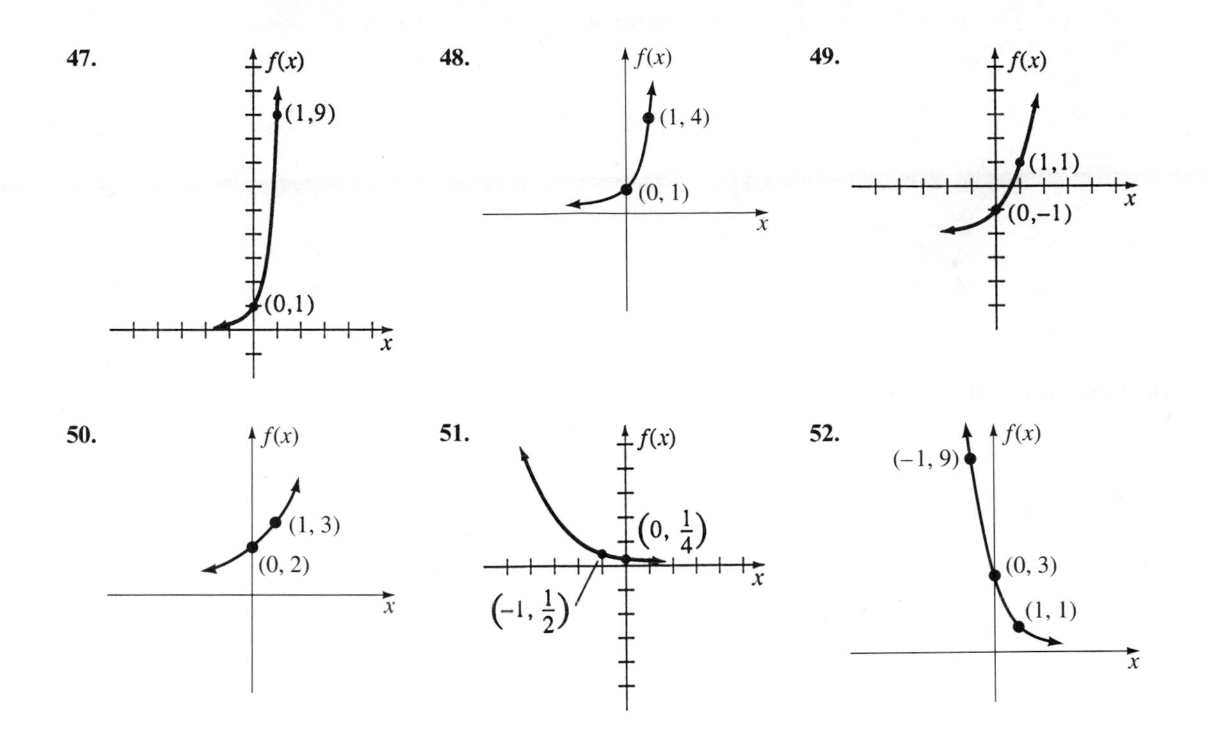

Problem Set 10.2 (page 506)

1. **(a)** \$.62 **(c)** \$2.06 **(e)** \$10,950 **(g)** \$658 **3.** \$283.70 **5.** \$865.84 **7.** \$1782.25
9. \$2725.05 **11.** \$16,998.71 **13.** \$22,553.65 **15.** \$567.63 **17.** \$1422.36
19. \$8963.38 **21.** \$17,547.35 **23.** \$32,558.88

25.

	1 yr	5 yrs	10 yrs	20 yrs
compounded annually	\$1120	1762	3106	9646
compounded semiannually	1124	1791	3207	10,286
compounded quarterly	1126	1806	3262	10,641
compounded monthly	1127	1817	3300	10,893
compounded continuously	1127	1822	3320	11,023

27. Nora will have \$.24 more. **29.** 9.54% **31.** 50 grams; 37 grams

33. f(x) (0,2) x

34. f(x) (1, .7) (–1, –1.6) (0, –1) x

35. f(x) (0,2) x

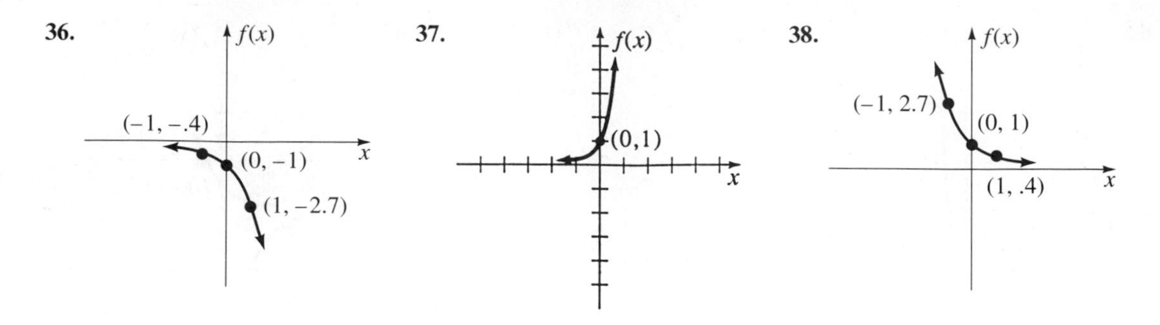

39. 2226; 3320; 7389 **41.** 2000 **43.** **(a)** 82,888 **(c)** 96,302
45. **(a)** 6.5 pounds per square inch **(c)** 13.6 pounds per square inch

Problem Set 10.3 (page 516)

1. $\log_2 128 = 7$ **3.** $\log_5 125 = 3$ **5.** $\log_{10} 1000 = 3$ **7.** $\log_2\left(\frac{1}{4}\right) = -2$ **9.** $\log_{10} .1 = -1$

11. $3^4 = 81$ **13.** $4^3 = 64$ **15.** $10^4 = 10000$ **17.** $2^{-4} = \frac{1}{16}$ **19.** $10^{-3} = .001$ **21.** 4

23. 4 **25.** 3 **27.** $\frac{1}{2}$ **29.** 0 **31.** -1 **33.** 5 **35.** -5 **37.** 1 **39.** 0

41. {49} **43.** {16} **45.** {27} **47.** $\left\{\frac{1}{8}\right\}$ **49.** {4} **51.** 5.1293 **53.** 6.9657

55. 1.4037 **57.** 7.4512 **59.** 6.3219 **61.** −.3791 **63.** .5766 **65.** 2.1531
67. .3949 **69.** $\log_b x + \log_b y + \log_b z$ **71.** $\log_b y - \log_b z$ **73.** $3\log_b y + 4\log_b z$
75. $\frac{1}{2}\log_b x + \frac{1}{3}\log_b y - 4\log_b z$ **77.** $\frac{2}{3}\log_b x + \frac{1}{3}\log_b z$ **79.** $\frac{3}{2}\log_b x - \frac{1}{2}\log_b y$ **81.** $\left\{\frac{9}{4}\right\}$

83. {25} **85.** {4} **87.** $\left\{\frac{19}{8}\right\}$ **89.** {9} **91.** {1}

Problem Set 10.4 (page 524)

1. 0.8597 **3.** 1.7179 **5.** 3.5071 **7.** −0.1373 **9.** −3.4685 **11.** 411.43
13. 90095 **15.** 79.543 **17.** 0.048440 **19.** 0.0064150 **21.** 1.6094 **23.** 3.4843
25. 6.0638 **27.** −0.7765 **29.** −3.4609 **31.** 1.6034 **33.** 3.1346 **35.** 108.56
37. 0.48268 **39.** 0.035994

41.

f(x)
(4, .6) (8, .9) (10, 1)
(2, .3)
(1, 0)
(.1, −1)
x

42.

f(x)
(2, .7) (4, 1.4) (8, 2.1) (10, 2.3)
(1, 0)
(.5, −.7)
(.1, −2.3)
x

43.

44. This graph is the same as Figure 10.5 on page 519.

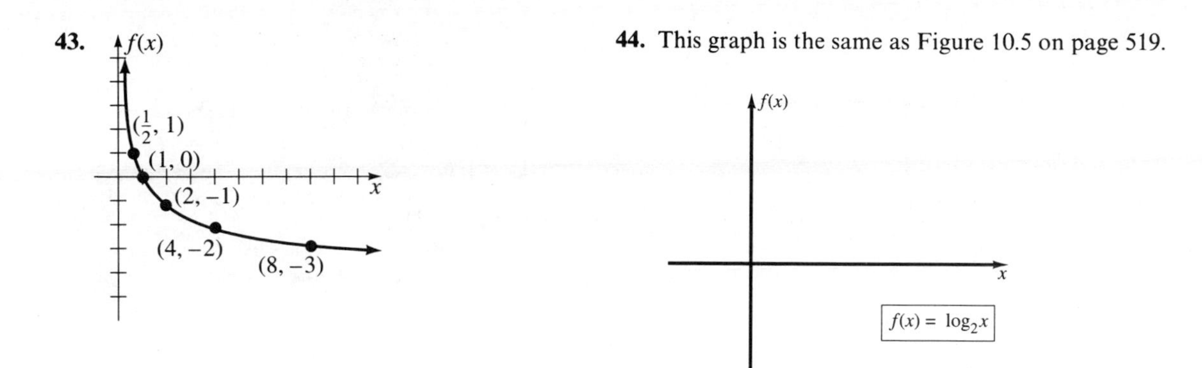

45.

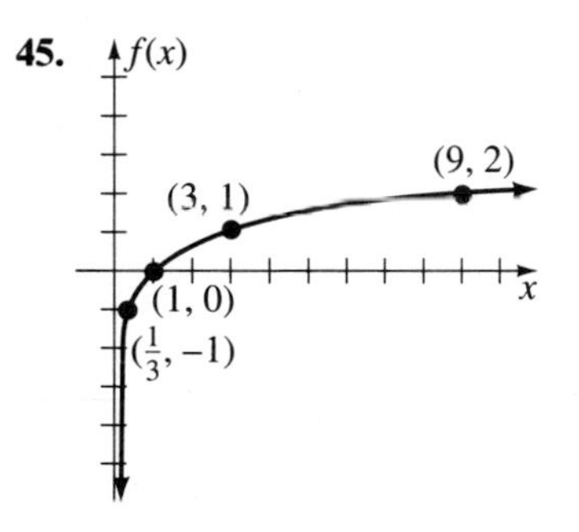

46.

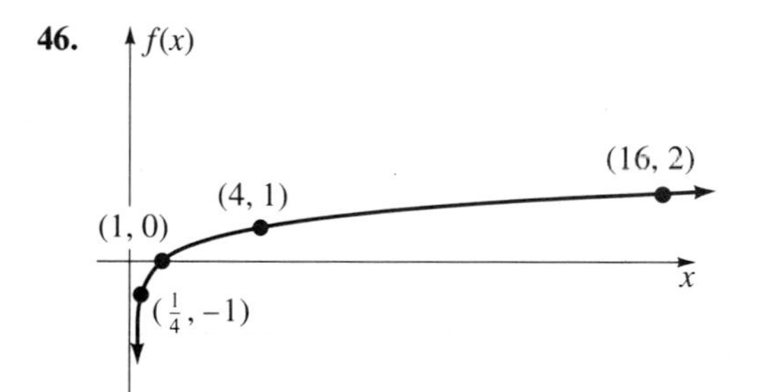

47.

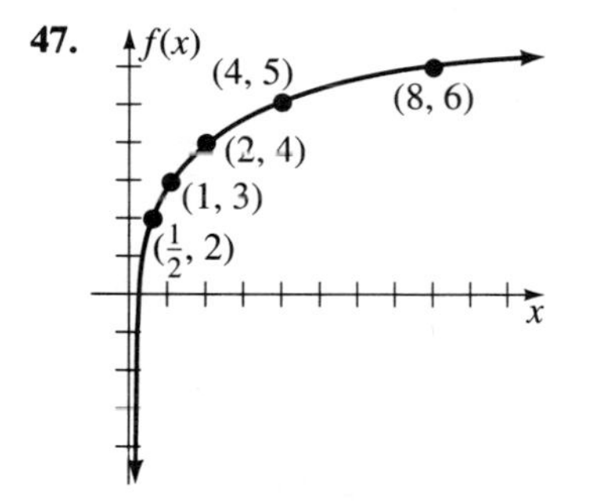

48.

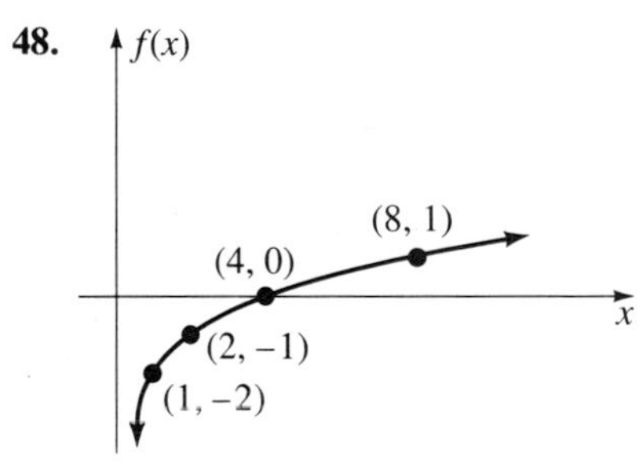

49.

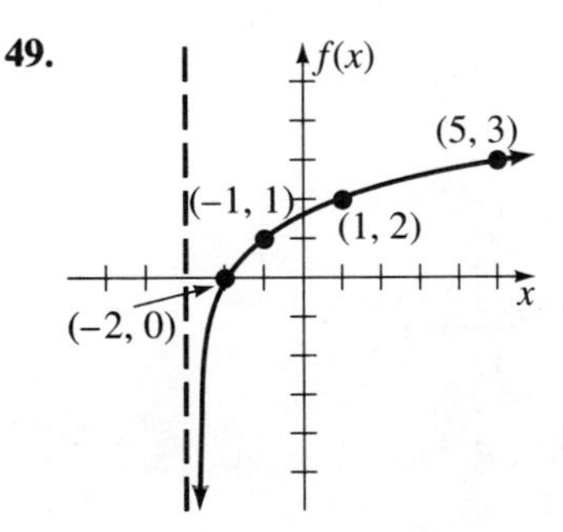

50.

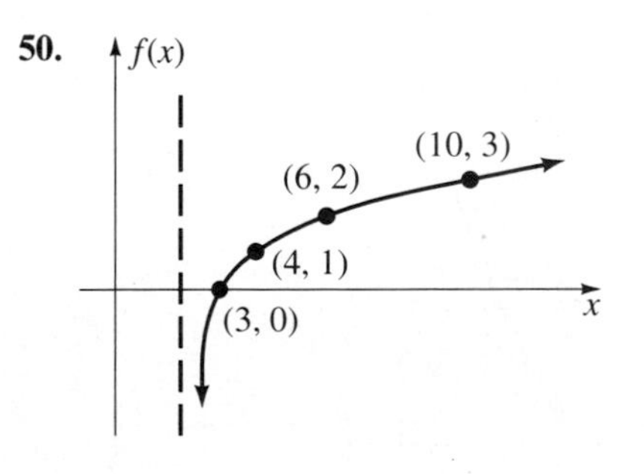

51. 52. 53.

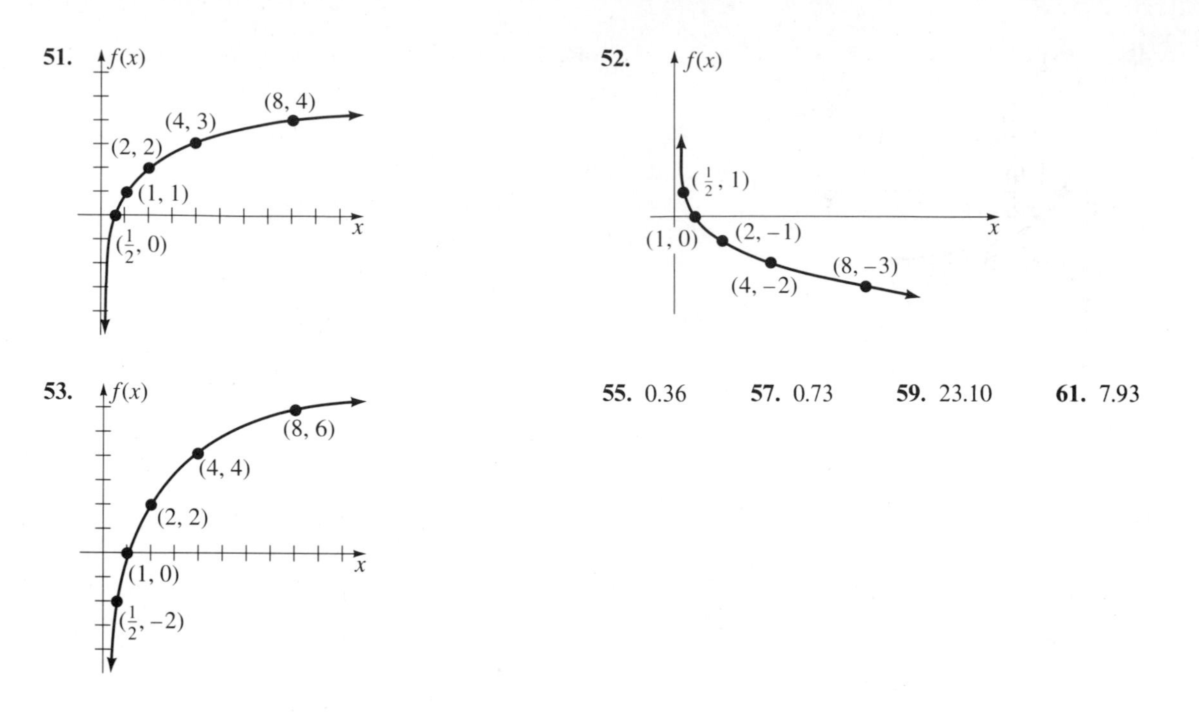

55. 0.36 **57.** 0.73 **59.** 23.10 **61.** 7.93

Problem Set 10.5 (page 532)

1. $\{3.15\}$ **3.** $\{2.20\}$ **5.** $\{4.18\}$ **7.** $\{0.12\}$ **9.** $\{1.69\}$ **11.** $\{4.57\}$ **13.** $\{2.46\}$
15. $\{4\}$ **17.** $\left\{\frac{19}{47}\right\}$ **19.** $\{1\}$ **21.** $\{8\}$ **23.** 4.524 **25.** −.860 **27.** 3.105
29. −2.902 **31.** 5.989 **33.** 2.4 years **35.** 5.3 years **37.** 6.8 hours **39.** 1.5 hours
41. 34.7 years **43.** 6.7 **45.** Approximately 8 times

Chapter 10 Review Problem Set (page 535)

1. 7 **2.** 3 **3.** 4 **4.** −3 **5.** 2 **6.** 13 **7.** 0 **8.** −2 **9.** $\{-4\}$
10. $\{3\}$ **11.** $\left\{-\frac{3}{4}\right\}$ **12.** $\{4\}$ **13.** $\{9\}$ **14.** $\left\{\frac{1}{2}\right\}$ **15.** $\{5\}$ **16.** $\{8\}$
17. 1.8642 **18.** 4.7380 **19.** −4.2687 **20.** −2.6289 **21.** 1.1882 **22.** 2639.4
23. .013289 **24.** .077197 **25.** $\{3.40\}$ **26.** $\{1.95\}$ **27.** $\{5.30\}$ **28.** $\{5.61\}$

29. **30.**

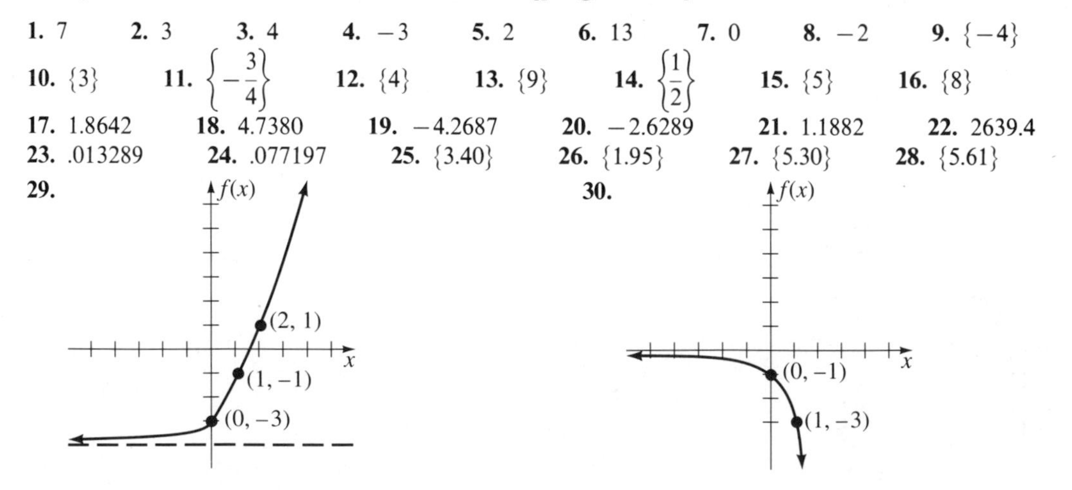

31.

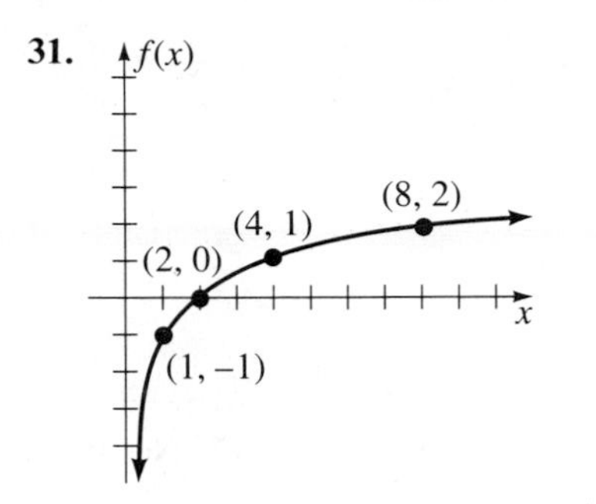

32.

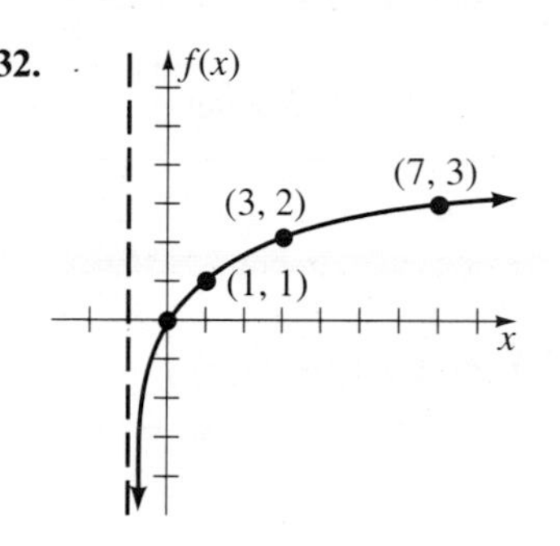

33. 2.842 **34.** \$2913.99 **35.** \$8178.72 **36.** \$5656.26 **37.** 133 grams
38. Approximately 5.3 years **39.** Approximately 12.1 years **40.** 61,070; 67,493; 74,591
41. Approximately 4.8 hours **42.** 8.1

CHAPTER 11

Problem Set 11.1 (page 542)

1. $-1, 2, 5, 8, 11$ **3.** $3, 1, -1, -3, -5$ **5.** $-1, 2, 7, 14, 23$ **7.** $0, -3, -8, -15, -24$
9. $-1, 5, 15, 29, 47$ **11.** $\frac{1}{2}, 1, 2, 4, 8$ **13.** $-\frac{2}{3}, -2, -6, -18, -54$ **15.** $a_8 = 54$ and $a_{12} = 130$
17. $a_7 = -32$ and $a_8 = 64$ **19.** $a_n = 2n - 1$ **21.** $a_n = 4n - 6$ **23.** $a_n = -2n + 7$
25. $a_n = -3n - 4$ **27.** $a_n = \frac{1}{2}(n + 1)$ **29.** 34 **31.** 78 **33.** -155 **35.** 106
37. 50 **39.** 170 **41.** -1 **43.** -13 **45.** 136 **47.** \$27,800 **49.** \$1170

Problem Set 11.2 (page 547)

1. 2550 **3.** 9030 **5.** -4225 **7.** 410 **9.** 8850 **11.** 5724 **13.** 5070
15. 346.5 **17.** 3775 **19.** 39,500 **21.** -8645 **23.** 122,850 **25.** 39,552
27. 76 seats, 720 seats **29.** \$10,000 **31.** \$357,500 **33.** 169 **35.** **(a)** $7 + 12 + 17; 36$
(c) $(-3) + (-5) + (-7) + (-9) + (-11) + (-13); -48$ **(e)** $3 + 6 + 9 + 12 + 15; 45$

Problem Set 11.3 (page 554)

1. $a_n = 3^{n-1}$ **3.** $a_n = 2(4)^{n-1}$ **5.** $a_n = \left(\frac{1}{3}\right)^{n-1}$ or $a_n = \frac{1}{3^{n-1}}$ **7.** $a_n = .2(.2)^{n-1}$ or $a_n = (.2)^n$
9. $a_n = 9\left(\frac{2}{3}\right)^{n-1}$ or $a_n = (3^{-n+3})(2)^{n-1}$ **11.** $a_n = (-4)^{n-1}$ **13.** 19,683 **15.** -512
17. $-\frac{6561}{256}$ **19.** 14,762 **21.** -9842 **23.** 1093 **25.** 511 **27.** 19,680 **29.** 2730
31. 1093 **33.** $1\frac{1023}{1024}$ **35.** $15\frac{31}{32}$ **37.** $\frac{1}{2}$ **39.** $0; -1$ **41.** 125 liters

43. \$102.40; \$204.75 **45.** $971.\overline{3}$ meters **47.** \$5609.66 **49.** 1,048,576 **51.** $\frac{4096}{531{,}441}$

53. $-\frac{177{,}147}{2048}$ **55.** (a) 126 (c) 31 (e) $1\frac{49}{81}$

Problem Set 11.4 (page 561)

1. 4 **3.** 1 **5.** 2 **7.** $\frac{2}{3}$ **9.** 9 **11.** No sum **13.** $\frac{4}{7}$ **15.** $\frac{16}{3}$ **17.** No sum

19. $\frac{81}{2}$ **21.** $\frac{4}{9}$ **23.** $\frac{47}{99}$ **25.** $\frac{5}{11}$ **27.** $\frac{427}{999}$ **29.** $\frac{7}{15}$ **31.** $\frac{72}{33}$ **33.** $\frac{47}{110}$

Problem Set 11.5 (page 565)

1. $x^8 + 8x^7y + 28x^6y^2 + 56x^5y^3 + 70x^4y^4 + 56x^3y^5 + 28x^2y^6 + 8xy^7 + y^8$
3. $81x^4 + 108x^3y + 54x^2y^2 + 12xy^3 + y^4$ **5.** $x^5 - 5x^4y + 10x^3y^2 - 10x^2y^3 + 5xy^4 - y^5$
7. $x^{10} + 10x^9y + 45x^8y^2 + 120x^7y^3 + 210x^6y^4 + 252x^5y^5 + 210x^4y^6 + 120x^3y^7 + 45x^2y^8 + 10xy^9 + y^{10}$
9. $64x^6 + 192x^5y + 240x^4y^2 + 160x^3y^3 + 60x^2y^4 + 12xy^5 + y^6$
11. $x^5 - 15x^4y + 90x^3y^2 - 270x^2y^3 + 405xy^4 - 243y^5$
13. $243a^5 - 810a^4b + 1080a^3b^2 - 720a^2b^3 + 240ab^4 - 32b^5$
15. $x^6 + 6x^5y^3 + 15x^4y^6 + 20x^3y^9 + 15x^2y^{12} + 6xy^{15} + y^{18}$
17. $x^7 + 14x^6 + 84x^5 + 280x^4 + 560x^3 + 672x^2 + 448x + 128$ **19.** $x^4 - 12x^3 + 54x^2 - 108x + 81$
21. $x^{15} + 15x^{14}y + 105x^{13}y^2 + 455x^{12}y^3$ **23.** $a^{13} - 26a^{12}b + 312a^{11}b^2 - 2288a^{10}b^3$
25. $462x^5y^6$ **27.** $-160x^3y^3$ **29.** $2000x^3y^2$

Chapter 11 Review Problem Set (page 567)

1. $-1, 1, 5, 13, 29$ **2.** (a) $a_n = 6n - 4$ (b) $a_n = -5n + 8$ (c) $a_n = 4(3)^{n-1}$
(d) $a_n = 5\left(\frac{1}{2}\right)^{n-1}$ **3.** (a) $\frac{1}{4374}$ (b) 155 **4.** (a) 3775 (b) 4092 (c) -1325
5. 90,000 **6.** $\frac{1365}{4096}$ **7.** 1375 **8.** 20,520 **9.** $5333.\overline{3}$ liters **10.** 3600 feet **11.** \$2.10
12. \$409.55 **13.** 32 **14.** $\frac{29}{99}$
15. $128x^7 + 448x^6y + 672x^5y^2 + 560x^4y^3 + 280x^3y^4 + 84x^2y^5 + 14xy^6 + y^7$ **16.** $252a^5b^5$

Practice Exercises for Appendix A (page A6)

1. $(x - 6)$ **2.** $x + 6$ **3.** $x + 6, R = 14$ **4.** $x - 2, R = -1$ **5.** $x^2 - 1$
6. $x^2 - 6x + 8$ **7.** $x^2 - 2x - 3$ **8.** $x^2 + 7x + 2, R = 1$
9. $2x^2 - x - 6, R = -6$ **10.** $3x^3 - 4x^2 + 6x - 13, R = 12$
11. $x^3 + 7x^2 + 21x + 56, R = 167$ **12.** $2x^3 - 4x^2 + 11x - 22, R = 47$

13. **(a)**

$$\begin{array}{r|rrrr} -1 & 1 & 6 & 11 & 6 \\ & & -1 & -5 & -6 \\ \hline & 1 & 5 & 6 & 0 \end{array}$$

Therefore, $(x + 1)$ is a factor.

$$\begin{array}{r|rrrr} -2 & 1 & 6 & 11 & 6 \\ & & -2 & -8 & -6 \\ \hline & 1 & 4 & 3 & 0 \end{array}$$

Therefore, $(x + 2)$ is a factor.

$$\begin{array}{r|rrrr} -3 & 1 & 6 & 11 & 6 \\ & & -3 & -9 & -6 \\ \hline & 1 & 3 & 2 & 0 \end{array}$$

Therefore, $(x + 3)$ is a factor.

(b)

$$\begin{array}{r|rrrr} 2 & 1 & -4 & -11 & 30 \\ & & 2 & -4 & -30 \\ \hline & 1 & -2 & -15 & 0 \end{array}$$

Therefore, $(x - 2)$ is a factor.

$$\begin{array}{r|rrrr} -3 & 1 & -4 & -11 & 30 \\ & & -3 & 21 & -30 \\ \hline & 1 & -7 & 10 & 0 \end{array}$$

Therefore, $(x + 3)$ is a factor.

$$\begin{array}{r|rrrr} 5 & 1 & -4 & -11 & 30 \\ & & 5 & 5 & -30 \\ \hline & 1 & 1 & -6 & 0 \end{array}$$

Therefore, $(x - 5)$ is a factor.

Practice Exercises for Appendix B (page A16)

1. 0.6362 **2.** 1.4382 **3.** 2.1103 **4.** 3.5473 **5.** $-1 + 0.9426$ **6.** $-2 + 0.8861$
7. $-3 + 0.7148$ **8.** $-4 + 0.4079$ **9.** 2.945 **10.** 62.67 **11.** 155.8 **12.** 5279
13. 0.3181 **14.** 0.08704 **15.** 20,930 **16.** 773.8 **17.** 5.749 **18.** 87.12
19. 1,549,000 **20.** 110,700 **21.** 614.1 **22.** 383.1 **23.** 0.9614 **24.** 5.041
25. 13.79 **26.** 48.49 **27.** 15.91 **28.** 1.116

Index

Rectangle

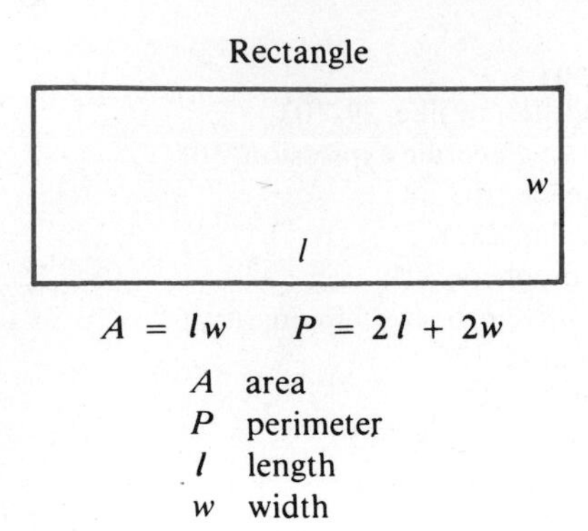

$A = lw \qquad P = 2l + 2w$

A area
P perimeter
l length
w width

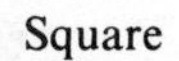

Square

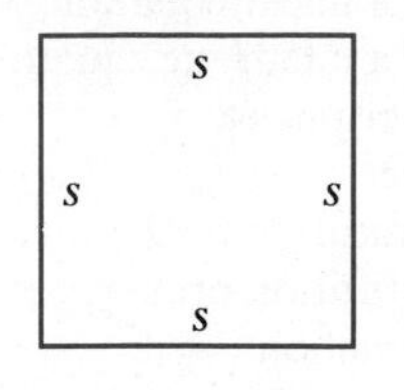

$A = s^2 \qquad P = 4s$

A area
P perimeter
s length of a side

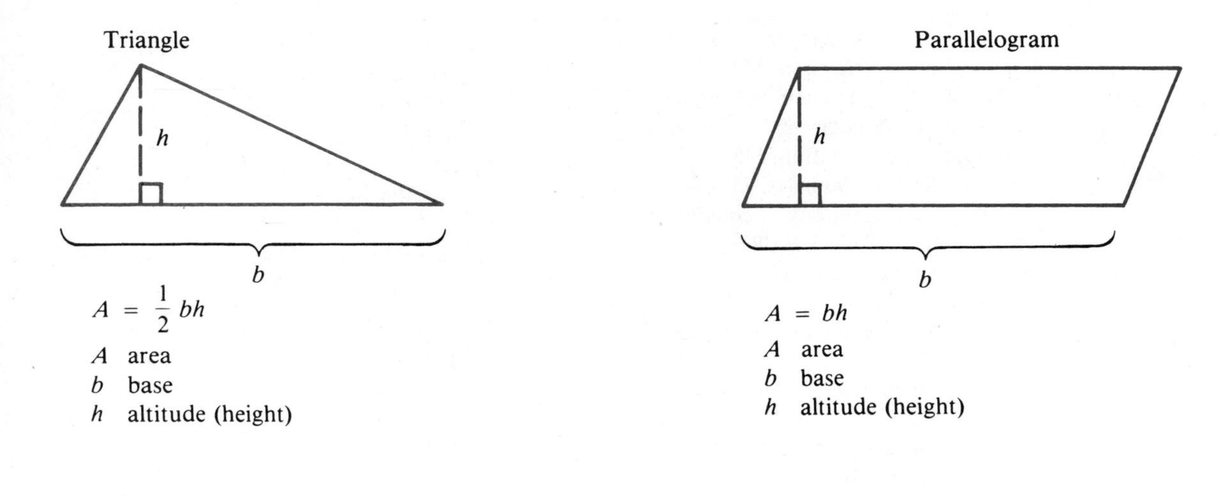

Triangle

$A = \frac{1}{2} bh$

A area
b base
h altitude (height)

Parallelogram

$A = bh$

A area
b base
h altitude (height)

Trapezoid

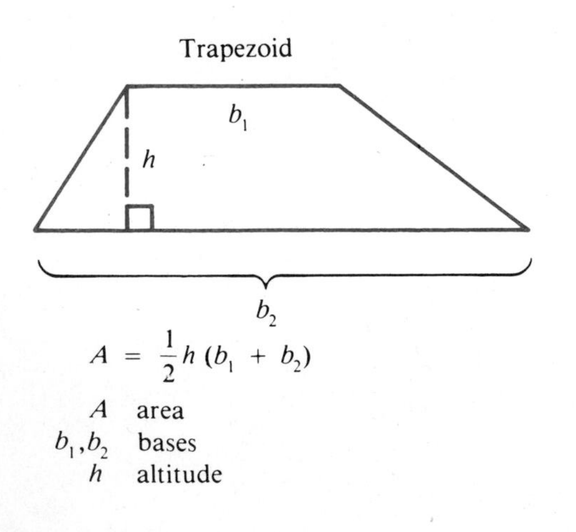

$A = \frac{1}{2} h (b_1 + b_2)$

A area
b_1, b_2 bases
h altitude

Circle

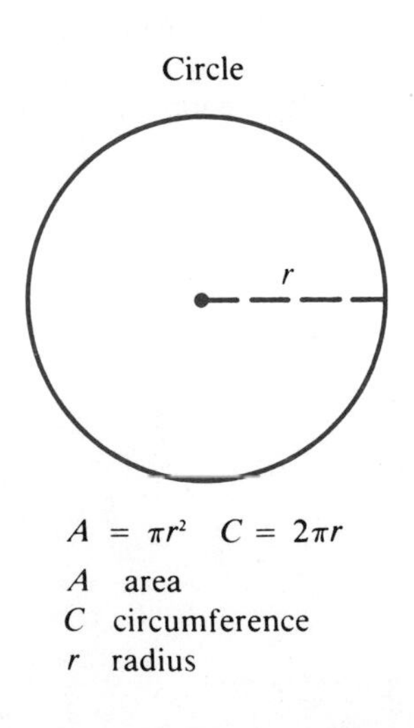

$A = \pi r^2 \quad C = 2\pi r$

A area
C circumference
r radius

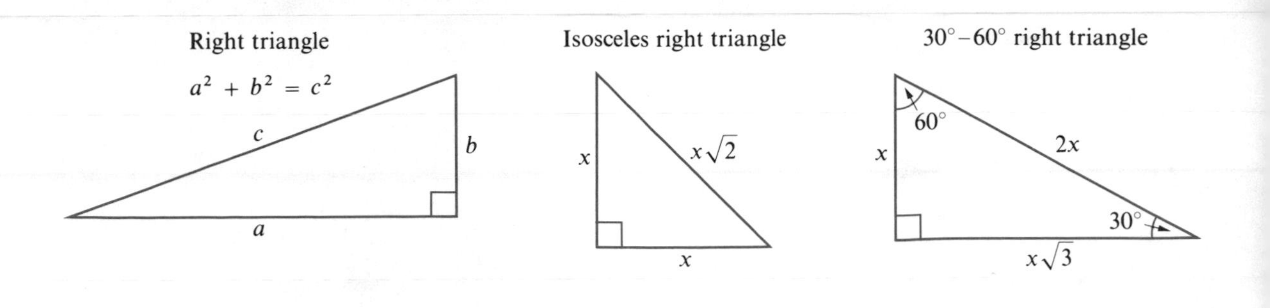

Sphere

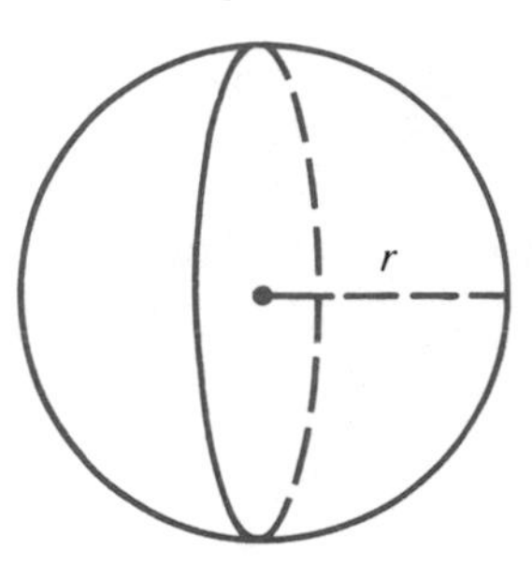

$S = 4\pi r^2 \quad V = \frac{4}{3}\pi r^3$

S surface area
V volume
r radius

Right circular cylinder

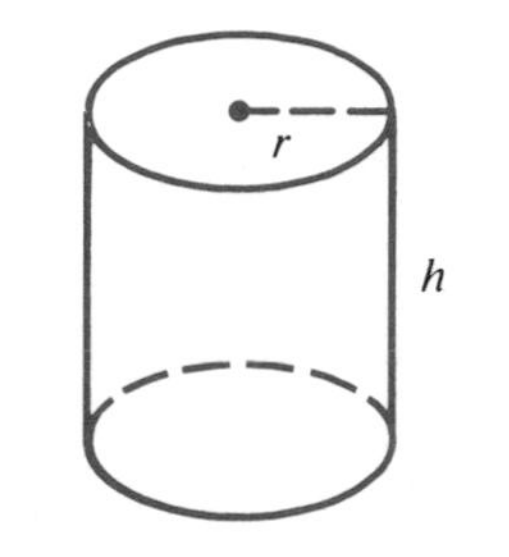

$V = \pi r^2 h \quad S = 2\pi r^2 + 2\pi rh$

V volume
S total surface area
r radius
h altitude (height)

Right circular cone

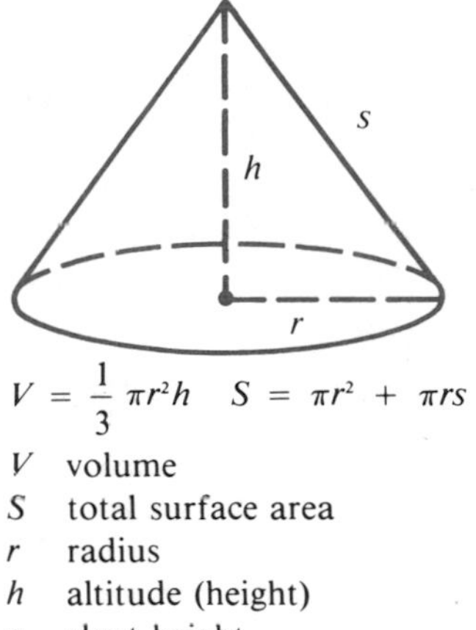

$V = \frac{1}{3}\pi r^2 h \quad S = \pi r^2 + \pi rs$

V volume
S total surface area
r radius
h altitude (height)
s slant height

Prism

Pyramid

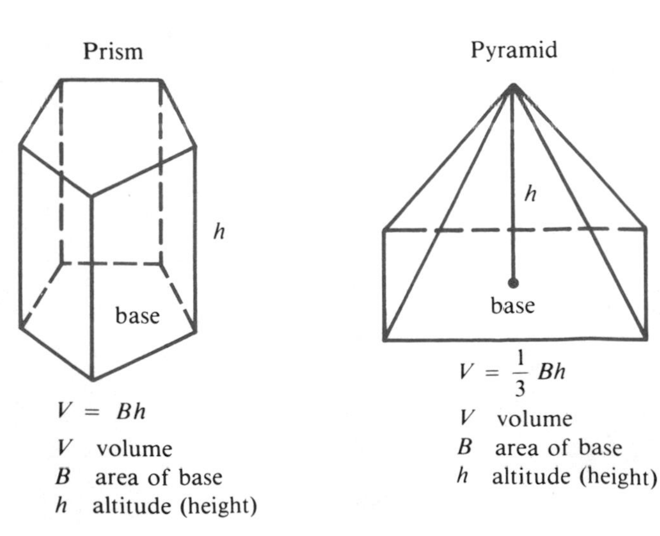

$V = Bh$

V volume
B area of base
h altitude (height)

$V = \frac{1}{3}Bh$

V volume
B area of base
h altitude (height)

Table of Common Logarithms

N	0	1	2	3	4	5	6	7	8	9
1.0	.0000	.0043	.0086	.0128	0.170	0.212	.0253	.0294	.0334	.0374
1.1	.0414	.0453	.0492	.0531	.0569	.0607	.0645	.0682	.0719	.0755
1.2	.0792	.0828	.0864	.0899	.0934	.0969	.1004	.1038	.1072	.1106
1.3	.1139	.1173	.1206	.1239	.1271	.1303	.1335	.1367	.1399	.1430
1.4	.1461	.1492	.1523	.1553	.1584	.1614	.1644	.1673	.1703	.1732
1.5	.1761	.1790	.1818	.1847	.1875	.1903	.1931	.1959	.1987	.2014
1.6	.2041	.2068	.2095	.2122	.2148	.2175	.2201	.2227	.2253	.2279
1.7	.2304	.2330	.2355	.2380	.2405	.2430	.2455	.2480	.2504	.2529
1.8	.2553	.2577	.2601	.2625	.2648	.2672	.2695	.2718	.2742	.2765
1.9	.2788	.2810	.2833	.2856	.2878	.2900	.2923	.2945	.2967	.2989
2.0	.3010	.3032	.3054	.3075	.3096	.3118	.3139	.3160	.3181	.3201
2.1	.3222	.3243	.3263	.3284	.3304	.3324	.3345	.3365	.3385	.3404
2.2	.3424	.3444	.3464	.3483	.3502	.3522	.3541	.3560	.3579	.3598
2.3	.3617	.3636	.3655	.3674	.3692	.3711	.3729	.3747	.3766	.3784
2.4	.3802	.3820	.3838	.3856	.3874	.3892	.3909	.3927	.3945	.3962
2.5	.3979	.3997	.4014	.4031	.4048	.4065	.4082	.4099	.4116	.4133
2.6	.4150	.4166	.4183	.4200	.4216	.4232	.4249	.4265	.4281	.4298
2.7	.4314	.4330	.4346	.4362	.4378	.4393	.4409	.4425	.4440	.4456
2.8	.4472	.4487	.4502	.4518	.4533	.4548	.4564	.4579	.4594	.4609
2.9	.4624	.4639	.4654	.4669	.4683	.4698	.4713	.4728	.4742	.4757
3.0	.4771	.4786	.4800	.4814	.4829	.4843	.4857	.4871	.4886	.4900
3.1	.4914	.4928	.4942	.4955	.4969	.4983	.4997	.5011	.5024	.5038
3.2	.5051	.5065	.5079	.5092	.5105	.5119	.5132	.5145	.5159	.5172
3.3	.5185	.5198	.5211	.5224	.5237	.5250	.5263	.5276	.5289	.5302
3.4	.5315	.5328	.5340	.5353	.5366	.5378	.5391	.5403	.5416	.5428
3.5	.5441	.5453	.5465	.5478	.5490	.5502	.5514	.5527	.5539	.5551
3.6	.5563	.5575	.5587	.5599	.5611	.5623	.5635	.5647	.5658	.5670
3.7	.5682	.5694	.5705	.5717	.5729	.5740	.5752	.5763	.5775	.5786
3.8	.5798	.5809	.5821	.5832	.5843	.5855	.5866	.5877	.5888	.5899
3.9	.5911	.5922	.5933	.5944	.5955	.5966	.5977	.5988	.5999	.6010
4.0	.6021	.6031	.6042	.6053	.6064	.6075	.6085	.6096	.6107	.6117
4.1	.6128	.6138	.6149	.6160	.6170	.6180	.6191	.6201	.6212	.6222
4.2	.6232	.6243	.6253	.6263	.6274	.6284	.6294	.6304	.6314	.6325
4.3	.6335	.6345	.6355	.6365	.6375	.6385	.6395	.6405	.6415	.6425
4.4	.6435	.6444	.6454	.6464	.6474	.6484	.6493	.6503	.6513	.6522
4.5	.6532	.6542	.6551	.6561	.6571	.6580	.6590	.6599	.6609	.6618
4.6	.6628	.6637	.6646	.6656	.6665	.6675	.6684	.6693	.6702	.6712
4.7	.6721	.6730	.6739	.6749	.6758	.6767	.6776	.6785	.6794	.6803
4.8	.6812	.6821	.6830	.6839	.6848	.6857	.6866	.6875	.6884	.6893
4.9	.6902	.6911	.6920	.6928	.6937	.6946	.6955	.6964	.6972	.6981
5.0	.6990	.6998	.7007	.7016	.7024	.7033	.7042	.7050	.7059	.7067
5.1	.7076	.7084	.7093	.7101	.7110	.7118	.7126	.7135	.7143	.7152
5.2	.7160	.7168	.7177	.7185	.7193	.7202	.7210	.7218	.7226	.7235
5.3	.7243	.7251	.7259	.7267	.7275	.7284	.7292	.7300	.7308	.7316
5.4	.7324	.7332	.7340	.7348	.7356	.7364	.7372	.7380	.7388	.7396